Smith, Currie & Hancock's COMMON SENSE CONSTRUCTION LAW:

A Practical Guide for the Construction Professional

Smith, Currie & Hancock's COMMON SENSE CONSTRUCTION LAW:

A Practical Guide for the Construction Professional

FOURTH EDITION

Edited by
Thomas J. Kelleher, Jr. and G. Scott Walters
Smith, Currie, & Hancock LLP

JOHN WILEY & SONS, INC.

This book is printed on acid-free paper. ♾

Published by John Wiley & Sons, Inc., Hoboken, New Jersey
Published simultaneously in Canada

For general information about our other products and services, please contact our Customer Care Department within the United States at (800) 762–2974, outside the United States at (317) 572–3993 or fax (317) 572–4002.

Wiley also publishes its books in a variety of electronic formats. Some content that appears in print may not be available in electronic books. For more information about Wiley products, visit our web site at www.wiley.com.

Library of Congress Cataloging-in-Publication Data:

Smith, Currie & Hancock LLP's common sense construction law : a practical guide for the construction professional/edited by Thomas J. Kelleher, Jr. —4th ed.
p. cm.
Includes index.
ISBN 978-0-470-23136-4 (cloth/cd : acid-free paper)
1. Construction contracts—United States. 2. Construction industry—Law and legislation—United States. I. Kelleher, Thomas J. II. Smith, Currie & Hancock. III. Title: Smith, Currie and Hancock LLP's common sense construction law. IV. Title: Common sense construction law.
KF902.S63 2009
346.73'078624—dc22

2008033284

Printed in the United States of America

10 9 8 7 6 5 4 3 2 1

In the first edition of *Common Sense Construction Law,* we recognized Overton A. Currie and Luther P. House, Jr., for their leadership and mentoring of the firm and its construction law practice. Consistent with the tradition established by the founders of this firm, we dedicated the second edition of *Common Sense Construction Law* to our clients, whose confidence and trust are essential to our practice and success. In the third edition of *Common Sense Construction Law* we recalled with fond memories our deceased partners, G. Maynard Smith (1907-1992), E. Reginald Hancock (1924–2004), and Bert R. Oastler (1933–2002), who devoted their professional lives to the practice of our profession at the highest level of excellence and set the standard for future generations of lawyers in the firm. In this fourth edition we honor the remarkable life and achievements of the late Overton A. Currie (1926–2005). Without him, the practice of construction law as we know it today would not be nearly as defined. Overton was truly a giant in, and has left an indelible mark upon, the construction industry and our firm.

CONTENTS

Preface xxv

Author Biographies xxvii

1 Legal Context of Construction 1

I. Introduction / 1
II. Contract Law / 1
III. Breach of Contract / 2
IV. Implied Contract Obligations / 2
V. Evolution of Construction Law / 4
VI. Torts / 5
  A. Intentional Torts / 5
  B. Negligence / 6
  C. Strict Liability / 7
VII. Statutory and Regulatory Laws Affecting Construction / 8
Points to Remember / 9

2 Alternative Contracting Methods 10

I. Traditional Approach to Construction: Advantages and Disadvantages / 11
II. Multiprime Contracting and Fast-Tracking / 12
III. Construction Management / 14
  A. Agency Construction Management / 15
  B. Program Management / 16
  C. Construction Manager/General Contractor / 16
IV. Design-Build Contracting: What Works to Avoid Disputes / 17
  A. The Design-Builder's Perspective / 17
  B. The Owner's Viewpoint / 19
  C. The Successful Design-Build Project / 24
  D. Design Professional Liability Issues in a Design-Build Project / 24

V. Design-Build Aspects of Traditional Construction / 25
A. Performance Specifications / 25
B. Shop Drawings / 27
C. Secondary Design Review / 28
D. The Interpretations Clause / 28
E. Identification of Patent Defects / 29
F. Compliance with Permits, Codes, and Regulations / 29
VI. Contractor Liability Issues / 30
VII. Engineer-Procure-Construct / 31
A. Limitations of Liability / 32
B. Liquidated Damages / 33
VIII. Private Public Partnerships / 34
Points to Remember / 35

3 Preparing to Work in a New State: Preproposal and Performance Considerations **36**

I. Qualifying to Do Business / 36
II. State Registration Requirements: Bonds to Secure Payment of Taxes / 38
III. State Licensing and Qualifications / 38
IV. Public Construction Awards / 40
V. State Statutes and Policies Affecting Contract Terms and Conditions / 42
A. Subcontractor/General Contractor Relationships / 42
B. Contract Award Preference Regulations / 43
C. Public Policy Limitations on Contract Clauses / 45
D. Impact of New Legislation / 47
VI. Preservation of Lien/Bond Rights / 48
A. Preconstruction Knowledge of Lien Law Peculiarities Is Essential / 48
B. Lien Law Protections and Procedures Vary Greatly from State to State / 49
C. "Almost Right" Is Almost Always Not Good Enough / 49
D. A Valid Claim of Lien Does Not Guarantee Payment / 50
E. Arm Yourself Early with Accurate Lien Law Information / 50
VII. Risk Assessment Checklists / 50
VIII. Information Sources / 53
Points to Remember / 54

4 Competing for the Contract **55**

I. Introduction: Sealed Bids to Negotiated Best-Value Awards / 55
II. General Considerations in Bids for Public Contracts / 55
III. The Responsible Bidder / 56
IV. The Responsive Bidder / 57
V. The "Lowest And Best" Bidder / 61
VI. Negotiated "Best Value" Selection Process / 63
VII. Electronic Bids / 66
VIII. Reverse Auctions / 67
IX. General Considerations When Competing on Private Contracts / 68
X. Effect of Past Performance Evaluations on Award Process / 69
 A. Past Performance Evaluation Procedures / 70
 B. Challenges to Past Performance Evaluations / 72
XI. Contractor Bid Mistakes / 73
 A. Elements for Relief from Bid Mistake / 74
 B. Owner's Duty If Bid Mistake Is Suspected / 77
 C. Withdrawal versus Reformation of Bid Mistake / 77
XII. Bid Protests / 78
 A. Protests on Federal Government Contracts / 79
 B. Protests before or after Receipt of Bids or Offers / 79
 C. Protests to the Contracting Agency / 80
 D. Protests to the Comptroller General / 82
 E. Bid Protests in Court / 84
 F. Bid Protests on State and Local Public Contracts / 85
XIII. Bid Bonds / 86
XIV. "Bid Shopping": What Is the Obligation of the Prime Contractor to the Subcontractor Submitting the Lowest Price? / 88
XV. Holding Subcontractors/Vendors to Their Bids / 91
 A. Promissory Estoppel / 91
 B. Elements of Promissory Estoppel / 92
XVI. Statute of Frauds Issues / 94
XVII. Damages / 94
 Points to Remember / 95

5 Intepreting the Contract **98**

I. The Importance of Contract Interpretation / 98
II. What Is a "Contract"? / 98
III. The Goal of Contract Interpretation / 99

IV. Defining Contract Terms / 100
- A. Terms Defined by the Parties / 100
- B. Technical Terms / 100
- C. Generally Accepted Definitions / 101

V. Interpreting the Contract's Wording / 101
- A. The Contract Must Be Considered as a Whole / 101
- B. Specific Terms versus General Terms / 101
- C. Handwritten, Typed, and Preprinted Terms / 102
- D. Order-of-Precedence Clause / 102

VI. The Facts and Circumstances Surrounding Contract Formation / 103
- A. Discussions and Conduct / 103
- B. The Parties' Prior Dealings / 104
- C. Industry Custom and Usage / 105
- D. Limitations on the Use of Facts and Circumstances Surrounding the Contract / 106

VII. Resolving Ambiguities / 106
- A. Construing the Contract against the Drafter / 107
- B. Duty to Request Clarification / 107

VIII. Implied Contractual Obligations / 108
- A. Duty of Good Faith and Fair Dealing / 108
- B. Duty to Cooperate / 109
- C. Warranty of Plans and Specifications— The *Spearin* Doctrine / 111

IX. Contractual Obligations Arising by *Operation* of Law / 112

Points to Remember / 113

6 The Uniform Commercial Code and the Construction Industry 114

I. The Uniform Commercial Code / 114

II. Applicability of the UCC to Construction / 115

III. Determining When Article 2 Applies / 115

IV. Modifying UCC Obligations / 116

V. Contract Formation under the UCC / 117
- A. Agreement on All Terms and Conditions Is Not Required / 117
- B. Methods of Acceptance / 117
- C. Requirement for a Written Contract / 118
- D. Withdrawal of an Offer / 119
- E. Filling Gaps in Essential Contract Terms / 119

VI. Risk of Loss / 120

VII. Inspection, Acceptance, Rejection, and Revocation of Acceptance / 121
A. Inspection of Goods / 121
B. Rejection of Goods / 122
C. Acceptance and Notice of Breach / 122
D. Seller's Right to Cure / 123
E. Revocation of Acceptance in Whole or in Part / 123
VIII. Warranties under the UCC / 124
A. Warranty Disclaimers / 125
B. Warranty Limitations / 125
IX. Statute of Limitations and Commencement of the Warranty Period / 126
X. Performance Issues / 128
A. Anticipatory Repudiation/Adequate Assurance of Performance / 128
B. Dealing with the Other Party's Insolvency / 129
C. Excuse of Performance by Failure of Presupposed Conditions / 130
D. Buyer's Remedies / 131
E. Seller's Remedies / 131
Points to Remember / 132

7 Authority And Responsibility of the Design Official 134

I. Overview / 134
II. Standard of Care and Professional Responsibility / 135
III. Authority of the Design Professional / 137
A. Actual Authority / 137
B. Implied Authority / 138
C. Apparent Authority / 139
D. Ratification of the Design Professional's Authority / 141
IV. Supervisory and Administrative Functions of the Design Professional / 141
A. Interpretation of the Plans and Specifications / 142
B. Review and Approval of Shop Drawings and Submittals / 144
C. Inspections and Testing / 147
D. Issuance of Certificates of Progress or Certificates of Completion and Certificates for Payment / 151
E. Resolution of Disputes between the Owner and the Contractor / 155
V. Other Duties of the Design Professional / 158

VI. The Design Professional's Liability to the Contractor / 159
A. Negligence and the "Economic Loss Rule" / 159
B. Intentional Torts / 163
C. Third-Party Beneficiary Theory / 163
D. Professional Liability Coverage / 166
VII. Statutes of Repose / 167
VIII. Effects of Contractual Limitations on Design Professional Liability / 168
IX. Assumption of Design Liability by the Contractor / 170
X. Recent Standard-Form Contract Efforts Regarding Shared Responsibility and Risk / 173
XI. The Design Professional's Copyright for Design Documents / 175
Points to Remember / 178

8 Subcontract Administration and Dispute Avoidance 180

I. Dispute Avoidance Begins at the Bidding Stage / 180
A. The Importance of the Low Price / 180
B. Know the Other Parties / 181
C. Problem Areas in Subcontract Bidding / 184
II. Preparation of the Subcontract Agreement / 185
A. "Flow-Down" Obligations / 188
B. Scope of the Work / 188
C. Payment Obligations / 189
D. Subcontractor Termination / 194
E. No Damages for Delay, Except as Paid by the Owner / 198
F. Changes / 199
G. Subcontractor's Indemnification of the Contractor / 201
H. Labor Affiliation / 203
I. Disputes Procedures / 204
J. Federal Government Projects / 206
III. Should Subcontractors Be Bonded? / 207
IV. Dispute Avoidance by Diligent Project Administration / 208
A. General Contractor's Duty to Coordinate the Work / 208
B. Implied Duty to Cooperate / 209
C. Implied Duty to Coordinate / 209
D. Limitation of Liability / 210
E. Pay Applications and Partial Lien Waivers / 211
F. Prime Contractor Financing of Subcontractors / 211
G. Remedies for Defective Performance / 212
H. Remedies for Delayed Performance / 213

I. Relationship between Subcontractor and the Owner: Can the Prime Contractor Assert the Subcontractor's Rights against the Owner? / 214
J. Prime Must Be Liable to the Subcontractor for the Pass-Through Claim: The *Severin* Doctrine / 215
K. States Have Adopted the *Severin* Doctrine / 216
L. Conclusion / 216
Points to Remember / 217

9 Contract Changes **219**

I. What Is a Changes Clause? / 219
II. Recovery under the Changes Clause / 225
A. Is There a Change to the Contract Work? / 226
B. Deductive Changes / 227
III. Who Is Authorized to Order Changes? / 228
A. Implied Authority / 229
B. Apparent Authority / 230
C. Alternatives to Demonstrating Authority / 231
IV. Written Documentation of Changes / 231
A. Written Directives / 231
B. Requirements for Written Notice of a Change / 234
V. Constructive Changes / 236
A. Informal Extra Work Directives / 237
B. Defective Plans and Specifications / 237
C. Misinterpretation of Plans and Specifications by the Owner / 238
D. Acceleration / 239
VI. Cardinal Changes / 242
VII. The Impact of Numerous Changes on Unchanged Work / 244
VIII. Impossibility/Impracticability / 245
Points to Remember / 247

10 Differing Site Conditions **249**

I. "Differing Site Condition" Defined / 249
II. Responsibility for Differing Site Conditions / 249
III. Standard Differing Site Conditions Clauses / 250
A. Federal Government Contracts / 251
B. ConsensusDOCS / 251
C. Other Standard Forms / 253
IV. Types of Conditions Covered / 257
A. Type I and Type II Changed Conditions / 257
B. Notice Requirements / 258

V. Operation of the Differing Site Conditions Clause / 258
A. Recovery for a Type I Changed Condition / 259
B. Recovery for a Type II Changed Condition / 263
VI. Stumbling Blocks to Recovery / 266
A. Site Investigations / 266
B. Exculpatory Clauses / 268
C. Notice Requirements / 269
VII. Relief in the Absence of a Contract Provision / 271
A. Misrepresentation / 271
B. Duty to Disclose / 272
C. Breach of Implied Warranty / 272
D. Mutual Mistake / 273
Points to Remember / 274

11 Schedules, Delays, and Acceleration **276**

I. Developments in Contractual Risk Allocation / 276
A. "Time Is of the Essence" Clause / 277
B. Contract Commencement and Completion Dates / 277
C. Substantial Completion and Final Completion / 278
D. Typical Contract Time and Scheduling Clauses / 278
II. Use of Schedules In Project Management / 280
III. Types of Delays / 281
A. Excusable Delays versus Nonexcusable Delays / 281
B. Excusable Delays / 282
C. Compensable Excusable Delays / 283
D. Nonexcusable Delays / 283
IV. Typical Causes of Excusable Compensable Delay / 283
A. Defective Drawings or Specifications / 284
B. Failure to Provide Access and Improper Site Preparation / 284
C. Failure to Supply Materials or Labor / 285
D. Failure to Provide Plans/Approve Shop Drawings / 285
E. Failure to Coordinate Prime Contractors / 286
F. Failure to Give Timely Orders for Work / 286
G. Failure to Make Timely Payments to Contractors / 287
H. Failure to Inspect / 287
I. Suspensions / 287
J. Excessive Change Orders / 288
K. Failure to Accept Completed Work / 288

V. Concurrent Delay / 288
- A. Traditional View: No Recovery by Either Party / 289
- B. Modern Trend: Apportionment of Delay Damages / 289

VI. Noncompensable Excusable Delays / 290
- A. Weather / 290
- B. Acts of God / 291
- C. Labor Problems / 291

VII. Acceleration / 291
- A. Directed Acceleration / 292
- B. Constructive Acceleration / 293

VIII. Contractual Limitations to Recovery / 293
- A. Requirement for Written Notice / 293
- B. No-Damages-for-Delay Clauses / 294
- C. Trade-to-Trade Clauses / 295

IX. Delay Claims and the Use of CPM Schedules / 296

X. Delay Claims Supporting Documentation / 298

Points to Remember / 300

12 Inspection, Acceptance, Warranties, and Commissioning 303

I. Inspection / 304
- A. Introduction / 304
- B. Standard Inspection Clauses / 305
- C. Safety-Related Inspection Obligations / 307
- D. Costs of Inspection / 308
- E. The Owner's Right to Inspect / 309
- F. Rejection and Correction / 311
- G. Limitation on Owner's Inspections / 311
- H. Inspection by the Design Professional or Inspector / 315
- I. Inspection by the Contractor / 316

II. Acceptance / 317
- A. Overview / 317
- B. Types of Acceptance: Formal versus Constructive / 318
- C. Authority as an Element of Constructive Acceptance / 319
- D. Limitations on the Finality of Acceptance / 320
- E. Contract Provisions Related to the Finality of Acceptance / 321
- F. Substantial Completion / 322
- G. Final Completion / 323
- H. Revocation of Acceptance / 323

III. Contractual Warranties / 325
A. Express Warranties / 325
B. Implied Warranties / 327
C. Statutory Warranties / 329
IV. Project Commissioning and Post-acceptance Facility Operations / 329
A. Commissioning Programs / 330
B. Bonding Considerations / 333
Points to Remember / 333

13 Management Techniques to Limit Risks and Avoid Disputes 335

I. Construction: A Risk-Prone Business / 335
II. Qualifying the Project and the Participants / 336
A. Qualifying the Project / 336
B. Qualifying the Project Participants / 337
C. Qualifying the Site and Locale / 342
III. Defining Rights, Responsibilities, and Risks: Parties and Their Contracts / 344
IV. Contract Framework / 345
A. Standard Contract Forms / 345
B. Critical Contract Provisions / 350
V. Avoiding and Preparing for Disputes through Proper Management and Documentation / 350
VI. Prudent and Responsible Estimating / 351
VII. Establishing Standard Operating Procedures / 352
VIII. Establishing Lines of Communication / 352
IX. Project Documentation / 353
X. Electronic Communications on Construction Projects / 360
A. Industry Forms Addressing Electronic Communications / 360
B. Web-Based Project Management Systems / 363
XI. Building Information Modeling / 363
A. Collaborative Uses of BIM / 364
B. BIM's Legal Implications / 365
XII. Cost Accounting Records / 366
XIII. Monitoring the Work through Scheduling / 367
XIV. Preserving Electronically Stored Information / 368
XV. Conclusion / 369
Points to Remember / 369

Appendix 13.1: ConsensusDOCS 221—Contractor's Statement of Qualifications for a Specific Project / 371
Appendix 13.2: Checklist: Prebid/Proposal Environmental Considerations / 381
Appendix 13.3: Checklist: Contracts in Foreign States / 383
Appendix 13.4: Logs and Forms / 386

14 Payment Bonds **406**

I. Payment Bonds Required by Statute / 407
A. The Miller Act / 407
B. Little Miller Acts / 411
II. Payment Bonds on Private Projects / 412
III. Qualifying for Payment Bond Coverage / 413
A. "Subcontractors" and "Suppliers" / 413
B. Suppliers of Customized Materials / 414
C. "Substantiality and Importance" of Relationship with Prime Contractor / 415
D. "Dummy" Subcontractors, Alter Egos, and Joint Ventures / 416
E. Claimants on Private Payment Bonds / 416
IV. Work Qualifying for Payment Bond Coverage / 417
A. Labor and Materials / 417
B. Equipment Repairs and Rental / 418
V. Recovery under Payment Bonds for Extra Work, Delay Damages, or Lost Profits, and Other Costs / 419
A. Extra Work / 420
B. Damages for Delay and Lost Profits / 421
C. Attorneys' Fees and Other Costs / 422
VI. Distinguishing between Payment Bond Claims and Performance Bond Claims / 422
VII. Procedural Requirements / 422
A. Time of Notice / 422
B. Time of Lawsuit / 423
C. Surety Response to Notice of Claim / 424
VIII. Effect of Payment Bonds on Lien Rights / 425
IX. The Surety's Defenses to Payment Bond Liability / 425
A. Pay-If-Paid Clauses / 426
B. Misrepresented Status of Payments / 426
C. Claim or Lien Waivers / 427
Points to Remember / 428

15 Performance Bonds and Termination **429**

I. Surety Performance Bonds / 430
- A. Fundamentals of Suretyship Law / 430
- B. Extension of Performance Bond Surety's Liability to Third Parties / 435
- C. Increase of the Surety's Liability under a Performance Bond / 436
- D. Surety's Defenses to Performance Bond Liability / 438
- E. Surety Entitlement to Contract Funds / 442
- F. Industry Performance Bond Forms: A Comparison / 443
- G. Effect on the Surety of Arbitration of Construction Disputes / 444
- H. Alternatives to Bonds: "Subguard" Programs / 446

II. Termination / 447
- A. Overview / 447
- B. Termination for Default / 447
- C. Standard Form Contracts: Grounds for Default Termination / 448
- D. Defenses of the Contractor / 451
- E. Contractor Response to Termination Notices / 453
- F. Remedies of the Owner / 454
- G. Termination for Convenience / 456
- H. Convenience Termination Costs / 459

Points to Remember / 460

16 Proving Costs and Damages **461**

I. Basic Damage Principles / 461
- A. The Compensatory Nature of Damages / 461
- B. Categories of Damages / 462
- C. Causation / 465
- D. Cost Accounting Records / 465
- E. Mitigation of Damages / 465
- F. Betterment / 466

II. Methods of Pricing Claims / 466
- A. The Total Cost Method / 467
- B. Segregated Cost Method / 468
- C. Modified Total Cost Method / 468
- D. Quantum Meruit Claims / 469

III. Contractor Damages / 470
- A. Contract Changes and Extras / 470
- B. Wrongful Termination or Abandonment / 472
- C. Owner-Caused Delay and Disruption / 473

D. Owner-Caused Acceleration / 479
E. Defective Drawings or Specifications / 480
F. Inefficiency Claims / 480
IV. Owner Damages / 485
A. Direct Damages / 486
B. Consequential Damages / 487
C. Liquidated Damages / 488
Points to Remember / 492

17 Construction Industry Environmental and Safety Concerns **494**

I. Sources of Environmental Regulation and Liability / 495
A. Encountering Hazardous Materials on a Construction Site—CERCLA Liability / 495
B. Water Quality / 499
C. Air Quality / 500
II. Minimizing Environmental Risks / 503
A. Conduct a Prebid Environmental Review of the Contract Documents / 503
B. Contract Provisions and Indemnification / 505
C. Insurance / 509
III. Proper Management Techniques / 510
A. Management Review of Environmental Risks / 510
B. Have a Response Plan / 510
C. Immediately Stop Work in the Affected Area / 511
D. Provide Immediate Notice / 511
E. Do Not Resume Work without Proper Authorization / 512
IV. Mold: Developing a Program to Limit Liability / 512
V. Environmentally Friendly Construction: Green Buildings / 513
A. LEED Certification / 513
B. Design and Construction Issues Affected by Green Building Construction / 515
C. Bearing the Risk: Legal Issues Raised by LEED / 516
VI. Construction Safety / 516
A. Sources of Safety Requirements: OSHA-Specific Project Procedures / 517
B. Successfully Working with OSHA / 517
C. Who Is Responsible for Project Safety? / 518
D. Indemnification / 521
E. Workers' Compensation Statutes / 521
Points to Remember / 523

18 Construction Insurance **524**

I. Importance of Insurance Planning / 524
II. Types of Insurance / 525
- A. Commercial General Liability / 525
- B. Builder's Risk Insurance / 528
- C. Errors and Omissions Insurance / 528
- D. Alternatives to Traditional Insurance Programs / 529

III. Contract Requirements for Insurance / 530
- A. Standard Contract Clause / 530
- B. Waiver of Subrogation / 531
- C. Proof of Insurance / 532

IV. Prompt Action to Protect Potential Coverage / 533
- A. Sensitivity to Insurance Issues / 533
- B. Immediate Notice / 533

V. Insurer's Response to Claims / 534
- A. Reservation of Rights / 534
- B. Litigation with the Insurer / 535

VI. Routine Coverage Issues / 535
- A. CGL Coverage Issues / 536
- B. Builder's Risk Coverage Issues / 540

VII. Concurrent Causes / 541
VIII. Construction Insurance and Mold Claims / 542
- A. The Insurance Industry Responds / 542
- B. Mold Claims: Establishing whether Coverage is Available / 543

Points to Remember / 545

19 Labor and Employment Issues Affecting the Construction Industry **549**

I. Immigration Issues / 549
- A. Immigration Documentation and Control / 550
- B. Responses to Immigration Inquiries / 552
- C. Immigration Issues in Public Contracting / 553

II. Employee Safety and Health / 554
III. Wage and Hour Requirements / 555
- A. Basic Calculation / 556
- B. Overtime Calculation / 557
- C. Independent Contractor Status Considerations / 559
- D. Government Contracts / 559

IV. Employee Benefits: ERISA / 562

V. Employment Discrimination / 563
A. Equal Employment Opportunity Commission / 563
B. Title VII of the Civil Rights Act of 1964 / 564
C. Americans with Disabilities Act / 564
D. Age Discrimination in Employment Act / 566
E. Discrimination and Public Contracting / 566
F. Office of Federal Contract Compliance Programs / 567
G. Executive Order 11246 / 567
H. Uniform Service Employment and Re-Employment Rights Act / 568
VI. Family and Medical Leave Act / 569
VII. Union Activity / 570
A. Union-Organizing Tactics / 571
B. Appropriate Employer Responses to Union Activism / 572
VIII. Utilizing Employee Background Investigations / 574
Points to Remember / 576

20 Bankruptcy in the Construction Setting **578**

I. Introduction / 578
II. The Players / 578
III. Bankruptcy Code / 579
A. Chapter 11—Reorganization / 579
B. Chapter 7—Liquidation / 580
C. Automatic Stay / 580
D. Sanctions for Violation of Automatic Stay / 581
E. Relief from the Automatic Stay / 582
F. Preferential Transfers / 583
G. Exceptions to the Preferential Transfer Rule / 584
H. Discharge / 586
I. Nondischargeable Debts / 587
IV. Status of the Debtor's Contracts / 588
A. Executory Contracts / 588
B. Affirmance or Rejection / 588
C. Assignment / 590
D. Minimizing the Impact on Executory Contracts / 590
V. Status of Materials and Equipment / 592
A. Property of the Debtor's Estate / 592
B. Supplier's Right to Recover Goods / 593

C. Stored Materials / 593
D. Voiding Unperfected Security Interests / 594
VI. Status of Contract Funds / 595
A. Unearned Contract Funds / 595
B. Earned but Unpaid Contract Funds / 595
VII. Other Sources of Funds / 598
A. Performance and Payment Bond Claims / 598
B. Mechanic's Liens / 599
C. Guarantors / 600
Points to Remember / 600

21 Resolution of Construction Disputes 603

I. Early Claim Recognition and Preparation / 603
II. Early Involvement of Experts and Attorneys / 605
III. Demonstrative Evidence / 606
IV. Contemporaneous Records / 607
V. Components of a Well-Prepared Claim Document / 607
VI. Calculating and Proving Damages / 609
VII. Pursuing Negotiation and Settlement / 610
VIII. Alternatives to Litigation / 611
A. Dispute Review Boards / 611
B. Minitrials/Summary Jury Trials / 611
C. Mediation / 612
IX. Arbitration / 614
A. Time and Costs of Arbitration / 615
B. Selection of Arbitrators / 616
C. Informality and Limited Appeals in Arbitration / 617
D. Enforceability of Agreements to Arbitrate / 618
E. Special Problems Involving Multiple Parties to Arbitration / 620
F. Med-Arb / 622
G. Arbitration Agreements and Procedures / 622
H. American Arbitration Association Rules and Procedures / 622
I. Center for Public Resources Rules and Procedures / 623
J. International Chamber of Commerce Rules and Procedures / 624
K. Party-Drafted Arbitration Agreements: Federal and State Law Considerations / 624
X. Litigation / 625
A. Profile of the Construction Trial / 625
B. The Court System / 626

C. Federal Rules of Civil Procedure / 627
D. Discovery / 627
E. Judge or Jury? / 629
F. Alternative Dispute Resolution in the Courts / 630
G. Trial / 630
H. Live Testimony / 630
Points to Remember / 632

22 Federal Government Construction Contract Disputes **633**

I. Historical Overview / 633
II. Scope of the Contract Disputes Act / 635
III. The Freedom of Information Act: A Claim Preparation Tool / 637
IV. Contractor Claims / 639
V. When Must a Claim Be Submitted? / 639
VI. Notice Requirements / 640
VII. Who May Submit a Claim? / 641
VIII. What Constitutes a Claim? / 642
A. Written Submission to the Contracting Officer / 643
B. Elements of a Claim / 644
C. Certification Requirement / 647
D. Monetary Threshold for Certification / 648
E. Modification of Claim Amount / 649
F. Certification Language / 651
G. Supporting Data / 653
H. Who Can Certify the Claim? / 654
IX. Other Certification Requirements / 654
X. Certification of Subcontractor Claims / 655
XI. Government Claims / 656
XII. Contracting Officer's Decision / 658
A. Time Allowed for Issuing the Decision / 658
B. Contents of the Final Decision / 659
XIII. Appeal Deadlines / 661
XIV. Choosing a Forum: Board or Court of Federal Claims / 664
XV. Transfer and Consolidation of Cases / 666
XVI. ADR and Government Contract Disputes / 667
XVII. Recovery of Attorneys' Fees in Government Contract Claims / 669
XVIII. False or Inflated Contract Claims / 670
Points to Remember / 672

Index **673**

PREFACE

The attorneys of Smith, Currie & Hancock LLP have practiced construction law and government contracts law for more than four decades. During that time, we have conducted hundreds of construction and government contract law seminars for clients, trade associations, colleges and universities, and professional groups. Throughout those efforts, our consistent goal has been to provide a practical, common sense perspective to the legal issues affecting the construction industry. In many respects, this book reflects a culmination and refinement of those educational endeavors and the practical approach they entail.

Construction law and other legal concepts and issues are essential to the construction industry, as they provide the theories, principles, and generally established rules that contribute to the smooth running of the construction process. When that process falters and disputes arise involving the design or construction of a project, a variety of procedures can be used to resolve these differences. These procedures vary in their legal formalities; however, an appreciation of these processes and principles should not be limited to lawyers. Individuals in management and supervisory positions in the industry must know what the law requires of them and what they can expect and require from others. Nor can lawyers focus on legal rules and procedures to the exclusion of the *business* of construction and expect to effectively represent and assist their clients. Construction law and the business of construction are inextricably intertwined. We hope this book reflects this interrelationship in the topics that it covers and the various perspectives and approaches it employs.

Claims and disputes are addressed throughout this fourth edition of *Common Sense Construction Law*. They must be in any complete and competent analysis of the construction environment. This book, however, is about more than just preparing claims and resolving disputes once they arise. Rather, our goal is to help provide the kind of insight and understanding needed to *avoid* claims and disputes. Reasonable recognition of the contractual allocation of rights, risks, and legal responsibilities, coupled with a spirit of communication and teamwork in the execution of the work, is far more likely to culminate in a successful project than an atmosphere rife with confrontation and dispute. Practical knowledge of the general rules governing employment-related issues can have the same beneficial result. Of course, the possibility of claims and disputes cannot be ignored. Careful attention and planning is required to avoid disputes and to deal effectively with them when they become inevitable.

The fourth edition of *Common Sense Construction Law* is a general teaching tool and is not a substitute for the advice of your attorney. Specific concerns and problems require the timely attention of legal counsel familiar with construction law, government contracts, and employment law. Nevertheless, this book will help you to expand your knowledge and awareness of the issues in construction that may affect you at any time. It may not provide all the answers, but you will be well equipped to ask the right questions. To make these materials more useful to you, we have included checklists, sample forms, and summary "Points to Remember" for each chapter.

We thank our clients who have shared their insights and concerns and provided the opportunities for experience and learning that are shared in this book. We also owe much gratitude to the construction industry as a whole for allowing us as construction attorneys to participate in the challenges of the industry to avoid and resolve problems. We hope this work will contribute to the worthy goals of the industry. We also hope that this book helps its readers pursue their interests in and commitments to construction from concept to completion.

Smith, Currie & Hancock LLP

Atlanta, Georgia
January 2009

SMITH, CURRIE & HANCOCK LLP

A Firm Concentrating Its Practice on the Construction Industry

For more than four decades Smith, Currie & Hancock LLP, with offices in Atlanta, Georgia; Charlotte, North Carolina; Fort Lauderdale and Tallahassee, Florida; Las Vegas, Nevada; and Washington, DC, has developed a nationally recognized practice focused on the construction industry and the variety of legal issues facing that industry. Federal government construction contract law has also been a principle practice area since the firm was founded The firm represents private and public clients working or located in all fifty states, as well as Mexico, Canada, Central and South America, Europe, Asia, and Africa.

After developing construction and employment law practices in the context of a general service firm, G. Maynard Smith, Overton A. Currie, and E. Reginald Hancock formed this firm in 1965 to concentrate their practices in those areas in order to provide more effective service to the firm's clients. Having trained and practiced law in the culture created by those three outstanding attorneys, the current members of the firm remain committed to a tradition of providing quality, cost-effective legal services to clients ranging from small, family owned concerns to multibillion-dollar corporations.

In representing the many construction industry participants competing for and performing private construction work, as well as those working under local, state, and federal government construction contracts, we are necessarily involved in a wide variety of legal and business related issues. The breadth of those issues is reflected by the spectrum of topics addressed in this fourth edition of *Common Sense Construction Law* and in *Federal Government Construction Contracts—A Practical Guide for the Industry Professional.* The goal of this fourth edition of *Common Sense Construction Law* is to provide an informative and timely discussion of these topics for the construction professional without all of the specific details of a multi-volume legal treatise. To accomplish that task in a practical and meaningful manner, this book reflects the combined efforts of many attorneys drawing more than 500 years of collective experience in the areas of construction law and federal government contracts law.

Many of the firm's attorneys have engineering degrees in addition to their legal education, and several worked in the construction industry before pursuing their law degrees. Others joined this firm after military service as government contracts legal counsel or have extensive training in public procurement. Three members of Smith, Currie & Hancock LLP have served as Chair of the Section of Public Contract Law of the American Bar Association, and another partner has served as the Chair of the American Bar Association Forum Committee on the Construction Industry.

Smith, Currie & Hancock LLP has represented clients from the entire spectrum of the construction industry: contractors, subcontractors, construction managers, owners (public and private), architects, engineers, sureties, insurance companies, suppliers, lenders, real estate developers, and others. They include multinational and *Fortune* 500 companies and trade associations representing billion-dollar industries, as well as local and regional clients. Although our attorneys have appeared in numerous cases resulting in reported court decisions and even more arbitrations, our traditional goal has been to achieve resolution of differences by communication and agreement rather than formal dispute resolution. Consequently, in the last four decades we have assisted in the amicable resolution of many more matters than such reported decisions.

In addition to serving clients nationwide, Smith, Currie & Hancock LLP attorneys have published numerous articles in trade magazines and other periodicals and have authored or co-authored dozens of books on construction and public contract law. Our lawyers maintain a heavy schedule of lectures and seminars sponsored by various trade associations, colleges, and universities, including the U.S. Army Corps of Engineers, Georgia Institute of Technology, Auburn University, the American Bar Association, the Practicing Law Institute, the Associated General Contractors of America, and the Associated Builders & Contractors, Inc.

ATLANTA OFFICE

THOMAS E. ABERNATHY, IV. Vanderbilt University, B.A. 1963 and J.D. 1967; Fellow and past member of Board of Governors, American College of Construction Lawyers; Past Chair, American Bar Association, Section of Public Contract Law; American Arbitration Association National Roster of Arbitrators; listed in *Chambers USA Directory of America's Leading Business Lawyers* in the practice area of construction law (2003–2007); listed in the Top 100 of Georgia Super Lawyers; co-author, *Changed Conditions/Edition II, Construction Briefings*, West Group (2000); contributing author, *Construction Business Handbook* (2004).

CLIFFORD F. ALTEKRUSE: B.A., Reed College (Meritorious Senior Thesis); J.D., University of South Carolina School of Law; Law Review, Order of the Coif. Lecturer on Alternative Dispute Resolution, Emory University School of Law, Georgia State University School of Law; Lecturer on Alternative Dispute Resolution and Construction Law for various CLE programs and professional groups. Professional Service Memberships: State Bar of Georgia (Past Chair and Member, Alternative Dispute Resolution Section); Atlanta Bar Association (Member,

Construction Law and ADR Sections); American Bar Association (Member, Alternative Dispute Resolution Section and Forum on the Construction Industry); Consortium on Negotiation and Conflict Resolution.

WILLIAM L. BAGGETT, JR. A.B., Dartmouth College, 1984 (*magna cum laude*); Phi Beta Kappa; J.D., Vanderbilt University, 1987. Articles Editor, *Vanderbilt Law Review,* 1986–1987. Law Clerk to the Honorable Albert J. Henderson, Senior Circuit Judge, United States Court of Appeals for the Eleventh Circuit, 1987–1988. Member: American Bar Association; Tennessee Bar Association; State Bar of Georgia; State Bar of Tennessee.

PHILIP E. BECK: Univ. of Tennessee, B.S., 1978 (*high honors*), M.B.A., 1981, J.D. 1981; Order of the Coif; Moot Court Board; Omicron Delta Kappa; Beta Gamma Sigma. Law clerk to Hon. Houston Goddard, Tennessee Court of Appeals, 1980–81. Listed in the 2003–04 *Chambers USA Directory of America's Leading Business Lawyers* and in *Law & Politics* and *Atlanta Magazine*'s 2004 Georgia Super Lawyers, in the area of construction law. Frequent author and lecturer on various aspects of construction law; experienced litigator, arbitrator and mediator. Member, Associated General Contractors of America Board of Directors, 2002–2007.

HUBERT J. BELL, JR. B.A., Davidson College, 1966; M.A., Pacific Lutheran University, 1974; J.D., University of Georgia, 1981 (*cum laude*); Order of the Coif; Past Chair, American Bar Association, Section of Public Contract Law; Co-author, "The Economic Loss Rule; A Fair Balancing of Interests," *Construction Lawyer*, Volume II, Number 2, April 1991; *Alternative Clauses to Standard Construction Contracts* (second edition) (contributing author), John Wiley & Sons, 1998.

JAMES K. BIDGOOD, JR. B.S. Civil Engineering, Georgia Institute of Technology 1974; M.B.A. Phillips University 1980; J.D. Emory University, 1983 (with distinction); Order of Coif; Member, Emory Law Journal. Pilot, Instructor pilot, Captain, USAF 1974–1980. Experience in variety of construction and environmental related matters in arbitration, litigation, and mediation. Member, American Arbitration Association Panel of Neutrals. Frequent lecturer on topics of construction law.

JAMES F. BUTLER, III. A.B., Duke University, J.D., University of Kentucky. Member: State Bars of Arkansas, Georgia, Florida, Kentucky, New York, North Carolina, Texas, and Washington. His practice is concentrated in the negotiation, arbitration, and litigation of construction contracts and hazardous-waste issues. He has experience with a variety of projects and sites, including petrochemical plants, dams, power plants, tunnels, water/wastewater systems, airports, office buildings, casinos, semiconductor facilities, food-processing facilities, schools, condominiums, prisons, highways, hospitals, Superfund sites, and government facilities.

MARK B. CARTER. B.S., University of Alabama, 1992; J.D., University of Georgia (2003). 2006–2007 *Georgia Super Lawyers'* Rising Star, in the area of construction law. Member: State Bar of Georgia. Design-Build Institute of America (Board Member 2006–2007). Experienced in variety of construction matters in negotiation, arbitration, litigation, and mediation.

ROBERT C. CHAMBERS. Managing Partner of Smith, Currie & Hancock LLP, B.S.C.E., Georgia Institute of Technology, 1979 (high honors); M.S.C.E., 1980 (Study Emphasis—Geotechnical Engineering); J.D., University of Georgia, 1985 (*magna cum laude)*; Order of the Coif. Co-Author: "Changed Conditions," (1984); "Changed Conditions/Edition II," (2000), *Construction Briefings* Federal Publications. Frequent lecturer on environmental issues affecting the construction industry, including environmental liability.

IAN C. CLARKE. B.S., Electrical Engineering, Florida Agriculture and Mechanical University, 1995 (with Honors); J.D., Emory University Law School, 2007. Member: American Bar Association and State Bar of Georgia.

AUBREY L. COLEMAN, JR. Senior Partner of Smith, Currie & Hancock LLP. B.A., Tulane University, 1964 (*cum laude*); LL.B., Vanderbilt University, 1967; Order of the Coif; Managing Editor, *Vanderbilt Law Review*, 1966–67; Phi Beta Kappa. Co-author: Georgia Construction Law, Professional Education Systems, 1986. Author and lecturer on various aspects of construction law and construction claims, including participation in programs presented by Georgia State University, Professional Education Systems, Inc., and Federal Publications, Inc.

JAMES W. COPELAND. B.E., Vanderbilt University, 1988; J.D., Duke University, 1995. Member: American Bar Association, State Bar of Georgia.

JOSEPH J. DINARDO. B.S., Biology and Chemistry, Palm Beach Atlantic University, 1995 (*Summa Cum Laude*); J.D., Cumberland School of Law, 1999; LL.M., International Business Transactions, Emory University School of Law, 2000. Member of the State Bar of Georgia and the U.S. Patent and Trademark Bar.

KARL DIX, JR. B.S., Economics (*magna cum laude*), Wharton School of Finance and Commerce at the University of Pennsylvania, 1980; J.D., with Specialization (*cum laude*) in International Law from Cornell University, 1983. Captain, U.S. Army Corps of Engineers, Attorney Advisor (1983–1987); Chief Counsel's Honors Program. Member of the State Bars of Florida, Georgia, and New York, U.S. Court of Appeals for the 11th and Federal Circuits and U.S. Court of Federal Claims.

ROBERT O. FLEMING, JR. B.S. with Honors, North Carolina State University, 1972; J.D., with distinction, Duke University, 1977. Co-author: "Deposition Strategy, Law and Forms: Building Construction," Matthew-Bender, 1981. Approved Arbitrator for Construction Contract Disputes, American Arbitration Association. Member of the American Bar Associations; State Bar of Georgia.

PHILIP L. FORTUNE. B.A., University of North Carolina at Chapel Hill, 1967, Gamma Beta Phi; J.D., University of Toledo, 1970. Comments Editor, *University of Toledo Law Review.* Author and lecturer on construction law, bonds, liens and preparation of claims. Lecturer on various aspects of the Hazardous Waste and Environmental Protection laws and regulations including Superfund.

C. DAMON GUNNELS. B.A., Accounting, University of Tennessee, 2002; J.D., University of Tennessee 2006, Managing Editor, *Transactions, Tennessee Journal of Business Law.* American Bar Association - Forum Committee on the Construction Industry.

EUGENE J. HEADY. University of Hartford, BSE, 1981 (Kappa Mu Honorary Engineering Society); Texas Tech University, J.D., 1996 (*cum laude*); Editor-in-Chief, *Texas Tech Law Review*, 1995–96; Editor, *Texas Tech Legal Research Board*, 1995–96. Listed in *Who's Who in American Law*, 2000 and *Georgia Super Lawyers*, 2004. Contributing author to numerous professional publications including, Author, Chapters *in Alternative Clauses to Standard Construction Contracts*, Aspen Law & Business, 1998. Public Contract Law Section, Vice Chair, Region IV, Georgia, 1999–2004. State Bars of Colorado, Florida, Georgia, and Texas.

KIRK D. JOHNSTON: B.S., Business Management and Marketing, Brigham Young University, 1996 (Dean's List); J.D., Santa Clara University School of Law, 2000. Professional Service Memberships: State Bars of California and Georgia.

S. GREGORY JOY. B.A (with high distinction), Univ. of Virginia, 1981; Phi Beta Kappa, Raven Honor Society; J.D., University of Virginia, 1984; Editor or co-author of four books and over 50 seminar manuals and speaker at over 70 seminars on construction and government contract law issues, and author of articles on international and environmental law issues. Represented owners, contractors, subcontractors and/or suppliers regarding construction projects throughout the United States, in the Caribbean, Latin America and the Middle East. Member: American Bar Association; State Bar of Georgia; ABA Section on Public Contract Law; ABA Forum Committee on the Construction Industry.

RAMSEY KAZEM. B.S., Farleigh Dickinson University, 1996 (*magna cum laude)*; M.B.A., Goizueta Business School, 2000; J.D., Emory University School of Law, 2000. Member, State Bar of Georgia.

THOMAS J. KELLEHER, JR. A.B., Harvard College, 1965 (*cum laude*); J.D., University of Virginia, 1968. Fellow, American College of Construction Lawyers; Co-author, "Preparing and Settling Construction Claims," *Construction Briefings,* Federal Publications, Inc., December 1983; *Construction Litigation: Practice Guide with Forms*, Aspen Law & Business; "Development in Federal Construction Contracts," *Wiley Construction Law Update,* Wiley Law Publications (1992–1999). Member: American College of Construction Lawyers; American Bar Association, State Bars of Georgia and Virginia; Associated General Contractors of America, Board of Directors (1999–2003); Chair: AGC's Federal Acquisition Regulation Committee (1999–2003).

DAVID C. KING. B.A., Management, Maryville College, 1993 *(summa cum laude*); J.D., University of Georgia School of Law, 1996 (*cum laude*); Board Member: Georgia Bar Association Tort and Insurance Practice Section.

JOHN M. MASTIN, JR. B.S., Electrical Engineering, University of Alabama, 1976 (*cum laude*); J.D., University of Montana School of Law, 2005 (with Honors).

ANDREW R. McBRIDE. B.A., History, Rutgers College, 1992; J.D., Emory University School of Law, 1996.

JOHN E. MENECHINO, JR.: B.A., Political Science and Economics, University of South Carolina, 1986 (*magna cum laude,* Phi Beta Kappa); J.D., University of South Carolina College of Law, 1989 (American Inns of Court). Contributing author, *Construction Subcontracting: A Legal Guide*, John Wiley & Sons, 1991; *Alternative Clauses to Standard Construction Contracts*, John Wiley & Sons, 1991, supplements 1992, 1997, second edition 1999. Member, ABA Forum on the Construction Industry; ABA Tort and Insurance Practice Section, Fidelity and Surety Law Committee.

GARRETT E. MILLER. B.S., University of North Carolina at Chapel Hill; J.D., University of Mississippi School of Law, 2001; Moot Court Board; Staff Member of Journal of National Security Law; Law School Honor Council Representative.

ERIC L. NELSON. B.A., California Polytechnic State University at San Luis Obispo; J.D., Washington & Lee University. Co-author: "Delay Claims Against the Surety," *The Construction Lawyer,* July 1997. Lecturer on construction claims for various professional organizations. Member, American Bar Association, Section of Public Contract Law (Chair, Construction Division, Insurance and Bonding Subdivision, 2001-Present; Vice-Chair, Subcontracting Subdivision (2000–2001), Forum on the Construction Industry; Associated General Contractors of America, Inc. (Young Leadership Council 1998–2000); State Bar of Georgia.

HELEN H. POPE. B.S., Geography, University of South Carolina, 1997; M.S., Geography, University of Wisconsin at Madison, 2000; J.D., Tulane University Law School, 2005 *(cum laude).*

RONALD G. ROBEY. B.A., Centre College and University of Kentucky, 1974 (with distinction); J.D., University of Kentucky, 1977; Order of the Coif, Lead Articles Editor, *Kentucky Law Journal*, 1976–1977. Author: "Construction Management—Avoid Being a Fiduciary," *Contractor Profit News*, October 1986; co-author: "Winning Strategies in Construction Negotiations, Arbitration and Litigation," *Construction Contracts,* Practicing Law Institute, 1986. Chair of CLE Construction Seminars in Atlanta, Orlando, and Las Vegas; Member, American Bar Association, Forum on the Construction Industry and Section of Public Contract Law, State Bars of Georgia, Florida, and Kentucky.

CHARLES E. ROGERS. A.B., Duke University, 1989; J.D., Emory University, 1993. Lecturer on design, construction, and handicap accessibility for various CLE programs and professional organizations. Professional service memberships: American Bar Association, Litigation Section and Construction Industry Forum; Atlanta Bar Association, Construction Section and Litigation Section.

KENT P. SMITH. B.A., Northwestern Oklahoma State University, 1963; J.D., Washburn University School of Law, 1966 (*cum laude*); *Washburn Law Journal* Editorial Board, Alumni Fellow in Law, 1992. Law clerk to the U.S. Court of Appeals for the Tenth Circuit (1966–1967); attorney for the United States Atomic Energy Commission (1967–1968). Author: "Contractor Default: How to Recognize It, What to Do About it, and How to Finish the Project," The Florida Bar (5th Annual Construction Contract Litigation Seminar), 1981; "Changed Conditions in Dredging Contracts," *World Dredging & Marine Construction*, Vol. 14, No. 4, 1978; Guest lecturer, Georgia Institute of Technology, Legal

Aspects of Architecture, Engineering and Contracting. Member, State Bars of Georgia and Kansas.

JOSEPH C. STAAK. B.S. Civil Engineering, Georgia Institute of Technology, 1974; J.D., University of Georgia, 1981 (*cum laude*); Member, *Georgia Law Review* 1979–81. U.S.A.F. 1974–78 (Engineering Officer). Co-Author: *Georgia Construction Lien and Public Contract Bond Law,* Lorman Education Services; Co-Author: *Practical Construction Law for Florida Contractors.* Member, State Bars of Alabama, District of Columbia, Florida, and Georgia, American Bar Association Section of Public Contract Law. Associate Member, American Society of Civil Engineers (Georgia Section).

CHARLES W. SURASKY. B.A., University of South Carolina, 1973 (*cum laude)*; J.D., Georgia Washington University, 1978 (with Honors). *George Washington Law Review*, 1976–1978. Member, State Bars of Georgia, Florida, South Carolina, and the District of Columbia.

DOUGLAS L. TABELING. B.A., Economics and Political Science, University of Kentucky, 2000 (with Honors), Otis A. Singletary Scholar, Honors Program; J.D., University of Kentucky, 2006; Editor in Chief, Journal of Natural Resources & Environmental Law, 2005–2006; Honor Council; Law Clerk, United States Attorney's Office for the Eastern District of Kentucky, 2004–2006.

JAMES B. TAYLOR. B.A., Technical Writing, University of Tennessee, 2001; J.D., University of Tennessee (*magna cum laude*), 2004; Phi Eta Sigma; Tennessee Law Review; College of Law Scholar.

JOHN S. TOBEY. B.A., English, Emory University, 2002 (with Honors); J.D., University of Texas School of Law, 2007 (with Honors); Universidad de Salamanca, Spain, Cursos Internacionales, 2001; University College, University of Oxford, 2001; University of College London, Faculty of Laws, 2006.

G. SCOTT WALTERS. B.C.E. (Bachelor of Civil Engineering), Georgia Institute of Technology, 1985; Chi Epsilon Honor Fraternity, 1984–85; J.D., Widener University School of Law, 1996; External Managing Editor, *The Widener Law Symposium Journal,* 1995–1996. Professional Service Memberships: American Bar Association, Forum on the Construction Industry, Section of Environment, Energy and Resources; State Bar of Georgia: Environmental Law Section; Air and Waste Management Association-Georgia Chapter; State Bars of Georgia and the District of Columbia.

GEORGE D. WENICK. B.S., University of Pittsburgh, 1972; J.D., University of Pittsburgh, 1975. Senior Claims Attorney, Office of Chief Counsel, Pennsylvania Department of Transportation, 1975–1984. Member, American Bar Association, Section on Litigation, Section of Public Contract Law; Forum Committee on the Construction Industry; State Bars of Georgia and Pennsylvania.

JONATHAN Y. YI. B.A., Anthropology with minor in English, University of Florida, 2001 (Highest Honors); J.D., University of Chicago Law School, 2004. Member: State Bars of Georgia, Florida, and New Jersey.

REGINALD M. JONES. (See Washington D.C. office)

CHARLOTTE OFFICE

RYAN L. BEAVER. B.A., History and Philosophy, Wake Forest University, 2003 (*cum laude*); J.D., Wake Forest University School of Law, 2006.

MATTHEW E. COX. B.S., Brigham Young University, 1996; Mu Sigma Rho (Statistical National Honor Society); J.D., University of South Carolina, 1999; Recipient, 1999 SCTLA Mock Trial Competition, 1st Place; Law Clerk to the Honorable J. Derham Cole, 7th Judicial Circuit, State of South Carolina, 1999-2000; Member of South Carolina, North Carolina and American Bar Associations (Construction Litigation Section).

H. LEE FALLS, III. B.A., Duke University, 2003 (Dean's List with Distinction); J.D., Duke University School of Law, 2006.

ROBERT J. GREENE. B.A., University of Utah (*cum laude*), 1973; J.D., University of Utah, 1978. Member; American Bar Association (Construction Industry Forum) and North Carolina Bar Association (Construction Law Section and Litigation Section), former member of National Panel of Arbitrators of the American Arbitration Association; Certified Mediator for North Carolina Superior Court; founding member, The Arbitration Group of the Carolinas.

MARK S. WIERMAN. B.S., Naval Architecture, United States Naval Academy, 1988 (*cum laude*); M.B.A., Wake Forest, 1995 (with distinction); J.D., University of North Carolina, 2007.

TALLAHASSEE/FT. LAUDERDALE OFFICES

MAURA ANDERSON. B.B.A., University of Notre Dame, 1991; M.A.L.S., Valparaiso University, 1992; J.D., Emory University School of Law, 1996. Member of the American Bar Association, Atlanta Bar Association.

F. ALAN CUMMINGS: B.S., Auburn University; Alabama; J.D., Florida State University School of Law. During his three decades of practice, he has practiced in the area of contract litigation and counseling and has handled all aspects of construction disputes from preconstruction and post-award bid disputes to construction litigation on behalf of contractors, owners, subcontractors, design professionals and sureties. He has extensive experience in litigating delay claims on major construction projects throughout the Southeast. He also has authored numerous articles and is a frequent lecturer on the subjects of risk avoidance and construction claims preparation.

XAVIER A. FRANCO. B.S., Psychology, University of Florida, 2002; J.D., University of Miami School of Law, 2006. Professional Service Memberships: State Bar of Florida.

JASON D. KATZ. B.S., University of Virginia, 2001, J.D., Nova Southeastern University, 2005 (*summa cum laude*); *Law Review*, 2003–2005; *Moot Court*, 2003–2005.

S. ELYSHA LUKEN. B.A., University of Florida, 1996; J.D., *(magna cum laude)* Florida State University, 1999. Legislative Editor and Editorial Board Member, Florida State University Law Review. Member, American Bar Association and the State Bar of Florida.

LISA F. MILLER. B.A., University of Florida, 1999. J.D., University of Florida College of Law, 2002. Member of the State Bar of Florida.

LENNY N. ORTIZ. B.S., Florida International University, Miami, FL, major in Criminal Justice. J.D., Florida State University, 2001; Member of the Moot Court Team and Journal of Land Use and Environmental Law. Staff Attorney at the Florida Supreme Court, 2001–2002. Professional Service Memberships: American Bar Association; Broward County Bar Association; Broward County Bar—Construction Law Committee; Florida Bar—Construction Law Committee.

BRIAN A. WOLF. University of Florida, B.S.B.A. 1990. J.D. Stetson University College of Law 1993; Stetson Law Review; Intern, Honorable Judge Elizabeth A. Kovachevich, United States District Court M.D. Florida. Member: Construction Law Committee, Florida Bar R PPTL Section; Construction Law and Government Contracts Sections, Associated Builders & Contractors. Co-Author *Florida Lienlaw Online*, Construction Publications, Inc. Lecturer on various aspects of construction law and construction claims.

LAS VEGAS OFFICE

J. WILLIAM EBERT. B.S. Biology, University of Utah, 1976; J.D., University of Utah College of Law, 1980. Professional service memberships: Clark County Bar Association; The Supreme Court of the United States, The United States Court of Appeals, and Ninth and Tenth Circuits.

EVANGELIN LEE. A.B., University of Michigan, 2003; J.D., William S. Boyd School of Law, University of Nevada, 2006. Fluent in Korean and speaks conversational Japanese.

WASHINGTON, D.C. OFFICE

DOUGLAS P. HIBSHMAN. B.A., Anthropology and History, Bloomsburg University of Pennsylvania, 1993 (*magna cum laude*); J.D., University of Cincinnati College of Law, 1998; LL.M. Taxation, Georgetown University Law Center, 2006. Law Clerk to the Honorable Mary Ellen Coster Williams, U.S. Court of Federal Claims, 2007–2008. Major, U.S. Marine Corps, Judge Advocate, 1999–2007: Department Head, Judge Advocate Division, Headquarters Marine Corps, 2006–2007; Administrative Law Attorney, Judge Advocate Division, Headquarters Marine Corps, 2003–2006; Military Prosecutor, MCRD San Diego, 2001–2003. Professional Service Memberships: State Bars of the District of Columbia and Ohio, Bar of the U.S. Court of Federal Claims.

REGINALD M. JONES. B.A, College of William & Mary, 1991. J.D., University of Georgia, 1999 (*cum laude*); Editorial Board, *Georgia Law Review*, Omicron Delta Kappa; Captain, United States Army; Author, *Lost Productivity: Claims for the Cumulative Impact of Multiple Change Orders*, 31 Public Contract Law Journal 1 (Fall 2001); Author, *Update on Proving and Pricing Inefficiency Claims*, 23 The Construction Lawyer 19 (Summer 2003); Lecturer on Construction Delay, Acceleration & Inefficiency Claims for Federal Publications Seminars, LLC.

STEVEN L. REED. B.S., Physical Geography (emphasis in mathematics), University of Georgia, 1972 (*magna cum laude*); J.D., University of Georgia School of Law, 1977. Administrative Judge/Mediator, (Ret.), Armed Services Board of Contract Appeals, 2000–2006; Administrative Judge/Mediator/Hearing Examiner, U.S. Armed Corps of Engineers Board of Contract Appeals, 1988–2000; Division Trial Attorney, U.S. Army Engineer Division, Pacific Ocean, 1985–88; Division Trial Attorney, U.S. Army Engineer Division, Europe, 1983–1985; Contracts Trial Attorney, U.S. Army Engineer Division, Pacific Ocean, 1981–83; Litigator, U.S. Army Judge Advocate General's Corps, 1978; State Prosecutor, Atlanta Metro Area, 1978. Colonel, U.S. Army Judge Advocate General's Corps (Reserve) (Ret.), 2002; Legion of Merit, 2002; Chief Senior Military Judge, Senior Military Judge, Military Judge, U.S. Army Reserve, 1996–2002; U.S. Army Field Artillery Officer, 1972–1975. Professional Service Memberships: State Bar of Georgia; Association for Conflict Resolution; Board of Contract Appeals Judges Association; Board of Contract Appeals Bar Association.

S. GREGORY JOY. (See Atlanta Office)

THOMAS J. KELLEHER, JR. (See Atlanta Office)

1

THE LEGAL CONTEXT OF CONSTRUCTION

I. INTRODUCTION

Construction projects are complex and multifaceted. Likewise, the law governing construction is complex and multifaceted. Aside from questions of what one can do in construction, there looms also the question of what one may do—that is to say, what the law of construction allows. So what is the law of construction? What factors influence the evolution of construction law?

For practical purposes, the law applicable to construction projects falls into three major categories: contract, tort, and statutory/regulatory. Contract law may seem intuitively logical, at least on the surface. Tort law may not seem logical in application, but it is an omnipresent influence on any construction project. Statutory or regulatory law generally applies to construction simply because some governing body has said it should, whether the application is logical or not. This book discusses in detail these legal bases of construction law. In this first chapter, each theory is introduced in concept.

II. CONTRACT LAW

Contracts are the threads from which the fabric of commerce is woven. A contract may be as simple as an agreement to pay for food ordered in a restaurant, or so complicated that no legion of lawyers could hope to decipher the real intent, or somewhere in between. Whatever their character, contracts govern the transactions that permeate our existence.

Contracts and contract law dominate construction. What is a contract, and what is contract law? A contract has traditionally been defined as "a promise or set of promises, for the breach of which the law gives a remedy, or the performance of which the law in some way recognizes as a duty."[1] Thus, a contract is basically a set of promises made by one party to another party, and vice versa. In the United States, contract law reflects both the common law of contracts, as set forth in court decisions, and statutory law governing the terms of certain transactions. An example of the latter is the Uniform Commercial Code (UCC), adopted by every state except Louisiana. See **Chapter 6.**

Parties with capacity to contract may generally agree to whatever they wish, as long as their agreements do not run afoul of some legal authority or public policy. Consequently, an owner and a contractor generally are free to agree to an allocation of risks in the context of a construction project,[2] but they may not agree to gamble on the project's outcome. The former agreement reflects a policy of freedom of contract; the latter violates public policy prohibitions on certain gambling transactions.

III. BREACH OF CONTRACT

A "breach of contract" results when one party fails in some respect to do what that party has agreed to do, without excuse or justification.[3] For example, a contractor's failure to use the specified trim paint color, or its failure to complete the work on time, constitutes a breach of contract. An owner may likewise breach its contract obligations. Many contracts expressly provide, for example, that the owner will make periodic payments to the contractor as portions of the work are completed. If the owner unjustifiably fails to make these payments, this failure constitutes a breach. Similarly, an owner may be held in breach for failing to meet other nonfinancial contractual obligations, such as timely review and return of shop drawings and submittals. In short, any failure to live up to the promises that comprise the contract is a breach.

Whenever there is a breach of contract, the injured party has a legal right to seek and recover damages. In addition, if there has been a serious and "material" breach—that is, a breach that, in essence, destroys the basis of the parties' agreement—the injured party is justified in treating the contract as ended.[4]

IV. IMPLIED CONTRACT OBLIGATIONS

Express contract obligations are those that are spelled out in the agreement or contract. Less obvious than the express duties under a contract, but just as important, are those obligations that are implied in every contract. Examples of these duties include the obligations of good faith and cooperation (See **Chapter 5**).

[1] Samuel Williston and Richard Lord, Williston on Contracts § 1:1 (4th ed. 2007).
[2] *See Interstate Contracting Corp. v. City of Dallas, Tex.*, 407 F.3d 708 (5th Cir. 2005) (involving contract between city and contractor for execution of storm water detention tanks. The contract shifted the risk of defective plans to contractor. The court held that city contract was valid under Texas law because the contract language expressly and unambiguously placed that risk on the contractor).
[3] *See* Restatement (Second) of Contracts § 235.
[4] *See generally* 17 Am. Jur. 2d *Contracts* § 606 (2008).

In the context of a construction project, one of the most important of these implied duties is the obligation that each of the contracting parties cooperate with the other party's performance.[5] The fact that this obligation is implied rather than express is not reflective either of its importance or of the frequency with which it forms the basis for breach actions. Rather, the obligation to cooperate forms the very foundation of the agreement between the parties.

The implied obligations to coordinate and cooperate are reciprocal and apply equally to all contracting parties. By way of illustration, an owner owes a contractor an obligation to allow the contractor access to the site in order to perform its work; a prime contractor has a similar duty not to hinder or delay the work of its own subcontractors; and one prime contractor is obligated not to delay or disrupt the activities of other parallel prime contractors to the detriment of the owner. Each example demonstrates that a contracting party owes an obligation of cooperation to the other party, whether owner-contractor or contractor-subcontractor. See **Chapter 5 and Chapter 8.**

In addition to the obligation of cooperation, the owner and the contractor have other implied obligations, such as warranty responsibilities. The owner's implied warranty of the adequacy of plans and specifications furnished by the owner is of great importance to the contractor, and the breach of this warranty forms the basis for a large portion of contractor claims. The existence of an implied warranty in connection with owner-furnished plans and specifications was recognized in *United States v. Spearin.*[6] The so-called *Spearin* doctrine has become well established in virtually every American jurisdiction that has considered the question of who must bear responsibility for the results of defective, inaccurate, or incomplete plans and specifications. In layman's language, the doctrine states that when an owner supplies the plans and specifications for a construction project, the contractor cannot be held liable for an unsatisfactory final result attributable solely to defects or inadequacies in those plans and specifications. The key in this situation is the allocation of the risk of the inadequacies of the design to the contracting party, which furnished the design. Thus, in a design-build project, the design-build contractor, not the owner, may bear the risk for a design error or deficiency. See **Chapters 2, 5, and 9.**

Similarly, contractors have other warranty responsibilities with regard to the results of their performance. For example, when the owner of a newly built structure or the purchaser of construction-related goods or services is justifiably dissatisfied with the facility, goods, or services, the owner may have a cause of action against the general contractor based on a breach of construction warranties. The nature of that action and the remedies available to the owner would, in large measure, depend on the provisions of the contract.

Even where there are no express warranties in the contract, most courts, under applicable state law, will imply a warranty for workmanship and materials, provided there is no contract provision to the contrary.[7] This implied warranty may, in some

[5] *See* 73 Samuel Williston and Richard Lord, Williston on Contracts § 63:22 (4th ed. 2007).
[6] 248 U.S. 132 (1918).
[7] 13 Am. Jur. 2d *Building, etc. Contracts* § 10 (2008).

instances, be operative regardless of the presence in the contract of express warranties of limited duration that would appear to restrict the scope of the contractor's warranty liability.

V. EVOLUTION OF CONSTRUCTION LAW

The basic concept of contractual rights and duties and the legal principles governing the interpretation and enforcement of contracts have been established for centuries in English and American common law.[8] The fact that the common law of contracts is grounded in the precedent of prior decisions helps provide certainty in commercial transactions and may suggest that construction law does not evolve. The latter is not correct. Rather, over the last century, there have been significant developments that have materially altered the allocation of rights and responsibilities of the parties and the construction process. This evolution is driven by a combination of factors including practices and principles developed in the context of federal government contracting, the widespread adoption of standard-form contracts, and a shift from the traditional design-bid-build project delivery system.

Since World War II, the federal government has been the largest single buyer or construction services in the world. The procurement and administration of construction projects involve contracts that employ relatively standard terms and conditions. An array of administrative boards of contract appeals (boards) and special courts have operated for decades for the sole purpose of resolving disputes on federal contracts. These tribunals have generated a tremendous number of decisions that collectively provide the single largest body of law in the area of construction disputes. Many of the decisions noted in this book, which address substantive principles of construction law, involve the resolution of disputes and claims on federal government construction contracts. Numerous fundamental principles of construction law have their genesis in federal government construction contracts. It is impractical to speak of modern American construction law without the consideration of federal procurement law, and the discussion of that law is interlaced in each of the substantive chapters of this book. For example, the concept of a termination for convenience clause is clearly rooted in federal government contracts. While relatively unknown in nonfederal construction contracts 50 years ago, it has become relatively common place in standard form commercial contracts today.[9] The precedents developed in the context of federal government contracts provide examples of contract provisions and the law interpreting those provisions. Even if those precedents are not binding authority on other courts or tribunals, the analysis can be and often does serve as persuasive authority.

Similarly, the shift away from the traditional design-bid-build delivery system, which began in the second half of the twentieth century, has stimulated the adoption of contract forms that blur and often blend the role of designer and contractor.[10] This trend

[8]E. ALLAN FARNSWORTH, CONTRACTS, § 1:4–1:7 (3d ed., Aspen L. & Bus. 1999).
[9]*See e.g.,* ConsensusDOCS 200, ¶ 11.4 (2007 ed.) and AIA A201, §14.4 (2007 ed.)
[10]*See* **Chapter 7,** "Authority and Responsibility of the Design Professional."

has resulted in the development of new standard form agreements that seek to allocate duties and risks consistent with the parties' evolving roles. Even if the parties' roles and duties are substantially altered, however, the legal analysis of those duties may still seek to follow and apply basic principles of the common law of contracts.[11]

VI. TORTS

A tort is a "civil wrong, other than a breach of contract, for which the court will provide a remedy in the form of an action for damages."[12] A tort is not the same thing as a crime, although the two areas overlap to some extent. A crime is an offense against the public at large and is prosecuted by public officials. In contrast, a tort action is a civil action commenced and maintained by the injured party itself. Such an action seeks to recover compensatory and even punitive damages from the wrongdoer (tort-feasor).

The law of torts changes as societal norms change, because one important consideration is: Who must bear the burden of a loss? Torts are generally divided into three basic categories: intentional torts, negligent torts, and strict liability.

A. Intentional Torts

An "intentional" tort is just what the name suggests: a tort where the wrongdoer either expressly or by implication intended the act that resulted in the injury. Assault and battery falls into this category, although this tort rarely occurs in a construction dispute. Fraud and misrepresentation, by contrast, may appear in a construction case, as with a claim that the contract documents "misrepresented" some material fact or condition on which a contract was based.

In *Sherman R. Smoot Co. v. Ohio Department of Administrative Services,*[13] the Court of Appeals of Ohio discussed the application of the theory of misrepresentation in the context of a changed condition claim. The Ohio Department of Administrative Services supplied the contractor with inaccurate information concerning the subsurface conditions permitting the use of trench footings. The court noted that the contractor introduced uncontroverted evidence that soil borings, like the plans, indicated that the subsurface conditions would permit the use of trench footings. Yet the contractor eventually found the subsurface conditions to be unsuitable for trench footings. When the contractor asserted a changed condition claim on the basis of the misrepresentation in owner provided soil borings, the department contended that the contractor was not entitled to rely on the soil borings because the contract stated that they were for information purposes only. The court found, however, that it was apparent that the state intended the contractor to use the soil borings in preparation of its bid such that the contractor was entitled to rely on them.

[11] *See, e.g., United Excel Corp.*, VABCA No. 6937, 04-1 BCA ¶ 32,485 (traditional contract interpretation principles related to the resolution of ambiguities in the specifications applied to a design-build project).
[12] W. Page Keeton, Prosser and Keeton on the Law of Torts, § 1 (5th ed. 1984).
[13] 136 Ohio App. 3d 166, 736 N.E.2d 69 (2000).

Another example of an intentional tort is "conversion." Conversion is the interference with the property of another to the extent that the wrongdoer deprives the owner of possession. When the wrongdoer sufficiently deprives the rightful owner of his or her property, a court might remedy the harm by assessing damages in an amount equal to the fair market value of the property at the time of the conversion. Conversion may come into a construction project where a project participant alleged to have received but not distributed (i.e., kept to itself) contract proceeds for the work of a lower-tier contracting party. Economic relations are entitled to protection against unreasonable interference. Interference with economic relations, which has been regarded by the courts as a separate tort—and which is of particular interest to contractors—is inducement to breach of contract, or interference with contract rights.

The area where this tort comes into play most often is in labor disputes. Interference with contract rights was, at one time, a fertile field for labor union liability. The common law quite strictly curtailed union activities (such as secondary boycotts and picketing) that prevented the performance of existing contracts. Over the years, however, the tort liability of labor unions has been radically affected by federal legislation affecting industries involved in interstate commerce. The existence of a contract, in itself, is no longer the exclusive consideration. The effect of labor legislation (such as the Norris-LaGuardia Act, the National Labor Relations Act, the Wagner Act, and the Taft-Hartly Act) on common law liability of unions is beyond the scope of this section but should be analyzed in connection with the effects of strikes and boycotts on a contractor's performance.

B. Negligence

Negligence in the popular sense is the lack of due diligence or care.[14] It is a second branch of the law of torts, and may be distinguished from intentional torts by the fact that no specific intent need be proven for the imposition of liability. The traditional elements of a claim for negligence are:

(1) A duty, or obligation, recognized by the law, requiring the actor to conform to a certain standard of conduct, for the protection of others against unreasonable risks.

(2) A failure on its part to conform to the standard required.

(3) A reasonably close causal connection between the conduct and the resulting injury. This is what is commonly known as legal cause or proximate cause.

(4) Actual loss or damage resulting to the interests of another.[15] Typically, these would be the claimant's economic or monetary damages.

[14] *See generally* W. Page Keeton, Prosser and Keeton on the Law of Torts, § 30 (5th ed. 1984).

[15] *Velez Constr. Corp. v. United States*, 777 F.2d 1550 (Fed. Cir. 1985); *Leviton Mfg. Co., Inc. v. Fireman's Fund Ins. Co.*, 226 Fed. App. 671 (9th Cir. 2007); W. Page Keeton Prosser And Keeton on The Law of Torts, § 30, 165 (5th ed. 1984).

Negligence as a theory of recovery against the project architect or engineer (design professional) is important for contractors because their lack of contractual privity with the design professional makes a breach of contract action impossible. Third-party beneficiary arguments (i.e., where the contractor asserts that it is an intended beneficiary of the contract between the owner and the design professional, and is thus entitled to recover damages for negligent breach of contract) are accepted in some jurisdictions;[16] however, traditional negligence arguments probably stand a greater chance for success.[17]

The standard of conduct applied to the design professional is not that of the reasonable person but rather that of the "reasonable design professional." The usual requirement is that the design professional must have the skill and learning commonly possessed by members, in good standing, of its profession. It will be liable for the harmful results if it does not meet that standard.[18]

This duty of competence is owed to anyone, including the contractor, who could foreseeably be injured by the design professional's negligence. It is not dependent on any contractual relationship between the parties. The duty of competency also extends to both the design and supervisory functions of the design professional (to the extent that it is obligated by the contract with the owner to function in a supervisory capacity).

The contractor, like the design professional, may be held liable for negligence that results in injury to third parties—whether that negligence is attributable to unsafe construction methods while the work is in process or to defects in the completed structure. According to 13 AM. JUR. 2D *Contracts* § 139 (2008), during the course of construction, a contractor may be liable for its negligence that results in personal injury or property damage to persons rightfully on the premises, occupants of adjacent premises, and persons lawfully using a street or highway abutting the construction site. Subcontractors, likewise, are liable to third parties where their negligence results in personal injury or property damage. This liability extends to the owner when a subcontractor negligently damages the building during construction, even though there is no contractual privity between the owner and the subcontractor.

C. Strict Liability

Strict liability is liability without regard to fault. Contractors generally encounter strict liability when they become involved in highly dangerous activities, such as blasting or demolition. In general, if a party is aware of the abnormally dangerous condition or activity, and has voluntarily engaged in or permitted it, that party accepts that it will be liable for resulting damage even though it has taken every reasonable precaution, was not negligent, or was not at fault in any moral sense.[19]

[16]*See COAC, Inc. v. Kennedy Eng'rs*, 67 Cal. App. 3d 916, 136 Cal. Rptr. 890 (1st Dist. 1977) (California has been liberal in granting third-party beneficiary status in such contexts); *Shaw Constructors v. ICF Kaiser Eng'rs, Inc.*, 395 F.3d 533 (5th Cir. 2004) (holding that owner was third-party beneficiary).
[17]*Associated Indus. Contractors, Inc. v. Fleming Eng'g, Inc.*, 162 N.C. App. 405, 590 S.E.2d 866 (N.C. App. 2004); W. PAGE KEETON, PROSSER AND KEETON ON THE LAW OF TORTS, § 30, 167 (5th ed. 1984).
[18]*Associated Indus. Contractors, Inc. v. Fleming Eng'g, Inc.*, 162 N.C. App 405., 590 S.E.2d 866 (N.C. App. 2004). W. PAGE KEETON, PROSSER AND KEETON ON THE LAW OF TORTS, § 30, 161 & 185 (5th ed. 1984).
[19]*Hilltop Nyack Corp. v. TRMI Holdings, Inc.*, 694 N.Y.S.2d 717 (N.Y. App. Div. 1999). 57 AM. JUR. 2D *Negligence* § 377 (2007).

VII. STATUTORY AND REGULATORY LAWS AFFECTING CONSTRUCTION

The contract and tort bodies of law just discussed trace their origins to the historical dealings of peoples in organized society. They reflect, rather than dictate, the customs, values, and expectations of society's members. As society evolves, so do the customs, values, and expectations of its members, followed in short order by the development and evolution of the common law of contracts and of torts. This evolution occurs principally through the mechanisms of court decisions of actual cases and controversies between members of society. In short, contract and tort law reflect society.

Statutory and regulatory law, by contrast, does not necessarily evolve from any aspect of society. These laws become laws because some governing body with authority to do so declares that they should be laws. This is not to suggest that statutes and regulations are less significant or valuable than common law concepts of contract and tort; it is merely to observe that statutes and regulations derive from a different source and in a different manner. These statutes and regulations can materially affect how business is conducted in the construction industry. For example, the statutes and implementing regulations establishing ethical standards of conduct in business transactions and setting forth severe sanctions for practices such as false or inflated claims, kickbacks, collusive or bidding schemes, and the like, that deviate from the expected standards of ethical conduct have become commonplace in the context of public construction projects. These statutes provide a context for the measurement of ethical business conduct. With the adoption of the United States Sentencing Guidelines[20] and the Sarbanes-Oxley Act,[21] similar standards came to be applied to business conduct in general. Given the expense associated with defending against a claim that one of these statutes has been violated, prudent contractors should devote senior management attention to developing and implementing effective compliance plans with these standards of conduct.

Anyone familiar with construction and the construction process can name numerous statutes and regulations and may at some time have dealt with these statutes, such as:

- Licensing statutes for designers and contractors
- Statutes governing qualifications to conduct business in a particular jurisdiction
- Building codes
- Regulations governing the issuance of building permits
- Environmental laws and regulations
- Regulations governing the public procurement process, such as the Federal Acquisition Regulation and comparable state and local government contracting procedures and regulations
- Statutes governing workplace safety

[20]*See* False Claims Act (Federal), 31 U.S.C. § 3729, *et seq.*; California False Claims Act, CAL. GOV'T CODE § 12650, *et seq.*; Florida False Claims Act, FLA. STAT. § 68.081 *et seq.*; Illinois, 740 Ill. COMP. STAT. ANN. 175/1, *et seq.*; VA. CODE ANN. §§ 8.01–2.16.1, *et seq.*

[21]Sarbanes-Oxley Act of 2002, Pub. L. No. 107–204, 116 Stat. 7456.

- Statutes governing wages and benefits paid to construction workers
- Statutes governing social policies such as equal opportunity laws
- Statutes governing labor relations in construction
- Statutes governing the concept of prompt payment on public and even nonpublic projects
- Insurance statutes such as workers' compensation laws and regulations
- Lien and bond statutes
- Statutes addressing the rights and remedies available to owners of certain properties, such as condominiums
- Bankruptcy laws
- Dispute resolution procedures, such as those that prescribe the disputes process on contracts with the United States

The list could go on and on. This book, in subsequent chapters, will address each of these topics, which frequently present themselves in construction projects. In this effort, neither this book nor any book can be exhaustive; hence, it is imperative that those involved in the construction process obtain competent and timely counsel and advice about construction-related legal issues pertinent to their particular project. The objective here is simply to give the reader an overview of some of the more commonly occurrring legal topics, as well as the interpretation of principles governing the duties and obligations of the parties to a construction project.

POINTS TO REMEMBER

- The rights and obligations of the parties to a construction project are, in general, first defined by their contracts.
- In general, parties are free to contractually allocate the risks, duties, and obligations associated with a construction project, so long as the parties do not violate the law or public policy.
- An understanding of any contract often requires analysis of the written terms and conditions as well as the application of the common law of contracts and any statutes governing that particular transaction.
- Contract obligations and duties are also subject to implied obligations, such as the duties of good faith and cooperation.
- In addition to contract law principles, the conduct of the parties to a construction project will also be evaluated in light of the applicable law of torts.
- Tort liability can be based on conduct amounting to an intentional tort, negligence (breach of the applicable standard of care), or strict liability reflecting public policy.
- In every jurisdiction, the construction project and the parties' obligations are also subject to myriad federal, state, and local laws and regulations.

2

ALTERNATIVE CONTRACTING METHODS

Construction projects have traditionally been designed, bid, built, and paid for (collectively, "delivered") within a framework of strictly defined roles, relationships, and procedures. The traditional structure for construction contracting has proved effective overall, but perceived weaknesses or opportunities for improvement have led to consideration and use of alternative delivery methods. These alternative methods include multiprime contracting, construction management contracting, design-build contracting, and engineer-procure-construct contracting (EPC). Although these methods have provided many advantages, they raise questions about the altered roles, relationships, and procedures involved in these options.

Under some of these alternative approaches to construction, the classic relationship between the general contractor and its subcontractors can be dramatically changed or eliminated altogether. Although the subcontractor's actual work in the field may be precisely the same as under the traditional approach, the subcontractor's relationship with the owner, the design professional, and other subcontractors may be fundamentally changed. Indeed, the term "subcontractor" may no longer apply. A subcontractor in the traditional model may become a trade contractor under a construction management approach or a parallel prime contractor under a multiprime contracting approach. Fast-track construction introduces additional uncertainties and ambiguities about the roles of subcontractors and the rights and responsibilities of various parties in the altered contractual structures.

The manner in which these alternative contracting methods diverge from traditional practices and roles requires careful attention, to make certain that the advantages sought through their employment are not lost through unanticipated problems and disputes. Despite significant use of these alternative methods, questions remain about how they affect otherwise established theories of liability among project participants. Consequently, it is difficult and perhaps dangerous to make generalizations about the impact of the alternative contracting methods on the role of the project participants.

I. TRADITIONAL APPROACH TO CONSTRUCTION: ADVANTAGES AND DISADVANTAGES

The traditional project delivery method, like the family minivan, is dependable, practical, and familiar. Construction industry professionals are familiar with the traditional construction contract structure and the roles of the parties under that structure: owner, design professional, independent general contractor, and subcontractors. In the traditional structure, design and construction proceed sequentially, with construction commencing only after the design is complete.

The process is initiated by the owner's recognition of a need or an opportunity for construction. The owner then contracts with a design professional to transform the owner's general concept ultimately into a complete set of plans and specifications for the entire project—from site preparation to finishes. The prepared plans and specifications are then used to solicit bids or proposals from general contractors, who rely on the scope of work defined by the plans and specifications as the basis for their pricing and to solicit subcontractor quotes. Ultimately the construction contract is awarded to one general contractor, usually the one with the lowest price. Typically, the general contractor then procures subcontractors for those divisions of the work that the general contractor will not perform with its own forces. After the owner makes the site available and issues a notice to proceed, construction starts. The general contractor's work is financed by regular monthly progress payments from the owner. The designer generally maintains a review and inspection role throughout construction, but the owner continues to maintain control over both the design professional and the general contractor.

The traditional approach and sequencing for construction are reflected in and reinforced by industry customs and practices, by statutes, and by standard contract documents. Although not welcome, the problems that arise are fairly predictable and can be resolved through established procedures and remedies. It is well established, for example, that the owner will generally be liable to the general contractor for additional costs associated with defects in the project plans and specifications. The subcontractors are responsible to the general contractor for their work but also look to the general contractor, or through it to the owner, for resolution of problems. The familiarity and predictability arising from long use of this traditional approach to construction has generated standard procedures for obtaining insurance and bonding, dealing with unexpected conditions, making changes, and generally resolving unforeseen contingencies. A well-defined model of rights, duties, and remedies is in place.

The traditional mode affords many advantages. It provides the owner with a complete design and a stated maximum price before construction begins. Also, the owner maintains exclusive control over the design professional and the contractor throughout construction. The traditional approach to construction has generally proven reliable and satisfactory.

Conversely, there are certain disadvantages to the traditional approach. The sequence of completing the design before construction arguably is not the most effective use of time and money. Waiting for a complete design before obtaining price commitments from contractors or starting any construction may expose the owner to inflation and delayed occupancy. That long lead time also denies the owner the ability to react quickly to changing market conditions and revised needs. Use of completed plans and specifications to generate competition solely on a price basis may also be counterproductive. The practice tends to encourage contractors to employ the lowest acceptable standards, and it frequently generates disputes as to what is acceptable under the plans and specifications and what is an "extra" requiring additional payment. The manner in which the traditional approach to construction relies so heavily on the owner's management of the design professional and the general contractor often presumes far greater expertise than some owners may possess.

Certainly, none of these or other perceived shortcomings are fatal flaws. The general success and effectiveness of the traditional approach to construction is evident. Nonetheless, other approaches do suggest ways around those shortcomings. As alternative approaches are tested and applied, the roles and relationships of subcontractors and others can change dramatically.

II. MULTIPRIME CONTRACTING AND FAST-TRACKING

Multiprime contracting, sometimes referred to as parallel prime contracting, differs from the traditional method of construction by displacing the general contractor. The owner no longer contracts directly with one general contractor who, in turn, subcontracts out portions of the work to various subcontractors. In multiprime contracting, the owner contracts directly with a number of specialty or trade contractors that would otherwise be the first-tier subcontractors in the traditional system. Each trade contractor is responsible directly to the owner for the performance of a discrete portion of the work.

Cost saving is frequently cited as a justification for a multiprime contracting approach. One source of savings is the elimination of the general contractor's markup and profit on a large portion of the construction costs. Further savings may be available by employing multiple prime contractors in conjunction with other alternative approaches to construction, such as fast-track construction.[1] The hope is that the compressed time frames associated with these techniques will save borrowing costs, blunt the impact of inflation on the construction budget, or allow the owner to use the project sooner.

Multiprime contracting is often employed with fast-track construction. Fast-track construction phases the design and construction so that construction can begin on preliminary items of work, such as site work, foundations, and even the structure, as design of mechanical and electrical systems, interior partitions, and finishes continues. The goal of this phasing is to constrict the time required to complete a project from commencement of design to final completion by overlapping the end of the design process with the beginning of construction.

[1]There is some ambiguity in the literature and in the trade about whether fast-track construction is the same as phased construction. This chapter uses the terms synonymously.

In many instances, however, the use of multiprime contracting is unrelated to reducing the time of construction. Instead, the sequence of design and construction may be the same as under the traditional approach, with no bidding or construction on any of the work commencing until the entire design is complete. For example, North Carolina public projects, which can be delivered under the multiprime system, must be divided into at least four categories—(1) heating, ventilation, and air conditioning, (2) plumbing, (3) electrical, and (4) general work relating to erection, construction, alteration, or repair—and all aspects of the construction proceed largely contemporaneously.[2] Regardless of the lack of any time savings, this approach to multiprime contracting is promoted by subcontractors, who emphasize the cost benefit of increased competition on multiple contracts and the elimination of the general contractor's markup and profit.

Although there are certainly benefits to be gained by the use of multiprime contracting, experience counsels caution on the part of all parties. The most significant hazard is the coordination problem that multiprime contracting seems to create. Without a general contractor with clear responsibility for overall project coordination, some other party must fill that void and provide essential management, administrative, and scheduling functions. As the party bringing the trade contractors together on the project, the owner, by implication, assumes the coordination duties and the corresponding liability, just as the general contractor does with its subcontractors. Scheduling and coordinating trade contractors on a major project is a formidable task that some owners are ill equipped to handle. The owner that lacks the requisite expertise and resources must recognize the need to pay someone else to perform such services. By employing a competent construction manager, the owner can largely offset its own lack of ability and expertise.

Unfortunately, some owners refuse to acknowledge the need for a single person or entity with the expertise and authority to provide essential coordination in a multiprime approach. Instead, these owners simply seek to shift the responsibility and liability for coordination to unwary trade contractors, who may not recognize the scope of the undertaking. Owners have succeeded in shifting coordination responsibilities by designating one of the multiprime contractors as being responsible for project management and control[3] or simply by expressly requiring the trade contractors to coordinate with each other.[4] Owners are not always successful and can be exposed to significant liability if they fail even to make an effort to coordinate.[5]

The coordination problems that accompany multiprime contracting are not limited to the private sector.[6] Until recently, North Carolina required public entities to employ multiprime contracting as the default delivery system for public building contracts. This rigid system was unsatisfactory because many qualified general contractors, who would otherwise fill the "general work" division under the North Carolina model, wanted no

[2]N.C. GEN. STAT. § 143–128(a) (2007).

[3]*Broadway Maint. Corp. v. Rutgers*, 447 A.2d 906 (N.J. 1982).

[4]*Hanberry Corp. v. State Bldg. Comm'n*, 390 So. 2d 277 (Miss. 1980).

[5]*Id.; Broadway Maint. Corp. v. Rutgers*, 447 A.2d 906 (N.J. 1982).

[6]*See Sauer, Inc. v. Danzig*, 224 F.3d 1340 (Fed. Cir. 2000) (describing the coordination challenges faced on a federal government project).

part of the coordination problems that plagued the state's public projects. In order to attract qualified general contractors to participate in public projects, North Carolina amended its public procurement law and now allows separate-prime, single-prime, dual bidding, construction management at risk, and "alternative contracting methods."[7]

The amount of litigation over the issue of responsibility for coordination in a multiprime contracting approach to construction demonstrates that there is great uncertainty about where ultimate responsibility will rest. The litigation also suggests that uncertainty on a project about who is responsible for coordination will likely breed problems with coordination. Experience also shows that even if a particular party is designated as being responsible for coordination, that party must possess a "big stick" for enforcement, usually in the form of some authority or substantial influence on key management, scheduling, and payment issues. Without such enforcement mechanisms, a project can quickly descend into chaos.

III. CONSTRUCTION MANAGEMENT

Construction management departs from the traditional model of construction contracting by replacing the general contractor with a construction manager that typically offers diverse expertise in design, construction, and management. Construction management emerged as a method of project delivery in the late 1960s, but it did not gain industrywide approval or broad use and support until the late 1970s. Although there has been a tremendous growth in the use of construction management since its inception, it appears to have leveled off in recent years.

Ironically, although construction management is used very much as a term of art, it means different things to different people. The term actually describes a broad range of services and contractual frameworks that may be applied to a particular project. Consequently, great caution must be exercised when discussing construction management, either in abstract terms or in terms of its use and impact on a particular project, to make certain everyone is clear about the specific parameters of the concept being discussed.

The role of the construction manager (CM) will be as diverse and extensive as the expertise the individual or entity brings to the project, and based on the owner's needs on a particular project. The CM may be a general contractor hired to supervise and coordinate work of specialty contractors during construction. The CM may also be a multifaceted team—supervising design; having important input into site selection, financing, and accounting; providing expertise in the areas of cost control, constructability reviews, value analysis, contract interfacing, and quality control; and serving as supervisor and coordinator of construction activities. The CM may act exclusively as the owner's agent, having responsibility without risk, or may have to assume some of the risks borne by the general contractor in the traditional model, such as guaranteeing a maximum price for the project. Within these two extremes, there are

[7]N.C. Gen. Stat. § 143–128(a1) (2007).

any number of variations or combinations of services provided, or roles filled by the CM. Because general principles regarding construction management are lacking, and because someone operating under a title and position designated as CM performs many and varied roles, it is always necessary to examine the specific contractual obligations undertaken by the CM and the owner through the contract at issue.

Generally, the CM's role appears to depend on whether the CM most closely resembles the design professional or the general contractor in traditional construction. CMs are usually either design professionals (i.e., architects or engineers) or general contractors, but their approaches to construction management differ. The key distinction is whether the risk of completing the project on time and within budget has been shifted from the owner to the CM.

A. Agency Construction Management

The definition employed when a CM is described from a design professional's perspective usually involves construction management in its purest form—the CM acts solely as the owner's agent. In this pure form, generally referred to as agency construction management, the CM takes no entrepreneurial risk for costs, timeliness, or quality of construction, and all subcontractors contract directly with the owner.

The agency CM acts as the owner's agent in supervising and coordinating all aspects of the construction project from the beginning of design to the end of construction. Since the agency CM is providing management expertise, the role of the general contractor typically is eliminated. Because the agency CM assumes no financial risk of construction, however, the owner will directly enter into contracts with the trade contractors. The agency CM may execute those contracts, but does so only as the owner's agent.

As discussed in the context of multiprime contracting, the owner's direct contractual relationship with the specialty contractors required in the agency CM approach changes the owner's relationship to the construction process by creating increased owner responsibility for coordinating and solving problems among the contractors. Although the agency CM assumes some of this responsibility, the owner ultimately may be liable to the specialty contractors if lack of coordination generates claims. The owner may pursue rights against the agency CM for such extra costs based on the agency CM's failure to coordinate, but the owner's ability to succeed in such a claim is impeded by the fact that the agency CM generally does not guarantee a maximum price for the construction; thus the owner bears the risk of cost overruns.

Contract documents produced by ConsensusDOCS and the American Institute of Architects (AIA) provide some degree of standardization of the agency CM system. Under the ConsensusDOCS series of documents,[8] an owner's separate contracts with the architect, trade contractors, and agency CM consider the agency CM's role in assisting the owner in the design and construction phases. The AIA's CM Adviser

[8]ConsensusDOCS 801 (2007 ed.); ConsensusDOCS 802 (2007 ed.); and ConsensusDOCS 803 (2007 ed.).

Family of construction management documents is based on a similar premise, where the owner has direct privity with the architect, agency CM, and contractors.[9]

B. Program Management

Construction management continues to evolve. Program management is a rapidly emerging delivery model that arose out of the agency CM model. Program management is used most frequently where an owner, such as a school board, requires multiple projects to be constructed within a certain time period. The program manager's role is tailored to the needs of the owner and, as yet, is not widely defined by standard contract forms, although the ConsensusDOCS 800 (2007 ed.) provides a standard Owner/Program Manager agreement and general conditions. The program manager may provide design services; manage budgets; assist in selecting architects, engineers, contractors, and subcontractors; monitor and oversee multiple projects; and play a significant role in planning the owner's program. The program management model benefits from the institutional knowledge, experience, and synergies that result from overseeing an entire building program.

C. Construction Manager/General Contractor

At the other extreme from the agency CM or the program manager is the CM that offers its services during the design phase and then also acts as the general contractor during the construction phase—the construction manager/general contractor (CM/GC).[10] This variation makes the CM/GC's duties, rights, and risks ultimately resemble those of a typical general contractor.

Industry forms exist for parties to use in an owner and CM/GC contracting method. For example, the ConsensusDOCS Construction Management Contracts (500 Series) provides a standard set of documents that recognizes the collaborative relationship between the owner and CM/GC through all phases of the work.[11]

The CM/GC also will become involved early in the project design. Although a design professional remains solely responsible for the design in this construction management model, the CM/GC will offer its practical construction expertise to suggest more effective approaches to the design and methods to save costs. Such services often are referred to as value engineering or constructability review.

Frequently, the CM/GC immediately begins estimating and scheduling functions, both to expedite the design process and to provide the owner and design professional with feedback about the construction cost and time implications of the evolving design. Once construction is ready to commence, the CM/GC, very much like a general contractor, will enter into fixed-priced subcontracts on its own behalf and not

[9]The AIA Construction Manager-Adviser Family includes: AIA A101 CMa (1992 ed.); AIA A201 CMa (1992 ed.); AIA B141 CMa (1992 ed.); AIA B801 CMa (1992 ed.).

[10]AIA A121 (CM 2003) and AIA A131 (CM 2003) are used where the construction manager is also the constructor.

[11]The ConsensusDOCS 500 series includes: ConsensusDOCS 500 (2007 ed.), ConsensusDOCS 510 (2007 ed.), and ConsensusDOCS 525 (2007 ed.).

on the owner's behalf. These subcontract prices, together with the CM/GC's costs of performance and markups, will be the basis of the CM/GC's guaranteed maximum price (GMP). It is these two features—that the CM/GC binds itself to subcontracts and to a GMP—which are the principal distinctions from the agency CM approach.

Although the CM/GC may be the party entering into the subcontracts, it is common for the owner to be afforded considerable involvement in selecting subcontractors. In addition, cost-reporting requirements often found in the CM/GC's contract with the owner require the subcontractors to provide more job cost information and reporting than a fixed-price contract usually requires. Finally, consistent with the CM/GC's more extensive dealings with the design professional and owner, subcontractors may perceive that the CM/GC is much more closely aligned with the owner than is the case with a general contractor.

Once construction begins, the coordination duties owed to the subcontractors by the CM/GC are virtually identical to those owed by a general contractor. Indeed, as part of its contractually defined services, the CM/GC may assume even greater coordination and scheduling responsibility than would a typical general contractor, although these additional services would primarily be for the owner's benefit.

IV. DESIGN-BUILD CONTRACTING: WHAT WORKS TO AVOID DISPUTES

A. The Design-Builder's Perspective

The "master builder" is again becoming prominent on the construction scene. An increasing number of owners, both public and private, have turned to the design-build project delivery method to fast-track the project and reduce overall project costs. Although the traditional checks and balances that come with using a separate designer and builder are sacrificed to some extent, the design-build method provides the owner with a single point of responsibility for project design and construction. In addition, the owner is relieved from responsibility for potential delays and costs associated with design errors and omissions. To this extent, the method protects the owner; but where does it leave the design-builder?

The design-builder is liable for both design problems and construction defects. The design-builder warrants the adequacy of the design and agrees that the finished project will meet certain performance specifications. At first glance, this level of risk would seem unacceptable to most contractors and design professionals, but careful project selection and definition, contract formation, and project control enable the risks to be minimized and managed.

Obviously, projects must be selected that are within the expertise of the design-builder. There are specialized design-build firms, as well as contractors and design firms owned by the same companies, that work as a team on design-build projects. A contractor without design capabilities, however, should not automatically shy away from design-build opportunities. Design-builders take a variety of forms, and often the design professional part of the organization is a subcontractor or joint venture partner of the contractor. Similarly, the design professional can retain the contractor as a subcontractor, or both the design professional and the contractor can act as subcontractors to a CM.

Regardless of the form chosen by the design-builder, the single most important step in the project is to arrive at a mutually understood and agreed definition of the project with the owner. Once the project definition, parameters, and requirements are established, the contract documents must be prepared consistent with the mutual expectations of the owner and the design-builder.[12] The design-builder can then limit its risk with contract clauses limiting or fixing damages to a specific amount—say, the amount of the design-builder's fee—or excluding certain types of damages, such as lost revenues or consequential damages. In addition, a contingency fee can be used as a component of the GMP, to be used to absorb unanticipated cost growth. Cost overruns or savings can be addressed in such a way that all parties have an incentive to ensure cost-effective results.

If the design-builder is composed of different design and construction entities, it is important that the respective roles, responsibilities, and liabilities are clearly established. Often, breakdown of the design-builder costs and fees between designer and contractor will be necessary for licensing and insurance purposes. Design professionals will want to limit their risk to the design portion of the work, for which they can obtain professional liability (i.e., errors and omissions) insurance. Likewise, the contractor will need to restrict its exposure to completion of the project in accordance with the design so it can obtain any necessary bonds. Finally, design-build team members may wish to provide for cross-indemnification of each other for any claims arising out of the other team member's work.

The ConsensusDOCS 300 (2007) Tri-Party Agreement brings the parties even closer together. This agreement recognizes a collaborative project delivery method where the owner, designer, and constructor are all parties to the same agreement. In this agreement, many of the decisions are made by a management group consisting of a representative from each of the parties. Those decisions are designated as "safe harbor" decisions, with the parties releasing each other from any liability for those decisions, as long as there is no willful default of an obligation under the agreement. Beyond the safe harbor decisions, the parties are liable for their own negligence and breaches of contract. And finally, where allowed by law, the ConsensusDOCS 300 allows both the designer and the constructor to limit their total liability under the agreement.[13]

Careful contract formation with anticipation of the possible areas of exposure allows contractors or design professionals to again assume the role of master builder. An understanding of what design-build is, how it works, and how the various parties can protect their respective interests hopefully will result hopefully in more profitable opportunities.

[12]Industry standard forms can provide a starting point for drafting or tailoring appropriate design-build contract documents. The American Institute of Architects, the Engineers Joint Contract Documents Committee, and ConsensusDOCS all have comprehensive contract documents and related forms for project management for both traditional and alternative contracting methods.

[13]ConsensusDOCS 300, ¶¶ 3.8.2, 11.5, 11.6n (2007 ed.).

Every proposed design-build project should be methodically reviewed to assess the possible risks. Issues to be addressed by the design-builder include:

(1) Selection of the owner
(2) Project definition and performance expectations
(3) Qualifications and experience of team members
(4) Contractual relationship of team members
(5) Licensing concerns
(6) Insurance and bonding
(7) Responding to the request for proposals (RFPs)
(8) Innovativeness of the proposal
(9) Flexibility of contract with owner
(10) Design review
(11) Handling of tenant or user input
(12) Scheduling
(13) Trade contractors
(14) Cost control
(15) Quality control
(16) Changes to the work
(17) Differing site conditions
(18) Environmental remediation risks
(19) Contingencies
(20) Allowances
(21) Shared savings
(22) Responsibility for cost overruns
(23) Design errors and omissions
(24) Construction defects
(25) Limitations of liability
(26) Delay damages
(27) Preventing and resolving disputes

B. The Owner's Viewpoint

The design-builder also needs to understand the owner's approach and perspective on the design-build project. To the owner, there is no better method of project delivery than design-build. Since the design-builder is responsible for both design problems and construction defects, the owner can avoid the traditional trailer battles between its design professional and its contractor. Free at last of the dreaded *Spearin* doctrine and its implied warranty of fitness of the plans and specifications,[14] the owner can relax and enjoy a viewer's perspective on its project as it moves from design through construction to completion. Finally, design-build provides a method of designing and constructing a project that permits the owner to avoid the three dreaded plagues of construction projects: changes, claims, and disputes.

The design-build method is so enticing to owners that the dollar volume of design-build projects continues to grow every year. An increasing number of public and private owners have elected to go design-build because it allows for the fast-tracking of projects without the risk of cost and time impacts due to defective or untimely completion of design elements.[15] Design-build works! Sometimes.

Design-build works when the owner knows what the desired end product is and adequately communicates that information to the design-builder. The single most important aspect of a successful design-build project is the preparation of the project's scope of work. The owner must adequately define the project. If a comparable project can be identified, the owner should specify this in the scope of work in order to give the design-builder a better idea of the owner's project definition. Although this description is too general for construction, it allows the design-builder to understand the nature of the project.

If the owner does not have in-house capability or consulting professionals, the required clarity of scope may be lacking. The owner must have preestablished and definitive design criteria identifying the project requirements before the project can evolve toward design and construction. Adequate project definition at this stage represents the best opportunity for the owner to protect itself on the project.

If the owner is going to obtain competitive proposals on a design-build project, it must establish a clear program of requirements and performance specifications. This is commonly done through an RFP. Guidelines should be established that allow an apples-to-apples comparison of the proposals received. The RFP should set out the scope of work and the criteria to be used for selection of the design-builder. It helps both the owner and design-builder if the RFP includes, at a minimum:

- The size and character of the project
- Technical scope of work
- Budget and financial considerations
- Schedule requirements
- Requirements and timing for establishing the price
- Provisions for value engineering and alternates
- Performance standards and guarantees
- Quality control/quality assurance requirements
- Operations, maintenance, and life-cycle considerations

[14]*United States v. Spearin*, 248 U.S. 132 (1918).

[15]As of October 2007, 67 design-build bills were considered in 26 states of which 21 were signed into law. *See www.dbia.org.*

- Liability, warranty, licensing, and bonding requirements
- Clarification of the consequences of nonperformance or late delivery
- Clear guidelines for selection of the successful proposer

Thoroughness in the preparation of the RFP allows the owner to define the project and to develop overall priorities in terms of spatial and system requirements, cost, design excellence, size, construction quality, schedule, and life-cycle costs. The more specifically the scope of work and priorities are set out in the RFP, the better the completed project. Evaluation criteria should be clearly set out in the RFP. If the evaluation criteria are weighted or are set up in priority order, the proposal needs to be is structured to emphasize the owner's priority items. The owner should provide a clear scope of work to the design-builder.

Inadequate or erroneous information in the RFP is one of the more common sources of disputes and can be costly to the design-builder. In *United Excel Corp.,*[16] the Veterans Affairs Board of Contract Appeals (VABCA) rejected a design-builder's claim for $120,000 that it incurred as a result of conflicting heating, ventilation, and air conditioning (HVAC) specifications in the RFP. The design-builder, United Excel Corp. (UEC), knew that some portions of the specifications called for aluminum ceiling diffusers and other portions called for more expensive stainless steel diffusers. UEC did not raise the conflict with the Department of Veterans Affairs (VA) when UEC submitted its response to the RFP.

When it came time to install diffusers, the VA insisted that UEC install the more expensive stainless steel diffusers. UEC performed the work under protest and filed a claim seeking an equitable adjustment for the difference in cost between stainless steel and aluminum diffusers. The VA argued that UEC was barred from recovery because UEC knew of, but failed to inquire about, the aluminum/stainless steel discrepancy. The VA's defense was based on the *contra proferentum* rule, where ambiguous specifications are construed against the government unless the contractor knew of the ambiguity or the ambiguity was so glaring or obvious as to be a "patent" ambiguity. A corollary of this rule is that a contractor will be barred from recovery if the contractor knew or should have known of the ambiguity but failed to inquire about it. UEC argued that the *contra proferentum* rule did not apply in the design-build context. UEC argued that the RFP drawings and specifications established only "design parameters" and UEC was therefore entitled to choose aluminum diffusers as the most economical way to achieve the design intent.

The VABCA rejected UEC's argument, holding that UEC was not relieved of its obligation to inquire about the discrepancy. The VABCA was not persuaded by UEC's design-build argument, stating that "a design build contract shifts risks to a contractor that a final design will be more costly than the bid price" and that the traditional rules of contract interpretation apply to design-build contracts.

[16]VABCA No. 6937, 04-1 BCA ¶ 32,485.

✔ DESIGN-BUILD PROJECT CHECKLIST FOR OWNERS

The prudent owner should consider the following issues in evaluating whether to use a design-build project delivery system:

- Exculpatory language and risk-shifting clauses are potentially helpful to the owner. The best way for an owner to protect itself on a design-build project, however, is through complete and precise project definition. Develop that definition with input from design professionals, construction professionals, operational personnel, maintenance personnel, tenants, and users.
- If there are no statutory restraints, pre-qualify potential design-builders and develop a short list of the most qualified teams. Establish clear guidelines for selecting the design-builder and adhere to those guidelines.
- Structure the RFP so that the would-be design-builders can understand the project definition, including all important elements of the project.
- Include the contract documents in the RFP. Inform the candidates that selection will be based on qualifications, technical quality of proposal, and responsiveness to invitation (including proposed contract documents).
- Tailor the contract documents to the particular project. Each project is unique, and this should be recognized during contract formation. Do not use a standard-form contract without modification.
- There is no justifiable reason for the design professional to disclaim its professional responsibility to the owner on a design-build project. For example, AIA Document A141 purports to extinguish any professional obligation of the designer to the owner. Modify this standard-form language to specifically provide that the owner is an intended third-party beneficiary of all contracts for design or engineering services and all subcontracts, purchase orders, and other agreements between the design-builder and third parties.
- Limit the owner's obligations under the contract. Modify the contract, if practical, to require that the design-builder obtain all permits, conduct all geotechnical testing, and perform any environmental assessments. If the owner retains responsibility for the site conditions, the old problems—extra costs and delays—caused by inaccurate information can again plague the design-build project. The design-builder is supposed to be the single point of responsibility. Any responsibilities, aside from payment, that remain with the owner tend to vitiate the desired "one-stop shopping."
- Require adherence to the contract timetable, including owner established milestones. Eliminate any standard-form language inconsistent with the design-builder's obligation to complete its work in strict accordance with the contract requirements.
- Structure payment terms so that the owner has an adequate time to verify, process, and fund any applications for payment. Establish procedures that will be used for payment and allow for any anticipated slippage in payment due to lender, grantor, or third-party involvement. Also, specify the rate of interest that will be assessed for any late payments.

- Set out the owner's termination rights, specifically establishing the owner's right to terminate the design-builder for default or at the owner's election (convenience). Limit the owner's liability in the circumstances of a termination even if a default termination is later determined to have been improper.
- Require the design-builder to include all its costs within the GMP or lump-sum price. Eliminate separate reimbursable items, contingencies, or allowances that are not included within the contract price.
- Design-build projects are scope-driven. Tailor exculpatory language to the particular project. Include a no-damages-for-delay clause for general application but also specifically tie anticipated delays to any remaining owner responsibilities. Similarly, clearly establish the design-builder's responsibility for the performance of equipment, processes, and components. Require the design-builder to verify the appropriateness of any equipment, process, or component for achieving the desired performance criteria.
- Do not allow the design-builder's proposal to become a contract document unless it contains no qualifications, no exceptions, no ambiguities, no exclusions, no limitations, and no language contrary to the contract documents. Instead, set out the specific scope of work—that is, technical specifications; drawings by drawing number and date; and other pertinent equipment, material, component, and finishes information. Too often the proposal and "killer" contract documents that have so lovingly been created will be contradictory or create ambiguities in the owner's desired contractual scheme.
- Require the design-builder to provide all insurance, including a design professional project policy with an extended discovery period. The advantage of the project policy is that it reserves coverage for that particular project, so that coverage will not be reduced by other claims or be subject to cancellation when the project is completed.
- Require the design-builder to provide sufficient guarantees of performance of the work and payment to its subcontractors. The design-builder should provide performance and payment bonds or some alternative (and acceptable) form of security to the owner.
- Set out a procedure for any project change orders. Establish a strict timetable for notice to the owner, and require contemporaneous submission of cost and time impact documentation. Control the change order process with procedures that are actually implemented.
- Do not allow the design-builder to limit its liability, disclaim guarantees or warranties, or otherwise vitiate its responsibility as the single-source responsible party. If anything, the design-builder's responsibility (and liability) should be greater than the sum of the contractor's construction responsibility and the designer's design responsibility.
- Do not meddle with the design after the GMP has been established, unless absolutely necessary. After the GMP has been established, any design modification puts the owner at risk in terms of cost and time to complete.
- Do not allow the owner's program consultant, staff personnel, or users to alter or modify the scope of work. To the extent that the program consultant requires

any change in quantity, quality, means, methods, techniques, sequences, or procedures, the owner can be liable.

- Do not provide any equipment, materials, or components. If the owner does, then to the extent late deliveries are experienced, the single-source responsibility of the design-builder is lost, and potential exposure to changes, claims, and disputes returns.
- Once the owner has the design-builder indeed performing as the single-source responsible entity, the owner should provide some incentive to the design-builder. Shared savings, with a percentage going to the design-builder, or bonuses for early completion, should be considered. In such a way, owners can demonstrate that they are being fair.

C. The Successful Design-Build Project

What works to avoid disputes on design-build projects is what works to avoid disputes on traditional projects: a fair allocation of risk, reasonable interpretation of the contract, a clear scope of work, acknowledgment of responsibility, acceptance of change, and good-faith cooperation between the parties. The design-build method of project delivery is not a panacea for the perceived ills of the construction marketplace, nor is it a substitute for adequate design and sound construction management. The design-build method is simply an alternative manner of providing the owner with a high-quality project, on time and within budget—if the project participants will commit the necessary time and resources to project definition, definitization, and actualization.

D. Design Professional Liability Issues in a Design-Build Project

In many design-build projects, the architect or engineer enters into a subcontract with the design-builder. The fact that the design professional is one step removed from the owner does not mean the design professional is off the hook for design defects. In fact, the design professional must exercise a greater deal of caution when venturing into the less settled realm of design-build.

One source of unanticipated or increased risk to the design professional is the construction contract itself. For example, under the Subcontract between the Design-Builder and Designer, issued by the Design Build Institute of America (DBIA),[17] the design professional may be held to performance standards that exceed the industry standard of care. The DBIA subcontract also requires the design professional to make site visits to "determine if the construction is proceeding in accordance with the Construction Documents."[18] If this requirement is not altered, the design professional could face liability for defective work.

[17]DBIA Document 540, ¶ 2.2.1.
[18]*Id.* at ¶ 2.7.6.

The design professional must also be cognizant of tort liability. In many jurisdictions, the design professional may be liable not only to the party with whom it contracted but to third parties (such as the owner) who could foreseeably be injured by the professional negligence. Consequently, it is possible for an owner to bring a claim directly against the design professional even where there is no privity of contract. The net result is the same as under the traditional system: The design professional may be liable both to the owner and to the builder.

Licensing laws present another source of liability for architects or engineers involved in design-build projects. Licensing laws are a simpler matter in the traditional delivery system, where the functions of the contractor and the design professional are neatly compartmentalized. In design-build, however, the design professional may wear one of many hats. For example, the architect or engineer might form a joint venture with a contractor to act as the design-builder and, in turn, subcontract its design services to the joint venture. The design professional might lead the design-build team by contracting directly with the owner and then subcontracting the construction portion of the work to one or more contractors.

Under each of these scenarios, the design professional must check the state licensing laws governing architects, engineers, contractors, and subcontractors. State laws governing the formation and authority of business organizations, such as joint ventures and professional corporations, must also be checked if the design professional intends to form a joint venture or other entity as part of the design-build team. The language of the design-build contract should also be analyzed to ensure that the design professional does not promise to perform services that would violate licensing laws.

V. DESIGN-BUILD ASPECTS OF TRADITIONAL CONSTRUCTION

Even within the traditional design-bid-build approach, contractors must be aware of the extent to which they can assume design-build responsibility through performance specifications, the shop drawing process, and clauses that impose design review responsibilities on the contractor.

A. Performance Specifications

Specifications fall into two general categories: performance specifications and design specifications. Design specifications precisely describe the work the contractor is to accomplish, including dimensions, tolerances, and materials. The design specifications are the "recipe" the contractor is required to follow in constructing the work. Performance specifications do not tell the contractor how to accomplish the result but only dictate what the result must be. This distinction is of critical importance, because design liability under performance specifications is generally allocated as it would be if the entire project were design-build. The contractor is responsible for all the costs associated with achieving the end result described in the performance specifications.

Often a specification is not exclusively a performance specification. Instead, the desired end result is described, but at least some design information is offered. When problems result from the contractor's inability to meet the performance criteria using the design data provided, liability for the design defect is unclear.[19] Liability ultimately may be allocated on the basis of which party had superior knowledge. A contractor, on its own or through some specialty subcontractor or supplier, may be deemed to have sufficient knowledge to recognize the conflict between the design outlined and the performance required, so that the defect constituted a patent defect. In such a case, the contractor is, at a minimum, required to call the defect to the attention of the owner. In *Brunson Associates, Inc.*,[20] the owner's lack of superior knowledge was a factor in holding the design-build contractor liable for a design defect that caused two fabric structures to collapse simultaneously.

In *Regan Construction Co. & Nager Electric Co.*,[21] the project specifications included performance criteria for air-handling units. At trial, the contractor proved that the only air-handling unit on the market that met the performance criteria was the one provided by the contractor. Unfortunately, that unit did not fit into the space allotted for the air-handling unit in the overall design. Thus, although the contractor could establish that the performance specification was defective within the context of the overall design, the contractor was nonetheless found liable for the extra cost associated with accommodating the unit because the contractor should have discovered the conflict earlier and called it to the owner's attention.

Contractors need to know the risk of performance specifications that may take somewhat unconventional forms. *Florida Board of Regents v. Mycon Corp.*[22] involved what appeared to be a "brand name or equal" specification for architectural concrete. Under the specifications, the contractor was to "provide a skin plate with a smooth, non-corded 'true-radius' forming surface, equal to that manufactured by Symons." The contractor used the referenced Symons system not only for the "skin plate" but also throughout the project, thinking that a Symons forming system would provide a suitable result. The owner concluded, however, that the concrete work failed to achieve the required tolerances. The contractor argued that its use of the specified Symons system was subject to the owner's implied warranty of the adequacy of the specifications so that the contractor could not be liable if the finished product did not meet specifications. The Florida Court of Appeals disagreed.

The court initially noted that if only one brand of product was specified so that the contractor had no discretion but to use that one product, then the owner's implied warranty of the specifications would apply and the contractor would be entitled to relief. Moreover, the court recognized that if there were true "or equal" language in the contract, the contractor could meet the contract by proposing a system equal to the brand specified. In this case, however, the court pointed out that the "or equal"

[19] *See, e.g., M. A. Mortenson Co.*, ASBCA No. 39978, 93-3 BCA ¶ 26,189 (owner forced to pay for extra structural concrete and steel despite contract status as design-build project, because owner provided conceptual design for project).
[20] ASBCA No. 41201, 94-2 BCA ¶ 26,936.
[21] PSBCA No. 633, 80-2 BCA ¶ 14,802.
[22] 651 So. 2d 149 (Fla. Dist. Ct. App. 1995).

references in the specifications related solely to the "skin plate" portion of the work, not to the entire steel concrete forming system that would need to be used on the project. Although the court appeared to agree that there was an implied warranty of the plans by use of the Symons system in connection with the "skin plate," the contractor nevertheless had to meet the specific tolerances for other concrete surfaces, and there was no representation in the contract that the Symons forming system would be adequate for such other work.

B. Shop Drawings

Shop drawings are an essential element of construction, bridging the gap between the design set forth in the plans and specifications and the details and specifics necessary to fabricate material and install the work in the field. There is the ever-present risk that, in translating the design to shop drawings, the contractor may intentionally or unintentionally alter the design. If that occurs and the change results in a design defect impacting construction, the contractor assumes the liability for that defective design even if the design professional generally approved the shop drawing incorporating the change.

Most construction contracts contain a shop drawing clause which states that the design professional's approval does not relieve the contractor from responsibility for complying with the plans and specifications.[23] Unless there is some basis to argue an ambiguity in the plans and specifications, this clause likely will shift responsibility to the contractor for any changes to the design via the shop drawing process, at least as between the owner, contractor, and designer.[24] The design liability for such changes can, however, be shifted back to the owner and designer if the changes are clearly identified and called to the attention of the owner or the designer that reviews and approves the shop drawing.[25] Standard contract clauses state that changes in or deviations from the plans and specifications must be specifically identified in writing as changes and deviations and must be specifically approved. The ConsensusDOCS 300 Tri-Party Agreement (2007 ed.) strikes a more even balance, requiring the constructor to conduct constructability reviews in collaboration with the designer.[26]

The numerous factual issues involved in what constitutes a change or deviation, whether it was sufficiently highlighted as such to the designers and whether there was specific acceptance and approval, are all fertile ground for disagreement and litigation. Structural steel shop drawings, including the detailing of fabrication, erection plans, and welding details, as compared to information contained in the structural drawings and specifications, generate a tremendous number of disputes because the stakes are so high and the technical issues so complex. Steel fabricators and erectors frequently feel that an unreasonable amount of design responsibility is being shifted to them, as they perceive that structural drawings omit much critical detail is necessary to fabricate and install the structural steel frame.

[23]*See* AIA A201, § 3.12.6 § 3.12.8 (2007 ed.); EJCDC C-700, § 6.17 (2007 ed.).
[24]*See, e.g., Fauss Constr. Inc. v. City of Hooper,* 249 N.W.2d 478 (Neb. 1977).
[25]*See, e.g., Montgomery Ross Fisher & H. A. Lewis,* GSBCA No. 7318, 85-2 BCA ¶ 18,108.
[26]*See* ConsensusDOCS 300, ¶ 6.15 (2007 ed.).

The extent of the problems and pressures described by steel fabricators is not imagined. The seriousness of the problem is highlighted by the infamous 1981 Kansas City Hyatt disaster that killed 114 people and injured 186 others when two walkways collapsed. In one of the resulting lawsuits the Missouri Court of Appeals held that the design of a structural steel connection could not be delegated to the steel fabricator. To have done so was grounds for revoking the license of the professional engineer, who was deemed to be grossly negligent for making only a cursory review of the involved shop drawings.[27]

Another area of frequent dispute in the shop drawing process involves dimensional errors in the plans and specifications. Standard shop drawing clauses require the contractor to verify field dimensions when preparing shop drawings. A failure to verify existing field conditions can therefore transfer liability for dimensional errors on the plans from the owner to the contractor even if the owner's designer included the erroneous dimensions in the original design and approved the shop drawing that repeated the erroneous dimension.[28]

C. Secondary Design Review

Compliance with performance specifications and providing details in the shop drawing process are affirmative acts that, it is hoped. alert the contractor to potential design liability. There are other standard contract clauses, however, that seek to shift design responsibility to the contractor. These clauses are based not on the contractor's actions but on the contractor's failure to take affirmative action to identify and correct design defects. Such boilerplate language generally is unnoticed and often not even an issue. Unfortunately, if there is a major design bust and more than enough damages and blame to go around, the contractor is likely to hear arguments that assumed liability through these stealthy risk-shifting clauses.

D. The Interpretations Clause

Plans and specifications often lack all details and specifics necessary to translate the design into construction—hence the need for shop drawings and material and equipment submittals. The requirement for the contractor to detail the design for construction purposes may also require the contractor to fill in the gaps of the design. Recognizing that all details cannot be addressed, many construction contracts contain a catchall clause requiring the contractor not only to supply and construct the work specifically set forth in the plans and specifications but also to furnish all necessary labor and materials that may be "reasonably inferred" from the plans and specifications in order to achieve a complete and functional project.[29] Disputes over the coverage of this clause frequently involve responsibility for piping and control wiring and

[27] *Duncan v. Mo. Bd. for Architects, Prof'l Eng'rs & Land Surveyors,* 744 S.W.2d 524 (Mo. Ct. App. 1988).

[28] *KAM Elec. Enters.*, VABCA No. 2492, 89-1 BCA ¶ 21,558.

[29] *See* AIA A201, § 1.2.1 (2007 ed.); EJCDC C-700, § 3.01 (2007 ed.); and ConsensusDOCS 300, ¶ 3.7 (2007 ed.).

whether the contractor is responsible to provide them at no additional cost. The stakes can be considerably higher than simply the cost of the omitted detail, if the omission of the procurement or installation creates a major delay to the project or requires completed work to be torn out.

What the contractor must provide is tied to the circumstances of the project and what a reasonable contractor would do. If a reasonable contractor would have included the work in its bid, or at least should have inquired about it, then the contractor generally will be responsible for the cost of providing the work, regardless of the fact that the work was not detailed in the plans and specifications.

E. Identification of Patent Defects

Even the most carefully designed and engineered project is not going to be perfect. The plans and specifications are too voluminous to eliminate all errors. In recognition of this reality, many construction contracts attempt to impose on the contractor the responsibility to review the design and to call to the attention of the owner or design professional any errors or omissions it finds.[30] Failure to disclose such errors renders the contractor potentially liable for them. Under AIA A201, the contractor's liability extends only to those errors that the contractor knowingly fails to report. Similarly, under ConsensusDOCS 200, the contractor is not liable for errors, omissions, or inconsistencies in the contract documents unless the contractor "recognized and failed to timely report to the Owner any error, inconsistency, omission, or unsafe practice" discovered in the contract documents.[31] The ConsensusDOCS 300 Tri-Party Agreement does not relieve any party from performance in accordance with the terms of their respective standard of care.[32]

Apart from the ConsensusDOCS Tri-Party Agreement, these clauses basically shift the liability for defective design for patent or obvious design defects to the contractor, while maintaining the owner's liability for hidden defects. What is patent or obvious will depend on the specific circumstances of the project and the relative expertise of the contractor.

F. Compliance with Permits, Codes, and Regulations

The design professional must design the project in accordance with applicable building codes and regulations. In most construction contracts, the contractor also assumes a separate and additional duty to obtain necessary permits and comply with applicable building codes and regulations.[33] These clauses are intended primarily to hold the contractor liable for construction means and methods. But they can also be used to hold the contractor liable for design defects arising from conflicts with building codes. As a practical matter, however, the contractor will not assume such liability unless the design defect was patent and the contractor failed to notify the owner.

[30] *See, e.g.*, AIA A201, § 3.2.1 (2007 ed.); EJCDC C-700, § 3.03 (2007 ed.).
[31] ConsensusDOCS 200, ¶ 3.1.2 (2007 ed.).
[32] ConsensusDOCS 300, ¶ 6.15 (2007 ed.).
[33] *See, e.g.*, AIA A201, § 3.7 (2007 ed.); EJCDC C-700, § 6.09 (2007 ed.).

VI. CONTRACTOR LIABILITY ISSUES

In the traditional system, the design professional is responsible for design deficiencies. As the single point of contact for the owner, however, the design-build contractor may be required to bear not only its own costs incurred because of a defect but also those extra costs incurred by the owner or other parties arising out of the design defect.

Some owners are attracted to design-build contracts for the very reason that such contracts relieve the owner from being caught between its design professional and its contractor with respect to design disputes.[34] In addition to assuming the risks of defects and design, the design-build contractor also assumes the risk that construction will cost more than originally anticipated. As long as the performance criteria provided by the owner are not impracticable, a design-build contractor may be forced to bear the extra cost of its performance due to extended construction time.[35]

One way that a contractor performing under the traditional design-bid-build delivery method may also assume a degree of design liability is by the owner's use of performance specifications or criteria. Although performance specifications dictate the results to be achieved by the contractor, they do not tell the contractor how to accomplish the desired results. Consequently, the contractor is responsible for the costs associated with achieving the end result specified.

Conflict may exist between design and performance specifications. If the contractor recognizes (or should recognize) such a conflict among the performance specifications, the contractor must notify the owner and the design professional in order to avoid potential liability due to the conflict. For example, in *Regan Construction Co. & Nager Electric Co.,*[36] performance criteria provided by the owner for air-handling units could be met only by one particular unit on the market, which the contractor incorporated into the construction. The unit, however, did not fit into the space allotted for it by the design. As a result, the contractor incurred extra costs associated with accommodating the unit. In an action against the owner, the board of contract appeals held that the contractor assumed the risk of extra costs, reasoning that the contractor could have calculated that the specified unit would not fit in the space provided, based on the contract drawings, before it ordered the air-handling unit.[37]

Another specification issue that may give rise to contractor design liability involves the use of a "brand name or equal" specification by the owner. Use of the specified brand-name product by a contractor does not necessarily or automatically relieve the contractor of liability for noncompliance with applicable performance criteria. In other words, a design defect that results from a contractor's use of a brand name as

[34] *Mobile Hous. Env'ts v. Barton & Barton*, 432 F. Supp. 1343 (D. Colo. 1977) ("turn-key" contractor and replacement both responsible for design liability).

[35] *Appeal of Ruscon Constr. Co. Inc.*, ASBCA No. 39586, 90-2 BCA ¶ 22,768.

[36] PSBCA No. 633, 80-2 BCA ¶ 14,802.

[37] *See also Modern Cont'l S. v. Fairfax County Water Auth.*, 70 Va. Cir. 172 (Va. 2006) (firmly placing the ultimate responsibility for errors in the contract documents and the risks associated therewith on the contractor, as provided for in the contract); and *D.C. McClain, Inc. v. Arlington Co.*, 452 S.E.2d 659 (Va. 1995) (contract documents which did not allow adequate room for on-site installation of post-tensioning apparatus did not relieve the contractor of obligation to obtain easements necessary to install apparatus).

specified in a "brand name or equal" specification does not necessarily fall within the owner's implied warranty of the adequacy of the specifications. If the contractor has a choice of products, the contractor may bear the burden of compliance.[38]

As noted, one of the common duties of a design professional is to review and approve contractor submittals, including shop drawings. If shop drawings prepared by the contractor require a change to the original design, however, the contractor still may be held responsible for impacts to the overall construction caused by problems with that change even though the design professional has appoved the change.[39] It may, however, be possible for the contractor to effectively reduce the risk of this type of design liability by immediately bringing any shop drawing–related design modifications to the attention of both the owner and the design professional.

Sometimes the limits of risk shifting related to substitutions are established by statute so that an architect's approval of the contractor's suggested substitution may not relieve the contractor of some warranty obligations. In *Leisure Resorts, Inc. v. Frank J. Rooney, Inc.*,[40] a contractor suggested the substitution of an air-conditioning unit on a condominium project. The substitution was approved by the architect and the engineer. When many of the units failed to perform, several condominium unit owners filed a class-action lawsuit against the developer who, in turn, sought indemnity from the contractor in a third-party action. The court held that although developers are subject to statutory "warranties of fitness or merchantability for the purposes or uses intended," the contractor's statutory warranty is to provide work and materials that "conform with the generally accepted standards of workmanship and performance of similar work."[41] In short, although the contractor had a duty to provide acceptable materials, it did not have a duty to evaluate broader issues of suitability of those materials for the purpose intended. Presumably, that would fall to the developer and the developer's design professional.

Finally, contractors should be wary of standard contract clauses that expose them to potential liability for design defects. Many contracts contain catchall clauses, which, for example, could require the contractor to supply any and all labor and materials that can reasonably be inferred from the plans and specifications as being necessary to achieve a complete project. If the contractor fails to provide such labor or materials, and a defect results, the contractor may be held liable.

VII. ENGINEER-PROCURE-CONSTRUCT

Engineer-procure-construct (EPC) has emerged as a popular delivery method across many construction industry sectors. Indeed, the terms "EPC" and "design-build" often are used interchangeably. EPC is similar to design-build in that an owner contracts

[38]*Fla. Bd. of Regents v. Mycon Corp.*, 651 So. 2d 149 (Fla. Dist. Ct. App. 1995).
[39]*See generally Fauss Constr., Inc. v. City of Hooper*, 249 N.W.2d 478, 481 (Neb. 1977).
[40]654 So. 2d 911 (Fla. 1995).
[41]*Id.* at 914.

with a single entity for the entire project. Thus, the EPC contractor assumes the project's corresponding business risk. The parties to an EPC contract must consisder the same issues that apply to design-build contracts.

EPC is widely used in industrial projects (i.e., major utilities projects and manufacturing facilities). Because these projects may take several years to complete, scheduling and sequencing are key to the success of an EPC project. The owner normally contracts with a consultant and the EPC contractor. Generally the EPC contractor self-performs most of the construction work while subcontracting some of the trade construction work. The EPC contractor then contracts with material suppliers for the major equipment purchases. In some cases, the EPC contractor is a major equipment supplier that in turn subcontracts the construction and equipment installation work.

There are no generally accepted form EPC contracts. The contract forms from the International Federation of Consulting Engineers (FIDIC) are gaining in popularity but still are not widely used. As a result, contracts for each EPC project are different, and must be carefully negotiated in order to arrive at a middle ground. EPC contractors should be aware that owner-provided EPC contracts undoubtedly will favor the owner. The EPC contractor should review the owner generated contract carefully to ensure that it does not take on unnecessary risk.

Performance requirements provide the key difference between a design-build project and an EPC project. The technical scope of work and the project definition provide the primary risks associated with an EPC project. Many owners minimize the risks associated with the scope of work by using performance specifications or output/throughput criteria that require the EPC contractor to achieve certain performance standards through any means or process selected by the EPC contractor. EPC projects and design-build projects share many of the same contractual issues. But because of the high-tech nature of EPC projects—and thus the heightened importance of performance requirements—a couple of contract risks are influenced more heavily by such performance requirements. These provisions are limitations of liability and liquidated damages.

A. Limitations of Liability

Most EPC contractors will not bet their company's entire net worth on an owner's project. Although this is perplexing to owners who have been sold on an EPC contractor's abilities based on its proposal, glossy brochures, business development materials and portfolio of successful projects, and, most likely, financial resources, it is understandable that an EPC contractor in a position to earn a reasonable profit on a project does not want to have a downside risk which far exceeds its upside potential.

Limitiation of liability clauses are commonplace and expected by most experienced owners. A reasonable limitation of liability clause does not diminish an EPC contractor's commitment to the project, nor does it represent an attempt to evade its responsibilities for the project. Rather, it represents an arm's-length negotiated cap on the potential downside risk of the project. Owners are certainly willing, and eager, to negotiate an EPC contractor's fee down, so it is not unreasonable to look at the entire risk spectrum and negotiate reasonable limitations that afford each party appropriate protection on the project.

Often the limitation of liability is tied to limits of applicable insurance coverage, or to a percentage of the contract price or the EPC contractor's fee. In addition, EPC contractors commonly negotiate a limitation of their obligations in the event of design errors or omissions. Most of these clauses derive from the design professional's agreement that often limits the obligation to perform redesign work. While many contractors will not enter into an EPC contract without a limitation of liability clause, the validity of such a clause needs to be determined under the applicable state law.[42]

B. Liquidated Damages

Liquidated damages are the primary way an owner may minimize the risks associated with the scope of work. Liquidated damages clauses are common to construction contracts as a way for the owner to fairly approximate its actual damages if the contractor fails timely to deliver the completed project (see **Chapter 11**). Liquidated damages may be tied to a variety of criteria. In the EPC contract, the two most common criteria for imposing LDs are performance guarantees and project delivery date(s).

In EPC projects, time is money and money is time. High-tech owners simply do not have the time to complete the design, procure the contractor and materials, and then construct the project in a linear sequence. The time savings, or perceived time savings, often creates the greatest risk on the project. For example, every day that a power plant is not generating electricity is lost revenue to the owner. It is incredibly important that the EPC contractor hit the project delivery date so the owner can begin generating revenue. Most owners will put teeth into the EPC contract by including a very strict liquidated damages provision for delay. The owner may go so far as to include a clause that allows the owner to terminate the EPC contractor and obtain an assignment of all subcontracts and purchase orders for the project in the event the EPC contractor is extremely late in performing or completing the work. Owners also may tie liquidated damages to the project's performance specifications. Often the EPC contractor has no choice but to deliver a project that is operating at less than 100% of the performance specifications called for in the EPC contract. There are a myriad of reasons that performance specifications are not met. In the vast majority of cases, the owner finds the completed project perfectly acceptable but operating at a lower efficiency than expected. In those cases, it would be far too costly to tear out the offending materials or machinery and replace them. Liquidated damages for decreased performance are a common way to compensate an owner for a project that does not meet 100% of the performance guarantees the EPC contractor promised but is otherwise operational and acceptable.

The prudent EPC contractor, when reviewing a liquidated damages clause, needs to make sure that these damages take the place of, and are not in addition to, actual damages that the owner might incur. Likewise, the EPC contractor would be well served to include contract terms that expressly avoid (or waive) consequential damages. If the liquidated damages truly are a fair approximation of the damages that an owner will incur as a result of the EPC contractor's failure to timely deliver the work or to deliver the work at the specified performance levels, then these should be the only damages for such delays or inefficiences.

[42]*Compare Blaylock Grading Co. v. Smith,* 658 S.E.2d 680 (N.C. App 2008) (clause valid) *with Lanier at McEver, L.P. v. Planners & Eng'rs Collaborative, Inc.* 663 S.E.2d 240 (Ga. 2008) (clause invalid).

VIII. PRIVATE PUBLIC PARTNERSHIPS

A private public (PPP) is a contractual relationship between a federal, state, or local public agency and a private company. A PPP is not necessarily a construction project delivery method, but it merits discussion because it is increasing in popularity in the United States. Under this concept, the public and private sectors come together to deliver a facility (or service) for the use of the general public. PPPs continue to gain in popularity due to the fiscal realities facing today's public agencies. Public resources cannot keep up with public needs, and public agencies are increasingly unwilling to take on too much financial risk. With a PPP, the private sector shares the risks and benefits with the public sector.

PPPs are as much a political tool as they are a contractual relationship. PPPs are used as policy to help leverage private-sector resources for public purposes. The result is that a public agency can meet the infrastructure needs of the public in a politically feasible manner. Because PPPs are largely funded by private funds, a government can finance the PPP in many creative ways. Local government incentives are controlled by state law, and there is no standard toolbox for providing the tax revenue stream necessary to execute a PPP. A public entity may issue tax-exempt or taxable bonds or offer other incentives, such as tax credits. In concept, the PPP is a win-win situation because public agencies can leverage the private sector to accomplish the agencies' goals.

Georgia, as just one example, has passed legislation allowing the Georgia Department of Transportation (GDOT) to solicit and receive proposals for PPPs.[43] Georgia's PPP law allows for local, state, and federal funds to be combined with private-sector funds on a PPP project. Georgia's PPP law also exempts contracts from the public bid process. Still, its PPP law does provide for legal public notice of the PPP proposal and an opportunity for other potential bidders to submit competing proposals. Interestingly, Georgia's PPP law also allows the GDOT to consider unsolicited proposals if they are unique and innovative.

State PPP laws have survived legal challenges.[44] Nevertheless, most likely there will be issues that will need to be worked through until PPP becomes a fully accepted method of financing public projects.

[43]*See* GA. CODE ANN. § 32-2-78 through 32-2-80 (2007).

[44]*See Coastside Fishing Club v. Cal. Res. Agency*, 71 Cal. Rptr. 87 (Cal. Ct. App. 2008) (upholding the creation of a PPP to provide the resources necessary to comply with the Marine Life Protection Act); *Bd. of Dirs. of Indus. Dev. Bd. of City of Gonzalez, Louisiana, Inc. v. All Taxpayers*, 938 So. 2d 11 (La. 2006) (upholding the constitutionality of the Tax Increment Financing Act, which authorized the issuance of bonds to provide public funding for private projects deemed to create economic development).

➢ POINTS TO REMEMBER

- The traditional approach to construction of design-bid-build has many strengths but also weaknesses that have prompted pursuit of other approaches to the construction process.
- The manner in which alternative contracting methods divert from clearly defined and accepted practices and roles requires careful attention to avoid unanticipated problems and disputes.
- Multiprime (or parallel prime) contracting, particularly when employed with fast-track construction, can reduce the time and cost of construction.
- All parties to a multiprime project must recognize that with the elimination of the general contractor, another party with appropriate power should be designated to assume the coordination responsibilities traditionally fulfilled by the general contractor. Without some express disclaimer, the owner assumes the duty to coordinate in the multiprime setting.
- Construction management generally entails involving an entity with diverse expertise in design, construction, and management in the design and construction process. The precise role of a construction manager on any project, however, can be determined only by reference to specific contract language.
- The role of the construction manager can range from that of a traditional general contractor, which provides some estimating and constructibility input during the design phase but is still required to guarantee time and price of performance, to a traditional design firm, which simply provides a higher level of construction administration and scheduling services for a fixed fee, without any guarantee of time or cost of performance.
- Design-build contracting represents the most radical departure from the traditional approach to construction by vesting all design and construction responsibilities, and resulting liabilities, in one party. The dramatic alteration of the traditional roles of the parties in design-build requires special attention to make certain the contract sets out the mutually understood and specific rights and responsibilities of each party.
- Even in the traditional build-to-design approach, contractors can assume discrete design liability as the result of performance specifications, the shop drawing process, and where secondary design review responsibility is imposed by standard contract clauses.
- Many states have passed laws allowing alternative methods of project delivery. Design professionals, contractors, and subcontractors should still check state laws governing licensing and the formation of business entities before signing on the dotted line.
- EPC contracting is gaining in popularity particularly in industrial construction. There are no widely accepted form contracts, so the EPC contractor should review carefully any proposed EPC contract.
- Performance requirements are very important in EPC contracts. Performance requirements will influence two major areas of contract negotiation in EPC contracts; limitations of liability and liquidated damages.

3

PREPARING TO WORK IN A NEW STATE: PREPROPOSAL AND PERFORMANCE CONSIDERATIONS

Every new construction project presents challenges and risks as well as opportunities. When a construction company attempts to do business in a new state for the first time, or without adequate experience in that jurisdiction, the potential stumbling blocks to success are significant. In order to meet these challenges, construction firms must arm themselves with information about the political climate, the labor market, and the laws affecting construction projects in that new jurisdiction. To assume that the laws and the construction climate in a new jurisdiction are identical, or even similar, to those in a contractor's home state is dangerous.

This chapter identifies some fundamental issues a contractor should consider before doing business in a new state. These guidelines are a first step, not an all-encompassing road map, to avoiding legal land mines while doing business in an unfamiliar jurisdiction.

I. QUALIFYING TO DO BUSINESS

Many construction companies perform work in states outside the state in which they are established. When a company conducts business in another jurisdiction, it is deemed a "foreign" business entity for purposes of that state's laws and regulations governing business transactions. For example, if a Delaware corporation performs work in Florida, that company is considered a foreign corporation under Florida's laws. A business entity (such as a corporation, partnership, limited liability company, or limited liability partnership) that does business—or obtains, possesses, or disposes of property—in a foreign state must first obtain a certificate of authority from that state.[1] The performance of construction work generally constitutes "doing business" as the term is used in foreign business entity statutes.[2]

By requiring a foreign business entity to obtain a certificate of authority, a state can maintain a record of the foreign business entity, require the appointment of a registered agent to accept service of legal process, and obtain revenue, generally in the form of taxes. To ensure compliance with this requirement, most states will not allow a foreign business entity to bring disputes to the state's courts until it has obtained the necessary certificate of authority.[3] A foreign business entity's failure to acquire the certificate of authority, however, generally will not invalidate any contract entered into by the unregistered foreign entity.[4]

In most instances, the failure to obtain a certificate of authority is curable at any time. Nevertheless, timely compliance with this statutory requirement is essential as "action" to preserve legal claim rights may be time sensitive. Consequently, a foreign company that procrastinates in its efforts to obtain the required certificate of authority does so at it own peril in that a legal right may expire before the issuance of the certificate of authority. For example, a foreign contractor may lose its mechanics' lien rights if it fails to obtain the certificate of authority before the statutory deadline to foreclose the lien.[5] A certificate of authority is generally not required when the work constitutes an "isolated transaction" in the foreign state. A number of factors are considered when determining whether a contractor's work constitutes an "isolated transaction." As described by the Maryland Court of Appeals,[6] the analysis for determining whether a foreign entity is "doing business" in the state does not rest on a single factor. Rather, it focuses on the nature and extent of the business and activities that occur in the state. Because very little business activity is necessary to constitute "doing business," most construction work is subject to the certificate of authority requirement as a "single or isolated transaction".[7]

[1]*See e.g.*, ALA. CODE § 10–2b–15.01; GA. CODE ANN. § 14–2–1501; HAW. REV. STAT. § 414–431; KY. REV. STAT. ANN. § 271b.15–010; N.C. GEN. STAT. § 55–15–01; S.C. CODE ANN. § 33–15–101; WIS. STAT. ANN. § 180.1501.
[2]*A.H.L. Inc. of Del. v. Star Ins. Co.*, 10 F. Supp. 2d 1216, 1219 (D. Kan. 1998) (holding "the majority rule is that the performance of construction work by a foreign corporation in a state is generally considered doing business in the state. . . ."); *S&H Contractors v. A.J. Taft Coal Co.*, 906 F.2d 1507, 1510 (11th Cir. 1990) (holding that "a foreign corporation doing construction work within a state is held to be doing business in that state. . . .").
[3]*St. Paul Fire & Marine v. Paw Paw's Camper City, Inc.*, 346 F.3d 153 (5th Cir. 2003); *Northfield Ins. Co. v. Odom Indus. Inc.*, 119 F. Supp. 2d 631 (S.D. Miss. 2000).
[4]*Springwall, Inc. v. Timeless Bedding, Inc.*, 207 F. Supp. 2d 410 (M.D.N.C. 2002); *Quarles v. Miller*, 86 F.3d 55 (4th Cir. 1996).
[5]*Space Planners Architects, Inc. v. Frontier Town–Missouri, Inc.*, 107 S.W.3d 398 (Mo. Ct. App. 2003); *In re Branson Mall, Inc.*, 970 F.2d 456 (8th Cir. 1992) (holding unregistered architectural firm could not enforce mechanics' lien even though firm registered with board after work was complete).
[6]*Tiller Constr. v. Nadler*, 637 A.2d 1183 (Md. 1994). *See also S.A.S. Pers. Consult. v. Pat-Pen*, 407 A.2d 1139, 1143 (Md. 1979).
[7]67 WILLIAM M. FLETCHER, FLETCHER ENCYCLOPEDIA OF LAW OF CORPORATIONS, § 8469 (2007).

II. STATE REGISTRATION REQUIREMENTS: BONDS TO SECURE PAYMENT OF TAXES

A company performing construction work in a foreign jurisdiction must determine if a bond is required to provide security for any costs or taxes payable to the state. Many states require a bond from out-of-state contractors to ensure payment of sales and use taxes. The bond requirement may apply when the work is performed pursuant to a federal government contract.[8] Although a state may recover the tax from the contractor that posted the tax bond, it cannot, however, recover such taxes from the federal government, which may be a party to the construction contract.[9]

A nonresident contractor's surety that has provided a performance bond also can be held liable for the sales and use taxes.[10] This rule was first announced by the United States Supreme Court in 1827 and remains a tenet of law recognized by a majority of the states.[11]

The laws regarding the issuance and necessity of tax bonds vary by state. Consequently, prior to starting construction work in a foreign jurisdiction, a contractor must identify the specific requirements for a given jurisdiction. For example, Alabama requires every nonresident contractor to register with the Department of Revenue before engaging in any work.[12] Upon registration, the nonresident contractor must deposit with the Department of Revenue 5% of the contract amount or provide a surety bond to guaranty payment of applicable taxes.[13] Early identification of, and compliance with, state law requirements for bonds to secure payment of taxes will avoid costly problems in the future—especially since the failure to comply with a state's requirements may prevent the assertion of contract or claim rights in the courts of that state.

III. STATE LICENSING AND QUALIFICATIONS

In an effort to protect the public from potential problems associated with poor-quality construction, state and local governments have passed laws and regulations requiring licensure of certain construction professionals. To ensure strict adherence to these licensing laws, most states' laws preclude an unlicensed individual or company from recovering monies owed for services performed. In contrast to the certificate

[8] *See, e.g., United Pac. Ins. Co. v. Wyo. Excise Tax Div., Dep't of Revenue and Taxation*, 713 P.2d 217 (Wyo. 1986); *C.R. Frederick, Inc. v. State Bd. of Equalization*, 120 Cal. Rptr. 434, 440 *cert. denied*, 419 U.S. 1120 (1974).

[9] 4 U.S.C. § 107(a); *Wash. v. United States*, 460 U.S. 536 (1983).

[10] *United Pac. Ins. Co. v. Wyo. Excise Tax Div., Dep't of Revenue and Taxation*, 713 P.2d 217 (Wyo. 1986).

[11] *Ogden v. Saunders*, 25 U.S. 213 (1827).

[12] ALA. CODE § 39–2–14.

[13] *Id.*

of authority requirement, the failure to comply with licensing requirements—even when it is a mistake—is generally not curable.[14]

It is not wise for a contractor to delay the completion of the licensing process until after it is awarded the contract. In some jurisdictions, the failure to have the required license when a bid is submitted is a violation of the licensing requirements and thus could jeopardize the contract award. In addition, a contract entered into by an unlicensed contractor may be deemed invalid or void in some jurisdictions. In California, however, a licensing failure at the time of the contract signing will not invalidate the contract—at least where the contractor is fully licensed at all times during contract performance.[15]

Although states have a universal interest in regulating construction companies and professionals, each state adopts its own unique set of licensing requirements. For example, only about half the states require contractors to be licensed. In Florida, contractors must possess a license, and certain corporations involved in construction[16] and architecture[17] must maintain specific certifications. In Maryland, a nonresident contractor must obtain a construction license[18] and pay an additional fee to perform work in the state.[19] The fee requirement is waived where the nonresident contractor's home state does not require a license of Maryland contractors. In Kansas, while there is no state law mandating contractors to be licensed, a water well contractor must be licensed by the Kansas Department of Health and Environment.[20] The lesson to be learned, of course, is that it is unwise to make assumptions about the licensing requirements in a foreign jurisdiction. The requirements vary from state to state.

Even though all states do not require contractors to be licensed, all states have laws regulating the practice of design professionals.[21] For example, in West Virginia, architects must register with the West Virginia Board of Architects,[22] and engineers must register with the West Virginia Board of Registration for Professional Engineers.[23] In Missouri, the failure of a person involved in architecture or engineering to obtain the necessary license constitutes a criminal misdemeanor and renders all contracts entered into by the unlicensed architect or engineer void and unenforceable.[24]

Compliance with licensing requirements can be greatly influenced by the subjective interpretations of those who are charged with the enforcement of licensing provisions. Contractors are well advised to get to know not only the letter of the licensing laws but also the manner in which those laws are applied by the licensing board and staff members.

[14] *See, e.g.,* NEV. REV. STAT. § 624.700.
[15] *MW Erectors, Inc. v. Niederhauser Ornamental and Metal Works Co., Inc.*, 115 P.3d 41 (Cal. 2005).
[16] FLA. STAT. § 489.119.
[17] FLA. STAT. § 481.219.
[18] MD. CODE ANN., Bus. Reg. § 17–601 through § 17–603.
[19] MD. CODE ANN., Bus. Reg. § 17–603(b).
[20] KAN. STAT. ANN. § 82a–1206(a).
[21] *See, e.g.*, Ind. Code § 25–4–1–26.
[22] W. VA. CODE § 30–12–11.
[23] W. VA. CODE § 30–13–13.
[24] MO. REV. STAT. § 327.191; MO. REV. STAT. § 327.461.

IV. PUBLIC CONSTRUCTION AWARDS

When performing construction work in a new state, it is essential that a contractor know about the regulations governing construction projects in that jurisdiction. To ensure compliance with local construction rules, a contractor's knowledge base must extend beyond a mere understanding of the applicable code requirements. State legislatures, counties, and municipalities frequently impose additional obligations (e.g., license or permit requirements) for a specific scope of work or type of construction activity. For example, Baltimore County, Maryland, requires a special building permit for the removal and disposal of asbestos shingles.[25]

Statutes governing contracts awarded by state agencies may also contain requirements that add, or take away, contractual rights, even though the contract does not address the particular issue. A contractor doing business with a state agency, therefore, cannot rely solely on the express terms of its contract to fully define its contractual rights, obligations, and risks. A prudent contractor must also examine the statutes and regulations affecting contracts with the contracting state agency. In Georgia, for example, the statutory provisions governing Department of Transportation contracts purport to insulate the state from liability for delay damages, even when no such limitation is apparent from the face of the parties' contract.

In addition to the substantive obligations imposed by local regulations, a contractor must be familiar with the procurement laws of local government entities. The procedure to award a contract can vary depending on the type of construction or government entity involved. For example, the Georgia Department of Transportation has its own authority to plan, designate, improve, manage, control, construct, and maintain the state highway system.[26] Likewise, the Alaska procurement code exempts 41 categories of procurement, including certain contracts for the Alaska Marine Highway system and the Alaska Industrial Development and Export Authority,[27] from the code's requirements. The city of Anchorage also has its own municipal procurement regulations.[28]

State agencies, counties, and municipalities are implementing new procurement methods that offer more options and flexibility in obtaining construction services. The trend away from the traditional sealed bidding process is motivated by increased pressure on local government entities to accomplish more with less funding, to complete projects faster, and to eliminate as much risk as possible. To address these objectives, new procurement rules are giving local government agencies more discretion in evaluating bids and proposals.

To address funding restrictions, states have implemented alternative procurement systems aimed at lowering the project cost or assisting in project financing. Several states, for instance, have authorized direct negotiation with a low bidder

[25] *www.baltimorecountymd.gov/agencies/environment/asbestos-removal.html.*

[26] GA. CODE ANN. §§ 32–2–1 *et seq.*

[27] ALASKA STAT. § 36.30.850(b).

[28] ANCHORAGE MUN. CODE § 7.20.010 *et seq.*

when the low bid exceeds the state agency's cost estimate.[29] These direct negotiations allow the public owner to adjust or revise the scope of work in an effort to meet budget limitations. The "Reverse Bid Auction" is another system implemented by some states to achieve lower prices for public works projects.[30] This method for soliciting bids keeps a bidder informed of its relative position among all competing bidders and allows a firm to adjust its bid price during a designated bidding period. States have also enacted legislation authorizing the use of Private Public Partnership (PPP)[31] This alternative to the competitive sealed bid process allows public owners to consider and accept solicited and unsolicited proposals from private companies. PPP encourages the private sector to participate in the development, financing, and operation of public projects. Finally, public owners are also experimenting with privatization of public services and "Build-Operate-Transfer" projects.

In addition to alternate procurement systems, public owners are authorizing the use of alternative project delivery methods to meet more aggressive project schedules and to place more construction risk on the contractor. Design-build contracts, for example, shift design responsibility (and the corresponding risk) from the owner to the contractor. This project delivery system has increased in popularity, and many states have enacted legislation to address and promote its use. The Colorado legislature, for instance, recognized the advantages of design-build contracts and enacted specific laws to encourage its use for highway construction.[32] In Tennessee, design-build contracts are authorized as long as the invitation for proposals (or other bid information provided) addresses use of the design-build system on the project.[33] Some states, however, are wary of awarding both design and construction responsibility to the same entity and have limited design-build contracts to certain agencies.[34]

A second alternative project delivery system authorized in some states is the use of "guaranteed maximum price at-risk construction management."[35] Public owners benefit from this system as it limits its exposure to project cost overruns and price escalation. In addition, as compared to design-build contracts, the public agency retains more control over the design process. Finally, public owners may also include incentive or bonus provisions in their contracts to encourage the earliest possible completion of the project.

[29]*See, e.g.,* GA. CODE ANN. § 36–91–22; 62 PA. STAT. ANN. § 3911; and VA. CODE ANN. § 2.2–4318.

[30]*See, e.g.,* ARIZ. REV. STAT. § 41–2672; COLO. REV. STAT. § 24–103–208.

[31]*See, e.g.,* COLO. REV. STAT. §§ 43–1–1201 *et seq.*; F.S.A. § 334.30; GA. CODE ANN. §§ 32–2–78 through 32–2-80; OR. REV. STAT. § 383.005; WASH. REV. CODE ANN. §§ 47.46.010 *et seq.*

[32]COLO. REV. STAT. §§ 43–1–1401, *et seq.*

[33]TENN. CODE ANN. § 12–10–124(c).

[34]*See, e.g.,* WASH. REV. CODE § 39.10.051.

[35]*See, e.g.,* ARIZ. REV. STAT. § 41–2579; N.M. STAT. ANN. §§ 13–1–124.2 through 124.5; WASH. REV. CODE § 39.04.220; WIS. STAT. § 16–6–701; 30 Ill. COMP. STAT. §§ 500/33–55, KAN. STAT. ANN. §§ 76–786; 75–37,144; KY. REV. STAT. ANN. § 45A.045(11)(b); NEB. REV. STAT. § 79–2003; S.D. CODIFIED LAWS § 5–18–49; ARK. CODE ANN. § 19–11–801; GA. CODE ANN. § 36–91–20(c); MISS. CODE ANN. § 31–7–13.2; N.C.GEN. STAT. ANN. §§ 143–128; 143–128.1; TENN. CODE ANN. §§ 12–10–124(c)(4) through 12–10–124(d); TEXAS GOV'T CODE ANN. § 2166.2532; VA. CODE ANN. § 2.2–4308; 5 ME. CODE R. § 1743; N.H. REV. STAT. ANN. §§ 21–I:78; 228:1.

V. STATE STATUTES AND POLICIES AFFECTING CONTRACT TERMS AND CONDITIONS

Each state has specific statutes and policies affecting the enforcement and validity of contractual terms frequently included in construction contracts. The three primary areas addressed by these statutes include: (1) subcontractor/general contractor relationships, (2) contract award preference policies, and (3) public policy limitations on contract terms and clauses.

A. Subcontractor/General Contractor Relationships

Promissory estoppel can be a fundamental tenet of construction law in many states. Under this legal doctrine, a subcontractor's or supplier's quote cannot be changed or withdrawn when the general contractor has reasonably relied on the quote in preparing its bid to the owner.[36] When a firm refuses to provide the goods and services at the quoted price—even if the quote is incorrect because of a mathematical error or otherwise—the doctrine of promissory estoppel is triggered, and the subcontractor will be required to honor the quoted price. Application of the doctrine will vary depending on the laws of a particular state. For instance, in some states, promissory estoppel will not be triggered by the above scenario as these jurisdictions view a subcontractor's quote as simply an *offer* to perform work, not an act that creates a binding contract.[37] Accordingly, subcontractors doing business in a new jurisdiction should understand that state's view on promissory estoppel and its application to subcontractor bids.[38]

A common issue involving subcontractor bidding is whether a state has subcontractor listing requirements for contractors bidding on a project. Some states have specific statutes requiring the general contractor to list its intended subcontractors when submitting its bid to the owner.[39] Other states, such as Louisiana, do not have this statutory requirement. Nonetheless, the failure to provide this information can result in the disqualification of the bid or proposal ***if*** the owner's request for bids or request for proposals asks for it to be provided.[40]

Prompt payment to subcontractors is another issue that has received significant attention from state legislatures and courts. Several states have enacted legislation to ensure that subcontractors are paid timely.[41] These statutory provisions define a

[36]*Olson v. Synergistic Tech. Bus. Sys.*, 628 N.W.2d 142 (Minn. 2001).

[37]*Electro Lab of Aiken v. Sharp Constr. Co.*, 593 S.E.2d 170 (S.C. Ct. App. 2004); *Ark Constr. Co. v. Indian Constr. Servs.*, 848 P.2d 870 (Ariz. Ct. App. 1993).

[38]The subject of promissory estoppel is discussed in more detail in **Chapters 4 and 8**.

[39]*See, e.g.,* CAL. PUB. CONT. CODE § 4100 *et seq.* (requiring bidders to submit names of subcontractors whose work constitutes at least 12% of the total contract amount).

[40]*Boh Bros. Constr. v. DOT*, 698 So. 2d 675 (La. Ct. App. 1997) (*citing C.R. Kirby Contractors, Inc. v. City of Lake Charles*, 606 So. 2d 952 (La. Ct. App. 1992)).

[41]*See, e.g.,* ARIZ. REV. STAT. §§ 32–1129.02; 41–2577.B; ARK. CODE ANN. § 19–4–1411; FLA. STAT. §§ 218.70; 255.071; MONT. CODE ANN. § 28–2–2103; N.M. STAT. ANN. § 57–28–5; MICH. COMP. LAWS ANN. § 125.1561 through 1562; GA. CODE ANN. § 13–11–1, *et seq.*; OHIO REV. CODE ANN. § 4113.61.

contractor's obligation for payment to the subcontractor by specifying a time period for issuing such payments and imposing a penalty for noncompliance. States that have not adopted a prompt payment statute may nonetheless recognize a subcontractor's right to prompt payment by judicial decision. Courts in Kentucky and other jurisdictions, for example, have declared that a contractor must pay a subcontractor within a reasonable time period.[42]

Criminal and civil regulations may also impact the manner by which contractors and subcontractors perform their work. For example, some states have enacted "trust fund" statutes to discourage the misuse of construction funds paid to the contractor.[43] Construction proceeds paid to a contractor that are used for purposes other than the proper payment of construction services and materials may result in the assessment of civil and criminal penalties to the contractor. These penalties may also extend to individual members of the company who actively or passively permitted the misuse of construction proceeds.[44]

A state trust fund statute may provide that if a contractor or subcontractor fraudulently obtains an "advance" of contract payments with a promise to perform construction work, and fails to do so, the contractor, or the principals of the contracting firm, may be guilty of larceny.[45] Oklahoma law provides that managing officers of a corporate contractor can be found guilty of embezzlement for trust fund statute violations.[46] Wisconsin courts have allowed civil suits against officers of corporate contractors for diverting funds protected by trust fund statutes.[47]

B. Contract Award Preference Regulations

Many states provide preferential treatment to certain entities when awarding a contract. Those qualifying for preferred treatment generally are selected to further a socioeconomic or political interest. For example, to reward those contractors who contribute to the funding of public improvements through the payment of taxes, many states have a stated preference for resident contractors (i.e., contractors who reside in the state awarding the contract).[48] In some jurisdictions, however, this residency preference has been eliminated by court decisions finding that such preference violates objectives of competitive bidding and the "privileges and immunities" clause of the United States Constitution.[49]

[42]*Thomas J. Dyer Co. v. Bishop Int'l Eng'g Co.*, 303 F.2d 655 (6th Cir. 1962); *see also Midasco Inc. v. M.E. Hunter & Assocs.*, WL 452414 (E.D. Va. 2006); *Envirocorp Well Servs., Inc. v. Camp Dresser & McKee, Inc.*, 2000 WL 1617840 (S.D. Ind. 2000).

[43]*See, e.g., McMahon v. State*, 574 S.E.2d 548 (Ga. App. 2002).

[44]*Doyle Dickerson Co. v. Durden*, 461 S.E.2d 902 (Ga. App. 1995).

[45]VA. CODE ANN. § 18.2–200.1; *Holsapple v. Commonwealth*, 587 S.E.2d 561 (Va. 2003); *State v. Cohen*, 783 So. 2d 1269 (La. 2001).

[46]*See* OKLA. STAT. tit. 42, § 153; 21 OKLA. STAT. ANN. tit. 21, § 1451.

[47]*See Tri-Tech Corp. of Am. v. Americomp Servs., Inc.*, 646 N.W.2d 822 (Wis. 2002).

[48]41 U.S.C. §§ 10a through 10d.

[49]*See, e.g., C.S. McCrossan Constr. Co. v. Rahn*, 96 F. Supp. 2d 1238 (D.N.M. 2000).

Because "resident contractor" status has obvious benefits, the standard for determining whether an entity qualifies as a *resident* is critical. The standard is not universal and will vary from state to state. For example, states differ as to whether a joint venture contractor comprised of at least one nonresident entity should qualify for the preferential treatment afforded resident contractors. In Alaska, a joint venture contractor qualifies for "resident contractor" status so long as one of the coventure entities is an Alaska resident.[50] Louisiana, however, requires that all coventure entities be state residents.[51]

The federal government[52] as well as many state and local governments[53] provide statutory preferences for "small business" contractors. These preferences typically take the form of a "set-aside" program where a percentage of the public contract is designated specifically for contractors qualifying as a "small business." A "small business" is defined by statute and includes certain entities that fall below specified average employment/average annual revenue limitations.[54] A contractor bidding on a contract with such a set-aside provision must ensure that a sufficient amount of the contract work is performed by a qualified "small business."

State governments may also grant preferences to minority- or women-owned enterprises.[55] The definition of "minority" varies from state to state, but typically includes African Americans, Native Americans, and females.[56] States that do not have a specific statutory requirement providing such preferences may nonetheless encourage the use of minority- or women-owned contractors in its request for bids.

Statutory preferences for minority- or women-owned businesses have been challenged in courts across the United States on the basis that they constitute unlawful discrimination against nonqualifying entities. In an attempt to clarify when and how these preference policies can be used, the United States Supreme Court[57] set out four requirements that must be met when formulating these preferences:

(1) A high, "strict scrutiny" standard of review.
(2) The discrimination must be particularly linked to the market area of the implementing agency.
(3) The government organization must evaluate race- or gender-neutral remedies before adopting race-conscious requirements.
(4) The plan must be carefully tailored and must be in place only for the amount of time required to reverse the effects of past discrimination.

[50] *Irby-Northface v. Commonwealth Elec. Co.*, 664 P.2d 557 (Ala. 1983).
[51] *Bristol Steel & Ironworks, Inc. v. State Dept. of Transp. & Dev.*, 507 So. 2d 1233 (La. 1987).
[52] 48 C.F.R. §§ 19.5, *et seq.*
[53] 2000 Model Procurement Code for State and Local Governments § 11–101–11–301.
[54] 13 C.F.R. § 121.101 through 08; 13 C.F.R. § 121.201.
[55] *See, e.g.*, N.J. Stat. Ann. § 52:32–20.
[56] *See, e.g.*, N.C. Gen. Stat. § 143–128.2.
[57] *City of Richmond v. J.A. Croson Co.*, 488 U.S. 469 (1989).

Despite this guidance, the debate continues on whether race, ethnicity, and gender should be used as criteria for awarding contracts.

C. Public Policy Limitations on Contract Clauses

Risk is inherent in any construction project. A well-written contract will address the project risk and clearly allocate it among the contracting parties. Popular risk shifting provisions include: (1) no-damages-for-delay clauses; (2) liquidated damages clauses; (3) pay-when-paid versus pay-if-paid clauses; and (4) indemnity provisions. Recognizing that these provisions often are mandated by the party with superior negotiation leverage, legislatures in various jurisdictions have limited or nullified the enforceability of these provisions on public policy grounds. In an effort to gain the benefit of greater consistency with the interpretation of an organization's subcontract or purchase order forms, many contractors include provisions that stipulate that the contract will be interpreted under the laws of a particular state, usually that organization's home state. Many states that have nullified contract provisions on public policy grounds have also enacted laws addressing these choice of law clauses.[58]

The no-damages-for-delay clause is a contract provision that precludes a claimant from recovering monetary damages resulting from project delays. This clause effectively shifts the risk of project delay from the owner to the general contractor (or from general contractor to subcontractor). This clause, which is common in construction contracts, is the subject of frequent debate. The protection it provides to an owner or general contractor will vary by state. For example, some state legislatures prohibit the use of no-damages-for-delay clauses—at least in some contracts—as they believe it unfairly shifts a big-dollar construction risk to the other party.[59] Other states, however, adopt a more moderate approach by enforcing the parties' negotiated contract terms, even the no-damages-for-delay clause, as long as concealment, misrepresentation, or fraud is not at issue in the dispute.[60]

The cost of delay to a construction project's completion may also be addressed in terms of a liquidated damages provision. Typically, this contractual provision will establish a daily rate for delay damages that accrues after the specified contract completion date has passed. A liquidated damages provision is enforceable as long as it comports with the standards established in a given jurisdiction. In Colorado, for instance, a liquidated damages clause is enforceable when these prerequisites are satisfied: (1) it is established at a time when the delay damages are difficult to ascertain; (2) it represents an accurate estimate of the actual costs that may be incurred; and (3) it reflects an agreement by all parties.[61] In South Carolina, a liquidated damages

[58] *See e.g.*, FLA. STAT. ANN. § 685.101; N.C. GEN. STAT ANN. § 22B-2; N.Y. GEN. BUS. LAW § 757 (2003).

[59] N.C. GEN. STAT. § 143-134.2; OR. REV. STAT. § 279C.315.

[60] *See, e.g., Triple R. Paving, Inc. v. Broward County*, 774 So. 2d 50 (Fla. Dist. Ct. App. 2000); *Pasty & Fuhrman v. Hous. Auth. of City of Providence*, 68 A.2d 32 (R.I. 1949).

[61] *Klinger v. Adams County Sch. Dist.*, 130 P.3d 1027 (Colo. 2006); *Rohauer v. Little*, 736 P.2d 403 (Colo. 1987).

clause is enforceable when a party proves a breach of contract,[62] but only if the liquidated damages clause is not deemed to serve as a penalty for untimely completion.[63]

Pay-when-paid and pay-if-paid clauses have long been the topic of heated debate in the construction industry. Compared to traditional payment terms, which require payment to a subcontractor within a defined time period after performance of the work, these clauses attempt to place certain restrictions on a contractor's obligation to pay its subcontractor. A pay-when-paid clause typically requires payment by the contractor to the subcontractor within a defined time period after payment from the owner. A pay-if-paid clause makes payment by the owner to the general contractor an express condition precedent of the contractor's obligation to the subcontractor. General contractors insert these clauses into subcontracts to shift the risk of an owner's nonpayment to the subcontractor. Because an owner's refusal to pay the general contractor may be for reasons unrelated to the subcontractor's work, strict enforcement of these clauses is viewed with skepticism. Subcontractors, in particular, are critical of these clauses as they believe a general contractor should not be relieved of its payment obligations when the subcontractor has satisfactorily completed its work. Some courts agree with this perspective, holding that a pay-when-paid clause—while affording the general contractor a reasonable amount of time to make payment—establishes an *absolute* commitment to pay the subcontractor.[64] Other courts, however, apply a stricter interpretation of the clause, holding that when payment from the owner is a *condition precedent* to a general contractor's obligation to pay its subcontractors, the applicable contract language must clearly state this.[65] The latter interpretation, a pay-if-paid clause, will be enforced only when the court determines that the contract language is clear and unambiguous (e.g.: "Subcontractor acknowledges that payments will be made from money received from the Owner").

Subcontractors performing work on federal government projects subject to the Miller Act are afforded some protection against conditional payment provisions. (See **Chapter 14.**) This right, however, can be waived by the subcontractor if such waiver is clear and explicit, in writing, and signed by the subcontractor after the work has started.[66] Some courts have construed this waiver provision very narrowly.[67]

Another risk-shifting provision subject to much debate is the indemnity clause. This provision, which imposes "hold-harmless" obligations on one of the contracting parties, is common in construction contracts and can take various forms. Because many indemnity provisions are drafted very broadly, some state legislatures and courts have limited the effect and applicability of these provisions. For example, Colorado recently enacted an anti-indemnity statute, which states that any provision requiring a person to indemnify another for damages or injuries caused by the negligence or

[62]*Carolinas Cotton Growers Ass'n v. Arnotte*, 371 F. Supp. 65 (D.C.S.C. 1974).

[63]*Benya v. Gamble*, 321 S.E.2d 57 (S.C. Ct. App. 1984).

[64]*See, e.g., Koch v. Constr, Tech., Inc.*, 924 S.W.2d 68 (Tenn. 1996).

[65]*Printz Svcs. v. Main Elec.*, 949 P.2d 77 (Colo. Ct. App. 1997).

[66]40 U.S.C. § 3133(c) (formerly cited as 40 U.S.C. § 270b(c)).

[67]*See U.S. ex rel. Walton Tech., Inc. v. Weststar Eng'g, Inc.*, 290 F.3d 1199 (9th Cir. 2002) (finding that subcontractor's execution of settlement agreement containing a pay-if-paid provision was not a clear and explicit waiver of the subcontractor's Miller Act payment rights).

fault of that party "is void as against public policy and unenforceable."[68] California, Georgia, Kentucky, Montana, New Mexico, New York, and many other states have passed similar legislation.[69] Contractors should be cautious when working outside of their home state as the enforceability of indemnity provisions, as well as the other provisions discussed earlier, will vary by state and should be tailored to meet the standards established by the applicable laws and judicial decisions of a given jurisdiction.

Certain payment terms and conditions may also be impacted when a construction contract involves foreign entities. International construction projects present a unique risk when defining and valuing payment terms. For example, currency fluctuations during the project can alter the contract's value. A United States company receiving local currency for construction services provided in a foreign country runs the risk that, between the time the service is rendered and payment is made, the value of the local currency, relative to the U.S. dollar, may change. Consequently, to ensure fair value, it is critical for a contractor performing services abroad and receiving payment in a foreign currency to analyze the volatility of that currency as well as the U.S. dollar and devise a strategy to address this impact on the contract amount.

D. Impact of New Legislation

When preparing to work in a new state, contractors should consider the political climate of that jurisdiction for emerging trends that could impact their contract work. Public interest groups and media frequently champion issues of public concern to prompt lawmakers to enact new legislation. The proposed or new legislation may directly target the construction industry in which case all players in the industry should carefully review and analyze such provisions to determine: (1) the scope of any new rights and obligations; (2) the impact on pricing, schedule, or overall attractiveness of a project; and (3) the effective date of the legislation. Knowing the impact of new legislation on the front end may avoid potential and unnecessary problems during performance of the contract.

Legislation which is not directly targeted at the construction industry may nonetheless ascribe new obligations to the contractor. For example, in 2005, the Florida legislature enacted the "Jessica Lunsford Act," which is intended to protect children from sexual predators. The legislation mandates stiff minimum sentences for child abusers and requires offenders to wear global positioning satellite (GPS) monitors when released from prison. At first glance this legislation seems inapplicable to contractors performing construction work in Florida. But the legislation requires "contractual personnel"—defined as any vendor, individual, or entity under contract with a school or the school board—who are permitted on school grounds when students are present to undergo background screening and provide a complete set of fingerprints.[70] Any individual found to have been convicted of a crime of moral turpitude (e.g., terrorism, sexual

[68]Colo. Rev. Stat. § 13–21–111.5(6).
[69]*See, e.g.,* Cal. Civ. Code § 2782; Ga. Code Ann. § 13–8–2; Ky. Rev. Stat. Ann. § 371.180; Mo. Rev. Stat. § 434.100; N.M. Stat. Ann. § 56–7–1; N.Y. Gen. Oblig. Law. § 5–322.1.
[70]Fla. Stat. Ann. §§ 1012.32, 1012.465.

misconduct, murder, kidnapping, etc.) may not provide services on school grounds. Thus, after digging through the details, this legislation clearly has significant ramifications to a contractor performing construction work on school grounds because the contractor now must undergo a background check and provide fingerprints of employees. This obligation applies to all contractors, subcontractors, and their employees. Many jurisdictions have enacted legislation similar to the "Jessica Lunsford Act" and may impose additional obligations on contractors under certain circumstances.

Likewise, several jurisdictions have enacted legislation designed to crack down on illegal immigration. A prominent part of such legislation requires employers to verify a new employee's citizenship status. In Arizona, for example, no employer may "intentionally" or "knowingly" employ an illegal immigrant.[71] Moreover, illegal immigrants are explicitly prohibited from employment on public works projects.[72] Similarly, Colorado requires a contractor to certify that it does not knowingly employ an illegal immigrant (or contract with a subcontractor that knowingly employs illegal aliens) and that it has attempted to confirm the residency status of new employees.[73] Pennsylvania, Vermont, New Hampshire, Georgia, and Kansas have similar legislation prohibiting the knowing employment of illegal immigrants or imposing an obligation to verify the residency status of its employees.[74] Contractors must be aware of these emerging trends in legislation and devise a strategy so that any new obligations imposed are satisfied in accordance with the statute.

VI. PRESERVATION OF LIEN/BOND RIGHTS

The lien laws in "foreign" jurisdictions present significant challenges for contractors, subcontractors, suppliers, and owners involved in construction projects outside their resident state. Lien law requirements are cumbersome and confusing; nonetheless, the failure to meet these requirements precisely can be fatal to a company's legitimate lien rights.

A. Preconstruction Knowledge of Lien Law Peculiarities Is Essential

Typically, contractors (as well as owners) do not analyze and review the lien law requirements in a particular jurisdiction until there is an immediate need to file a claim of lien or, from the owner's perspective, after the lien has been filed. This approach can be dangerous as many jurisdictions require action from the contractor early in the project to preserve lien rights or, in the case of the owner, to limit lien obligations. If these initial obligations are not satisfied, those lien rights will be lost forever.

[71] ARIZ. REV. STAT. § 23–212.
[72] ARIZ. REV. STAT. § 34–301.
[73] COLO. REV. STAT. ANN. § 8-17.5–102.
[74] 43 PA. STAT. ANN. §§ 166.1 through 166.5; 21 VT. STAT. ANN. tit. 21, § 444a; N.H. REV. STAT. § 275–A:4–a; GA. CODE ANN. §§ 13–10–90; 13–10–91; KAN. STAT. ANN. § 21–4409.

In some states, affirmative action must be taken before the project even begins, or lien-related rights may be lost. In Alabama, for instance, a "full price" lien (as opposed to a lien on just the unpaid balance of the owner/general contractor contract) is available only to those who have a direct contract with the owner or to materialmen who give notice to the owner *before* furnishing materials.[75] An owner receiving notice from a material supplier may avoid responsibility for the supplier's materials by issuing a "Predelivery Notice" to that supplier. The notice should convey to the supplier that the owner will not be responsible for the price of the materials being furnished.[76]

In other jurisdictions, lien rights survive the start of construction but are in jeopardy if some action is not taken very soon after project work begins. For example, in Michigan, to preserve its lien rights a subcontractor or supplier must serve a "Notice of Furnishing" on the general contractor and on the owner's designated recipient within 20 days after *first* furnishing labor or materials to a construction project.[77] In addition, an owner must file and post a "Notice of Commencement."[78] Any delay in providing the owner with notice of the intended lien claim may result in a forfeiture of lien rights.

B. Lien Law Protections and Procedures Vary Greatly from State to State

A contractor must not assume that familiarity with its home state's lien laws will equip the company to protect its rights under the lien laws of another jurisdiction. Lien laws—perhaps more than any other statutory scheme affecting the construction industry—vary widely from state to state. For example, in North Carolina, punch list or warranty work typically will not extend the time for filing a lien claim. In neighboring South Carolina, however, such remedial or warranty work may qualify as a basis to extend the time for filing a claim of lien.[79]

C. "Almost Right" Is Almost Always Not Good Enough

In many areas of contract law, parties are not penalized when they are "almost right" or when they have "substantially complied" with the contract or other legal requirements.

For example, contractors and subcontractors that fail to follow the precise notice requirements prescribed by contract may be excused by the owner's actual knowledge of the circumstances requiring such contract notice. The lien laws are not so forgiving. The failure of a contractor, subcontractor, supplier, or owner to exactly satisfy lien law requirements is often fatal to that party's rights under the lien statute. Even the slightest noncompliance can be costly. For example, the Maryland lien statute

[75] ALA. CODE §§ 35–11–210 *et seq.*

[76] ALA. CODE §§ 35–11–210; 35–11–218.

[77] MICH. COMP. LAWS ANN. §§ 570.1101 *et seq.*; MICH. COMP. LAWS ANN. § 570.191.

[78] MICH. COMP. LAWS ANN. §§ 570.102; 570.1108 through 570.1301.

[79] N.C. GEN. STAT. ANN. § 44A–7 *et seq.*; S.C. CODE ANN. § 29–5–10 *et seq.*

provides that a "Notice of Intent to Claim a Lien" must be served within 120 days of the claimant's last work. This does not mean that the notice can be sent four months after the last work. There is a difference between 120 days and four months, and the failure to recognize this difference can be fatal to a company's lien rights.[80]

D. A Valid Claim of Lien Does Not Guarantee Payment

Even if a contractor successfully complies with the precise requirements of the lien laws in the "foreign" jurisdiction, the exercise of its mechanics' lien right may fall short of the payment guarantee it seeks. The inadequacy of lien rights as payment security may result from several causes: (1) the lien may be subordinate to other security interests that have a combined value greater than the property; or (2) the lien may not cover all of the damages or costs to which the contractor is entitled.

From the contractor's or supplier's perspective, the limited scope of the damages covered by most lien law statutes is a frequent problem. For example, indirect delay damages (the extended job site and home office overhead attributed to a project's delay) are one of the "big-dollar" consequences of a troubled construction project that cannot be included in a lien claim.

The existence of lien rights, however, is no substitute for the exercise of "good financial management," which includes systems and procedures to prequalify those with whom your company intends to do business. Obtain pre-project proof of adequate project financing, and avoid contracts that require your company to continue working during payment disputes.

E. Arm Yourself Early with Accurate Lien Law Information

A contractor's ability to exactly satisfy the lien requirements—and to avoid the many pitfalls that await it in the lien laws of foreign jurisdictions—begins with the gathering of reliable lien law information. A number of sources are available to the construction industry. For example, mechanics' lien filing services (as well as various contractor associations) have pamphlets, handouts, and other publications available that cover basic—and even more sophisticated—lien law information. Courts often have construed statutory language defining lien rights in ways that make a literal reading of the underlying lien statute unreliable. Similarly, statutorily mandated lien forms may be misleading to those who do not know how the lien statutes define the terms used in the lien form. Investing time in preparing oneself to take full advantage of the lien laws and other payment protections available to a contractor or subcontractor in any new project jurisdiction is priceless.

VII. RISK ASSESSMENT CHECKLISTS

The process of identifying general construction project risks, including risks unique to the state where construction is planned, is crucial to the success of a construction project. Owners, construction managers, designers, contractors, subcontractors, and materialmen

[80]Md. Code Ann., Real Prop. § 9–101.

cannot afford to wait until a problem arises before assessing the risks inherent in doing business in an unfamiliar state. Instead, this risk assessment process must start early, and it should be a systematic approach to risk identification. If a firm does not have such a systematic approach to risk identification, development of one should be a high priority. It is often possible to borrow from the numerous checklists and guidelines created by industry participants and adapt that system to meet the particular operational style of any company. A contractor should preserve the information gathered during the risk assessment so that it can be of benefit to the project management team. The checklist that follows can be used as a start to an early and thorough evaluation of the issues to be considered before performing work in a new state:

✔ RISK ASSESSMENT CHECKLIST

(1) Qualifying to Do Business

a. Unless the planned construction work constitutes an isolated transaction, obtain a certificate of authority.

b. Establish whether the state has any additional requirements for nonresident bidders—for example, filing with the state's Department of Revenue a statement of a contractor's tangible property.

c. Ascertain and satisfy required registration fees.

d. Establish a registered agent/office.

(2) State Revenue Department Requirements

a. Ascertain whether the state requires a bond to ensure payment of taxes

b. Identify the bond amount and renewal requirements

c. Identify any sales or use tax liability

d. Plan for ad valorem or other taxes

(3) Licensing Requirements

a. Identify any need for contractor or subcontractor licenses

b. Determine whether the state requires a contractor to obtain a license before bidding on a construction project or performing work

c. Verify compliance with architect and engineer licensing requirements

d. Determine whether a license is required for other construction professionals or specific trades, such as plumbers, electricians, fire alarm installers, individuals involved in asbestos or lead abatement, and others

e. Review unique requirements of city or county licensing or permit requirements

f. Review building code requirements

g. Identify any required environmental permits or requirements

(4) Evaluating the Effect of a State's Laws on Contracts Terms

a. Subcontractor issues:

(1) Must subcontractors be listed on the bid or identified before performance?

(2) Is there a state prompt-payment statute?

(3) Can parties contract for payment terms that vary from the prompt-payment act?

b. Contract award preference regulations:

(1) Determine whether the jurisdiction allows an award preference for "residents."

(2) Determine whether the state has a specified small business preference.

(3) Determine whether the state or bid requires a certain amount of the work to be performed by a minority-or a woman-owned business.

c. Contract term laws or regulations:

(1) Determine the existence/validity of no-damages-for-delay clauses.

(2) Determine the existence/validity of pay-if-paid clauses.

(3) Does a state statute affect the enforceability of indemnity clauses?

(4) Does the contract adopt the laws of another jurisdiction?

(5) Are there other public policy limits on exculpatory clauses?

(6) Does a state statute affect the validity of a provision specifying a choice of venue for court actions?

(5) Special Insurance Requirements

a. Identify the nature of required insurance coverages

b. Is an additional rider required for work in this jurisdiction?

c. Identify filing/reporting requirements.

(6) Preservation of Lien/Bond Rights

a. Is the project public or private?

b. Verify the existence of lien rights

c. Gather property legal description/owner information

d. Verify the existence and coverages of payment bonds

e. Identify any precontract filing requirements

f. Identify preliminary/early notice requirements

g. Verify postperformance notice requirements

h. Determine any limitations on lien/bond recovery

i. Make an early request for bond/lien information

j. Identify any need to use mandatory lien waiver forms

(7) Payment Security Verification

a. Verify the adequacy of project financing

b. Identify notice-to-lender requirements

c. Are lender approvals required for change orders or other actions?

d. Identify any extra work payment limitations

e. Identify change order authority limitations

f. Verify the existence and impact of a state prompt payment statute

g. Identify trust fund/retainage requirements

(8) Labor, Equipment, and Material Issues
 a. Verify the adequacy of available local labor
 b. Is the locale primarily a union or nonunion labor market?
 c. Determine materials/equipment availability
 d. Identify applicable sales or use taxes
 e. Verify applicable labor rates/fringes
 f. Is the project affected by project/union agreements?
(9) Political Climate Assessment
 a. Identify special political ties/circumstances
 b. Identify other business climate considerations

VIII. INFORMATION SOURCES

In the pursuit of obtaining a state's requirements for performing construction work, the Internet can provide a plethora of information. Each state maintains a Secretary of State Web site that provides information for obtaining a certificate of authority within that particular jurisdiction.[81]

For an overall review of a state's construction contracting requirements, the state chapters of the Associated General Contractors of America (AGC)[82] or the Associated Builders and Contractors (ABC)[83] trade organizations can provide valuable information. For example, on the AGC's Web site, a State Law Matrix may be purchased that offers a comprehensive listing and review of laws that influence the construction process in all 50 states, the District of Columbia, and Puerto Rico.

Some states have organizations unique to the state and the construction professional. For example, in New York, the General Building Contractors of New York State is an organization of more than 150 general contractors and construction managers.[84] Contacting these local organizations can also familiarize a contractor with each state's unique requirements.

Other construction industry trade associations (e.g., the Construction Financial Manager's Association) publish magazines, newsletters, and books with valuable information. Construction law seminars that focus specifically on the laws of a particular state can be found in most states. Several publishing companies sell books that detail construction law issues for a particular state or that focus on how all 50 states handle a particular construction law issue (e.g., LienLaw Online, a subscription Web service providing information regarding the lien laws in all 50 states and the District of Columbia).[85]

[81] *E.g., www.in.gov/sos/; www.sos.georgia.gov/.*
[82] *www.agc.org.*
[83] *www.abc.org.*
[84] *www.gbcnys.agc.org.*
[85] *www.lienlawonline.com.*

If a contractor is bidding on work for one of the state's governmental entities, it is important to review the requirements unique to each department, such as the Department of Transportation[86] or the Department of Administration,[87] as these requirements may vary from those at the state level.

➢ POINTS TO REMEMBER

- As construction work generally constitutes "doing business" in a state, a contractor preparing to work in a new state must first obtain a certificate of authority from the Secretary of State's office.
- Most states require a bond from out-of-state contractors to ensure prompt and full payment of taxes that become due as a function of performing construction work in the state.
- Although not every state requires that a contractor be licensed before bidding or performing construction work, all states require that architects, engineers, and other design professionals be licensed.
- State lien and bond statutes often impose early filing or notice requirements. Identify those requirements before you start work on a new project.
- The manner by which public construction projects are awarded varies by state and should be understood and followed to ensure proper consideration of the bid.
- State and local regulations regarding such things as general contractor relationships with subcontractors, contract award preferences, and the enforceability of contract terms vary greatly from one jurisdiction to another. Do not assume that the laws in a foreign jurisdiction will be identical, or even similar, to those in your home jurisdiction.
- Perform comprehensive risk assessment before bidding on a construction project, and preserve and revisit that assessment throughout the project duration to minimize potential problems.

[86] *www.state.hi.us/dot.*
[87] *www.idoa.in.gov/doa.*

4

COMPETING FOR THE CONTRACT

I. INTRODUCTION: SEALED BIDS TO NEGOTIATED BEST-VALUE AWARDS

The bidding process is often the general contractor's first exposure to the particular construction project and to the owner of that project. Bidding procedures vary widely depending on whether the owner is public or private. Most federal, state, and municipal contracts are awarded pursuant to public bidding. Competitive bidding, although less prevalent in the private sector, is used by many private owners as a means of obtaining an advantageous price while maintaining some control over the quality of the successful awardee. Although private owners are rarely, if ever, "required" to use competitive bidding procedures, for those private owners that do choose to solicit competitive bids for a project, both the owner and potential bidders should pay close attention to general principles of public bidding, because many of the same considerations are present whether the project is public or private.

II. GENERAL CONSIDERATIONS IN BIDS FOR PUBLIC CONTRACTS

The essential characteristics of public contracts at all levels of government are similar. Public contracts generally are awarded to the contractor presenting the "lowest responsible and responsive" offer or the "lowest and best offer." In other words, a successful bid usually will be the one that represents the lowest price from a responsible bidder and that meets the specifications provided by the public owner or contracting agency. The next sections discuss in greater detail the elements of the typical successful bid.[1]

[1]Many of the principles (responsiveness, responsibility) related to competitive bidding have developed in the context of federal government contracting. Therefore, reference is made to the decisions in federal government procurement that help define the concepts and illustrate their application.

III. THE RESPONSIBLE BIDDER

The terms "lowest responsive and responsible" and "lowest and best" bidder, as used in statutes, are not confined to the lowest-dollar bid. Indeed, the words "responsible" and "best" are often as important as the word "lowest." Responsibility determinations focus on whether the contractor has the necessary technical, managerial, and financial capability and integrity to perform the work. A "responsible" bidder is a contractor capable of satisfactorily undertaking and completing the work. Public contracting authorities will consider a number of factors in determining whether a contractor is responsible. These factors, falling into two general categories, are also relevant to a private owner's evaluation of prospective contractors.

First, a contractor must be able to perform the work required by the solicitation. In determining the contractor's ability to perform, public contracting authorities will consider the contractor's financial resources, facilities and equipment, experience, and licenses and permits. The second general category of responsibility standards addresses the contractor's desire and reliability to perform the contract. In this category, procurement officials will consider the ethical integrity of the contractor and the contractor's ability to perform and complete previous (and typically similar) projects.

In federal procurement, Federal Acquisition Regulation (FAR) § 9.104–1 specifies that a contractor must demonstrate these qualifications in order to be considered responsible:

(1) Have adequate financial resources to perform the contract, or the ability to obtain them.

(2) Be able to comply with the required or proposed delivery or performance schedule, considering all existing commercial and governmental business commitments.

(3) Have a satisfactory performance record.

(4) Have a satisfactory record of integrity and business ethics.

(5) Have the necessary organization, experience, accounting and operational controls, and technical skills, or the ability to obtain them.

(6) Have the necessary production, construction, and technical equipment and facilities, or be able to obtain them.

(7) Be otherwise qualified and eligible for award under applicable laws and regulations.

In general, under the principles governing federal contracting, the determination of responsibility is based on the contractor's ability to perform the specified work when such work is to commence, not at the time of bidding. Therefore, a bidder does not have to demonstrate the ability to perform at the time the bid is submitted. Rather, a bidder will be deemed responsible if the bidder has or can obtain the apparent ability to perform the work as of the date work is to begin.[2]

[2]Comp. Gen. B-176227, 52 Comp. Gen. 240.

All available facts, whether submitted with the bid or not, should be submitted by the contractor and considered by the government to resolve responsibility questions. Since a public official is making a quasi-judicial decision when determining whether a bidder is "responsible" within the meaning of the governing statute, the bidder is entitled to the prerequisites of due process. Therefore, a finding by a public body that a bidder is not responsible should be supported by a record establishing (1) the facts on which the decision was based, (2) details of the investigation that disclosed these facts, and (3) the opportunity that was offered to the bidder to present its qualifications.

Generally, the contractor will be entitled to an informal hearing on the matter, although a formal hearing may be prescribed by statute. The contractor is normally entitled to present evidence in its own behalf. Once a decision is made, however, the courts generally will not reverse the agency's decision and require award of the contract, except where there has been a clear violation of the law. In *Ward La France Truck Corp. v. City of New York,*[3] for example, the court found that requirements of a fair hearing had not been observed. The court, however, did not order award of the contract to the affected bidder; instead, it returned the matter to the public body for reconsideration.

In the federal procurement setting, the Comptroller General of the United States (Comptroller General) has limited challenges to responsibility determinations to those involving serious concerns that the contracting officer failed to consider available relevant information or otherwise violated a statute or regulation. The result has been the acceptance of agency determinations in the vast majority of cases.[4] Where an agency determination of nonresponsibility is so unreasonable as to be arbitrary and capricious, however, the affected bidder will be permitted to recover its bid preparation costs.[5]

IV. THE RESPONSIVE BIDDER

A public contract bid usually will not be accepted unless the bid is "responsive"—that is, the bid must comply with all material requirements of the solicitation. Responsiveness differs from responsibility. Responsiveness focuses on whether the bid, as submitted, is an offer to perform the exact tasks spelled out in the bid invitation and whether acceptance will bind the contractor to perform in strict conformance with the invitation.[6]

Failure of a contractor to carefully comply with the requirements for competitive bidding may result in the bid being declared "nonresponsive" or, if an award has been made, may render the contract voidable or prevent the contractor from recovering full compensation for work performed. In determining the responsiveness of a bid, the bidder's intent must be clearly ascertainable from the face of the bid.[7] In order

[3]160 N.Y.S.2d 679 (N.Y. Sup. Ct. 1957).

[4]4 C.F.R. § 21.5(c); *Impresa Construzioni Geom. Domenico Garufi v. United States,* 238 F.3d 1324 (Fed. Cir. 2001).

[5]*PGBA, LLC v. United States,* 60 Fed. Cl. 196 (2004).

[6]*Johnson v. United States,* 15 Cl. Ct. 169 (1988).

[7]*Jarke Corp.,* Comp. Gen. B-231858, 88–2 CPD ¶ 82.

to rise to the level of nonresponsiveness, however, the deviation must be considered material.[8]

A deviation is considered material if it gives one bidder a substantial competitive advantage that prevents other bidders from competing equally. A deviation is also material if it goes to the substance of the bid or works prejudice on other bidders. The deviation goes to the substance of the bid if it affects price, quantity, quality, or delivery of the items offered.[9] A contractor bidding on public work generally must take the contract as presented. Thus, any qualification of a bid that limits or changes one or more of the terms of the proposed contract subjects the contractor to the risk of its bid being deemed nonresponsive. For example, in *Fire-Trol Holdings, LLC,*[10] a contractor's bid was found to be nonresponsive where it offered to provide fire retardants that were not on an approved products list as required by the solicitation.[11] Obviously, such a qualification, if accepted by the owner, could have given the contractor an unfair price advantage over other bidders.[12]

The contractor's inclusion of reservations or conditions in its bid generally renders the bid nonresponsive. According to the FAR, a bid is nonresponsive if it includes conditions such as:

(1) Protecting against future changes in conditions, such as increased costs, if total possible costs to the government cannot be determined.
(2) Failing to state a price and indicating that price shall be "price in effect at time of delivery."
(3) Stating a price but qualifying it as being subject to "price in effect at time of delivery."
(4) Conditioning or qualifying a bid by stipulating that it is to be considered only if, before date of award, the bidder receives (or does not receive) award under a separate solicitation when that is not authorized by the invitation.
(5) Requiring the government to determine that the bidder's product meets applicable government specifications.
(6) Limiting rights of the government under any contract clause.[13]

Most rules concerning bid responsiveness are aimed at preventing a contractor from having "two bites at the apple." In other words, the concept of "bid responsiveness" is used to guard against a low bidder having the opportunity, after bids are opened and all prices are revealed, to accept or reject an award based on some contingency that the bidder created itself, and that applies only to, and works to the advantage of, that bidder.

[8]*R.D. Brown Contractors, Inc. v. Bd. of Educ. of Columbia County,* 626 S.E.2d 471 (Ga. 2006).
[9]FAR § 14.404–2.
[10]*Fire-Trol Holdings, LLC. v. United States,* 68 Fed. Cl. 281 (2005).
[11]*See also Kipp Constr. Co.,* Comp. Gen. B-181588, 75–1 CPD ¶ 20.
[12]*See also Chemtech Indus., Inc.,* B-186652, 76–2 CPD ¶ 274.
[13]FAR § 14.404–2(d).

The prohibition against a bidder having "two bites at the apple" applies most often when a defect in the bid or an ambiguity in a solicitation subjects the intended bid to differing interpretations. For example, in *Caprock Vermeer Equipment, Inc.,*[14] a bidder for an equipment supply contract included in its bid descriptive literature on which the word "optional" was written. The Comptroller General found that there were two interpretations of the bid, at least one of which rendered the bid nonresponsive. The Comptroller General, therefore, upheld the government's rejection of the bid.

A bid will also be considered nonresponsive if the bidder attempts to make the bid contingent upon some act or event. In *Hewlett Packard,*[15] the Comptroller General found a bid to be nonresponsive where the bidder sent a transmittal letter stating that the bid was contingent on the removal of a contract clause. The Comptroller General found that the contingency rendered the bid nonresponsive because the bidder sought to change the terms of the contract to the bidder's sole advantage.

A bid will also be considered nonresponsive where the bidder deviates from the bidding requirements by failing to acknowledge addenda (i.e., changes to the bid documents issued before the bids become due), particularly where the addenda contain a statutorily required provision.[16] Also, an oral, rather than written, acknowledgment of an amendment is unacceptable.[17]

According to the Comptroller General, nonresponsive bids also include bids that fail to acknowledge an amendment that would impose a new legal obligation (even if it would not affect price)[18] and that fail to certify that a federally mandated small business concern will be utilized.[19]

A determination that a bid was nonresponsive because it was materially unbalanced was upheld by the Comptroller General even though the bidder contended the "unbalancing" resulted from allocated technical evaluation and preproduction costs to first articles.[20] The Comptroller General said that, where costs necessary to produce the first articles also are a necessary investment in the production quantity, the costs should be amortized over the total contract rather than allocated solely to the first articles. The reason for rejecting "front-loaded" bids is that the greatly enhanced first-article prices will provide funds to the firm in the early period of contract performance, and will be, in essence, an interest-free loan to which the contractor is not entitled.[21]

Rejection of a bid for nonresponsiveness may also be proper when the principal on the bid bond submitted by the bidder is not the same legal entity as the offeror on the bid form. Generally, a surety can be obligated on a bid bond only if the principal named in the bond fails to execute the contract after award to that bidder. The refusal of another entity to contract with the awarding authority does not result in a forfeiture of the bid bond. Defective bid bonds constitute a substantial deviation, ordinarily

[14]Comp. Gen. B-217088, 85–2 CPD ¶ 259.
[15]Comp. Gen. B-216530, 85–1 CPD ¶ 193.
[16]*Integrated Bus. Solutions, Inc. v. United States,* 58 Fed. Cl. 420 (2003).
[17]*Alcon, Inc.,* Comp. Gen. B-228409, 88–2 CPD ¶ 114.
[18]*Am. Sein-Pro,* Comp. Gen. B-231823, 88–2 CPD ¶ 209.
[19]*Delta Concepts, Inc.,* Comp. Gen. B-230632, 88–2 CPD ¶ 43.
[20]*M. C. Gen., Inc.,* Comp. Gen. B-228334, 87–2 CPD ¶ 572.
[21]*Fid. Techs. Corp.,* Comp. Gen. B-232340, 88–2 CPD ¶ 511.

requiring rejection of the bid as nonresponsive, because such bonds do not protect the public body and enable bidders to get out of contracts with impunity.[22]

Although many deviations such as those just noted may be considered "material," the awarding authority may waive minor irregularities.[23] This long-established policy permitting waiver of minor irregularities or informalities preserves the focus of competitive bidding on lowest price by discouraging questions over matters not affecting the substance of the bid.[24]

The basic rule observed in connection with minor irregularities is that the defect or variation in the bid must have trivial or negligible significance when contrasted with the total cost or scope of the invitation for bids. Deviations affecting price, quantity, quality, delivery, or completion are generally material and merit especially stringent standards to protect against any bidder obtaining a competitive advantage.[25] For bid irregularity to be waived, it must be so inconsequential or immaterial that the bidder does not gain a competitive advantage after all bids have been exposed. Thus, a minor irregularity may be found where the bidder fails to initial a price change in its bid before bid opening;[26] fails to mark its bid envelope with the solicitation number, date, and time of bid opening;[27] or fails to provide incidental information requested by the invitation.[28]

Determining what constitutes a minor informality generally is left to the discretion of the contracting officer,[29] and courts often recognize the contracting officer's broad discretion to determine what constitutes a minor irregularity. For example, a Maryland court held that a bidder's failure to furnish a bid bond was not a material irregularity necessitating bid rejection.[30] Specifically, the court characterized the governing considerations in this way:

> Of course, bidders should make every effort to comply as strictly as possible with specifications. On the other hand, it is the duty of an administrative agency to secure the most advantageous contracts possible for the accomplishment of its work. A bidder's variation from specifications will not exclude him from consideration for the award of the contract unless it is so substantial as to give him a special advantage over the other bidders. In judging whether or not the omission or irregularity in a bid is so substantial as to invalidate it, the court must be careful not to thwart the purpose of competitive bidding by declaring the lowest bid invalid on account of variations that are not material.[31]

[22]*See Yank Waste Co.,* Comp. Gen. B-180418, 74–1 CPD ¶ 190.

[23]*R.D. Brown Contractors, Inc. v. Bd. of Educ. of Columbia County,* 626 S.E.2d 471 (Ga. 2006); *Interstate Constr. Inc.,* Comp. Gen. Feb. B-281,465, 99–1 CPD ¶ 31.

[24]Comp. Gen. B-148624, 41 Comp. Gen. 721; *Faist v. Hoboken,* 60 A. 1120 (N.J. Sup. 1905).

[25]*Tel-Instrument Elecs. Corp. v. United States,* 56 Fed. Cl. 174 (2003); *Bishop Contractors, Inc.,* Comp. Gen. B-246526, 91–2 CPD ¶ 555. *See* FAR § 14.405.

[26]*Werres Corp.*, Comp. Gen. B-211870, 83–2 CPD ¶ 243.

[27]*Bond Transfer & Storage Co.*, Comp. Gen. B-210251, 83-1 CPD ¶ 97.

[28]*Indus. Design Laboratories, Inc.*, Comp. Gen. B-215162, 84-2 CPD ¶ 413.

[29]*Interstate Rock Prods., Inc. v. United States,* 50 Fed. Cl. 349 (2001).

[30]*Bd. of Educ. of Carroll County v. Allender,* 112 A.2d 455 (Md. 1955).

[31]*Id.* at 460 (citing *George A. Fuller Co. v. Elderkin,* 154 A. 548 (Md. 1931)(emphasis supplied.)

As noted previously, however, the court's conclusion that a bid without a bid bond did not constitute a "special advantage over the other bidders" is inconsistent with the great majority of such decisions.[32]

Another far-reaching example of a procuring agency's exercise of discretion to award to the low bidder despite a potential bid irregularity arose in *Pullman Inc. v. Volpe.*[33] There the court upheld "clarifications" made by the low bidder after bid opening on the ground that the clarifications demonstrated that the bid conformed to and did not alter the specifications.

In contrast to the two cases just noted, a number of Comptroller General decisions may help illustrate what are more typically considered to be minor informalities:

(1) The omission of unit prices under circumstances where they could be calculated by dividing total prices by estimated quantities[34]
(2) The insertion of the wrong solicitation number on a bid bond[35]
(3) The omission of a principal's signature on a bid bond when the bond is submitted with a signed bid[36]
(4) An ambiguous bid price if the bid is low under all reasonable interpretations[37]
(5) A failure to include required information on affiliates[38]
(6) A failure to acknowledge an amendment to the solicitation that would not have a material impact on price[39] or only a trivial impact on price[40]
(7) A failure to acknowledge an amendment reducing the quantity of items to be ordered where the amendment imposed no obligations not already in the original invitation and had no impact on the bid price[41]
(8) A failure to provide equipment description information when the solicitation did not make it clear such failure would result in bid rejection[42]

V. THE "LOWEST AND BEST" BIDDER

Some who engage in competitive bidding assume that an owner must award the construction contract to the lowest bidder. This belief, however, is only partially accurate in public contracting and generally is not true at all in private contracting. A contractor's

[32]*See, e.g., Thorp's Mowing,* Comp. Gen. B-181154, 74–2 CPD ¶ 37; *George Harms Constr. Co. v. Ocean County Sewerage Auth.,* 394 A.2d 360 (N.J. Super. Ct. App. Div. 1978).
[33] 337 F. Supp. 432 (E.D. Pa. 1971).
[34]*GEM Eng'g. Co.,* Comp. Gen. B-231605.2, 88–2 CPD ¶ 252.
[35]*Kirila Contractors, Inc.,* Comp. Gen. B-230731, 88–1 CPD ¶ 554.
[36]*P-B Eng'g. Co.,* Comp. Gen. B-229739, 88–1 CPD ¶ 71.
[37]*NJS Dev. Corp.,* Comp. Gen. B-230871, 88–2 CPD ¶ 62.
[38]*A & C Bldg. & Indus. Maint. Corp.,* Comp. Gen. B-229931, 88–1 CPD ¶ 309.
[39]*Adak Commc'ns Sys., Inc.,* Comp. Gen. B-228341, 88–1 CPD ¶ 74.
[40]*Star Brite Constr. Co.,* Comp. Gen. B-228522, 88–1 CPD ¶ 17 ($2,000 out of a $118,000 difference between low and second low bid).
[41]*Automated Datatron, Inc.,* Comp. Gen. B-231411, 88–2 CPD ¶ 137.
[42]*Houston Helicopters, Inc.,* Comp. Gen. B-231122, 88–2 CPD ¶ 149.

bid, in a legal sense, is simply an offer to perform work; until the owner accepts that offer, no enforceable obligation arises.[43] Highlighting this idea that the bid itself is only an offer to perform, the owner's Invitation for Bids (Invitation) usually contains a provision to the effect that "the owner reserves the right to reject any or all bids." This provision puts contractors on notice in the bidding documents that the owner may choose not to "accept" any of the bid "offers."[44]

Although the federal government must consider factors other than price in the evaluation process, the relative pricing of bids is of great importance in federal procurement. Federal agencies must accept the bid that is the most financially advantageous to the federal government, so long as that bid is responsive to the terms of the solicitation and the bidder is a responsible contractor. According to one federal statute, agency heads must award contracts "to the responsible bidder whose bid conforms to the solicitation and is most advantageous to the United States, *considering only price and the other price-related factors included in the solicitation*" (emphasis added).[45]

The federal concern for securing the most financially advantageous bid is well illustrated by the evaluation of bids responding to solicitations with multiple items. Bids presented on this type of solicitation generally must be evaluated based both on the unit price offered by a bidder as well as the total price. In *S. J. Groves & Sons Co.,*[46] the Comptroller General stated that "where awards on a combination of schedules is contemplated the award made must result in the lowest cost to the Government to carry out the mandate of 10 U.S.C. § 2305(c). . . ."[47] As a result, a federal agency actually may be required to make multiple awards if different bidders offer the lowest unit price on different specific items or if the value of prices of individual items is lower than the total price offered by another bidder.

Even though the relative prices offered by bidders are of great importance to federal procurement officials, a contractor submitting the low bid generally has no vested interest in the award of that contract.[48] Nevertheless, that contractor cannot be denied the contract unless the denial would be in the public's best interest. A further qualification to the importance of price is that even if the bidder on a federal contract

[43]As stated in Samuel Williston & Richard Lord, Williston on Contracts (4th ed. 2007): Often tenders or bids are advertised for by public corporations, municipalities, counties or states, or private corporations. The rules governing such bidding are analogous to the rules governing auction sales. Thus, an ordinary advertisement for bids or tenders is not itself an offer, but the bid or tender is an offer that creates no right until accepted.

[44]Some courts, however, have found that a "reservation of right to reject all bids" clause will not permit an arbitrary rejection of bids, particularly when the bids meet the terms of the solicitation and the rejection is intended primarily to avoid a bid protest. *See Pataula Elec. Membership Corp. v. Whitworth,* 951 F.2d 1238, 1243–44 (11th Cir. 1992), *rev'd on other grounds, sub nom. Flint Elec. Membership Corp. v. Whitworth,* 68 F.3d 1309 (11th Cir. 1995), *corrected decision,* 77 F.3d 1321 (11th Cir. 1996); *Cianbro Corp. v. Jacksonville Transp. Auth.,* 473 So. 2d 209 (Fla. Dist. Ct. App. 1985).

[45]10 U.S.C. § 2305(b)(1)(3).

[46]Comp. Gen. B-184260, 76–1 CPD ¶ 205.

[47]*Id.* at 23.

[48]Courts in some jurisdictions, however, have found that the contractor submitting the lowest, responsible responsive bid does have a protected interest under state or local law in contract award. *See Pataula Elec. Membership Corp. v. Whitworth,* 951 F.2d 1238, 1243–44 (11th Cir. 1992), *rev'd on other grounds, sub nom. Flint Elec. Membership Corp. v. Whitworth,* 68 F.3d 1309 (11th Cir. 1995), *corrected decision,* 77 F.3d 1321 (11th Cir. 1996).

does not present the lowest bid, the bid is entitled to fair consideration.[49] Where a federal agency fails to consider adequately a bidder's proposal, the bidder may be entitled to recover bid preparation costs.[50]

Most state and local governments must also award their contracts to the lowest and best bidder. Most states and many localities have enacted statutes or ordinances mandating award to the lowest bidder whose bid meets the solicitation's technical requirements.[51] Some courts addressing state and local public competitive bidding statutes, regulations, ordinances, and guidelines have held that such provisions can create a protected property interest in favor of the lowest responsible, responsive bidder in the expectation of award, if the solicitation is subject to competitive bidding requirements protected by federal or state law.[52] Other courts interpreting state or local competitive bidding requirements have found that the lowest responsible, responsive bidder does not have a protected property interest in the award of the contract.[53] Determining whether a protected property interest exists turns on the nature of the state or local law, regulation, or ordinance.[54]

If a protest is contemplated, the contractor should first determine whether a competitive bidding statute or regulation governs the state's or locality's procurement, and, if so, what such regulation permits or requires.

VI. NEGOTIATED "BEST VALUE" SELECTION PROCESS

Historically, the preferred method of contracting in the federal government system was through sealed bidding with the government issuing an Invitation. Bids were received on the date specified in the Invitation, and award was made to the lowest-dollar bid from a responsive, responsible bidder. Negotiated procurement was another method of contracting but was used only under specific circumstances set forth in statute and regulations. Under negotiated procurement procedures, the federal government issued a request for proposals (RFP). Generally, once proposals were received, the government and the offerors entered into negotiations under the terms of the solicitation, with award being made to the offeror that ultimately submitted the lowest negotiated price.

The negotiated "best value" procurement process is authorized under 10 U.S.C. § 2304 and the FAR.[55] Under this procedure the government is not required to award

[49]*Concept Automation, Inc. v. United States,* 41 Fed. Cl. 361 (1998).

[50]*Coflexip & Servs., Inc. v. United States,* 961 F.2d 951 (Fed. Cir. 1992).

[51]*See, e.g.,* FLA. STAT. ANN. §§ 255.29, 337.11; GA. CODE ANN. § 36–91–21.

[52]*See, e.g., Harris v. Hays,* 452 F.3d 714 (8th Cir. 2006); *Club Italia Soccer & Sports Org., Inc. v. Charter Twp. of Shelby, Mich.,* 470 F.3d 286 (6th Cir. 2006); *Enertech Elec. v. Mahoning County Comm'rs,* 85 F.3d 257 (6th Cir. 1996).

[53]*See, e.g., Interior Contractors, Inc. v. Bd. of Trs. of Newman Mem'l County Hosp.,* 185 F. Supp. 2d 1216 (D. Kan. 2002); *John Gil Constr., Inc. v. Riverso,* 72 F. Supp. 2d 242 (S.D.N.Y. 1999).

[54]*See, e.g., Pataula Elec. Membership Corp. v. Whitworth,* 951 F.2d 1238, 1243–44 (11th Cir.), *rev'd on other grounds, sub nom. Flint Elec. Membership Corp. v. Whitworth,* 68 F.3d 1309 (11th Cir. 1995), *corrected decision,* 77 F.3d 1321 (11th Cir. 1996).

[55]*See* FAR Part 15.

to the lowest-priced offeror; but through a trade-off evaluation process as specified in the RFP, the government makes an award to the offeror whose proposal provides the "best value" to the government. In this regard, FAR § 15.101–1 provides:

(a) A tradeoff process is appropriate when it may be in the best interest of the Government to consider award to other than the lowest priced offeror or other than the highest technically rated offeror.
(b) When using a tradeoff process, the following apply:
  (1) All evaluation factors and significant subfactors that will affect contract award and their relative importance shall be clearly stated in the solicitation; and
  (2) The solicitation shall state whether all evaluation factors other than cost or price, when combined, are significantly more important than, approximately equal to, or significantly less important than cost or price.
  (3) This process permits tradeoffs among cost or price and non-cost factors and allows the Government to accept other than the lowest priced proposal. The perceived benefits of the higher priced proposal shall merit the additional cost, and the rationale for tradeoffs must be documented in the file in accordance with FAR § 15.406.

The negotiated best-value procurement process has been the subject of numerous appeals to the Comptroller General as well as to the United States Court of Federal Claims and the United States Court of Appeals for the Federal Circuit. The Comptroller General and federal courts will give substantial deference to the discretion of the government in determining which offer is the most advantageous to the government. For example, in *Park Tower Mgmt., Ltd. v. United States,*[56] the agency selected a maintenance services contractor after an extended negotiated procurement process. The disappointed bidder filed a protest claiming that the agency's best-value determination was unreasonable because, in part, the award went to a contractor with a lower technical evaluation score. The *Park Tower* court upheld the agency's decision, noting that the best-value approach gives the agency substantial discretion, even more so than when using a sealed bid approach.

The protestor's burden when seeking to overturn an agency's decision becomes heavier as the agency's level of discretion increases. In addition, evaluation scores are merely guides to be used by the agency, which must exercise judgment in determining how differences in technical evaluation scores will impact overall performance on the contract. The ultimate standard or test applied by both the Comptroller General and the federal courts is (a) whether the government complied with the FAR requirements for negotiated best-value procurements; and (b) whether the agency followed the selection and evaluations procedures set forth in the RFP.[57]

The negotiated best-value method is also widely used in state and local government procurement.[58] Even when state statutes mandate the use of sealed bidding, the use

[56]67 Fed. Cl. 548 (Fed. Cl. 2005).
[57]*Microdyne Outsourcing, Inc. v. United States,* 72 Fed. Cl. 694 (2006); *Widnall v. B3H Corp.,* 75 F.3d 1577 (Fed. Cir. 1996). *See also Banknote Corp. of Am., Inc. v. United States,* 56 Fed. Cl. 377 (2003).
[58]*See, e.g.,* CAL. EDUC. CODE § 17250.15; Colo. Rev. Stat. § 24–93–106; FLA. STAT. ANN. § 287.055.

of best value might be possible when a project's overall costs can be reduced while still maintaining public policy objectives or when sealed bidding is not feasible or practical.[59] For example, when substantial cost savings could be achieved without encouraging favoritism and without substantially diminishing competition for public contracts, an Oregon court upheld an agency's decision to forgo the use of competitive sealed bidding.[60] Sealed bidding may not be feasible or practical in cases of insufficient competition, lack of detailed specifications, or under emergency situations.

Although many states have permitted their procuring agencies to use the best-value approach, especially for design-build projects (see **Chapter 2**), there is little similarity between the details of various state plans. The major differences lie primarily in the evaluation criteria used to determine best value and the process by which the winning bid is selected.

The process used to select the best value may involve specific requirements for creating the solicitation, scoring and ranking the submitted proposals, making an initial determination of the most qualified bidders, or rules regulating negotiations with bidders. In general, however, the process may be reduced to a two-step procedure consisting of a qualification stage and a best-value evaluation stage. In the first stage the procuring agency receives proposals and creates a short list of the best candidates meeting some specified qualification threshold, based on appropriate selection criteria. The most qualified candidates receive an invitation to negotiate. For example, Florida agencies seeking architectural or engineering services narrow the list of bidders to not less than three candidates, based on the ability to furnish the required services.[61] In the second stage, the procuring agency typically conducts an in-depth evaluation of the qualified proposals and selects the bidder that presents the best value to the project.

States following the best-value method have also created different guidelines for using the evaluation criteria to award contracts. In some states, statutes enumerate evaluation factors that the procuring agency must use. Colorado's integrated project delivery statute, for instance, requires the use of price, technical approach, past performance, and project management and craft labor capabilities as factors.[62] In other states, the procuring agency itself decides what factors determine best value. For example, a California school district may independently determine appropriate selection criteria for best-value design-build projects.[63] Regardless of the guidelines used to create best-value criteria, state procurement regulations are subject to the same types of challenges as seen in the federal arena. That is, the somewhat subjective nature of the best-value method gives considerable discretion to the agency, and, therefore, the agency must comply with the applicable statutory or regulatory requirements as well as the evaluation and selection procedures set forth in the RFP.[64]

[59] *Associated Builders and Contractors, Inc. v. Tri-County Metro. Transp. Dist. of Or., (Tri-Met) Bd. of Dirs.,* 12 P.3d 62 (Or. Ct. App. 2000); *Graydon v. Pasadena Redevelopment Agency,* 164 Cal. Rptr. 56 (Cal. Ct. App. 1980).

[60] *Associated Builders and Contractors, Inc. v. Tri-County Metro. Transp. Dist. of Or., (Tri-Met) Bd. of Dirs.,* 12 P.3d 62 (Or. Ct. App. 2000).

[61] FLA. STAT. ANN. § 287.055.

[62] COLO. REV. STAT. § 24–93–106.

[63] CAL. EDUC. CODE § 17250.15.

[64] *See, e.g., Nachtigall v. N.J. Turnpike Auth.,* 694 A.2d 1057 (N.J. Super. Ct. App. Div. 1997).

VII. ELECTRONIC BIDS

Electronic commerce has been implemented in virtually all of the states and the federal government, but there is no predominant pattern among the states. Many have passed enabling legislation allowing state officials to establish procurement programs for electronic commerce, while others provide only electronic solicitation notices. Some states allow online bidding as an alternative method, and still others will accept electronic bids for selected agencies. The federal government utilizes electronic procurement practices. Section 15 of the FAR, which governs the conduct of negotiated procurement, has been rewritten to allow federal agencies to use online auction technology. Still prohibited, however, is conduct that "favors one offeror over another," "reveals an offeror's technical solution," or "knowingly furnishes source selection information" contrary to regulatory or statutory requirements.[65]

Four decisions from the Comptroller General illustrate problems that e-bidding for federal contracts may present. In *PM Tech, Inc.,*[66] a contractor bidding online attempted to take the necessary steps to timely submit its bid but failed to submit anything other than the cover sheet for its proposal by the time responses to the solicitation were due. PM Tech provided the information missing from its original submittal one day after solicitations were due; however, the bid was still rejected as untimely. PM Tech protested to the Comptroller General. The Comptroller General rejected PM Tech's arguments and stated in its decision: "We view it as an offeror's responsibility, when transmitting its proposal electronically, to insure the proposal's timely delivery by transmitting the proposal sufficiently in advance of the time set for receipt of proposals to allow for timely receipt by the agency."[67] In a decision similar to those involving late delivery of bids or late delivery of faxes, the Comptroller General concluded that PM Tech assumed the risk of submitting an incomplete bid when it waited until only 13 minutes before the deadline for receipt of proposals to begin to transmit its proposal.

Similarly, in *Sea Box, Inc.,*[68] the Comptroller General held that although the bid was received by the federal government's e-mail system before the deadline, it was untimely because it failed to reach the specified e-mail box by the stated deadline. In the *Sea Box* decision, the Comptroller General pointed out that the protestor sent its bid, consisting of seven e-mails, approximately 11 minutes before the stated deadline. The seven e-mails reached the agency's initial point of entry for e-mail before the deadline, were held for a period of time, then were sent to a virus-scanning server and subsequently arrived 7 to 24 minutes late at the specified e-mail box for receipt of bids. The agency rejected the bid as late, and the Comptroller General agreed.

In *Tischman Construction Corp.,*[69] the prospective contractor submitted identical bids electronically and on paper as was required by the RFP for construction

[65]FAR § 15.306(e).
[66]Comp. Gen. B-291082, 2002 CPD ¶ 172.
[67]*Id.* at 3.
[68]Comp. Gen. B-291056, 2002 CPD ¶ 181.
[69]Comp. Gen. B-292097, 2003 CPD ¶ 94.

management services. The paper version of the bid was received by the contracting entity approximately 73 minutes after the stated deadline. The electronic version of the bid was received approximately 50 minutes before the bid deadline. The agency rejected the bid as untimely. The Comptroller General disagreed and sustained Tischman's protest, relying on a prior decision in which a bidder submitted its bid at two separate locations, as required by the solicitation, but only one was timely.[70] In that earlier decision, the Comptroller General found that the agency had received a complete copy of the bid in a timely manner and no competitive advantage was obtained by the contractor. Under those circumstances, the late bid at the second location was considered a minor informality by the Comptroller General.

Finally, in *VSA Information Sys. Inc.,*[71] the Comptroller General denied a protest where the solicitation materials were available only on the Internet. The procuring agency had posted an amendment to the solicitation with a short response time and did not specifically advise the protestor of the amendment. The Comptroller General found that the protestor had not taken reasonable steps to make itself aware of amendments. The agency cited registering for e-mail notification or checking the Internet site as reasonable precautions to take to ensure receipt of all relevant amendments. Because of its failure to take these reasonable steps, the protestor had insufficient time to protest the terms of the solicitation.

The strict and longstanding rules regarding timeliness for the submission of bids or proposals in proper format should be adhered to by firms submitting bids electronically.

VIII. REVERSE AUCTIONS

The advancement of technology has also given rise to the use of online "reverse auctions." A reverse auction is a live, Internet-based auction in which prospective contractors bid down the price to provide the products or services sought by the purchaser. A typical reverse auction starts with the owner selecting a list of prequalified bidders. Specifications and drawings are provided to the bidders before the auction with sufficient lead time to prepare bids. Once the auction starts, the contractors bid simultaneously and can evaluate their relative standing, although they do not know the identity of the other bidders and may not have a competitor's specific quote or bid. Bids are made in multiple rounds, and the duration of each round is relatively short. The auction ends at a predetermined time, and the lowest bidder is selected.

Reverse auctions may offer many potential benefits to owners. When the products being sought are commodities or when owners can clearly define the scope, specifications, and conditions of the desired products and services, a reverse auction may help to get to the bottom-line price. This is especially true when there are many qualified bidders. The speed of a reverse auction may also decrease procurement time as there is less need for negotiations between the owner and potential contractors.

[70]*ABT Associates, Inc.,* Comp. Gen. B-226063, 87–1 CPD ¶ 513.
[71]Comp. Gen. B-291488, 2002 CPD ¶ 205.

Finally, reverse auctions leverage technology, and owners may easily outsource their auctions to third-party providers. This may lower the transactional costs of procurement and may ensure that a detached and neutral party administers the process.

The use and efficacy of reverse auctions, however, is hotly contested in the construction industry.[72] One of the chief complaints lodged by contractors and suppliers is that a reverse auction is heavily focused on price to the exclusion of other relevant factors. Since construction products and services seldom can be characterized as commodities, the focus on price may lead owners to ignore the inherent variability found in construction projects. That is, the winning bidder's price may simply be a bare-bones bid that does not fully consider the numerous unknown issues likely to be encountered. This may lead to disagreements over the necessity of change orders and the subsequent pricing of such changes. Also, important qualitative factors such as schedule, quality, and performance are likely to be discounted by the reverse auction process. When these factors are deemphasized, the resulting true cost of the project is likely to be higher than anticipated.

Reverse auctions may also lead to bid gaming, in which the owner fails to obtain the lowest bid that otherwise would have been offered. Since the contractors understand that they may simply offer a lower bid during a later stage of the reverse auction, they may lack the motivation to put forth their truly lowest bid. Instead, contractors hold back their lowest bids and monitor the progress of the other bidding. As a consequence, the winning bidder may win the auction with a bid that exceeds the price it otherwise would have been prepared to offer. In sealed bidding, by contrast, there is no opportunity to view the other bids, and contractors realize that there is but one opportunity to offer a bid. Sealed bidders are, therefore, motivated to provide the lowest acceptable bid when the bid is initially submitted.

IX. GENERAL CONSIDERATIONS WHEN COMPETING ON PRIVATE CONTRACTS

Few rules exist in private contract bidding that obligate the owner to award to a particular contractor. Although a private owner is not bound by statute to operate under the "lowest responsible, responsive bidder" approach, the private owner likely will consider many of the same factors when reviewing bids. The private owner may select a contractor based on any criteria it deems appropriate. The private owner is generally under no obligation to make award to the lowest bidder, and might attempt to negotiate an even lower price once the low bid has been determined. Moreover, private owners are generally under no duty to disclose the bases for their decisions. Private owners may open bids outside the presence of all bidders. Private owners may require the bidders to provide a wide variety of information concerning the contractor's ability to perform the project. In addition, private owners may or may not require the contractor to provide a bid bond to guarantee its bid or proposal, or payment and performance bonds or other security, to guarantee performance once the work starts.

[72]*See, e.g.,* Associated General Contractors of America, White Paper on Reverse Auctions for Procurement of Construction (2003).

As with public bidding, the contractor should present itself as a responsible business that is responsive to the owner's needs and that will perform at a reasonable price. The contractor should also take care to determine whether the owner can meet payment obligations and provide promised logistical support. For the most part, a private owner presents a much greater financial risk than a public body with its tax base. Additionally, the contractor should carefully review all specifications and plans presented by the owner, because the private owner may not have the resources to provide the review necessary to find and eliminate potential, and costly, design problems.

X. EFFECT OF PAST PERFORMANCE EVALUATIONS ON AWARD PROCESS

In 1994, Congress passed the Federal Acquisitions Streamlining Act (FASA).[73] In so doing, Congress acknowledged that a contractor's past performance should be evaluated during a current acquisition process to determine whether that contractor should receive future work. Section 1091 of FASA provides that "past contract performance of an offeror is one of the relevant factors that a contracting official of an executive agency should consider in awarding a contract." Under FASA, all federal departments and agencies must have procedures to record contractor performance contemporaneous with that performance on all contracts over $100,000 and to use the contractor's past performance during future acquisition processes. The federal government believes that the use of past performance evaluations will sufficiently motivate contractors to perform at the highest standards or, to the extent they are not performing at such level, to improve their performance before they are again rated by an agency or department. Past performance evaluations are a significant factor in the agencies' determinations of a best-value selection. Procuring agencies can better predict the quality of the contractor's performance and, therefore, the associated customer's satisfaction.

The primary rationale supporting the evaluation of contractor performance and its use in future procurements is twofold. First, Congress believes that an active dialogue between the contractor and the government during performance will result in better current performance by the contractor. Second, because contractors are made aware that their performance evaluations will be used by other procuring agencies on future procurements, procuring agencies will be better able to select high-quality contractors for new contracts.

The federal government has limited the scope and content of past performance evaluations to matters related to a contractor's actual contract performance. In April 2002, the Office of Federal Procurement Policy issued a government-wide memorandum providing that contractors could not be given "downgraded" past performance evaluations for availing themselves of their rights to file protests and claims or for deciding not to use alternative dispute resolution (ADR). That same memorandum also provided that contractors could not be given more "positive" past performance evaluations for refraining from filing protests and claims or agreeing to use ADR.

[73]Pub. L. No. 103–355 (Oct. 13, 1994) (codified in scattered sections of 10 U.S.C. and 41 U.S.C.).

Several Comptroller General decisions have addressed the relevance of past performance evaluations and their use in current procurements. In *C. Lawrence Construction Company, Inc.*,[74] the Comptroller General found that the requirement of the United States Army Corps of Engineers (Corps) that offerors provide at least five past performance references for relevant contracts, which were defined in the RFP as similar construction contracts in the $5 to $10 million range, was not unduly restrictive and was consistent with the Corps' position that evaluating at least five projects provided a "comfort zone" with respect to a prospective contractor's overall performance trends.

The Comptroller General has also ruled that it is reasonable for an agency to consider the *specific* experience of a contractor to which award was being made more favorably than the *general* experience of the protesting company.[75] Similarly, agencies can look at the contractor's performance on contracts other than those similar to the one for which the past performance is being evaluated and can, in fact, look to see how a firm has carried out its contractual obligations. Such evaluation can provide the agency with a prediction of whether the contractor will satisfactorily perform the new contract.[76] A smaller dollar value does not render the previous contract performance irrelevant for past performance purposes.[77] Agencies may also consider negative past performance comments even though such comments have not been documented contemporaneously with the occurrence of the past performance.[78]

A. Past Performance Evaluation Procedures

As part of the overall evaluation process for a current procurement, the procuring agency should confirm the accuracy of any prospective offeror's past contract information and assign a performance risk rating. Final past performance ratings may be reflected in a color, adjective rating, a number, or some other means, depending on the particular agency policy for indicating the relative ranking of offerors. In a prospective contractor's performance risk assessment, the number and severity of problems in its overall work record should be considered, as should demonstrated effectiveness of corrective action taken. In instances where good or poor performance is noted, it should be related to a specific solicitation requirement. When a prospective contractor's past performance reflects problems, the procuring agency must evaluate the extent to which the government played a part in that poor performance. Naturally, the procuring activity should look for the areas of performance that are most critical to the procurement and seek instances of excellent or exceptional performance by contractors in related areas on previous procurements.

When a procuring agency is drafting past performance evaluation criteria, the Office of Federal Procurement Policy has provided the following factors as guidance for drafting a past performance evaluation:

[74]Comp. Gen. B-289341, 2002 CPD ¶ 17.

[75]*M&W Constr. Corp.*, Comp. Gen. B-288649.2, 2002 CPD ¶ 30.

[76]*The Standard Register Co.*, Comp. Gen. B-289579, 2002 CPD ¶ 54.

[77]*Dan River, Inc.*, Comp. Gen. B-289613, 2002 CPD ¶ 80.

[78]*Kathpal Techs., Inc.*, Comp. Gen. B-291637.2, 2003 CPD ¶ 6.

(1) *Use Past Performance as a Distinct Factor.* The past performance factor should be distinct and identifiable in order to reduce the chances of its impact being lost within other factors and to ease the evaluation process. But if integrating past performance with other noncost/price factors provides a more meaningful picture, each agency should use its own discretion. The key is not to dilute the importance or impact of past performance when determining the best-value contractor.

(2) *Choose Past Performance Subfactors Wisely.* Tailor the subfactors to match the requirement and to capture the key performance criteria in the statement of work. Carefully consider whether subfactors add value to the overall assessment, warrant the additional time to evaluate, and enhance discernment among the competing proposals.

   a. *Quality of Product or Service.* The offeror will be evaluated on compliance with previous contract requirements, accuracy of reports, and technical excellence, to include quality awards/certificates.

   b. *Timeliness of Performance.* The offeror will be evaluated on meeting milestones, reliability, responsiveness to technical direction, deliverables completed on time, and adherence to contract schedules, including contract administration.

   c. *Cost Control.* The offeror will be evaluated on its ability to perform within or below budget; use of cost efficiencies; relationship of negotiated costs to actuals; submission of reasonably priced change proposals; and timely providing current, accurate, and complete billing.

   d. *Business Relations.* The offeror will be evaluated on its ability to provide effective management, meet subcontractor and SDB [Small Disadvantaged Business] goals, cooperative and proactive behavior with the technical representative(s) and contracting officer, flexibility, responsiveness to inquiries, problem resolution, and customer satisfaction. The offeror will be evaluated on satisfaction of the technical monitors with the overall performance and final product and services. Evaluation of past performance will be based on consideration of all relevant factors and circumstances. It will include a determination of the offeror's commitment to customer satisfaction and will include conclusions of informed judgment. The basis for the conclusions of judgment should be substantially documented.

(3) *Subcontractor, Teaming and Joint Venture Partner's Past Performance.* For the purpose of evaluation of past performance information, offerors shall be defined as business arrangements and relationships such as joint ventures, teaming partners and major subcontractors. Each firm in the business arrangement will be evaluated on its performance under existing and prior contracts for similar products or services.[79]

[79]Executive Office of The President, Best Practices for Collecting and Using Current and Past Performance Information, Office of Federal Procurement Policy, Office of Management & Budget, 19–20 (2000).

Procuring activities using the foregoing guidelines in evaluating past performance of prospective contractors should first determine how well a prospective contractor has performed and how relevant that performance is to the instant procurement. That final rating is then used along with other rated evaluation factors, such as price, in a comparative assessment to determine the most highly rated offeror. That offeror is most likely to be awarded the contract. If the procuring activity identifies negative findings that translate to high performance risk ratings, the agency should determine the extent to which the government may have played a part in or contributed to the negative finding and the extent of that contribution. Finally, procuring activities are required by the Office of Federal Procurement policy to be cautious not to downgrade or penalize offerors "for the judicious use of the contract claims process."[80]

B. Challenges to Past Performance Evaluations

FAR § 42.1503 provides that contractors can respond to performance evaluations after the contracting officer has signed the final assessment. The initial assessment of the contractor's performance is made by the program and contracting office and typically is signed by the program office person most familiar with the contractor's performance. The contracting officer initials the initial assessment and signs the final assessment. Once the contracting officer signs the final assessment, it should be sent to the contractor for the contractor's comments. The contractor has at least 30 days to respond. The contracting officer may extend the response time on an as-needed basis. The contracting officer is encouraged to call the contractor and initiate discussions on the performance assessment, even if the contractor fails to respond by the established deadline. If the contractor fails to respond at all, then the government's comments will stand alone. If, however, the contractor provides a rebuttal for a part of, or all of the ratings, then the contracting officer and lead assessor must work with the contractor to see if an agreement can be reached on the contractor's ratings. When the contracting officer and lead assessor meet with the contractor to determine whether the ratings can be agreed on, these meetings should be face to face. If no agreement can be reached, the contractor may seek review at least one level above the contracting officer.

Any rebuttal statements provided by the contractor and agency review thereof must be attached to the performance evaluation and provided to any acquisition officials evaluating the contractor for future contracts. Contracting officers must send a copy of the government's assessment to the contractor once it is complete, and in no event later than when the assessment is input into any automated Past Performance Information System (PPIS). Any interim assessment should be retained until contract completion and included as part of the final assessment in the contract file, but not entered in the PPIS. Finally, no assessments may be retained in any PPIS longer than three years after contract completion. Any assessment storage system should provide an individual contractor access to only its own assessments.

[80] *Id.* p. 21.

Other than as just discussed, challenges to a contractor's past performance are normally in the context of challenges following the government's terminating a contract for default based on a contractor's unacceptable performance or alleged failures of the government to comply with the Code of Federal Regulations (C.F.R.), FAR, or agency regulations regarding use of past performance information in bid evaluations. (See **Section IX** of this chapter.)[81]

There is a split between the boards and the Court of Federal Claims regarding a contractor's right to challenge a past performance evaluation. The boards, which do not have bid protest jurisdiction, have held that the Contract Disputes Act (41 U.S.C. §§ 601–613) does not give them jurisdiction to hear an appeal by a contractor challenging a past performance evaluation.[82] The Court of Federal Claims does have jurisdiction to hear bid protests and claims.[83] The Court of Federal Claims has held that the Tucker Act vested the court with jurisdiction to review a contractor's *claim* challenging a past performance evaluation if the applicable *claim submission* requirements of the Contract Disputes Act were satisfied.[84]

XI. CONTRACTOR BID MISTAKES

Bid mistakes occur often in the rush of competitive bidding situations. Contractors do not work under ideal conditions in the scramble to meet the deadline for submitting bids. Most courts recognize that honest, sincere people, even in the exercise of ordinary care, can make mistakes of such a fundamental character that holding the contractor to the bid would be fundamentally unfair.[85]

The rationale behind allowing a contractor to withdraw an erroneous bid lies in the principle that, in most cases where performance has not yet commenced, the owner can be returned to the status quo and will suffer no injury if withdrawal of the bid is permitted. The mere fact that the owner will have to accept a bid at a higher price generally is not viewed as the type of "injury" that would justify holding the contractor to its mistaken bid. If the contractor is low due solely to a mistake, the owner would receive an unearned windfall if it could require the contractor to perform at the mistakenly low price.[86]

[81]*Microdyne Outsourcing, Inc. v. United States,* 72 Fed. Cl. 694 (2006); *Widnall v. B3H Corp.,* 75 F.3d 1577 (Fed. Cir. 1996). *See also Banknote Corp. of Am., Inc. v. United States,* 56 Fed. Cl. 377 (2003); *Aerospace Design & Fabrication, Inc.,* Comp. Gen. B-278896.3, 98–1 CPD ¶ 139 (upholding a protest based on agency's improper evaluation of proposals).

[82]*See TLT Constr. Corp.,* ASBCA No. 53769, 02–2 BCA ¶ 31,969.

[83]28 U.S.C. §§ 1491–1509.

[84]*Record Steel and Constr., Inc. v. United States,* 62 Fed. Cl. 508 (2004). See **Chapter 22** for a discussion of CDA claim submission requirements.

[85]*Info. Intern. Assocs. Inc. v. United States,* 74 Ged. Cl. 192 (2006); *Kenneth E. Curran, Inc. v. State,* 215 A.2d 702 (N.H. 1965).

[86]Comp. Gen. B-137155, 38 Comp. Gen. 218; Comp. Gen. B-167649, 1969 WL 3519.

A. Elements for Relief from Bid Mistake

Despite the general availability of equitable relief, not every bid mistake entitles a contractor to retract its bid. To justify the court's intervention, a contractor generally must satisfy four criteria:

(1) The mistake is of such consequence that enforcement would be unconscionable.
(2) The mistake must relate to the substance of the consideration, that is, a "material" feature.
(3) The mistake must have occurred regardless of the exercise of ordinary care.
(4) It must be possible to place the other party in status quo.[87]

This test has been adopted in a number of cases.[88] For example, the Arizona Court of Appeals held, in *Marana Unified School District No. 6 v. Aetna Casualty & Surety Co.,*[89] that a contractor could withdraw its bid containing a bid mistake in excess of $300,000 without forfeiting its bid bond because the mistake: (1) was large; (2) related to a material matter; (3) resulted from an honest mathematical error; and (4) did not seriously prejudice the owner.

Other cases suggest a fifth requirement: timely notice. When there has been a unilateral mistake, prompt notification to the other party concerning the error may be crucial in determining whether the contractor can withdraw its bid.[90] Relief from a bid mistake is most easily obtained where notification of the mistake is given before the bid has been accepted.[91]

Even if award has already been made, relief may be available if the contractor acts promptly. In *School District of Scottsbluff v. Olson Construction Co.,*[92] a public owner sued a contractor and its bonding company because the contractor refused to enter into a school facilities contract. The contractor had submitted a bid of $177,000. Its bid had been accepted and the award made. The only other bid was $203,700.

When the low bidder's vice president learned of the discrepancy in the two bids, he immediately suspected an error. He examined the estimate sheets and found that a clerk, although experienced in preparing bids, had inadvertently entered an amount of $2,628 instead of $26,289 for the structural steel, resulting in an error of approximately $23,000. The contractor notified the owner at once of the error, but the owner insisted on compliance with the bid. Only four days had elapsed between the bid

[87] *Oce N. Am., Inc. v. Caputo,* 416 F. Supp. 2d 1321 (S.D. Fla. 2006).
[88] *See, e.g., Real Estate Value Co. v. USAir, Inc.,* 979 F. Supp. 731 (N.D. Ill. 1997); *Patel v. Northfield Ins. Co.,* 940 F. Supp. 995 (N.D. Tex. 1996); *Dick Corp. v. Associated Elec. Coop., Inc.,* 475 F. Supp. 15 (W.D. Mo. 1979); *Clinton County Dep't of Pub. Works v. Am. Bank & Trust Co.,* 268 N.W.2d 367 (Mich. Ct. App. 1978), *rev'd on other grounds,* 276 N.W.2d 7 (Mich. 1979); *Puget Sound Painters, Inc. v. State,* 278 P.2d 302 (Wash. 1954).
[89] 696 P.2d 711 (Ariz. Ct. App. 1984).
[90] *James Cape & Sons Co. v. Mulcahy,* 700 N.W.2d 243 (Wis. 2005).
[91] *Id.*
[92] 45 N.W.2d 164 (Neb. 1950).

opening and notification of the mistake. The court held that the contractor was entitled to retract the bid. Although the contractor's bid had been accepted and the contract awarded to it, the contract had not been signed and performance had not started.

If a contractor waits too long to request relief from a bid mistake, courts may be reluctant to allow the mistake to be corrected. For example, a contractor that made a $317,000 mathematical error in its $15.8 million Florida state highway bid was not entitled to an increase in its contract price, despite the Florida Department of Transportation's knowledge of a possible error. The contractor did not sue for correction of the mistake until 21 months after bid opening, when the project was 75% complete.[93]

Courts regularly emphasize the requirement that the bid mistake must not have resulted from any culpable negligence by the contractor.[94] The contractor has the burden to establish the absence of such negligence in bid preparation.[95] Often courts also distinguish between (l) errors of a mathematical, typographical, or clerical nature (e.g., incorrect transposition of figures on the bid sheet); and (2) errors in judgment (e.g., incorrectly estimating the amount of steel required for the project). Generally, courts will grant relief from clerical errors but not for errors in judgment.[96]

Application of this concept can be found in a Comptroller General decision, *Matter of Continental Heller Corp.*[97] Continental Heller submitted the apparent low bid on a contract for construction of a Navy training building. Before contract award, Continental Heller notified the Navy it had made a mistake in its bid by including only $50,000 for profit rather than the intended $500,000. In support of its claim, Continental Heller submitted, among other things, its bid preparation worksheets showing the $500,000. The Comptroller General sustained Continental Heller's bid protest because its bid preparation documents provided clear and convincing evidence of a clerical error and the specific amount of the intended bid. Continental Heller was thus allowed to receive the contract at the corrected price.

A contractor that realizes before award that the bid contains a material error should take immediate steps to safeguard its chances of obtaining relief. First, the contractor should notify the owner or public contracting official awarding the contract of the error, preferably in writing; but if time will not permit, then orally, followed promptly by a written request for modification or withdrawal with documentation supporting the assertion of a mistake.[98]

[93]*Dep't of Transp. v. Ronlee, Inc.*, 518 So. 2d 1326 (Fla. Dist. Ct. App. 1987). *See also Chris Berg, Inc. v. United States,* 426 F.2d 314 (Ct. Cl. 1970); *Dick Corp. v. Associated Elec. Coop., Inc.*, 475 F. Supp. 15 (W.D. Mo. 1979); *C.N. Monroe Mfg. Co. v. United States,* 143 F. Supp. 449 (E.D. Mich. 1956).

[94]*See Oce N. Am., Inc. v. Caputo,* 416 F. Supp. 2d 1321 (S.D. Fla. 2006); *Liebherr Crane Corp. v. United States,* 810 F.2d 1153 (Fed. Cir. 1987).

[95]*See, e.g., James Cape & Sons Co. v. Mulcahy,* 700 N.W. 2d 243 (Wis. 2005); *State Bd. of Control v. Clutter Constr. Corp.*, 139 So. 2d 153 (Fla. Dist. Ct. App. 1962); *Ex parte Perusini Constr. Co.,* 7 So. 2d 576 (Ala. 1942).

[96]*See, e.g., Info. Intern. Assocs. Inc. v. United States,* 74 Fed. Cl. 192 (2006); *Balaban-Gordon Co., Inc. v. Brighton Sewer Dist.*, 342 N.Y.S.2d 435 (N.Y. App. Div. 1973).

[97]Comp. Gen. B-230559, 88–1 CPD ¶ 571.

[98]*To The Sec'y of the Army,* Comp. Gen. B-137155, 1958 CPD ¶ 91; *To Giltron Assocs., Inc.*, Comp. Gen. B-167649.

Once the contractor has notified the owner of the bid mistake, it should then request an informal conference with the owner to discuss the bid mistake. At the conference, the contractor should present evidence supporting its request for modification or withdrawal. This evidence should include statements (preferably affidavits) explaining the mistake, copies of the bid, original worksheets, subcontractor quotations, published price lists, and any other data used in preparing the bid that establishes: (1) the existence of the error; (2) the manner in which it occurred; and (3) the intended bid. The contractor should gather this information as quickly as possible and secure it until presentation to the owner.[99]

If the contractor does not detect the error until after award, it faces greater difficulty in modifying or withdrawing its bid. The contractor should notify the owner of an error discovered after award immediately upon discovering the error, just as if the error were discovered before award. The likelihood of successfully reforming the bid at this later time, however, is much more remote.[100]

Some states have enacted laws regarding the withdrawal of bids containing mistakes. For example, in Georgia, a bid on a public project (other than a Georgia Department of Transportation project) containing a bid mistake may be withdrawn after bid opening without forfeiture of the bid bond if: (1) the bidder has made an appreciable error in the calculation of the bid that can be documented by clear and convincing evidence; (2) such errors can be clearly shown by objective evidence drawn from inspection of the original work papers or other materials used in preparing the bid; (3) the bidder serves written notice on the public entity that invited the proposals before award of the contract and not later than 48 hours after the opening of bids, excluding Saturdays, Sundays, and legal holidays; (4) the bid was submitted in good faith and the mistake was due to calculation or clerical error, an inadvertent omission, or a typographical error, as opposed to an error in judgment; and (5) withdrawal of the bid will not result in undue prejudice to the public entity or other bidders by placing them in a materially worse position than if the bid had never been submitted.[101] If the bid is withdrawn, the remaining bids will be considered as if the withdrawn bid had not been submitted.[102] If the project is relet for bids after the bid is withdrawn, the withdrawing bidder cannot submit a bid on the resolicitation, and the withdrawing bidder cannot supply any material or labor to the project for compensation and cannot subcontract work on that project.[103]

In private contracting, a contractor has no statutory right to withdraw an erroneous bid, and no statutory protocol is available to outline the procedure for such withdrawal. A contractor may be able to retract its bid when equitable criteria are met and when such action is not affected by a statutory prohibition or restriction against withdrawal of bids.[104] Similarly, when equitable conditions are satisfied, a bidder may be entitled to recover its bid deposit.[105]

[99]*Matter of Cont'l Heller Corp.*, Comp. Gen. B-230559, 88–1 CPD ¶ 571.

[100]*James Cape & Sons Co. v. Mulcahy,* 700 N.W.2d 243 (Wis. 2005).

[101]GA. CODE ANN. § 36–91–52; *see also* GA. CODE ANN. § 32–2–69.

[102]GA. CODE ANN. § 36–91–52(c).

[103]GA. CODE ANN. § 36–91–52(c)-(d).

[104]*See Jobco, Inc. v. Nassau County,* 514 N.Y.S.2d 108 (N.Y. App. Div. 1987).

[105]*See James Cape & Sons Co. v. Mulcahy,* 700 N.W.2d 243 (Wis. 2005); *Rushlight Automatic Sprinkler Co. v. Portland,* 219 P.2d 732 (Or. 1950).

B. Owner's Duty If Bid Mistake Is Suspected

The owner (or prime contractor) may also have a duty to notify the contractor (or subcontractor) that there may be an error in the bid if the circumstances are sufficient to place the owner on notice of a possible error. For example, in *I & R Mechanical, Inc. v. Hazelton Mfg. Co.,*[106] a heating and cooling equipment supplier submitted a bid to a contractor for three boilers as part of a school construction project. The supplier's bid was nearly 40 percent lower than the next lowest bid. The contractor attempted to hold the equipment supplier to its bid. The court found the facts sufficient to have put the contractor on notice and to impose a duty to attempt to verify the accuracy of the bid before allowing reliance on the bid.

As a general rule, the existence of a wide range between the low bid price and the other bid prices is generally sufficient to put the bid recipient on notice of a possible error. Similarly, a substantial variation in the bid from the owner's estimate of the approximate cost can also put the owner on notice of a possible error.[107]

C. Withdrawal versus Reformation of Bid Mistake

Most state laws recognized the right to withdraw a mistaken bid. But the right to correct the bid is often not available. Some state courts recognize that an erroneous bid should be reformed where the bid was the lowest bid before and after reformation.[108] Other courts have held that a bid is properly rescinded where the bidder makes a unilateral mistake.[109]

The right to withdraw and the opportunity to correct a mistaken bid are better established under federal law. FAR § 14.407 details the procedures for dealing with bid mistakes and the circumstances in which a bid may be withdrawn or corrected. To guard against mistakes, the contracting officer must review all bids and must notify a bidder if an apparent mistake is found or if there is reason to believe the bid contains a mistake.[110]

Even if the bidder verifies the accuracy of its bid, relief from mistake may still be allowed where the government's request for verification does not alert the bidder to the basis for the government's belief that the bid is erroneous.[111] If the contracting officer knows, or should have known, of a contractor's error, then award of a contract is tantamount to bad faith and constitutes overreaching. Under such circumstances, relief may be appropriate if the contracting officer does not properly verify the contractor's price.[112]

[106]817 N.E.2d 799 (Mass. App. Ct. 2004).
[107]*See, e.g., Jensen & Reynolds Const. Co. v. State Dept. of Transp. & Public Facilities,* 717 P.2d 844 (Alaska 1986).
[108] *E.H. Oftedal and Sons, Inc. v. State ex rel. Mont. Transp. Com'n,* 40 P.3d 349 (Mont. 2002).
[109]*See, e.g., Naugatuck Valley Dev. Corp. v. Acmat Corp.,* 523 A.2d 924 (Conn. App. Ct. 1987); *Clinton County Dep't of Pub. Works v. Am. Bank & Trust Co.,* 268 N.W.2d 367 (Mich. Ct. App. 1978), *rev'd on other grounds,* 276 N.W.2d 7 (1979); *Wil-Fred's, Inc. v. Metro. Sanitary Dist. of Greater Chi.,* 372 N.E.2d 946 (Ill. App. Ct. 1978); *Baltimore County v. John K. Ruff, Inc.,* 375 A.2d 237 (Md. 1977).
[110]FAR § 14.407–1.
[111]*P. T. Serv. Co.,* GSBCA No. 7589, 85–3 BCA ¶ 18,430.
[112]*Chemtronics, Inc.,* ASBCA No. 30883, 88–2 BCA ¶ 20,534.

In federal procurement, bidders can alter their bids to correct mistakes so the government can enjoy the cost benefits of a downward correction.[113] If a downward correction would displace the low bid, the bidder must present clear and convincing evidence, ascertainable substantially from the invitation and the bid itself,[114] establishing both the existence of the mistake and the dollar amount of the bid actually intended by the contractor seeking alteration.

Where the low bidder seeks to increase its bid and still remain low, it must also present clear and convincing evidence of the mistake and its actual intended bid. The contractor may not, however, be limited to the solicitation and the bid itself in the evidence it presents. The contractor may include statements, a file copy of the bid, original worksheets, subcontractor quotations, and published price lists.[115] But even if there is evidence to support such a correction, the correction may be disallowed if it brings the low bidder too close to the second low bidder.

Finally, federal procurement law expressly provides for relief from mistakes that are alleged after award. The FAR authorizes such relief under specific factual circumstances. Initially, such requests are addressed to the agency contracting officer and are processed in accordance with FAR § 14.407–4. If the matter is not resolved at the contracting officer level, any dispute about the contractor's entitlement to relief or the nature of the relief is handled under FAR Subpart 33.2, Disputes and Appeals, in accordance with the Contract Disputes Act. Under the regulations, the contracting officer can rescind a contract or modify the contract by deleting the item involved in the mistake or increasing the price.

As noted previously, some state laws permit the withdrawal of bids in limited circumstances, although reformation of bids after bid opening is rarely allowed.[116]

XII. BID PROTESTS

The federal government, as well as virtually all other governmental entities, primarily uses either sealed bidding or competitive best-value proposals in the award of public construction contracts. Historically, governments prefer the competitive bid system because independently submitted sealed bids result in the lowest cost to the owner and best protect the public interest. Bid protests arise almost exclusively in connection with sealed bids (Invitations) or competitive proposals (RFPs)—predominantly at the federal level, but to an increasing degree at the state and local government levels. In bidding or proposing on projects for public owners, contractors must rely not only on their own evaluation of the Invitation but also on a proper application of the owner's competitive bidding procedures. Increased competition for contracts, combined with economic cycles that reduce the total volume of contracts available

[113] FAR § 14.407–3; *P K Contractors, Inc.*, Comp. Gen. B-205482, 82–1 CPD ¶ 368.

[114] FAR § 14.407–3.

[115] FAR § 14.407–3(g)(2); *Coleman Indus. Constr. Co.*, Comp. Gen. B-207682, 82–2 CPD ¶ 213.

[116] *See* GA. CODE ANN. § 36–91–52; *see also* GA. CODE ANN. § 32–2–69.

for bidding, have forced both contractors and their attorneys to become knowledgeable about the rules governing competitive procurements, the enforcement of rules governing sealed bids or competitive proposals, and related bid protest rules and regulations. Whether the procurement is conducted using sealed bids or competitive proposals, challenges to the public agency's conduct of a particular procurement action are labeled "bid protests."

The nature and degree of formality of competitive procurement procedures varies greatly among states and localities and with the type of owner involved. That said, some general guidelines, primarily developed from federal competitive bidding procedures, can apply in many situations. Increased use of competitive bidding at state and local levels has resulted in the development of the American Bar Association Model Procurement Code for state and local governments. This Model Code attempts to further standardize competitive bid procedures for public construction contract awards.

The guiding principle in bid protests is quick action since most regulations specify that such action must be taken within a few days. In addition, under some circumstances, the contracting authority may award the contract while the protest is still pending. Once the award is made, the reviewing body will be reluctant to reverse the decision of the awarding authority. Thus, obtaining a favorable ruling on the merits of a bid protest may prove to be an empty victory if the contract itself has already been awarded to another bidder.

A. Protests on Federal Government Contracts

When a contractor submitting a bid or proposal on a public contract believes that the rules and regulations applicable to the particular procurement have not been followed, it may want to protest award of the contract. The protesting contractor generally will base the protest on the successful offeror's or bidder's failure to meet one or more of the requirements previously discussed: lowest bid, responsibility, responsiveness to the solicitation, or the agency's failure to evaluate the offer consistent with the solicitation's evaluation criteria.

In the federal contract arena, a protester has many options for filing a bid protest. A bid protest may be filed with:

(1) The contracting agency responsible for the procurement
(2) The Comptroller General of the Government Accountability Office (GAO), formerly the General Accounting Office
(3) The United States Court of Federal Claims

B. Protests before or after Receipt of Bids or Offers

A disappointed bidder may initially submit a bid protest to the agency that is involved in the procurement. Federal regulations do not discuss in detail the procedures for agency protests, but some general guidelines are discussed at FAR Subpart 33.1.

Agency protests may be filed before or after contract award by a bidder/offeror or a prospective bidder/offer if its direct economic interest is affected by the award.[117] The protester must submit its protest in writing.[118] The agency receiving the protest must respond using a method that provides evidence of receipt.[119] The interested parties and agency should use their best efforts through open and frank discussion to try resolving concerns raised by an interested party before the submission of a protest.[120]

There are time limits for agency protests. Protests based on alleged apparent improprieties in the solicitation shall be filed before bid opening or the closing date for the receipt of proposals.[121] In all other cases, protests must be filed not later than 10 days after the basis for the protest is known or should have been known, whichever is earlier.[122] An agency can consider an untimely protest if the agency determines that the protest raises issues significant to the agency's acquisition system.[123] In general, the protester should submit a protest as soon as it becomes aware of the basis for the protest, because the likelihood of success declines significantly once the award has been made.

C. Protests to the Contracting Agency

Protests to the contracting agency must be addressed to the contracting officer or other official designated to receive such protests.[124] The protester should also consider submitting its protest at a level within the contracting agency or contracting entity higher than the contracting officer, if allowed by the agency's rules. Often decisions of the contracting officer may be the very decisions in question. The agency must provide a procedural mechanism for the protester to request an independent review above the contracting officer level, either as an initial review of the protest or as an appeal from the contracting officer's decision.[125]

The protest must be concise and logically presented and must include: (1) the name, address, and fax and telephone numbers of the protester; (2) the solicitation or contract number; (3) a detailed statement of the legal and factual grounds for the protest, including a description of the resulting prejudice to the protester; (4) copies of relevant documents; (5) a request for ruling by the agency; (6) a statement of the form of relief requested; (7) all information establishing that the protester is an interested party; and (8) all information establishing the timeliness of the protest.[126]

[117]FAR § 33.101.
[118]*Id.*
[119]FAR § 33.103(h).
[120]FAR § 33.103(b).
[121]FAR § 33.103(e).
[122]*Id.*
[123]*Id.*
[124]FAR § 33.103(d)(3).
[125]FAR § 33.103(d)(4).
[126]FAR § 33.103(d)(2).

Upon receipt of a protest before award, the contract may not be awarded pending agency resolution of the protest unless the award is justified in writing for urgent and compelling reasons, or it is determined in writing to be in the government's best interest.[127] This justification must be approved at a level above the contracting officer.[128] If the award is withheld pending agency resolution of the protest, the contracting officer must notify the other bidders/offerors whose bids or offers may become eligible for award and, if appropriate, request that the bidders or offerors extend the time for acceptance.[129] If an extension cannot be obtained, the agency should consider proceeding with the award.[130] Upon receipt of a protest within 10 days after contract award or within 5 days after a debriefing date offered to the protester under a timely debriefing request (whichever is later), the contracting officer shall immediately suspend performance pending resolution of the protest by the agency, including any review at a higher agency level, unless continued performance is justified, in writing, for urgent and compelling reasons, or it is determined in writing to be in the government's best interests.[131] Such justification or determination must be approved at a level above the contracting officer, or by another official pursuant to agency procedures.[132] An agency protest will not extend the time for obtaining a stay from the GAO; agencies must include in their protest procedures a voluntary suspension period when the agency protest is denied and the protester subsequently protests to the GAO.[133] Agencies must use their best efforts to resolve protests within 35 days after the protest is filed.[134]

Bidders are often reluctant to protest to the contracting agency because the agency is being asked to judge the actions of its own employees. Protests to the GAO or in the United States Court of Federal Claims provide more of an opportunity for a "neutral, third-party review" of the agency's actions and positions. Furthermore, agency protests are likely to be less successful than protests to the GAO if the basis for the protest involves an unusual issue or one that is not a clear violation of applicable laws or regulations. The contracting agency is likely to ratify the actions of its employees if there is any basis for such actions.

There are circumstances in which an agency protest may be advantageous to a bidder. An agency protest may be less costly and time-consuming if the protest is unquestionably valid because the agency may act quickly to correct any deficiency so as not to delay commencement of work on the particular project. Additionally, agency protests may be useful where the protest is filed well before bids are to be received. In such a situation, the protester may have the opportunity to obtain a quick decision from the contracting agency; and, if that decision is adverse to the protester's

[127]FAR § 33.103(f)(1).
[128]*Id.*
[129]FAR § 33.103(f)(2).
[130]*Id.*
[131]FAR § 33.103(f)(3).
[132]*Id.*
[133]FAR § 33.103(f)(4).
[134]FAR § 33.103(g).

position, the protester may then seek relief in another forum, such as the GAO. The protester should be mindful that if the agency protest is unsuccessful, the protester has 10 days from the actual or constructive knowledge of the initial agency decision to file a protest with the GAO.[135]

D. Protests to the Comptroller General

This section summarizes the important initial time constraints and factors to consider in filing a protest with the GAO. Beware that these regulations can change (including the initial deadline for filing a protest), and the current bid protest regulations as published in the Federal Register and the C.F.R. must be followed to assure a valid protest.[136]

Any "interested party" may file a protest with the GAO alleging an irregularity in the solicitation or award of a federal government contract to which a federal agency is a party. This "interested party" is an actual or prospective bidder or offeror whose direct economic interest would be affected by the award of a contract or by the failure to award a contract.[137] In several cases, protesters who did not even submit offers were nevertheless held to be interested parties. For example, protesters precluded from submitting a proposal because of short response time and restrictive specifications,[138] or otherwise denied the opportunity to compete,[139] have been deemed to have standing to protest.

Protests based on alleged improprieties in a solicitation that are apparent before bid opening must be filed before bid opening or the time set for receipt of initial proposals.[140] As a result, the Comptroller General has held that protests regarding improprieties that were apparent before bid opening are untimely even if submitted with a bid[141] or proposal.[142] Generally, other protests must be filed no later than 10 calendar days after the basis of the protest is known or should have been known, whichever is earlier, except protests challenging a procurement conducted on the basis of competitive proposals under which a debriefing is requested and, once requested, is then required. If the debriefing is required and the basis for the protest is known before or as a result of the debriefing, the protest must be filed after the debriefing but within 10 calendar days of the debriefing.[143] The GAO must notify the contracting agency of the protest within 1 working day after receipt of a protest.[144]

Under most circumstances, a contracting agency cannot award a contract after the agency has received notice of the protest and while the protest is pending.[145] The head of an agency may authorize award of a contract notwithstanding a protest upon a finding

[135] 4 C.F.R. § 21.2(a)(2); FAR § 33.103(d)(4).
[136] *See* 4 C.F.R. § 21.0, *et seq.; see also* 31 U.S.C. § 3553.
[137] 4 C.F.R. § 21.0(a).
[138] *Vicksburg Fed. Bldg. Ltd. P'ship,* Comp. Gen. B-230660, 88–1 CPD ¶ 515.
[139] *Afftrex, Ltd.,* Comp. Gen. B-231033, 88–2 CPD ¶ 143; *REL,* Comp. Gen. B-228155, 88–1 CPD ¶ 25.
[140] 4 C.F.R. § 21.2 (a)(1).
[141] *Fredrico Enter., Inc.,* Comp. Gen. B-230724.2, 88–1 CPD ¶ 450.
[142] *Darome Connection,* Comp. Gen. B-230629, 88–1 CPD ¶ 461.
[143] 4 C.F.R. § 21.0(e); 4 C.F.R. § and 21.2(a)(2).
[144] 4 C.F.R. § 21.3(a).
[145] 4 C.F.R. § 21.6; 31 U.S.C. § 3553(c)(1).

that: (1) urgent and compelling circumstances that significantly affect the interests of the United States will not permit waiting for the Comptroller General's decision (as long as the award was otherwise likely to occur within 30 days after the making of the finding); and (2) after the Comptroller General is advised of the finding.[146]

If a protest is appropriate and is filed within 10 calendar days after the award (or 5 calendar days after the date offered for a required debriefing), and the agency is notified of the protest, the contracting officer may not authorize performance to begin while the protest is pending, or the contracting officer shall immediately direct the contractor to cease performance and suspend related activities.[147] The head of the contracting agency can authorize performance of the contract notwithstanding the protest: (1) upon a written finding that the performance is in the best interests of the United States, or urgent and compelling circumstances that significantly affect the interests of the United States will not permit waiting for the Comptroller General's decision on the protest; and (2) after the Comptroller General has been notified of the finding.[148]

Under 4 C.F.R. § 21.1(b), a protest must be in writing and addressed to General Counsel, Government Accountability Office, 441 G Street, NW, Washington, D.C. 20548, Attention: Procurement Law Control Group. The protest must include the name, address, and telephone number of the protester and be signed by the protester or its representative. It must also identify the contracting activity and the solicitation or contract number, or both, and include a detailed statement of the legal and factual grounds of the protest, including copies of all relevant documents. Finally, the protest must set out all information establishing that the protester is an interested party, set out the information establishing the timeliness of the protest, specifically request a ruling by the Comptroller General, and state the form of relief requested.[149]

A protest shall not be deemed filed unless it is actually received by the GAO within the time for filing and is accompanied by a certificate that a copy of the protest, together with relevant documents not issued by the contracting agency, was concurrently served on the contracting agency and the contracting activity (a contracting activity is the subelement of the contracting agency that actually issued the solicitation or contract, or both).[150] No formal briefs or other technical forms of pleadings are required.[151]

The GAO's regulations permit consideration of untimely protests raising significant issues. For example, the Comptroller General invoked its discretion to consider an untimely protest under the "significant issue" exception in *Reliable Trash Service Co.*,[152] since the record clearly indicated that bids could not have been evaluated on a common basis. In *Associated Professional Enterprises, Inc.*,[153] however, the

[146]31 U.S.C. § 3553(c)(2) – (3).
[147]31 U.S.C. § 3553(d)(3)(A).
[148]31 U.S.C. § 3553(d)(3)(C).
[149]4 C.F.R. § 21.1 (c).
[150]4 C.F.R. § 21.0(g); 4 C.F.R. § 21.1(e).
[151]4 C.F.R. § 21.1(f).
[152]Comp. Gen. B-234367, 89–1 CPD ¶ 535.
[153]Comp. Gen. B-235066.2, 89–1 CPD ¶ 480.

Comptroller General held that the "good cause" exception to the timeliness rule will be limited in future cases to circumstances in which a compelling reason beyond the protester's control prevents timely filing.

The protest initially may be filed with the contracting agency and a subsequent protest to the GAO may be filed within 10 calendar days of actual or constructive knowledge of initial adverse agency action (unless the contracting agency imposes a more stringent time for filing).[154]

In most cases, the agency to which the protest is submitted must issue a report regarding the bases for the protest within 30 days of telephone notice of the protest to the agency from the GAO.[155] The protester may submit comments to the GAO on the agency report within 10 calendar days of receipt of the report, with a copy to the agency and other participating parties.[156]

The Comptroller General must render a decision within 100 days after the protest is filed.[157] If the Comptroller General finds that the protested solicitation, termination of the contract, proposed award, or award does not comply with statute or regulation, it can direct the agency to refrain from exercising options under the contract, terminate the contract, recompete the contract, issue a new solicitation, award a contract consistent with the law, or make other recommendations as deemed appropriate.[158] If the Comptroller General determines that applicable statutes or regulations have not been followed, it may find the protester entitled to bid preparation costs, protest costs, and reasonable attorneys' fees.[159] Despite this, in *Princeton Gamma-Tech, Inc.*[160] the Comptroller General held that costs incurred in connection with the agency-level protest cannot be reimbursed under the GAO's rule permitting reimbursement of costs for a prevailing protester.

Comptroller General decisions are technically advisory only due to constitutional separation of powers considerations.[161] The Competition in Contracting Act requires an agency to provide a full report on any refusal to follow a Comptroller General decision.[162]

E. Bid Protests in Court

One of the most important aspects of bid protests is quick action. From the standpoint of the protester, it is imperative to prevent the award or performance of the disputed contract to go any further than it has when the decision to pursue a protest is made. The farther along the award or performance on the contract, the less likely that a court will be willing to grant injunctive relief to suspend any further performance.

[154] 4 C.F.R. § 21.2(a)(3).
[155] 4 C.F.R. § 21.3(c).
[156] 4 C.F.R. § 21.3(i).
[157] 4 C.F.R. § 21.9(a).
[158] 4 C.F.R. § 21.8(a).
[159] 4 C.F.R. § 21.6.
[160] Comp. Gen. B-228052.5, 89–1 CPD ¶ 401.
[161] *Advanced Sys. Dev., Inc. v. United States,* 72 Fed. Cl. 25 (Fed. Cl. 2006); *Ameron, Inc. v. U.S. Army Corps of Eng'rs,* 809 F.2d 979, 995 (3d Cir. 1986), *cert. granted,* 485 U.S. 958, *cert. dismissed,* 488 U.S. 918 (1988).
[162] 31 U.S.C. § 3554(e)(1).

Often a contractor will want to obtain a temporary restraining order and injunction to prevent the award to, or commencement of the contract by, another bidder. This is particularly true on state and local projects where award is not automatically stayed and in federal contracts where the agency has indicated it intends to award despite the automatic stay regulation,[163] or because the agency claims award and performance is in the agency's best interests and urgent and compelling reasons exist to start performance.

An injunction suspends any further activity on the contract, whether award or performance, while the court or appropriate agency has the opportunity to decide the merits of the protest. Although the automatic stay provisions of the Competition in Contracting Act diminish the need for injunctive relief in federal government procurements, there still will be situations in which such relief is required. In addition, although federal courts give substantial deference to Comptroller General decisions,[164] a protester may want the court involved to review the administrative decision rendered by the Comptroller General. Most important, court action may provide the only vehicle to pursue a protest beyond "agency level" in state and local government procurements.

The U.S. Court of Federal Claims is the only court for the protesting bidder seeking injunctive relief against a federal agency or activity. The Federal Court Improvement Act of 1988 gives the Court of Federal Claims exclusive jurisdiction to decide preaward federal contract disputes.[165] Previously, only federal district courts had jurisdiction over bid protests. The district courts' jurisdiction was based on the *Scanwell* doctrine, named for the first case in which the federal courts took jurisdiction over a federal bid protest.[166]

Until 2001, federal district courts had concurrent jurisdiction over post-award bid protests.[167] As of January 1, 2001, Congress ended the federal district courts' concurrent jurisdiction over protests regarding federal government procurements,[168] leaving the Court of Federal Claims as the exclusive judicial forum for such actions.

A contractor's primary relief in the form of money damages in a court action is its bid or proposal preparation costs, but **not** anticipated profits.[169] The protester may also recover bid protest costs under the Equal Access to Justice Act.[170] Additionally, a protester can be required to post a bond for security if a restraining order is issued.[171]

F. Bid Protests on State and Local Public Contracts

For state or local government projects, absent specific procedures, a bid protest similar to that just described should be delivered as soon as possible to the awarding authority and other involved parties. For example, in connection with federally

[163]31 U.S.C. § 3553(2)(B).
[164]*Patriot Contract Servs. v. United States,* 388 F. Supp. 2d 1010 (N.D. Cal. 2005).
[165]28 U.S.C. § 1491(a).
[166]*Scanwell Lab., Inc. v. Shaffer,* 424 F.2d 859 (D.C. Cir. 1970).
[167]*In re Smith & Wesson, Inc.,* 757 F.2d 431 (1st Cir. 1985); *Coco Bros. v. Pierce,* 741 F.2d 675 (3d Cir. 1984); *see also Am. Dist. Tel. v. Dep't of Energy,* 555 F. Supp. 1244 (D.D.C. 1983).
[168]Pub. L. No. 104–320, 110 Stat. 3870.
[169]*Gentex Corp. v. United States,* 61 Fed. Cl. 49 (2004).
[170]28 U.S.C. § 2412(d)(1)(A).
[171]FED. R. CIV. P. 65(c).

funded projects at the local level (such as federal Environmental Protection Agency [EPA] sewage treatment projects), notice of the protest should be sent to the grantor as well as the grantee, usually the local awarding authority. Also, the protester should consult any pertinent federal statutes or regulations governing protests of federally funded state or local contracts (such as the EPA Construction Grant Regulations for EPA treatment projects). Thereafter, it is essential to prepare as quickly as possible a comprehensive statement with supporting documentation stating the detailed basis for the protest. This presentation should be submitted to all concerned parties.

In some states, a disappointed bidder may be able to seek bid protest relief on public (state or local) contracts through administrative procedures or court. Many state and local agencies have administrative protest procedures.[172] Additionally, judicial relief may be available for disappointed bidders in some states.[173] Federal courts, however, have been reluctant to entertain bid protests on state or local government contracts on the basis of a violation of procedural due process rights under 42 U.S.C. § 1983.[174]

XIII. BID BONDS

In connection with public contracts in particular, agencies frequently require that each bidder provide security with the bid that will guarantee execution of the contract if the bidder is awarded the contract.[175] In some instances a cash deposit is required. More commonly, the security called for is in the form of a "bid bond."[176] These bonds commonly provide that if the contractor does not execute the contract, the surety will be liable to the owner only to the extent of the difference between the contractor's bid and the lowest amount for which the owner may be able, in good faith, to award the contract within a reasonable time. A further limit generally is the penal sum of the bond.[177]

Federal contracts usually require that a bid bond (also known as a bid deposit) be provided to guarantee that the lowest responsive and responsible bidder will enter into a contract with the government. As the Court of Claims noted in *Anthony P. Miller, Inc. v. United States,* "[i]t is well established that in the event of a default by a bidder the United States may retain the bid deposit as liquidated damages unless the amount is so large or disproportionate as to constitute a penalty."[178]

[172]*See, e.g.,* Georgia Department of Administrative Services, Georgia Vendor Manual § 3.8; Code of Miami-Dade County § 2–8.4; N.C. Gen. Stat. § 143–53.
[173]*See, e.g., Amdahl Corp. v. Ga. Dep't of Admin. Servs.,* 398 S.E.2d 540 (Ga. 1990).
[174]*See, e.g., Flint Elec. Membership Corp. v. Whitworth,* 68 F.3d 1309 (11th Cir. 1995), *corrected decision,* 77 F.3d 1321 (11th Cir. 1996).
[175]*See Diamond Int'l Corp.,* Comp. Gen. B-180426, 74–2 CPD ¶ 139.
[176]*See, e.g., All Seasons Constr., Inc. v. United States,* 55 Fed. Cl. 175 (2003); *Bolivar Reorganized Sch. Dist. No. 1 v. Am. Sur. Co.,* 307 S.W.2d 405 (Mo. 1957).
[177]*See Powder Horn Constructors, Inc. v. City of Florence,* 754 P.2d 356 (Colo. 1988); *Bd. of Educ. of Union Free Sch. Dist. No. 3, Town of Cheektowaga v. Md. Cas. Co.,* 98 N.Y.S.2d 865 (N.Y. App. Div. 1950) (in the context of the difference in cost exceeding the penal sum of the bond).
[178]161 Ct. Cl. 455, 468 (1963).

State or local governments may require bid bonds on their construction contracts as a matter of law or public policy. Although not all states have statutory provisions requiring bid bonds, the power of public agencies to require such bonds is not generally dependent on statute. Thus, it is generally conceded that a state, city, county, or other public agency or body empowered to let a contract may require the contractor to furnish a surety bond, even though there is no express statutory authority for such a requirement.[179] Although the practice is perhaps more common in connection with public contracts, bids on private contracts can be required to be accompanied by bid bonds.

The extent of liability under a bid bond is usually a fixed sum specified in the bond or the difference between the bid submitted by the defaulting contractor and the next-lowest bidder or the price at which the owner is forced to contract.[180] For example, in *Board of Education of Community United School District No. 303 v. George S. Walker Plumbing & Heating, Inc.,*[181] summary judgment was granted to the owner for the difference between the lowest bidder's bid and the second bidder's contract price because the lowest bidder, whose bid was accepted, refused to enter into a contract. With respect to the contractor's risks, the surety also received summary judgment against the contractor for indemnification. Indemnification between the contractor and surety usually arises out of the express obligations in the bond itself as well as the general agreement of indemnity executed by one or more of the contractor's principals in favor of the surety. (See **Chapter 14** and **Chapter 15**.)

Courts have given various treatments to the limit of the surety's liability to the owner pursuant to the bid bond. In *A. J. Colella, Inc. v. Allegheny County,*[182] the amount of recovery was limited to the penal sum of the bid bond, although the difference between the amount of the bid and the amount at which the county ultimately let a contract greatly exceeded the bond's penal sum.

Other cases have held that the penal amount of the bid bond is, in effect, a liquidated damages provision. Thus, upon failure of the low bidder to enter into a contract, the owner could recover the penal sum of the bond, even though the difference between the defaulting contractor's bid and the eventual contract price was less.[183]

Several defenses have been argued by a defaulting contractor and its bid bond surety when the contractor has refused to enter into a contract and there has been a call on the bond. Under appropriate facts, the contractor may successfully argue that there has been a material change in the contract upon which the bid was submitted. Such was the case in *Northeastern Construction Co. v. City of Winston-Salem,*[184] when, after bid opening, the municipal authority eliminated approximately 15% of the work.

[179]*Union Indem. Co. v. State,* 114 So. 415 (Ala. 1927); *Foster v. Kerr & Houston,* 179 A. 297 (Mass. 1907); *Sw. Portland Cement Co. v. Williams,* 251 P. 380 (N.M. 1926).
[180]*Brown v. United States,* 152 F. 964 (2d Cir. 1907).
[181]282 N.E.2d 268 (Ill. App. Ct. 1972).
[182]137 A.2d 265 (Pa. 1958).
[183]*See City of Merrill v. Wenzel Bros., Inc.,* 277 N.W.2d 799 (Wis. 1979); *Bellefonte Borough Auth. v. Gateway Equip. & Supply Co.,* 277 A.2d 347 (Pa. 1971).; *City of Lake Geneva v. States Improvement Co.,* 172 N.W.2d 176 (Wis. 1969).
[184]83 F.2d 57 (4th Cir. 1936).

Furthermore, the contractor and its surety may also be released when the contractor made a material mistake in its bid as well as when bid conditions are not met.[185]

In some circumstances, contractors may have claims against their sureties if, after execution of the bid bond and before award, the surety refuses to execute the necessary payment and performance bonds for the contractor. Although furnishing a bid bond ordinarily does not obligate the surety to furnish additional bonds, some contractors have claimed against their bid bond surety for the surety's refusal to issue the other bonds necessary for the contractor to accept award.

The contractor might assert an estoppel argument against the surety if facts can be presented to prove that words or conduct of the surety were reasonably relied on by the contractor to the contractor's detriment.[186] Even when the surety is not obligated to furnish payment and performance bonds upon the furnishing of the bid bond, the contractor may show that the writing of payment and performance bonds is customary after the writing of the bid bond in a claim against the surety.[187]

In addition to requiring bid bonds, some states require posting of separate bonds as part of the process to qualify to conduct business or obtain licenses. These bonds may exist to secure payment of taxes (sales, workers' compensation, etc.) and are distinct from those associated with a particular project's bid bond. (See **Chapter** 3.) The State Law Matrix of the Associated General Contractors of America *DocuBuilder* publication identifies state laws addressing bidding, requirements for bid bonds, as well as other bonds that contractors may be obligated to post in order to conduct business in that jurisdiction. A summary of the major provisions of all state laws concerning public bonds may also be found in the *Credit Manual of Commercial Laws,* published by the National Association of Credit Management and updated annually.

XIV. "BID SHOPPING": WHAT IS THE OBLIGATION OF THE PRIME CONTRACTOR TO THE SUBCONTRACTOR SUBMITTING THE LOWEST PRICE?

One of the most emotional issues connected with the prime contractor/subcontractor relationship involves "bid shopping" by the prime contractor. "Bid shopping" refers to actions taken by the prime contractor after award of the prime contract to reduce subcontractor prices by "shopping" the lowest bid in a particular trade from subcontractor to subcontractor. One court has referred to such postaward conduct as "bid chiseling," and defined "bid shopping" or "bid peddling" as those actions occurring before award of the prime contract.[188]

[185]*See* David B. Harrison, Annotation, *Right of Bidder for State or Municipal Contract to Rescind Bid on Ground That Bid Was Based upon His Own Mistake or That of His Employee*, 2 A.L.R. 991 (1980).
[186]*See generally Conner Bros. Constr. Co., Inc. v. United States,* 65 Fed. Cl. 657 (2005); *Reynolds v. Gorton,* 213 N.Y.S.2d 561 (N.Y. Sup. Ct. 1960).
[187]*Commercial Ins. Co. v. Hartwell Excavating Co.,* 407 P.2d 312 (Idaho 1965).
[188]*People v. Inland Bid Depository,* 44 Cal. Rptr. 206 (Cal. Ct. App. 1965).

Although "bid shopping" has been described as "the purest form of competition," virtually all subcontractors and suppliers categorically reject the possibility of any virtue in the practice.[189] Bid shopping has been condemned by at least one construction industry trade organization.[190] Yet it remains present in the industry, and the difficult position of the subcontractor is magnified by the lack of available legal theories to bind the prime.

Much of the attention is focused on allegations of such conduct by general contractors. Still, the practice can occur at any tier in the chain of subcontracts: for example, subcontractor and sub-subcontractor, subcontractor and supplier, and so on. Although the issues in this section are addressed in the context of general contractors and their prospective subcontractors, the analysis has equal application to all tiers of the contracting process.

As a result of the various criticisms that have been leveled at bid shopping, several measures have been taken to eliminate, or at least minimize, the practice. The first of these measures has been the establishment of local bid depositories for subcontractor bids. A number of depository (by law) provisions necessary for the effective operation of bid depositories, however, have been held to violate state and federal antitrust laws because of their restrictive effect on competition among both subcontractors and general contractors. In *Mechanical Contractors Bid Depository v. Christiansen,*[191] treble damages were awarded the plaintiff for antitrust violations by use of a bid depository.[192]

A second measure to combat bid shopping (addressing any postaward bid shopping) has been the statutory or contractual requirement that the prime contractor list in its bid the subcontractors it intends to use.[193] Where listing is required, there may be a statutory prohibition on changing subcontractors.[194] Only when a subcontractor is unable or unwilling to perform will substitution be permitted, and the prime contractor must seek consent to substitute from the awarding authority.[195] The requirement of subcontractor listing has resulted in actions by subcontractors against prime contractors in instances where the contractor later refused to contract with the listed subcontractor.[196]

Absent organized measures such as bid depositories or subcontractor listing requirements, what measures are available to subcontractors to counteract the negative

[189]*See* William H. Orrick, Jr., *Trade Associations Are Boycott Prone—Bid Depositories as a Case Study*, 19 HASTINGS L.J. 505 (1968).

[190]*See* Code of Ethical Conduct of the Associated General Contractors of America (1947, reprinted 1970).

[191]352 F.2d 817 (10th Cir. 1965), *cert. denied,* 384 U.S. 918 (1965).

[192]*Id. But see Cullum Elec. & Mech., Inc. v. Mech. Contractors Ass'n of S.C.,* 436 F. Supp. 418 (D.S.C. 1976), *aff'd,* 569 F.2d 821 (4th Cir.), *cert. denied,* 439 U.S. 910 (1978) (four-hour bid plan approved).

[193]*See, e.g.,* New Mexico Subcontractors Fair Practices Act, N.M. STAT. ANN. § 13–4–31, *et seq.*

[194]*Id.* at § 13–4–36.

[195]*See R.J. Land & Assocs. Constr. Co. v. Kiewit-Shea,* 81 Cal. Rptr. 2d 615 (Cal. Ct. App. 1999); *S. Cal. Acoustics Co. v. C. V. Holder, Inc.,* 456 P.2d 975 (Cal. 1969).

[196]*Id.*

effects of bid shopping? Subcontractors bringing suits against "shopping" prime contractors generally base their actions on two theories of recovery:

(1) That the prime contractor through its acts (principally by using the subcontractor's bid in its own bid to the owner) has "accepted" the bid of the subcontractor, and has thereby entered into a subcontract.
(2) That the conversations and negotiations between the subcontractor and the general contractor have created an oral agreement for performance of the subcontracted work.

Such subcontractor actions have generally not met with great success. For example, the first theory of recovery set out above was rejected in *Merritt-Chapman & Scott Corp. v. Gunderson Bros. Engineering Corp.*[197] The *Merritt-Chapman* court found that there had never been the requisite "meeting of minds" in negotiations between the prime and subcontractor to create a contract, even though the subcontractor's bid had been listed as a part of the bid on the prime contract. The naming of Gunderson, the subcontractor, in the prime contractor's bid did not affect the result. Correspondingly, the argument for an oral contract generally is defeated by the statute of frauds, which requires that certain types of contracts be in writing in order to be enforceable.

The general lack of success of subcontractors in proving that a contract exists with the prime contractor before execution of a formal written subcontract or purchase order should not, however, leave the impression that there is no possibility of a prime contractor being contractually obligated to a subcontractor under appropriate circumstances. One of the common problems in proving a contract is to establish that there is a sufficiently specific agreement between the parties.

A sufficiently specific agreement is necessary for enforcement. Detailed items, however, such as the precise scope of the work to be performed, the period of performance, and so on, are subjects that may not be discussed in preliminary negotiations. Nonetheless, many such details can be supplied from various sources.

For example, certain items may be contained in the prime contract's general and special conditions, specifications, and drawings, all of which are typically included in the Invitation or RFP. These details may also be incorporated by reference. Contract terms can also be proved by reference to prior dealings between the parties or custom and usage in the particular industry. Terms defining exactly what work is included in the contract, and establishing a price for that work, may be confirmed in writing without an elaborate document. For example, a one-page letter or memorandum may succinctly encompass the few remaining details sufficient for a contract, and, if signed by the general contractor, might serve as an enforceable contract.

[197]305 F.2d 659 (9th Cir. 1962), *cert. denied,* 371 U.S. 935 (1962). *But see Elec. Constr. & Maint. Co. v. Maeda Pac. Corp.,* 764 F.2d 619 (9th Cir. 1985) (finding that a subcontractor was successful, when compelling facts were present).

XV. HOLDING SUBCONTRACTORS/VENDORS TO THEIR BIDS

Prime contractors can be faced with the unfortunate situation of having relied on a low subcontractor bid only to discover after award of the prime contract that the subcontractor will not perform at the price originally bid. Although the prime contractor's formal, written bid has been accepted, and the prime contractor is bound to the owner, the prime contractor has no contract with the subcontractor and is faced with making up the shortfall. Similar to the bid shopping discussion in the **Section XIV** of this chapter, analysis of a contractor's ability to hold a subcontractor or vendor to its bid applies to any tier in the contracting process. Under the doctrine of promissory estoppel, however, the subcontractor may be bound to the prime contractor as if a subcontract had actually been executed.[198]

A. Promissory Estoppel

The doctrine of promissory estoppel is set forth in Section 90 of the Restatement (Second) of Contracts (1981):

> A promise which the promisor should reasonably expect to induce action or forbearance on the part of the promisee (or a third person) and which does induce such action or forbearance is binding if injustice can be avoided only by the enforcement of the promise. The remedy granted for breach may be limited as justice requires.

The leading case regarding the application of the doctrine of promissory estoppel to subcontractors' bids is *Drennan v. Star Paving Co.*[199] Drennan, a general contractor, was preparing to bid on a school project. Star Paving Company submitted a bid to Drennan to perform paving work on the school job. Star's bid was submitted by telephone on the day that Drennan had to submit its bid to the school district. Because Star's bid was the lowest received on the paving work, Drennan included Star's price in its computation of the overall costs of the project and submitted the total as its bid to the school district. Drennan was awarded the contract, as it had submitted the lowest general bid.

The next day Star informed Drennan that Star was revoking its bid. Drennan indicated that it had used the bid in computing its own overall bid and that it expected Star to perform the work in accordance with the terms of Star's bid. Star refused to perform. Drennan was forced to obtain a contract for the paving work from a different company at a higher price. Drennan sued Star for the amount of the price increase. The court applied the doctrine of promissory estoppel to rule that Star had become bound to the terms of its offer as a result of Drennan's detrimental reliance on the promises contained in Star's offer.

[198]*See* O. Currie and N. Sweeney, Construction Briefings, Holding Subcontractors to Their Bids, No. 86–3 (Feb. 1986).
[199]333 P.2d 757 (Cal. 1958).

Indeed, subcontractors may be bound not only by the price given in their quote but also to subcontract terms deemed to be standard in the industry. For example, in *Crook v. Mortenson-Neal,*[200] a subcontractor whose bid was accepted, balked at the scheduling and bonding requirements in the subcontract prepared by the general contractor. The court stated that:

> At the time [the subcontractor] bid on its subcontract, it should have expected to be bound by reasonable additional terms governing standard conditions implicit in the relationship between subcontractor and general contractor. Both industry custom, as expressed in standard form subcontracts, and the circumstances surrounding the particular project, dictate the kinds of provisions [the subcontractor] should reasonably have expected in its final subcontract.

Most courts now hold that a contractor may enforce a subcontractor's bid under a promissory estoppel theory.[201] Yet, other courts have rejected the *Drennan* reasoning. For example, in *Home Electric Co. v. Underdown Heating & Air Conditioning Co.,*[202] a North Carolina court ruled as a matter of law, in that state, that a subcontractor was not bound to its bid. The court was disturbed by the one-sided arrangement caused by promissory estoppel, since the prime can enforce the subcontractor's price while the subcontractor has no recourse if it is not awarded the subcontract. The court reasoned that prime contractors can avoid the problem by securing a contract with a subcontractor at the outset, conditioned on a successful bid to the owner.[203]

B. Elements of Promissory Estoppel

In order to hold a subcontractor to its bid under the doctrine of promissory estoppel, the contractor must show all four of the following:

(1) A clear and definite offer by the subcontractor to perform the work at a certain price
(2) A reasonable expectation by the subcontractor that the prime contractor will rely on the subcontractor's price in preparing the prime contractor's bid
(3) Reasonable reliance by the prime contractor on the subcontractor's bid
(4) Detriment to the general contractor as a result of reliance on the subcontractor's bid and the subcontractor's subsequent refusal to perform[204]

[200]727 P.2d 297 (Alaska 1986).
[201]*See Allen M. Campbell Co. Gen. Contractors, Inc. v. Va. Metal Indus., Inc.,* 708 F.2d 930 (4th Cir. 1983); *Hoel-Steffen Constr. Co. v. United States,* 684 F.2d 843 (Ct. Cl. 1982); *Jenkins & Boller Co. v. Schmit Iron Works, Inc.,* 344 N.E.2d 275 (Ill. App. Ct. 1976); *E. A. Coronis Assocs. v. M. Gordon Constr. Co.,* 216 A.2d 246 (N.J. 1966); *James King & Son, Inc. v. De Santis Constr. No. 2 Corp.,* 413 N.Y.S.2d 78 (N.Y. Sup. Ct. 1977).
[202]358 S.E.2d 539 (N.C. Ct. App. 1987), *aff'd,* 366 S.E.2d 441 (N.C. 1988).
[203]*See also Anderson Constr. Co. v. Lyon Metal Prods., Inc.,* 370 So. 2d 935 (Miss. 1979).
[204]*Preload Tech., Inc. v. A.B. & J. Constr. Co.,* 696 F.2d 1080 (5th Cir. 1983); *E. A. Coronis Assocs. v. M. Gordon Constr. Co.,* 216 A.2d 246 (N.J. 1966).

1. *Clear and Definite Offer*

Generally, a lower-tier subcontractor or supplier submits its price to the upper-tier firm as an offer to do the work at a specific price. But the requirement that there be a clear and definite offer to perform a certain part of the work for a particular price prevents the application of the doctrine of promissory estoppel where a subcontractor or supplier offers only an estimate of the cost of the work without intending to make a definite offer to perform the work at that price.[205] Likewise, where the bid or proposal is made expressly revocable or subject to revision, there is no clear and definite offer.[206]

2. *Subcontractor Expects Reliance*

In order for the doctrine of promissory estoppel to apply to a subcontractor's bid, it must be reasonable for the subcontractor to have expected the firm receiving the bid or proposal to rely on the subcontractor's bid or proposal in the preparation of the prime contractor's overall bid or proposal. In order to prove that this was the expectation of the subcontractor, or else that it reasonably should have been its expectation, the courts have, in cases like *Constructors Supply Co. v. Bostrum Sheet Metal Works, Inc.,*[207] allowed the prime contractor/upper-tier firm to introduce testimony as to the ordinary customs and practices of the construction industry in this regard. Of course, evidence may also be presented to demonstrate a subcontractor's actual knowledge and expectation that its quotation would be used in submitting a bid or proposal for the overall project.[208]

3. *Reliance Must Be Reasonable*

Another fundamental requirement for the application of promissory estoppel is that the reliance on the subcontractor's bid or proposal must have been reasonable. Ordinary customs and practices of the construction industry can be relied on to establish the reasonableness of such reliance. Where the allegedly mistaken bid or proposal does not differ substantially from the other bids or proposals received, and there is no obvious mathematical error in the bid or proposal, the reliance normally will be considered reasonable.

It may be necessary in some cases to refute the contention of the subcontractor that its bid or proposal reflected an obvious mistake that should have been evident to the general contractor or upper-tier firm and, therefore, that its reliance on the bid or proposal was unreasonable. If the subcontractor's bid or proposal is much lower than other quotes received on the same work, any reliance on that bid or proposal may be unreasonable.[209] Reasonable reliance is not established where the general contractor

[205]*See N. Litterio & Co. v. Glassman Constr. Co.,* 319 F.2d 736 (D.C. Cir. 1963).
[206]*Preload Tech., Inc. v. A.B. & J. Constr. Co.,* 696 F.2d 1080 (5th Cir. 1983).
[207]190 N.W.2d 71 (Minn. 1971).
[208]*Debron Corp. v. Nat'l Homes Constr. Corp.,* 493 F.2d 352 (8th Cir. 1974).
[209]*Edward Joy Co. v. Noise Control Prods., Inc.,* 443 N.Y.S.2d 361 (N.Y. Sup. Ct. 1981); *Anderson Constr. Co. v. Lyon Metal Prods., Inc.,* 370 So. 2d 935 (Miss. 1979).

misleads an inexperienced subcontractor into believing that the subcontract work can be performed at a price suggested by the general contractor when that price underestimates the true cost of performance.[210]

If a prime contractor or upper-tier contractor is confronted with an unusually low bid or proposal that it did not suggest, and verifies this quote with the subcontractor or supplier, the resulting reliance may be shown to be reasonable.[211]

Even if it can be established that the subcontractor's bid or proposal was mistaken, the subcontractor still may be bound to it. In *Constructors Supply Co. v. Bostrum Sheet Metal Works, Inc.,*[212] the court found that a bid that was 10 percent to 11 percent lower than the other subcontract bids was not self-evidently mistaken, and the general contractor was reasonable in relying on it.[213]

XVI. STATUTE OF FRAUDS ISSUES

It is common in the construction industry for a general contractor to receive subcontractor bids or proposals on the same date that the general contractor must submit its bid or proposal. Often these subcontractor bids or supplier quotes will be given over the telephone at the last minute because the subcontractors wish to prevent general contractors from engaging in preaward bid shopping. As a result, the general contractor is often in a position where it must rely on these oral bids with the possibility of substantial financial exposure should the subcontractors refuse to perform for the price bid. The question therefore arises whether a general contractor that is invoking the doctrine of promissory estoppel to enforce a subcontractor's bid is barred by the statute of frauds if the subcontractor's or supplier's bid is oral.

Courts have split as to whether the statute of frauds defense is overcome by the promissory estoppel argument. In numerous cases the statute of frauds has been held not to apply to subcontractor bids enforced under the doctrine of promissory estoppel.[214] But in *Anderson Construction Co. v. Lyon Metal Products, Inc.,*[215] the court ruled that a prime contractor was barred from recovering damages by the statute of frauds for a subcontractor's failure to honor its oral bid.

XVII. DAMAGES

Assuming that the contractor has established all the elements of an action based on promissory estoppel, what is the measure of damages it may receive from the subcontractor or supplier? In most cases, this measure has been the difference between

[210] *Architects & Contractors Estimating Serv. Inc. v. Smith,* 211 Cal. Rptr. 45 (Cal. Ct. App. 1985).
[211] *H.W. Stanfield Constr. Corp. v. Robert McMullan & Son, Inc.*, 92 Cal. Rptr. 669 (Cal. Ct. App. 1971); *Preload Tech., Inc. v. A.B. & J. Constr. Co.*, 696 F.2d 1080 (5th Cir. 1983).
[212] 190 N.W. 2d 71 (Minn. 1971).
[213] *See also Saliba-Kringlen Corp. v. Allen Eng'g Co.*, 92 Cal. Rptr. 799 (Cal. Ct. App. 1971).
[214] *See, e.g., Allen M. Campbell Co. Gen. Contractors, Inc. v. Va. Metal Indus., Inc.*, 708 F.2d 930 (4th Cir. 1983); *Ralston Purina Co. v. McCollum,* 611 S.W.2d 201 (Ark. Ct. App. 1981).
[215] 370 So. 2d 935 (Miss. 1979).

the price at which the original subcontractor bid the work and the price the general contractor had to pay to obtain a replacement subcontractor.[216]

A somewhat different formulation of damages was used in *Constructors Supply Co. v. Bostrum Sheet Metal Works, Inc.*[217] In that case the court found that the general contractor had not attempted to bid shop the defendant subcontractor's quote. As a result, the general contractor was entitled to recover its damages from the subcontractor. The court found, however, that the general contractor had engaged in bid shopping on other subcontracts for the same project. As a result the contractor had saved money. The damages awarded to the general contractor for the difference in cost between the defendant subcontractor's bid and the cost of obtaining a replacement subcontractor were then reduced by the amount that the general contractor had saved by bid shopping on other subcontractors. This case illustrates the extent to which some courts disdain the practice of bid shopping and the danger that bid shopping poses to a general contractor that uses promissory estoppel to obtain damages from a subcontractor that has withdrawn its bid and refused to perform.

➢ POINTS TO REMEMBER

Competitive Bid Public Contracts

- Generally awarded to low bidder that is financially, technically, and historically responsible to perform the work.
- Past performance or evaluations may be critical to the determination of a bidder's responsibility.
- The bid must be responsive to the solicitation.
- Minor irregularities in the bid can be waived.
- Material irregularities may not be waived.
- Online offerors or bidders will be held to the same strict rules regarding timely submission.

Competitive Proposals on Public Contracts

- Generally awarded on a best-value trade-off basis.
- Low price may not be the controlling factor.
- Past performance evaluations may be a significant factor in the technical evaluation of the offeror's proposal.

Bidding on Private Contracts

- Owners may restrict eligible bidders.
- Owners generally have broad discretion in selecting the contractor.

[216]*See, e.g., C & K Eng'g Contractors v. Amber Steel Co.*, 587 P.2d 1136 (Cal. 1978); *James King & Son, Inc. v. De Santis Constr. No. 2 Corp.*, 413 N.Y.S.2d 78 (N.Y. Sup. Ct. 1977).
[217]190 N.W.2d 71 (Minn. 1971).

Elements of a Mistaken Bid Entitling the Bidder to Relief

- The error must be large in comparison to overall bid price.
- The mistake must relate to an important or "material" aspect of the work.
- The mistake cannot be the product of the bidder's own negligence.
- Relief is more likely for mathematical error.
- Relief is unlikely for estimating/subjective error.
- The owner must be capable of being returned to the status quo.
- Timely notice of the mistake must be given.

Potential Relief from Bid Mistakes (Maximizing the Mistaken Bidder's Potential for Success)

- Notify the owner or public contracting official immediately upon discovery of the mistake.
- Request modification or withdrawal of bid in writing.
- Request immediate conference with owner or public contracting official to discuss the mistake.
- Compile evidence of the mistake to present to owner or public contracting official, including:
    - Affidavits of those individuals involved in bid preparation explaining the mistake
    - Copies of the bid
    - Original worksheets
    - Subcontractor quotes
    - Published price lists or trade catalogs
    - Any other data used to prepare the bid
    - Gear presentation to prove:
    - The existence of the mistake
    - How the mistake happened
    - The bid that was intended

Efforts to Relieve Bid Shopping

- Bid depositories—limited success.
- Subcontractor listing requirements—infrequently used.

Holding the Subcontractor to Its Bid—Summary of Promissory Estoppel Principles:

- There is a clear and definite offer from the subcontractor.
- Subcontractor or lower-tier sub-subcontractor expects the recipient of the bid to rely on it.

- There is actual and reasonable reliance on the bid from the upper-tier firm.
- A party may not rely on obviously mistaken bid.
- A contractor may not mislead a lower-tier bidder to submit an unreasonably low price.
- Damages, if subcontractor refuses to perform in accordance with its bid, may be measured by the difference between the subcontractor's bid and ultimate price paid to a substitute subcontractor for that scope of work bid by the original bidding subcontractor.

5

INTERPRETING THE CONTRACT

I. THE IMPORTANCE OF CONTRACT INTERPRETATION

The contract is the foundation of virtually every relationship in the construction industry. "Contract interpretation" is the process of determining exactly what the parties agreed to in their contract. It involves deciding the meaning of words, filling in gaps, and resolving conflicts. Familiarity with the basic rules of contract interpretation will increase a party's chances of avoiding many problems that can arise during the negotiation and performance of construction contracts.

II. WHAT IS A "CONTRACT"?

A "contract" may be defined succinctly as a set of promises. If a contract is enforceable, the law requires the performance of these promises and provides a remedy if they are not performed. Every contract must satisfy these five conditions to be enforceable:

(1) There must be a real agreement between the parties—that is, a true "meeting of the minds" on the contract's subject matter.
(2) The subject matter of the contract must be lawful.
(3) There must be valid consideration for each promise.[1]

[1]*See Regal Homes, Inc. v. CNA Ins.,* 171 P.3d 610, 617 (Ariz. Ct. App. 2007); *Leesburg Cmty. Cancer Ctr. v. Leesburg Reg'l Med. Ctr.,* Inc. 972 So. 2d 203, 206 (Fla. Dist. Ct. App. 2007). A party can supply consideration either by giving a benefit or incurring a detriment. *See Sunflower Bank, N.A. v. Kindsvater,* 144 P.3d 81 (Kan. Ct. App. 2006); *Christian v. Gouldin,* 804 A.2d 865 (Conn. App. Ct. 2002).

(4) The parties must have the legal capacity to contract.

(5) There must be compliance with legal requirements regarding the form of the contract (e.g., some contracts must be made in writing).[2]

The law will not enforce a contract that fails to meet one or more of these requirements.

Not all construction contracts need be in writing to be valid and enforceable. Many construction businesses have entered into an agreement on a handshake. Modern times, however, require recognition of practical considerations in the complex world of construction. Even if an oral contract is enforceable as a matter of law, reliance on an oral agreement is a risky proposition at best.

The promises that constitute a contract may do more than just impose duties on the promisor and grant rights to the promisee. They may also operate to allocate certain risks that ultimately would make performance by one party more difficult or expensive. Thus, a court, when interpreting a contract, may speak not only in terms of contractual duties and rights, but also in terms of which party assumed the risk of certain contingencies.

III. THE GOAL OF CONTRACT INTERPRETATION

Contracts require interpretation to determine and enforce the parties' intent at the time of contracting. It is rarely possible to determine what was in the minds of the parties when they executed the contract, especially if a dispute later arises. Therefore, courts rely on the objective manifestations of the parties.

A primary contract interpretation rule, which has several facets, is that the reasonable, logical meaning of the contract language will be presumed to be the meaning intended by the parties. This reasonable and logical meaning rule overrides all other rules of contract interpretation.[3]

According to this rule, contract language is interpreted as it would be understood by a reasonably intelligent and logical person familiar with the facts and circumstances surrounding the contract. Courts use two sources of information in determining this objective intent: (1) the language used by the parties in the contract and (2) the facts and circumstances surrounding contract formation.

[2]A "statute of frauds" attempts to prevent false contract claims, typically by requiring that the contract be in writing. Not all contracts are subject to a statute of frauds. Most jurisdictions have statutes of frauds for government contracts, contracts of marriage, contracts that cannot be performed within one year, contracts for the sale of land, contracts over $500 for the sale of goods, contracts to guaranty the debt of another, and other types of contracts. *See, e.g.* GA. CODE ANN. §13–5–30; IOWA CODE ANN. § 622.32; OKLA. STAT. ANN. tit. 15, § 136.

[3]*Lobo Painting, Inc. v. Lamb Constr. Co.*, 231 S.W.3d 256 (Mo. Ct. App. 2007); *Sprucewood Inv. Corp. v. Alaska Hous. Fin. Corp.*, 33 P.3d 1156 (Alaska 2001); *Alvin Ltd. v. U.S. Postal Serv.*, 816 F.2d 1562 (Fed. CIR. 1987).

IV. DEFINING CONTRACT TERMS

Contract interpretation starts by defining the contract's terms. These definitions come from three sources:

(1) The parties may have defined the terms within the contract.
(2) Technical terms are given their meaning within various industries and trades.
(3) General terms are given their widely accepted meanings.

These sources of definitions are discussed in the following sections.

A. Terms Defined by the Parties

Frequently parties will define the terms they use in a contract. These agreed-on definitions are the clearest manifestation of the parties' intent. Therefore, courts generally will abide by the parties' definitions.[4] Examples of some of the more common terms that are frequently defined in a construction contract include the:

- Owner
- Design professional (architect or engineer)
- Contractor
- Contract documents
- Work
- Contract price
- Contract time (or time for performance)
- Change order
- Substantial and/or final completion[5]

B. Technical Terms

Terms may acquire nonstandard or technical meanings in certain industries or trades. These meanings may differ substantially from the meanings generally associated with those terms. Technical meanings will override generally accepted meanings when circumstances indicate that the parties intended to use the technical meaning of the term.[6]

[4]*See, e.g., Hayden Corp. v. Glacier Park Co.,* 896 P.2d 604 (Or. Ct. App. 1995) (warranty definition); *Guy F. Atkinson Co.,* ENGBCA No. 4891, 86–1 BCA ¶ 18,555.
[5]*See, e.g.,* EJCDC C-700, § 1.01 (containing "Defined Terms").
[6]*See, e.g., L.K. Comstock & Co., Inc. v. Becon Constr. Co.,* 932 F. Supp. 948 (E.D. Ky. 1994); *P.J. Dick Contracting, Inc.,* PSBCA No. 1097, 84–1 BCA ¶ 17,149.

C. Generally Accepted Definitions

Terms will be given the meanings generally ascribed to them unless the parties have defined a term otherwise or intended for a term to have a technical meaning.[7]

V. INTERPRETING THE CONTRACT'S WORDING

The following sections examine the legal rules generally applicable to interpreting the language of a contract once its individual terms have been defined in accordance with the principles discussed in the prior sections.

A. The Contract Must Be Considered as a Whole

A fundamental principle of contract interpretation is that a contract must be considered as a whole, giving effect to all of its parts.[8] If, for example, a proposed interpretation makes some part of a written agreement meaningless or illogical, that proposed interpretation usually will be rejected as unreasonable if there is another reasonable interpretation that gives effect to every part of the contract.[9] Each part of the agreement should be examined with reference to all other parts, because one clause may modify, limit, or illuminate another.[10]

Similarly, where several documents form an integral part of one transaction, a court may read these together with reference to one another even where the documents involved do not specifically refer to one another. A similar rule applies to documents annexed to the contract or incorporated by reference. Therefore, an interpretation that leaves portions of the contract meaningless will generally be rejected. Likewise, because the contract is to be read as a whole, its provisions should, if possible, be harmonized.

B. Specific Terms versus General Terms

Sometimes parties to a contract will agree on terms that appear inconsistent with each other. In such a case, if one written term specifically addresses a particular issue, and the other term is a general term, the specific term usually will be viewed as

[7]*Lodge Corp. v. Assurance Co. of Am.*, 775 N.E.2d 1250 (Mass. App. Ct. 2002); *Atlas R.R. Constr. Co.*, ENGBCA No. 5972, 94–3 BCA ¶ 26,997; *see, e.g., Sauter Constr. Co.*, ASBCA No. 22338, 78–1 BCA ¶ 13,092.

[8]*New Valley Corp. v. United States*, 119 F.3d 1576 (Fed. Cir. 1997); *Downey v. Clauder*, 811 F. Supp. 338 (S.D. Ohio 1992); *McDevitt Mech. Contractors, Inc. v. United States*, 21 Cl. Ct. 616 (1990).

[9]*See Medlin Constr. Group, Ltd. v. Harvey*, 449 F.3d 1195 (Fed. Cir. 2006) (rejecting interpretation that made part of the specifications meaningless).

[10]*T. Brown Constructors, Inc. v. Pena*, 132 F.3d 724 (Fed. Cir. 1997); *Plaza Dev. Serv. v. Joe Harden Builders, Inc.*, 365 S.E.2d 231 (S.C. Ct. App. 1988).

creating an exception to the general term and usually will be given precedence over the more general term.[11]

C. Handwritten, Typed, and Preprinted Terms

Many times, written contracts will be on preprinted forms, such such as the American Institute of Architects (AIA), ConsensusDOCS, or Engineers Joint Contract Documents Committee (EJCDC) families of contract document forms.[12] To accommodate issues or concerns unique to the specific construction project, the parties often will modify these preprinted terms with typed and sometimes even handwritten modifications. Often the preprinted, typewritten, and handwritten terms will conflict with each other.

As a general rule, handwritten terms take priority over conflicting typewritten or preprinted terms. Likewise, typewritten contract terms take precedence over preprinted contract terms. These general rules of interpretation will apply unless circumstances show that the parties intended otherwise.[13]

D. Order-of-Precedence Clause

In some circumstances, it may be impossible to interpret a contract without resolving a conflict between different terms. Construction contracts are complex and frequently contain numerous sections drafted by different people or organizations. When two or more conflicting provisions cannot be harmonized, the rules of contract interpretation establish an order of precedence that may resolve the conflict.

The general conditions of many contracts include an order-of-precedence clause expressly stating which provisions control over others in case of conflict.[14] For example, the order-of-precedence clause may state that the specifications generally take precedence over the drawings, special conditions take precedence over general conditions, and so on. In federal government contracts, provisions required by law generally cannot be altered by such a clause.

In the absence of an order-of-precedence clause, general common law rules of precedence will apply. For example, it is a basic rule of contract interpretation that general terms and provisions in a contract yield to specific ones.[15] It is also a general rule of contract interpretation that when specific requirements or definitions are itemized and

[11] *T. Brown Constructors, Inc. v. Pena,* 132 F.3d 724, 729 (Fed. Cir. 1997) ("the more specific test data trumps the more general test results.") (citations omitted); *Smith Barney, Inc. v. Sarver,* 108 F.3d 92 (6th Cir. 1997).

[12] See **Chapter 13** for a detailed list of the ConsensusDOCS, AIA, and EJCDC construction documents released in 2007.

[13] *Benedict v. Snead,* 271 Ga. 585, 586 (Ga. 1999).

[14] *See, e.g., Gen. Eng'g & Mach. Works v. O'Keefe,* 991 F.2d 775 (Fed. Cir. 1993) (citing *Hensel Phelps Constr. v. United States,* 886 F.2d 1296 (Fed. Cir. 1989).

[15] This rule is generally known as *ejusdem generis. See Smith Barney, Inc. v. Sarver,* 108 F.3d 92 (6th Cir. 1997). *See also* 17A C.J.S. *Contracts* § 313.

spelled out, anything not expressly included is deemed to be excluded.[16] Additionally, handwritten terms take precedence over typewritten terms, and typewritten terms take precedence over printed terms.[17]

VI. THE FACTS AND CIRCUMSTANCES SURROUNDING CONTRACT FORMATION

Courts frequently interpret a contract based on the facts and circumstances surrounding the contract's formation. This evidence comes in three forms: (1) evidence of discussions and conduct; (2) evidence of the parties' prior dealings; and (3) evidence of custom and usage in the industry.

A. Discussions and Conduct

The parties' discussions and conduct can be persuasive when interpreting a contract. For example, a contractor may become aware of a possible ambiguity at a prebid conference and request a clarification. Such a clarification may serve as proof that the parties resolved a possible ambiguity and agreed on a common interpretation of the contract.

Similarly, one of the parties may make its interpretation of the contract known to the other party. This can be done through express discussions or may arise impliedly by the party's conduct. If the other party, knowing this interpretation, remains silent or does not object, this interpretation will be binding.

1. Parol Evidence

Evidence of the parties' discussions and conduct before, and at the time, a written contract is signed (parol evidence) may not, however, be admissible as evidence to resolve a dispute regarding the interpretation of a contract. Reducing a contract to writing has legal consequences. Traditionally, the law has imposed rules that limit the use of external or parol evidence to vary or contradict the terms of a written contract. This is commonly called the parol evidence rule.[18]

The first question that must be answered to determine the admissibility of parol evidence is whether the contract is a final and complete expression of the parties' agreement. Parol evidence may be used to make this determination. If the contract is final and complete, normally parol evidence cannot be used to vary or contradict

[16]This rule of contract law is generally known as *expressio unius est exclusio alterius* (the express mention of the one is the exclusion of the other). *See* 17A C.J.S. *Contracts* § 312.

[17]*Authentic Architectural Millworks v. SCM Group,* 586 S.E.2d 726 (Ga. Ct. App. 2003); *Patellis v. 100 Galleria Parkway Assoc.,* 447 S.E.2d 113, 115 (Ga. Ct. App. 1994); *see also Wood River Pipeline Co. v. Willbros Energy Servs. Co.,* 738 P.2d 866 (Kan. 1987).

[18] *S & B Mining Co. v. N. Commercial Co.,* 813 P.2d 264 (Alaska 1991); *see also Lower Kuskokwim Sch. Dist. v. Alaska Diversified Contractors, Inc.,* 734 P.2d 62 (Alaska 1987).

its unambiguous terms.[19] For example, a contractor was not allowed to rely on an oral, prebid extension of time to establish an acceleration claim when that evidence would have contradicted the express provisions of the written contract.[20] If a term is ambiguous, however, courts may admit extrinsic evidence concerning the parties' negotiations to ascertain the intent of the parties at the time of contracting.[21]

2. *Merger Clauses*

Most construction contracts address the parol evidence issue by including a clause indicating that the writing is a complete and final statement of all the terms of the contract. This type of contract clause is known as a merger or integration clause. For example, the first sentence of Paragraph 24.7 ConsensusDOCS 300 is an integration clause, which states: "This Agreement is solely for the benefit of the Parties, represents the entire and integrated agreement between the Parties, and supersedes all prior negotiations, representations or agreements, either written or oral." A merger or integration clause is intended to exclude evidence of contrary meaning or interpretation of contract terms that are within the "four corners" of the contract, or evidence of additional contract terms.[22] Such "extrinsic" evidence is typically in the form of precontract negotiations, prior dealings between the parties, or standard industry practices.[23]

B. The Parties' Prior Dealings

If parties have dealt with each other previously, courts may look at their earlier behavior and practices to help interpret their current contract. Although evidence of an established pattern of prior dealings may be offered to aid a court, it cannot be used to vary or modify the clear, express terms of a written contract. The parol evidence rule prevents such a use of extrinsic evidence.

The admission of prior dealings serves the purpose of showing what the parties intended by the language in the contract. For example, a Pennsylvania court interpreted an ambiguity in a contract as to the meaning of the term "positive shielding" in favor of the contractor based on the "conduct of the parties" throughout a prior project.

[19]*Fid. & Deposit Co. of Md. v. ROTEC Indus., Inc.*, 2004 WL 432513 (N.D. Ill. 2004); *Rothlein v. Armour and Co.*, 377 F. Supp. 506, 510 (W.D. Pa. 1974). *See also Fuller Co. v. Brown Minneapolis Tank & Fabricating Co.*, 678 F. Supp. 506 (E.D. Pa. 1987).

[20]*See Lower Kuskokwim Sch. Dist. v. Alaska Diversified Contractors, Inc.*, 734 P.2d 62 (Alaska 1987).

[21]*Hickman v. Kralicek Realty and Constr. Co.*, 2003 WL 22723484 (Ark. Ct. App. 2003); *Teleport Commc'ns Group, Inc. v. Barclay Fin. Group*, 176 F.3d 412 (7th Cir. 1999); *Judge v. Wellman*, 403 S.E.2d 76 (Ga. Ct. App. 1991).

[22]*McAbee Constr. Inc. v. United States*, 97 F.3d 1431, 1434 (Fed. Cir. 1996).

[23]The presence of a merger or integration clause will not, however, prevent a party from using extrinsic evidence of precontract negotiations, past dealings, or industry custom to supply terms or meaning to a written agreement if the written agreement is obviously incomplete or imprecise. *See Wood v. Phoenix Ins. Co.*, 34 S.E.2d 688 (Ga. 1945).

The court allowed extrinsic evidence as to the parties' conduct to determine the parties' intent in using the term.[24]

As previously stated, extrinsic evidence is not generally admissible to show intent entirely different from what is clearly stated in the contract. A prior course of dealing may, however, show that the contract is not the final and complete agreement of the parties.[25]

Prior conduct may also amount to waiver or estoppel. A party may be prevented from enforcing an explicit contract requirement if in its prior dealings it did not require compliance with the requirement.[26]

C. Industry Custom and Usage

Evidence of customs within a particular industry may be used to show that the parties intended for an ordinary word to have a specialized meaning.[27] However, courts are divided on the role of such evidence.[28] One line of cases holds that evidence of trade practice and custom may be admitted to show the meaning of an ambiguous contract term but not to override a seemingly unambiguous term.[29] The second line of cases maintains that evidence of trade practice and custom may be introduced to show that a term, which appears on its face to be unambiguous, has, in fact, a specialized meaning other than that of its ordinary meaning.[30]

A party seeking to assert a trade custom or practice must present evidence that the custom is well established.[31] One method of establishing trade custom is to show the interpretations of other bidders on the contract.[32]

Similarly, a technical word will be given its ordinary meaning in the industry unless it is shown that the parties intended to use it in a different sense. The appropriate meaning of ambiguous technical terms may also be clarified by the introduction of extrinsic evidence. For example, in a classic case, a Texas appellate court allowed the introduction of evidence of custom to establish the intended meaning of the contract term "working days" as it related to the owner's right to assess liquidated damages for

[24]*Dep't of Transp. v. IA Constr. Corp.*, 588 A.2d 1327 (Pa. Commw. Ct. 1991).

[25]Restatement (Second) of Contracts § 65 (1981).

[26]*James v. Zurich-Am. Ins. Co. of Ill.*, 203 F.3d 250 (3d Cir. 2000); *Sperry Flight Sys. v. United States*, 548 F.2d 915 (Ct. Cl. 1977).

[27]*See, e.g., Rosenberg v. Turner*, 98 S.E.763 (Va. 1919).

[28]*Metric Constructors v. Nat'l Aeronautical & Space Admin.*, 169 F.3d 747 (Fed. Cir. 1999).

[29]*R. B. Wright Constr. Co. v. United States*, 919 F.2d 1569 (Fed. Cir. 1990); *George Hyman Constr. Co. v. United States*, 564 F.2d 939 (Ct. Cl. 1977); *WRB Corp. v. United States*, 183 Ct. Cl. 409 (1968). *See also Dubois Constr. Co. v. Monda Constr. Co., Inc.*, 907 So. 2d 855 (La. Ct. App. 2005) (holding that in absence of agreed-upon rate for removing concrete slab, and where subcontractor did not offer any more specific proof, the reasonable value of work could be determined by the rate the subcontractor had charged for similar work on prior projects).

[30]*W. States Constr. Co., Inc. v. United States*, 26 Cl. Ct. 818 (1992); *Haehn Mgmt. Co. v. United States*, 15 Cl. Ct. 50 (1988), *aff'd*, 878 F.2d 1445 (Fed. Cir. 1989); *Gholson, Byars & Holms Constr. Co. v. United States*, 351 F.2d 987 (Ct. Cl. 1965).

[31]*W. G. Cornell Co. v. United States*, 376 F.2d 299 (Ct. Cl. 1967).

[32]*See Eagle Paving*, AGBCA No. 75–156, 78–1 BCA ¶13,107.

delay in completion.[33] More recently, the United States Claims Court relied on trade practice to interpret patently ambiguous pipe-wrapping requirements in a federal government construction contract.[34]

D. Limitations on the Use of Facts and Circumstances Surrounding the Contract

When a contract is reduced to writing, the three outside sources of meaning just discussed can be used only to (1) supply terms or meanings for the purpose of completing a written agreement that is incomplete;[35] (2) show the correct interpretation for an agreement that is complete but susceptible to more than one reasonable meaning;[36] or (3) show that the parties actually have more than one contract.

Evidence of the parties' discussions and conduct, prior dealings, or industry customs usually cannot be relied on to contradict the plain language of the contract. Recognizing this rule, the construction industry professional who enters into a contract relying on a prebid "clarification" or other precontract assurance differing from the plain language of the final written agreement runs the risk of having such inconsistent provisions declared unenforceable.[37]

VII. RESOLVING AMBIGUITIES

The rules of contract interpretation previously discussed may not resolve every contract ambiguity. If an ambiguity remains, courts will apply one of the two risk-allocation principles (or rules of last resort) to resolve the conflict: (1) the ambiguity should be construed against its drafter; or (2) the ambiguity should be construed against the party that failed to request a clarification of the ambiguity. These principles are discussed next.

[33] *Lewis v. Jones,* 251 S.W.2d 942 (Tex. App. 1952).

[34] *W. States Constr. Co., Inc. v. United States,* 26 Cl. Ct. 818 (1992).

[35] *Stamford Wrecking Co. v. United Stone Am., Inc.,* 912 A.2d 1044 (Conn. App. Ct. 2007) (written agreement did not state how much work was allocated to subcontractor).

[36] *Eric A. Carlstrom Constr. Co. v. Indep. Sch. Dist. No. 77,* 256 N.W.2d 479 (Minn. 1977) (industry custom used to supply meaning to "means and methods" clause).

[37] *Ala. Elec. Coop., Inc. v. Bailey's Constr. Co., Inc.,* 950 So. 2d 280 (Ala. 2006) (pre-contract statement about providing insurance coverage could not be enforced because statement was inconsistent with the plain language of the final written agreement); *Baroid Equip., Inc. v. Odeco Drilling, Inc.,* 184 S.W.3d 1 (Tex. App. 2005) (lessor's prelease assurances were unenforceable, because the assurances were inconsistent with the plain language of the written lease agreement); *Neal & Co., Inc. v. Ass'n of Vill. Council Presidents Reg'l,* 895 P.2d 497 (Alaska 1995) (architect's oral pre-bid clarification about temporary power was unenforceable because the clarification was inconsistent with plain language of the bid documents and not added by written addendum).

A. Construing the Contract against the Drafter

The risk of ambiguous contract language generally belongs to the party responsible for drafting the ambiguity.[38] For example, one court found that a contractor was not entitled to assess a percentage markup for overhead on certain change order allowances on the basis that the contract stated that these allowances included the contractor's "actual costs."[39] Construing the term "actual costs" against the contractor that drafted the applicable contract provision, the court concluded that this term was ambiguous because it was "capable of more than one meaning when viewed objectively by a reasonably intelligent person who has examined the context of the entire integrated agreement."[40]

This rule of contract interpretation applies unless the nondrafting party knew of, or should have known of, the ambiguity and several requirements must be met for this principle to apply:

(1) There must truly be an ambiguity—that is, the contract must have at least two reasonable interpretations. A nondrafting party's interpretation need not be the only reasonable interpretation for this principle to apply.[41]

(2) One of the two parties must have drafted or chosen the ambiguous contract language.

(3) The nondrafting party must demonstrate that it relied on its interpretation.[42]

B. Duty to Request Clarification

An obviously (or "patently") ambiguous contract provision will not be construed against its drafter if the nondrafting party fails to seek clarification of that ambiguity before submitting its proposal or bid.[43] Ambiguities are either patent or latent. A patent ambiguity is one that is readily apparent from the wording of the contract.[44] In contrast, language containing a latent ambiguity initially appears to be clear and unambiguous but actually contains an underlying ambiguity that becomes apparent only after a close examination or presentation of extrinsic facts.[45]

[38]The technical name for this interpretive rule is *contra proferentem*. *See Freeman & Co. v. Bolt,* 968 P.2d 247 (Idaho Ct. App. 1998); *United States v. Turner Constr. Co.,* 819 F.2d 283 (Fed. Cir. 1987).

[39]*All Star Constr. Co., Inc. v. Koehn,* 741 N.W.2d 736 (S.D. 2007).

[40]*Id.* at 744.

[41]*Fry Comm'ns Inc.* 22 Cl. Ct. 497 (1991); *Bennett v. United States,* 371 F.2d 859 (Ct. Cl. 1967); *Gall Landau Young Constr. Co.,* ASBCA No. 25801, 83–1 BCA ¶ 16,359.

[42]*Interstate Gen. Gov't Contractors v. Stone,* 980 F.2d 1433 (Fed. Cir. 1992); *Fruin-Colnon Corp. v. United States,* 912 F.2d 1426 (Fed. Cir. 1990).

[43]*Kiska Constr. v. Wash. Metro. Area Transit Auth.,* 321 F.3d 1151 (D.C. Cir. 2003); *Triax Pac., Inc. v. West,* 130 F.3d 1469 (Fed. Cir. 1997).

[44]*See Big Chief Drilling Co. v. United States,* 15 Ct. Cl. 295 (1988).

[45]*See AWC, Inc.,* PSBCA No. 1747, 88–2 BCA ¶ 20,637.

A bidder or offeror generally has an obligation to seek clarification of patent ambiguities or inconsistencies that appear in the bid or proposal documents.[46] Typically, government construction bid documents contain an express provision imposing an affirmative duty on a contractor to seek clarification of patent ambiguities.[47] But the lack of such a provision does not relieve a contractor of its duty to request clarification of obvious ambiguities. For example, the United States Court of Federal Claims held that, when a provision in the solicitation conflicts directly and openly with a provision in a referenced handbook, a contractor must identify the obvious ambiguity in order to obtain clarification.[48] Because the contractor in that case did not alert the contracting officer to the glaring discrepancy, the court barred the contractor from recovering any damages caused by the conflicting provisions within the solicitation and handbook.[49]

The difficulty arises in determining whether an ambiguity was obvious before bidding. One factor used to make this determination is whether other bidders requested a clarification before bidding.[50] Ultimately, whether the ambiguity was obvious will depend on "what a reasonable man would find to be patent and glaring."[51]

VIII. IMPLIED CONTRACTUAL OBLIGATIONS

The law includes implied obligations in every contract, in addition to the obligations that are expressly agreed on by the parties.[52] Implied obligations can allocate risk and responsibility between contracting parties with the same force as terms that are negotiated and reduced to writing in a signed written agreement. Three of the most important implied obligations arising in construction contracts are briefly discussed in the next three sections.

A. Duty of Good Faith and Fair Dealing

A general duty of good faith and fair dealing is implied in virtually every contract in almost every state.[53] The implied duty of good faith and fair dealing has been explained as an "implied covenant...that imposes obligations on both contracting parties that include the duty not to interfere with the other party's performance and

[46] *White v. Edsall Constr. Co.,* 296 F.3d 1081 (Fed. Cir. 2002); *Newsom v. United States,* 676 F.2d 647 (Ct. Cl. 1982).
[47] *Blount Bros. Constr. Co. v. United States,* 346 F.2d 962 (Ct. Cl. 1965).
[48] *Nielsen-Dillingham Builders, Joint Venture v. United States,* 43 Fed. Cl. 5 (1999).
[49] *Id. See also Big Chief Drilling Co. v. United States,* 15 Ct. Cl. 295 (1965).
[50] *See W.M. Schlosser Co.,* VABCA No. 1802, 83–2 BCA ¶ 16,630.
[51] *Max Drill, Inc. v. United States,* 427 F.2d 1233 (Ct. Cl. 1970).
[52] *See, e.g.,* Restatement (Second) of Contracts § 205 (1981).
[53] A few states do not recognize an implied covenant of good faith and fair dealing in contracts, including Texas and Indiana. Teri J. Dobbins, *Losing Faith: Extracting the Implied Covenant of Good Faith from (Some) Contracts*, 84 Or. L. Rev. 227 (2005).

not to act so as to destroy the reasonable expectations of the other party regarding the fruits of the contract."[54] For contracts for the sale of goods among merchants, the Uniform Commercial Code (UCC) defines good faith as "honesty in fact in the conduct or transaction concerned and the observance of reasonable commercial standards of fair dealing in the trade."[55]

B. Duty to Cooperate

A duty to cooperate is implied in every contract wherever one party's cooperation is necessary for the other party to perform.[56] From the contractor's perspective, the owner's implied duty to cooperate is probably the most important of all the implied contractual obligations. In one old but often cited case where a contractor obtained damages for owner-caused delays, the United States Court of Claims reviewed several generally accepted legal authorities supporting the duty to cooperate and concluded:

> [I]t is however, an implied provision of every contract, whether it be one between individuals or between an individual and the Government, that neither party to the contract will do anything to prevent performance thereof by the other party or that will hinder or delay him in its performance.[57]

Owners can breach their duty to cooperate in a number of ways. Some more common examples are:

(1) Failure to coordinate the work of separate contractors within their control[58]

(2) Late delivery of the completed design for which the owner is responsible[59]

[54]*Centex Corp. v. United States,* 395 F.3d 1283 (Fed. Cir. 2005).

[55]*See, e.g.,* GA. CODE ANN. § 11–2–103(1) (b) and GA. CODE ANN. § 11–1–201(19). The U.C.C. also states that "the obligations of good faith, diligence, reasonableness, and care. . .may not be disclaimed by agreement but the parties may by agreement determine the standards by which the performance of such obligations is to be measured if such standards are not manifestly unreasonable." Ga. Code Ann. § 11–1–102(3). For additional information on application of the U.C.C. in the construction industry, see **Chapter 6**.

[56] *Renda Marine, Inc. v. United States,* 66 Fed. Cl. 639, 649 (2005); *Case Corp. v. Hi-Class Bus. Sys. of Am., Inc.,* 184 S.W.3d 760 (Tex. App. 2005); *Hamlin v. Steward,* 622 N.E.2d 535 (Ind. Ct. App. 1993).

[57]*George A. Fuller Co. v. United States,* 69 F. Supp. 409 (Ct. Cl. 1947). *See also Coatesville Contractors v. Borough of Ridley,* 506 A.2d 862 (Pa. 1986).

[58]*See Clifford R. Gray, Inc. v. City Sch. Dist. of Albany,* 716 N.Y.S.2d 795 (N.Y. App. Div. 2000) (owner's construction manager failed to coordinate and supervise the parallel prime contractors); *Snyder Plumbing & Heating Corp. v. State,* 198 N.Y.S.2d 600, 604 (N.Y. Ct. Cl. 1960); *Scocollo Constr., Inc. v. City of Renton,* 9 P.3d 886 (Wash. Ct. App. 2000) (finding breach of implied duty for failing to coordinate the work of other contractors).

[59]*See Great Lakes Aircraft Co., Inc. v. City of Claremont,* 608 A.2d 840 (N.H. 1992) (jury question as to whether city breached implied duty to cooperate with airport builder by failing to timely obtain required FAA waiver of terms of airport layout plan on builder's behalf). *But see Whittaker Elec. Sys. v. Dalton,* 124 F.3d 1443 (Fed. Cir. 1997) (no breach of implied duty where government required contractor to submit formal rather than redline drawings).

(3) Late or incomplete review of submittals[60]
(4) Delayed issuance of, or refusal to issue, change orders[61]
(5) Untimely or incomplete inspections[62]
(6) Delay in obtaining necessary permits and approvals[63]
(7) Taking other actions that unreasonably hinder or interfere with the contractor's progress[64]

The owner's implied duty to cooperate includes an implied duty to exercise care to coordinate the activities of the various entities with which the owner has contracts, such as parallel prime contractors.[65] Although sophisticated owners often attempt to avoid their duty to coordinate through the use of contract clauses that obligate parallel prime contractors to coordinate directly with each other, courts have recognized that when an owner retains the right of control over its contractors, it cannot absolve itself of all responsibility for the coordination of their work.[66]

When an owner fails to provide the necessary coordination, it generally must respond by paying delay or disruption damages to the impacted contractor.[67] These principles also apply to general contractors, which have implied duties to cooperate with and coordinate the work of the subcontractors and trade contractors under their control.[68]

[60]*See Sterling Millwrights, Inc. v. United States,* 26 Cl. Ct. 49 (1992) (government failed to timely review erection drawings).

[61]*See Orlosky, Inc. v. United States,* 68 Fed. Cl. 296 (2005) (government required contractor to incur expense in submitting request for equitable adjustment on termination for convenience and then withdrew the termination for convenience).

[62]*See Allied Fire & Safety Equip. Co. v. Dick Enters., Inc.,* 886 F. Supp. 491 (E.D. Pa. 1995); *Crawford Painting & Drywall Co. v. J.W. Bateson Co, Inc.,* 857 F.2d 981 (5th Cir. 1988).

[63]*See Precision Pine & Timber, Inc. v. United States,* 50 Fed. Cl. 35 (2001) (finding that government breached its implied duty to cooperate where it failed to consult with U.S Fish and Wildlife on timber contracts as required by Endangered Species Act). *But see Hunt Constr. Group, Inc. v. United States,* 48 Fed. Cl. 456 (2001) (finding that government did not breach implied duty where it had refused to sign agency agreement that would have allowed contractor to obtain state sales tax exemption on permanent construction materials purchased).

[64]*CEMS, Inc. v. United States,* 59 Fed. Cl. 168 (2003). *See also Jones v. Calpine Corp.,* 172 Fed. Appx. 186 (9th Cir. 2006) (finding plaintiff stated a cause of action for breach of implied duty to cooperate based on defendant's actions in preventing completion construction of power plant in order to avoid payment obligation).

[65]*Baldwin-Lima-Hamilton Corp. v. United States,* 434 F.2d 1371 (Ct. Cl. 1970); *Snyder Plumbing & Heating Corp. v. State,* 198 N.Y.S.2d 600, 604 (Ct. Cl. 1960).

[66]*Amp-Rite Elec. Co., Inc. v. Wheaton Sanitary Dist.,* 580 N.E.2d 622 (Ill. App. Ct. 1991); *Shea-S&M Ball v. Massman-Kiewet-Early,* 606 F.2d 1245 (D.C. Cir. 1979); *Eric A. Carlstrom Constr. Co. v. Indep. Sch. Dist. No. 77,* 256 N.W.2d 479 (Minn. 1977).

[67]*Freeman Contractors, Inc. v. Cent. Sur. & Ins. Corp.,* 205 F.2d 607 (8th Cir. 1953); *L. L. Hall Constr. Co. v. United States,* 379 F.2d 559 (Ct. Cl. 1966); *Gasparini Excavating Co. v. Pa. Turnpike Comm'n,* 187 A.2d 157 (Pa. 1963).

[68]*United States f/b/o Wallace v. Flintco, Inc.,* 143 F.3d 955 (5th Cir. 1998); *Allied Fire & Safety Equip. Co. v. Dick Enters ., Inc.,* 886 F. Supp. 491 (E.D. Pa. 1995); *Crawford Painting & Drywall Co. v. J.W. Bateson Co,* 857 F.2d 981 (5th Cir. 1988).

C. Warranty of Plans and Specifications—The *Spearin* Doctrine

Another important implied obligation in construction contracts is that the party responsible for furnishing the completed design impliedly warrants the adequacy and sufficiency of that design.[69] This rule was first pronounced by the United States Supreme Court in *United States v. Spearin*:

> [I]f the contractor is bound to build according to plans and specifications prepared by the owner, the contractor will not be responsible for the consequences of defects in the plans and specifications.[70]

Often referred to as the *Spearin* doctrine, this implied duty has been recognized by the courts in nearly every state. The owner impliedly warrants the adequacy and sufficiency of the completed plans and specifications to the contractor, even when the design is prepared by the owner's independent architect or engineer.[71] Generally, in federal construction contracts, all delays caused by defective design specifications are compensable.[72]

In "performance specifications," an owner specifies the performance characteristics that are to be obtained by the contractor and leaves the details of the design to the contractor's judgment. Although the *Spearin* doctrine does not apply to performance specifications,[73] an owner still can be liable for a contractor's unanticipated difficulties under a performance specification if the contractor shows that the owner-furnished performance specification was impossible or commercially impracticable to achieve.[74] A performance specification is commercially impracticable if it can be performed only at an excessive and unreasonable cost.[75]

In design-build contracts, the designer-builder has the implied duty to furnish an adequate and sufficient design, instead of the owner, because the designer-builder is responsible for furnishing the design.[76]

The *Spearin* doctrine can serve as both a shield and a sword for a party that is not responsible for furnishing the design. When the owner furnishes the design, the contractor will not be liable to the owner for an unsatisfactory final result if the contractor performs in accordance with owner-furnished plans and specifications. If an inadequacy in an owner-furnished design results in delay, disruption, or additional cost

[69] *Big Chief Drilling Co. v. United States,* 26 Cl. Ct. 1276 (1992); *Ordnance Research, Inc. v. United States,* 609 F.2d 462 (Ct. Cl. 1979); *State Highway Dep't v. Hewitt Contr. Co.,* 146 S.E.2d 632 (Ga. 1966).

[70] 248 U.S. 132, 136 (1918).

[71] *Greenhut Constr. Co.,* ASBCA No. 15192, 71–1 BCA ¶ 8845.

[72] *Daly Constr. Inc. v. Garrett,* 5 F.3d 520 (Fed. Cir. 1993); *Am. Line Builders, Inc. v. United States,* 26 Cl. Ct. 1155 (1992); *Chaney & Jones Constr. Co. v. United States,* 421 F.2d 728 (Ct. Cl. 1970).

[73] A design specification details the materials to be used and how the work is to be performed. A performance specification specifies the results to be obtained, and leaves it to the contractor to determine how to achieve those results. *Neal & Co., Inc. v. United States.,* 945 F.2d 385, 389 (Fed. Cir. 1991).

[74] *Int'l Elec. Corp. v. United States,* 646 F.2d 496 (Ct. Cl. 1981).

[75] *Oak Adec, Inc. v. United States,* 24 Cl. Ct. 502 (1991); *W. F. Magann Corp. v. Diamond Mfg. Co.,* 775 F.2d 1202 (4th Cir. 1985); *Keller Constr. Corp. v. George W. Coy & Co.,* 119 So. 2d 450 (La. 1960).

[76] *Mobile Hous. Env't v. Barton & Barton,* 432 F. Supp. 1343 (D. Colo. 1977).

to the contractor, the contractor may use the *Spearin* doctrine as the basis for claims for additional time and recovery of the additional costs incurred.[77] For example, in a recent case arising from the construction of two helicopter hangars, the Federal Circuit Court of Appeals affirmed a board decision awarding a contractor additional costs under the *Spearin* doctrine, where the government's design called for the hangar doors to be constructed and rigged with three "pick" points, which would not work.[78] In another recent case, the Seventh Circuit enforced the *Spearin* doctrine under Illinois law, holding that a city impliedly warranted the suitability of a specified quarry to produce adequate armor rock.[79]

A contractor cannot recover under the *Spearin* doctrine if it knew, or through the exercise of reasonable care should have known, of the defective nature of the design before submitting its bid or proposal.[80] Thus, when the solicitation documents direct the bidders to conduct a prebid site inspection, the contractor will be deemed to have prior knowledge of any inaccuracies in the design that could have been discovered through a reasonable site inspection, even if the contractor never inspects the site.[81] In addition to site inspection clauses, sophisticated owners often attempt to avoid liability under the *Spearin* doctrine by including clauses in their contracts that disclaim the sufficiency of the plans and specifications. When owners provide design information with the intent that contractors rely on it when formulating their bids, however, the *Spearin* doctrine should prevail over a disclaimer clause, unless it expressly and specifically shifts the risk of the design flaw to the contractor.[82]

IX. CONTRACTUAL OBLIGATIONS ARISING BY OPERATION OF LAW

In addition to the implied obligations discussed in the previous section, many jurisdictions will read certain obligations into a contract or void certain obligations that are contrary to that jurisdiction's statutes, regulations, or stated public policy. For example, local building codes usually provide a minimum standard of care for construction contracts by operation of law. Likewise, the UCC can supply important terms to contracts for the sale of goods.[83]

The law also nullifies certain express contract terms. For example, many states have laws that nullify terms that indemnify a construction owner or contractor against

[77] *USA Petroleum Corp. v. United States,* 821 F.2d 622 (Fed. Cir. 1987); *Felton Constr. Co.,* AGBCA No. 406–9, 81–1 BCA ¶ 14,932; *R. M. Hollingshead v. United States,* 111 F. Supp. 285 (Ct. Cl. 1953).
[78] *White v. Edsall Constr. Corp.,* 296 F.3d 1081 (Fed. Cir. 2002).
[79] *Edward E. Gillen Co. v. City of Lake Forest,* 3 F.3d 192 (7th Cir. 1993).
[80] *Blount Bros. Constr. Co. v. United States,* 346 F.2d 962 (Ct. Cl. 1965).
[81] *Stuyvesant Dredging Co. v. United States,* 834 F.2d 1576 (Fed. Cir. 1987); *Johnson Controls, Inc. v. United States,* 671 F.2d 1312 (Ct. Cl. 1982); *Allied Contractors, Inc. v. United States,* 381 F.2d 995 (Ct. Cl. 1967).
[82] *See United States v. Spearin,* 248 U.S. 132 (1918); *White v. Edsall Constr. Co, Inc.,* 296 F.3d 1081 (Fed. Cir. 2002); *Sherman R. Smoot Co. of Ohio v. Ohio Dep't of Admin. Servs.,* 736 N.E.2d 69 (Ohio Ct. App. 2000); *Morris, Inc. v. State ex rel. S.D. Dept. of Transp.,* 598 N.W.2d 520 (S.D. 1999).
[83] *Gen. Eng'g & Mach. Works v. O'Keefe,* 991 F.2d 775, 779–81 (Fed. Cir. 1993). *See* U.C.C. § 2–204(3).

its own sole negligence, or prospectively waive mechanic's lien rights, or penalize a contracting party for its failure to perform.[84] Some states have also enacted legislation providing that "No Damages for Delay" clauses are void and unenforceable. [85]

➢ POINTS TO REMEMBER

- Construction industry professionals should enter into and perform their contracts with an awareness and understanding of the principles of contract interpretation.
- Construction industry contracts should be written.
- Construction industry should fully allocate risks between the parties and embody the parties' entire agreement.
- Important construction industry contract terms should be defined.
- A construction industry contract should include an order-of-precedence clause.
- Reliance on prior or contemporaneous discussions and conduct, prior dealings, and industry usages is risky. The terms of the contract itself should address these issues explicitly.
- The drafter of the construction industry contract must be aware that when both parties to such a contract submit a reasonable interpretation of an ambiguous provision, in most cases the contract language will be construed against the drafter.
- Contractors must inform owners of patent ambiguities or inconsistencies in the bid or proposal documents.
- Virtually every construction industry contract imposes on both parties an implied duty of good faith and fair dealing.
- A construction industry contract imposes an implied duty of cooperation on the contracting parties.
- Owners have an implied duty to coordinate the work of their contractors, and contractors have an implied duty to coordinate the work of their subcontractors.
- The party responsible for furnishing the completed project design has an implied duty to furnish an adequate and sufficient design.
- Many jurisdictions impose additional contract obligations, or void certain express contract provisions, by operation of law.

[84]*See, e.g.* GA. CODE ANN. § 13–8–2; 770 Ill. COMP. STAT. ANN. 60/21.

[85]ARIZ. REV. STAT. § 41–2617; CAL. PUB. CONT. CODE § 7102; COLO. REV. STAT. § 24–91–103.5(1) (a); MASS. GEN. LAWS, ch. 30, § 390; MO. REV. STAT. § 34.058; N.J. STAT. ANN. § 2A:58B-3 (2002); N.C. GEN. STAT. § 143–134.3; OHIO REV. CODE ANN. § 4113.62; R.I. GEN. LAWS § 37–2–42; VA. CODE ANN. § 2.2–4335; WASH. REV. CODE § 4.24.360. See **Chapter 11** for a more detailed explanation of the "no damages for delay" clause.

6

THE UNIFORM COMMERCIAL CODE AND THE CONSTRUCTION INDUSTRY

I. THE UNIFORM COMMERCIAL CODE

The Uniform Commercial Code (UCC or Code) is a set of rules governing various business transactions and commercial instruments. These include the sale of goods, negotiable instruments, bulk transfers, letters of credit, and some credit transactions involving security interests.

The Code was developed to lend uniformity to state laws governing commercial matters. Uniformity is important because commercial transactions often involve parties in two or more states. Forty-nine states, the District of Columbia, and the Virgin Islands have fully adopted the UCC with only minor variations, resulting in considerable uniformity. The sole exception is Louisiana, which has not adopted the uniform version of UCC Article 2 but has revised the Louisiana Civil Code law of Sales to parallel UCC Article 2.[1] The UCC is codified by the individual states through their respective statutes.[2]

[1]*See* La. Civ. Code Ann. art. 2601 *et seq.*

[2]The U.C.C. can be found in these state statutes: Ala. Code §§ 7–1–101 *et seq.*; Alaska Stat. §§ 45.01.101 *et seq.*; Ariz. Rev. Stat. §§ 47–1101 *et seq.*; Ark. Code. Ann. §§ 4–1–101 *et seq.*; Cal. Com. Code §§ 1101 *et seq.*; Colo. Rev. Stat. §§ 4–1–101 *et seq.*; Conn. Gen. Stat. §§ 42a–1–101 *et seq.*; Del. Code Ann. tit. 6, § 1–101 *et seq.*; D.C. Code Ann.§ 28:1–101 *et seq.*; Fla. Stat. §§ 670.101 *et seq.*; Ga. Code Ann. §§ 11–1–101 *et seq.*; Haw. Rev. Stat. §§ 490:1–101 *et seq.*; Idaho Code §§ 28–1–101 *et seq.*; 810 Ill. Comp. Stat. Ann. 5/4-101 *et seq.*; Ind. Code 26–1–1–101 *et seq.*; Iowa Code §§ 554.1101 *et seq.*; Kan. U.C.C. Ann. 84–1–101 *et seq.*; Ky. Rev. Stat. Ann. § 355.1–101 *et seq.*; La. Rev. Stat. Ann. § 10:1–101 *et seq.*;

II. APPLICABILITY OF THE UCC TO CONSTRUCTION

Article 2 of the UCC, which governs the sale of "goods," is the article most relevant to the construction industry. The provisions of Article 2 will be called on often in construction transactions or disputes relating to the sale of materials or equipment. In 2003, after a decade-long process, the National Conference of Commissioners on Uniform State Laws approved amendments to Article 2 of the UCC (the 2003 Amendments).[3] As of January 1, 2008, only three state legislatures (Kansas, Nevada, and Oklahoma) had even considered bills proposing to enact the 2003 Amendments to UCC Article 2. None of these bills was enacted.

III. DETERMINING WHEN ARTICLE 2 APPLIES

Article 2 defines "goods" as "all things (including specially manufactured goods) that are movable at the time of identification to the contract for sale."[4]

Because UCC Article 2 broadly defines goods, it might seem that the UCC would apply to nearly all construction contracts. This is not the case, however. The UCC is inapplicable to many construction contracts because the UCC does not cover contracts for the provision of labor or other services. Many construction contracts involve a mixture of goods and services (or "hybrid" contracts). In these cases, determining UCC applicability can be difficult.

For hybrid contracts, there are several approaches for determining whether the UCC applies. Most courts adhere to the primary or predominant purpose rule. That is, if the

ME. REV. STAT ANN. tit. 11, § 1–101 *et seq.*; MD. CODE ANN., COM. LAW, §§1–101 *et seq.*; MASS. GEN. LAWS. ANN. ch. 106, § 2–101 *et seq.*; MICH. STAT. ANN. § 440.1101 *et seq.*; MINN. STAT. §§ 336.1–101 *et seq.*; MISS. CODE ANN. §§ 75–1–101 *et seq.*; MO. REV. STAT. §§ 400.1–101 *et seq.*; MONT. CODE. ANN. §§ 30–1–101 *et seq.*; NEB. REV. STAT. U.C.C. §§ 1–101 *et seq.*; NEV. REV. STAT. §§ 104.1101 *et seq.*; NEV. REV. STAT. ANN. 104.2101 *et seq.*; N. J. STAT. ANN. 12A:1–101 *et seq.*; N.M. STAT. ANN. §§ 55–1–101 *et seq.*; N.Y.U.C.C. LAW §§ 1–101 *et seq.*; N.C. GEN.STAT. §§ 25–1–101 *et seq.*; N.D. CENT. CODE §§ 41–01–02 *et seq.*; OHIO REV. CODE ANN. §§ 1301.01 *et seq.*; OKLA. STAT. ANN. tit. 12A, § 1–101 *et seq.*; OR. REV. STAT. §§ 71.1010 *et seq.*; 13 PA. CONST. STAT. ANN. § 2101 *et seq.*; R.I. GEN. LAWS §§ 6A–1–101 *et seq.*; S.C. CODE ANN. §§ 36–1–101 *et seq.*; S.D. Codified Laws §§ 57A–1–101 *et seq.*; TENN. CODE ANN. §§ 47–1–101 *et seq.*; Tex. BUS. & COM. CODE ANN. 1.101 *et seq.*; UTAH CODE ANN. §§ 70A–1a–101 *et seq.*; VT. STAT. ANN. tit. 9A, § 2–101 *et seq.*; 11A V.I. CODE ANN. § 2–101 *et seq.*; 1–101 *et seq.*; VA. CODE ANN. §§ 8.1A–101 *et seq.*; Wash. Rev. Code §§ 62A.1–101 *et seq.*; W. VA. CODE §§ 46–1–101 *et seq.*; WIS. STAT. §§ 401.101 *et seq.*; WYO. STAT. ANN. §§ 34.1–1–101 *et seq.*

[3]These amendments range from changes to make Article 2 gender neutral to those deemed necessary to reflect the use of electronic commerce in lieu of traditional writings. Article 2 was initially drafted in a period when nearly all business forms and contracts were "in writing." Hence, Article 2 often refers to written documents or writings. In the twenty-first century, business transactions are often conducted electronically, thereby making it necessary to adapt mid-twentieth-century concepts to a new medium to conduct business. Since one or more of these amendments may be adopted by the states, potentially significant changes proposed in the 2003 Amendments are noted in this chapter.

[4]U.C.C. § 2–105 (1999). The 2003 Amendment continues with this concept.

contract's primary purpose is the sale of goods, with labor or other services incidentally involved, then the UCC will apply.[5] Otherwise, the UCC will not apply.

Other courts apply the UCC to that portion of the hybrid contract pertaining to the sale of goods but not to the rest of the contract. The UCC is applied even if the contract's primary or predominant purpose is to provide services.[6] This approach is more likely if the goods are supplied separately and the buyer could have purchased them off the shelf.[7]

Even under the predominant purpose approach, no litmus test is available for determining UCC applicability because of the difficulty in identifying a contract's predominant purpose. These rules of thumb can provide guidance: The UCC is more likely to apply to a hybrid construction contract that does not require the performance of significant labor or other services at the job site. The UCC is also more likely to apply where the hybrid contract is a purchase order rather than a subcontract. Conversely, the UCC is less likely to apply to a prime contract with an owner, which courts usually find to be a contract primarily for services. Because of the many exceptions, these rules of thumb cannot predict with certainty whether the UCC will apply to a particular transaction.

IV. MODIFYING UCC OBLIGATIONS

Buyers and sellers may alter most UCC provisions and obligations by agreement, except where specifically limited. In addition, other Code sections expressly authorize parties

[5] *Trident Constr. Co., Inc. v. Austin Co.,* 272 F. Supp. 2d 566 (D.S.C. 2003); *J. Lee Gregory, Inc. v. Scandinavian House,* 433 S.E.2d 687 (Ga. Ct. App. 1993) (a lump-sum contract for the purchase of replacement windows was for the sale of goods even though a substantial amount of service was necessarily involved); *Pittsley v. Houser,* 875 P.2d 232 (Idaho Ct. App. 1994) (a lump-sum contract for the purchase and installation of carpet was considered a contract for the sale of goods when the particular carpet was the focus of the contract and not the installation; *Plantation Shutter Co., Inc. v. Ezell,* 492 S.E.2d 404 (S.C. Ct. App. 1997) (a contract for the sale of custom-built and installed shutters was predominantly a contract for the sale of goods); *Gulf Coast Fabricators, Inc. v. Mosley,* 439 So. 2d 36 (Ala. 1983) (the sale and erection of prefabricated metal building was governed by Article 2); *Port City Constr. Co. v. Henderson,* 266 So. 2d 896 (Ala. Civ. App. 1972) (a contract for the sale of concrete and "all labor to pour and finish" was a sale of goods); *Colo. Carpet Installation, Inc. v. Palermo,* 668 P.2d 1384 (Colo. 1983) (installation of carpet a sale of goods); *Mennonite Deaconess Home & Hosp., Inc. v. Gates Eng'g Co.,* 363 N.W.2d 155 (Neb. 1985) (installation of one-ply roof membrane governed by Article 2); *Meyers v. Henderson Constr. Co.,* 370 A.2d 547 (N.J. 1977) (contract to supply and install overhead doors governed by Article 2).

[6] *See TK Power, Inc. v. Textron, Inc*., 433 F. Supp. 2d 1058 (N.D. Cal. 2006) (court would apply the common law to that portion of the contract to which the U.C.C. did not apply); *Stephenson v. Frazier,* 399 N.E.2d 794 (Ind. Ct. App. 1980) (finding that in a contract to purchase a modular home with a single purchase price, the U.C.C. was inapplicable to the part of the contract relating to the construction of the foundation and installation of the septic system because they were not "goods" but rather constituted services. The modular home was covered by the U.C.C.); *Foster v. Colo. Radio Corp*., 381 F.2d 222 (10th Cir. 1967) (a contract for the sale of assets of a radio station which included both goods as well as the license, goodwill, real estate, studios, and transmission equipment was viewed in two parts—the sale of goods and nongoods—the U.C.C. being applicable to the former but not latter).

[7] *Anthony Pools v. Sheehan,* 455 A.2d 434 (Md. 1983) (supply of diving board governed by U.C.C. even though construction of swimming pool was not).

to "contract out" of that section. Some obligations cannot be limited or disclaimed even by agreement. These unwaivable obligations include the implied duties to act in good faith, with diligence, due care, and in a reasonable manner. The parties may, however, agree to the standards by which performance of these obligations will be measured.

V. CONTRACT FORMATION UNDER THE UCC

The traditional, common law rules of contract formation provide that a contract is formed when there is an offer and an acceptance of that offer. Under those traditional rules, an offer is rejected unless the acceptance is its "mirror image." For example, if the response to an offer includes terms that add to or differ from the terms in the original offer, then the response is considered to be a rejection of the original offer and a counteroffer, based on the additional or different terms. If the parties proceed to perform, a contract is formed that includes the terms of the counteroffer.

Article 2 ignores the traditional, common law rules and brings the process of contract formation for the sale of goods into line with the perceived realities of business practices. The basic principles governing the formation of a contract for the sale of goods are outlined next.

A. Agreement on All Terms and Conditions Is Not Required

Under Article 2, a contract is formed when the parties, through their words or conduct, manifest a sufficient intent to form a contract.[8] An acceptance is not required to be a "mirror image" of the offer.[9] Instead, an acceptance may contain different or additional terms, so long as it is coupled with a definite expression of acceptance or a written confirmation sent within a reasonable time, unless the acceptance is expressly made conditional upon the original offeror's acceptance of the additional terms. Additional terms are treated as proposals for additions to the contract. These additional terms become part of the contract unless the offeror limited acceptance to its original terms or the offeror timely objects to the additional terms. Finally, the parties' conduct may be sufficient to establish a contract despite unresolved items in the written offer and acceptance. Complete agreement on all terms is not required. In the absence of complete agreement, the contract will consist of the terms on which the parties agreed (price, time, method of shipment, place of delivery, etc.) along with the implied terms provided under other provisions of Article 2.[10]

B. Methods of Acceptance

Unless the parties have indicated otherwise, an offer to contract for goods is construed as inviting acceptance in any manner and by any medium reasonable under the

[8]U.C.C. § 2–207(3) (1999).
[9]U.C.C. § 2–207(1) (1999).
[10]The 2003 Amendments refer to terms that appear in the writing or record of electronic information exchange of both parties on terms, whether or not in a writing or record of information exchange to which both parties agree.

circumstances.[11] For example, an offer to purchase goods for prompt shipment may be accepted by a promise to ship or prompt shipment of the goods.[12] If nonconforming goods are shipped, however, this will create a breach of the contract unless the seller notified the buyer that the goods were being sent as an accommodation. Where the beginning of a requested performance is a reasonable mode of acceptance, an offeror who is not notified of acceptance within a reasonable time may treat the offer as having lapsed before acceptance.[13]

C. Requirement for a Written Contract

Contracts for the sale of goods with a value of more than $500 are unenforceable unless there is a written document that is "sufficient to indicate that a contract for sale has been made between the parties and signed by the party against whom enforcement is sought."[14] This rule has several exceptions and can be satisfied in several ways. For example, the requirement of a written document may be satisfied where one party, within a reasonable time, sends a written confirmation of the contract to the other party, and the receiving party has reason to know of the contents of the confirmation and fails to give written notice of objection to its contents within 10 days after receipt.[15] The writing requirement does not apply when:

(1) The goods are specially manufactured for the buyer, and the seller has started manufacturing or procuring the goods;

(2) The opposing party admits in court papers the existence of the contract; or

(3) The buyer has received and accepted the goods or made payment that the seller has accepted.[16]

The best way to ensure compliance with the requirement for a written contract is to formalize the contract in a document signed by both parties. This may not be customary or it may be impractical, such as where one is ordering supplies or materials by telephone or e-mail.[17] In such cases, either party may be able to back out if none

[11] U.C.C. § 2–206(1) (a) (1999).
[12] U.C.C. § 2–206(1) (b) (1999).
[13] U.C.C. § 2–206(2) (1999).
[14] U.C.C. § 2–201 (1999). The 2003 Amendments, if adopted by a state, would increase this threshold from $500 to $5,000.
[15] *See Atlas R. Constr. Co. v. Commercial Stone Co.*, 33 Pa. D. & C. 3d 477 (Ct. Comm. Pl. 1984) (contractor's lawyer's demand letter sent to stone supplier three and one-half months after breach of oral contract held timely written confirmation of oral contract); *but see Starry Constr. Co., Inc. v. Murphy Oil USA, Inc.*, 785 F. Supp. 1356 (D. Minn. 1992) (six-month delay in sending of letter confirming oral modification of oil sales contract was unreasonable and thus ineffective as confirmation for purposes of "merchant exception" of § 2–201(2)); *Bureau Serv. Co. v. King*, 721 N.E.2d 159 (Ill. App. Ct. 1999) (Under the circumstances of an action for breach of an oral contract to deliver grain, a delay of over eight months in sending a confirmatory memorandum was unreasonable in view of the grain market.).
[16] U.C.C. § 2–201(3) (1999).
[17] To the extent that a writing is required, many courts have held that electronic communication can be treated as a "writing." *See, e.g., Int'l Casings Group, Inc. v. Premium Standard Farms, Inc.*, 358 F. Supp. 2d 863 (W.D. Mo. 2005) (Lack of a traditional signature on electronic communications should not permit a party to escape responsibility for promises made merely because no handwritten signature existed).

of the just-mentioned exceptions apply. If a contract that is required to be in writing is later modified, the modification must also be in writing.[18]

D. Withdrawal of an Offer

In another departure from the common law, the Code allows a buyer or seller to hold an offer open even if the other party does not pay to keep it open. A written offer to buy or sell goods that gives assurances that the offer will be held open is irrevocable for the stated period or, if no period is stated, for a reasonable period not exceeding three months.[19]

E. Filling Gaps in Essential Contract Terms

Ideally, parties to a contract for the sale of goods will agree in writing to all essential contract terms after careful negotiations, during which the terms are openly presented and discussed. The UCC recognizes, however, that a contract may be formed without such negotiations and that gaps in the contract may result. To resolve problems arising from such gaps, the UCC provides that a contract will not be unenforceable because of indefiniteness, merely because the parties left one or more terms open.[20] This applies only, however, where the parties clearly intend to make a contract, and a reasonably certain basis exists for formulating a remedy in the event of a breach.

Under the UCC, the quantity of goods bought and sold is usually the only essential contract term. Except in the case of "requirements" or "output" contracts, if the quantity of goods is missing, the courts will not fill in the "gap" by implying a "reasonable" quantity of goods.[21] If the parties agree on quantity, however, a court will enforce a contract even if they have to imply a reasonable price.[22] Likewise, if the delivery date is missing, courts will imply a reasonable date based on the circumstances.[23]

[18]U.C.C. § 2–209(3) (1999).
[19]U.C.C. § 2–205 (1999).
[20]U.C.C. § 2–204(3) (1999).
[21]A "requirements" contract is one in which the buyer expressly or implicitly agrees to obtain all of the requirements for a certain kind of goods from the seller. In certain circumstances, a "requirements" contract may be useful to a contractor purchasing construction materials. Where it would be difficult to estimate accurately the quantity of material required, the supplier may simply agree to meet the contractor's requirements—that is, to supply all the material the contractor needs to perform the contract. Should the quantity exceed that which was reasonably contemplated, the courts limit the quantity, at the contract unit price, to a reasonable amount. If the buyer does not in fact require any of the goods, there is no obligation under a requirements contract to take or pay for any quantity. An "output" contract calls for the buyer to purchase all of the seller's output of a certain kind of goods. In the construction setting, the "output" contract is rarely encountered. *Brem-Rock, Inc. v. A. C. Warmack,* 624 P.2d 220 (Wash. Ct. App. 1981) (sand and gravel requirements contract upheld despite being "harsh bargain"); *Atl. Track & Turnout Co. v. Perini Corp.,* 989 F.2d 541 (1st Cir. 1993) (sale of salvaged materials on railroad rehabilitation project); *R. A. Weaver & Assocs., Inc. v. Asphalt Constr., Inc.,* 587 F.2d 1315 (D.C. Cir. 1978) (subcontract to supply and install crushed limestone).
[22] *See, e.g., Mathis v. Exxon Corp.,* 302 F.3d 448 (5th Cir. 2002); *Neugent v. Beroth Oil Co.,* 560 S.E.2d 829 (N.C. Ct. App. 2002).
[23] *Ewanchuk v. Mitchell,* 154 S.W.3d 476 (Mo. Ct. App. 2005) (because the parties failed to agree on the terms of delivery, the UCC cured this omission for them by specifying both the time and place of delivery).

A court enforcing a contract may also take notice of customary industry practices. Express terms of the contract may be interpreted in light of the parties' conduct, course of dealing, trade usage, or course of performance.[24] The Code recognizes that the parties' actions under their agreement are the best indication of what was meant by that agreement.

VI. RISK OF LOSS

The Code provides a comprehensive scheme for allocating risk of loss between the parties while the goods are in transit or storage. This provides greater certainty and enables the party that bears the risk of loss to arrange for appropriate insurance.

Where neither party is in breach of contract, the risk of loss is allocated in this way:

(1) If shipment is by a carrier, risk of loss passes to the buyer upon delivery to the carrier, unless the contract requires delivery at a particular destination, in which case the risk of loss passes to the buyer when the carrier tenders delivery at that destination.[25]

(2) If a third party is to hold the goods in a bailment, the risk of loss passes on tender of documents of title or when the bailee acknowledges the buyer's right to possession. Thus, the risk of loss may pass to the buyer for construction materials that are stored off-site in a warehouse over which the buyer has no control.[26]

(3) In all cases not covered by (1) and (2) above, the risk of loss passes to the buyer upon receipt of the goods if the seller is a merchant, or on tender of delivery if the seller is not a merchant. For example, where a supplier ships in its own trucks, the risk of loss passes upon the buyer's receipt of the goods.[27]

A contract that specifies "F.O.B. [free on board] place of shipment," also known as a "shipment" contract, places the risk of loss on the buyer as soon as the goods are placed in the possession of a common carrier.[28] Conversely, a contract that specifies "F.O.B. place of destination," also known as a "destination" contract, leaves the risk of loss on the seller until the delivery of the goods is tendered at the named destination.[29] The Code allows any of these provisions to be changed by agreement of the parties.[30]

[24]U.C.C. § 2–208 (1999).

[25]U.C.C. § 2–509(1)(a) (1999).

[26]U.C.C. § 2–509(2) (1999).

[27]U.C.C. § 2–509(3) (1999). The 2003 Amendments, if adopted by a state, would alter this allocation of risk by providing that risk of loss passes on the buyer's receipt of goods in all cases, not just when the seller is a merchant.

[28]U.C.C. § 2–319(1)(a) (1999). The 2003 Amendments would no longer define delivery terms such as "FOB" on the grounds that such terms are no longer current and any attempted scheme of statutory definitions would not reflect evolving domestic and international practices. If the delivery terms found in Article 2 are not used, the parties need to carefully address risk of loss, insurance obligations, and so on in the sales agreement.

[29]U.C.C. § 2–319(1)(b) (1999).

[30]In a typical purchase order or sales order form, the risk of loss is covered by delivery terms, such as FOB.

VII. INSPECTION, ACCEPTANCE, REJECTION, AND REVOCATION OF ACCEPTANCE

A. Inspection of Goods

Where the seller is required or authorized to send the goods to the buyer, the Code gives the buyer the right, before acceptance, to inspect the goods at any reasonable time and place and in any reasonable manner.[31] The parties may agree to modify the time, place, or even the right of inspection by inserting appropriate terms in the contract.[32]

Inspection and payment are linked under the Code, although the parties may agree otherwise by contract. Where the buyer is to pay on or after delivery of the goods, the buyer has the right to inspect before payment. Absent an opportunity to inspect, the buyer has no obligation to pay. The buyer may lose the right of inspection, however, by agreement or by conduct showing waiver.

The buyer may contract away its right of inspection by agreeing to cash on delivery (C.O.D.) payment terms or other similar terms. C.O.D. effectively requires payment before delivery, which precludes inspection.[33] Where the contract requires payment before inspection, payment does not necessarily constitute acceptance or defeat the buyer's right to a later inspection and assertion of its rights and remedies.[34] Denial of the right of inspection is a breach of contract. Consequently, if the UCC-governed contract does not allow the seller to ship C.O.D., the seller breaches the contract by shipping C.O.D. because this denies the buyer its right to inspect.[35]

The buyer may waive the right of inspection by unreasonably delaying inspection.[36] Whether an inspection delay is reasonable or so unreasonable as to amount to a waiver depends on the facts and circumstances of the transaction. The buyer must also give notice of any defect within a reasonable time after inspection, or it will likewise waive any remedy.[37]

The right of inspection includes the right to test the goods, which may include tests that must destroy a small amount of the goods. Use of the goods for destructive or nondestructive testing does not constitute acceptance of the goods as long as the testing is necessary and reasonable. Initially, the buyer bears the expenses of inspection and testing. It may recover those expenses from the seller if the buyer properly rejects the goods as a result of the inspection and testing.[38]

[31]U.C.C. § 2–513(1) (1999).
[32]*Id.*
[33]U.C.C. § 2–513(3) (1999).
[34]U.C.C. § 2–512 (1999).
[35]4 Anderson, Uniform Commercial Code § 2–513:13 (1983).
[36]*EPN-Delaval, S.A. v. Inter-Equip., Inc.*, 542 F. Supp. 238 (D.C. Tex. 1982); *Trinity Indus., Inc. v. McKinnon Bridge Co., Inc.*, 77 S.W.3d 159 (Tenn. Ct. App. 2001).
[37]U.C.C. § 2–607 (1999); *see* discussion in **Section VII.C** of this chapter.
[38]U.C.C. § 2–513(2) (1999).

B. Rejection of Goods

If the goods themselves or the seller's manner of tendering their delivery do not meet the contract requirements, the buyer has the option of accepting or rejecting the goods or any part of them.[39] If the buyer rejects some or all of the goods, the buyer must do so within a reasonable time and give notice to the seller.[40] If the buyer fails to notify the seller of the reason for rejection, the buyer cannot rely on that reason to justify rejection if the seller could have cured the defect after reasonable notice.[41] After rejecting goods in its possession, a buyer must hold them with reasonable care for a sufficient time to permit the seller to remove them. [42]

C. Acceptance and Notice of Breach

A buyer "accepts" goods by failing to make an effective rejection, by knowingly taking nonconforming goods, or by "doing any act inconsistent with the seller's ownership."[43] For example, a buyer may accept construction materials by using them to perform work because such use is inconsistent with the seller's continued ownership of those materials.

The buyer must pay for all accepted goods.[44] This may seem obvious because, in the absence of problems with quantity, quality, or timeliness, the buyer usually pays for the goods in the ordinary course of business. Where the buyer contends that there are problems with the goods but it fails to give the seller notice or otherwise protect its interests, the buyer still may have to pay for the goods because it may be deemed to have accepted them.[45] Also, the buyer cannot sit idly by while defective goods are being used. The buyer is required to give the seller notice of any breach within a reasonable time to preserve its remedies.[46] Failure to give notice bars any remedy.[47]

[39]U.C.C. § 2–601 (1999).
[40]U.C.C. § 2–602(1) (1999).
[41]U.C.C. § 2–605(1) (1999).
[42]U.C.C. § 2–602 (1999).
[43]U.C.C. § 2–606(1)(c) (1999). *See Hooten Equip. Co. v. Trimat, Inc.*, 2004 WL 444134 (Ohio Ct. App. 2004) (construction company's use of range and hood on project was act inconsistent with seller's ownership and thus acceptance); *Moore & Moore Gen. Contractors, Inc. v. Basepoint, Inc.*, 485 S.E.2d 131 (Va. 1997) (contractor's installation of nonconforming cabinets constituted act inconsistent with subcontractor's ownership and amounted to acceptance of goods under the U.C.C.); *Meland v. Intermountain Sys., Inc.*, 712 P.2d 1295 (Mont. 1985) (buyer accepted goods by erecting entire building after discovering incorrect lengths in parts of prefabricated metal building). Under the 2003 Amendments, use of goods after a rightful rejection may not, in all cases, constitute acceptance of those goods. If adopted, this change could be significant unless the parties' agreement addresses the buyer's use of rejected goods.
[44] U.C.C. § 2–607(a) (1999).
[45]*Econ. Forms Corp. v. Kandy, Inc.*, 391 F. Supp. 944 (N.D. Ga. 1974), *aff'd*, 511 F.2d 1400 (5th Cir. 1975) (general contractor was obligated to pay for concrete forms despite alleged defects).
[46]*Smith-Wolf Constr., Inc. v. Hood,* 756 P.2d 1027 (Colo. Ct. App. 1988) (subcontractor gave reasonable notice of breach of warranty of methods and rates of application of waterproofing product).
[47]*Hitachi Elec. Devices (USA), Inc. v. Platinum Techs., Inc.*, 621 S.E.2d 38 (S.C. 2005) (buyer who failed to give reasonable notice of breach was barred from any remedy). The 2003 Amendment introduces the concept of prejudice to the seller as the extent to which lack of notice bars or diminishes the seller's remedy for nonconforming goods.

For the notice to be "reasonable," the buyer must give it within sufficient time to afford the seller a chance to remedy the breach and minimize the resulting damages.[48]

D. Seller's Right to Cure

In a single delivery contract, with few exceptions, the buyer can reject goods for any defect in the goods or the tender. This is the Code's "perfect tender rule," which does not require a material breach of the contract for the buyer to reject the goods. The seller does, however, have several rights to cure where it has failed to deliver perfect tender.

Where a buyer has rejected goods because of defects and time remains in the original contract for performance, the seller may give reasonable notice of its intention to do so and make a new tender of conforming goods.[49] The seller normally has no time to cure beyond the original contract time. If, however, the buyer rejects a tender of goods that the seller reasonably believed would be acceptable, the seller, upon reasonable notification to the buyer, will have further reasonable time to tender conforming goods.[50] A reasonable belief that the tender would be acceptable can come from trade practices between the parties or ignorance as to the defect on the part of the seller.

E. Revocation of Acceptance in Whole or in Part

The buyer may revoke its acceptance of a lot or commercial unit whose nonconformity substantially impairs its value to the buyer in two situations. First, if the buyer accepted on the reasonable assumption, which was not realized, that the nonconformity would be cured; and second, if the nonconformity was not discovered and the buyer's acceptance was reasonably induced either by the difficulty of discovery before acceptance or by the seller's assurances.[51] A buyer's revocation of acceptance must occur within a reasonable time after the buyer discovers or should have discovered the grounds for it and before any substantial change in the condition of the goods.[52] If the goods are changed in condition such change must only be a result of the goods' own defects in order for revocation of acceptance to be valid.[53] Revocation of acceptance is not effective until the buyer notifies the seller of such revocation.[54] Upon a proper revocation of acceptance, a buyer has the same rights and duties toward the goods as if it had initially rejected the goods.[55]

[48] *Metro Inv. Corp. v. Portland Rd. Lumber Yard, Inc.*, 501 P.2d 312 (Or. 1972) (notice of siding defects not untimely although given two years after the defect was discovered).
[49] U.C.C. § 2–508(1) (1999).
[50] U.C.C. § 2–508(2) (1999). The 2003 Amendments, if adopted, significantly revise the seller's right to cure and obligation if it elects to exercise its right to cure.
[51] U.C.C. § 2–608(1)(2003).
[52] U.C.C. § 2–608(2)(2003).
[53] *Id.*
[54] *Id.*
[55] U.C.C. § 2–608(4)(2003). The 2003 Amendments provide an additional provision that "if a buyer uses the goods after a rightful rejection or justifiable revocation of acceptance, the following rules apply: (a) Any use by the buyer that is unreasonable under the circumstances is wrongful as against the seller and is an acceptance only if ratified by the seller. (b) Any use of the goods that is reasonable under the circumstances is not wrongful as against the seller and is not an acceptance, but in an appropriate case the buyer is obligated to the seller for the value of the use to the buyer."

Several remedies are available to a buyer after proper revocation of acceptance.[56] A buyer can cancel the order and, regardless of whether the buyer chooses to cancel, the buyer can recoup as much of the purchase price as has been paid.[57] A buyer may "cover" and recover the difference between the cost of the substituted goods and the contract price.[58] If a buyer does not cover, then, in addition to the purchase price paid, the buyer may recover the difference between the market price and the contract price.[59] Provable incidental and consequential damages are always recoverable.

VIII. WARRANTIES UNDER THE UCC

The Code identifies four types of warranties to protect buyers of goods.

(1) The warranty of title and against infringement warrants that the title to goods conveyed is good and their transfer rightful. The seller also warrants that the transfer "shall not unreasonably expose the buyer to litigation because of any colorable claim to or interest in the goods."[60]

(2) The express warranty is an affirmation of fact or promise made by the seller to the immediate buyer which relates to the goods' capabilities or uses. It is not necessary to the creation of an express warranty that the seller use the words "warranty" or "guarantee."[61]

(3) The implied warranty of merchantability is the warranty that the goods will be fit for ordinary use.[62] Unless excluded or modified by the parties, a warranty that the goods will be merchantable is implied in contracts for their sale if the seller is a merchant with respect to goods of that kind.[63] Most construction contracts that might be governed by the UCC expressly exclude this warranty.

[56] U.C.C. § 2–711(2003).
[57] *Id.*
[58] U.C.C. § 2–711(1)(a)(2003).
[59] U.C.C. § 2–713(2003).
[60] U.C.C. § 2–312 (1999).
[61] U.C.C. § 2–313 (1999). Although the U.C.C. warranty provisions are generally written in the context of the buyer and seller, the context of warranty enforcement on a construction project involves multiple parties (owner, contractor, subcontractors, and suppliers) and multiple tiers of contracts. Owners often specify warranty periods that are triggered by substantial completion or even final completion/acceptance of the entire project and may obligate the contractor to enforce the warranties on behalf of the owner. Contractors and subcontractors need to carefully evaluate these warranties and their enforcement when soliciting the initial bids or proposals. In that context, there may be little or no additional cost in obtaining the necessary warranty protection for the ultimate buyer (the owner) or obtaining acceptance of a warranty period commencing well after the delivery or initial operation of equipment. In contrast, filling gaps in warranty coverage after the project is under way can be very costly.
[62] U.C.C. § 2–314 (1999).
[63] U.C.C. § 2–314 (1999).

(4) The implied warranty of fitness for the intended purpose. This warranty is also part of a contract, whether or not written in the contract, but only where the seller has or should have knowledge of the buyer's specific needs and sells goods to satisfy those needs. In that situation, the seller warrants that the goods are suitable for meeting such specific needs.[64] Again, most construction contracts that might be governed by the UCC expressly exclude this warranty.

A. Warranty Disclaimers

The Code includes rules governing a seller's attempt to limit or disclaim warranty coverage.[65] Generally, a seller's disclaimer language must conform to the wording suggested by the Code. For example, the Code states that "language to exclude warranties of fitness for a particular purpose in a consumer contract must state: "The seller assumes no responsibility that the goods will be fit for any particular purpose for which you may buying these goods, except as otherwise provided in the contract."[66] Notwithstanding this, the Code allows all implied warranties to be excluded by expressions such as "as is" or "with all faults."[67] Case law has also upheld warranty disclaimers which stated that "there is no implied warranty of merchantability or any other implied warranty that extends beyond the express warranty included in this contract." The law addressing this subject has developed to the point that sellers have little trouble including enforceable disclaimer provisions in their contracts.[68]

B. Warranty Limitations

Because warranties are easily disclaimed, contractual limitations on remedies for breaches such as delivery delays and nonconforming goods and materials are often more important to buyers.[69] Sellers often seek to limit their exposure to buyers by limiting the buyers' damages or substituting an alternative remedy. For example, a seller may seek to limit the buyer's remedy to a return of the goods and repayment of the purchase price, or to repair and replacement of the defective goods.[70] To be effective in barring consequential damages—that is, business losses other than those incurred to obtain conforming goods—the seller must provide that the

[64]U.C.C. § 2–315 (1999).
[65]U.C.C. § 2–316 (1999).
[66]U.C.C. § 2–316(2) (1999).
[67]U.C.C. § 2–316(3)(a) (1999).
[68]*See, e.g., Haight v. Dales Used Cars, Inc.*, 87 P.3d 962 (Idaho Ct. App. 2003); *Hou-Tex, Inc. v. Landmark Graphics*, 26 S.W.3d 103 (Tex. App. 2000).
[69]U.C.C. § 2–718 (1999) and U.C.C. § 2–719 (1999).
[70]*Mitsubishi Corp. v. Goldmark Plastic Compounds, Inc.*, 446 F. Supp. 2d 378 (W.D. Pa. 2006) (exclusive remedy clause that precluded damages otherwise available under the U.C.C. and the common law was enforceable because the parties clearly agreed that the remedy was exclusive and the limitation was not unconscionable); *In Re Access Cardiosystems, Inc.*, 361 B.R. 626 (Bankr. D. Mass. 2007) (excluding remedies available under the U.C.C. and limiting damages to price paid for product was an enforceable remedy limitation).

alternative remedy is exclusive of all other remedies. If the alternative remedy is not expressly exclusive, then the stated remedy will merely be one of many forms of relief.[71] A UCC disclaimer may not be effective to eliminate tort liability—that is, for negligence or strict liability in tort—although a seller may obtain a waiver of tort claims through a properly drafted contract provision.

Finally, the buyer has recourse to all other Code remedies when the exclusive, limited remedy fails of its essential purpose.[72] In other words, the buyer is not restricted to an exclusive, limited remedy that fails to work. The most common limited remedy is repair or replacement of the defective goods. If the defect cannot be repaired or replaced within a reasonable time, however, that remedy may fail of its essential purpose and the buyer may be able to sue for damages.[73]

IX. STATUTE OF LIMITATIONS AND COMMENCEMENT OF THE WARRANTY PERIOD

Tender of delivery is "an offer of goods under a contract as if in fulfillment of its conditions even though there is a defect when measured against the contract obligation."[74] Tender of delivery generally triggers the start of the warranty period for the goods. It also gives rise to the cause of action or right to sue for delivery of defective goods. The statute of limitations begins to run when the cause of action arises.[75] In most, but not all states, the statute of limitations under the Code is four years from tender of delivery.[76]

There are several exceptions to these general principles. The most important exception permits the parties to alter by agreement the timing of both the commencement of the

[71] *Id.*

[72] U.C.C. § 2–719(2) (1999).

[73] *Atwell v. Beckwith Mach. Co.*, 872 A.2d 1216 (Pa. Super. 2005) (repair or replacement limited remedy failed of its essential purpose when the claimed breach was that the seller supplied the wrong tractor); *Bishop Logging Co. v. John Deere Indus. Equip. Co.*, 455 S.E.2d 183 (S.C. Ct. App. 1995) (a repair or replacement limited warranty failed in its essential purpose when the seller was unable to repair the equipment after having a reasonable opportunity to do so); *Chatlos Sys., Inc. v. NCR Corp.*, 635 F.2d 1081 (3d Cir. 1980) *remanded to* 670 F.2d 1304 (3d Cir. 1982), *cert. dismissed*, 457 U.S. 1112 (1982) (ineffective repair of computer system over one-and-one-half-year period was failure of repair remedy); *Garden State Food Dist., Inc. v. Sperry Rand Corp.*, 512 F. Supp. 975 (D.N.J. 1981); *But see Kaplan v. RCA Corp.*, 783 F.2d 463 (4th Cir. 1986) (repair remedy did not fail where seller replaced defective antenna immediately); *Middletown Eng'g Co. v. Climate Conditioning Co.*, 810 S.W.2d 57 (Ky. Ct. App. 1991) (126 days found to be reasonable; summary judgment granted for the seller). Even if the limited remedy fails and the buyer is entitled to sue for damages, a contract clause excluding consequential damages still may be effective.

[74] *Kittitas Reclamation Dist. v. Spider Staging Corp.*, 27 P.3d 645 (Wash. Ct. App. 2001).

[75] *Superior, Inc. v. Behlen Mfg. Co.*, 738 N.W.2d 19 (N.D. 2007) (buyer's claim for breach of sales contract regarding nonconforming bolts accrued, and four-year limitation period began to run, upon tender of delivery of bolts).

[76] U.C.C. § 2–725 (1999). The 2003 Amendments proposed a revision seeking to address a situation when the breach is discovered near the end of the four-year period. The revision would establish a one-year period to assert a claim but no longer than five years after the cause of action accrued. The cause of action for defective or nonconforming goods typically accrues at the time of delivery. As of 2008, no state has adopted this revision.

warranty period and the running of the statute of limitations. For example, the parties can agree to make both events contingent upon acceptance of the goods by the buyer after a reasonable period for inspection.

A second exception arises when the seller attempts repairs that ultimately prove unsuccessful. Sometimes such repair attempts are considered to "toll," or suspend, the running of the statute of limitations. The rules governing this exception vary significantly from state to state. Some states hold that repair efforts do not suspend the running of the limitation period.[77] Other states hold that the limitation period does not run during the seller's repair efforts.[78] The key factor in determining whether a seller's repair efforts toll the running of the statute is often whether the seller made any promises or assurances to the buyer that the defects could be repaired. A court usually will examine such promises or assurances to see if they caused the buyer to forbear filing suit.[79] A third exception arises where the seller makes a warranty of future performance. Some parties, seeking to avoid the running of the limitation period, have argued that this exception applies merely because they could not discover the defect until some point after tender of delivery. Courts usually have rejected this argument unless the warranty required more in the way of future performance than merely meeting the contract specifications.[80] One example of a construction warranty that extends to future performance is the typical 10-year or 20-year roofing warranty. Express language such as "bonded for up to 20 years" may provide a warranty of future performance.[81] In contrast, a statement that concrete forms will withstand specified loads was not a warranty of future performance, and breach of that warranty occurred upon tender of delivery, not upon the use and failure of the forms.[82]

[77] *Holbrook, Inc. v. Link-Belt Constr. Equip. Co.*, 279, 12 P.3d 638 (Wash. Ct. App. 2000) (manufacturers' efforts to repair hydraulic log loader did not toll four-year statute of limitations applicable to buyer's breach of warranty claims). *See, e.g., K/F Dev. & Inv. Corp. v. Williamson Crane & Dozer Corp.*, 367 So. 2d 1078 (Fla. Dist. Ct. App. 1979), *cert. denied*, 378 So. 2d 350 (Fla. 1979).

[78] *Keller v. Volkswagen of Am., Inc.*, 733 A.2d 642 (Pa. Super. Ct. 1999) (fact question existed regarding whether statute of limitations was tolled pursuant to repair doctrine, thus precluding summary judgment for vehicle manufacturer). *See, e.g., Ontario Hydro v. Zallea Sys., Inc.*, 569 F. Supp. 1261 (D. Del. 1983).

[79] *See, e.g., Mills v. Forestex Co.*, 134 Cal. Rptr. 2d 273 (Cal. Ct. App. 2003) (where the plaintiffs produced no evidence of statements or conduct by defendant after repair efforts ceased that could have induced them to forbear filing suit, the tolling ended when the defendant quit working and failed to return to the job).

[80] *See Safeway Stores, Inc. v. Certainteed Corp.*, 710 S.W.2d 544, 547 (Tex. 1986) (containing an extensive listing of cases supporting the "universal rule in other jurisdictions that an implied warranty does not fall under the exception in the Code because, by its very nature, it cannot explicitly extend to future performance") (citations omitted).

[81] *Poli v. DaimlerChrysler Corp.*, 793 A.2d 104 (N.J. Super. Ct. App. Div. 2002) (holding that a sales contract that contained a "seven year/seventy thousand mile-repair" power train warranty constituted a promise of future performance rather than a mere representation of the car's condition at the time of delivery.); *Little Rock Sch. Dist. of Pulaski County v. Celotex Corp.*, 574 S.W.2d 669 (Ark. 1978). *See Mittasch v. Seal Lock Burial Vault, Inc.*, 344 N.Y.S.2d 101 (N.Y. 1973) (warranty that a burial vault would provide "satisfactory service at all times" was a warranty of future performance).

[82] *Raymond-Dravo-Langenfelder v. Microdot, Inc.*, 425 F. Supp. 614 (D. Del. 1976); *W. Recreational Vehicles, Inc. v. Swift Adhesives, Inc.*, 23 F.3d 1547 (9th Cir. 1994) (where the seller promised that its adhesive would work on the buyer's new Filon vehicles, this promise, while implicitly touching on the future performance of the adhesive, did not specifically refer to a future time and therefore did not explicitly extend to future performance).

A different problem exists with regard to the statute of limitations where the purchase agreement provides that the seller will indemnify and hold the buyer harmless from claims and losses arising from the sale of, or the buyer's use of, the goods. In this context, where defective goods might trigger the indemnity obligation, a breach claim would arise if the seller failed to honor its indemnity obligation. The statute of limitations generally does not begin to run on a breach of an indemnification agreement until the seller refuses to honor it, even though delivery of the goods occurred much earlier.[83] Contractors and subcontractors need to appreciate the distinction between a warranty obligation and an indemnity obligation and ensure that the latter is addressed in any lower-tier subcontracts and purchase orders.

X. PERFORMANCE ISSUES

A. Anticipatory Repudiation/Adequate Assurance of Performance

When one party to a contract has reasonable grounds for feeling insecure about the other party's ability or desire to fully perform in the future, the aggrieved party has a right to require the other party to provide adequate assurances of full performance.[84] The adequacy of such assurances is measured by commercial standards. The term "commercial standards," as well as the notion of insecurity, may be defined by agreement of the parties to avoid many of the problems of interpretation inherent in this section.

The aggrieved party may make a demand on the other party, who then has the duty to provide adequate assurance of performance within a reasonable time but not longer than 30 days. The failure to provide adequate assurances within a reasonable time is a repudiation of the contract. In case of repudiation, the aggrieved party is entitled to take action to mitigate its damages.[85] The aggrieved party may: (1) await performance by the repudiating party for a reasonable time; (2) resort to any remedy for breach even though the aggrieved party has stated it will await performance; or (3) simply suspend its own performance.[86] The aggrieved party cannot demand more than "adequate" assurances of performance. Otherwise, the aggrieved party has none of these rights, even if the other party refuses to meet those demands.

The aggrieved party may also ask the other party to provide additional security, but not so much as to modify the essential terms of the contract. For instance, the seller cannot demand payment in advance when the contract expressly or impliedly

[83]*Abbott Lab., Inc. v. Gen. Elec. Capital*, 765 So. 2d 737 (Fla. Dist. Ct. App. 2000) (the statute of limitations applicable to an equipment seller's agreement to indemnify a finance company for satisfying liability to a customer began to run when the seller refused to pay and breached the indemnity clause). *See, e.g., Tolar Constr. Co. v. GAF Corp.*, 267 S.E.2d 635 (Ga. Ct. App. 1980), *rev'd on other grounds*, 271 S.E.2d 811 (Ga. 1980).
[84]U.C.C. § 2–609 (1999).
[85]U.C.C. § 2–610 (1999).
[86]U.C.C. § 2–610 (1999).

provides for a 30- to 60-day billing cycle and other security is offered. Verbal assurances by a buyer that payment eventually will be made is not adequate, however, when the seller is demanding that payment be made on time. Examples of adequate assurances by a buyer would include presentation of statements from its banker, submission of a letter of credit, establishment of an escrow account for future payments, or even offering a mortgage on other property.

B. Dealing with the Other Party's Insolvency

Under the UCC, a party is insolvent if it has ceased to pay its debts in the ordinary course of business, cannot pay its debts as they become due, or is deemed insolvent under federal bankruptcy laws.

Upon discovering the buyer's insolvency, the seller may:

(1) Stop delivery if the goods are still in transit;
(2) Refuse delivery unless the buyer pays cash for the delivered goods and all prior deliveries; or
(3) Reclaim the goods:
    (a) Within 10 days, if the buyer received the goods on credit while insolvent; or
    (b) At any time, if the buyer misrepresented its solvency in writing to the seller within three months prior to delivery.

The seller's right to reclaim the goods is an important right under the Code, but it has certain limitations. The original seller's right to reclaim is lost if the original buyer has resold the goods to a good-faith purchaser. For example, in the construction context, a supplier may not recover goods from an insolvent subcontractor, even if the goods have not yet been incorporated into the project, if the prime contractor or owner has already paid the subcontractor for the goods. A prime contractor or owner should, however, be cautious about attempting to use this exception to defeat a supplier's right to reclaim goods. Payment to an insolvent subcontractor is generally unwise unless the subcontractor has already performed the work for which it is entitled to be paid, as opposed to merely purchasing supplies that it will later incorporate into the project. As this exception requires good faith by the prime contractor or owner, it may not apply if the subcontractor's insolvency is obvious. Unless the subcontract requires prompt payment and the contractor has made prior prompt payments for materials, a sudden hasty payment to a subcontractor may show that the prime contractor knew of the subcontractor's insolvency and was attempting in bad faith to impair the seller's ability to reclaim. In these circumstances, the better course of action would be to work directly with the seller to obtain possession of and title to the goods.

The Code also addresses the opposite situation in which a buyer is faced with an insolvent seller. A buyer may take possession of goods for which it has contracted, where two conditions apply:[87]

[87]U.C.C. § 2–502 (1999).

(1) The goods in question must be specially manufactured goods or goods specifically identified as intended for the buyer.

(2) The buyer must have made at least partial payment for the goods, and the seller became insolvent within 10 days after the first payment.

Absent these special circumstances, the buyer may not obtain possession but may acquire a security interest in the goods under Article 9 of the Code.

C. Excuse of Performance by Failure of Presupposed Conditions

Purchase orders and sale of goods agreements usually include a force majeure clause. A typical force majeure clause excuses nonperformance by either the buyer or the seller that is caused by an act of God, war, strike by a common carrier, and other causes beyond the control of the parties. Under the Code, a force majeure clause is lawful and enforceable.

Where the contract does not include a force majeure clause, the Code provides that a late delivery or nondelivery will be excused if the seller's performance has been made commercially impracticable by the occurrence of a contingency, the non-occurrence of which was a basic assumption of the contract, or by good-faith compliance with a domestic or foreign governmental regulation.[88] Where the seller remains capable of partial performance, the seller must allocate production and deliveries among its customers.[89] In addition, the seller must notify the buyer of delay or non-delivery and the amount of any allocation of available deliveries.[90] If the impracticability affects the agreed manner of delivery, and a reasonable substitute is available, the buyer must accept the reasonable substitute.[91]

"Impracticability" under this Code section means "commercial impracticability." This is a more lenient standard for excusing performance than is available under the common law that applies to most construction contracts.[92] Consequently, contractors should bear this Code section in mind when purchasing goods, and should seek to impose by contract a greater obligation on the seller than the Code imposes.[93] Otherwise, the contractor may be required to perform while its suppliers—which are providing goods and, therefore, are governed by the Code—will be excused from performance.[94]

Increased costs do not constitute commercial impracticability unless they result from an unforeseen contingency. Mere unexpected difficulties and expenses do not excuse performance unless they are so extreme that they are outside the contemplation

[88]U.C.C. § 2–615(a) (1999).

[89]U.C.C. § 2–615(b) (1999).

[90]U.C.C. § 2–615(c) (1999).

[91]U.C.C. § 2–614(a) (1999).

[92]*Helms Constr. & Dev. Co. v. State,* 634 P.2d 1224 (Nev. 1981).

[93]U.C.C. § 2–615 (1999).

[94]*Am. Laminates, Inc. v. J.S. Latta Co.*, 980 S.W.2d 12 (Mo. Ct. App. 1998) (subcontrator's loss of project did not excuse its performance on purchase orders placed with manufacturer of laminated casework under doctrine of commercial impracticability).

of the parties.[95] A severe shortfall of raw materials or supplies due to a war embargo or unforeseen shutdown of a major source of supply that either causes a large increase in costs or prevents the seller from procuring basic supplies, however, may constitute commercial impracticability.[96]

D. Buyer's Remedies

If the seller delivers nonconforming goods or fails to make delivery, the buyer may either purchase replacement goods ("cover") or, in appropriate circumstances, obtain specific performance by forcing the seller to make delivery of conforming goods.

The measure of damages in the event of cover is the difference between the contract price and the amount actually paid for the substitute goods, plus incidental and consequential damages.[97] An alternate measure of damages for the buyer is "hypothetical cover." Hypothetical cover is the difference between the contract price and the market price, plus incidental and consequential damages.[98] Hypothetical cover may be used to establish that the buyer's actual cover price was unreasonably high or low. Where the two measures of damages differ only marginally, however, the party seeking damages may recover by using the more favorable amount.

A buyer may obtain specific performance where the goods are otherwise unavailable or the price of replacement goods has increased so dramatically that it would be a poor business practice to cover. Specific performance is an extraordinary remedy that may be available from a court on an expedited basis. The buyer must provide the seller with prompt notice of rejection or revocation of acceptance. Failure to provide prompt notice will almost always constitute wrongful rejection/revocation by the buyer.[99] Prompt notice is important because the seller has the right to cure the defects and tender the repaired goods as conforming to the contract if the time for performance has not yet expired.[100] Although the seller's right to cure does not extend to major repairs, the definition of "major" in the construction context will vary, depending on the cost and nature of the goods.

E. Seller's Remedies

Where the buyer wrongfully rejects goods, wrongfully revokes acceptance of goods, fails to make payment, or repudiates the contract, the seller may stop delivery, resell

[95]*Sachs v. Precision Prods. Co.,* 476 P.2d 199 (Or. 1970); *Res. Inv. Corp. v. Enron Corp.,* 669 F. Supp. 1038 (D. Colo. 1987).

[96]*Leon County v. G.J. Gluesenkamp,* 873 So. 2d 460 (Fla. Dist. Ct. App. 2004) (holding that if the performance of a duty is made impracticable by having to comply with a domestic or foreign governmental regulation or order, further performance will be excused); *Swift Textiles, Inc. v. Lawson,* 219 S.E.2d 167 (Ga. Ct. App. 1975) (holding merchant liable for the difference between contract and market prices when unable to deliver on its contract with manufacturer even though its inability was due to farmers' repudiating their contracts with merchant due to increase in cotton prices).

[97]U.C.C. § 2–712 (1999). *Egerer v. CSR West, LLC,* 67 P.3d 1128 (Wash. Ct. App. 2003).

[98]*Allied Canners & Packers, Inc. v. Victor Packing Co.,* 209 Cal. Rptr. 60 (Cal. Ct. App. 1984).

[99]UCC. § 2–601 (1999); U.C.C. § 2–714(1) (1999).

[100]U.C.C. § 2–508 (1999).

the goods and sue for losses from the resale, recover damages, or cancel the contract.[101] A seller's damages may also include incidental costs incurred in transportation, care, and custody of the goods involved in stopping delivery and any resale.[102]

➢ POINTS TO REMEMBER

- Article 2 of the Uniform Commercial Code (UCC) applies to some extent virtually to every construction project.
- Although Article 2 of the UCC does not apply to contracts for the sale of construction services, it does govern transactions involving the sale of "goods," which are defined as "all things . . . which are movable at the time of identification to the contract for sale." These concepts typically apply to purchase orders for materials, equipment, or systems incorporated into the construction project.
- Each state that has adopted the UCC has done so through its state statutes. There may be slight variances in these statutes. The state's version of the UCC that governs a particular contract or project should be reviewed in analyzing UCC questions, not the version of the UCC adopted by the National Conference of Commissioners on Uniform State Laws.
- Where a contract involves the sale of both goods and services, most courts will look to the predominant nature of the contract to determine whether the UCC applies.
- Most of the terms imposed by the UCC can be modified by agreement of the parties.
- Some of the traditional common-law rules governing contract formation do not apply under the UCC. It is easier to establish the existence of an enforceable contract under the UCC.
- The UCC provides for four kinds of warranties for the protection of buyers of goods: warranty of title, express warranties, the implied warranty of merchantability, and the implied warranty of fitness for the goods' intended purpose.
- Parties may disclaim or limit express or implied warranties, but the Code imposes limitations on the warranties that can be disclaimed and may require special language.
- The UCC grants certain rights to the seller in the event that the buyer is insolvent.
- The UCC contains a comprehensive scheme for allocating the risk of loss between the contracting parties.
- Other issues expressly addressed by the UCC include: the buyer's right to inspect the goods; the circumstances under which a buyer may reject the goods; the buyer's obligations and rights upon acceptance of the goods; the requirements governing notice of a breach; and a party's rights upon learning that the other party is not likely to perform ("anticipatory repudiation").

[101] U.C.C. § 2–703 (1999).
[102] U.C.C. § 2–710 (1999).

- Most purchase orders and sale of goods agreements contain a force majeure clause; in the absence of such a clause, the UCC provides that late delivery or nondelivery will be excused if the seller's performance has been made impracticable by the occurrence of a contingency, the nonoccurrence of which was a basic assumption of the contract, or by good-faith compliance with a governmental regulation.
- The UCC sets forth the remedies that are available to a buyer or seller, should the other party breach the agreement.

7

AUTHORITY AND RESPONSIBILITY OF THE DESIGN PROFESSIONAL

I. OVERVIEW

Design is an integral part of the construction process, and many construction disputes have their genesis in real or perceived design problems. In addition to design responsibility, the project architect or engineer (or design professional)[1] often plays a pivotal role in the administration of the construction project, with responsibility for contract interpretation, certifications of the contractor's applications for payment, inspections, contract compliance, and acceptance of the work.[2] Therefore, proper recognition of the relative rights and responsibilities of the design professional is essential to an understanding of the legal relationships involving the parties to construction projects and to the proper allocation of responsibilities and liabilities among those parties.

Professional services performed by architects and engineers overlap to such an extent that in many cases it is difficult to draw a distinction between them.[3]

[1]For the purposes of this chapter, the two professionals will be considered interchangeable (or collectively as "design professionals"), as the principles discussed are in most cases applicable to both. *See Redies v. Attorneys Liab. Prot. Soc'y,* 150 P.3d 930 (Mont. 2007) (defining A/E as design professional); *Cramer Hill Residents Ass'n, Inc. v. Primas,* 928 A.2d 61 (N.J. Super. Ct. App. Div. 2007). Unless otherwise noted, the term "design professional" will be used in this chapter to refer to licensed architects and engineers.

[2]Although there is no such thing as a "typical" construction project, for the sake of brevity, comments in this chapter anticipate, unless otherwise noted, project delivery by what Professor J. Sweet calls the "eternal triangle" (i.e., an owner contracting with a design professional and a contractor, each acting as an independent party). JUSTIN SWEET, LEGAL ASPECTS OF ARCHITECTURE, ENGINEERING, AND THE CONSTRUCTION PROCESS 85 (Marc M. Schneier ed., Thompson 2004).

[3]*Baltimore County v. RTKL Assocs., Inc.,* 846 A.2d 433 (Md. 2004); *Rosen v. Bureau of Prof. & Occupational Affairs, State Architects Licensure Bd.,* 763 A.2d 962 (Pa. Commw. Ct. 2000). Architects and engineers generally are licensed individually (not as corporations or partnerships) and must meet the standards for their particular profession.

Federal agencies use the term "architect/engineer" to refer to both types of services.[4] Frequently the architect or engineer is primarily responsible for the design and contract documents. That entity or person, in turn, subcontracts specific responsibility to independent mechanical, electrical, structural, and plumbing engineers or other consultants. Many design firms provide both architectural and engineering services.

II. STANDARD OF CARE AND PROFESSIONAL RESPONSIBILITY

The scope of the design professional's authority and responsibility on a particular project often is defined in the contract between the design professional and the owner.[5] Contractual privity is the direct relationship between parties to a contract.[6] Contractual privity allows the owner to sue the design professional directly for damages attributable to a breach of contract or the design professional to sue the owner if the owner breaches the contract.

The design professional and the owner need to clearly identify the terms of their contract, particularly regarding the scope of the professional's standard of care, to avoid unintended consequences. In *Chesapeake Paper Products Co. v. Stone & Webster Engineering Corp.,*[7] the owner and engineer exchanged alternative contract forms, neither of which was signed by the other party. The owner's form was sent after the engineer's form, however, and was not rejected by the engineer. The owner's contract form contained a higher standard of care, providing "all materials and articles covered by this order . . . will be free from defects in material and/or workmanship, and merchantability." The owner and engineer later signed an amendment modifying the owner's contract form, even though that contract had not been signed by the engineer. When a dispute later arose concerning the engineer's standard of care, the jury found that the engineer had accepted the owner's contract and the higher standard of care as evidenced by the contract amendment that modified the owner's contract form.[8]

Even if the design contract is between the owner and a design corporation, individual design professionals can be personally liable to the owner, where their actions are negligent and the professional reasonably knew or should have known that a client would be injured by the negligent performance of their services.[9] As a general rule, all construction-related contracts contain an implied obligation and a duty to perform the contract with the ordinary skill, care, and diligence commensurate with

[4]*See Redies v. Attorneys Liab. Prot. Soc'y,* 150 P.3d 930 (Mont. 2007) (defining A/E as design professional).
[5]The former AGC Doc. 240 (2000 ed.) included a definition of the standard of care to be applied. The current ConsensusDOCS 240 (2007 ed.) eliminated such definition from the agreement and leaves the standards to the local governing boards.
[6]For sample standard form contracts between architects and owners, *see* ConsensusDOCS 240 (2007 ed.) and American Institute of Architects (AIA) Forms AIA B101 (2007 ed.), AIA B103 (2007 ed.) to AIA B105 (2007 ed.). For sample standard form contracts between engineers and owners, *see* ConsensusDOCS 240 (2007 ed.) and Engineers Joint Contract Documents Committee (EJCDC) E-520 (2002 ed.).
[7]51 F.3d 1229 (4th Cir. 1995).
[8]*Id.*
[9]*Smith v. N. Am. Stainless, L.P.,* 158 Fed. Appx. 699 (6th Cir. 2005).

that of members in the profession.[10] Unless the agreement states otherwise, the design professional must meet that duty even if it is not specifically stated in the contract.[11]

In the traditional design-bid-build project delivery system, contractual privity is usually absent between the contractor and the design professional. Generally, a contractor cannot sue a design professional engaged directly by the owner for breach of contract. A contractor seeking to recover directly against a design professional must resort instead to lawsuits based on legal theories such as negligence, third-party beneficiary, and breach of implied warranty, which are discussed later in this chapter. In a design-build delivery system, the design professional and the contractor are more likely in privity of contract. In such arrangement, a breach of contract action against the design professional by the contractor is an available remedy.

Design professionals, like other professionals, are not required to be perfect. When a claim for professional negligence or malpractice is asserted by the project owner, the contractor, or some third party, the performance of the design professional will be measured against the applicable professional standard of care. As one court explained:

> Architects, doctors, engineers, attorneys, and others deal in somewhat inexact sciences and are continually called upon to exercise their skilled judgment in order to anticipate and provide for random factors which are incapable of precise measurement. The indeterminate nature of these factors makes it impossible for professional service people to gauge them with complete accuracy in every instance. Thus, doctors cannot promise that every operation will be successful; . . . and an architect cannot be certain that a structural design will interact with natural forces as anticipated. [Instead,] the law has traditionally required, not perfect results, but rather the exercise of that skill and judgment which can be reasonably expected from similarly situated professionals.[12]

Therefore, the design professional is not liable for the occasional inconsequential error in judgment, and for a plaintiff to recover on a professional liability claim, the offense typically must rise to the level of a failure on the part of the design professional to exercise reasonable care and professional skill. If, before engagement, a representative of the design professional represents to the owner that he or she possesses specialization and experience in a particular area, that design professional may be held to a higher standard of care.[13] In addition, the description of the design professional's tasks or objectives may be so specific as to the final performance of the completed project that those representations can amount to an express warranty of the design (with the resulting contractual liability applied) in lieu of a professional negligence standard.[14]

[10]*E. Steel Constructors, Inc. v. City of Salem,* 549 S.E.2d 266 (W. Va. 2001).

[11]*Parkette, Inc. v. Micro Outdoors Adver., LLC,* 617 S.E.2d 501 (W. Va. 2005); *Flintkote Co. v. Dravo Corp.,* 678 F.2d 942, 949 (11th Cir. 1982).

[12]*Russo v. NCS Pearson, Inc.,* 462 F. Supp. 2d 981, 995 (D. Minn. 2006).

[13]*See generally* Justin Sweet, Legal Aspects of Architecture, Engineering, and the Construction Process 234 (Marc M. Schneier ed., Thompson 2004).

[14]*Collins Co., Ltd. v. Carboline Co.,* 532 N.E.2d 834 (Ill. 1988).

III. AUTHORITY OF THE DESIGN PROFESSIONAL

The design professional normally assumes one of three distinct roles on the construction project delivery team: (1) independent contractor, (2) an agent of the owner, or (3) an interpreter/arbiter of disputes.[15] How much control and authority the design professional has over a particular project will determine its role.[16]

During the design development phase, the design professional generally acts as an independent contractor to the owner. An independent contractor performs work but is not controlled in the manner or the methods in which the work is achieved.[17] As an independent contractor, the design professional may not be deemed to be an agent of the owner, and, in such cases, the design professional may not have the power to bind the owner with regard to third parties. (Similarly, design consultants to the project architect or engineer are typically independent contractors, not agents of the project architect or engineer.)[18] Once the design is completed, a design professional may deliver ownership of all documentation to the project owner or maintain ownership in the copyright of the design. If the design professional maintains ownership, it must then grant a license to the contractor and subcontractors to allow the further use of the design documents.[19]

When the design professional assumes inspection and supervisory responsibilities during construction (as often happens), that person becomes a "special" agent of the owner with limited authority to act on behalf of the owner.[20] The parties may, by their conduct during the course of a project, depart from the written terms of their contract with the result being a modification of the parties' respective authority or responsibilities. In such circumstances, the limits of the design professional's job-site supervisory authority may be governed by application of the common law agency principles of actual, implied, or apparent authority.

A. Actual Authority

"Actual authority" refers to that authority that an owner (i.e., the principal) expressly confers on its agent (commonly the design professional) and that the agent accepts. As a "special agent," the design professional's actual authority usually is outlined in its contract with the owner and is limited to specific functions. For example, on a typical construction project (and in the absence of contract terms to the contrary), the design professional does not have the authority to make or modify contracts on behalf of the owner/principal[21] or to materially change the scope of the work, the contract price, or the contract time.[22] The owner can, however, grant to the design professional

[15] S Philip Lane Bruner & Patrick J. O'Connor, Bruner & O'Connor on Construction Law § 17:4 (2007).
[16] Restatement (Third) of Agency § 1.01 (2006).
[17] Restatement (Third) of Agency § 7.07 (2 & 3) (2006).
[18] *Lagerstrom v. Beers Constr. Co.*, 277 S.E.2d 765 (Ga. App. 1981).
[19] *See* ConsensusDOCS 200, ¶ 2.3.1 (2007 ed.) (allowing the parties the flexibility to decide who owns the final delivered construction documents and who should grant the contractors and subcontractors the license to use the documents); EJCDC E-505, § 6.03 E (2004 ed.) (granting owner a license to use but allowing the engineer to retain ownership of the documents).
[20] *See generally* J. Acret, Architects and Engineers § 8.01 (Shepard's/McGraw-Hill 1993).
[21] *Crown Constr. Co. v. Opelika Mfg. Corp.*, 480 F.2d 149, 151 (5th Cir. 1973).
[22] *See, e.g.*, AIA A201, § 4.1.2 (2007 ed.); EJCDC C–700, § 9.04 (2007 ed.).

the actual authority to take these actions by so providing in the applicable contracts. The ConsensusDOCS 240, Standard Form of Agreement between Owner and Architect/Engineer (2007 ed.) (ConsensusDOCS 240), and ConsensusDOCS 245, Short Form Agreement Between Owner & Architect/Engineer (2007 ed.) (ConsensusDOCS 245), address the authority of the architect to act on behalf of the owner, as does the American Institute of Architects' (AIA) Standard Form of Agreement Between Owner and Architect, Form B101 (2007 ed.) (AIA B101).[23]

The design professional's authority may also be established in documents to which the design professional is not a party. For example, it is common for the contract between the owner and the contractor to specify the extent of the design professional's authority to act as agent for the owner.

The ConsensusDOCS 200, Standard Form of Agreement and General Conditions Between Owner and Contractor (2007 ed.) (ConsensusDOCS 200), provides that the project owner will provide professional services through the project architect/engineer and that the owner can direct the contractor to submit various documents to the project architect/engineer, including applications for payment, notice of unknown or concealed site conditions, shop drawings, samples, and submittals.[24]

The most recent edition of the standard-form AIA A201, General Conditions of the Contract for Construction (2007 ed.) (AIA A201), describes with some specificity the authority of the architect to act in the owner's behalf.[25] Similarly, AIA B101 also addresses the architect's authority to act on the owner's behalf.[26] But the limits set forth in the owner/architect agreement do not typically bind the contractor. In most cases, the contractor has not even seen the owner/architect agreement. Furthermore, during precontractual negotiations, the parties often modify these standard forms. To the extent that the various contracts and incorporated documents are not coordinated, disputes may arise due to the imposition of inconsistent obligations on the parties.

Language pertaining to a design professional's authority often can be found in different sections of the general conditions, in the specifications, and even on the plans. Many disputes arise because one or more of the parties failed to understand those limits. The design professional must, therefore, know the extent of its authority to act on behalf of the owner. Every participant in the construction process should be familiar with the design professional's actual authority as expressed in its contract with the owner and in the contract documents. Any inconsistencies should be resolved by the parties before disputes arise.

B. Implied Authority

The doctrine of "implied authority" is closely related to that of actual authority. Implied authority gives the design professional the means to act in ways that are incidental to the exercise of actual authority. In other words, implied authority allows the design

[23]ConsensusDOCS 240, ¶ 3 (2007 ed.); ConsensusDOCS 245 (2007 ed.); AIA B101 (2007 ed.).

[24]*See, e.g.,* ConsensusDOCS 200, ¶¶ 3.15, 3.14, 3.16, 9.2.1 (2007 ed.).

[25]*See, e.g.,* AIA A201, Art. 4 (2007 ed.); *see also* EJCDC C-700, § 9.09 (2007 ed.) (detailing the limitations of the engineer's authority and responsibility).

[26]*See, e.g.,* AIA B101, §§ 2.6.1.6, 2.6.2 (2007 ed.).

professional to do those things that are considered reasonable and necessary for the exercise of the design professional's actual authority. The design professional acting under implied authority may bind the owner to those acts even though the authority to perform such acts is not expressly set forth in the contract. For example, if the contract makes the design professional responsible for the review and approval of pay requests, by implication, the design professional is also authorized to reject pay requests.[27] As this example illustrates, determination of the reasonableness and necessity of incidental acts is often a matter of common sense. But this is not always the case.

Consider hypothetically the situation in which a contract expressly grants the design professional the actual authority to decide whether to require the contractor to post payment and performance bonds. Is then the design professional also granted the implied authority to represent to a potential subcontractor that a payment bond will be required? At least one court considering that question held that the design professional had the implied authority to make that representation, and thus the design professional's representation was binding on the owner.[28]

C. Apparent Authority

"Apparent authority" differs from actual and implied authority. In construction, apparent authority may arise when an owner acts in a way that leads a contractor (or other third party) to reasonably believe that the design professional has authority beyond the actual authority expressed in the contract. If the owner, by its actions (or inactions), leads a third party to reasonably believe that the design professional is authorized to act on the owner's behalf, then the owner likely will be bound by such acts of the design professional. Thus, by their conduct, parties may unwittingly expand the authority of the design professional to act on the owner's behalf.

Suppose a contract expressly authorizes a design professional to issue change orders only if a written change order is signed by the owner. If a design professional later verbally approves extra work without issuing a signed written change order (but with the owner's knowledge and consent), the owner likely will be bound by these change orders because the design professional had the apparent authority to issue them.[29] A contractor must be particularly careful, however, when relying on the apparent authority of a design professional to approve changes or additional work without the owner's consent.[30] It is always best to obtain the owner's signature on change orders or other contract modifications potentially impacting cost, time, or third parties.

[27]*See Cullum Mech. Constr., Inc. v. S.C. Baptist Hosp.*, 544 S.E.2d 838 (S.C. 2001) (architect generally does not have a duty to ensure payments to subcontractors, but special conditions in the contract may give rise to a special relationship with subcontractors and therefore a duty of care arises).

[28]*Bethlehem Fabricators, Inc. v. British Overseas Airways Corp.*, 434 F.2d 840 (2d Cir. 1970).

[29]*V.L. Nicholson Co. v. Transcon Inv. & Fin. Ltd., Inc.*, 595 S.W.2d 474 (Tenn. 1980) (architect's oral approval was binding on developer based on developer's actions and knowledge).

[30]*Hussey, Gay & Bell v. Ga. Ports Auth.*, 420 S.E.2d 50 (Ga. Ct. App. 1992) (owner held not liable to subcontractor because engineer did not have express nor apparent authority to agree to pay for repair work).

The dilemma of apparent authority with respect to a design professional's issuance of change orders has been addressed in the evolution of the standard form contract general conditions. Before 1976, the design professional had the authority under the AIA A201 to issue a written change order without the owner's signature if the design professional had written authority that could be furnished to the contractor on request. The amendments then allowed change orders to be signed jointly by the owner and the architect,[31] change orders prepared by the architect and signed by the owner, architect, and contractor;[32] and finally change orders that could be modified only by the owner or with the owner's consent.[33] With the exception of minor changes in the work that do not involve adjustment of the contract sum or the contract time, the AIA A201 General Conditions limit the authority of the design professional to issue changes without the owner's signature. In more recent revisions of the general conditions, the design professional is now responsible for making interim determinations of amounts to be paid for the purposes of the certification of the contractor's monthly pay applications.[34]

Likewise, the ConsensusDOCS 200 does not provide for the project architect or engineer to issue changes, reserving that responsibility for the owner.[35] This ConsensusDOCS contract provides that the owner can, however, direct the contractor to submit its pay applications to the architect/engineer.[36]

It has long been generally established that a design professional does not have authority to modify the contract documents unilaterally. In *Smith v. Board of Education,*[37] an architect directed the original contractor to dispense with the specified wainscoting in the corridors of a building. Upon completion of the building, the owner engaged a replacement contractor to install the omitted wainscoting and then deducted this cost from the amount paid to the original contractor. In the lawsuit that followed, the court held in favor of the owner, reasoning that the original contractor's reliance on the architect's unilateral directive to omit the specified wainscoting was unreasonable. This principle generally applies today, unless the contractor can establish apparent authority of the design professional to alter the plans and specifications through the actions of the owner.

Disputes frequently arise in connection with a design professional's verbal orders for extra work or changes when the contract requires a written change order. Generally, the owner is not bound by such oral authorizations unless the owner waives the written change order requirement. In one case, for example, the design professional verbally instructed the contractor to use a more expensive material than was originally specified. The contract provided that no changes resulting in added cost could be authorized by the architect and that any such changes could be made only by a written order of the owner signed by the architect. The court held that the contractor

[31]*See* AIA A201 (1976 ed.).

[32]*See* AIA A201, § 7.3 (1987 ed.).

[33]*See* AIA A201, § 7.4 (1997 ed.).

[34]*See* AIA A201, § 7.3.9 (2007 ed.). *Compare* EJCDC C-700, Art. 14 (2007 ed.).

[35]*See* ConsensusDOCS 200, ¶ 8 (2007 ed.).

[36]*See* ConsensusDOCS 200, ¶ 9.2.1 (2007 ed.).

[37]85 S.E. 513, 515 (W. Va. 1915).

was not entitled to recover any additional cost because the design professional did not have the authority to direct the change.[38]

A design professional that, without authority, issues orders or directives risks exposure to potential liability for the consequences. Some courts have demonstrated an increased willingness to hold design professionals responsible for their actions and not let them hide behind concepts of agency, especially when a contractor has acted reasonably in relying on the design professional.[39]

D. Ratification of the Design Professional's Authority

Contractual limits on a design professional's authority to act on an owner's behalf may afford some protection to owners where an agent acts outside the limits of its authority. Still, a party—for example, an owner or contractor—may undermine such protection by conduct or representations indicating a ratification of the agent's acts or the owner's acceptance of the benefits related to them.

For example, in one case, the general contractor's superintendent routinely signed delivery tickets presented by a concrete supplier. The back of those tickets contained a preprinted broad indemnification provision in favor of the concrete supplier. After a job-site accident, the concrete supplier sought enforcement of the indemnification provision. The general contractor defended by asserting that the superintendent lacked the authority to agree on behalf of the general contractor to indemnify the supplier. The court held that this defense, although it had some merit, did not overcome the fact that the general contractor had, without objection to the indemnity clause, routinely paid invoices based on the delivery tickets. Such conduct was deemed to have ratified the initially unauthorized conduct of the superintendent, and that ratification defeated the general contractor's agency-related defense.[40]

IV. SUPERVISORY AND ADMINISTRATIVE FUNCTIONS OF THE DESIGN PROFESSIONAL

Sometimes, particularly when the federal government is involved, responsibility for the supervision or administration of various aspects of the construction process is placed with the owner or an independent construction manager. Many contracts, however, still require the design professional to perform supervisory and administrative functions during the construction phase. Five frequently occurring aspects of the design professional's construction-phase supervisory and administrative roles include:

(1) Interpretation of the plans and specifications

(2) Review and approval of shop drawings and submittals

[38] *Iowa Elec. Light & Power Co. v. Hopp*, 266 N.W. 512 (Iowa 1936); *see also C.B.I. Na-Con, Inc. v. Macon-Bibb Water & Sewerage Auth.*, 421 S.E.2d 111 (Ga. Ct. App. 1992).

[39] *E. Steel Constrs., Inc. v. City of Salem*, 549 S.E.2d 266 (W. Va. 2001); *Prichard Bros., Inc. v. The Grady Co.*, 436 N.W.2d 460, 464–65 (Minn. Ct. App. 1989).

[40] *Pioneer Concrete Pumping Servs., Inc. v. T&B Scottdale Contractors, Inc.*, 462 S.E.2d 627 (Ga. Ct. App. 1995) (citing for general principle in different context).

(3) Inspection and testing
(4) Issuance of certificates of progress or completion in connection with the contractor's applications for payment
(5) Resolution of disputes between the owner and the contractor

A. Interpretation of the Plans and Specifications

The plans and specifications commonly generate questions during construction. Minor issues often arise because a specific element of work is inadequately detailed or specified in the contract documents. It is rare for every detail of the work to be expressly set forth in the plans and specifications. Thus, parties often are forced to interpret the contract requirements based on their prior dealings, industry practice, code requirements, and the conduct of the parties leading up to the point of dispute. Simple contract interpretation is often inadequate to resolve disputes resulting from ambiguous or defective plans and specifications or oral change directives. During the project, the parties' interests may conflict to the degree that it is difficult or impossible to distinguish between "interpretation" of the contract documents and "remediation" of those that are defective. Fortunately, most construction industry participants historically have employed reasonable judgment and open communication to resolve "informally" the great majority of interpretation issues and other disputes, relying on dispute review boards, arbitrators, or the courts only when necessary to resolve intractable problems related to defective design documents.

Under most standard-form contracts and under many project-specific "custom" contracts, the supervising design professional has the authority to interpret the plans and specifications. This authority theoretically gives the design professional some degree of control over the work as it progresses and establishes an initial framework within which to solve problems and maintain consistent project administration. The responsibility that accompanies this authority, however, exposes design professionals to liability to owners for negligent interpretation of the plans and specifications. This may be true even when an owner uses the design of a separate design consultant. In one case, the architect was found liable for a faulty design that had been recommended to the owner by an independent engineer. The architect alleged that it was entitled to rely on the engineer's professional advice. The court viewed the architect's act of affixing its professional stamp on the plans as an implicit acceptance of an affirmative duty to review the plans to ensure that they were technically sound. In deciding against the architect, the court reasoned that, if the architect did not have the expertise to review the engineer's plans, the architect should have abstained from stamping the engineer's documents or should have withdrawn from the project.[41]

[41]*S.D. Bldg. Auth. v. Geiger-Berger Assocs.*, 414 N.W.2d 15, 24 (S.D. 1987); *see also Kerry, Inc. v. Angus-Young Assocs., Inc.*, 694 N.W.2d 407 (Wis. Ct. App. 2005) (architects held liable for damages when failing to identify and notify owner of missing information in engineer's report); *Nicholson & Loup, Inc. v. Carl E. Woodward, Inc.*, 596 So. 2d 374 (La. Ct. App. 1992).

In general, a design professional that seals drawings is assuming design responsibility for those plans.[42]

The design professional's authority to interpret the contract requirements does not include the authority to place additional burdens or liabilities on the owner or the contractor by modifying contract terms. For example, a dispute arose on one project over a design professional's interpretation of the contractor's obligation to provide additional fill material. The court held that the architect did not have the power, express or implied, to modify the contractual rights or liabilities of either party under the contract.[43] Although the plans and specifications contained no reference to additional fill material, the design professional determined that the contractor had provided it without additional compensation.

In addition to the literal wording of the contract documents, the design professional's discretion in interpreting the plans and specifications may be limited by construction industry general practices and trade customs. For example, a court held that a subcontractor was entitled to additional compensation for performing certain painting work beyond normal custom, despite the architect's contrary interpretation of the subcontract. The court stated that although the architect had the authority to make final decisions regarding the meaning of plans and specifications, the contract implied that the architect would render a reasonable decision that was consistent with industry custom or trade usage.[44]

This same principle applies in the federal arena. A contractor on a government project may be entitled to rely on industry custom and practice in the absence of any contractual provision to the contrary. In one case, the United States Court of Claims awarded damages to a contractor for costs incurred in meeting higher standards than required under a reasonable interpretation of the specifications. In that case, the contractor was directed to construct wooden concrete forms within tolerances that were not set forth in the specifications and that were more restrictive than necessary to meet the specified tolerances for the finished concrete. The contract merely specified that the "forms shall be true to line and grade." The court applied common trade practices to support an award to the contractor, reasoning that the federal government must state in clear and unambiguous language any intention to alter industry custom or trade usage with respect to required tolerances.[45]

[42]*Ambassador Baptist Church v. Seabreeze Heating & Cooling Co.,* 184 N.W.2d 568 (Mich. Ct. App. 1970); *see also State Bd. of Architects v. Clark,* 69 A.2d 1247 (Md. Ct. Spec. App. 1997) (where architect's sealing of drawings by unlicensed draftsman did not violate professional responsibility).

[43]*Tomlinson v. Ashland County,* 173 N.W. 300, 303–04 (Wis. 1919); *But see Fontaine Bros., Inc. v. City of Springfield,* 617 N.E.2d 1002 (Mass. App. Ct. 1993) (architect did not improperly alter the contract when deciding the meaning of certain specifications).

[44]*John W. Johnson, Inc. v. J.A. Jones Constr. Co.,* 369 F. Supp. 484 (E.D. Va. 1973); *see also Batson-Cook Co. v. Loden & Co.,* 199 S.E.2d 591 (Ga. Ct. App. 1973) (general contractor liable for enforcement of architect's unreasonable rejection of slightly flawed bricks; rejection was inconsistent with industry practices but not technically violative of express contract terms); *City of York v. Turner-Murphy Co., Inc.,* 452 S.E.2d 615 (S.C. Ct. App. 1994) (city failed to prove that engineer's performance did not conform with industry standards).

[45]*Kenneth Reed Constr. Corp. v. United States,* 475 F.2d 583 (Ct. Cl. 1973); *see also Robert E. Moore Constr.,* AGBCA No. 85–262–1, 90-2 BCA ¶ 22,803.

The design professional's power to interpret contract requirements also carries with it the duty to render these interpretative decisions honestly and in good faith. Practically speaking, the design professional is working for the owner. Thus, it may be difficult at times for the design professional to exercise independent judgment. Furthermore, the design professional often is placed in the awkward position of being forced to judge the adequacy of the very plans and specifications that it produced. A sense of professional pride (combined with a reluctance to present the owner with a change order to correct an omission or mistake in the design) sometimes can influence a design professional's objectivity in the field. Courts faced with this situation often find that the failure of a design professional to perform its administrative duties honestly and in good faith diminishes or nullifies the authority to interpret contract requirements or arbitrate contract disputes.[46]

Owners and contractors must also be careful to avoid giving the design professional's opinion more weight than the contract or law allows. For example, the Colorado Department of Transportation was found liable when its engineer contended that its engineer's design characterized as a "final decision" triggered a 30-day judicial review deadline or waiver of the claim.[47] The court held that the contract between the parties did not give the engineer the power to make binding decisions which could be admitted in a legal proceeding.[48] Similarly, an architect's decision to refuse to certify final payment due to artistic defects was held not to be a final decision binding on the parties.[49] Thus, the contractor was liable for delay when it did not fully comply with the terms of the contract.

B. Review and Approval of Shop Drawings and Submittals

The design professional generally reviews data submitted by the contractor, such as samples, shop drawings, product data, and schedules. The primary purpose of shop drawings and submittals is for the owner and the design professional to obtain an understanding of how the contractor intends to perform certain aspects of the work.[50] These submittals also inform the design professional and the owner of the processes, materials, and methods to be used by the contractor, and reflect the contractor's understanding of the contract requirements.

The design professional's review of submittals has legal consequences. One question that arises frequently is whether the design professional's approval of these processes and methods relieves the contractor of its responsibility to comply with the specifications. Generally, it does not.[51]

[46]*See, e.g., State Highway Dep't v. Knox-Rivers Constr. Co.,* 160 S.E.2d 641 (Ga. App. 1968) (engineer's refusal to certify release of funds withheld did not excuse owner's payment where reason for refusal was based on engineer's erroneous interpretation of documents).

[47]*Castle Rock Constr. Co. v. Dep't of Transp.,* 74 P.3d 491 (Colo. Ct. App. 2003).

[48]*Id.*

[49]*NSC Contractors, Inc. v. Borders,* 564 A.2d 408 (Md. 1989).

[50]*See* Justin Sweet, Legal Aspects of Architecture, Engineering, and The Construction Process 399 (Marc M. Schneier ed., Thompson 2004).

[51]*See Cmty. Sci. Tech. Corp.,* ASBCA No. 20244, 77–1 BCA ¶ 12,352; *see also D.C. McClain, Inc. v. Arlington County,* 452 S.E.2d 659 (Va. 1995) (county's approval of shop drawings did not relieve the

In one case, for example, a contractor with a design-build contract hired a design professional to perform the design work and review shop drawings.[52] The contractor later awarded work to a subcontractor that was required to prepare and submit shop drawings to the contractor for review and approval. The contractor then forwarded the shop drawings to the design professional. The design professional approved them even though they contained errors. After the work had been performed in accordance with the shop drawings, problems were discovered and the subcontractor was forced to correct them. The subcontractor subsequently sued the design professional for its alleged negligence in approving the defective shop drawings. The court held in favor of the design professional, stating that the design professional owed no duty to the subcontractor, that the subcontractor was responsible for meeting all of the requirements of its subcontract, and that the subcontractor had no right to rely on the design professional's approval of its shop drawings because the design professional was hired to provide services only to the contractor, not to subcontractors.

When a design professional's performance of this review function goes beyond mere review and comment and the design professional attempts to change the contractor's scope of work, courts likely will deem such an attempt beyond the scope of its authority. In that situation and absent some ratifying act, the owner may not be bound by such approval.[53]

The 1987 edition of the AIA A201 was specific as to the design professional's liability for approving contractor submittals that change the specifications. This version of the AIA A201 relieved the contractor of responsibility for deviations from the contract documents contained in submittals, if and only if "the Contractor has specifically informed the Architect in writing of such deviation at the time of submittal and the Architect has given written approval to the specific deviation." Additional language in the 1997 edition of the AIA A201 specifically excluded from the architect's indicated approval those revisions made by the contractor, its subcontractors, or suppliers to shop drawings and other submittals resubmitted without a specific graphic indication of each revised element.[54] This change arguably shifted responsibility from the architect to the contractor for revisions in submittal and shop drawings that were not specifically graphically highlighted by the contractor. The 2007 edition of AIA A201 does not change the parties' responsibilities for submittals from those set forth in the 1997 version.

Subparagraph 3.14.1 of ConsensusDOCS 200 contains requirements for the contractor's written identification of deviations, substitutions, or changes and express written approval from the "Owner" authorizing the deviation, substitution, or change. Subparagraph 3.14.5 of ConsensusDOCS 200 contains an additional specific requirement for the approval of substitutions. One difference in the two industry forms is

contractor of obligation to properly construct project); *McDevitt Mech. Contractors., Inc. v. United States,* 21 Cl. Ct. 616 (1990) (government approval of defective shop drawings did not relieve contractor of responsibility for errors in shop drawings under terms of contract).

[52]*Lutz Eng'g Co. v. Indus. Louvers, Inc.*, 585 A.2d 631 (R.I. 1991).

[53]*Fauss Constr., Inc. v. City of Hooper,* 249 N.W.2d 478, 481 (Neb. 1977) (enforcing contract terms limiting design professional's authority).

[54]AIA A201, § 3.12.9 (1997 ed.).

that the ConsensusDOCS form initially reserves to the owner, not the design professional, the right to make such decisions or approvals.

Courts can and do impose liability on design professionals for approving shop drawings that deviate from the contract documents. For example, in one case, the contract documents called for 10-gauge steel to be used for stair pans. The contractor submitted shop drawings showing 14-gauge steel, which the design professional approved. Workmen later walked onto the pans and were injured when the landing pan collapsed through the framework. The court held in favor of the workmen and against the design professional, finding that the contract language, ostensibly limiting the scope of the architect's review of shop drawings to "conformance with the design concept of the Project and for compliance with the information given in the Contract Documents," did not relieve the design professional of liability for its negligence in supervising the shop drawing process as evidenced by the architect's failure to detect the error.[55]

Perhaps the most notable example highlighting professional responsibility related to the shop drawing process arose out of the 1981 Kansas City Hyatt Regency disaster in which a multistory interior hotel bridge collapsed, killing more than 100 persons and injuring 186. The cause of the collapse was a faulty structural steel detail that had been introduced into the process through shop drawings. The court discounted the engineer's attempt to avoid liability for negligence by asserting a defense that industry "custom and practice" allowed engineers to rely on fabricators "to design certain structural steel connections." The court noted that, by affixing his professional stamp to the documents, the certifying engineer assumed responsibility for the entire engineering project and that his professional duties were "nondelegable."[56]

The design professional generally has an obligation to review and act on a shop drawing or submittal within a reasonable time. To the extent the contractor is delayed or hindered by the design professional's failure to timely approve or reject a submittal, the owner, and perhaps even the design professional, may be liable to the contractor.[57] The AIA A201 (2007 ed.) attempts to limit an owner's (or design professional's) liability for submittal review delays. Together, §§ 3.10.2 and 4.2.7 of this industry contract document require the contractor to prepare a submittal schedule promptly after award of the contract and submit the schedule for the architect's approval. The submittal schedule must be coordinated with the construction schedule and allow the architect "reasonable time" to review submittals. If the contractor fails to submit the submittal schedule, then the contractor will not be entitled to any increase in contract sum or an extension of time based on the time required for the review of submittals by the design professional.

[55]*Jaeger v. Henningson, Durham & Richardson,* 714 F.2d 773, 776 (8th Cir. 1983).

[56]*Duncan v. Mo. Bd. for Architects, Prof'l Eng'rs & Land Surveyors,* 744 S.W.2d 524, 536–37 (Mo. Ct. App. 1988); *see also Henningson, Durham & Richardson, Inc. v. Swift Bros. Constr. Co.,* 739 F.2d 1341 (8th Cir. 1984) (architect's disclaimer on its shop drawing stamp did not relieve the architect of its affirmative duty to review shop drawings and, therefore, language requiring indemnity of the architect by the shop drawing preparer was not lawful or enforceable in light of architect's negligence).

[57]*See, e.g., E. C. Ernst, Inc. v. Manhattan Constr. Co. of Tex.,* 551 F.2d 1026 (5th Cir. 1977), *cert. denied,* 434 U.S. 1067 (1978); *Prichard Bros. Inc. v. The Grady Co.,* 436 N.W.2d 460 (Minn. Ct. App. 1989).

Moreover, the recent trend toward Building Information Modeling (BIM), which provides a means for members of the project delivery team to collaboratively share project design and construction data, could present significant risks as relates to a design professional's review and approval of shop drawings and submittals. Because BIM is collaborative in nature, multiple parties will provide input into, and may exercise control over, such review processes. As such, the design professional involved in BIM should ensure that responsibility for review and approval of such submittals is well-defined. For a more detailed discussion of BIM, including its legal implications, see **Chapter 13**.

C. Inspections and Testing

Frequently, the owner engages the design professional to inspect work for conformity to contract requirements. In performing these inspection duties, the design professional acts as the owner's agent. The design professional, under these circumstances, typically has the implied authority to reject defective work (i.e., work not in compliance with the contract documents). Disputes often arise when the design professional rejects work that the contractor believes complies with the contract. Factual disputes in this area cover every aspect of construction and generally turn on issues of contract interpretation, workmanship, and sufficiency of materials.

In *United States Fidelity & Guaranty Co. v. Stanley Contracting, Inc.*,[58] a suit arose from a dispute between the owner and the contractor where the owner relied on the recommendation of an engineer and terminated the contractor for failing to comply with the contractual requirements. The parties used a standard Engineers Joint Contract Documents Committee (EJCDC) contract that contained a provision granting the engineer sole discretion to determine if a particular product was an appropriate substitution or conformed to the contract's "or-equals"clause.[59] When the contractor attempted a substitution, the engineer inspected the product but recommended that the owner reject it. The owner allowed the substitution; however, the engineer rejected the first four sets of drawings submitted. These rejections significantly delayed the work. Ultimately, the owner terminated the contractor for breach. The court held that the owner was justified in relying on the engineer's recommendations because the contractor had agreed to the contract's or-equals clause.[60] The EJCDC document afforded the engineer great discretion to determine the substitution procedures, and the engineer did not exercise bad faith or poor judgment in rejecting the contractor's proposed substitution or implementing the review and approval procedures.[61]

Improper rejection of work generally will not expose the design professional to liability to the contractor, because the design professional usually is acting on the owner's behalf.[62] But incorrect or negligent inspections that result in improper

[58] 396 F. Supp. 2d 1157 (D. Or. 2005).
[59] *Id.* at 1160–61.
[60] *Id.* at 1171.
[61] *Id.*
[62] *See, e.g., Nannis Terpening & Assocs., Inc. v. Mark Smith Constr. Co.,* 318 S.E.2d 89 (Ga. Ct. App. 1984).

rejection of the work or in the owner's liability to the contractor for damages may generate legal action by the owner against the design professional. Additionally, problems often arise when a design professional fails to inspect work or fails to timely detect nonconforming work as the work proceeds. Contractors, believing their work to be satisfactory, may cover up or build on nonconforming work, making corrections or remedial work more difficult. Who should be responsible when corrections to nonconforming work are more expensive because the problem is discovered too late: the contractor that improperly constructed the work or the design professional that failed to timely catch the error?

The AIA A201 (1987 ed.), and many other construction contract forms, limit the scope of the owner's and design professional's responsibility to what can generally be described as inspection and contract compliance obligations.[63] The contractor's former and more general "study, compare, and report" obligations were considerably expanded in the 1997 version of AIA A201, and these expansions continue in the 2007 edition of AIA A201. The contractor must take timely field measurements "related to that portion of the work" and to "observe any conditions at the site affecting it." Arguably, this imposes on the contractor an ongoing task-specific duty to confirm the constructability of the design. Reports of inconsistencies or omissions must be in a form "as the Architect may require." The AIA A201 (1997 ed.) contains no model forms on which a contractor is to report these problems. The prudent contractor, therefore, must determine exactly what form the architect requires, since these requirements have an onerous "club" attached. If a contractor fails to timely and properly observe, study, compare, and confirm the contract documents and field conditions, or to report potential conflicts and discrepancies, that contractor may be required to pay those resulting costs of correction that could have been avoided had the contractor done so. Similarly, under Subsections 3.2.3 and 3.2.4 of AIA A201 (2007 ed.), if a contractor fails to report promptly a discovered violation or deviation of the contract documents from "applicable laws, statutes, ordinances, codes, rules and regulations, or lawful orders of public authorities" the contractor may be liable for the related avoidable costs.

The scope of a design professional's potential liability for job-site accidents caused by unsafe working conditions continues to trouble many courts. Generally, the design professional's obligation to monitor the quality of the work is weighed against the contractor's responsibility to control the means and methods of its performance. The courts usually resolve this issue by closely examining the language of both the design professional's contract and the construction contract.[64] Generally, the contractor retains the primary responsibility for contract compliance and for control of construction means, methods, and techniques. Section 3.1.2 of ConsensusDOCS 200 (2007 ed.) and § 3.3.1 of AIA A201 (2007 ed.) exemplify standard-form means and methods clauses.

[63]*See generally D.C. McClain, Inc. v. Arlington County,* 452 S.E.2d 659 (Va. 1995) (contract required contractor to verify dimension at the site before commencing construction; owner not liable for elevation discrepancies discovered in contract documents during construction process).

[64]*Shepherd Components, Inc. v. Brice Petrides-Donohue & Assocs. Inc.,* 473 N.W.2d 612 (Iowa 1991); *Moore v. PRC Eng'g, Inc.,* 565 So. 2d 817 (Fla. Dist. Ct. App. 1990); *Marshall v. Port Auth. of Allegheny County,* 568 A.2d 931 (Pa. 1990).

In addition, the design professional is not deemed to be responsible for supervision or inspection of construction when the contract does not require it to do so.[65] For example, in a Texas case, a supervising engineering firm was found not liable for a contractor's failure to use adequate supports when pouring a concrete slab that resulted in the death of a construction worker after the slab collapsed.[66] In its decision, the court relied on the engineer's contract, which specified that the firm was not responsible for the contractor's construction means and methods or for safety at the job site.[67]

A means and methods clause, however, does not always absolve the design professional of liability. In another Texas case, the court held an architect liable for failing to adequately inspect a project despite exculpatory contract language stating that the architect would "not be responsible for the Contractor's failure to carry out the work in accordance with the contract documents." The court reasoned that an architect that is paid to protect the owner from construction defects should be held responsible for failing to detect defective work.[68]

In another case, a court considered whether a design professional could be held liable for a fire caused by the contractor's deficient performance.[69] The design professional had entered into a standard AIA design contract that provided:

> The Architect shall visit the site at intervals appropriate to the stage of construction . . . to become generally familiar with the progress and quality of the Work and to determine in general if the Work is proceeding in accordance with the Contract Documents. However, the Architect shall not be required to make exhaustive or continuous on-site inspections to check the quality or quantity of the Work. On the basis of such on-site observations as an architect, the Architect shall keep the Owner informed of the progress and quality of the Work, and shall endeavor to guard the Owner against defects and deficiencies in the Work of the Contractor.

The fire was apparently caused by the contractor's improper placement of insulation around light fixtures that the design professional had not detected. The court rejected the argument that the design professional was not obligated to catch such a defect. Despite the general disclaimers in the AIA form contract, the court denied summary judgment to the design professional, reasoning that the design professional

[65] *Goette v. Press Bar & Café, Inc.*, 413 N.W.2d 854, 856 (Minn. Ct. App. 1987); *see also C.L. Maddox, Inc. v. Benham Group, Inc.*, 88 F.3d 592 (8th Cir. 1996) (design professional not responsible to guard contractor against construction defects and deficiencies despite contract requirement for design professional to keep contractor informed of quality and progress of the work because the contractor, not designer, was responsible for construction means and methods. Design professional was responsible to make site visits but not to ensure that contractor made no errors).

[66] *Rodriquez v. Universal Fastenings Corp.*, 777 S.W.2d 513, 516 (Tex. App. 1989).

[67] *Id. See also Block v. Lohan Assocs., Inc.*, 645 N.E.2d 207 (Ill. App. Ct. 1993) (architect not liable for failure to supervise and coordinate work when subcontractor's employee fell from ladder; architect's contract was limited to design and design conformance, not worker safety); *Davis v. Lenox Sch.*, 541 N.Y.S.2d 814 (N.Y. App. Div. 1989).

[68] *Hunt v. Ellisor & Tanner, Inc.*, 739 S.W.2d 933 (Tex. App. 1987).

[69] *Diocese of Rochester v. R-Monde Contractors, Inc.*, 562 N.Y.S.2d 593 (N.Y. Sup. Ct. 1989).

had a duty to be generally knowledgeable of the quality and quantity of work to evaluate payment applications. The design professional was not relieved from liability simply because the contractor had failed to perform its work correctly. While the design professional was not a guarantor of the contractor's performance, it still had the responsibility to keep abreast of the quality of work.[70]

The design professional is also obligated to take action when it discovers or should have discovered the contractor's defective workmanship. The design professional can be held liable to the owner for inspection and approval of a contractor's faulty or nonconforming work. For instance, an architect was found to have breached its duty to guard the owner against defects and deficiencies in the work, and was therefore liable to the owner for the cost of repairing interior wall coverings damaged because of inadequate waterproofing observed and approved by the architect.[71]

In another case, a homeowner sued an architect for habitually neglecting its duties to inspect and approve the contractor's work.[72] The architect argued that while it was contractually obligated to perform "general supervision of the construction work," its fee did not include "the cost of superintendence by a full-time inspector or Clerk of the Works." The court, however, held for the homeowner, noting that:

> The term "general supervision," as used in the instant agreement, must mean something other than mere superficial supervision. Obviously, there can be no real value in supervision unless the same be directed toward securing a workmanlike adherence to specifications and adequate performance on the part of the contractor.[73]

In *Central School District No. 2 v. The Flintkote Co.,*[74] the court required a trial on the issue of whether the design professional had properly inspected a roof. The owner had employed a "clerk of the works" to make inspections, and the design professional contended that this eliminated its duty to inspect the work. The court rejected this argument, holding that if the architect could rely solely on the clerk of the works in this regard, then:

> [T]he owner would be deprived of the professional judgment [which] he had the right to expect. The owner's retainer of a "Clerk of the Works" for full-time, on-site services, constituted a protection that is an addition to and not a substitute for the contractual and professional obligations of the architect.[75]

From the contractor's standpoint, a design professional's periodic inspection of construction work and failure to object to nonconforming materials and workmanship

[70]*Id.* at 596. *But see C.L. Maddox, Inc. v. Benham Group, Inc.,* 88 F.3d 592 (8th Cir. 1996).
[71]*Dan Cowling & Assocs., Inc. v. Bd. of Educ.,* 618 S.W.2d 158 (Ark. 1981).
[72]*Pancoast v. Russell,* 307 P.2d 719 (Cal. Ct. App. 1957).
[73]*Id.* at 722.
[74]391 N.Y.S.2d 887 (N.Y. App. Div. 1977).
[75]*Id.* at 888.

may, under the right circumstances, establish a constructive acceptance of the work and preclude an owner from refusing to pay the contractor. Although no set rule applies in these situations, the contractor is more likely to prevail if it can show either a reasonable reliance on the design professional to detect nonconforming work or fault on the part of the design professional or owner.

For example, an architect (that in this case also happened to be the owner and the contractor) failed to inspect a subcontractor's repairs to work that the architect had originally found inadequate. The court held that the architect's failure to inspect the remedial work and to specify any additional corrections constituted a waiver of a condition precedent that work be performed to the satisfaction of the architect before payment and that the owner-contractor was estopped from raising the condition precedent as a defense in a breach of contract suit brought by the subcontractor.[76]

D. Issuance of Certificates of Progress or Certificates of Completion and Certificates for Payment

Most construction projects are performed over an extended time. Because of this fact, construction contracts frequently contemplate interim (or progress) payments to the contractor at specified intervals, usually monthly, throughout the project. Extended construction projects also contemplate a final payment upon final completion and acceptance of the work. Often the owner engages the design professional to issue certificates of progress or completion, which generally are related to the owner's duty to pay the contractor for work performed. This type of contract requirement is valid, despite the fact that the design professional acts as the owner's agent.[77]

The AIA A201 (2007 ed.) provides seven grounds for the architect to withhold approval on a certificate for payment. These are:

(1) defective Work not remedied;
(2) third-party claims filed or reasonable evidence indicating probable filing of such claims unless security acceptable to the Owner is provided by the Contractor;
(3) failure of the Contractor to make payments properly to Subcontractors for labor, materials or equipment;
(4) reasonable evidence that the Work cannot be completed for the unpaid balance of the Contract Sum;
(5) damage to the Owner or a separate contractor;
(6) reasonable evidence that the Work will not be completed within the Contract Time, and that the unpaid balance would not be adequate to cover actual or liquidated damages for the anticipated delay; or
(7) repeated failure to carry out the Work in accordance with the Contract Documents.[78]

[76] *Hartford Elec. Applicators of Thermalux, Inc. v. Alden,* 363 A.2d 135, 138 (Conn. 1975).
[77] *Friberg v. Elrod,* 296 P. 1061 (Or. 1931).
[78] AIA A201, § 9.5.1 (2007 ed.).

ConsensusDOCS 200 provides that the owner, rather than the architect, has the authority to adjust pay requests and sets forth the bases entitling an owner to adjust or reject a contractor's pay request as follows:

> 9.3 ADJUSTMENT OF CONTRACTOR'S PAYMENT APPLICATION The Owner may adjust or reject a payment application or nullify a previously approved payment application, in whole or in part, as may reasonably be necessary to protect the Owner from loss or damage based upon the following, to the extent that the Contractor is responsible therefore under this Agreement:
>
> 9.3.1 the Contractor's repeated failure to perform the Work as required by the Contract Documents;
>
> 9.3.2 loss or damage arising out of or relating to this Agreement and caused by the Contractor to the Owner or to Others to whom the Owner may be liable;
>
> 9.3.3 the Contractor's failure to properly pay Subcontractors and Material Suppliers following receipt of such payment from the Owner;
>
> 9.3.4 rejected, non-conforming or defective Work not corrected in a timely fashion;
>
> 9.3.5 reasonable evidence of delay in performance of the Work such that the Work will not be completed within the Contract Time; and
>
> 9.3.6 reasonable evidence demonstrating that the unpaid balance of the Contract Price is insufficient to fund the cost to complete the Work; and
>
> 9.3.7 third party claims involving the Contractor or reasonable evidence demonstrating that third party claims are likely to be filled unless and until the Contractor furnishes the Owner with adequate security in the form of a surety bond, letter of credit or other collateral or commitment which are sufficient to discharge such claims if established.
>
> No later than seven (7) Days after receipt of an application for payment,the Owner shall give written notice to the Contractor, at the time of disapproving or nullifying all or part of an application for payment, stating its specific reasons for such disapproval or nullification, and the remedial actions to be taken by the Contractor in order to receive payment. When the above reasons for disapproving or nullifying an application for payment are removed, payment will be promptly made for the amount previously withheld.

Although contract language may provide the design professional with a broad range of bases on which it can rely to withhold a payment certification, the refusal to issue a certificate must be exercised carefully. Common issues arising from this duty involve the acceptance of the work upon the design professional's approval of pay requests and the liability of the design professional for the improper failure to approve progress payments. As with issues relating to inspection of the work, the analysis of questions regarding progress payments is almost always fact- and contract-specific.

A number of cases confirm the design professional's duty to use the care of one skilled in that profession when issuing payment certificates. As such, the design professional can be held liable for damages resulting from a failure to act accordingly. For example, in one California case, the owner asserted that the architect was negligent in issuing certificates of payment without first determining whether the contractor had fully paid its subcontractors and materialmen. The court held that the architect, under its contract, should have protected the owner by making certain that all bills were paid or by insisting on lien waivers from suppliers.[79]

The wrongful withholding of a payment certificate can also result in a design professional's liability directly to the contractor. For example, a Mississippi court found that an engineer that reduced a contractor's pay request contrary to an audit of the contractor's work, without consulting the contractor and without adequately checking the extent of the work, was directly liable to the contractor and was liable for bad-faith damages to the contractor.[80] Similarly, an engineer's failure to properly construe the contract requirements that resulted in withholding payment from the contractor entitled the contractor to release of the contract balance.[81]

The design professional's negligence in approving pay requests and issuing progress certificates may also expose it to liability to third parties with a financial interest in the project, such as banks or sureties. For example, in *National Surety Corp. v. Malvaney,*[82] an architect was found liable to the contractor's surety for the negligent issuance of a substantial completion certificate. Under the contract, the architect could require the contractor to present evidence that payroll and material bills had been paid before releasing retained funds. The architect, however, failed to enforce this requirement. Consequently, the contractor collected its retainage and then defaulted, leaving the burden of completing the work and satisfying claims to the surety.

The court rejected the architect's argument that its lack of privity of contract with the surety insulated it from liability to the surety. The court relied on a negligence theory, ruling that the design professional owed a duty to determine whether the contractor had paid the bills to both the owner and the surety for whose mutual benefit and protection contractor funds were retained. The surety's failure to take steps to ascertain whether outstanding bills were being paid by the contractor was not held to constitute contributory negligence, because the surety had a right to assume that the retainage would not be released until the contract was fully performed.

Furthermore, in *Peerless Insurance Co. v. Cerny & Associates, Inc.,*[83] the court determined that a surety was subrogated to any claim that the owner would have

[79] *Palmer v. Brown,* 273 P.2d 306 (Cal. Ct. App. 1954); *see also Newton Inv. Co. v. Barnard & Burk, Inc.,* 220 So. 2d 822 (Miss. 1969).

[80] *City of Mound Bayou v. Roy Collins Constr. Co.,* 499 So. 2d 1354 (Miss. 1986); *see also Magnolia Constr. Co. v. Miss. Gulf S. Eng'rs, Inc.,* 518 So. 2d 1194 (Miss. 1988) (recognizing that the contractor had a claim against the project engineer where engineer did not recommend that the owner pay the contractor).

[81] *Textor Constr. Co. v. Forsyth R-III Sch. Dist.,* 60 S.W.3d 692 (Mo. Ct. App. 2001).

[82] 72 So. 2d 424 (Miss. 1954).

[83] 199 F. Supp. 951 (D. Minn. 1961).

against the architect for negligent certification of progress payments.[84] Similarly, in *Westerhold v. Carroll,*[85] the court confirmed that an architect can be held liable to the indemnitor of a contractor's surety for certifying payments in excess of the value of work performed. In *Boren v. Thompson & Associates,*[86] a subcontractor on a public project was allowed to sue the project architect on claims that the architect negligently allowed the disbursal of funds without first verifying whether the contractor had provided a payment bond.

Although the contractor generally is entitled to receive periodic payment as work progresses, this right must be balanced against the owner's right to verify that the contractor is performing the work in accordance with the contract documents. To address this, construction contracts usually state that periodic payment to the contractor does not constitute acceptance of the work and does not waive the owner's right to demand contract compliance. Because the design professional frequently is vested with the responsibility to certify payment based on progress, industry contract forms attempt to foreclose the implication that the contractor has performed in accordance with the contract documents based solely on the architect's approval of the contractor's payment applications. For example, AIA A201 (2007 ed.) provides:

> The issuance of a Certificate for Payment will constitute a representation by the Architect to the Owner, based on the Architect's evaluation of the Work and the data comprising the Application for Payment, that to the best of the Architect's knowledge, information and belief, the Work has progressed to the point indicated and that the quality of the Work is in accordance with the Contract Documents . . . [and] that the Contractor is entitled to payment in the amount certified.[87]

Despite such wording to the contrary, the design professional's certification of progress or completion still may benefit the contractor. In one case, the owner brought a breach of contract claim against the contractor alleging failure to comply with the plans and specifications.[88] The court held that the design professional's certificate of completion was an effective defense to the owner's claim of noncompliance unless there was proof of fraud, gross negligence, or bad faith on the part of the design professional in issuing the certificate.

The design professional's certification of progress payments has also been used by subcontractors as support for claims against the general contractor. In one case, the prime contractor terminated a subcontractor for deficient performance.[89] The subcontractor, however, argued that the general contractor had sought and obtained progress

[84]*Id. See also Unity Tel. Co. v. Design Serv. Co.,* 201 A.2d 177 (Me. 1964); *But see Fireman's Fund Ins. Co. v. SEC Donohue, Inc.,* 679 N.E.2d 1197 (Ill. 1997) (surety's subrogation rights not sufficient to allow surety to sue project engineer; privity of contract required).

[85]419 S.W.2d 73 (Mo. 1967).

[86]999 P.2d 438 (Okla. 2000).

[87]AIA A201, § 9.4.2 (2007 ed.).

[88]*Fuchs v. Parsons Constr. Co.,* 111 N.W.2d 727, 734 (Neb. 1961).

[89]*Sweeney Co. of Md. v. Eng'rs-Constrs., Inc.,* 823 F.2d 805 (4th Cir. 1987).

payments for the subcontractor's work and that the design professional's certification of the subcontractor's work was strong evidence that the work was satisfactory. The general contractor was placed in the difficult position of arguing either that the payment applications were not representations that the work was acceptable or that the design professional's approval of the payments did not constitute a binding acceptance of the work. The court of appeals held that the trial court did not err in allowing the admission of the applications into evidence and in considering that evidence in the court's determination of the subcontractor's satisfactory performance.[90]

The requirement that the design professional issue a certificate before issuing payment to the contractor may be excused or waived by the parties under certain circumstances. In one case, the court found that neither the owner nor the contractor had relied on a design professional's certificate as a basis for making or receiving progress payments for completed excavation work. Consequently, the contract provision requiring a certificate before final payment was deemed waived so that neither party could insist on strict compliance with the contract requirement.[91]

Furthermore, an architect cannot delegate its responsibility for certifying progress payments to another architect and thereby escape liability for a defective building. In *Sheetz, Aiken & Aiken, Inc. v. Spann, Hall, Ritchie, Inc.*,[92] the court found that the project architect's duty as inspecting architect was to verify that the contractor had performed the work in accordance with the plans and specifications, while the duty of the architect hired to certify progress was merely to determine the amount owed to the contractor for direct construction costs and not to ensure a defect-free building.

E. Resolution of Disputes between the Owner and the Contractor

Another important function that may be performed by the design professional pursuant to a contract provision is the evaluation of claims and disputes between the owner and contractor. Many of the older standard construction documents designated the "Architect" as the initial arbiter of disputes and evaluator of claims between the owner and contractor. For example, Section 4.4.1 of the AIA A201 (1997 ed.) provided:

> Claims, including those alleging an error or omission by the Architect but excluding those arising under Sections 10.3 through 10.5, shall be referred initially to the Architect for decision. An initial decision by the Architect shall be required as a condition precedent to mediation, arbitration or litigation of all Claims between the Contractor and Owner arising prior to the date final payment is due, unless 30 days have passed after the Claim has been referred to the Architect with no decision having been rendered by the Architect. The

[90] *Id.* at 808–11; *see also Magnolia Constr. Co. v. Miss. Gulf S. Eng'rs, Inc.*, 518 So. 2d 1194 (Miss. 1988) (recognizing that third parties are entitled to rely on an architect's performance of its contractual obligations to the owner, including the architect's certification of work performed).

[91] *Palmer v. Watson Constr. Co.*, 121 N.W.2d 62 (Minn. 1963).

[92] 512 So. 2d 99 (Ala. 1987).

> Architect will not decide disputes between the Contractor and persons or entities other than the Owner.

The 2007 version of the AIA A201 allows for an "Initial Decision Maker" designated by the owner, which may not be the architect. The owner may select another person or entity as initial decision maker, but the architect will be the initial decision maker if no other person is selected. Section 15.2.1 of the AIA A201 (2007 ed.) provides in part:

> Claims, excluding those arising under Sections 10.3, 10.4, 11.3.9, and 11.3.10, shall be referred to the Initial Decision Maker for initial decision. The Architect will serve as the Initial Decision Maker, unless otherwise indicated in the Agreement. Except for those Claims excluded by this Section 15.2.1, an initial decision shall be required as a condition precedent to mediation of any Claim arising prior to the date final payment is due, unless 30 days have passed after the Claim has been referred to the Initial Decision Maker with no decision having been rendered. Unless the Initial Decision Maker and all affected parties agree, the Initial Decision Maker will not decide disputes between the Contractor and persons or entities other than the Owner.

ConsensusDOCS 200 goes further in removing the design professional from the dispute resolution process. Article 12, "Dispute Mitigation and Resolution," calls for the owner's and contractor's representatives initially to conduct direct discussions "in good faith" and, if those fail, to have the senior executives of the parties meet to try to resolve the dispute.[93] If those procedures fail, then the parties can elect to have their disputes submitted to a "project neutral" or a dispute review board.[94] If one of those procedures is not used or if it is used but does not resolve the dispute, the parties then can proceed to mediation and ultimately to arbitration or litigation.[95] There is no decision-making role for the design professional in the ConsensusDOCS 200 dispute procedure.

In contracts requiring disputes to be referred to the design professional for decision, questions may arise as to whether the design professional's decision is binding, and if not, at what point a party can initiate arbitration or litigation. Courts are reluctant to recognize final and binding power of a design professional to resolve disputes unless the contract clearly and unambiguously assigns that power.[96] This reluctance may be based on a feeling that design professionals may be partial to the owner, or perhaps because such recognition would result in the waiver of a party's right to its day in court or arbitration.

More difficult issues tend to arise with contracts that require claims to be submitted to the design professional for decision before a party commences litigation. How long must a contractor wait for a decision from the design professional before

[93]ConsensusDOCS 200, ¶ 12.2 (2007 ed.).

[94]*Id.* at § 12.3.

[95]*Id.* at §§ 12.3 through 12.5.

[96]*Beers Constr. Co. v. Pikeville United Methodist Hosp. of Ky., Inc.*, 129 Fed. Appx. 266 (6th Cir. 2005); *see also Global Constr., Inc. v. Mo. Highway & Transp. Comm'n*, 963 S.W.2d 340 (Mo. Ct. App. 1997); *Top Line Constr. Co. v. J.W. Cook & Sons, Inc.*, 455 S.E.2d 463 (N.C. Ct. App. 1995).

litigation may be commenced? What if the design professional repeatedly asks for additional documentation to support the claim, thus delaying a decision? Courts typically will not consider initial referral of a claim to the design professional to be a condition precedent to litigation unless the contract clearly states that it is.[97]

If the contract requires the design professional to render a decision within a certain period of time, the commencement of litigation would be appropriate after the time for a decision has elapsed. For example, Section 15.2.2 of AIA A201 (2007 ed.), provides that the Initial Decision Maker will within 10 days of receipt of a claim (1) request additional supporting data, (2) reject the claim in whole or part, (3) approve the claim, (4) suggest a compromise, or (5) advise the parties that the Initial Decision Maker cannot resolve the claim if it does not have sufficient information or if it would be inappropriate for the Initial Decision Maker to resolve the claim.

If no time is specified, a reasonable time often is assumed. Some courts have held that the design professional's role as arbitrator of disputes between owner and contractor ends once the design professional is no longer responsible for supervising the contractor's performance.[98]

When a design professional assumes the role of an impartial arbitrator and the contract language deems the decision of the design professional to be "final," some design professionals have enjoyed limited immunity from arbitration or litigation brought by owners and contractors resulting from the professional's decision[99] because of the performance of a quasi-judicial function.[100] Such immunity is limited to those acts performed in a "judicial" capacity. The crucial question in such situations is whether the challenged act was within the scope of the design professional's decision-making duties.[101]

For example, in *Blecick v. School District No. 18,* the architect was not liable for its refusal to issue a final certificate because that refusal was within the architect's capacity as arbitrator. As such, the refusal did not amount to a failure to exercise care in the issuance of a final certificate.[102] In a contrasting case, after the contractor pointed out defects in the plans and specifications, the design professional that prepared them allegedly became enraged, intentionally and maliciously tried to bankrupt the contractor, and attempted to interfere with the contract between the contractor and the owner. The contractor also claimed that the design professional issued contradictory instructions, changed the plans and specifications without

[97]*Shook of W. Va., Inc. v. York City Sewer Auth.,* 756 F. Supp. 848 (M.D. Pa. 1991); *see also Aberdeen Golf & Country Club v. Bliss Constr., Inc.,* 932 So. 2d 235 (Fla. Dist. Ct. App. 2005).

[98]*HIM Portland, LLC v. DeVito Builders, Inc.* 211 F. Supp. 2d 230 (D. Me. 2002).

[99]*Lane v. Geiger-Berger Assocs.,* 608 F.2d 1148 (8th Cir. 1979); *But see RPR & Assocs. v. O'Brien/Atkins Assocs., P.A.,* 24 F. Supp. 2d 515 (M.D.N.C. 1998) (architect denied immunity as arbitrator where claims were based on architect's negligent performance as an architect and project supervisor); *MCI Constr., LLC v. Hazen and Sawyer, P.C.,* 2003 WL 22061226 (M.D.N.C. 2003) (court denied arbitral immunity and stated that design professional could be deposed) (unpublished opinion).

[100]*Lundgren v. Freeman,* 307 F.2d 104, 117 (9th Cir. 1962); *see also Phoenix Servs. Ltd. P'ship v. Johns Hopkins Hosp.,* 892 A.2d 1185 (Md. Ct. Spec. App. 2006).

[101]*RPR & Assocs. v. O'Brien/Atkins Assocs., P.A.,* 24 F. Supp. 2d 515 (M.D.N.C. 1998); *HBS Contractors, Inc. v. Nat'l Fire Ins. Co. of Hartford,* 501 S.E.2d 372 (N.C. Ct. App. 1998).

[102]406 P.2d 750 (Ariz. Ct. App. 1965).

regard to added cost or delay, interfered with the contractor's coordination of its subcontractors, and improperly withheld progress payments. The court found that the design professional's claim of immunity based on its status as an arbitrator was an inadequate defense because so many of the challenged acts were clearly outside the architect's role as arbitrator.[103]

Distinguishing the design professional's protected functions from other duties can be difficult. The deliberate or negligent misuse of the design professional's powers as arbitrator may excuse a party's disregard of a design professional's decision. In one case, a state engineer's unreasonable behavior voided his authority to classify and quantify a contractor's excavation work.[104] The contractor had agreed to construct a bridge and reconstruct a section of state highway. All "unclassified excavation" was to be performed at one price while rock excavation was to be performed at five times that price. When the contractor unexpectedly struck rock and notified the state in accordance with the contract, the department's engineer, who was designated as the final arbiter of contract disputes, was noncommunicative and delayed visiting the job site to measure the extent of the rock encountered. Although the court recognized the engineer's authority under the contract to determine the quantity of rock, it found that the parties were not bound by an engineer's decision "manifestly arbitrary or rendered in bad faith." On the basis of the engineer's intentional failure to conduct a timely inspection, the court awarded the contractor damages for the uncompensated rock excavation. The engineer's failure to exercise its powers as arbitrator in good faith rendered the resulting decision effectively meaningless.[105]

V. OTHER DUTIES OF THE DESIGN PROFESSIONAL

The administrative duties of the design professional just discussed are found in many private construction contracts. They illustrate the design professional's basic responsibilities and obligations on a construction project. These general duties of the design professional often are supplemented by the inclusion in the contract documents of more specific duties, such as the following:

(1) If soil conditions are such that, in the Architect's opinion, yard hydrants cannot be secured, use bridles.

(2) Defects in the extension of existing gas lines must be repaired to the Architect's satisfaction.

(3) The entire jacking and shoring procedure is subject to the Architect's review.

[103] *Craviolini v. Scholer & Fuller Associated Architects,* 357 P.2d 611, 614 (Ariz. 1960); *see also Laurel Race Course, Inc. v. Regal Constr. Co.,* 333 A.2d 319 (Md. 1975) (design professional's immunity lost when design professional fails to use reasonable care and good faith when acting as an arbitrator subjecting professional to liability to owner or contractor).

[104] *Brezina Constr. Co. v. S.D. Dep't of Transp.,* 297 N.W.2d 168 (S.D. 1980).

[105] *Id.*

(4) Set manhole castings in fresh bed of mortar and carefully adjust elevation as directed by the Architect.

(5) Concrete surfaces shall be finished to a true and even plane well within limits of best trade practice and to the Engineer's satisfaction.

VI. THE DESIGN PROFESSIONAL'S LIABILITY TO THE CONTRACTOR

Under certain circumstances, the design professional's failure to perform properly its supervisory or administrative responsibilities may result in liability for resulting damages to the owner. These same acts or omissions may also justify an action by the contractor either against the owner or directly against the design professional.

Design professionals are increasingly being held directly liable to contractors under various legal theories, with the most prevalent involving the design professional's negligent breach of a duty owed to the owner or to the contractor.

A. Negligence and the "Economic Loss Rule"

The negligent breach of a duty by one person often injures another person. Such harm may include a bodily injury, an injury to property, or an injury to a financial or business interest of some type. The latter often is referred to as an "economic loss." It has long been established that a person may sue in negligence for personal injuries and property damage even where the person has no contractual relationship with the defendant.[106] Historically, courts have been reluctant to allow parties to recover in negligence for purely economic loss where there is no contractual privity between the parties. This limitation on negligence actions is called the "Economic Loss Rule."[107]

The Economic Loss Rule is still recognized in many states[108] and was unanimously upheld in 1986 by the United States Supreme Court in an admiralty case.[109] The rule is less rigidly applied in many jurisdictions, however, particularly regarding design professionals.[110] Third parties have been allowed to recover damages for

[106]*MacPherson v. Buick Motor Co.,* 111 N.E. 1050 (N.Y. 1916).

[107]*City Express, Inc. v. Express Partners,* 959 P.2d 836 (Haw. 1998) (holding that plaintiff may not recover purely economic losses against architect for professional negligence where plaintiff contracted with architect); *2314 Lincoln Park W. Condo. Ass'n v. Mann, Gin, Ebel & Frazier, Ltd.,* 555 N.E.2d 346 (Ill. 1990) (holding that economic loss rule bars cause of action in tort for professional malpractice against architect).

[108]*See, e.g., Franklin Grove Corp. v. Drexel,* 936 A.2d 1272 (R.I. 2007); *EBWS, LLC v. Britly Corp.,* 928 A.2d 497 (Vt. 2007); *City of Cairo v. Hightower Consulting Eng'rs, Inc.,* 629 S.E.2d 518 (Ga. Ct. App. 2006); *Fleischer v. Hellmuth, Obata & Kassabaum, Inc.,* 870 S.W.2d 832 (Mo. Ct. App. 1993); 2314 *Lincoln Park W. Condo. Ass'n v. Mann, Ginn, Ebel, & Frazier, Ltd.,* 555 N.E.2d 346 (Ill. 1990); *Floor Craft Floor Covering, Inc. v. Parma Cmty. Gen. Hosp. Ass'n,* 560 N.E.2d 206 (Ohio 1990).

[109]*E. River S.S. Corp. v. Transamerica Delaval, Inc.,* 476 U.S. 858 (1986).

[110]*See, e.g., Dufficy & Sons, Inc. v. BRW, Inc.,* 74 P.3d 380 (Colo. Ct. App. 2002); *Moransais v. Heathman,* 744 So. 2d 973 (Fla. 1999); *Prichard Bros., Inc. v. The Grady Co.,* 428 N.W.2d 391 (Minn. 1988); *Shoffner Indus., Inc. v. W. B. Lloyd Constr. Co.,* 257 S.E.2d 50 (N.C. Ct. App. 1979); *A. R. Moyer, Inc. v. Graham,* 285 So. 2d 397 (Fla. 1973).

a design professional's (1) negligent performance of duties;[111] 2) negligent failure to measure removal of rock and boulders;[112] (3) intentional and negligent preparation of defective plans and specifications;[113] (4) negligent performance of a water percolation test;[114] and (5) negligent evaluation of soil permeability.[115]

Some courts recognize a duty on the part of the design professional toward the contractor based on the design professional's inherent power over the contractor. For example, in *A.R. Moyer, Inc. v. Graham,*[116] both architect and engineer were held liable to the contractor for negligently issuing contrary instructions to the contractor and for failing to revise and coordinate modified corrective plans and specifications they had issued to the contractor. The court allowed the contractor to recover damages for delay directly from the architect, stating:

> Altogether too much control over the contractor necessarily rests in the hands of the supervising architect for him not to be placed under a duty imposed by law to perform without negligence his functions as they affect the contractor. The power of the architect to stop the work alone is tantamount to a power of economic life or death over the contractor.[117]

If a design professional has agreed to perform supervisory tasks on a construction project, the contractor on the project may have a right to rely on the competence of that supervision.[118] In *HBS Contractors, Inc. v. National Fire Insurance Corp. of Hartford,*[119] the contractor brought a claim against the architect for negligence and bad faith. The architect had agreed to provide architectural services and supervision at the project. The contractor claimed that the architect failed to properly design an erosion control plan, failed to grant an extension for time due to weather delays, and improperly nullified valid pay requests. The court found that the architect was negligent in its administration of the contract.[120] The court stated that an architect with general supervisory power that determines final compliance of a contract between the owner and contractor, and negligently performs a contractual duty, may be held

[111] *Ins. Co. of N. Am. v. Town of Manchester,* 17 F. Supp. 2d 81 (D. Conn. 1998); *Davidson & Jones, Inc. v. New Hanover County,* 255 S.E.2d 580 (N.C. Ct. App. 1979).

[112] *Forte Bros., Inc. v. Nat'l Amusements, Inc.* 525 A.2d 1301 (R.I. 1987); *COAC, Inc. v. Kennedy Eng'rs,* 136 Cal. Rptr. 890 (Cal. Ct. App. 1977).

[113] *McElwee Group LLC v. Mun. Auth. of Borough of Elverson,* 476 F. Supp. 2d 472 (E.D. Pa. 2007); *Dufficy & Sons, Inc. v. BRW, Inc.,* 74 P.3d 380 (Colo. Ct. App. 2002); *Bilt-Rite Contractors, Inc. v. The Architectural Studio,* 866 A.2d 270 (Pa. 2005).

[114] *S.E. Consultants, Inc. v. O'Pry,* 404 S.E.2d 299 (Ga. Ct. App. 1991).

[115] *City of Cairo v. Hightower Consulting Eng'rs, Inc.,* 629 S.E.2d 518 (Ga. App. 2006); *Robert & Co. v. Rhodes-Haverty P'ship,* 300 S.E.2d 503 (Ga. 1983).

[116] 285 So. 2d 397 (Fla. 1973).

[117] *A.R. Moyer, Inc. v. Graham,* 285 So. 2d 397, 401 (Fla. 1973) quoting *U.S. ex rel. L.A. Testing Lab. v. Rogers & Rogers,* 161 F. Supp. 132, 136 (S.D. Cal. 1958).

[118] *See Tommy L. Griffin Plumbing & Heating Co. v. Jordan, Jones & Goulding, Inc.,* 463 S.E.2d 85 (S.C. 1995); *Day v. Nat'l U.S. Radiator Corp.,* 128 So. 2d 660 (La. 1961) (dictum).

[119] 501 S.E.2d 372 (N.C. Ct. App. 1998).

[120] *Id.*

liable to the contractor.[121] The court recognized that each participant in a construction project must be able rely on the performance of others, thus obligating each to perform its duties with due care.

In *Davidson & Jones, Inc. v. New Hanover County,*[122] a general contractor submitted a bid partially based on a soil report furnished by the design professional on behalf of the city. The bid package contained language that stated that before submitting a proposal, the bidder shall:

(a) carefully examine the drawings and specifications;
(b) visit the site of the work and fully inform himself of existing conditions and limitations;
(c) rely entirely upon his own judgment in preparing his proposal, and include in his bid a sum sufficient to cover all items required by the contract.[123]

The general contractor engaged subcontractors to excavate and install foundation pilings. During construction, damage occurred to a city library on the adjacent site. The city declared the contractor at fault and required that the library be repaired. The contractor made the repair and then filed suit against the city and a third-party complaint against the design professionals. The court held that a contract provision in the agreement between the contractor and the owner stating that the contractor "was to rely entirely upon its own judgment in submitting its bid and was to conduct a site inspection to have a complete understanding of all existing conditions relating to the work" was not enough to shield the design professional from liability for negligently preparing the report.[124]

Some courts permitting recovery for economic losses by parties not in privity with one another under negligence theories have not distinguished between a design professional's duties that are "supervisory" and those that are "administrative." For instance, in *Dickerson Construction Co. v. Process Engineering Co.,*[125] the project architect's duties were defined by the contract as administrative. The court concluded that the use of the word "supervision" in the jury instructions did not connote a higher or lower standard of duty than the use of the word "inspection." The court further held that the basic question was whether the design professional had exercised reasonable care in making the observations and inspections at the project. In another case, the architect admitted during testimony that "supervision" and "administration" at the job site constitute the same set of obligations.[126]

Even in jurisdictions rejecting the Economic Loss Rule, design professionals have successfully defended against claims for economic damages by alleging contributory negligence by the contractor. A third party asserting the negligence claim against a

[121]*Id.* at 377 (dictum taken from *Shoffner Indus., Inc. v. W.B. Lloyd Constr. Co.,* 257 S.E.2d 50, 55 (N.C. Ct. App. 1979)).
[122]255 S.E.2d 580 (N.C. Ct. App. 1979).
[123]*Id.* at 581.
[124]*Id.* at 585.
[125]341 So. 2d 646 (Miss. 1977).
[126]*Kleb v. Wendling,* 385 N.E.2d 346, 349 (Ill. App. Ct. 1979).

design professional must be able to prove that the design professional's negligence proximately caused the injury or damage. For instance, a Florida court held that an architect was not liable to a subcontractor for negligently advising the owner to reject the subcontractor's request for a substitution. According to the court, the owner made the final decision on the subcontractor's request, and the architect therefore was not the "proximate cause" of the subcontractor's damages.[127]

In some cases, courts may not allow a contractor to sue the design professional if the contractor cannot prove its claim against the owner. For example, a contractor sued an owner for delay damages allegedly caused by the design professional's negligence. The contract between the contractor and the owner contained a "no-damage-for-delay" clause. The court stated that an independent contractor bears responsibility for its own actions absent any interference by owner or owner's representative. The contractor was unable to prove interference by the design professional and thus could not prevail on its claim against the owner.[128] A contractor cannot circumvent the restrictions of a no-damages-for-delay clause by suing a design professional for negligence where the no-damages-for-delay clause is broad enough to include claims against the design professional.[129]

In jurisdictions recognizing a contractor's right to sue the design professional in negligence for economic loss, contractors have a difficult burden to establish that the design professional is liable. In general, it is much easier to establish owner liability for defective design under contract and warranty theories than it is to prove professional negligence on the part of the design professional. Negligence is determined by evaluating the acts or omissions of the alleged wrongdoer according to a specific standard of care. Design professionals are required to exercise that degree of skill and diligence ordinarily exercised under like circumstances by design professionals in good standing in the same or similar communities.[130] This standard of care may vary according to local practice.[131] Under some circumstances and in some jurisdictions, however, the "locality rule" does not apply.[132] A design professional's deviation from established standards of care as announced in trade publications or industry manuals published by professional organizations such as the AIA, the National Society of Professional Engineers (NSPE), and the American Consulting Engineers Council

[127]*McElvy, Jennewein, Stefany, Howard, Inc. v. Arlington Elec., Inc.*, 582 So. 2d 47 (Fla. Dist. Ct. App. 1991). *See generally* Hubert J. Bell, Jr. & Luther P. House, Jr., Current Status of Economic Loss Rule: 1992 Wiley Construction Law Update 239 (Overton A. Currie & Neal J. Sweeney eds., John Wiley & Sons, Inc. 1992).

[128]*Guy M. Cooper v. E. Pa. Sch. Dist.*, 903 A.2d 608 (Pa. Commw. Ct. 2006).

[129]*Bates & Rogers Constr. Corp. v. Greeley & Hansen*, 486 N.E.2d 902 (Ill. 1985).

[130]*John Day Co. v. Alvine & Assocs., Inc.*, 510 N.W.2d 462 (Neb. Ct. App. 1993); *Overland Contractors, Inc. v. Millard Sch. Dist., No. 17, Douglas County*, 369 N.W.2d 69 (Neb. 1985).

[131]*See* James Acret & Annette Davis Perrochet, Architects and Engineers, Liability For Negligence: Standard of Practice § 1:3 (2008); *Handex of Carolinas, Inc. v. County of Haywood*, 607 S.E.2d 25 (N.C. Ct. App. 2005); *Hobson v. Waggoner Eng'g, Inc.*, 878 So. 2d 68 (Miss. Ct. App. 2003); *see also Collins Co. v. City of Decatur*, 533 So. 2d 1127 (Ala. 1988).

[132]*See, e.g., E. Steel Constructors, Inc. v. City of Salem*, 549 S.E.2d 266 (W. Va. 2001); *Georgetown Steel Corp. v. Union Carbide Corp.*, 806 F. Supp. 74, 78 (D. S.C. 1992); *McMillan v. Durant*, 439 S.E.2d 829 (S.C. 1993); *Doe v. Am. Red Cross Blood Servs.*, 377 S.E.2d 323 (S.C. 1989).

(ACEC) can be considered evidence of negligence. Such evidence is not conclusive, however, and proof of professional malpractice usually requires expert testimony by someone in the same field.[133] Moreover, some jurisdictions require parties suing a design professional for negligence to file with their complaints an affidavit from a professional in the same field substantiating the malpractice claim.[134]

B. Intentional Torts

Although less common than actions for negligence, lawsuits for intentional torts can also present problems for design professionals. The intentional tort theory is used much less frequently, however, because negligence theories now are more widely recognized as grounds for recovery of economic loss.

On occasion, lawsuits for an intentional tort against a design professional can be successful. In one case, a court held that an architect's unjustified and therefore arbitrary refusal to issue a certificate of performance subjected it to intentional tort liability.[135] In another case, an appellate court sent the matter back to the trial court for a determination of whether the architect was acting within its arbitral role or whether the architect had willfully and wrongfully interfered with a school construction contract by inducing the school district to default the contractor.[136] The court further held that the contractor could recover punitive damages in addition to compensatory damages if the contractor could demonstrate that malicious conduct on the part of the architect was unrelated to any arbitral contractual function.[137]

A design professional will not be held liable for tortious interference with contractual relations where the design professional has a bona fide interest in the project.[138] Furthermore, if the design professional's unfair recommendation to terminate a subcontract is not followed by a contractor, the design professional will not be liable to a subcontractor.[139]

C. Third-Party Beneficiary Theory

The contractor might also argue that it is a "third-party beneficiary" of the contract between the owner and the design professional. This theory, for the most part, occasionally has been successful in the construction context.

[133] *James Corp. v. N. Allegheny Sch. Dist.*, 938 A.2d 474 (Pa. Commw. Ct. 2007); *McCabe v. Schoenwald*, 166 P.3d 450 (Kan. Ct. App. 2007).

[134] *See, e.g.*, GA. CODE ANN. § 9–11–9.1 (Georgia statute requiring an affidavit by a licensed professional averring at least one act of professional malpractice). This affidavit requirement applies to individual design professionals as well as design professional business organizations. *See Sembler Atlanta Dev. I, LLC v. URS/Dames & Moore, Inc.*, 601 S.E.2d 397 (Ga. Ct. App. 2004).

[135] *Unity Sheet Metal Works, Inc. v. Farrell Lines, Inc.*, 101 N.Y.S.2d 1000 (N.Y. Sup. Ct. 1950).

[136] *Lundgren v. Freeman*, 307 F.2d 104, 119 (9th Cir. 1962).

[137] *See also Craviolini v. Scholer & Fuller Associated Architects*, 357 P.2d 611 (Ariz. 1960).

[138] *RPR Assocs. v. O'Brien/Atkins Assocs., P.A.*, 24 F. Supp. 2d 515 (M.D. N.C. 1998).

[139] *See John W. Johnson, Inc. v. Basic Constr. Co.*, 292 F. Supp. 300, 304 (D. D.C. 1968), *aff'd*, 429 F.2d 764 (D.C. Cir. 1970) (where unfair and unjust recommendation by an architect to terminate a subcontract was not heeded by the contractor, subcontractor's cause of action against architect was properly dismissed).

A "third-party beneficiary" is a party that is not one of the original contracting parties but that may bring an action to enforce the provisions of the contract.[140] To do so, however, the third party must have been an "intended beneficiary" of the parties to the contract. In other words, the parties to a contract must intend to confer a direct or immediate benefit on the third party that claims to be a beneficiary.[141]

In the absence of a written expression, that intent is difficult to prove because courts require a clear expression of intent to recognize a third-party beneficiary.[142] When a third party benefits by an agreement, but that benefit is not the specific object of the intended contractual obligation, the third party is referred to as an "incidental beneficiary." Incidental beneficiaries do not have a legal right to bring an action to enforce the provisions of the contract.[143]

As a practical matter, a contractor is more likely to succeed in a tort action than through a third-party beneficiary theory under the contract between the owner and the design professional. For example, in one case, a contractor claimed to be a third-party beneficiary to the design professional/owner contract.[144] The court found that a contractor could assert a claim in negligence against the architect but reasoned that generally the provisions of the owner-architect contract are for the benefit of the owner and the design professional, and the contractor is only an incidental beneficiary, not a third-party beneficiary. In a South Carolina case, the contractor sued a project engineer under both tort and third-party beneficiary theories. The contractor alleged that the engineer wrongfully stopped work based on false allegations, erroneously interpreted contract, and made disparaging remarks to contractor's bonding company. The court found the tort claims supportable but dismissed the contractor's third-party beneficiary claims.[145]

Courts typically determine whether a contractor is a third-party beneficiary by examining the apparent intent of the contracting parties.[146] If there is an intent to make a third party a beneficiary, some contracts will specifically state the intent of the parties, such as EJCDC C-700, Standard General Conditions of the Construction Contract (2007 ed.) [EJCDC C-700]. For example, in the context of describing certain insurance policies, this form contract states that:

[140]*See Keel v. Titan Constr. Corp.*, 639 P.2d 1228 (Okla. 1981).

[141]*Kane Enters. v. MacGregor (USA), Inc.*, 322 F.3d 371 (5th Cir. 2003).

[142]*VP Bldgs., Inc. v. No-Co Constr., Inc.*, 202 Fed. Appx. 733 (5th Cir. 2006); *Peter Kiewit Sons' Co. v. Iowa S. Utils. Co.*, 355 F. Supp. 376, 392–93 (S.D. Iowa 1973).

[143]*Richardson Assocs. v. Lincoln-Devore, Inc.* 806 P.2d 790 (Wyo. 1991); *Edward B. Fitzpatrick, Jr. Constr. Corp. v. County of Suffolk,* 525 N.Y.S.2d 863 (N.Y. App. Div. 1988); *Valley Landscape Co. v. Rolland,* 237 S.E.2d 120, 124 (Va. 1977); *Engle Acoustic & Tile, Inc. v. Grenfell,* 223 So. 2d 613, 620 (Miss. 1969).

[144]*A. R. Moyer, Inc. v. Graham,* 285 So. 2d 397, 402-03 (Fla. 1973); *see also E. Steel Constructors, Inc. v. City of Salem*, 549 S.E.2d 266 (W. Va. 2001) (owner was not a third-party beneficiary of architect-engineer contract but could assert claim against engineers based on negligence theory).

[145]*Tommy L. Griffin Plumbing & Heating Co. v. Jordan, Jones & Gouldings, Inc.*, 463 S.E.2d 85 (S.C. 1995).

[146]*See* 17A Am. Jur. 2d *Contracts* § 440; *see also Malta Constr. Co. v. Henningson, Durham & Richardson, Inc.*, 694 F. Supp. 902, 908 (N.D. Ga. 1988) (upholding third-party beneficiary theory by a general contractor against supplier of shop drawings under contract with a subcontractor); *Plourde Sand & Gravel v. JGE E., Inc.*, 917 A.2d 1250 (N.H. 2007); *Thompson v. Espey Huston & Assocs., Inc.*, 899 S.W.2d 415 (Tex. App. 1995).

> Owner and Contractor intend that all policies purchased . . . will protect Owner, Contractor, Subcontractors, and Engineer, and all other individuals or entities identified in the Supplementary Conditions as loss payees (and the officers, directors, members, partners, employees, agents, consultants, and subcontractors of each and any of them) in such policies and will provide primary coverage for all losses and damages caused by the perils or causes of loss covered thereby.[147]

But there may be circumstances in which the design professional assumes extra responsibilities to the contractor by its conduct. An example is where the architect has no contractual duty to the owner to make extensive and thorough on-site inspections but nevertheless does make such inspections.[148] Therefore, in some cases, damages that result from improper performance of assumed duties may be recovered through third-party beneficiary actions.

Furthermore, the New York Court of Appeals found in *Ossining Union Free School District v. Anderson LaRocca Anderson, et al.*,[149] that an owner could assert a claim against two engineering consultants who had been retained by the architect even though there was no direct contractual relationship between the engineers and the owner if the relationship between the engineers and the owner was so close as to approach that of privity. The court indicated that it was not enough that it was "foreseeable" that the owner might have relied on negligently prepared reports. Rather, the court found there was a bond "so close as to approach that of privity" (1) if there was awareness that the reports were to be used for a particular purpose or purposes, (2) there was reliance by one party or parties in furtherance of that purpose, and (3) some conduct by the defendant's engineers linked them to the owner and showed that the engineers should have understood that the owner would rely on the reports. The court found that the engineers knew that the reports were being prepared for the owner school district and would be relied on by the school district. It was clear that the engineers were retained to visit the school district's property, examine its buildings, and prepare reports and findings regarding those buildings. The court also noted that retention of the engineers was specifically authorized in the contract between the school district and the architect, and the engineers were informed that in seeking compensation, the architect had indicated that it was hired by the school district. The court found that the engineers' reports were prepared with the objective of shaping the owner's conduct, and, therefore, the engineers owed a duty of diligence to the school district.[150]

By carefully drafting the contract, a design professional can reduce or avoid third-party beneficiary claims. An example of this is found in Section 1.1.2 of AIA A201 (2007 ed.), which provides:

> The Contract Documents shall not be construed to create a contractual relationship of any kind (1) between the Contractor and the Architect or the Architect's

[147]EJCDC C-700, § 5.07 (2007 ed.).
[148]*Krieger v. J. E. Greiner Co.*, 382 A.2d 1069, 1081 (Md. 1978) (Levine, J., concurring).
[149]539 N.E.2d 91 (N.Y. 1989).
[150]*Id.*

consultants, (2) between the Owner and a Subcontractor or a Sub-subcontractor, (3) between the Owner and the Architect or the Architect's consultants or (4) between any persons or entities other than the Owner and the Contractor.

Likewise, EJCDC C-700, § 9.09 (A) (2007 ed.) states:

Neither Engineer's authority or responsibility . . . nor any decision made by Engineer in good faith either to exercise or not exercise such authority or responsibility or the undertaking, exercise, or performance of any authority or responsibility by Engineer shall create, impose, or give rise to any duty in contract, tort, or otherwise owed by Engineer to Contractor, any Subcontractor, any Supplier, any other individual or entity, or to any surety for or employee or agent of any of them.[151]

ConsensusDOCS 200, § 2.2, states:

This Agreement is solely for the benefit of the Parties, represents the entire and integrated agreement between the Parties, and supersedes all prior negotiations, representations or agreements, either written or oral. This Agreement and each and every provision is for the exclusive benefit of the Owner and Contractor and not for the benefit of any third party except to the extent expressly provided in this Agreement.

Such clauses are usually judicially enforced.[152] A court might be persuaded by a third-party beneficiary argument where it is needed to salvage a cause of action. In most cases, however, the same act that would constitute the basis for a third-party beneficiary action would also constitute a negligence action, which is a more likely basis for relief.[153]

D. Professional Liability Coverage

Often design firms are of relatively smaller size or financial strength than their clients and the contractors with whom they work. Consequently, contractors and third parties with large claims often obtain a more substantive recovery by suing the owner rather than the design professional. Professional liability insurance available to architects and engineers increases a party's prospects for recovery against a design professional while at the same time reducing the potential impact on the design professional of a large judgment in a negligence action.

[151]EJCDC C-700, § 9.09 A (2007 ed.); *see also* EJCDC E-505, § 6.07 C.1 (2004 ed.).

[152]*Watson, Watson, Rutland/Architects, Inc. v. Montgomery County Bd. of Educ.*, 559 So. 2d 168 (Ala. 1990); *Sheetz, Aiken & Aiken, Inc. v. Spann, Hall, Ritchie, Inc.*, 512 So. 2d 99 (Ala. 1987); *Mich. Abrasive Co. v. Poole*, 805 F.2d 1001 (11th Cir. 1986).

[153]*Richardson Assocs. v. Lincoln–Devore, Inc.*, 806 P.2d 790 (Wyo. 1991); *Edward B. Fitzpatrick, Jr. Constr. Corp. v. County of Suffolk*, 525 N.Y.S.2d 863 (N.Y. App. Div. 1988).

Every professional liability insurance policy contains language that limits or defines the scope of coverage. Questions often arise involving interpretation of the scope of coverage to determine whether the design professional is protected from the particular claims alleged.[154] Professional liability insurance policies are very specific as to the negligent acts that are covered. For instance, if a design professional performs services outside its covered range of services, it may well risk exposure to a negligence claim arising out of those services that would not be covered under the typical malpractice liability policy.[155] Also, some design professional liability policies expressly exclude from coverage certain acts that are not considered "insurable risks" because they involve conscious and deliberate conduct on the part of the professional.[156] Such exclusions commonly include a design professional's failure to complete drawings or specifications on time or a professional's commission of intentional torts. Because professional liability insurance coverage is so limited, professional liability insurance policies often exclude coverage of claims for a design professional's general negligence in the performance of its duties.[157]

VII. STATUTES OF REPOSE

Often injury and resulting liability relating to construction projects do not surface until many years after the work has been completed and accepted. With the increased scope of liability for design professionals and the prospect of perpetual liability, courts and legislatures have set statutes of limitations and statutes of repose to bar claims for damages against a design professional after a set time period.

Statutes of limitations establish the time period within which a suit can be filed upon the discovery of the act or omission giving rise to the claim. (See **Chapter 6**.) Conversely, statutes of repose establish an outer time limit beyond which the design professional cannot be held liable for design and construction defects after the completion or substantial completion of a project.

A party that discovers its claim after the expiration period provided in the statute of repose is barred from asserting that claim. The language of statutes of repose varies from state to state. Often these statutes are broadly written, so as to bar claims against "any person performing or furnishing the design, planning, supervision, or observation of construction or the construction or repair of the improvement" that are brought after the specified time.[158] Therefore, a contractor suing a design professional for damages sustained in connection with the project must institute an action that complies with the applicable statute of repose. Most, if not all, statutes of repose apply to tort claims, such as those brought under negligence or intentional

[154]*Stine v. Cont'l Gas Co.*, 349 N.W.2d 127 (Mich. 1984).
[155]*Bell Lavalin, Inc. v. Simcoe & Erie Gen. Ins. Co.*, 61 F.3d 742 (9th Cir. 1995).
[156]*Id.*
[157]*Doucet v. Huffine Roofing & Constr.*, 841 So. 2d 916 (5th Cir. 2003); *Westfield Ins. Co. v. Weis Builders, Inc.*, 2004 WL 1630871 (D. Minn. 2004).
[158]*See, e.g.*, ARK. CODE ANN. § 16–56–112 (West 2007).

tort theories. Many of the statutes also time-bar contract claims. Thus a party suing a design professional on a third-party beneficiary theory must pay close attention to the applicable statute of repose.

Statutes of repose usually take one of two forms. Most set a maximum time after completion of the construction, beyond which no claims may be brought relating to the project's design or construction.[159] The second type of statute also sets a maximum time after project completion for bringing a claim. This latter variety of statutes, however, may also specify a shorter time period after the cause of action accrues within which such action must be brought.[160] Statutes of repose will specify when the statute begins to run. Depending on the jurisdiction, the statute may begin to run upon completion of the project, substantial completion of the project, termination of the design of the improvement, or termination of the construction services.[161] Furthermore, a few statutes can be triggered by other events, such as the date of actual possession by the owner of the improvement, the date of the issuance of a certificate of occupancy, or the date of completion or termination of the contract between the design professional or the contractor and the owner.[162] Unless a statute is triggered by the issuance of a particular document, such as a certificate of substantial completion or a certificate of occupancy, judicial interpretation of the facts and the statute may ultimately be required.

VIII. EFFECTS OF CONTRACTUAL LIMITATIONS ON DESIGN PROFESSIONAL LIABILITY

Design professionals may try to limit their liability through contractual terms. For example, the AIA has published Document B503–2007, Guide for Amendments to AIA Owner—Architect Agreements. Paragraph 15 of AIA B503 permits two options for limiting liability. One option is to limit liability to the amount of the architect's contract compensation or to a specified dollar amount. That provision states:

> Neither the Architect, Architect's consultants, nor their agents or employees shall be jointly, severally or individually liable to the Owner in excess of the compensation to be paid pursuant to this Agreement or _______ Dollars ($_______), whichever is greater, by any reason of any act or omission, including breach of contract or negligence not amounting to a willful or intentional wrong.

[159] *See, e.g.,* Mo. Ann. Stat. § 516.097 (West 2007).

[160] *See, e.g.,* Mass. Gen. Laws Ann. ch. 260, § 2B (West 2007).

[161] *See, e.g.,* Fla. Stat. Ann. § 95.11(3)(c) (West 2007); Ga. Code Ann. § 9–3–51(a) (Michie Supp. 1998) (eight years after substantial completion); Mo. Ann. Stat. § 516.097 (Vernon Supp. 2007) (10 years after improvement complete); Mont. Code Ann. § 27–2–208 (2007) (10 years after completion of improvement to real estate); N.J. Stat. Ann. § 2A: 14–1.1 (West 2007) (10 years after substantial completion).

[162] *See, e.g.,* Fla. Stat. Ann. § 95.11(3)(c) (West 2007); Colo. Rev. Stat. § 13–80–104 (2007).

Some courts have accepted such limitations.[163] Courts in other states have taken a dim view of such limitations. In *Moransais v. Heathman,*[164] the Florida Supreme Court indirectly addressed the issue of whether a design professional can contractually limit its liability through contract. The case involved a homeowner who had contracted with an engineering firm for design services. Apparently, because the contract with the firm contained a very strict limitation of professional liability, the homeowner brought the negligence action against the individual professionals employed by the firm that worked on the project. The Florida Supreme Court held that professionals owe a duty to those who use their services that is independent of any obligation owed under a contract.[165] As the court stated:

> [T]he mere existence of such a contract should not serve per se to bar an action for professional malpractice. Further, the mere existence of a contract between the professional services corporation and a consumer does not eliminate the professional obligation of the professional who actually renders the service to the consumer or the common law action that a consumer may have against the professional provider. While the parties to a contract to provide a product may be able to protect themselves through contractual remedies, we do not believe the same may be necessarily true when professional services are sought and provided. Indeed, it is questionable whether a professional, such as a lawyer, could legally or ethically limit a client's remedies by contract in the same way that a manufacturer could do with a purchaser in a purely commercial setting. In any case, we conclude that the principles underlying the economic loss rule are insufficient to preclude an action for professional malpractice under the circumstances presented here.[166]

If a state has enacted legislation barring broad form indemnity clauses in contracts relating to construction, that legislation may provide a basis to challenge a broadly worded limitation of liability clause. For example, Georgia's anti-indemnity statute, found at GA. CODE ANN. § 13-8-2, provides as follows:

> A covenant, promise, agreement, or understanding in or in connection with or collateral to a contract or agreement relative to the construction, alteration, repair, or maintenance of a building structure, appurtenances, and appliances, including moving, demolition, and excavating connected therewith, purporting to indemnify or hold harmless the promisee against liability for damages arising out of bodily injury to persons or damage to property caused by or resulting from the sole negligence of the promisee, his agents or employees, or indemnitee is against public policy and is void and unenforceable, provided

[163] *1800 Ocotillo, LLC v. WLB Group, Inc.,* 176 P.3d 33 (Ariz. Ct. App. 2008); *Blaylock Grading Co., LLP v. Smith,* 658 S.E.2d 680 (N.C. CT. APP. 2008); *Markborough Cal., Inc. v. Superior Court,* 277 Cal. Rptr. 919 (Cal. Ct. App. 1991).
[164] 744 So. 2d 973 (Fla. 1999).
[165] *Id.*
[166] *Id.* at 983.

that this subsection shall not affect the validity of an insurance contract, workers' compensation, or agreement issued by an admitted insurer.

In *Lanier at McEver, L.P. v. Planners & Engineers Collaborative, Inc.*, the Georgia Supreme Court applied this statute to determine the validity of a limitation of liability clause contained in a contract between a developer and an engineer related to the design of a storm-water drainage system for an apartment complex. The limitation of liability clause at issue provided as follows:

> In recognition of the relative risks and benefits of the project both to [Lanier] and [PEC], the risks have been allocated such that [Lanier] agrees, to the fullest extent permitted by law, to limit the liability of [PEC] and its sub-consultants to [Lanier] and to all construction contractors and subcontractors on the project or any third parties for any and all claims, losses, costs, damages of any nature whatsoever [,] or claims expenses from any cause or causes, including attorneys' fees and costs and expert witness fees and costs, so that the total aggregate liability of PEC and its subconsultants to all those named shall not exceed PEC's total fee for services rendered on this project. It is intended that this limitation apply to any and all liability or cause of action however alleged or arising, unless otherwise prohibited by law.[167]

Distinguishing discussion in other jurisdictions, the Georgia Supreme Court ruled that the language extending the limitation of liability to "any and all liability or cause of action" and to "any third parties to any and all claims" violated the intent of Georgia's anti-indemnity statute and rendered the contractual limitation of liability at issue in *McEver* unenforceable.

IX. ASSUMPTION OF DESIGN LIABILITY BY THE CONTRACTOR

Traditionally, the design professional is responsible for defects and other problems associated with the project design. There are several ways, however, that a contractor may assume design liability. When this happens, a contractor may be required to bear not only the costs it incurred because of the defect, but also those extra costs incurred by the owner or other parties related to the defect.

The most common situation where a contractor assumes the risks of design defects is on design-build projects. A design-build contractor performs tasks ordinarily assigned to the design professional in addition to its construction duties.[168] Some owners are attracted to design-build contracts because they ostensibly create a single source of responsibility. These contracts also relieve the owner from being caught between its design professional and its contractor with respect to design disputes.[169]

[167] 663 S.E.2d 240, 241-2 (Ga. 2008).

[168] See **Chapter 2** for a more detailed discussion of design-build projects.

[169] *Mobile Hous. Env'ts v. Barton & Barton,* 432 F. Supp. 1343 (D. Colo. 1977) ("turn-key" contractor and replacement both responsible for design liability).

In addition to assuming the risks of defects in design, the design-build contractor also assumes the risk that construction will cost more than originally anticipated. As long as the performance criteria provided by the owner are not impracticable, a design-build contractor may be forced to bear the extra cost of its performance due to extended construction time.[170]

Owners are not completely immunized from design liability by entering into a design-build contract. An owner may be subject to liability for design defects, depending on the level of the owner's technical expertise, the owner's active direction of the design process, or the information provided by the owner at the project's outset. Because the design-build project delivery regime deviates from the traditional allocation of design risks, owners, contractors, and design professionals must scrutinize the contract language carefully.

A contractor performing under the traditional design-bid-build delivery scheme may assume a degree of design liability by the owner's use of performance specifications or criteria. Although performance specifications dictate the results to be achieved by the contractor, they do not tell the contractor how to accomplish the desired results. Consequently, the contractor is responsible for the costs associated with achieving the end result specified. Section 3.12.10 of the AIA A201 also invites the design professional to allocate design responsibility to the contractor. The implications of such allocation could prove problematic if they are deemed as an attempt to shift professional responsibilities in violation of state professional licensing and registration laws.

Standard-form documents, however, allow for such delegation. Thus, a contractor must take great care to (1) competently perform any design tasks assumed; (2) account for the added costs and exposure related to assumption and performance of professional duties; and (3) make sure that the risks represented by such exposure are insured or otherwise managed.

Conflicts may exist between design and performance specifications. If the contractor recognizes (or should recognize) such a conflict among the performance specifications, the contractor must notify the owner and the design professional to avoid potential liability due to the conflict.

For example, in *Regan Construction Co. & Nager Electric Co.,*[171] performance criteria provided by the owner for air-handling units could be met only by one particular unit on the market, which the contractor incorporated into the construction. The unit, however, did not fit into the space allotted for it with the design. As a result, the contractor incurred extra costs associated with accommodating the unit. In an action against the owner, the board of contract appeals held that the contractor assumed the risk of extra costs, reasoning that the contractor could have calculated that the specified unit would not fit in the space provided based on the contract drawings before it ordered the air-handling unit.[172]

[170] *Ruscon Constr. Co.,* ASBCA No. 39586, 90–2 BCA ¶ 22,768.

[171] PSBCA No. 633, 80–2 BCA ¶ 14,802.

[172] *Id. See also D.C. McClain, Inc v. Arlington County,* 452 S.E.2d 659 (Va. 1995) (contract documents not allowing adequate room for on-site installation of post-tensioning apparatus did not relieve contractor of obligation to obtain easements necessary to install apparatus).

Another specification issue that may give rise to contractor design liability involves the use of a "brand name or equal" specification by the owner. Use of the specified brand-name product by a contractor does not necessarily or automatically relieve the contractor of liability for noncompliance with applicable performance criteria. In other words, a design defect that results from a contractor's use of a brand name as specified in a "brand name or equal" specification does not necessarily fall within the owner's implied warranty of the adequacy of the specifications. If the contractor has a choice of products, the contractor bears the burden of compliance.[173]

As discussed above, one of the common duties of a design professional is to review and approve contractor submittals, including shop drawings.[174] If shop drawings prepared by the contractor require a change to the original design, however, the contractor may be held responsible for impacts to the overall construction caused by problems with that change even though the change has been approved by the design professional.[175]

Where the contract requires the contractor to determine field conditions, the contractor may be liable for defective submittals or shop drawings that fail to comply with the actual field conditions even if the submittals are approved by the owner or architect. For example, where the contract required the contractor to determine the field elevations, and the drawings contained no representations regarding the elevations, the contractor was still liable for shop drawings that would not work with the actual field conditions, despite the fact that the owner had approved the shop drawings.[176] The court reasoned that the contractor had the responsibility to obtain the information on which the shop drawings were to be based and the contractor's failure to obtain the proper field of measurements meant the contractor was responsible for deficiencies in the shop drawings despite the owner's approval.[177] Similarly, the contractor can be responsible for deficiencies in the shop drawings that result from the lack of field verification of measurements where the contract provides that (a) the contractor is responsible for determining dimensions and that (b) the owner's/design professional's review of shop drawings is only for conformance with the design concept of the project and compliance with the information given in the contract documents.[178] Still, the contractor may be able to reduce its risk for this type of design liability by immediately bringing any shop drawing–related design modifications to the attention of both the owner and the design professional.

Sometimes the limits of risk shifting related to substitutions are established by statute so that an architect's approval of the contractor's suggested substitution may not relieve the contractor of some warranty obligations. In *Leisure Resorts, Inc. v. Frank J. Rooney, Inc.*,[179] a contractor suggested the substitution of an air-conditioning unit on a condominium project. The project architect and engineer both approved

[173]*Martin K. Eby Constr. Co., Inc. v. Jacksonville Transp. Auth.*, 436 F. Supp. 2d 1276 (M.D. Fla. 2005).

[174]*See, e.g.*, EJCDC C-700, § 6.17 (2007 ed.).

[175]*See generally Fauss Constr., Inc. v. City of Hooper*, 249 N.W.2d 478, 481 (Neb. 1977).

[176]*McDevitt Mech. Contractors, Inc. v. United States*, 21 Cl. Ct. 616 (1990).

[177]*Id.*

[178]*Joseph Co.*, ENGBCA No. 5887, 92–3 BCA ¶ 25,075.

[179]654 So. 2d 911 (Fla. 1995).

the substitution. When many of the units failed to perform, several condominium unit owners filed a class-action lawsuit against the developer, which, in turn, sought indemnity from the contractor in a third-party action. The court held that while developers are subject to statutory warranties of merchantability or fitness for the purposes or uses intended, the contractor's statutory warranty is only to provide work and materials that "conform to the generally accepted standards of workmanship and performance of similar work."[180] In short, although the contractor had a duty to provide acceptable materials, it did not have a duty to evaluate broader issues of suitability of those materials for the purpose intended. Presumably, such duty would fall to the developer and the developer's architect or engineer.

Finally, contractors should be wary of standard contract clauses that expose the contractor to potential liability for design gaps or omissions. Many contracts contain catchall clauses that, for example, could require the contractor to supply any and all labor and materials that can be reasonably inferred from the plans and specifications as being necessary to achieve a complete project. If the contractor fails to include such labor or materials in its proposal or bid, the contractor may be required to fill the "gap" and furnish the missing scope of work with no increase in the contract price.

X. RECENT STANDARD-FORM CONTRACT EFFORTS REGARDING SHARED RESPONSIBILITY AND RISK

The tension regarding the responsibilities of the design professional, the owner, and the contractor for project design and administration has led some industry groups to propose triparty agreements to encourage collaboration and minimize risks and disputes. In 2007, a consortium of contractor, owner, subcontractor, and surety groups issued ConsensusDOCS 300, Standard Form of Tri-Party Agreement for Collaborate Project Delivery. At the heart of this document is the concept of a collaborative relationship among the project design team, the owner, and the contractor, all party to one contract. Section 3.4 of ConsensusDOCS 300 provides:

> The Parties each accept the relationship of mutual trust, good faith, and fair dealing established by this Agreement and covenants with each other to cooperate and exercise their skill and judgment in furthering the interests of the Project. The Designer and Constructor each represents that it possesses the requisite skill, expertise, and, as applicable, licensing to perform the required services. The Owner, Constructor, Designer and all members of the CPD [Collaborative Project Delivery] Team agree to adhere to principals of collaboration based on mutual trust, confidence, good faith and fair dealing. Within the scope of their respective expertise, the Parties shall together actively and continually pursue collaboration in the best interests of the Project. The Parties shall endeavor to promote harmony and collaboration among all Project participants.

[180] *Id.* at 914.

ConsensusDOCS 300 continues at Section 3.6 regarding the designer's responsibilities:

> The Designer shall furnish or provide all of the design and engineering services necessary to design the Project in accordance with the Owner's objectives, as outlined in the Owner's Program and other relevant information defining the Project. Consistent with the collaborative approach set forth in this Agreement, the Designer shall draw upon the assistance of the Constructor and others in developing the Project design, but the Designer shall retain overall responsibility for all design decisions as required by applicable state laws. Cost and schedule are design criteria and the Designer, in collaboration with the CPD Team, shall ensure that the design fully considers cost and schedule implications. The Designer represents that it is an independent contractor and that in its performance of the Services it shall act as an independent contractor. The Designer's duties, responsibilities and limitations of authority shall not be restricted, modified or extended without written consent of the Management Group.

ConsensusDOCS 300 further attempts to assign risk in line with the goal of collaboration. Section 3.8.2 provides that for project risks arising from collaborative decisions mutually agreed on by the parties, the parties agree to release each other from liability from any nonnegligent act, omission, mistake, or error in judgment (regardless of negligence), acting in good faith in performing their obligations under the agreement except to the extent the act of omission constitutes a willful default of a duty under the agreement. Furthermore, the agreement provides for possible limitations on the designer's liability. Section 3.8.2.2.(a) provides:

> To the fullest extent permitted by law, and notwithstanding any other provision of this Agreement, the total liability, in the aggregate, of Designer to Owner or Constructor or anyone claiming by, through or under the Owner or Constructor, for any and all claims, losses, costs or damages of whatsoever kind arising out of, resulting from or in any way related to the Project or the Agreement from any cause or causes including but not limited to the negligence or breach of contract of Designer shall not exceed an amount equal to ($_____), unless such claims, losses, costs or damages are reimbursed pursuant to a policy of professional liability insurance maintained by Designer or Designer's Consultants.

The triparty agreement further provides at Section 3.8.3 that the parties mutually waive consequential damages against each other.

This collaborative process encouraged by ConsensusDOCS 300 is a result, in part, of contracts used by groups such as Sutter Health based in Sacramento, California.[181] Sutter Health, a system of nonprofit hospitals and doctors' groups in Northern California, has created multiparty contracts involving the owner, designer,

[181] Nadine M. Post, *A New Process Paradigm*, Engineering News Record 80 (Nov. 26, 2007).

and contractor to encourage cooperation and minimize risks and disputes.[182] The collaborative approach used by Sutter Health has also been used by the Lean Construction Institute (LCI).[183] LCI has a "relational contract" (also known as an integrated form of agreement, or IFOA).[184]

The collaborative process emphasized by ConsensusDOCS 300 has apparently been successful on multiple projects. The growth of standard-form documents may encourage a trend toward collaboration among all contracting parties through the contract terms to minimize risks and liabilities to design professionals as well as contractors and owners. Only time will tell whether this process becomes more prevalent and actually minimizes risks and liabilities of design professionals as well as other parties to the construction project.

XI. THE DESIGN PROFESSIONAL'S COPYRIGHT FOR DESIGN DOCUMENTS

Increasingly, design professionals are taking legal action to protect their proprietary designs from unauthorized use. Recognizing this reality, industry standard forms seek to address the parties' respective rights. For example, the ConsensusDOCS 240 provides some flexibility with regard to the ownership of project documents. Paragraph 10.1 provides that:

> OWNERSHIP OF TANGIBLE DOCUMENTS The Owner shall receive ownership of the property rights, except for copyrights, of all documents, drawings, specifications, electronic data and information (hereinafter "Documents") prepared, provided or procured by the Architect/Engineer or by consultants retained by the Architect/Engineer and distributed to the Owner for this Project, upon the making of final payment to the Architect/Engineer or in the event of termination under Article 8, upon payment for all sums due to Architect/Engineer pursuant to Paragraphs 8.1 and 8.2.
>
> 10.1.1 COPYRIGHT The Parties agree that Owner ______ shall/ ______shall not (indicate one) obtain ownership of the copyright of all Documents. The Owner's acquisition of the copyright for all Documents shall be subject to the making of payments as required by Paragraph 10.1 and the payment of the fee reflecting the agreed value of the copyright set forth below:
>
> If the Parties have not made a selection to transfer copyright interests in the Documents, the copyright shall remain with the Architect/Engineer.

ConsensusDOCS 240 further provides, at Section 10.1.3, that after completion of the project, the owner may reproduce, reuse or make derivative works from the architect's documents solely for the purpose of maintaining, renovating, remodeling

[182] *Id.*
[183] Nadine M. Post, *Relational Contract Already Used as a Model*, ENGINEERING NEWS RECORD 47 (Jan. 7/14, 2008).
[184] *Id.*

or expanding the project at the work site. This language provides the owner certain rights even if the book copyright interest has not been transferred to the owner.

In contrast, other contracts provide that all copyright rights remain with the design professional. The EJCDC specifically grants the design professional ownership and property interest in all of the documents related to a particular project, regardless of whether the project is completed.[185] The AIA A201 (2007 ed.) more specifically provides that the architect retains control of the copyrights for its design documents, stating:

> The Architect and the Architect's consultants shall be deemed the authors and owners of their respective Instruments of Service, including the Drawings and Specifications, and will retain all common law, statutory and other reserved rights, including copyrights. The Contractor, Subcontractors, Sub-subcontractors, and material or equipment suppliers shall not own or claim a copyright in the Instruments of Service. Submittal or distribution to meet official regulatory requirements or for other purposes in connection with this Project is not to be construed as publication in derogation of the Architect's or Architect's consultants' reserved rights.
>
> The Contractor, Subcontractors, Sub-subcontractors and material or equipment suppliers are authorized to use and reproduce the Instruments of Service provided to them solely and exclusively for execution of the Work. All copies made under this authorization shall bear the copyright notice, if any, shown on the Instruments of Service. The Contractor, Subcontractors, Sub-subcontractors, and material or equipment suppliers may not use the Instruments of Service on other projects or for additions to this Project outside the scope of the Work without the specific written consent of the Owner, Architect and the Architect's consultants.[186]

Under federal copyright laws,[187] design professionals' copyrights can apply to drawings, specifications, and nearly even the buildings themselves. The damages available for infringements of such copyrights can be substantial. They can include an award of all the profits of the infringer. Under federal law, the design professional generally is required only to show the accused infringer's gross revenues. The burden then shifts to the accused infringer to prove costs unrelated to the infringement.[188]

Furthermore, damages can be assessed at many times the gross fees or revenues realized by a particular infringer. In *Johnson v. Jones,*[189] an architect was awarded damages of over $107,000 against a replacement architect and a contractor who

[185]EJCDC E-505, § 6.03.A (2004 ed.).
[186]AIA A201, § 1.5.1 (2007 ed.). *See also* AIA C106—Digital Data Licensing Agreement (2007 ed.) (AIA licensing agreement that can be incorporated into contracts to address the use and transmission of information, communications, drawings, or designs created or stored for the project in digital form).
[187]*See, e.g.,* 17 U.S.C. § 101 *et seq.*
[188]17 U.S.C. § 504(b).
[189]149 F.3d 494 (6th Cir. 1998).

copied the original architect's drawings even though the original architect charged only approximately $16,000 for the original house design. The contractor had given the original drawings to the replacement architect, who used those plans despite not having the copyright or license from the original architect.

In *Shine v. Childs,*[190] a federal court in New York found that the architect and his firm that had designed the "Freedom Tower" (proposed to be built where the World Trade Center Towers had stood) may be liable to a former architecture student, Shine. While a student, Shine, had prepared two similar designs, one of which was called the "Olympic Tower." Shine had proposed his designs for a Yale University class on skyscrapers in 1999. Shine's final review jury at Yale included Childs, a consulting partner at Skidmore, Owings & Merrill. When the "Freedom Tower" design was unveiled in 2003, Shine believed there were similarities in the unique design features of his "Olympic Tower." Shine registered his designs in 2004 with the U.S. Copyright Office and, shortly thereafter, sued Childs for copyright infringement. The court would not dismiss the lawsuit, finding that, to succeed, Shine needed to show ownership of a valid copyright and copying of the original components of Shine's work. The court found the registration of the copyright was strong evidence of a valid intellectual property right. The court then found that to prove copying, Shine had to show that Childs had access to Shine's design and that there were substantial similarities between the two designs. Childs had acknowledged that he had seen Shine's design so the only issue remaining was whether the designs were substantially similar.[191]

Determining the true owner of intellectual property rights can be extremely complicated. Even if an architect agrees to transfer the copyright for its drawings or to grant a license to its design to an owner, that transfer or license may not be effective as to the copyrights of consulting engineers to drawings or specifications prepared by the engineers. For example, in *Guillot-Vogt Associates, Inc. v. Holly & Smith,*[192] a contract between an owner and its architect providing that the architect conveyed to the owner copyrights related to contract documents was ineffective to transfer copyrights of consulting engineers to the documents they prepared. Even though a state law appeared to create the conveyance to the owner of the design on state projects by operation of law, the court found that the statute and the owner-architect contract were ineffective and overruled by federal law (17 U.S.C. § 201(e)) to allow an involuntary transfer of an exclusive copyright.

In some circumstances, copyrighted architectural works may be used or reused without the copyright owner's explicit permission.[193] Some states have enacted statutes allowing public agencies to reuse the plans they purchase for one project on later projects.[194] Copyright law recognizes a "fair use" exception regarding copyrighted

[190]382 F. Supp. 2d 602 (S.D.N.Y. 2005).

[191]*Id.*

[192]848 F. Supp. 682 (E.D. La. 1994).

[193]*See Jeffrey A. Grusenmeyer & Assocs., Inc. v. Davison, Smith & Certo Architects, Inc.*, 212 Fed. Appx. 510 (6th Cir. 2007).

[194]*See, e.g.*, FLA. STAT. ANN. § 235.211 (regarding school plans); LA. REV. STAT. ANN. § 38: 2317 (regarding state projects in general); N.M. STAT. ANN. § 13–1–123 (permitting agency to reuse plans if plan preparer is held harmless for future use by the agency.

materials. Under this doctrine, copyrighted materials may be reproduced and distributed for limited purposes, such as news reporting, teaching, scholarship, or research.[195] Factors for determining whether the "fair use" exception applies include: (a) whether the materials are being used for profit or nonprofit purposes; (b) the nature of the copyrighted materials; (c) the extent of the copyrighted work used in relation to the whole work; and (d) the effect on the value or potential market for the copyrighted material.[196] The fair use doctrine generally does not apply to the unauthorized reproduction of copyrighted materials on the construction project. Therefore, a party wishing to use copyrighted plans and specifications still will need to obtain permission to reproduce the copyrighted documents which may be permitted through the project contracts.[197]

➢ POINTS TO REMEMBER

- Parties can avoid many construction contract disputes by understanding the authority, responsibility, and duties of the project's design professional.
- Owners, design professionals, and contractors need to review carefully their contract documents to ensure that the authority, responsibilities, and standard of care of the design professionals are clearly stated.
- In any given case, a design professional may have authority to do certain things on the job site because:
  - It is expressly stated in the contract documents (actual authority).
  - It is incidental or necessary to the exercise of the design professional's actual authority (implied authority).
  - The acts of the owner would lead a reasonable person to believe the design professional had such authority (apparent authority).
- During construction, the project design professional is generally responsible for:
  - The initial interpretation of the plans and specifications.
  - Reviewing and approving shop drawings and other submittals.
  - Inspection.
  - Reviewing the contractor's pay requests and certifying progress and completion.
  - Arbitrating certain disputes between the owner and the contractor.
- On most jobs, the design professional does not have authority to:
  - Make changes to the work.
  - Direct the operations of the contractor.
  - Deviate from the requirements of the contract.

[195] 17 U.S.C. § 107.
[196] *Marcus v. Rowley*, 695 F.2d 1171 (9th Cir. 1983).
[197] *See, e.g.*, ConsensusDOCS 240, ¶ 10.1.1 (2007 ed.).

- Some contracts, such as the AIA standard-form contract documents, may emphasize the design professional's role as "administrative" and otherwise seek to insulate the design professional from potential liability to the contractor or the owner for interpreting, reviewing, inspecting, approving, and performing other functions set out in the contract documents.
- The design professional may be liable in some states to the contractor for economic losses suffered due to negligent performance of the design professional's duties and responsibilities, even though there is no contract between the design professional and the contractor. The applicable law on this issue must be reviewed in each case to determine if a cause of action is available and, if so, what damages are recoverable.
- The contractor may assume design liability in certain situations. To avoid such liability, the contractor should always notify the owner and the design professional about inconsistencies between the design and the performance specifications as well as any changes to the design resulting from the shop drawings and submittals.
- The design professional generally will control the copyright to its design work, particularly the unique components of its design, and anyone who wants to use that design will need to obtain legitimate control of the copyright or a license from the design professional. Otherwise, the user may find itself liable to the design professional for substantial damages for copyright infringement.

8

SUBCONTRACT ADMINISTRATION AND DISPUTE AVOIDANCE

Successful project performance depends on the legal and business relationship among the prime contractor and subcontractors and suppliers as much as any other factor. Although the prime contractor is responsible for the satisfactory and timely completion of the project, much of the work is performed by subcontractors and suppliers.

A subcontractor's unsatisfactory performance often generates disputes among the prime contractor, that subcontractor, and the owner, and possibly other subcontractors. Disputes also arise between the prime contractor and the subcontractor from problems for which the owner may be responsible. Similarly, whenever the prime contractor fails to perform, everyone involved with the project is affected. These disputes spawn lawsuits or arbitration. Unfortunately, formal dispute resolution, such as litigation or arbitration, is costly, time consuming, and uncertain, and rarely solves the problem when it arises.

This chapter focuses on dispute avoidance and contract administration among the prime contractor, subcontractors, and suppliers. To help identify and prevent problems from the outset, this chapter also concentrates on areas where prime contractor and subcontractor disputes often originate.

I. DISPUTE AVOIDANCE BEGINS AT THE BIDDING STAGE

A. The Importance of the Low Price

Price is one important consideration in selecting a subcontractor. Yet the lowest-cost subcontract may turn out to be the most expensive one if the subcontractor is unwilling or unable to do the work.

A prospective subcontractor should consider the prime contractor's overall pricing strategy. Unfortunately, a subcontractor may never be completely sure how the prime contractor is pricing the project during the bid or proposal stage. History, however, may provide a workable indication as to whether the prime contractor can be expected to underprice the contract with the expectation of negotiating even lower subcontract prices after the award or of making up deficiencies on extras.

B. Know the Other Parties

Dealing with irresponsible or unscrupulous contractors very likely will result in disputes. Avoidance of problems begins at the bidding or proposal stage, when both prime contractors and subcontractors identify potential contractors with whom they will bid and enter into contracts for the work. As such, all contractors should consider four factors before deciding whether to contract with a particular company:

(1) Reputation
(2) Financial resources
(3) Experience and qualifications
(4) Union versus nonunion status

1. Reputation

A firm can avoid numerous headaches and possible losses simply by investigating the past performance record of a potential prime contractor or subcontractor. Unreliable contractors can be avoided at the outset by dealing with reputable individuals with a proven ability to perform, by running credit checks, and by inquiring about the experiences of other owners and contractors with that firm.

Even a cursory investigation may yield clues to potential problems. For example, engaging a subcontractor with a reputation for shoddy or defective work may force the prime contractor to remedy an unsatisfactory work product at its own expense. A subcontractor may find that a prime has a history of slow performance at the start of a project, forcing the subs to encounter extended schedules, frequent payment delays, or end-of-project acceleration. A particularly litigious contractor may refuse to negotiate a settlement in the event of a disagreement, disrupting the project schedule by refusing to work and possibly forcing arbitration or a court battle, and associated extra expense.

2. Financial Resources

A contractor that lacks adequate working capital brings myriad problems to the project, even without a default. For example, suppliers concerned about the contractor's ability to pay may hold deliveries until they receive payment. Laborers may refuse to work because they are not timely paid. These problems can impact the project.

For most subcontractors, the prime contractor's financial resources are the ultimate concern because the prime often must finance part of the performance. For instance, the contractor may be forced to finance corrective work for defects caused by other subcontractors or to continue performance in the face of differing site conditions or changes in the work while negotiations or litigation resolves ultimate entitlement to payment. If the prime contractor lacks sufficient financial resources to continue performance and pay its subcontractors and suppliers in the face of delayed payments from the owner, there is the potential of a prime contractor failure, affecting everyone involved with the project. That prospect warrants considerable attention to the reputation and resources of the prime contractors with whom subcontractors and suppliers do business.

Similarly, a subcontractor with insufficient working capital may be encouraged to front-end load pay requests so that payment may be made for more than the work actually performed is worth or for work not yet performed. If an overpaid subcontractor defaults, the cost to complete the work, combined with the sum already paid to the defaulting subcontractor, may be substantially greater than the amount originally allocated by the prime contractor for the work. The difference, of course, may come out of the prime contractor's pocket.

3. *Experience and Qualifications*

State and local laws may require licensing of general contractors and certain subcontractor trades. (See **Chapter 3**.) Such licensing requirements provide minimal assurance of the contractor's competence, but the investigation should not stop there.

Inquiry should be made into the contractor's experience on projects of similar type and size to the one being bid or proposed. The best residential electrical subcontractor in the county may not be the best subcontractor for a high-rise hotel project. That subcontractor's reputation may be built on its personal integrity and quality of its employees' work, which are laudable achievements. But they do not necessarily assure that the subcontractor has the experience to manage the work of 15 other electricians and coordinate with multiple other trades—other likely requirements for a subcontractor's work on a high-rise hotel project.

The subcontractor may find through inquiry of other subcontractors and suppliers that a prime contractor has considerable experience with one type of project, but that experience is different from the requirements of the prospective project. The subcontractor may also learn that the prime contractor is already overcommitted on existing contracts, which may affect the prime contractor's performance on a new project.

Similarly, a prime contractor must explore the subcontractor's experience in light of special performance requirements for the particular project. For example, if the specifications give only performance criteria, the subcontractor must be able to produce engineering and design details to timely meet the criteria. If a subcontractor proposes to sub-subcontract out the engineering or design work, it should investigate the credentials and experience of the proposed design professional.

Some building product and component suppliers condition the warranty of their products on installation or application by an approved (certified) subcontractor. In some cases, such requirements are not necessary and may serve to inflate the cost of the work. The applicable warranty and the legal basis to enforce that warranty must be reviewed before dismissing a supplier's or manufacturer's requirements. The prime contractor may find that its subcontractor is insolvent and that the supplier has no warranty obligation for failure of a roof or wall system because its product was installed by an unauthorized subcontractor. By contrast, if the work is performed by an approved installer, the prime contractor may be protected by both the supplier's periodic inspections and the warranties of the product and the installation.

4. Union versus Nonunion Status

A prime contractor's choice of a union or nonunion subcontractor generally depends on its own union status. If a union subcontractor and a nonunion prime contractor contract for a project, the subcontract should contain provisions dealing with labor relations on the work site and give the prime contractor the right to terminate the subcontractor if labor problems delay or impede the progress of the work.

A subcontractor should carefully consider its contractual right to time extensions in the event the prime contractor or another subcontractor has labor problems that affect its work. One option available to contractors is to operate "double-breasted."[1] In general, a double-breasted contractor is a union contractor that forms a second, separate company which operates "open shop." In essence, there are two employers, one union, one nonunion. In order to establish a legally correct double-breasted operation, the contractor must conclusively establish that both entities (i.e., the union operation as well as the open-shop operation) are distinct and separate business concerns and not merely convenient distortions, or alter egos, of each other. This form gives the contractor the advantage of being able to operate without a union where this is feasible while at the same time continuing operation of the union company where the use of union labor sources is desirable or required.

Of course, double-breasted operations do not guarantee freedom from labor problems, and the cost of maintaining two separate companies may be prohibitive for some organizations. Most important, the use of a double-breasted operation also requires careful adherence to very specific and technical labor law requirements. For example, the National Labor Relations Board (NLRB) has enumerated four criteria used to determine whether the two operations are truly independent in every conceivable aspect of their operation. These are: (1) common management and supervision of operations; (2) common control of personnel policies and labor relations; (3) interrelation of operations; and (4) common ownership or financial control.[2] Of these, the NLRB has held the most important factor to be the control that one party maintains over the other party's labor relations.[3]

Double-breasted operations can be an effective alternative for a contractor that wants to compete with open-shop contractors in some areas while maintaining good relations with the union where union labor predominates. Contractors should seek the

[1]*Gerace Constr., Inc.*, 193 N.L.R.B. 645 (1971).
[2]*Canned Foods, Inc.*, 332 N.L.R.B. 1449 (2000).
[3]*Id.*

advice of a labor attorney before attempting to establish a double-breasted operation so as to ensure compliance with the requirements of applicable laws and regulations.

C. Problem Areas in Subcontract Bidding

When is there an enforceable agreement between a prime contractor and subcontractor? The simple answer is only when both parties have signed the subcontract agreement. But this simple answer may also be wrong.

For example, a prime contractor that has relied on a subcontractor's telephone bid may be entitled to enforce that bid even though the parties have not signed an agreement. A letter of intent, or a telephone conversation, may give rise to binding obligations such that a refusal to execute a subcontract agreement following a letter of intent or a telephone bid may itself constitute a breach.

Oral agreements can be binding, although they may not be enforceable in certain transactions. When a transaction involves primarily the sale of goods, a binding contract may arise under the Uniform Commercial Code (UCC). The UCC requires that the parties agree to material terms, even though disagreement may still exist with respect to other important terms of the "contract."[4]

1. Enforcement of Subcontractor Bids

Practically every contractor has had a subcontractor or supplier revoke its bid or proposal after the contractor has entered into a contract with a third party in reliance on the subcontractor's bid. This issue has generated as much litigation between prime contractors and subcontractors as almost any other problem facing the construction industry.

Many early cases held that the subcontractor was not bound to the prime contractor; or, similarly, the supplier was not bound to the subcontractor, unless the parties had actually signed a contract.[5] Therefore, bids and proposals could be withdrawn up until the point where the parties were bound by formal documents.

In recent years, many courts (but not all) have held that a subcontractor is bound to the prime, when notified within a reasonable time after award that the prime relied on the bid in obtaining the prime contract award and intends to use the subcontractor for the project.[6] What constitutes a reasonable time for acceptance of an offer depends on the circumstances.[7]

[4]See **Chapter 6** for a more detailed discussion on the U.C.C. and the construction industry.

[5]*See, e.g., Anderson Constr. Co. v. Lyon Metal Prods.,* 370 So. 2d 935 (Miss. 1979). *But see Arango Constr. Co. v. Success Roofing, Inc.,* 730 P.2d 720 (Wash. Ct. App. 1986) (refusing to follow the *Anderson Constr. Co.* Decision).

[6]*See, e.g., Allen M. Campbell Co. v. Va. Metal Indus., Inc.,* 708 F.2d 930 (4th Cir. 1983); *Drennan v. Star Paving Co.,* 333 P.2d 757 (Cal. 1958); *Allied Grape Growers v. Bronco Wine Co.,* 249 Cal. Rptr. 872 (Cal. Ct. App. 1988); *SKB Indus., Inc. v. Insite,* 551 S.E.2d 380 (Ga. Ct. App. 2001); *But see Home Elec. Co. v. Hall & Underdown Heating & Air Conditioning Co.,* 366 S.E.2d 441 (N.C. 1987) (North Carolina Supreme Court refuses to follow *Campbell*).

[7]*See Piland Corp. v. REA Constr. Co.,* 672 F. Supp. 244 (E.D. Va. 1987) (subcontractor not bound by telephone bid where general contractor failed to notify it of acceptance within industry's customary 30-day period); *Cf. ATACS Corp. v. TransWorld Commcy'ns, Inc.,* 155 F.3d 659 (3d Cir. 1998).

The trend toward binding subcontractors to the prime contractor based on the bidder's proposal, even without a contract document, is a response to the hardship a contractor faces when it is awarded a contract based on the subcontractor's proposal, and then the subcontractor is not obligated to proceed with performance. Many contractors seek to avoid this problem by requiring that bidders also agree that their proposal will not be revoked for the same period of time allowed for award of the prime contract in the owner's bid documents.

2. Subcontractor Rights against the Prime Contractor

Although subcontractors may be responsible to prime contractors that rely on their bid, prime contractors are less frequently bound to the subcontractors whose bids were relied on by the prime.[8] The fact that the prime contractor used the subcontract bidder's proposal in its bid to the owner does not ordinarily provide the subcontractor with rights equivalent to those that the prime contractor may have to bind the subcontractor to a contract. But each set of circumstances must be judged on its own merits.

An argument has successfully been made that if the prime contractor sends the subcontractor a letter of intent or a clear statement that it intends to have the subcontractor perform the proposed work for a mutually acceptable price, and the subcontractor begins performance in reliance on the prime's request, then there is a basis to find that the parties have entered into a contract, even though it is not completely reduced to writing. Under such circumstances, a subcontractor may be entitled to recover damages for breach of contract if the prime contractor later withdraws its commitment.

The fact that the parties have not executed a formal contract is not necessarily controlling. If the circumstances indicate that a subcontractor intends to perform a specifically identified scope of work on mutually agreeable basic terms of price and time, and if the subcontractor begins performance pursuant to such terms without objection from the prime contractor, there is support for a finding that the parties are bound to each other. Thus, the subcontractor may be entitled to a remedy for breach when it is later not allowed to complete the work.

II. PREPARATION OF THE SUBCONTRACT AGREEMENT

The "subcontract," for purposes of this discussion, refers to the contract between parties at any level below the contract to which the owner is a party (usually referred to as the prime contract). The subcontract includes the contract between the prime contractor

[8]*See Williams v. Favret,* 161 F.2d 822 (5th Cir. 1947). *But see Elec. Constr. & Maint. Co., Inc. v. Maeda Pac. Corp.,* 764 F.2d 619 (9th Cir. 1985) (reversing trial court's holding that general contractor not bound to accept subcontractor's low bid). *See generally Steeltech Bldg. Prods., Inc. v. Edward Sutt Assocs., Inc.,* 559 A.2d 228 (Conn. App. Ct. 1989) (enforcing oral agreement against owner for benefit of contractor).

and a subcontractor, and contracts between a subcontractor and sub-subcontractor at all levels. Contracts with vendors that only supply materials or equipment and do minimal or no work on the project site are often subject to the principles of law in the Uniform Commercial Code, which are discussed further in **Chapter 6**.

At the outset, a firm should decide whether to draft its own subcontract or use a standard subcontract form. Form subcontracts developed by industry trade organizations such as ConsensusDOCS, the American Institute of Architects (AIA) contract documents, and the Engineers Joint Contract Documents Committee (EJCDC) documents are widely utilized in the industry. Form subcontracts have the advantage of being relatively inexpensive and fairly widely accepted.

In addition, the contract form developed by an industry group may offer the advantage of providing a series or family of documents that address a variety of needs. For example, the ConsensusDOCS 700[9] Series of subcontracting documents (2007 ed.) includes a variety of forms, such as:

ConsensusDOCS 751 Standard Short Form Agreement Between Contractor and Subcontractor (Where Contractor Assumes Risk of Owner Payment)

ConsensusDOCS 705 Invitation to Bid/Subbid

ConsensusDOCS 706 Subcontract Performance Bond

ConsensusDOCS 707 Subcontract Payment Bond

ConsensusDOCS 710 Subcontractor's Application for Payment

ConsensusDOCS 750 Standard Form of Agreement Between Contractor and Subcontractor

ConsensusDOCS 750.1 Standard Form Rider Between Contractor and Subcontractor for Storage of Materials at Subcontractor's Yard

ConsensusDOCS 760 Subcontractor Bid Bond

The AIA contract document A401 (2007 ed.) (Standard Form of Agreement Between Contractor and Subcontractor) is the only construction services form contract in the AIA documents.[10] Also, AIA A401 expressly incorporates AIA A201 into the subcontract.[11] Some other AIA G series, Architect's office and project (or project administrative forms) forms, such as the bond forms (AIA A312 [1984 ed.]); the lien

[9]ConsensusDOCS were released in 2007 as part of a collaborative effort between the AGC and numerous other designer, owner, contractor, subcontractor, and surety organizations to produce a series of documents that allocate risk among all parties to a contract in an effective and efficient manner. The new ConsensusDOCS are meant to supplement prior documents, such as a number of AGC Contract Documents. It should be noted that at the time of printing this reference, some of the original AGC 600 Series Documents have retained their 600 series prefixes. Thus, former AGC 601 Contractor/Subcontractor Agreement (Federal Construction) is now ConsensusDOCS 752 (Federal Construction). For a complete listing of the ConsensusDOCS, see **Chapter 13**.

[10]EJCDC does not publish construction services documents between contractor and subcontractor.

[11]AIA A401, § 2 (2007 ed.).

waiver form (AIA G706A); change order forms (AIA G701 [2001 ed.]); and application and certification for payment form (AIA G702 [1992 ed.]) are commonly used by contractors and subcontractors. Care should be taken in using these or any other industry forms in the contractor-subcontractor relationship where such forms are designed for the owner-prime contractor relationship.

Whenever standard forms are used, they still should be carefully reviewed, as these forms may not address specific items of concern to the contracting parties, or may contain terms unnecessary for the particular project or contractual relationship. For that reason, some contractors may choose to develop their own forms; however, the software packaging of the modern standard forms also allows for modification or adaptation to address particular concerns or needs.

Regardless of whether an industry standard or project specific document is used, an adequate subcontract form should, at a minimum, accomplish the following:

(1) Define the exact undertakings of the subcontractor so that there can be no dispute about the scope of work the subcontractor is to perform, including standard of quality and performance time.

(2) Detail the terms and conditions of payment, including the method of computing progress payments, terms of payment for stored material, due dates for pay requests, handling of interim retainage, and final payment.

(3) Anticipate areas of trouble or dispute, and define the consequences if work is not properly or timely performed.

(4) Specify the relief that the innocent party is entitled to receive.

(5) Avoid favoring one party to such an extent that the other party will decline to sign the proffered form or refuse to perform under its terms. A subcontract should be tailored to fit the needs of a particular project.

A collection of uncoordinated provisions from various unrelated form contracts can result in a monster that may create substantial unanticipated problems for both parties. Such monsters typically are created in one of two ways. The first is a "cut-and-paste" operation by which various provisions are borrowed from different subcontracts and brought together in one document without being reconciled or analyzed for consistency. The second is an "inclusion" operation, by which the contractor includes (either expressly or by reference) in its subcontract the general and special conditions of the prime contract, terms and conditions that it has prepared specially, and perhaps some other standard-form document, such as the contractor's project procedures. The justification for an inclusion contract rests on the theory that more is better. The result is often a hodgepodge of documents that are not consistent or coordinated.

Instead, a subcontract form should be prepared by someone competent for the task, and modified as needed for the conditions of each project. Also, it should be reviewed periodically in light of lessons learned through experience and changes in the law. Unlike terms of many public contracts, which are set by law or regulation, there are no "standard" subcontract terms for private contracts. The following points should be considered in drafting the terms of a subcontract.

A. "Flow-Down" Obligations

A flow-down clause, which contractually ties, for instance, the subcontractor to the prime in the same manner as the prime is bound to the owner, is imperative.[12] One example of a basic, yet enforceable, flowdown clause reads: "[C]ontractor shall have the same rights and privileges as against the Sub-contractor . . . as the Owner in the General Contract has against the Contractor."[13] Since subcontractors at every level are bound together on the project by a series of contracts, all subcontracts should contain flow-down clauses. The absence of a flow-down clause can leave the contractor exposed to liability but unable to demand of the subcontractor the performance that it is required to undertake.

In return for this protection, the prime contractor will generally provide the subcontractor with the same rights that the prime has against the owner up the contract ladder, including making claims, protesting, and providing notices to the owner. In other words, the rights and duties flow both ways—upward to the prime contractor as well as downward to the subcontractors at each level. This tends to keep the parties on an even basis, even though there is no privity of contract between the owner and the subcontractors.

B. Scope of Work

Every subcontract must have a clear and defined scope of work. From a practical standpoint, it is very difficult to define all of a subcontractor's work. Therefore, the subcontract form should refer to all contract documents between the owner and prime contractor.[14]

The general contract bid items are often too general to incorporate by reference into the subcontract as the definition of the scope of work. This practice should be avoided. Instead, the work description must explicitly define responsibilities for the scope of work and must also define the responsibilities each contractor assumes for its own activities.

The scope of work for a project is not simply the items of performance that are subcontracted out for a particular project. A scope of work provision also includes the responsibilities for the facilities and for contract administration. Therefore, a detailed description of the actual work to be performed by the subcontractor on a project, including incorporation of the owner's description of the work from the general contract, must be supplemented with revisions that address other contingencies.

[12] *See generally,* E. Sanderson Hoe et al., Flow-Down Clauses in Subcontracts, Briefing Paper No. 85–5 (1985).

[13] *L & B Constr. Co. v. Regan Enters., Inc.,* 482 S.E.2d 279 (Ga. 1997).

[14] *See S. Leo Harmonay, Inc. v. Binks Mfg. Co.,* 597 F. Supp. 1014, 1024 (S.D.N.Y. 1984) (holding that incorporation by reference clause applies only to specifications and actual statement of work), *aff'd,* 762 F.2d 990 (2d Cir. 1985); *Turner Constr. Co. v. Midwest Curtainwalls, Inc.,* 543 N.E.2d 249, 251–52 (Ill. App. Ct. 1989) (holding that incorporation by reference clause binds the subcontractor to all terms and conditions of the prime contract); *Gibbons-Grable Co. v. Gilbane Bldg. Co.,* 517 N.E.2d 559, 563 (Ohio Ct. App. 1986) (discussing incorporation by reference clauses with specific example). *See also* Overton A. Currie, et al., Construction Subcontracting: A Legal Guide for Industry Professionals § 3.11 (Overton A. Currie, et al., eds., John Wiley & Sons, Inc. 1991); Overton A. Currie et al., 1994 Wiley Construction Law Update § 9.9 (1994).

✔SCOPE OF WORK CHECKLIST

(1) An explanation of items listed in the subcontract scope of work that may differ from the plans or technical specifications; for example:
   (a) Items of work that appear in the specifications may be excluded by the subcontractor in its proposal;
   (b) Items of work may be specifically listed to clarify a possible ambiguity as to which trade is responsible for performance; and
   (c) Items may not be listed in the plans or technical specifications but are necessary for completion of work on the project and are included in the scope of work to be performed by the subcontractor.

(2) A clause clarifying that the subcontractor agrees to furnish, without extra charge, all work, labor, materials, and equipment not mentioned or shown in the contract documents that are nonetheless generally included under a particular class of subcontract or fairly implied by the subcontract as necessary for the satisfactory completion of the project (including work necessary to conform to applicable laws, ordinances, orders, rules, regulations, and requirements).

(3) A clause providing a method of interpretation of the subcontract document. This may include an agreement to allow the design professional to resolve any conflicts between various subcontract provisions, plans, and specifications. The form of the contract may require the design professional to resolve differences of interpretation under that contract, so a drafter may choose to carry that responsibility through into the subcontract.

If a general contractor divides a certain item of work between subcontractors, two additional considerations arise. First, extreme care must be used to define each subcontractor's scope of work to ensure that no aspect of the work is omitted. Second, the contractor must realize that responsibility for coordinating the efforts of these subcontractors will usually fall on the contractor that has divided the work. Efforts to avoid this responsibility through the use of subcontract disclaimers generally are ineffective because the necessary authority is not clearly and expressly delegated to make the assignment of responsibility effective.

C. Payment Obligations

The prime contractor's obligation to make progress payments or pay retainage to the subcontractor before the prime receives corresponding payment from the owner can be fraught with problems and lead to significant disputes. At the outset, the terms and conditions of payment should be clearly and unambiguously set out in the subcontract agreement. Industry form subcontracts may provide a model for such payment terms.

Section 11.3 of the AIA Standard Subcontract Form of Agreement (AIA Document A401 (2007 ed.)) provides that if the prime contractor does not receive payment from the owner for any reason that is not the fault of the subcontractor, the prime contractor shall, upon demand, pay the subcontractor the amount computed for the progress payment in accordance with the subcontract. This provision places the risk of nonpayment by the owner on the prime contractor. Paragraph 8.2.5 TIME OF PAYMENT of ConsensusDOCS 750 (Contractor/Subcontractor Agreement) similarly places the risk of nonpayment by the owner on the prime (general) contractor. Both of these industry documents reflect an allocation of the risk of nonpayment consistent with the rationale that the prime (general) contractor is in the best position to evaluate both the owner's willingness and ability to pay for the work within a reasonable period of time.

In contrast, "pay-when-paid" and "pay-if-paid" clauses seeks to place the risk of nonpayment for work on the subcontractor or supplier. Case law and statutory law vary significantly from jurisdiction to jurisdiction regarding the interpretation and subsequent enforceability of contractual "pay-when-paid" and "pay-if-paid" clauses. In drafting or reviewing contracts, it is imperative to be familiar with the law in the applicable jurisdiction regarding treatment of these types of payment clauses. It is also important to understand the differences between these payment provisions.

1. Pay-When-Paid Clauses

A pay-when-paid clause generally obligates the prime contractor to make payment to the subcontractor when it is paid by the owner, or within a reasonable time period after receipt of corresponding payment from the owner. One example of a pay-when-paid clause is: "The total price to be paid to subcontractor shall be $ . . . no part of which shall be due until five (5) days after owner shall have paid contractor therefore.. . ."[15] The variations on this theme are almost unlimited. In spite of this clause, the prime contractor in the *Thomas J. Dyer* case was not released from its payment obligation even though the owner filed bankruptcy before final payment. The court in that case held that the payment clause failed to express that the parties' intent was to shift the risk of the owner's nonpayment from the general contractor to the subcontractor.

a. The Case against the Pay-When-Paid Clause There are a number of reasons why the pay-when-paid clause should not be applied to excuse payment under all circumstances. First, the intent of the parties must be examined. In most cases, the subcontractor is not in privity of contract with the owner, and usually has no opportunity to deal directly with the owner in the proposal or bid phase or at any time during performance. The subcontractor looks to the general contractor for coordination with other trades and for use of site facilities, for scheduling, for interpretation of the plans and specifications, and for payment for the work. If the parties intend that the subcontractor will look only to the owner for payment, that intention must be clearly expressed in the subcontract.

Second, the payment clause in the subcontract must be examined. Look for specific circumstances mentioned in the subcontract, such as the delay of progress payments because of default of the subcontractor in the AIA A401 contract, or a reference to the owner's insolvency as a condition excusing nonpayment. If the payment clause

[15] *Thomas J. Dyer Co. v. Bishop Int'l Eng'g Co.,* 303 F.2d 655 (6th Cir. 1962).

deals with the amount, timing, method of payment, and other essential provisions, without reference to specific conditions excusing payment, then a court will likely decide that payment to the subcontractor is not conditioned on payment from the owner under all circumstances.[16]

Some courts have found that wording which fails to explicitly state that the parties to the agreement meant to shift the risk of the owner's nonpayment to the subcontractor simply means that the prime contractor must pay the subcontractor in a reasonable period of time.[17] As one court stated, "if such was the intention of the parties it could have been so expressed in unequivocal terms."[18] Accordingly, many courts have held that pay-when-paid provisions are unenforceable as against public policy when they require a subcontractor to wait indefinitely or forego payment under circumstances that the subcontractor cannot control.[19] Finally, a small but growing number of states have enacted statutes that declare these clauses to be unenforceable under mechanic's lien laws or as a matter of public policy.[20]

b. The Case for the Pay-When-Paid Clause Some states have taken a stance more favorable to the prime contractor and upheld the subcontract payment provision making the owner's payment to the prime contractor a condition precedent to the prime's obligation to pay the subcontractor for work performed.[21]

As an example, a prime contractor was excused from the obligation to pay until it had received payment when the subcontract provided: "[P]ayments will be made [to the subcontractor] from money received from the owner only and divided pro rata [among] all approved accounts of subcontractors, labor and materials."[22] This payment clause set up a fund consisting of money received from the owner, out of which subcontractors were to be paid, and made the fund the only source from which subs were to be

[16]*See Mrozik Constr., Inc. v. Lovering Assocs., Inc.,* 461 N.W.2d 49 (Minn. Ct. App. 1990).
[17]*See, e.g., Watson Constr. Co. v. Reppel Steel & Supply Co., Inc.,* 598 P.2d 116 (Ariz. Ct. App. 1979); *OBS Co., Inc. v. Pace Constr. Corp.,* 558 So. 2d 404 (Fla. 1990) (holding that unless parties' intent to shift risk of owner non-payment to subcontractor is unambiguously expressed, contractor is liable for payment to subcontractors); *Power & Pollution Serv., Inc. v. Suburban Power Piping Corp.,* 598 N.E.2d 69 (Ohio Ct. App. 1991).
[18]*Power & Pollution Serv., Inc.,* 598 N.E.2d 69, 70 (Ohio Ct. App. 1991) (citing *Thomas J. Dyer Co. v. Bishop Int'l Eng'g Co.,* 303 F.2d 655 (6th Cir. 1962)).
[19]*See William R. Clarke Co. v. Safeco Ins. Corp.,* 938 P.2d 372 (Cal. 1997); *S. States Masonry, Inc. v. J.A. Jones Constr. Co.,* 507 So. 2d 198 (La. 1987); *Kawall Corp. v. Capolino Design & Renovation,* 388 N.Y.S.2d 346 (N.Y. App. Div. 1976); *West-Fair Elec. Contractors v. Aetna Cas. & Sur. Co.,* 661 N.E.2d 967 (N.Y. 1995); *Certified Fence Corp. v. Felix Indus., Inc.,* 687 N.Y.S.2d 682 (N.Y. App. Div. 1999).
[20]*See, e.g.,* N.C. Gen. Stat. § 22C-2 (1991) and 770 Ill. Comp. Stat. Ann. 60/21 (2003); Wis. Stat. Ann. § 779.135 (2007).
[21]*See Midamerica Constr. Mgmt., Inc. v. Mastec N. Am., Inc.,* 436 F.3d 1257 (10th Cir. 2006) (holding Texas and New Mexico courts will enforce a contingent payment clause if it is clearly a condition precedent to payment under the subcontract); *Sasser & Co. v. Griffin,* 210 S.E.2d 34 (Ga. Ct. App. 1974); *A. A. Conte, Inc. v. Campbell-Lowrie-Lautermilch Corp.,* 477 N.E.2d 30 (Ill. App. Ct. 1985). *See also* AGC Document No. 655, ¶ 8.2.5 (containing a pay-when-paid provision, which states in part: "Receipt of payment by the Contractor from the Owner for the Subcontract Work is a condition precedent to payment by the Contractor to the Subcontractor. The Subcontractor hereby acknowledges that it relies on the credit of the Owner, not the Contractor for payment of Subcontract Work.").
[22]*Sasser & Co. v. Griffin,* 210 S.E.2d 34 (Ga. Ct. App. 1974).

paid. This provision was enforced as a pay-when-paid clause, relieving the prime contractor of the obligation to advance monies for progress payments to the subcontractor, because it showed the intent of the parties that the subcontractors' payments were to come only from the fund created by payments from the owner. The subcontractors were not relying on the prime contractor's credit for payment.[23]

2. *Pay-If-Paid Clauses*

In an effort to avoid making payments to subcontractors even though payment has not been received from the owner, some prime contractors now include a provision in their subcontracts making payment by the owner a "condition precedent" to the prime's obligation to pay its subcontractors.[24] Such a provision is referred to as a "pay-if-paid" clause, under the theory that the contractor's payment obligation only arises **if** it is paid by the owner. To be enforceable, such a contract payment provision must clearly show that the express condition is the mutual intent of the parties and that the parties recognize that payment may not be received from the owner so that the subcontractor assumes this risk as a part of the negotiated contract.[25] The arguments for or against a "pay-if-paid" clause are essentially the same as those set forth above in the discussion of "pay-when-paid" clauses.

Several courts recently have held that a subcontractor still may seek payment from a prime contractor, even with a valid pay-if-paid clause, when the contractor prevented the condition precedent from occurring. This is known as the Prevention doctrine.[26] For example, in the *Northeast Drilling* case, the prime contractor's subcontract contained a valid pay-if-paid clause. The prime contractor failed to submit change order requests for additional work that the subcontractor had performed. As a result, the owner never paid the prime contractor for the additional work the subcontractor had performed, and the prime contractor never paid the subcontractor. The court found that the prime contractor's actions in failing to submit the change order requests prevented the prime contractor from using the pay-if-paid defense during the subsequent nonpayment action brought by the subcontractor.

✔ PAYMENT CLAUSE CHECKLIST

Given the various and confusing approaches courts have taken in regard to contingency payment clauses, these items should be considered in drafting or reviewing contract language on the terms of payment.

[23]*See Wilson v. Post-Tensioned Structures, Inc.*, 522 So. 2d 79 (Fla. Dist. Ct. App. 1988).

[24]*Associated Mech. Corp., Inc. v. Martin K. Eby Constr. Co., Inc.*, 67 F. Supp. 2d 1375 (M.D. Ga. 1999); *St. Paul Fire & Marine Ins. Co. v. Ga. Interstate Elec. Co.*, 370 S.E.2d 829 (Ga. Ct. App. 1988); *see also* AGC Document No. 655 ¶ 8.2.5 (1998 ed.).

[25]*Galloway Corp. v. S. B. Ballard Constr. Co.*, 464 S.E.2d 349 (Va. 1995); *see also Imagine Constr., Inc. v. Centex Landis Constr. Co.*, 707 So. 2d 500 (La. Ct. App. 1998); *Sheldon Pollack Corp. v. Falcon Indus., Inc.*, 794 S.W.2d 380 (Tex. App. 1990).

[26]*Ne. Drilling, Inc. v. Inner Space Serv. Inc.*, 243 F.3d 25 (1st Cir. 2001); *Moore Bros. Co. v. Brown & Root, Inc.*, 207 F.3d 717 (4th Cir. 2000).

(1) *Conditions for payment.* Which conditions precedent must be met before the subcontractor is entitled to payment? For instance, the prime contract may require: (a) that all work meet the approval of the owner or its designated representative before payment is considered; and (b) proof that all invoices from subcontractors and suppliers for prior work have been paid. Before there is any entitlement to payment, the contract may also require that bonds or certificates of insurance be submitted and progress schedules and submittals be current.

(2) *Computing the amount.* The method of determining the amount of progress payments should be clearly described. The contract terms should describe how to determine the basis of payment, such as a percentage of the total estimated value of labor, materials, and equipment incorporated into the work, and a percentage of the value of materials suitably stored, less the aggregate of previous payments.

(3) *Pay-when-paid or Pay-if-paid:* Is payment to the subcontractor conditioned on payment by the owner for the same work? If this is the parties' agreement, and it is not against state law or public policy, state that the owner's payment is a "condition precedent"; that the subcontractor expressly accepts the risk that the owner may not pay the contractor; that the subcontractor relies for payment on the credit and ability of the owner to pay; and that the contractor's payment bond surety, if any, will be obligated only to the same extent as the prime contractor.

(4) *Final payment terms.* After the subcontract is fully completed, when will the subcontractor receive its final payment, and what are the conditions to be met before final payment will occur? These conditions may include:

    (a) the owner and design professional accept the subcontractor's work in writing;

    (b) the contractor is fully paid by the owner; and

    (c) a release (including a release of lien rights) is executed by the subcontractor for the contractor.

3. Changes to the Payment Terms Must Also Account for the Surety

After the subcontract is signed, there remains a potential for problems in the implementation of the agreement. The obvious problems are caused by a delay in payment or a dispute over the amounts earned for progress payments or for extra work. Careful draftsmanship and thoughtful consideration in advance will provide the parties with contract terms to measure and guide their performance. The subcontractor and prime contractor may find, however, that unexpected circumstances during performance may require a change in their expected payment procedures. Unexpected results may follow if the parties' actions to accommodate the new circumstances are not carefully considered. For example, the prepayment of a subcontractor may discharge the subcontractor's performance bond surety from liability. If a subcontractor falls behind schedule and is financially unable to increase its workforce to regain the lost time, the prime contractor may consider prepaying the subcontractor to finance the additional crews needed to regain the schedule while expecting to deduct the amounts advanced from later progress payments.

In one case, a prime contractor agreed to this arrangement but failed to obtain the approval of the subcontractor's payment and performance bond surety.[27] When it sued the surety for the amounts prepaid to the subcontractor, the surety defended on the basis that the prepayments materially altered the subcontract and operated to discharge the surety from its obligations.[28] The prime contractor's payment bond surety may take the same position if the prime contractor is in default of its payment obligations after agreeing to alter the payment terms of its contract without obtaining the surety's prior consent.

Some courts require that a surety show injury or prejudice in addition to a material alteration of the contract in order to be discharged from bond obligations.[29] The lesson for all parties is that payment terms must be considered carefully when initially entering into the contract, and any change in those terms must be reviewed for its impact on the present obligations between the parties and the rights and obligations of sureties and other payment guarantors that may be affected.

D. Subcontractor Termination

1. *The Right to Terminate*

a. Express Termination Rights (Subcontractor Default Clause) Virtually all construction contracts expressly recognize the right to terminate the contract upon the default of either party under certain circumstances. These provisions also recognize the right of the innocent party to recover damages flowing from the default and resulting termination. For example, subparagraph 10.1.2 of ConsensusDOCS 750 (2007 ed.) sets forth the following basic steps regarding the termination process:

> TERMINATION BY CONTRACTOR If the Subcontractor fails to commence and satisfactorily continue correction of a default within three (3) business Days after written notification issued under Subparagraph 10.1.1, then the Contractor may, in lieu of or in addition to the remedies provided for in Subparagraph 10.1.1, issue a second written notification, to the Subcontractor and its surety, if any. Such notice shall state that if the Subcontractor fails to commence and continue correction of a default within seven (7) Days of the written notification, the Agreement will be deemed terminated.[30]

[27] *Southwood Builders, Inc. v. Peerless Ins. Co.,* 366 S.E.2d 104 (Va. 1988).

[28] *See also United States v. Freel,* 186 U.S. 309, 316 (1902); *Chas. H. Tompkins Co. v. Lumbermens Mut. Cas. Co.,* 732 F. Supp. 1368, 1377–78 (E.D. Va. 1990); *Brunswick Nursing & Convalescent Ctr., Inc. v. Great Am. Ins. Co.,* 308 F. Supp. 297 (S.D. Ga. 1970); *In re Liquidation of Union Indem. Ins. Co.,* 632 N.Y.S.2d 788 (N.Y. App. Div. 1995).

[29] *United States v. Reliance Ins. Co.,* 799 F.2d 1382 (9th Cir. 1986); *Reliance Ins. Co. v. Colbert,* 365 F.2d 530 (D.C. Cir. 1966); *Mergentime Corp. v. Wash. Metro. Area Transit Auth.,* 775 F. Supp. 14 (D.D.C. 1991).

[30] ConsensusDOCS 750, ¶ 10.1.2 (2007 ed.).

Similarly, Section 7.2.1 of AIA Document A401 (2007 ed.) provides that the contractor may terminate the subcontractor upon 10 days notice if the subcontractor "repeatedly fails or neglects to carry out the Work in accordance with the Subcontract Documents or otherwise to perform in accordance with this Subcontract."[31] Both of these industry form subcontracts also expressly set forth the grounds for nonpayment that entitle the subcontractor to terminate the contract with the prime contractor.

b. Implied Termination Rights Even in the absence of an express termination provision, there generally exists an implied right to terminate a contract and sue for damages if the other party has materially breached the contract. It is nonetheless important that the prime contractor have the right to terminate a subcontract (and that a subcontractor be able to terminate a sub-subcontractor's right to proceed) under certain circumstances, and this right must be spelled out in detail, with notice requirements and cure rights specifically delineated.

c. Alternatives to Termination The subcontract termination clause should provide flexibility to the prime contractor, so that the extreme act of termination is not the contractor's only option in dealing with a subcontractor's default. The subcontract must expressly reserve the prime's right to pursue alternatives in lieu of termination, such as paying the subcontractor's suppliers and subcontractors or supplementing the subcontractor's workforce. A termination for default is the construction industry equivalent of capital punishment. It is an option not to be invoked lightly, and only after careful consideration of other options.

Although the discussion that immediately follows focuses on the prime contractor/subcontractor relationship, the principles stated apply equally to a subcontractor/sub-subcontractor agreement. Likewise, some of the options to termination are discussed in the Default Clause Checklist, in **Section II.D.2** of this chapter.

d. Consequences of Improper Termination Whether the subcontractor was actually in default at the time of termination is usually a fact question. The facts will determine the propriety of the termination. If the termination was proper, then the prime contractor will be entitled to damages, including the cost of completion; if improper, then the subcontractor may recover as damages the profit it would have received if allowed to complete the work. If there would be no profit pursuant to the subcontract, the subcontractor may recover damages based on an equitable theory of value of goods or services provided (*quantum meruit*) basis. If the subcontract contains only a "termination for default" option, the prime contractor may be exposed to significant subcontractor claims if the termination is later deemed improper, even if the prime contractor acted with a good-faith belief that the termination was justified.

[31] AIA A401, § 7.2.1 (2007 ed.).

2. *Considerations in Drafting Subcontract Default Clause*

Some of the considerations that should be addressed when drafting or reviewing a subcontract default clause are set out in the following checklist. Use this checklist to determine if and how the subcontract addresses major issues in a default situation:

✔ DEFAULT CLAUSE CHECKLIST

(1) *Obligation to proceed.* In the event of a dispute, can the subcontractor be required to carry out its work as directed by the prime contractor? Even so, if the subcontractor's refusal to proceed under protest is "reasonable" (e.g., where paving in cold weather might subject the subcontractor to later liability on its express warranty), it is possible that the subcontractor may successfully argue that its refusal to follow the prime's direction was justified. Accordingly, any resulting default termination was improper.[32]

(2) *Notice.* Does the subcontract specifically outline requirements for notice to the subcontractor if the subcontractor fails to comply with the contractor's instructions or the terms of the subcontract? In the absence of an express notice provision, some courts may find an implied duty to notify the subcontractor before termination and to give the subcontractor an opportunity to cure the default.[33]

(3) *Cure right.* The time that the subcontractor will be given the chance to cure a default (the cure period) should be specified. Some definition of an effective cure (e.g., a complete correction of the default, or an approved, written program for curing the default and reasonable progress in implementing such a program) is also warranted. Subcontractors should review the clause to verify that the contract expressly provides a right to cure a default.

(4) *Contractor options.* A well drafted default clause will give a contractor several options in dealing with a potential subcontractor default. If there is no cure within the specified time period, these options might include:
   (a) Supplementing the subcontractor's workforce;
   (b) Having a second subcontractor take over a portion of the subcontractor's work;
   (c) Making direct payments to subcontractor's suppliers, subcontractors, and laborers;
   (d) Accelerating the work; or
   (e) Subcontract termination (i.e., taking possession of the subcontractor's materials, tools, and equipment, and completing the subcontractor's work with another subcontractor).

[32] *See Wilson v. Kapetan, Inc.*, 595 A.2d 369 (Conn. App. Ct. 1991).
[33] *See McClain v. Kimbrough Constr.* Co., 806 S.W.2d 194 (Tenn. Ct. App. 1990).

The subcontract should clarify that the decision to first pursue one or more of these options does not prejudice the prime's right to later terminate or to claim damages against the subcontractor or its surety.

(5) *Payment rights.* What effect will a default have on payment to the subcontractor? This clause might include provisions giving the contractor the right to deduct all monies expended and costs incurred, direct and indirect (including attorneys' fees), as a result of the subcontractor's default, from all amounts otherwise owed to the subcontractor. Also, if the amounts owed to the subcontractor are not enough to cover such expenses, the termination clause would obligate the subcontractor to promptly pay upon demand the full amount of the difference.

(6) *Good-faith defense to wrongful termination.* If the prime contractor is mistaken about the subcontractor's default or failure to cure such default, but acts on a good-faith belief, will the contractor be liable for additional subcontractor damages (e.g., anticipated profits)? To minimize this risk, the termination clause might convert the "wrongful" termination to a "termination for convenience" and limit the contractor's exposure to the value or cost of the work in place, plus a reasonable profit and overhead markup on that work. The clause could expressly exclude the subcontractor's right to recover unearned profit, unexpended overhead, attorneys' fees, and other damages.[34]

(7) *Proof of termination costs.* What information from the contractor will be sufficient to demonstrate the amount of money owed the contractor as a result of the default?

(8) *Default-triggering events.* How does the subcontract define those subcontractor acts and omissions that will justify prime contractor action under the termination clause? Careful drafting is required to anticipate the potential problems that might give rise to a contractor's need to invoke the termination clause. In the absence of an express provision justifying a subcontract termination, a contractor must prove that the subcontractor "materially breached" the subcontract.

(9) *Use of subcontractor resources.* Does the termination clause effectively preserve the contractor's right to take over the subcontractor's materials, tools, and equipment in order to complete the subcontract work? Does the clause preserve the contractor's right of access to subcontractor information (e.g., as-built information) and the option to accept an assignment of subcontractor's contracts with suppliers, equipment dealers, and sub-subcontractors?

(10) *Flow-down limitations.* Does the subcontract anticipate that the termination of the subcontract may result from a termination of the prime contract by the owner? In a properly drafted flow-down clause, a prime contractor may attempt to limit its exposure to termination damages and expenses

[34] *But* see ConsensusDOCS 750, ¶ 10.7 (2007 ed.).

to whatever damages or costs the contractor is able to collect on the subcontractor's behalf from the owner. In federal government contracting, the government obligates the prime contractor to make a good-faith effort to negotiate the inclusion of a termination for convenience flow-down clause in all subcontracts.

3. Termination for Convenience

To protect the prime contractor in circumstances involving an improper default termination or where the prime contractor's contract is terminated for the convenience of the owner, the inclusion of a provision for the "termination for convenience" of the sub is recommended. This provision allows the prime to terminate the subcontractor with or without cause. If one party terminates another for convenience, the terminated party has the right to an equitable adjustment that will return that party to its precontract condition.

Although a typical termination for convenience clause appears to give one party the absolute right to terminate another for whatever reason, there is a conflict as to whether the termination must be made "in good faith." At least one court has held that there are circumstances in which a termination, even under a seemingly unrestricted termination clause, may not be allowed where done in bad faith.[35] Other courts have been hesitant to hold that there is a good-faith requirement for the exercise of a termination for convenience right.[36] The question of whether there is a good-faith requirement for termination without cause remains unresolved in many jurisdictions.[37] It is therefore a point that may be argued by any contractor that may have been terminated "in bad faith." In federal government contracting, the government's right to terminate a contract for its convenience and limit the contractor's recovery in accordance with the Termination Clause is very broad.[38] To avoid the limitations on recovery under that clause, the contractor must prove "bad faith" or "abuse of discretion." Either standard is very difficult to establish.[39]

E. No Damages for Delay, Except as Paid by the Owner

A contractor may limit delay claim exposure to its subcontractors in proportion to the owner's liability and actual payment to the prime. In *Dyser Plumbing Co. v. Ross Plumbing & Heating, Inc.,*[40] a Florida appellate court upheld a contract clause that limited a subcontractor's delay damages to a percentage of an amount paid by the owner. The court ruled that the subcontract's terms clearly and unambiguously established the extent of any recovery by the subcontractor for delay damages.

[35]*See RAM Eng'g. & Constr., Inc. v. Univ. of Louisville,* 127 S.W. 3d 579 (Ky. 2003).

[36]*See Niagara Mohawk Power Corp. v. Graver Tank & Mfg. Co.,* 470 F. Supp 1308 (N.D.N.Y. 1979), 806 S.W. 2d 194 (Tenn. Ct. App. 1990).

[37]*Id.*

[38]*Praecomm, Inc. v. United States,* 78 Fed. Cl. 5 (2007).

[39]*C.F.S. Air Cargo, Inc.,* ASBCA No. 36113, 91–1 BCA ¶ 23,583, *aff'd without opinion,* 944 F.2d 913 (Fed. Cir. 1991).

[40]515 So. 2d 250 (Fla. Dist. Ct. App. 1987).

If a general contractor hopes to limit a subcontractor's claim for delays to those amounts actually paid by the owner, the subcontract should expressly set out this limitation. Many of the considerations outlined earlier in this chapter apply to the determination of the subcontractor's rights when confronted with such a subcontract provision. Obviously, these same considerations apply when a subcontractor is drafting its sub-subcontracts and purchase orders.

F. Changes

One of the sources of frequent disputes in construction is the contract's changes clause. (See **Chapter 10**.) For the owner and contractor, it is important that the subcontract specify a procedure for allowing changes to the contract. The subcontractor's primary concern is that it be guaranteed the right to be paid for "extra" work. In addition, the subcontractor frequently is concerned that it not be required to finance, for an extended time period, a significant amount of "extra" work while waiting for (1) a determination of whether the work is "extra" and, if so, (2) a determination of the work's reasonable value. The parties' competing interests lead to frequent disputes.

In judging the completeness and fairness of the subcontract changes clause, contractors and subcontractors should consider these items.

✔ SUBCONTRACT CHANGES CLAUSE CHECKLIST

(1) *The right to order changes.* The contractor should verify that the subcontract clearly states the contractor's right at any time, on written order, without notice to the surety and without invalidating the subcontract, to make changes in the work contracted for and the subcontractor's duty to proceed with the work as directed by the contractor's written order. There is no implied right to order contract changes.

(2) *Notice requirements.* If the subcontractor is faced with an "informal" direction to perform additional work, the subcontract should set out express notice requirements and stated consequences for the failure to observe those notice requirements.

(3) *Specific timetable for change order resolution.* The contractor and subcontractor both should be concerned that the subcontract specify a timetable and specific steps for the resolution of the adjustment of the subcontract price and schedule. The contractor does not want to surprise the owner with subcontractor claims that remain unknown or unquantified until the end of the job. The subcontractor does not want to finance extra work for a prolonged period. The subcontract should detail the rights and duties of the parties in the event the work is performed before reaching an agreement on price.

(4) *Pricing limitations.* The parties should be aware of any limitations on the allowable adjustment to the subcontract price—for example, the amount allowed by the owner; the amount of the subcontractor's initial estimate; the amount determined by the design professional or some other "final decision" maker; overhead and profit limitation clauses; or other conditions precedent to a price adjustment.

(5) *Pricing proof.* The subcontract should identify the subcontractor's duty to provide cost data, certifications, and any other information necessary to support its proposal. The subcontract form should require the subcontractor to submit claims for additional costs in a manner and time sufficient to allow the general contractor to submit this information, if need be, to the owner under the terms of the prime contract. A simple flow-down clause may not be sufficient to provide for the timing requirements between the two contracts.

(6) *Binding effect of the change order.* Frequently, the subcontract will grant audit rights to either the owner or the general contractor, so as to hold open the possibility of reopening a change order in the event of defective or inadequate pricing by the subcontractor.

(7) *Emergency changes.* The subcontract should provide alternate change order procedures, rights, and duties in the event of the need to implement a change during an emergency that endangers life or property.

(8) *Oral change orders.* What is the effect of oral orders or directives that are believed to be changes, and what are the parties' duties and obligations under such circumstances? Does the subcontract require that the subcontractor obtain a written work order before performing extra work? Such requirements are common and can be enforceable.

(9) *Working under protest.* From the contractor's perspective, it is important that the subcontractor be obligated to proceed with the directed work, even if the subcontractor disputes the value of the work or if the contractor denies that the work is "extra." Such a subcontract clause appears to be enforceable. For example, in *Keyway Contractors, Inc. v. Leek Corp.,*[41] a Georgia court enforced a contract provision requiring that the work continue despite pending disputes. Following a dispute over payment, the subcontractor abandoned the work site. The court found that the subcontractor's action constituted a material breach of contract. The subcontractor confronted with such a clause must look for some express or implied limit to the obligation to perform work for which it may not be paid. Also, the subcontractor must be certain that it has the right to proceed "under protest" and that it understands the procedural requirements for preserving that right.

(10) *Change order authority.* Does the subcontract contain limitations on the authority of the contractor's representatives to direct changes? Limitations as to who has the authority to direct changes and limits on the amount of change order authority entrusted to a contractor representative are common.

(11) *Alternative methods for fixing change order amounts.* Does the subcontract provide unit prices for change order work, or pricing limitations in connection with "force account" work? Since it is likely that many change order amounts will not be agreed on in advance, the subcontract should provide alternate routes to fixing the change order amount.

[41] 376 S.E.2d 212 (Ga. Ct. App. 1988).

(12) *Time extensions.* The changes clause should require that the subcontractor quantify the time impact, as well as the cost impact, of changes to the work. Subcontractors must be aware of subcontract clauses that would deprive them of time extensions made necessary by changes unless the time extension request is presented in a specified format, with specified support, or in a certain timeframe.

G. Subcontractor's Indemnification of the Contractor

Traditionally, subcontracts have imposed very broad hold-harmless obligations on subcontractors. Some courts have been reluctant to enforce overly broad indemnification clauses as a matter of public policy. Some state legislatures have also taken steps to limit the effects of indemnity agreements.[42] For example, Section 13–8–2 of the Official Code of Georgia (2007) invalidates any attempt to require a contracting party to hold another party harmless from liability arising solely from the latter's own acts. Accordingly, while a broad, general indemnification clause is important, it should be drafted after consideration of the applicable state laws and decisions. Similarly, a clause that is enforceable in one state may be void in a neighboring state. Therefore, prime contractors should be cautious when doing work outside of their home state. Subcontractors should look for a governing law clause which provides that the contract is to be governed by the laws of a particular state.

Subcontract indemnity clauses come in many shapes and sizes. The clauses are typically long and cumbersome to read and difficult to understand. Consider the following potential problem areas in drafting or reviewing a subcontract indemnity clause.

✔ SUBCONTRACT INDEMNITY CLAUSE CHECKLIST

(1) *Scope of indemnity.* What types of claims and damages are covered? Is the indemnity clause limited to damages arising out of any injury to any person, or any death at any time resulting from such injury, or any damage to any property, which may arise (or which may be alleged to have arisen) out of or in connection with the work covered by the subcontract? Do the damages covered go beyond those connected with property damage or personal injury? Do the damages covered include attorneys' fees and consequential damages? Care in defining the scope of the damages covered is critical for the contractor and subcontractor alike.

(2) *Persons protected.* Is the indemnity obligation owed only to the contractor, or does it include other named parties or groups such as the design professional? How broadly is the protected class defined?

(3) *Conduct creating the indemnity obligation.* Is the indemnity obligation limited to damages that flow solely from the conduct of the subcontractor? Does the indemnity obligation arise if the subcontractor is only partially responsible for

[42] *See, e.g.,* 770 ILCS 60/21 (2003).

the damage? Does the subcontract attempt to place the indemnity obligation on the subcontractor even if the damages do not result in any part from the subcontractor's negligence?

(4) *Insurance availability.* Some types of risks that are assigned to subcontractors through indemnity clauses may be mitigated through insurance policies. Contractors and subcontractors, however, frequently fail to assess and compare the risks presented by indemnity clauses with their insurance coverage.

(5) *Public policy limits.* As previously mentioned, many states have taken the position that broad-form indemnity agreements (i.e., those obligating one party to indemnify another for all loss, even if caused solely by the indemnitee's own negligence) are void as against public policy. Illinois, Georgia, New York, Minnesota, California, Michigan, and many other states have such "anti-indemnity" statutes.[43] Depending on the jurisdiction involved, however, these anti-indemnity statutes may not preclude the contractor from requiring that the subcontractor obtain insurance to cover the negligent acts of the contractor.[44] Specifically, a court may view as acceptable the long-standing practice in the construction industry whereby parties to a subcontract agree that one party is to purchase insurance that protects "others" involved in the performance of the construction project. Depending on the jurisdiction, a subcontract provision requiring such insurance coverage may or may not be viewed as an unenforceable anti-indemnity agreement. A party contemplating the use of such a provision should check to determine if the protection afforded by the broad subcontract indemnity clause is not taken away by applicable laws and the court's interpretation of such laws.

[43]30 Alaska Stat. § 45.45.900; Ariz. Rev. Stat. Ann. § 34–226; Cal. Civ. Code § 2782; Conn. Gen. Stat. § 52–572k; Fla. Stat. Ann. § 725.06; Ga. Code Ann. § 13–8–2; Haw. Rev. Stat. § 431: 10–222; Idaho Code Ann. § 29–114; 740 Ill. Comp. Stat. 35/1; Ind. Code Ann. § 26–2–5–1; La. Rev. Stat. Ann. § 38–2216(D); Md. Code Ann., Cts. & Jud. Proc. § 5–401; Mass. Gen. Laws Ann. ch. 149, § 29C; Mich. Comp. Laws § 691.991; Minn. Stat. § 337.01–05; Miss. Code Ann. § 31–5–41; Neb. Rev. Stat. § 25–21–187; N.M. Stat. Ann. §§ 56–7–1; N.Y. Gen. Oblig. Law § 5–322.1; N.C. Gen. Stat. § 22B-1; N.D. Cent. Code § 9–08–02.1; Ohio Rev. Code Ann. § 2305.31; Or. Rev. Stat. § 140; R.I. Gen. Laws § 6–34–1; S.C. Code Ann. § 32–2–10; S.D. Codified Laws §§ 56–3–18; Tenn. Code Ann. § 62–6–123; Utah Code Ann. § 13–8–1; Va. Code Ann. § 11–4.1; Wash. Rev. Code Ann. § 4.24.115; W. Va. Code § 55–8–14. Several states have taken the position that broad-form indemnity agreements cannot be used to protect architects, engineers, and other design professionals. *See, e.g.,* Del. Code Ann. tit. 6, § 2704 (statute is designed to limit indemnity clauses in contracts of preconstruction designers and planners); N.H. Rev. Stat. Ann. § 338-A:1 (prohibiting indemnification agreements by architects, engineers, and surveyors); 68 Pa. Cons. Stat. § 491 (prohibiting indemnification agreements by architects, engineers, and surveyors); Tex. Civ. Prac. & Rem. Code Ann. §§ 130.001–005 (architects and engineers are not protected against injuries or damages caused by defects in their plans, designs, or specifications prepared for a construction project); Utah Code Ann. § 13–8–2 (subcontract cannot limit design professional's liability to subcontractor).

[44]*See ESI, Inc. of Tenn. v. Westpoint Stevens, Inc.,* 562 S.E.2d 198 (Ga. Ct. App. 2002) (holding Georgia anti-indemnity statute did not preclude contractual requirement for insurance coverage); *Jokich v. Union Oil Co. of Cal.,* 574 N.E.2d 214 (Ill. App. Ct. 1991); *Holmes v. Watson-Forsberg Co.,* 488 N.W.2d 473 (Minn. 1992) (holding that a subcontract provision obligating the subcontractor to maintain general liability insurance coverage was not an unenforceable indemnification agreement under Minnesota statutes but rather was a valid agreement to provide specific insurance coverage as allowed by statute).

(6) *Clarity of the parties' intent.* Since indemnity provisions tend to reverse the generally preferred rule that wrongdoers should pay for their own mistakes, courts often strictly construe the indemnity language. Loose language used in defining the indemnity obligation may result in a restrictive reading of the scope of the parties' indemnity agreement. Therefore, it is essential to understand the scope of the risk that is intended to be the subject of the indemnity clause and to define it clearly.

(7) *Collection rights.* Typically, the subcontract may give the contractor the right to withhold from any payment otherwise due the subcontractor such amounts as may be "reasonably necessary" to protect it against damages covered by the indemnity clause. From the subcontractor's perspective, it is desirable to place some safeguards on the process and timetable for determining indemnity losses, and some limits (e.g., indemnity payments will be limited to the amount of available insurance) on the exposure assumed under the subcontract.

H. Labor Affiliation

The prime contractor should have the right to determine whether a project will be open shop or union and to require, insofar as permitted by applicable law, its subcontractors to abide by this policy. The prime should retain the right to require a subcontractor to perform, or be terminated, even if the subcontractor's workers refuse to work as a result of a dispute that involves the general contractor and its employees (not the subcontractor's). In some instances, the owner may dictate whether the project will be union or open shop. In these cases, the general contractor should be sure to include language in the prime contract that specifically exempts the general contractor from responsibility for work stoppage as a result of labor strikes. A list of some labor issues that should be addressed in any subcontract follows.

✔ LABOR AFFILIATION CHECKLIST

(1) *Advice regarding labor practices.* The parties should define the subcontractor's obligation to keep itself and the prime contractor fully advised of all pertinent local and regional labor agreements and practices, including any local labor union contract negotiations occurring during the term of the subcontract.

(2) *Collective bargaining agreements.* If the subcontractor has a collective bargaining agreement either locally or nationally with a labor union engaged in local negotiations, or if the subcontractor will be affected, either directly or indirectly, by the outcome of local negotiations, is the subcontractor obligated to join such negotiations, if legally permissible, and participate or associate itself with the local contractor or contractors involved in the negotiations in an endeavor to resolve the labor dispute?

(3) *Labor harmony.* The subcontractor should be obligated to commit that all labor used throughout the work shall be acceptable to the owner and the contractor and of a standing or affiliation that will permit the work to be carried on harmoniously and without delay to the project and that will in no case or under

any circumstances cause any disturbance, interference, or delay to the progress of the building, structures, or facilities, or any other work being carried on by the owner or the contractor in any other town or city in the United States.

(4) *Labor agreements.* The subcontractor must recognize and comply with all agreements of the contractor with local building trade councils or separate unions concerning labor and working conditions and otherwise applicable to the work insofar as these agreements do not conflict with or violate any local, state, or federal laws or properly constituted orders or regulations.

(5) *Termination rights.* The contractor has the right to terminate the subcontract and proceed in accordance with the provisions of the subcontract if the subcontractor's work or the contractor's work is stopped or delayed or interfered with by strikes, slowdowns, or work interruptions resulting from the acts or failure to act of the employees of the subcontractor in concert, or if the subcontractor breaches other applicable labor provisions in the subcontract.

(6) *Work stoppages.* The contractor's remedies when the subcontractor's employees engage in a work stoppage solely as a result of a labor dispute involving the contractor or others and not in any manner involving the subcontractor should be clearly defined.

I. Disputes Procedures

1. In General

It is generally desirable that the prime contractor tie the subcontractor to the same remedies and disputes process that binds the prime contractor to the owner. Whether in court, an administrative disputes process (board hearing), or an alternative dispute resolution procedure (e.g., mediation, arbitration, or both) the subcontractor should be required to pursue diligently and exhaust those remedies and to be bound by the determination of its claims or rights under the specified disputes procedure. Consider the following dispute resolution issues when drafting or reviewing a subcontract.

✔ DISPUTES CLAUSE CHECKLIST

(1) *Flow-down effect of dispute resolution.* To what extent is the determination of any issue under the terms of the contract between the owner and the prime contractor binding on the rights of the subcontractor? From the prime contractor's perspective, the ability to "flow down" the impact of a binding decision under the owner/contractor agreement is critical.

(2) *Subcontractor participation in owner/contractor disputes procedures.* Is the subcontractor obligated to assist the contractor in the prosecution or defense of a proceeding under the terms of the general contract? Is the subcontractor obligated to reimburse the prime contractor for any portion of the costs and legal fees incurred? Even if the subcontractor is not a named party to the dispute proceedings, if the subcontractor's rights will be affected by the outcome, it will want the right to participate in the proceedings.

(3) *Decision-making authority.* Does the subcontractor have any control over the prosecution and defense of any proceeding under the terms of the general contract that involves or relates to the subcontract? Does the subcontractor have any protection against an unreasonable compromise of its dispute position?

(4) *Obligation to continue working.* Is the subcontractor obligated to proceed with the work as directed by the contractor pending resolution of any dispute under the terms of the general contract or any dispute between the contractor and its subcontractor?

(5) *Disputes not involving the owner.* Does the subcontract contain provisions, such as an arbitration clause controlling the resolution of disputes between the contractor and the subcontractor that are not related to the owner or controlled by the terms of the general contract?

(6) *Third-party claims.* Does the subcontract address the subcontractor's right or duty to proceed against third parties to recover claims for damages? Such clauses are increasingly popular, not only in subcontracts, but also in owner/contractor agreements.

(7) *Mandatory pursuit of administrative remedies.* Does the subcontract require the subcontractor to exhaust any contractual disputes process before instituting or prosecuting any statutory remedy or action against the contractor or the contractor's surety?

(8) *Mandatory alternative dispute resolution (ADR) procedures.* In recognition of the high cost and uncertainty of litigation and other "formal" disputes procedures, many in the construction industry are searching for less costly, less divisive, and quicker means for resolving disputes, such as mediation. The ability to control, at least to some extent, the dispute process is also a consideration. Any ADR procedure should be detailed in the subcontract. For example, Subparagraph 11.5.1 of ConsensusDOCS 750 provides that the parties shall first seek to resolve a dispute by direct discussions. If those do not resolve the matter, the subcontract requires the parties to participate in mediation under the Construction Industry Mediation Rules of the American Arbitration Association before resorting to any other form of binding dispute resolution. In the revised version of AIA Document A401, Section 6 requires mandatory mediation prior to binding dispute resolution. This clause allows the party drafting the contract to specify binding dispute resolution as either arbitration or litigation. In the absence of an expressed selection, the default is litigation. Similarly, EJCDC Documents allow the drafting party to decide between mediation and binding arbitration.

(9) *Appointment of final decision makers.* Contracts and subcontracts often attempt to designate some individual as having the authority to make "final and binding" decisions with respect to certain types of disputes. These contract clauses may also attempt to place short time limits for appeal or protest of a decision, before it becomes "final and binding." Identify such attempts before contracting and be aware of their impact.

2. *Arbitration Considerations*

The determination to accept or include an arbitration provision in a contract or subcontract involves both practical and legal considerations. This decision must be made before the subcontract is executed. If the subcontract does not contain an arbitration provision, disputes arising thereafter will be arbitrated only if both parties so agree. Such an agreement may not be easy to obtain after the dispute has arisen. If the subcontract includes an arbitration agreement, however, it probably will be enforceable under federal and perhaps state laws. One party typically cannot avoid arbitration at that point over the other party's objection.

Some subcontracts contain provisions requiring arbitration at the sole election of the contractor. Such unilateral provisions have been enforced by some courts.[45]

J. Federal Government Projects

If the prime contract is with a federal government agency, certain provisions in the contract become more important, and there are other provisions—perhaps unnecessary in private work—that need to be included in the subcontract form. The same may be true with respect to state governments and their agencies. Any contractor who works with a particular governmental agency or agencies would be well advised to consult an attorney concerning what, if any, additional provisions should be included in its standard subcontract form due to the particular way in which that agency procures construction services.

The following examples of contract provisions may be useful additions to subcontracts on federal government projects:

✔ FEDERAL GOVERNMENT PROJECT CHECKLIST

(1) *Termination for convenience.* The government's exercise of its rights under this clause has a significant substantive and procedural effect on the contractor's rights and remedies, including express directions to the contractor concerning its subcontractors and suppliers. The federal government's duty to reimburse the contractor for certain subcontractor/supplier claims may depend on the negotiations between the contractor and subcontractor/supplier concerning the effect of a termination for convenience.[46]

(2) *Cost and pricing data.* The subcontract/purchase order should address the subcontractor's/supplier's obligation to provide "cost or pricing" data or other cost and claim certifications required by or necessary under the terms of the prime contract. To the extent that the contractor relies on or is liable to the federal government due to data, information, or certificates provided by the subcontractor or supplier, the subcontract or purchase order should address the issues of indemnity, costs of defense, and so on, related to any government

[45] *See, e.g. TechnoSteel LLC v. Beers Constr. Co.*, 271 F.3d 151 (4th Cir. 2001).
[46] *See* FAR § 49.108–5(a).

claim. In addition, the clause should address the extent and duration of audit rights by any party and any other needed flow-down requirement.

(3) *Flow-down consistency.* Particular care must be given to flow-down clauses needed to promote the consistency of the prime contract and the subcontracts for proper administration of a subcontract or purchase order under a government contract. In addition, numerous social and economic clauses must be incorporated into specific subcontracts and purchase orders and further flowed down to lower-tier contracts. Satisfaction of these requirements dictates the careful drafting of appropriate flow-down provisions and a review of the terms of the government contract. More information on these issues can be found in **Chapter 22**.[47]

III. SHOULD SUBCONTRACTORS BE BONDED?

A performance bond is a written guaranty by a third party (based on the terms of the bond) that the principal will perform its contract obligations. In the context of a subcontractor principal, such a bond usually is intended to provide that if the subcontractor defaults or fails to complete the project, the surety will complete performance or pay damages up to the limit of the penal sum of the bond.

A labor and material payment bond (or, simply, payment bond) helps ensure that labor and materials used in the course of the subcontractor's work will be paid for by the surety if not by the subcontractor. A payment bond creates alternatives to the filing of liens on the project property and helps protect the prime contractor from liability to the various tiers of sub-subcontractors and suppliers. (See **Chapters 14 and 15** for a more complete discussion of payment and performance bonds.)

In the past, it was not unusual for prime contractors to require bonds of subcontractors and for subcontractors to require bonds of sub-subcontractors. The general contractor may decide at the outset that the subcontracted work will be bonded, no matter which subcontractor is chosen to do the work. In recent years, however, these bonds have become more expensive and sometimes more difficult to obtain. Today, whether subcontractors should provide bonds has become somewhat more complicated. Consequently, the decision to require bonds may need to be made on a case-by-case basis. Other alternatives may provide some assurance of the subcontractor's performance without involving the expense of a bond. For example, the subcontractor may be willing to post a letter of credit in lieu of a bond; another entity, perhaps a parent company or an individual principal of the company, may be willing to provide a guarantee of performance; or the subcontractor may allow the prime to hold an increased level of retainage from progress payments.

The prime contractor that decides to have its subcontractors bonded must be aware of the exact language of the bonds. No statutory requirements exist for subcontractors'

[47]*See also* Thomas J. Kelleher, Jr., Thomas E. Abernathy IV & Hubert J. Bell, Jr., Federal Construction Contracts—A Practical Guide for the Industry Professional (John Wiley & Sons 2008).

bonds. They are considered to be common law bonds, and their wording controls the obligations of the surety.

Subcontractor performance bonds can be worded so that they protect or benefit only the prime contractor and do not provide a direct right of action by third-party subcontractors and suppliers. Alternatively, the performance bond may expressly state that subcontractors and suppliers can maintain a direct action against the surety. Otherwise, a court may well find that the bond does not create such a right of action in third parties.

If bonds are to be provided, separate payment and performance bonds should be required. The contractor obtains greater protection by requiring two separate bonds. If the bonds are not separate, the penal sum provides the maximum liability of the surety for both the payment and performance bonds. For example, if a payment and performance bond is in the total penal amount of $1 million, the surety is generally liable for no more than $1 million, even where the cost to complete the subcontractor's performance together with the claims of materialmen and suppliers might exceed $1 million. If, however, a performance bond is obtained in the amount of $1 million and a payment bond is obtained in the same amount, the general contractor obtains potential protection of $2 million, with $1 million for performance and $1 million for payment.

It is essential that the surety providing the bonds be financially viable and responsible, since bonds from a deficient surety offer little or no protection. Consequently, the subcontract form should contain the limitation that the bonds will be provided by an "acceptable surety."

IV. DISPUTE AVOIDANCE BY DILIGENT PROJECT ADMINISTRATION

Disputes often can be avoided or minimized through effective project management by both the prime contractor and the subcontractors. An understanding of some important obligations and rights is critical to dispute avoidance or resolution.

A. General Contractor's Duty to Coordinate the Work

Subcontractors normally are engaged by the general contractor to perform various "trade" portions of the work. For the project to proceed smoothly and without unreasonable interference, the activities of the general contractor and of each subcontractor must be sequenced and coordinated so that all can promptly and efficiently perform their work.

The general contractor usually assumes responsibility for coordination, which is similar to the responsibility of the owner in the coordination of parallel prime contractors. Since the general contractor is customarily the only party in direct privity with all of the subcontractors, that contractor generally has the duty to coordinate.

B. Implied Duty to Cooperate

The duty to cooperate with subcontractors and coordinate their work is implied, even though the subcontract may not expressly provide that the prime or general contractor must respond in damages for any delay or extra cost incurred by a subcontractor due to a breach of this duty. Subcontract provisions, such as a "no damages for delay" clause, may, however, modify, limit, or shift this implied responsibility or the damages for breach of this duty.[48]

The implied duties to cooperate and coordinate require the general contractor to take reasonable measures to protect a subcontractor from delays or interference by the general contractor or other subcontractors. For example, in *Johnson v. Fenestra, Inc.*,[49] a subcontractor contracted to install panels on the exterior of a building under construction, and the prime agreed to furnish these panels. The prime contractor delivered defective panels. Several months elapsed before the correct panels were provided, and the delay prevented the subcontractor from completing work before winter weather set in. The court held that the subcontractor was entitled to damages for being forced into winter work and that the subcontract necessarily implied that the prime would supply panels in time and in quantities that were consistent with the parties' understanding that there would be a "prompt beginning and an early completion" of the job.

The obligation to reimburse for delay and extra cost may arise from provisions that require the prime to perform some act preparatory to the subcontractor's work. In *Manhattan Fire Proofing Co. v. John Thatcher & Son*,[50] the prime agreed that the building under construction would be made available to the concrete subcontractor so that the concrete work could be done "in one operation" and completed by a stated date. Because the prime delayed the job and was late in turning over the building, however, the subcontractor was required to divide its work into several phases. The court, in awarding the subcontractor delay damages, held that the subcontract provision that the work could be "done in one operation" required the prime contractor to protect the subcontractor from the type of disruption delay experienced on the job.

C. Implied Duty to Coordinate

It is not enough for the general contractor to ensure that its own forces do not interfere with the subcontractors. The duty to coordinate includes an additional requirement that the general contractor take reasonable steps to avoid interference between subcontractors. This requires the contractor to (1) schedule the work of all subcontractors so that the work of different trades can be sequenced as contemplated at the time of bidding, and (2) take reasonable steps during performance to maintain the schedule.

[48]*See Crawford Painting & Drywall Co. v. J. W. Bateson Co.*, 857 F.2d 981 (5th Cir.1988).
[49]305 F.2d 179 (3d Cir. 1962).
[50]38 F. Supp. 749 (W.D.N.Y. 1941).

A general contractor's responsibility for coordination of all subcontractors is illustrated in *J. J. Brown Co. v. J. L. Simmons.*[51] There the plastering subcontractor claimed damages for the prime's failure to provide heat in buildings where the plastering work was to be performed. The prime contended that other subcontractors were responsible for the failure to provide heat. The court held that the prime had a duty "to keep the work in such state of forwardness as to enable the subcontractor to perform within a limited time" and that the delays of other subcontractors were the prime's responsibility and would serve as no defense to the plastering subcontractor's suit.[52]

D. Limitation of Liability

A prime contractor may be able to limit liability by disclaiming the responsibility of coordination to subcontractors, as illustrated by *Crawford Painting & Drywall Co. v. J. W. Bateson Co., Inc.*[53] Bateson was the prime contractor on a hospital building project for the Army Corps of Engineers. Bateson's subcontract with Crawford included a clause in which Bateson disclaimed liability for any delay or disruption to Crawford's work.

After excessive change orders were issued, the work was significantly delayed. Crawford sued Bateson, and at trial introduced evidence that Bateson withheld from its subcontractors the actual project schedule and extensions to that schedule that Bateson had negotiated with the Corps. Crawford alleged that Bateson took advantage of the subcontractors in administering the project and in negotiating claim settlements. The jury awarded Crawford $7 million, but a federal court of appeals reversed the jury award and dismissed the lawsuit. The court held that Bateson's actions amounted to nothing more than a failure to coordinate the subcontractors, and Bateson had disclaimed this responsibility in its subcontract. Crawford could not recover.

Although the general contractor can be held answerable for its interference with or failure to coordinate subcontractors, the general may not be liable for delays or interference over which it has no control and that were unforeseeable at the time of contracting. In *Southern Fireproofing Co. v. R. F. Ball Construction Co., Inc.,*[54] the court held that a contractor was not liable for damages caused to a subcontractor by delays attributable to subsurface conditions that were unknown to the parties at the time of subcontracting. A similar result was reached when a prime contractor was held not liable for damages caused by delays that were due to inclement weather.[55]

Success in coordination depends largely on open and accurate communication between the general contractor and all subcontractors. The general must establish

[51]118 N.E.2d 781 (Ill. App. Ct. 1954). *See also Ragan Enters. v. L & B Constr. Co.,* 492 S.E.2d 671 (Ga. Ct. App. 1997).

[52]*J.J. Brown Co. v. J.L. Simmons Co.,* 118 N.E.2d 781, 785 (Ill. App. Ct. 1954). *See also United States ex. rel. Wallace v. Flintco, Inc.,* 143 F.3d 955 (5th Cir. 1998); *Allied Fire & Safety Equip. Co. v. Dick Enter., Inc.,* 886 F. Supp. 491 (E.D. Pa. 1995); *Cleveland Wrecking Co. v. Cent. Nat'l Bank of Chi.,* 576 N.E.2d 1055 (Ill. App. Ct. 1991); *Unis v. JTS Constructors/Managers, Inc.,* 541 So. 2d 278 (La. Ct. App. 1989).

[53]857 F.2d 981 (5th Cir. 1988).

[54]334 F.2d 122 (8th Cir. 1964).

[55]*Ben Agree Co. v. Sorensen-Gross Constr. Co.,* 111 N.W.2d 878 (Mich. 1961).

some method to both discover what is going on at the job site (e.g., what problems the subcontractors are having and the problems they anticipate) and to relay suggestions, recommendations, and requirements to the subcontractors to avoid or solve the problems. Efforts to develop these communications must begin at the preconstruction conference and continue throughout the contract performance period.

E. Pay Applications and Partial Lien Waivers

When a general contractor receives a pay application from a subcontractor, it generally should require a partial waiver of liens covering work for which payment is requested. At the end of a project, total (or final) lien waivers should be required. Subcontractors should obtain similar lien waivers from their lower-tier subcontractors and suppliers. Lien laws vary greatly from state to state, so advice of counsel familiar with the applicable lien law should always be obtained. Generally, waivers of lien rights are strictly construed, must be in writing, and must clearly express the potential lien claimant's unambiguous intent to waive lien rights.

Contractors must not confuse a lien waiver with a release of claims. A lien waiver, unlike a release, generally benefits only the property owner by protecting it from the claim of lien. It does not absolve the contractor from its contractual obligations to the lien claimant.[56] The general contractor should consider using a form that both waives lien rights and releases the contractor from subcontractor's claims, at least to the extent of payment actually received.

F. Prime Contractor Financing of Subcontractors

The question of whether and to what extent a prime should provide financial aid to an ailing subcontractor usually arises when an unbonded subcontractor is on the verge of default.

A prime contractor has several obvious reasons for wanting to help a faltering subcontractor. First, the prime is usually obligated to complete the project by a specified time. A work slowdown before default, and delay while obtaining a completion subcontractor, may extend the project beyond the completion date. Second, the prime must keep the work in a state of readiness so that other subcontractors can proceed efficiently. A failure to coordinate the work may subject the prime to delay damage claims by these subcontractors. Third, the prime may not be able to recover damages from the defaulting subcontractor because of the latter's financial predicament. In short, the subcontractor may be "judgment proof."

Other advantages may be gained by working out some form of financial assistance for a troubled subcontractor. The value of maintaining continuity on the job may justify the cost of financial assistance. The subcontractor's personnel have a basic familiarity with the project, and work experience would eliminate the need for another learning curve for replacement workers. The defaulting subcontractor may

[56]*See, e.g., Hampshire Homes, Inc. v. Espinosa Constr. Serv., Inc.,* 655 S.E.2d 316 (Ga. Ct. App. 2007); *David Shapiro & Co., Inc. v. Timber Specialties,* 233 S.E.2d 439 (Ga. Ct. App. 1977).

have executed fixed-price purchase orders that could be lost if the sub is terminated, causing the completing subcontractor's cost (and, correspondingly, the cost to the general contractor) to increase. Also, there may be a shortage of available subcontractors in particular trades and geographic areas at any given time, making it difficult or even impossible for the prime to find a suitable replacement subcontractor.

All these considerations suggest that it may be prudent for a prime contractor to consider providing financial assistance to an unbonded subcontractor on the verge of default. This assistance may come in the form of advanced payments, reduced retainage, guarantees of payment to sub-subcontractors or suppliers, and the like. Although these techniques may help the subcontractor complete the job, they also create potential risks by reducing the prime contractor's financial security in the subcontractor's performance or creating an obligation for the prime contractor that is larger than anticipated.

Different considerations are involved when the faltering subcontractor is bonded, and these problems are reviewed elsewhere. (See **Chapter 14 and Chapter 15**.) As a general matter, any payments the surety makes to a distressed subcontractor do not result in a credit to the surety against the penal sum of the performance bond.[57] But the prime contractor that lends financial assistance runs a risk of having the surety later claim to be discharged by reason of a material increase in the bonded risk.

G. Remedies for Defective Performance

Contract law generally provides that where a builder's performance is defective, the agreed contract price is subject to a deduction for the costs to remedy the defects. The same rule applies between prime contractors and subcontractors.[58] If a subcontractor fails to perform its work, the prime contractor may perform the work, backcharge the subcontractor, and deduct the backcharge from the subcontractor's progress payments.

The measure of the prime contractor's damages generally is the reasonable cost of remedying the defects—that is, the cost of making the subcontractor's work conform to the contract.[59] Where the prime contractor retains a portion of the subcontract price, the measure of damages is the reasonable cost of completing the contract and repairing the subcontractor's defective performance less the part of the contract price still unpaid.[60]

A prime contractor's decision to perform the subcontractor's work assumes that the subcontractor is in default. That assumption carries with it some risk. The contractor runs the risk of guessing incorrectly as to whether the subcontractor has substantially performed in the first instance and whether the backcharge is valid in the

[57]*See Caron v. Andrew,* 284 P.2d 544 (Cal. Ct. App. 1955); *Employers Mut. Cas. Co., v. United Fire & Cas. Co.,* 682 N.W.2d 452 (Iowa Ct. App. 2004)

[58]"[T]he builder. . .may be entitled to a deduction or set-off against such compensation [as is owed the subcontractor] for expenses incurred or damages sustained by reason of the subcontractor's failure to comply with his or her contract." 17A C.J.S., *Contracts,* § 392 (2008).

[59]*Sorensen v. Robert N. Ewing,* 448 P.2d 110 (Ariz. Ct. App. 1968).

[60]*Id.* at 114.

second. In making such an error, the contractor may, itself, be committing a material breach of the subcontract.[61]

H. Remedies for Delayed Performance

When a subcontractor fails to complete the work covered by its subcontract within the time specified, and the delay in completion is not excusable, the prime contractor is entitled to recover any resulting, foreseeable damages. A delay is excusable if it results from an event or omission for which the subcontractor is not responsible. Under certain circumstances, the subcontractor's unexcused delay may warrant subcontract termination.

Termination of a sub for late performance is proper only if the delay is not excusable. When a subcontractor is delayed because of the prime contractor's failure to schedule or coordinate other trades on the project, the delay is excusable and does not warrant a default termination.

In *Tribble & Stephens Co. v. Consolidated Services,*[62] the court held that a subcontractor was wrongfully terminated since the delay in prosecuting its work was excusable. In *Tribble,* the court relied on evidence showing that the contractor had ordered the subcontractor to relocate electrical junction boxes as a result of another subcontractor's work. In addition, a delay in completing the soffits, which was not in the subcontractor's scope of work, prevented the subcontractor from installing the light fixtures. Finally, certain owner-furnished material arrived late and impeded the subcontractor's wiring in certain areas of the project.

Typically, the liquidated damages assessed by the owner against the prime contractor constitute the owner's measure of damages for delays, although the owner may be able to recover actual damages. A prime contractor may be able to protect itself from liquidated damages when the delay is caused by a subcontractor and the subcontract contains an appropriately worded subcontract provision. For example, in *Taos Construction Co. v. Penzel Construction Co.,*[63] the court ruled that a subcontract's "pass on" of liquidated damages from the prime to its subcontractor was valid.

Krauss v. Greenberg[64] also illustrates the prime contractor's rights to recover damages caused by a subcontractor's delay. Krauss, the subcontractor, sued Greenberg for the value of material furnished in connection with work under a government contract that contained a liquidated damages clause. The prime contractor had been charged liquidated damages by the owner for delays that were attributable to the subcontractor. Therefore, the prime counterclaimed against the subcontractor, raising the question of whether consequential damages could be recovered for breach of a subcontract. The damages were allowed because the court found that assessment of liquidated damages by the owner in the event of late completion was within the contemplation of the subcontracting parties. Unless the subcontract language controlling

[61]*Howard S. Lease Constr. Co. v. Holly,* 725 P.2d 712 (Alaska 1986).
[62]744 S.W.2d 945 (Tex. App. 1988).
[63]750 S.W.2d 522 (Mo. Ct. App. 1988).
[64]137 F.2d 569 (3d Cir. 1943).

the subcontractor's liability for delay damages is carefully written, a prime contractor's recovery for delays caused by the subcontractor may be limited to the liquidated damages paid to the owner.[65]

I. Relationship between Subcontractor and Owner: Can the Prime Contractor Assert the Subcontractor's Rights against the Owner?

There is normally no "privity of contract" between the owner and any subcontractor. As a result, the owner and subcontractor generally cannot sue each other for breach of contract. In addition, the owner's only payment obligation is to the prime contractor under the terms of their contract. The prime contractor in turn has a corresponding contractual obligation to pay its subcontractors.[66]

In connection with federal government contracts, it is permissible for a prime contractor to bring an action against the government on behalf of a subcontractor that has been injured by government action or inaction. This "sponsorship" rule applies to appeals before the various agency boards of contract appeals as well as before the United States Court of Federal Claims. This right was confirmed in *United States v. Blair,*[67] where the United States Supreme Court stated that Blair (the prime contractor) "was the only person legally bound to perform his contract with the Government and he had the undoubted right to recover from the Government the contract price for the tile, terrazzo, marble, and soapstone work whether that work was performed personally or through another."[68]

This same theory may also be available as a basis for an action by a prime contractor on behalf of its subcontractor against a nonfederal government owner, where the prime, in effect, acts as a conduit for the cause of action.[69]

The subcontractor may be entitled to recover from both the prime and the owner when both are found to be responsible for damages. In *Rome Housing Authority v. Allied Building Materials, Inc.,*[70] the court held that the trial court was correct in awarding delay damages to a subcontractor apportioned equally between the contractor and the owner as a result of their failure to resolve disputes in a timely manner. In *Mobile Chemical Co. v. Blount Bros. Corp.,*[71] a general contractor, acting as construction manager, and an owner were found equally liable for damages to subcontractors resulting from their joint decision to accelerate. The court admonished the general contractor because its "first construction schedule was prepared by an

[65]*Indus. Indem. Co. v. Wick Constr. Co.,* 680 P.2d 1100 (Alaska 1984).
[66]*Marion Mach. Foundry & Supply Co. v. Colcord,* 294 S.W. 361 (Ark. 1927); *Utschig v. McClone,* 114 N.W.2d 854 (Wis. 1962).
[67]321 U.S. 730 (1944).
[68]*Id.* at 737.
[69]*See Robert E. McKee v. City of Atlanta,* 414 F. Supp. 957 (N.D. Ga. 1976); *St. Paul Dredging Co. v. State,* 107 N.W.2d 717 (Minn. 1961).
[70]355 S.E.2d 747 (Ga. Ct. App. 1987).
[71]809 F.2d 1175 (5th Cir. 1987).

employee innocent of the ability to prepare such a schedule for a complex project."[72] The message is clear: The prime contractor's right to insist that subcontractors perform in accordance with the project schedule carries with it a corresponding obligation to ensure that the schedule is accurate, realistic, and consistent with the parties' agreement.

In *Wexler Construction Co. v. Housing Authority of Norwich,*[73] the general contractor was not allowed to sue on behalf of a subcontractor. In that case the court implied, however, that the *Blair* doctrine was inapplicable only because there existed an implied contract between the subcontractor and the owner, and thus there was no reason why the subcontractor could not bring the action directly.

J. Prime Must Be Liable to the Subcontractor for the Pass-Through Claim: The *Severin* Doctrine

Many courts have held that a prime contractor can maintain a suit on behalf of one of its subcontractors only if it has reimbursed its subcontractor for the latter's damages or remains liable for such reimbursement in the future. Courts typically refer to this as the *Severin* doctrine, which was first articulated in *Severin v. United States.*[74] In *Severin,* the contract between the prime contractor and the subcontractor contained an exculpatory clause holding the prime contractor harmless from any claim caused by the actions of the government/owner. The court, therefore, held that it had no jurisdiction to hear any claim based upon damages to the subcontractor.

Exceptions to the *Severin* doctrine have been made since the decision was first rendered. In 1965, the Court of Claims held that the doctrine does not apply when subcontractors' claims are asserted as an equitable adjustment under the provisions of a prime contract with the federal government.[75] Additionally, the Court of Claims has held that in federal contract cases, the government bears the burden of proving that the prime contractor is not liable to the subcontractor.[76] Moreover, the Armed Services Board of Contract Appeals has held that a prime contractor's claim may proceed against the government even if the prime's liability to its subcontractors is contingent on the prime's successful appeal.[77]

[72] *Id.* at 1177.
[73] 183 A.2d 262 (Conn. 1962).
[74] 99 Ct. Cl. 435 (1943), *cert. denied,* 322 U.S. 733 (1944). *See also J. L. Simmons Co. v. United States,* 304 F.2d 886 (Ct. Cl. 1962).
[75] *Blount Bros. Constr. Co. v. United States,* 348 F.2d 471 (Ct. Cl. 1965). *See also Metric Constructors, Inc. v. United States,* 314 F.3d 578 (Fed. Cir. 2002) (government did not meet its burden of establishing that subcontractor executed iron-clad release sufficient to trigger application of the *Severin* doctrine). Boards and courts have limited the *Severin* doctrine out of reluctance to leave subcontractors with valid claims out in the cold. Numerous cases have recognized these limitations. *See, e.g., Ball, Ball & Brosamer, Inc.,* IBCA No. 2841, 97–1 BCA ¶ 28,897; *E.R. Mitchell Constr. Co. v. Danzig,* 175 F.3d 1369 (Fed. Cir. 1999); *Roof-Techs Int'l, Inc. v. State,* 57 P.3d 538 (Kan. Ct. App. 2002); *Morrison Knudsen Corp. v. Fireman's Fund Ins. Co.,* 175 F.3d 1221 (10th Cir. 1999).
[76] *Ace Constructors, Inc. v. United States,* 70 Fed. Cl. 253 (2006).
[77] *Oconto Elec., Inc.,* ASBCA No. 45856, 94–3 BCA ¶ 26958.

K. States Have Adopted the *Severin* Doctrine

Some state courts have adopted the *Severin* doctrine as their own. For instance, the Georgia Supreme Court held that a general contractor may not recover on behalf of its subcontractors absent proof of liability to those subcontractors.[78] Thus, if a subcontract absolves a general contractor of liability to its subcontractor for delay damages, the general contractor is precluded from bringing an action to recover for delays on behalf of the subcontractor.

In *Warren Bros. Co. v. North Carolina Department of Transportation,* the North Carolina Court of Appeals created an absolute bar to subcontractor claims against the North Carolina Department of Transportation even if the claim is brought by the general contractor and regardless of whether the general contractor is or will ultimately be liable to the subcontractor for the damages in question.[79] In reaching this decision, the court specifically disregarded considerations of the *Severin* doctrine. Rather, the court based its decision upon the language contained in § 108–6 of the Standard Specifications for Roads and Structures issued by the North Carolina DOT, providing that a subcontractor will not have any claim against the department by reason of the approval of the subcontract by the department, interpreting this section to preclude even claims by general contractors on behalf of a subcontractor. Exceptions to the "*Warren* doctrine" have been recognized, where the prime contractor included the subcontractor's damages as a subset of its own damages. Otherwise, the courts reason, there would be no means of recovering the subcontractor's damages.[80]

L. Conclusion

Avoidance of subcontract disputes demands diligent subcontract administration and constant vigilance by both the prime contractor and subcontractor.

Dispute avoidance begins at the bidding or negotiations stage of the project. Prime contractors should investigate their potential subcontractors in light of the subcontractor's ability to contribute to the overall project success. Subcontractors should be equally diligent in qualifying the prime contractor and owner.

The subcontract document should accurately express the parties' agreement. Whether a form subcontract or a specially drafted document is used, the subcontract should be consistent with the requirements of the general contract and address problems that may affect the progress and performance of the work. Finally, diligent project administration should achieve early detection of problems, which will help avoid disputes or, at least, minimize their consequences.

[78]*Dep't of Transp. v. Claussen Paving,* 273 S.E.2d 161 (Ga. 1980). *See also Univ. of Alaska v. Modern Constr., Inc.,* 522 P.2d 1132 (Alaska 1974); *D.A. Parrish & Sons v. County Sanitation Dist. No. 4,* 344 P.2d 883 (Cal. 1959); *Kinsington Court v. Dep't of State Highways,* 253 N.W.2d 781 (Mich. Ct. App. 1977); *Buckley & Co. v. State,* 356 A.2d 56 (N.J. 1975); *Tully & DiNapoli, Inc. v. State,* 272 N.Y.S.2d 667 (N.Y. Ct. Cl. 1966).

[79]307 S.E.2d 836 (N.C. Ct. App. 1983).

[80]*Metric Constructors, Inc. v. Hawker Siddeley Power Eng'g, Inc.,* 468 S.E.2d 435 (N.C. Ct. App. 1996); *Bolton Corp. v. T.A. Loving Co.,* 380 S.E.2d 796 (N.C. Ct. App. 1989).

➢ POINTS TO REMEMBER

SUBCONTRACT DEFAULT CLAUSE

- Before contracting, utilize a default clause checklist to identify the terms and verify the adequacy of a subcontract termination for default clause. The terms of the subcontract should be consistent with the performance and payment provisions of the prime contract.
- Some courts may recognize actual or constructive notice of default and the right to cure, even in the absence of express language in the default clause.
- The right to terminate a subcontract for the contractor's convenience may be restricted in some courts by a "good-faith" requirement in the exercise of the termination right.

NO DAMAGES FOR DELAY, EXCEPT AS PAID BY THE OWNER

- A subcontract flow-down provision may limit a subcontractor's right to recover delay damages.
- Subcontractors should review the owner/general contractor contract documents carefully to determine, before contracting, the risk assumed by the subcontractor through subcontract flow-down clauses.
- Subcontractors must be aware that their right to recover may depend on compliance with documentation and notice requirements in the prime contract.

CHANGES

- Before contracting, utilize a changes clause checklist to verify the adequacy and fairness of the subcontract changes clause.
- Clauses that attempt to limit the type and amount of additional costs (e.g., overhead, profit, indirect losses) recoverable as a result of contract changes are increasingly popular and potentially dangerous.
- Take the sample checklists outlined in this chapter, expand and tailor them to suit particular practices and industry needs, and require their use in prescreening each potential contract.
- At the beginning of each job, outline the contract procedures (e.g., notice requirements, time limits, cost support and record requirements, etc.) that will govern and limit change order rights.

SUBCONTRACT INDEMNIFICATION

- In some states, clauses requiring one party to hold another party harmless from liability arising solely from the latter party's own acts are void, as against public policy.

- Use a subcontract indemnity clause checklist to determine, in advance of contracting, the exact scope of any indemnity obligation created in the subcontract.

LABOR AFFILIATION

- Ensuring labor harmony on the project should be one of the goals of a properly drafted subcontract.
- Use the labor affiliation checklist to help identify the manner in which the subcontract treats labor issues.

SUBCONTRACT DISPUTES PROCEDURES

- The contractor should be careful that its subcontract disputes procedure is consistent with any disputes procedure required in the owner/contractor agreement.
- If the subcontractor is to be bound by determinations made as a result of owner/contractor disputes, the subcontract should spell out the subcontractor's rights or duties of participation in that disputes procedure.
- Alternative dispute resolution procedures such as arbitration or mediation can be less expensive and less time consuming; consider such alternative dispute procedures before contracting, in lieu of or as a preliminary step before resorting to litigation or arbitration.

FEDERAL AND STATE GOVERNMENT PROJECTS

- Standard subcontract forms utilized in private work are often inadequate for problems and requirements inherent to public contracting.
- Before doing public contract work, have your subcontract reviewed for any necessary inclusion of contract clauses (e.g., cost or pricing data requirements, notice, scope of work, claims and certification requirements) peculiar to government contracting.

9

CONTRACT CHANGES

I. WHAT IS A CHANGES CLAUSE?

Changes are inevitable on any significant construction project. Projects are rarely built exactly as they were originally designed. A multitude of variables can generate the need for a modification to the design, schedule, or performance.

The "changes" clause in a construction contract allows one party to implement changes in the work while the project is being constructed. A changes clause may also come into play when issues arise involving defective specifications, differing site conditions, impossibility of performance, acceleration, inspection, acceptance, and warranties. In short, a changes clause is often an umbrella provision, involving numerous aspects of performance under the contract. No other clause more clearly illustrates the uniqueness and complexity of a construction contract.

A changes clause in a construction contract departs from the established principle of contract law that a contract requires the mutual agreement of the contracting parties. Under the common law, a contract is created when two parties reach an agreement on an identified undertaking. Each party agrees to perform specific contract duties in exchange for some performance by the other party. Once the agreement is reached, the terms of the contract define and limit the obligations of each party, and no one party can unilaterally change or modify the contract. Generally, if changes or modifications are necessary, both parties have to reach a separate agreement incorporating any negotiated changes.[1]

[1]Occasionally a construction contract will be entered into without having a changes clause. In that event, the common law rule applies and a contractor cannot be required to perform changes unless the changes are separately agreed to. In one case, a general contractor entered into a subcontract for the removal, storage, and replacement of existing equipment on a renovation job. The subcontract did not contain a changes clause. When the owner decided to pay to have all new equipment, the general contractor deleted the requirement that the subcontractor reinstall the existing equipment. The general contractor took a credit for this deleted work. The subcontractor sued, arguing that the agreement could not be changed unless there was mutual agreement. The court agreed with the subcontractor and held the general contractor liable for this deleted work. *Werner v. Ashcraft Bloomquist, Inc.*, 10 S.W.3d 575 (Mo. Ct. App. 2000).

Rigid application of this common law rule would create practical problems in construction. Requirements and needs often change during the course of a long construction project. Small details left out of the scope of work can affect the practical use of the entire project and must be timely addressed. Problems would be caused by the common law rule due to the dynamic nature of construction, the ever-changing needs of the owner, or frequent encounters with unexpected developments. The changes clause grew out of the practical need for a contract tool to address changes required after contract formation.

Although a changes clause is found in construction contracts more often than in other types of contracts, it is not correct to presume that the clause violates principles of contract law. Through a changes clause, the contracting parties agree in advance that one of the parties has the right to revise the work in some way under defined terms. The requirement for legal consideration to support contract promises is satisfied by the provision requiring that the contract sum, contract time, or both be adjusted if the change requires extra work or an extended period of performance.

There are many variations in the standard industry wording of the changes clauses,[2] but most widely used changes clauses address these topics:

- Description of authorized changes and the party authorized to make changes.
- Process to determine adjustment, if any, to contract price, time for performance, or both.
- Procedure in the event the parties do not agree that a directive can control a change to the contract.
- Notice requirements related to contract changes.

Article 8 of ConsensusDOCS 200 (2007 ed.) illustrates many of the typical elements of a construction contract changes clause. That clause provides:

ARTICLE 8—CHANGES

Changes in the Work that are within the general scope of this Agreement shall be accomplished, without invalidating this Agreement, by Change Order, and Interim Directed Change.

8.1 CHANGE ORDER

8.1.1 The Contractor may request or the Owner may order changes in the Work or the timing or sequencing of the Work that impacts the Contract Price or the Contract Time. All such changes in the Work that affect Contract Time or Contract Price shall be formalized in a Change Order. Any such requests for a change in the Contract Price or the Contract Time shall be processed in accordance with this Article 8.

[2] *Compare* ConsensusDOCS 200, Standard Agreement and General Conditions Between Owner and Contractor, Article 8 (2007 ed.); AIA A201, General Conditions of the Contract for Construction, Article 7 (2007 ed.); EJCDC C-700, Standard General Conditions of the Construction Contract (2007 ed.); *and* Article 10; Federal Acquisition Regulation (FAR) Changes (June 2007) FAR § 52.243–4.

8.1.2 The Owner and the Contractor shall negotiate in good faith an appropriate adjustment to the Contract Price or the Contract Time and shall conclude these negotiations as expeditiously as possible. Acceptance of the Change Order and any adjustment in the Contract Price or Contract Time shall not be unreasonably withheld.

8.2 INTERIM DIRECTED CHANGE

8.2.1 The Owner may issue a written Interim Directed Change directing a change in the Work prior to reaching agreement with the Contractor on the adjustment, if any, in the Contract Price or the Contract Time.

8.2.2 The Owner and Contractor shall negotiate expeditiously and in good faith for appropriate adjustments, as applicable, to the Contract Price or the Contract Time arising out of an Interim Directed Change. As the Changed Work is performed, the Contractor shall submit its costs for such work with its application for payment beginning with the next application for payment within thirty (30) Days of the issuance of the Interim Directed Change. If there is a dispute as to the cost to the Owner, the Owner shall pay the Contractor fifty percent (50%) of its estimated cost to perform the work. In such event, the Parties reserve their rights as to the disputed amount, subject to the requirements of Article 12.

8.2.3 When the Owner and the Contractor agree upon the adjustment in the Contract Price or the Contract Time, for a change in the Work directed by an Interim Directed Change, such agreement shall be the subject of a Change Order. The Change Order shall include all outstanding Interim Directed Changes on which the Owner and Contractor have reached agreement on Contract Price or Contract Time issued since the last Change Order.

8.3 DETERMINATION OF COST

8.3.1 An increase or decrease in the Contract Price or the Contract Time resulting from a change in the Work shall be determined by one or more of the following methods:

8.3.1.1 unit prices set forth in this Agreement or as subsequently agreed;

8.3.1.2 a mutually accepted, itemized lump sum;

8.3.1.3 costs calculated on a basis agreed upon by the Owner and Contractor plus ___________% Overhead and ___________% profit; or

8.3.1.4 if an increase or decrease cannot be agreed to as set forth in Clauses .1 through .3 above, and the Owner issues an Interim Directed Change, the cost of the change in the Work shall be determined by the reasonable actual expense and savings of the performance of the Work resulting from the change. If there is a net increase in the Contract Price, the Contractor's Overhead and profit shall be adjusted accordingly. In case of a net decrease in the Contract Price, the Contractor's Overhead and profit shall not be adjusted unless ten percent (10%) or more of the Project is deleted. The Contractor shall maintain a documented, itemized accounting evidencing the expenses and savings.

8.3.2 If unit prices are set forth in the Contract Documents or are subsequently agreed to by the Parties, but the character or quantity of such unit items as originally contemplated is so different in a proposed Change Order that the original unit prices will cause substantial inequity to the Owner or the Contractor, such unit prices shall be equitably adjusted.

8.3.3 If the Owner and Contractor disagree as to whether work required by the Owner is within the scope of the Work, the Contractor shall furnish the Owner with an estimate of the costs to perform the disputed work in accordance with the Owner's interpretations. If the Owner issues a written order for the Contractor to proceed, the Contractor shall perform the disputed work and the Owner shall pay the Contractor fifty percent (50%) of its estimated cost to perform the work. In such event, both Parties reserve their rights as to whether the work was within the scope of the Work, subject to the requirements of Article 12. The Owner's payment does not prejudice its right to be reimbursed should it be determined that the disputed work was within the scope of Work. The Contractor's receipt of payment for the disputed work does not prejudice its right to receive full payment for the disputed work should it be determined that the disputed work is not within the scope of the Work.

8.4 CLAIMS FOR ADDITIONAL COST OR TIME Except as provided in Subparagraph 6.3.2 and Paragraph 6.4 for any claim for an increase in the Contract Price or the Contract Time, the Contractor shall give the Owner written notice of the claim within fourteen (14) Days after the occurrence giving rise to the claim or within fourteen (14) Days after the Contractor first recognizes the condition giving rise to the claim, whichever is later. Except in an emergency, notice shall be given before proceeding with the Work. Thereafter, the Contractor shall submit written documentation of its claim, including appropriate supporting documentation, within twenty-one (21) Days after giving notice, unless the Parties mutually agree upon a longer period of time. The Owner shall respond in writing denying or approving the Contractor's claim no later than fourteen (14) Days after receipt of the Contractor's claim. Any change in the Contract Price or the Contract Time resulting from such claim shall be authorized by Change Order.

There are many similarities in commonly used changes clauses, but they are not identical. For example, the American Institute of Architects (AIA), ConsensusDOCS, and Federal Acquisition Regulation (FAR) clauses all describe the extent of the authorized changes by reference to the owner's right to order changes to the "general scope" of the contract (work). In addition to using the "general scope" language, the changes claused by the federal government's changes clause is unique in identifying specific categories of authorized changes. These are:

- Specifications (including drawings and design)
- Method or manner of perfomance of the work
- Government furnished facilities, equipment services, or site
- Directing acceleration of the work

Possibly reflecting the fact that Engineers Joint Contract Documents Committee (EJCDC) are used more often for horizontal construction or civil projects on which unit price work is common, the EJCDC changes clause omits the "general scope" phrase but authorizes "additions, deletions, or revisions" in the contract work.[3]

Given the variety of circumstances that result in the need to issue a change to the contract, contractors and owners should review the changes clause in their contract or subcontract as well as any related contract provisions and consider potential issues such as:

- Can major quantity changes be directed?
- Can the project schedules be accelerated?
- Can the project be shifted to a different site?
- Can a particular method for performing the work be directed?
- Does the contract or subcontract specifically exclude the performance of certain work (e.g., abatement of hazardous materials)?

Although the parties to a private commercial project have very broad rights to agree to modifications that are beyond the "scope" of the original work, the basic question is whether such a change can be imposed by one party. Moreover, in the public arena with competitve bidding laws, there is the question of whether an "out of scope" change violates applicable laws or ordinances requiring competitive proposals or bids.

Related to the question of the extent or type of permissible changes is the necessity to obtain the consent of a performance or payment bond surety to those contractual modifications or changes. Most standard-form contracts and bond forms provide that modifications or changes to the contract may be made with no notice to or consent by the surety.[4] If the change or modification, however, is outside the scope of the contract and the surety did not consent to it, that modification may operate to discharge the surety completely[5] or to the extent that it suffered prejudice.[6] This principle requires consideration of the scope of permissible changes and the purpose to obtain the surety's consent to a particular modification.[7]

In summary, although questions of the extent or type of changes contemplated by a contract and the need to obtain the contract consent of any surety are addressed in

[3]EJCDC C-700, § 10 (2007 ed.).

[4]*See, e.g.,* FAR § 52.243.4(a) ("without notice to the sureties"); ConsensusDOCS 200, ¶ 8 (2007 ed.) ("Changes in the Work that are within the general scope of this Agreement shall be accomplished, invalidating this Agreement, by Change Order, and Interim Directed Change."); AIA A312 Performance Bond and Payment Bond, § 8 (1984 ed.) (surety waives notice of any changes); EJCDC C-700, § 10.01 (2007 ed.) (without notice to any surety).

[5]*United States v. Freel,* 186 U.S. 309, 317 (1902) (modification changing location for a dry dock discharged the surety); *Nat'l Sur. Corp. v. United States,* 118 F.3d 1542, 1547 (Fed. Cir. 1997) (improper release of retainage); *Mergentine v. Wash. Metro. Area Transit Auth.,* 775 F. Supp. 14, 23 (D.D.C. 1991) (prepayment of retainage); *In re Liquidation of Union Indem. Ins. Co. of New York,* 632 N.Y.2d 788 (N.Y. App. Div. 1995) (work added to subcontract by change order that was specifically excluded under the original contract).

[6]*Mergentine v. Wash. Metro. Area Transit Auth.,* 775 F. Supp. 14, 20 (D.D.C. 1991). Attempting to avoid this notice issue, EJCDC C-700, § 10.04 (2007 ed.) expressly places the responsibility of giving notice to the surety on the contractor.

[7]*See Trinity Universal Ins. Co. v. Gould,* 258 F.2d 883, 886 (10th Cir. 1958).

the various standard industry form contracts, these forms may not completely eliminate the need for surety consent under unusual circumstances. More important, to the extent that the parties utilize custom contract or subcontract forms, or detail specific exclusions from the scope of work, there exists an enhanced possibility that it would be prudent, if not essential, to obtain the surety's consent to contract modifications.[8]

Depending on the language of the changes clause and related provisions, typical changes clauses (in a general construction contract) may allow an owner to order extra work, delete work, or make changes to the quantity, timing, sequence, or methods of any contract work activity, with commensurate adjustments in the contract price and performance time. Subcontract forms contain similar provisions allowing for the adjustment of contract obligations at the subcontract level.

A "change order" provision, as the changes clause sometimes is called, gives the owner the contractual power to adapt actual conditions to the end product sought—for example, to correct errors in the plans and specifications or to take advantage of newly developed construction techniques and materials. A "changes" clause gives the owner contractual power to "order" the contractor to make changes, so long as the owner follows the terms of the clause and compensates the contractor consistent with the terms of that clause. This typically includes the owner's right to issue deductive changes that presents unique issues often more challenging than changes increasing the scope of work.

Changes clauses often specify the method for determining the amount of additional compensation or performance time that is due the contractor for changed or extra work. Payment for changed work often is accomplished based on an agreed or negotiated lump-sum, unit prices, or time-and-materials (cost-plus) basis. The changes clauses used in the ConsensusDOCS 200, Paragraph 8, AIA A201, § 7, and EJCDC C-700, § 10, list all of these methods. In contrast, the federal government changes clause at FAR § 52.243–4 calls for an equitable adjustment to the contract based on the "increase or decrease in the Contractor's cost of . . . the performance of any part of the work . . . ."

Although owners often prefer to negotiate the extra compensation the contractor is to receive before the changes are implemented, it is common for changes clauses to provide that the contractor may be required to proceed with the work without first reaching agreement on the added (or reduced) compensation and time due. Having the power to direct the contractor to proceed without reaching agreement on compensation protects the owner against possible work stoppages or delays that might otherwise result while negotiations are under way. For example, the Standard Agreement and General Conditions Between Owner and Contractor, ConsensusDOCS 200, provides for an "Interim Directed Change," which is signed by the owner and requires the contractor to proceed with changed work.[9] An Interim Directed Change becomes the subject of a "change order" when (if) the owner and contractor agree on an adjustment to the contract price or contract time or both.

[8]*See In re Liquidation of Union Indem. Ins. Co. of N.Y.,* 632 N.Y.2d 788 (N.Y. App. Div. 1995).

[9]ConsensusDOCS 200, ¶ 8.2.1 (2007 ed.) ("The Owner may issue a written Interim Directed Change directing a change in the Work prior to reaching agreement with the Contractor on the adjustment, if any, in the Contract Price or the Contract Time."). AIA A201 provides a similar procedure known as "Construction Change Directives." *See* AIA A201, § 7.3 (2007 ed.).

This type of provision requiring changed work to proceed pending an agreement on the compensation can be just as important for the general contractor to ensure that subcontractors do not stop if questions arise about changes. Thus, a provision requiring subcontractors to continue work during change order negotiations should be included in all subcontracts.

When a contractor has to proceed with changed work before the price is negotiated, it should protect its interests by keeping complete documentation, including records of all expenditures and impacts caused by the change. Costs incurred for the changed work should be documented using separate cost codes if possible. The ultimate cost generally includes an allowance for the contractor's profit and overhead. Profit and overhead markups either will be defined by the changes clause, or a reasonable markup will be negotiated. Change order costs may need to be certified or approved by the design professional, making record keeping even more critical.

The contractor may also be entitled to an appropriate extension of the contract time if the contract sets out a specific completion date and the changed work impacts the time of project completion. Regardless of whether the contract provides for liquidated damages, the contractor is well served to consider if an appropriate adjustment in the time of performance is due. Time issues generated by a change normally are negotiated at the same time that money issues are resolved and agreements documented in the final written change order form. It is also critical for the contractor to document the extension of performance time required by the change if no agreement can be reached.

The changes clause is also the contract mechanism that courts often will employ when the parties have a dispute over the scope of work or interpretation of the contract documents, or if there is an allegation of defective specifications. For whatever reason, a contractor may find that it has to perform work it did not originally envision. In such circumstances, the contractor believes it is being required to perform a change and therefore is entitled to be paid for this "extra." The owner, however, believes that the work is part of the original scope of the contract, so requiring the contractor to perform it would not be a change and would not give rise to revising the contract price.

Although these types of disputes always will be decided by their individual facts, the changes clause is the proper mechanism to employ in the event that the contractor's position is correct. Thus, the changes clause will come into play in circumstances where the parties disagree over whether there has been a change at all. These types of situations are commonly referred to as "constructive changes" and are covered in more detail later in this chapter.

II. RECOVERY UNDER THE CHANGES CLAUSE

There are four primary factors to consider in assessing the contractor's rights to recover under the changes clause:

(1) Is there a change to the contract work?

(2) Who is authorized to order changes?

(3) When are written change orders required?

(4) When is written notice of a claim for additional compensation arising from extra work required?

Each factor is discussed in detail in the next sections.

A. Is There a Change to the Contract Work?

As mentioned, the contracting parties may not be able to agree whether certain orders from the owner constitute "changes" under the changes clause. Determining the answer to this question depends on the plans, specifications, and contract provisions for each individual job. As a general rule, a contractor only is required to perform in accordance with its original agreement. If the owner orders work that is different from the requirements of the original agreement, the changes clause will entitle the contractor to additional compensation.[10]

The rules of contract interpretation discussed in **Chapter 5** often play an important role in a court's resolution of a dispute over whether certain work ordered by the owner is changed or extra work. The contract scope of work is defined by the contract documents. The interpretation of those documents often will decide whether a specific item of work is or is not included in the scope of the project. One of the more important rules of contract interpretation is that the written contract will be construed most strongly against the party who drafted it.[11] In other words, when a contract can have two or more reasonable interpretations (i.e., the contract is ambiguous), it will be construed against the author since it had the discretion or control in describing the requirement. Typically, the owner or its agent (the project architect/engineer) drafts the construction contract. Thus, in order for a contractor to recover for extra work, it is not generally necessary that its interpretation of the contract be the only reasonable interpretation or even the most reasonable interpretation. Rather, if the contractor's interpretation is reasonable, if the contractor can show that it relied on its reasonable interpretation, and if the ambiguity was not "patent" (or obvious), then the contractor's interpretation should prevail over an equally reasonable interpretation advanced by the party that drafted the ambiguous contract documents.[12]

[10] *See, e.g., E.C. Ernst, Inc. v. Koppers Co., Inc.,* 476 F. Supp. 729 (W.D. Pa. 1979), *modified,* 626 F.2d 324 (3d Cir. 1980); *Jones v. Pollock,* 208 P.2d 1031 (Cal. Ct. App. 1949), *vacated,* 215 P.2d 733 (Cal. 1950); *W.R. Ferguson, Inc. v. William A. Berbusse, Inc.,* 216 A.2d 876 (Del. 1966); *See also Town of Palm Beach v. Ryan Inc. E.,* 786 So. 2d 665 (Fla. Dist. Ct. App. 2001) (affirming judgment in favor of contractor where additional work was necessary to bring the project into conformity with the overall intent of the contract). *See also Emulsfield Asphalt, Inc. v. Transp. Comm'n of Wyo.,* 970 P.2d 858, 865 (Wyo. 1998) (rejecting state's argument that change order, which was missing "essential term of how to determine amount of consideration," must be considered nothing more than an agreement to agree).

[11] *Lytle v. Freedom Int'l. Carrier, S.A.,* 519 F.2d 129, 134 (6th Cir. 1975); *Cincinnati Bengals, Inc. v. Bergey,* 453 F. Supp. 129, 149 (S.D. Ohio 1974); *The Powell Co. v. The McGarey Group, LLC,* 508 F. Supp. 2d 1202 (N.D. Ga. 2007); *L&L Builders Co. v. Mayer Associated Servs., Inc.,* 46 F. Supp. 2d 875 (N.D. Iowa 1999).

[12] *See Bennett v. United States,* 371 F.2d 859 (Ct. Cl. 1967); *Turner Constr. Co., v. United States,* 367 F.3d 1319 (Fed. Cir. 2004).

B. Deductive Changes

It is common to discuss contract modifications that increase the cost of or time for performance of the work, but a changes clause also applies to changes that decrease the cost of or time required for performance. For example, if the owner deducts certain work activities from a contract, a changes clause will entitle the owner to a decrease in the contract price or time.[13] Deductive changes often are not specifically addressed in changes clauses, and these can generate unique questions over the contract adjustment or credit due. An owner may want to use the numbers in a bid schedule or schedule of values in valuing deleted work. The contractor, however, may have reasons why such pricing is not appropriate. Moreover, different questions arise when dealing with deductive changes from when dealing with extra work changes. Does one price deleted work based on the cost savings to the contractor or the value of the deleted work to the owner? How should overhead and profit be addressed in a deductive change? How should discounts be considered? The issues related to the pricing of credits can present some difficult questions of contract interpretation.[14]

Interesting issues arise when a contract contains both a partial termination for convenience clause and a changes clause. In such cases, the issue becomes whether the deduction of work should be categorized as a deductive change order or as a partial termination. The distinction will determine how the contractor is compensated for the deducted scope of work, as seen in *J.W. Bateson Co. v. United States.*[15] In *Bateson,* the government had awarded the contractor separate contracts for construction of two housing projects, one at Camp Pickett in Virgina and one at Camp Breckinridge in Kentucky. The government deleted 349 out of the 430 houses at the Kentucky camp. The government ordered the contractor to ship the prefabricated building materials for the deleted Kentucky camp to the Virginia camp. The resulting change order provided a 10% profit on the materials that were shipped to the Virginia camp and the remaining work completed under both contracts. The government did not, however, reduce the contractor's profit under the Kentucky contract for the deleted work. Subsequently, the government filed suit to recover what it termed as "double profit," arguing that the

[13]*Ragnar Benson, Inc. v. Bechtel Power Corp.*, 651 F. Supp. 962 (M.D. Pa. 1986) (subcontractor required to give credit for savings realized by the use of a more economical cleaning method).

[14]Deductive change cases can address unusual questions that may arise when parties disagree on how to value deleted work. For example, in *M. J. Paquet, Inc. v. N.J. Dep't of Transp.*, 761 A.2d 122 (N.J. Super. Ct. App. Div. 2000), *aff'd in part, rev'd in part,* 794 A.2d 141 (N.J. 2002), the contractor submitted a lump-sum bid on a bid schedule that required a breakdown of the price into work components. The contractor listed the price of a prebid quote in the breakdown for painting work. Immediately prior to bid, the contractor got a lower paint quote and reduced its lump-sum bid accordingly but did not adjust the breakout (contractor used a single line item to make cut/adds in the minutes before bid). When the owner decided to delete the painting, it wanted to use the breakout price rather than what the contractor actually had in its bid. The appellate court had to decide how to value this deductive change and ultimately ruled that the value should be based on the breakout number listed; otherwise, the breakout would be meaningless. On appeal, the New Jersey Supreme Court reversed on this point and held that the contractor was entitled a contract adjustment based on the amount it actually had included for the painting work in its bid price. The Supreme Court based its rationale on the concept that the contractor was entitled to an "equitable adjustment."

[15]308 F.2d 510 (5th Cir. 1962).

reduction in the Kentucky contract was essentially a "partial terminiation" that entitled the government to recover the contractor's anticipated profit. The contractor argued that under the changes clause used in the contract, it was entitled to retain profit that it would have earned on the deleted work. Upholding the government's claim, the court ruled that cancellation of the Kentucky work was a partial termination, not a change. More important, the court viewed the case as boiling down to unjust enrichment and stated that as Bateson had received a profit on prefabricated materials it furnished but did not erect at Camp Breckinridge, it received a second profit on the identical material when it erected it into housing units at Camp Pickett.[16]

III. WHO IS AUTHORIZED TO ORDER CHANGES?

The changes clause in most construction contracts identifies who can order changes to the work. Typically, the owner reserves to itself the right to order changes to the work, especially if the change requires a price adjustment. Many contracts will allow the design professional to order minor changes that do not affect price. In federal government contracts, the contracting officer is the only person authorized to order changes on behalf of the government.[17]

The individuals having authority to order changes should be specified in the contract. Yet contract documents do not always clearly establish who can order changes. This can be a significant problem if it is not feasible for a contractor to wait for a final change order to be processed before beginning additional work. As a result, the contractor must rely on assurances by the design professional or other owner representatives that additional costs will be paid. These persons may or may not be found to have the legal authority to alter the plans and specifications or impose additional obligations. Risks are created when a contractor proceeds with extras without a fully executed change or without a contractually authorized change directive.[18]

If the contract is not clear as to who has authority to direct changes, an understanding should be reached as early as the preconstruction conference as to the individuals authorized to order changes and the extent of their authority. These understandings should be confirmed in writing. If there are any questions regarding who may bind the other party, these need to be resolved at the inception of the project.

A lack of clear direction in the contract as to who is authorized to direct compensable changes can cause harsh results if the person directing the change is found to lack the necessary authority. This is particularly true when dealing with public

[16] *Id.* at 514.

[17] *See Winter v. Cath-dr/Balti Joint Venture,* 497 F.3d 1339 (Fed. Cir. 2007).

[18] *See, e.g., Smith v. Bd. of Educ. of Parkersburg Dist.,* 85 S.E. 513 (W. Va. 1915) (where the contractor was subsequently held responsible for the installation of an item that was specified in the contract and that had been omitted pursuant to the instructions of the architect); *County of Brevard v. Miorelli Eng'g, Inc.,* 703 So. 2d 1049 (Fla. 1997) (sovereign immunity barred recovery for extra work); *C.O.B.A.D. Constr. Corp. v. Sch. Bd. of Broward County,* 765 So. 2d 844 (Fla. Dist. Ct. App. 2000) (no recovery for extra work not authorized by a change order).

bodies. For example, in *Nether Providence Township School Authority v. Thomas M. Durkin & Sons, Inc.*,[19] the contract required changes be approved in writing by the school board. During performance, a disagreement arose as to whether certain work was required under the contract. Two members of the school board signed a letter to the contractor ordering the contractor to proceed with the disputed work and stating that the disagreement would be resolved at a later time. The court ruled that the letter was neither valid authorization under the terms of the contract nor a waiver of those terms. The court concluded that the contractor was not entitled to additional compensation because it had never been given instructions by the school board, the only party with such authority under the contract.

The *Durkin & Sons* decision is an unfortunate example of what can happen when a contractor does not follow the requirements of the changes clause. That case involved a public body spending tax dollars, and the court felt compelled to strictly follow the contract requirements. Although there are several ways (discussed below) that the contractor in *Durkin & Sons* might have recovered, the case is a good example of the substantial risk a contractor accepts when it does not strictly follow the requirements of the changes clause or does not understand the limits of those acting on behalf of another party, particularly a public entity.

A contractor performing extra work at the direction of someone other than the person designated in the contract has two primary legal theories for recovering under the changes clause. The lack of express authority may be overcome by: (1) the implied or apparent authority of the person who requested the extra (the design professional), or (2) the ratification by the owner of the actions of the person (the design professional) requesting the extra. These two theories are discussed next.

A. Implied Authority

In preparing plans and specifications for a traditional design-bid-build project, the design professional generally acts as an independent contractor to the owner. When providing supervisory functions on a project, however, including processing change orders for extra work, the design professional typically functions as an agent of the owner. (See **Chapter 7** for a general discussion of the design professional's authority.) In an extra work situation, the question often becomes whether the design professional has the "implied" authority to order extra work on the owner's behalf, even though the contract does not give the design professional the "express" authority.

"Implied authority" is the authority of an agent to do whatever acts are incidental to, or necessary, usual, or proper for, the exercise of the express authority delegated to that agent by the principal. For example, does an engineer who has the express authority to direct changes in the grade of a railroad also have the implied authority to enter into a supplemental contract to accomplish a grade change? In *Lafayette Railway Co. v. Tucker*,[20] the Alabama Supreme Court held that the engineer did have such implied

[19]476 A.2d 904 (Pa. 1984).
[20]27 So. 447 (Ala. 1900).

authority.[21] In federal government contracts, the concept of implied "actual" authority of representatives of contracting officers is applied very narrowly.[22]

B. Apparent Authority

In contrast to implied authority, the term "apparent authority" refers to the situation where the owner's representative (the design professional or some other person) acts in a way that leads the contractor to reasonably believe that the representative has authority beyond that actually possessed. For example, if a prudent contractor forms a reasonable belief from actions or statements made by the owner that the architect has a certain authority, then the owner may be bound by the architect's acts if the architect takes action within this "apparent" authority.

It is often difficult for a court to determine the scope of an agent's authority. Nonetheless, an error in this determination can result if substantial financial losses. Such was the case in *Foster Wheeler Enviresponse, Inc. v. Franklin County Convention Facilities Authority,*[23] which involved a contract for removal and disposal of hazardous waste. The public bid for the work required a base price to remove and dispose of an estimated amount of contaminated waste and for a unit price, in the event the quantity of waste deviated from the estimated amount. The final contract also required written authorization for any changes in the scope of work.

Once excavation began, the contractor immediately encountered hazardous waste and advised the owner that there was more waste than originally estimated. The owner advised the contractor to obtain direction for the work from the environmental consultant hired to oversee the project. With the oversight of the consultant's on-site representative signing off on truck manifest forms, the contractor continued to remove and transport the waste.

The contractor faxed a letter to the consultant confiming the total amount of waste beyond the original estimate that had been removed and an estimate for additional waste still to be removed. A few days later the consultant told the contractor to cease all work. Shortly thereafter, the contractor submitted a written request for clarification, seeking payment in the amount of nearly $1 million for extra waste that was removed. This request was denied for failure to obtain written authorization for work beyond the base bid. The court concluded that the consultant did not have the authority to waive the requirement for written orders and denied contractor's entire claim.

[21] *But see Albert Steinfield & Co. v. Broxholme,* 211 P. 473 (Cal. Ct. App. 1922) (holding that architect that was told to get other bids did not have authority to enter into a contract on behalf of the owner).

[22] *Winter v. Cath-dr/Balti Joint Venture,* 497 F.3d 1339 (Fed. Cir. 2007) (Department of Defense regulation expressly precluded delegation of change order authority to a person designated as the contracting officer's representative). *See also H. Landau & Co. v. United States,* 886 F.2d 322, 324 (Fed. Cir. 1989) (implied actual authority requires showing that contracting authority is an integral part of assigned duties); *Leonardo v. United States,* 63 Fed. Cl. 552, 557 (2005) (agency regulations must not grant such authority to other agency employees).

[23] 678 N.E.2d 519 (Ohio 1997).

Although the *Foster Wheeler* decision illustrates the type of factual analysis that may be made in a case involving implied authority, the concept of apparent authority does not apply to contracts with the federal government. The United States Supreme Court has expressly rejected the concept of apparent authority when dealing with a representative of the United States. Rather, the contractor or prospective contractor must ascertain the actual authority of the person ordering the work.[24]

C. Alternatives to Demonstrating Authority

Although the design professional (or other owner representative) may not have express authority to order extra work, it may nevertheless be possible to establish that the actions of the design professional directing extra work were "ratified" or approved by the owner after the fact. As emphasized in the case of *Kirk Reid Co. v. Fine,*[25] however, it is critical that the owner have actual knowledge of the change that is claimed to have been ratified.

Under certain circumstances, a contractor might recover for extra work performed without following contract procedures under equitable theories, such as *quantum meruit.* As discussed in *Trinity Products, Inc. v. Burgess Steel, L.L.C.,*[26] this typically requires proof that the work was not contemplated by the parties when they entered into the contract and the extra work is not controlled by the contract.

The best and safest course of action for any contractor is to not perform extra or changed work without a valid order from a person authorized by the contract to order such work. Failure to obtain proper authorization under the contract changes clause can jeopardize the contractor's claim for extra compensation.

IV. WRITTEN DOCUMENTATION OF CHANGES

A. Written Directives

Problems can arise when oral orders are given for extra work where the contract specifically requires that they be in writing. Most construction contracts today provide that the contractor shall not proceed with any extra or changed work until a written change order has been issued. For example, Article 8 of ConsensusDOCS 200 requires either a written change order signed by both parties or a written Interim Directed Change issued by the owner before the contractor may proceed with extra work.[27] Such provisions should be considered valid and binding on the parties, and, if not satisfied, may

[24]*Fed. Crop Ins. Corp. v. Merrill,* 332 U.S. 380 (1947); *Daly Constr., Inc. v. United States,* 5 F.3d 520, 521 (Fed. Cir. 1993); *Mil-Spec Contractors, Inc. v. United States,* 835 F.2d 865, 867 (Fed. Cir. 1987).

[25]139 S.E.2d 829, 834–35 (Va. 1965); *see also Winter v. Cath-dr/Balti Joint Venture,* 497 F.3d 1339, 1347 (Fed. Cir. 2007) (ratification requires knowledge of material facts and approval of activity by one with authority).

[26]486 F.3d 325 (8th Cir. 2007).

[27]*See also* AIA A201, § 7 (2007 ed.); EJCDC C-700, § 10 (2007 ed.).

prevent the contractor from recovering compensation for changed or extra work.[28] For this reason, the contractor should always obtain a written change order, or a written work directive, before performing changed or extra work. An instructive example is found in *Environmental Utilities. Corp. v. Lancaster Area Sewer Authority,*[29] where the contractor was not allowed to recover for extra work performed because it did not have written orders for the work. Similarly, in *County of Brevard v. Miorelli Engineering,*[30] the Florida Supreme Court denied recovery for changed work without a written change order, rejecting the contractor's argument that the county-owner had waived the written change order requirement by issuing oral directives to proceed with extras. The court based its decision on sovereign immunity, stating that the sovereign could be left with potentially unlimited liability if the doctrines of waiver and estoppel could be used to defeat the express terms of the contract.[31]

Obtaining a written change order before performing any changed work is always desirable, since it protects both owner and contractor, but this is frequently ignored in practice. Often it is not practical for the contractor to stop work while a formal change order is processed. Realities make it necessary for the contractor to rely on the assurances of the owner's representative or the design professional that a change order will be forthcoming. In recognition of this reality, ConsensusDOCS 200 provides that the contractor may proceed with changes directed by the owner in an Interim Directed Change before there is an agreement on the pricing and time associated with the change.[32]

A contractor sometimes can obtain relief when it proceeds with extras without an appropriate written directive or executed change order. A court sometimes will look for a way to avoid enforcing the written work order requirement when it leads to a particularly inequitable result. Various methods and theories have been used for this purpose.

Some courts will make a distinction between an "extra," or work that is not required at all under the contract, and "additional work," which itself is not precisely required but is a necessary extension of other work that is specified. Where the contractual written change order requirement covers only "extra" work, a court might characterize the work as "additional" work for which the contractor could recover despite not having a written order.[33]

A contractor also can sometimes avoid the harsh results of the written change order requirement by showing that the owner waived, modified, or abandoned the written change order requirement. Courts (focusing on the ultimate question of recovery) often

[28] *See, e.g., Plumley v. United States,* 226 U.S. 545 (1913); *see generally* 13 AM. JUR. 2D *Building and Construction Contracts* § 23 (2008).
[29] 453 F. Supp. 1260 (E.D. Pa. 1978).
[30] 703 So. 2d 1049 (Fla. 1997).
[31] *Id.* at 1051. *But see W&J Constr. Corp. v. Fanning/Howey Assocs.,* 741 So. 2d 582 (Fla. Dist. Ct. App. 1999) (reversing summary judgment ruling in favor of a school board where issue of fact existed as to whether contract required the contractor to install upgraded fire protection system).
[32] ConsensusDOCS 200, ¶ 8.2.1 (2007 ed.). *See also Id.* at ¶ 12.1 (providing that the contractor "shall continue the Work and maintain the Schedule of the Work during any dispute mitigation or resolution proceedings."). AIA A201 contains similar provisions addressing "Construction Change Directives"; *see* AIA A201, § 7.3 (2007 ed.); *see also* EJCDC C-700, § 10.01 (2007 ed.) (addressing "Work Change Directives").
[33] *See B&G Crane Serv. Inc. v. Lamastus,* 323 So. 2d 515 (La. Ct. App. 1975); *Roff v. S. Constr. Corp.,* 163 So. 2d 112 (La. Ct. App. 1964).

have used these theories interchangeably. The decisions are not altogether consistent as to what particular acts or conduct are sufficient to waive or modify the requirement for a written change order. The general principle is that the written change order requirement can be waived or modified not only by express words, but also by acts or conduct that by implication waive or otherwise derogate the writing requirement.[34] A waiver or modification of the written order requirement is most easily established where the owner has made progress payments for some of the extra work without having a written order in place.[35]

Similarly, repeated verbal directives from the owner to perform extra work can create a modification or waiver of the written change order requirement. For example, in *Consolidated Federal Corp. v. Cain,*[36] an owner who verbally ordered several changes could not contend that the contractor breached the contract by requesting additional compensation for those changes. The court recognized that the contract required written change orders but found that the owner had waived that requirement by repeatedly ordering changes verbally.[37]

A waiver or modification also has been found where the owner made verbal statements promising to pay for extra work and the contractor proceeded based on those promises, as illustrated by the decision in *Udevco, Inc. v. Wagner.*[38] In that case, a contractor was allowed to recover where the owner had made an express oral waiver of the writing requirement and the contractor relied on the statement in doing the work. Yet in a Texas case, a subcontractor lost a claim against its prime contractor because of the absence of a written change order, even though the prime's representative promised "they would take care of it down the line."[39] These cases clearly establish that while a contractor may have arguments why it should recover, there is always risk when either party departs from the terms of the contract.

The owner's knowledge that extra work is being performed, without objection, may be an excuse to the written change order requirement.[40] Mere knowledge on the part of the owner, however, often will not be enough to convince a court that the writing requirement has been waived. In many instances, the owner will not know

[34]*See* C.P. Jhong, Annotation, *Effect of Stipulation, in Private Building or Construction Contract, That Alterations or Extras Must Be Ordered in Writing,* 2 A.L.R. 3d 620 (1965).

[35]*See Cardinal Dev. Co. v. Stanley Constr. Co., Inc.,* 497 S.E.2d 847 (Va. 1998) (developer could not rely on contract clause requiring written authorization where developer made payments for additional work for over a year without requiring written orders); *Union Bldg. Corp. v. J&J Bldg. & Maint. Contractors, Inc.,* 578 S.W.2d 519 (Tex. App. 1979); *See also Safer v. Perper,* 569 F.2d 87 (D.C. Cir. 1977); *Custom Builders, Inc. v. Clemons,* 367 N.E.2d 537 (Ill. App. Ct. 1977) (where the owner only refused to pay for some selected extras without a written order); *W.E. Garrison Grading Co. v. Piracci Constr. Co., Inc.,* 221 S.E.2d 512 (N.C. Ct. App. 1975).

[36]394 S.E.2d 605 (Ga. Ct. App. 1990).

[37]*See also Allen & O'Hara, Inc. v. Barrett Wrecking, Inc.,* 898 F.2d 512 (7th Cir. 1990); *Eastline Corp. v. Marion Apartments, Ltd.,* 524 So. 2d 582 (Miss. 1988).

[38]678 P.2d 679 (Nev. 1984); *see also Winn-Senter Constr. Co. Katie Franks, Inc.,* 816 S.W.2d 943 (Mo. Ct. App. 1991); *Meadows v. Kinser,* 603 S.W.2d 624 (Mo. Ct. App. 1980); *D.K. Meyer Corp. v. Bevco, Inc.,* 292 N.W.2d 773 (Neb. 1980).

[39]*Austin Elcon Corp. v. Avco Corp.,* 590 F. Supp. 507, 513 (W.D. Tex. 1984).

[40]*See* C.P. Jhong, Annotation, *Effect of Stipulation, in Private Building or Construction Contract, That Alterations or Extras Must Be Ordered in Writing,* 2 A.L.R. 3d 620 (1965).

whether certain work is an extra. The owner may argue that it had a right to assume the work was covered by the contract unless there was specific notice and a change order was worked out.

Generally, a contractor can argue that extra work is compensable without a written change order where:

(1) The work was orally ordered or authorized by the owner (as opposed to the design professional or construction representative); or
(2) The owner has orally agreed or promised to pay additional compensation for the work in question; or
(3) The parties to the contract, throughout their performance, have entirely or repeatedly disregarded the writing requirement.

Although the legal theories of waiver, modification, and estoppel may, in many instances, provide relief where the contractor has not obtained a written change order, these theories present significant proof problems. There are always two sides to a story when parties argue over significant sums of money and where adequate documentation does not exist. The safer course always is to insist on written authorization (if not a formal change order) before proceeding with the work, rather than relying on the oral assurances of the design professional or the owner that the contractor will be compensated for its effort.

A contractor can get some protection by confirming in writing to the owner that the work being performed is in addition to that required under the contract and that the contractor anticipates additional compensation in money and time for the extra work.

B. Requirements for Written Notice of a Change

Owners typically include provisions in their construction contracts requiring written notification within a certain time if the contractor intends to file a claim for work considered to be beyond contract requirements. For example, ConsensusDOCS 200 requires written notice given to the owner of a claim within 14 days after the "occurrence" giving rise to the claim or within 14 days after the contractor first recognizes the condition giving rise to the claim, whichever is later.[41] Additionally, ConsensusDOCS 200 also requires written notice to be provided before commencing the work (except in emergency situations).[42] The AIA A201 (2007 ed.), requires written

[41] *See* ConsensusDOCS 200.
[42] *Id. See* Section 1 of this **Chapter.**

notice to the architect within 21 days after the occurrence of the event giving rise to the claim or within 21 days after the condition giving rise to the claim is first recognized, whichever is later.[43]

Provisions requiring written notice of claims protect the owner by allowing the owner an opportunity to evaluate the situation before the work is performed and the costs are incurred. Such provisions in subcontracts provide similar protection to the general contractor.

The contractor's failure to give the required notice of a claim for additional compensation may bar the recovery of additional compensation, especially where the owner can show some prejudice for not being notified.[44] Provisions requiring advance notification of a claim for extra work may be waived or modified, however, similar to the written change order requirement. For example, in *Wiscch & Vaughan Construction Co. v. Melrose Properties Corp.,*[45] the contractor on a hotel project submitted a claim for compensation for extra work. The owner denied the claim because the contractor failed to comply with contract provisions requiring notice and changes be "in writing." The court found the course of conduct on the project included the owner's oral agreement and partial payment for some of the extra work performed by contractor. Accordingly, the court held that the owner had waived the contractual notice and written change order requirements and upheld the contractor's claim.

In *Macri v. United States,*[46] a general contractor was to construct the foundation upon which a subcontractor was to erect certain tanks. The subcontractor complained that defects in the foundation would delay the subcontractor's performance and increase its costs. After unsuccessfully attempting to repair the foundation, the general contractor ordered the subcontractor to proceed. The court allowed the subcontractor to assert a claim for extras, even though the subcontractor had not clearly placed the general contractor on notice after being ordered to proceed. The court

[43]AIA A201, § 15.1.2 (2007 ed.) provides: "Claims by either the Owner or Contractor must be initiated by written notice to the other party to the Initial Decision Maker with a copy sent to the Architect, if the Architect is not serving as the Initial Decision Maker."

[44]*See, e.g., Linneman Constr., Inc. v. Mont.-Dakota Utils. Co., Inc.,* 504 F.2d 1365 (8th Cir. 1974); *Associated Mech. Contractors, Inc. v. Martin K. Eby Constr. Co., Inc.,* 983 F. Supp. 1121 (M.D. Ga. 1997), *aff'd in part, rev'd in part,* 271 F.3d 1309 (11th Cir. 2001); *A.H.A. Gen. Constr., Inc. v. N.Y. City Hous. Auth.,* 699 N.E.2d 368 (N.Y. 1998); *Buchman Plumbing Co., Inc. v. Regents of Univ. of Minn.,* 215 N.W.2d 479, 486 (Minn. 1974). *But see Ronald Adams Contractor, Inc. v. Miss. Transp. Comm'n,* 777 So. 2d 649, 653–54 (Miss. 2000) (contractor's failure to provide notice pursuant to the contract of a differing site condition was "irrelevant" and the contractor could recover damages where the state Transportation Commission had actual notice of the differing site condition); *New Pueblo Constructors, Inc. v. State,* 696 P.2d 185 (Ariz. 1985) (strict enforcement of a contractual notice provision was not required where the state Department of Transportation was aware of the changed conditions for which the contractor was seeking additional compensation and was not prejudiced by the lack of formal notice of such claims).

[45]21 S.W.3d 36 (Mo. Ct. App. 2000); *See also Transpower Constructors, A Div. of Harrison Int'l Corp. v. Grand River Dam Auth.,* 905 F.2d 1413 (10th Cir. 1990).

[46]353 F.2d 804 (9th Cir. 1965).

found that the general contractor had knowledge of the subcontractor's claim from the earlier communications and could not claim surprise or prejudice due to the subcontractor's failure to comply strictly with the written notice provision.[47]

In summary, a failure to give prompt written notice may prevent a contractor from recovering compensation to which it otherwise would be entitled. Prompt written notice should be given to the owner or to its authorized representative when any action (or inaction) on the part of the owner (or its representatives) would increase cost or time. Following this simple procedure can minimize disputes later. See **Chapter 13** for a discussion of a project administration procedure to address notice requirements.

V. CONSTRUCTIVE CHANGES

Under the "constructive change" doctrine, informal actions or inactions by the owner that are not initially acknowledged by the owner to be a change in the scope of work—but that require extra work by the contractor—may nonetheless constitute a change in the scope of the work. One author has stated that the "constructive change" doctrine recognizes that an informal requirement for performance of additional work on a construction project equates to a formal directive for such work and should, therefore, be governed by similar principles. Accordingly, this author has described the "constructive change" concept as "[owner] conduct which is not a formal change order, but which has the effect of requiring the contractor to perform work different from that prescribed by the original contract, but in theory, which could have been ordered under the changes clause."[48]

Although the term "constructive change" first arose in the context of federal government contracts,[49] the general concept is now well known in connection with private construction contracts. Courts simply use different terminology and legal theories to achieve the same result. Three such legal theories typically are employed in the private owner context:

(1) Directed work (which is most analogous to pure "constructive change" analysis)
(2) Breach of contract
(3) Implied contract[50]

[47]353 F.2d at 807. *See also Gen. Specialties Co. v. Nello L. Teer Co.,* 254 S.E.2d 658 (N.C. Ct. App. 1979) (court characterized the owner's oral agreement to an extra as a waiver of the formal notice requirement); *Nat Harrison Assocs., Inc. v. Gulf States Utils. Co.,* 491 F.2d 578, *reh'g denied,* 493 F.2d 1405 (5th Cir. 1974) (party to whom written notice was to be given waived the written notice requirement because it had knowledge of the extra work and did not object to it).

[48]Robert C. Gusman, *Constructive Change—A Theory Labeled Wrongly*, 6 PUB. CONT. L. J. 229 (January, 1974).

[49]*See Indus. Research Assocs., Inc.,* DCAB WB-5, 68–1 BCA ¶ 7069 at 32,685–86.

[50]*See, e.g., Denton Constr. Co. v. Mo. State Highway Comm'n,* 454 S.W.2d 44 (Mo. 1970) (directed work); *Udevco, Inc. v. Wagner,* 678 P.2d 679 (Nev. 1984) (breach of contract); *V. L. Nicholson Co. v. Transcon Inv. & Fin., Ltd., Inc.,* 595 S.W.2d 474 (Tenn. 1980) (implied contract).

The underlying concept is essentially the same whether the owner is a public or private entity, and similar rules apply in both spheres.

"Constructive changes" typically fall into one of four general categories:

(1) An informal extra work directive or order.

(2) The drawings or specifications are defective and, as a result, the contractor is required to expend extra effort.

(3) The owner or its representative misinterprets the contract—for example, where work that actually satisfies contract requirements is erroneously rejected or where an unreasonably high standard of performance is required.

(4) The owner denies the contractor a justified time extension, requiring compliance with the original completion schedule, and thereby forces the contractor to accelerate performance.[51]

Each of these four categories is discussed next.

A. Informal Extra Work Directives

An informal work directive occurs any time an authorized representative of the owner directs the contractor—either orally or in writing, but without a formal change order—to perform work beyond the original scope of the contract. Such conduct constitutes a "constructive change" entitling the contractor, under certain circumstances, to reimbursement of any additional costs incurred in performing the changed work and, possibly, an extension of the contract time. These informal orders or directives differ from advice, comments, or suggestions that may be offered by technical representatives of the owner.[52] The specific underlying circumstances must be examined on a case-by-case basis but generally include proper notice that the directive is considered to be a change and documentation of costs.

B. Defective Plans and Specifications

A "constructive change" arising in connection with defective plans and specifications has its basis in what is referred to as the *Spearin* doctrine[53] (See **Chapter 5**). This doctrine provides that, when an owner supplies the plans and specifications for a construction project, the contractor cannot be held liable for an unsatisfactory final result attributable solely to defects or insufficiencies in those plans and specifications. The *Spearin* doctrine assumes the absence of any negligence on the contractor's part and that the contractor made no express warranty with regard to the suitability of the plans and specifications. Under this principle, an implied warranty exists for

[51] *Dep't of Transp. v. Anjo Constr. Co.*, 666 A.2d 753, 757 (Pa. Commw. Ct. 1995); *See also Sherman R. Smoot Co. v. Ohio Dept. of Admin. Servs.*, 736 N.E. 2d 69 (Ohio Ct. App. 2000).

[52] *Indus. Research Assocs., Inc.*, DCAB WB-5, 68–1 BCA ¶ 7069.

[53] *United States v. Spearin*, 248 U.S. 132 (1918).

owner-furnished plans and specifications that if the contractor complies with them, a satisfactory product will result.[54] The delivery of defective plans and specifications is therefore a breach of the implied warranty, absolving the contractor from liability for unsatisfactory results or delays in completion.

The second aspect of the *Spearin* doctrine, the right of a contractor to recover its additional costs when defective plans and specifications necessitate extra or remedial work, has found similar acceptance. The general principle has been stated in this way:

> Where defects in the plans and specifications, the sufficiency of which is not warranted by the contractor, necessitate extra work or materials to complete the contract, the contractor may recover therefor from the owner.[55]

For example, in *Fairbanks North Star Borough v. Kandick Construction, Inc.,*[56] the plans and specifications on a roadway project understated the amount of excavation of material to be removed and disposed of away from the project site. The court found that the owner had breached its implied warranty of the adequacy of the plans and specifications, and allowed the contractor to recover its extra costs since the contractor had reasonably relied on the defective plans and specifications. In *APAC Carolina, Inc. v. Town of Allendale, S.C.,*[57] the court held that the owner's implied warranty of the sufficiency of the plans and specifications carried over to the subcontractor, entitling the subcontractor to recover from the general contractor for additional subcontract work required as a result of defects in project specficiations. The contractor's liability applied despite the fact that the contractor did not draft the plans nor held itself out as having special knowledge. In *Adams v. Tri-City Amusement Co.,*[58] the walls of a building collapsed because the plans and specifications did not make allowances for wet soil conditions. The court held that the contractor was entitled to recover the reasonable value of the work to reconstruct the wall.

C. Misinterpretation of Plans and Specifications by the Owner

The third category under the "constructive change" concept relates to misinterpretation of the plans and specifications by the owner or its representatives. This type of constructive change arises from the owner's implied duty not to hinder or delay the contractor in the performance of its work, which is an implied obligation contained in every contract.[59]

[54]*See, e.g., Ace Constructors, Inc. v. United States,* 499 F.3d 1357 (Fed. Cir. 2007); *A.G. Cullen Constr., Inc. v. State Sys. of Higher Educ.,* 898 A.2d 1145, 1156–58 (Pa. Commw. Ct. 2006) (explanation of difference between design and performance specifications to application of the *Spearin* doctrine).
[55]13 Am. Jur. 2d *Building and Construction Contracts* § 31 (2008).
[56]795 P.2d 793 (Alaska 1990).
[57]41 F.3d 157 (4th Cir. 1994); *See also Keller Constr. Corp. v. George W. McCoy & Co.,* 119 So. 2d 450 (La. 1960) (implied warranty of owner directed plans carried down to subcontracts, entitling subcontractor to recover from general contractor. General contractor entitled to indemnity from owner for amounts paid to subcontractor).
[58]98 S.E.647 (Va. 1919).
[59]*See Ajax Paving Indus., Inc. v. Charlotte County,* 752 So. 2d 143 (Fla. Dist. Ct. App. 2000). *See also* **Chapter 5.**

This type of "constructive change" arises where, for example, the contract specifies a particular method of performance or allows the contractor to select the method, but the design professional requires a different, more expensive method from that contemplated by the contractor when it prepared its bid. For example, in *H. I. Homa Co.,*[60] a constructive change was found where the contracting officer rejected a bar-type progress chart that satisfied the contract's progress of work clause and instead required the contractor to provide a critical path method (CPM) schedule.

In *Charles Meads & Co. v. City of New York,*[61] a contract to build a public library gave the contractor the option of performing the work in a certain manner. When the architect required the contractor to use a more complicated and expensive method to attain the same result, the court held the city liable for the contractor's extra costs. A similar principle was applied in *S. Hanson Lumber Co. v. Moss,*[62] where the contractor was entitled to additional compensation when the owner wrongfully refused to allow the use of the material specified in the contract and instead required a more expensive type of material.

This type of constructive change can also arise from the owner's interpretation of a contract ambiguity in its favor. For example, in *Julian Speer Co.,*[63] the contract for construction of a science building at Ohio State University required the installation of science laboratory fume hoods. The contract required the prime contractor to purchase the fume hoods for the project. The project specifications, however, omitted any reference to internal piping from the fume hoods. At the direction of the owner and the architect, the contractor supplied and installed internal piping for the fume hoods at additional cost of approximately $45,000. In upholding the contractor's claim, the court noted that the owner did not provide proper plans because the specifications seemed to reflect that all contractor would need to complete the project would be to "rough-in" the final plumbing connections, not install piping internal to the fume fixtures themselves.

Interpretations of contract specification requirements are a repetitive source of "constructive change" claims and disputes. Typically, the contractor must establish that its interpretation fell within the scope of reasonableness and that it relied on its interpretation during the bid/proposal phase. Failure to prove the latter element can defeat an otherwise valid claim for additional compensation.[64]

D. Acceleration

The fourth category of "constructive change" involves acceleration of the work. Constructive acceleration occurs in the absence of an owner-directed acceleration,

[60]ENGBCA Nos. PCC-41, PCC-42, 82–1 BCA ¶ 15,651.

[61]181 N.Y.S. 704 (1920).

[62]111 N.W.2d 681 (Iowa 1961).

[63]*Julian Speer Co. v. Ohio State Univ.,* 680 N.E.2d 254 (Ohio Ct. Cl. 1997).

[64]*See Seringetti Constr. Co. v. City of Cincinnati,* 553 N.E.2d 1371, 1376 (Ohio Ct. App. 1988) (court denied contractor's claim for extra costs incurred by using more expensive chemical-water-pressure method of cleaning building interior and noted it was incumbent upon the contractor to make a pre-bid determination of the method to be employed where the contract did not specify a method).

such as where the owner has refused a valid request for time extensions or threatened other action that requires the contractor to accelerate its work to avoid liquidated damages or other loss or risk of loss. The classic case is when a request for a time extension for excusable delay is denied and the contract provides liquidated damages for late completion. The law construes this as a constructive order by the owner to complete performance within the originally specified completion date, a shorter period of time, and potentially at a higher cost than otherwise would have been required. The constructive acceleration doctrine allows recovery for additional expenses where the owner refuses to give the contractor a time extension to which the contractor is contractually entitled, thereby forcing the contractor to "accelerate" its work efforts in an attempt to maintain the original work schedule.[65]

Court decisions have identified five elements normally required to establish a claim for constructive acceleration.[66] Those elements are:

(1) An excusable delay must exist.
(2) Timely notice of the delay and a proper request for a time extension should have been given.
(3) The time extension must have been postponed or refused.
(4) The owner must have ordered in some way (either by coercion, direction, or some other manner) the project be completed within its original performance period.
(5) The contractor must make efforts to accelerate its performance, thereby incurring additional costs.

The contractor should give appropriate notice of the delay and request an extension of time in order to recover its costs of acceleration.[67] Notice is important, especially when the contractor believes acceleration is occurring without a specific order to accelerate. Such notice will establish that the acceleration is not being undertaken voluntarily and, further, that the contractor expects the owner to pay the additional costs incurred. This notice to the owner of "forced" acceleration will assist the contractor greatly in recovering damages. This notice is not absolutely necessary if (a) the acceleration has been expressly directed; (b) the owner has indicated no time extensions will be permitted; or (c) the owner has waived the need for notice.[68] Moreover, if the owner has specific knowledge of excusable delays and unequivocally

[65] *See Dep't of Transp. v. Anjo Constr. Co.,* 666 A.2d 753 (Pa. Commw. Ct. 1995); *Howard J. White, Inc. v. Varian Assocs.,* 2 Cal. Rptr. 871 (Cal. Ct. App. 1960); *Siefford v. Hous. Auth. of City of Humboldt,* 223 N.W.2d 816, 820 (Neb. 1974).

[66] *Sherman R. Smoot Co. v. Ohio Dep't of Admin. Servs.,* 736 N.E.2d 69 (Ohio Ct. App. 2000); *Envirotech Corp. v. Tenn. Valley Auth.,* 715 F. Supp. 190 (W.D. Ky. 1988); *Nat Harrison Assocs., Inc. v. Gulf States Utils. Co.,* 491 F.2d 578 (5th Cir. 1974); *Natkin & Co. v. George A. Fuller Co.,* 347 F. Supp. 17 (W.D. Mo. 1972).

[67] *Envirotech Corp. v. Tenn. Valley Auth.,* 715 F. Supp. 190 (W.D. Ky. 1988).

[68] *Nat Harrison Assocs., Inc. v. Gulf States Utils. Co.* 491 F.2d 578 (5th Cir. 1974) (finding owner waived notice requirements where owner was advised that contractor was accelerating its work and would be incurring costs because of owner's failure to grant time extensions).

orders the contractor to finish work on the contract completion date without regard to excusable delays, the notice requirement is satisfied. Additionally, it is important that the contractor give detailed information about the delay to the owner, so that the owner can determine the reasonableness of the time extension.

The contractor must carefully and fully document all delays and other factors that will aid in proving its entitlement to damages. Courts and boards are willing to find for the contractor on constructive acceleration claims, but if not carefully documented, the damages may not be recovered.

The refusal to grant a time extension can be expressed either by a clear rejection of a time extension or by the postponement of a decision concerning the request. The owner owes a duty to the contractor to timely respond and grant or deny the request.[69] By failing to respond to the request, or even putting off the decision until the completion of the contract, the owner puts the contractor in a precarious position. If the contractor acts as if the time extension will be granted and continues at a pace that will complete the work after the completion date, and the time extension is not granted, liquidated damages may be assessed. If the contractor acts as if the time extension will not be granted and accelerates in order to complete the work by the established completion date, the contractor likely will incur additional expenses that might not be recovered. The contractor's updated schedule showing the contractor's reasonable expectations at the time of the request becomes especially important to support the decision to accelerate. Damages are recoverable even if acceleration is attempted only to avoid the risk of liquidated damages. If placed on appropriate notice, it is possible that the owner eventually will pay the cost created by postponing its decision on time extension.

In *James Corporation v. North Allegheny School District,*[70] the court stated that a contractor may recover for increased costs as a result of accelerating its performance where:

(1) Its own delays in performance are excusable;
(2) The contractor was ordered to accelerate; and
(3) The contractor did so and sustained extra costs.

The *James Corporation* court upheld the trial court's award of acceleration damages where the school district did not obtain the required permits before soliciting bids, the construction manager did not issue a comprehensive schedule until four months after the project began, and the school district refused to adjust the project completion date.

A constructive change also can arise when an owner's incorrect interpretation of the contract requires the contractor to accelerate. For example, in *Rogers Excavating,*[71] an earthwork contract required the contractor to start work within four days

[69]*Dep't of Transp. v. Anjo Constr. Co.,* 666 A.2d 753, 758 (Pa. Commw. Ct. 1995) (owner's granting of a 64-day extension did not preclude holding that owner constructively ordered contractor to accelerate project where owner waited months to grant the requested extension of time).
[70]938 A.2d 474, 483 (Pa. Commw. Ct. 2007).
[71]AGBCA No. 79–180, 83–2 BCA ¶ 16,701.

after receipt of the Notice to Proceed and finish all work in 90 days. The contractor submitted a proposed schedule showing mobilization to start within four days but actual excavation not starting until 40 days after Notice to Proceed. The owner refused to accept this schedule and required the contractor to start excavation work within the four-day period. The court found this to be an acceleration justifying additional compensation because mobilization was found to be "work" as it was defined in the contract. The contractor, however, ultimately lost the case because it had failed to give the required notice that it considered the owner's action to create a claim situation.

The term "acceleration" is best known in federal government contracts.[72] The same result is achieved, however, usually under a breach of contract theory, where the private owner fails to grant an extension of time promptly or properly. As the court stated in *Wallace Process Piping Co. v. Martin-Marietta Corp*:[73]

> The order to complete additional work without an extension of time, which made necessary a workweek in excess of 50 hours, was a change within the meaning of Paragraph 4, the Changes Clause of the contract. The refusal of Martin upon demand timely made by Wallace for an equitable adjustment constituted a breach of contract.

VI. CARDINAL CHANGES

The changes clause of a construction contract does not give the owner an unrestricted right to order extra work. Changed work or extra work must be within the general scope of the original contract.

As a simple example, an owner that contracts for the construction of a house cannot require the contractor, by change order, to build a second house. Such extra work is totally beyond the scope of the original agreement. This would be an example of a "cardinal change." A change order requiring the addition of a room or the finishing of the basement, however, probably would be valid under the changes clause. The difficult questions, of course, involve those cases that fall somewhere between these two extremes.

The term "cardinal change" refers to a change or changes ordered by the owner that are beyond the scope of the contract and therefore constitute a material breach of contract.[74] If a change is a cardinal change, the owner is in breach of its contract and the contractor can either refuse to perform or can perform and be paid the reasonable value for the work. But if a contractor refuses to perform a proper change, incorrectly thinking it to be a cardinal change, it will be in breach for refusing to perform.

If a contractor is confronted with an undertaking substantially different from that originally contemplated due to the extensive changes ordered by the owner or dictated by the owner's actions, then a cardinal change may exist. The contractor has

[72] *See, e.g. Ensign-Bickford Co.*, ASBCA No. 6214, 60–2 BCA ¶ 2817.
[73] 251 F. Supp. 411, 418–19 (E.D. Va. 1965).
[74] *Keeter Trading Co., Inc. v. United States*, 79 Fed. Cl. 243 (2007).

the right to disregard the contract agreement in a cardinal change situation and seek compensation for the reasonable value of all services and materials provided.

In *Becho, Inc. v. United States,*[75] the court discussed the concept of cardinal change in this way:

> Of course, the government may not, through a contracting officer's decision, impose obligations on a contractor far exceeding any contemplated by their contract. If the government orders a "drastic modification" in the performance required by the contract, the order is a considered a "cardinal change" that constitutes a material breach of the contract.

The court added that "each case must be analyzed on its own facts and in light of its own circumstances, giving just consideration to the magnitude and quality of the changes ordered and their cumulative effect upon the project as a whole."[76]

In an 1883 decision, the Illinois Supreme Court,[77] without coining the term "cardinal change", analyzed the concept in this way:

> Obviously, under a contract to construct a framed building at stipulated prices, a party could not be required to construct a stone or brick building, at prices to be fixed by the architect of the other party, by the use of these words in the contract. Nor could a party, by virtue thereof, contracting to build a small and inferior brick or stone building, be required to construct a large and superior stone or brick building. The mere combination of proportions and quantities, even of materials of the same class or grade, may be so different in different buildings of the same dimensions, that a party would not make the same bid, or be able, without financial loss, to construct them all for the same price.
>
> The terms stated in the writing were, we think, the controlling inducement to the contract, and the "changes, additions and alterations" therein provided for must have been contemplated and intended to be but such as were incidental to the complete execution of the work as described in the plans and specifications, and therefore of only minor and trifling importance, for otherwise some definite mode of determining what prices should be paid for them would also have been prescribed by the writing. We think any material departure from the plans and specifications with reference to which the contract was made, which resulted in a new and substantially different undertaking, cannot be regarded as within the meaning of this language.
>
> We cannot admit that a party entering into a contract to do a given work at stipulated prices, can, by the use of these words in the written contract, be made to do a different and more expensive work at prices to be named altogether, or in large part, by the architect of the other party.[78]

[75]47 Fed. Cl. 595, 600 (2000).
[76]*Id.* at 601.
[77]*County of Cook v. Harms,* 108 Ill. 151 (Ill. 1883).
[78]*Id.* at 159–60.

In *C. Norman Peterson Co. v. Container Corp. of America,*[79] a contractor brought an action for breach of contract and for the reasonable value of the labor and materials used after experiencing significant cost overruns at a paper mill project. The court found that the owner, by imposing hundreds of changes on the contractor, so altered the scope of the work under the original contract that the owner had abandoned the original contract. Accordingly, the guaranteed maximum cost provision of the contract was inapplicable and the contractor was entitled to recover the reasonable cost of its work. The court justified basing the contractor's damages on the total cost method because the owner was to blame for preventing the contractor from making a detailed showing as to how the damages claimed were caused by the breach.

A cardinal change may also arise from physical conditions encountered by the contractor that were not expected and that fundamentally changed the scope of the work. For example, in *Clark-Fitzpatrick, Inc./Franki Foundation Co. v. Gill,*[80] a contractor was awarded a contract by the Rhode Island Department of Transportation (RIDOT) for the construction of a replacement for the Jamestown Bridge connecting the island of Jamestown with the mainland at North Kingstown. After the work began, the contractor encountered a significant problem as a result of unexpected behavior of the soils at the bottom of the bay. The RIDOT was forced to change from a friction-pile design to a more expensive composite-pile design because of the soil difficulties. The RIDOT estimated the additional cost at $12 million while the contractor estimated the additional cost at $30 million. The contractor was ordered to continue working and subsequently filed suit claiming the change in the pile design amounted to a cardinal change of the contract. At trial, the contractor was awarded approximately $22 million, which was later reduced by $3 million on appeal.

The question of whether a particular change (or group of changes) is sufficient to constitute a cardinal change is a matter of degree—and often is very subjective. The basic tests for a cardinal change are:

(1) Whether the type of work was within the contemplation of the parties when they entered into the contract.
(2) Whether the job as modified is still the same basic job.

VII. THE IMPACT OF NUMEROUS CHANGES ON UNCHANGED WORK

Contractors occasionally may encounter a project where the owner makes multiple changes to a construction job, and no single change could be considered a cardinal change. After agreeing to many of these changes, however, the contractor realizes that the agreed prices for the changes are not adequate to compensate the contractor for the disruption and inefficiency created by the large volume of changes. This situation can occur when the contractor prices an individual change based on the direct

[79]218 Cal. Rptr. 592 (Cal. Ct. App. 1985).
[80]652 A.2d 440 (R.I. 1994).

costs to perform the work associated with that one change and does not account for the adverse impact the change could have on performing unchanged work.

Generally, a contractor is bound by the terms of the change orders it signs. Thus, if the signed change orders have broad release language or state that they include all the compensation the contractor is entitled to receive for the changed work—covering both direct and indirect costs—the contractor may have little legal recourse.

For example, in *Vanlar Construction, Inc. v. County of Los Angeles,*[81] a contractor brought an action against the county to recover impact costs arising out of the cumulative effect of numerous change orders. The court held that Vanlar was not entitled to impact costs, emphasizing that all of the change orders and supplemental agreements the contractor had entered included direct and indirect costs. The court further pointed out that if Vanlar contemplated a future claim for impact costs, it should have requested that a reservation clause be inserted in each change order and supplemental agreement.

A contractor concerned about possible indirect impacts numerous changes may have on a job should consider adding a clause to proposed change orders stating that the change order covers only the direct costs of the changed work and that the contractor reserves the right to claim impact costs later.

VIII. IMPOSSIBILITY/IMPRACTICABILITY

The legal theory of "impossibility" can provide relief in appropriate situations from the usual common law principle that a party must either perform its contract or respond in damages, even though performance proves to be more onerous or expensive than anticipated. The impossibility doctrine permits a contractor to walk away from a contract without penalty if performance is impossible or is so impracticable as to be virtually impossible. A party must generally show three elements to excuse performance of a contract based on impossibility: (1) a contingency—something unexpected—must have occurred; (2) risk of the unexpected occurrence must not have been allocated by agreement or by custom; and (3) occurrence of the contingency must have rendered performance commercially impracticable.[82]

The theory of commercial impracticability is related to the doctrine of impossibility. In the leading case of *Mineral Park Land Co. v. Howard,*[83] the court described the concept in this way:

> A thing is impossible in legal contemplation when it is not practicable; and a thing is impracticable when it can only be done at an excessive and unreasonable cost. . . . We do not mean to intimate that the defendants could excuse themselves by showing the existence of conditions which would make the

[81]217 Cal. Rptr. 53 (Cal. Ct. App. 1985) (unpublished opinion).

[82]*Island Dev. Corp. v. Dist. of Columbia,* 933 A.2d 340 (D.C. 2007); *see also Fraught v. Platte Valley Pub. Power & Irrigation Dist.,* 51 N.W. 2d 253 (Neb. 1952).

[83]156 P. 458 (Cal. 1916).

> performance of their obligations more expensive than they had anticipated, or which would entail a loss by them. But where the difference in cost is so great as here, and has the effect, as found, of making performance impracticable, the situation is not different from that of a total absence of earth and gravel.

In *Mineral Park Land,* the contractor was excused from performing a gravel excavation contract when the cost of performance proved to be 12 times more than originally anticipated.

Under traditional legal contract analysis, an unexpected rise in the price of materials is not an excuse for nonperformance. Courts and federal contract appeals boards consistently rule that a contractor signing a lump-sum contract accepts the cost risks associated with performance, including abnormal risks. Therefore, in all but the rarest of cases, the unexpected increase in the cost of goods or construction materials is a risk borne by the contractor.

It is generally stated that the doctrine of impracticability requires objective impracticability. Objective impracticability refers to the contemplated performance being impractical no matter who attempts performance. Subjective impracticability, or impracticability tied to the particular circumstances of the contractor in question, is not sufficient.[84] Thus, a contractor must show that performance would be impractical for the "reasonable contractor"[85] rather than merely impractical for it because of its particular abilities or circumstances.[86] For example, in *Piasecki Aircraft Corp. v. United States,*[87] the court refused to apply the doctrine, cautioning that the doctrine of impracticable performance may be invoked "only when the [contractor] has exhausted all its alternatives" and when all methods of performance are "commercially senseless."

Although the concept of impossibility can serve as a defense to excuse a contractor's entire contract performance, it also may prompt revisions or alterations in performance that bring the changes clause into play. For example, when the plans and specifications require the use of certain materials that subsequently prove impossible or impracticable to obtain, the contractor is necessarily forced to find a substitute. This substitution will be a compensable change if it increases the contractor's cost of performance. In *McIntyre v. United States,*[88] the contract required approved Colorado marble, but marble satisfying the specifications was unavailable in Colorado. The court held that the contractor was entitled to the extra costs incurred in obtaining marble from Tennessee, even though government permission to obtain it from the alternate source was subject to the condition that there would be no additional cost chargeable to the government.

[84]*See, e.g., Luminous Neon, Inc. v. Parscale,* 836 P.2d 1201 (Kan. Ct. App. 1992).
[85]*Guy F. Atkinson Co.,* ENGBCA No. 4771, 88–2 BCA ¶ 20,714 (impossible to achieve required moisture content); *Blount Bros. Corp. v. United States,* 872 F.2d 1003 (Fed. Cir. 1989) (specified material not available).
[86]*Koppers Co. v. United States,* 405 F.2d 554 (Ct. Cl. 1968).
[87]667 F.2d 50 (Ct. Cl. 1981), *cert. denied*, 444 U.S. 898 (1981).
[88]52 Ct. Cl. 503 (1917).

A similar rule may apply with regard to methods of performance.[89] The federal government contract appeals boards and the U.S. Court of Federal Claims,[90] however, have held that a method of performance dictated by the contract must be followed, even though performance may be more difficult (though short of impossible) or less efficient than anticipated. For example, *Natus Corp. v. United States*[91] involved an $8.5 million contract to produce a portable steel airplane landing mat. The contractor argued that production in accordance with government specifications was commercially "impracticable" and that, as a result, it was entitled to an equitable adjustment. Despite the problems the plaintiff encountered in developing a suitable production process, the court found that an alternative process, although less economical, appeared to be workable. In the words of the court:

> The law excuses performance (or, in the case of government contracts, grants relief through a change order) where the attendant costs of performance bespeak commercial senselessness; it does not grant relief merely because performance cannot be achieved under the most economical means.[92]

➢ POINTS TO REMEMBER

- Change order disputes arise when the contracting parties disagree about:
  - The original scope of the work, or
  - Compliance with notice requirements, or
  - Who has authority to order changes, or
  - Pricing issues
- Authority issues arise when a person who has no actual express authority orders a change for which the owner must pay because:
  - The person had implied authority, or
  - The person had apparent authority, or
  - The owner ratifies the change order.
- The lack of the written notice/written change order often required by a contract does not necessarily defeat a claim for extra compensation if the requirement for a writing has been waived by agreement or conduct of the parties. Under certain circumstances, a claim for *quantum meruit* may be available to recover compensation where there is no written change order.

[89]*See, e.g., Hol-Gar Mfg. Corp. v. United States,* 360 F.2d 634 (Ct. Cl. 1966); *Hobbs Constr. & Dev., Inc.,* ASBCA No. 34890, 91–2 BCA ¶ 23,755.
[90]Previously the United States Court of Claims and the United States Claims Court.
[91]371 F.2d 450 (Ct. Cl. 1967).
[92]*Natus Corp. v. United States,* 371 F.2d 450, 457 (Ct. Cl. 1967).

- Constructive changes occur when the owner, without issuing a written order, directs the contractor to perform additional work to complete or remedy defective plans and specifications, to meet a higher standard of performance, to change or accelerate its schedule, or to alter the sequence of work from that which the contractor had otherwise planned.
- "Cardinal changes" are those changes or accumulations of changes that so greatly exceed the scope of the contract as to create a material breach of contract.
- "Impact costs" refer to the cumulative or "ripple" cost effect of numerous changes, and can be recovered in addition to the cost of the underlying changes if the contractor has appropriately reserved its right, in each signed change order, to recover later-incurred impact costs.
- Impossible or commercially impracticable conditions can excuse a contractor from performance; contractors generally must show that performance by a reasonable contractor was impracticable in order to prove that it was impossible or commercially impracticable to perform.

10

DIFFERING SITE CONDITIONS

One of the more common risks on a construction project arises when a contractor encounters conditions that materially differ from those reasonably contemplated when the contract was bid. Unanticipated site conditions often generate extra costs and also can substantially delay and disrupt the project. Since these delays usually occur at the front end of a job, they can have a greater overall impact on the project.

I. "DIFFERING SITE CONDITION" DEFINED

A "differing site condition"—or "changed condition," as it is sometimes called—is a physical condition encountered in performing the work that was not visible and not known to exist at the time of bidding and that materially differs from the condition believed to exist at the time of pricing the contract. Often this condition could not have been discovered by a reasonable site investigation. Examples of changed conditions or differing site condition problems include soil with inadequate bearing capacity to support the building being constructed; soil that cannot be reused as structural fill; unanticipated groundwater (static or percolating); quicksand; muck; rock formations (or excessive or insufficient quantities of rock); and artificial (man-made) subsurface obstructions.

II. RESPONSIBILITY FOR DIFFERING SITE CONDITIONS

Under a traditional contract risk allocation analysis, a prudent contractor would be expected to protect itself against unforeseen conditions by including a contingency factor in its bid. The basic flaw in this approach is that a contractor cannot accurately value a true unknown. Even if included, the bid contingency may end up being totally inadequate or, alternatively, grossly inflated. The one constant is that including any contingency increases bid prices and thus works to the detriment of the owner if adverse conditions are not encountered. In other situations, the contingency may prove wholly inadequate to cover the contractor's actual increased costs.

To alleviate some of the risks associated with unexpected site conditions, differing site condition clauses have become a common feature in many construction contracts. The reason for this widely used provision has been explained by many courts. The United States Court of Claims (now the United States Court of Federal Claims) explained it in *Foster Construction C.A. & Williams Bros. Co. v. United States* as follows:[1]

> The purpose of the changed conditions clause is thus to take at least some of the gamble on subsurface conditions out of bidding.
>
> Bidders need not weigh the cost and ease of making their own borings against the risk of encountering an adverse subsurface condition, and they need not consider how large a contingency should be added to the bid to cover the risk. *There will be no windfalls and no disasters.* The Government benefits from more accurate bidding, without inflation for risks which may not eventuate. It pays for difficult subsurface work only when it is encountered and was not indicated in the logs. (Emphasis added.)

Despite this logic, some private and public owners do not include a changed conditions clause in their contracts. In fact, some owners go further and include clauses that purport to place all possible risks of differing site conditions on the contractor. These "exculpatory" clauses or disclaimers invite bidders to include contingencies in their bids.

The absence of a differing site conditions clause in the contract, however, does not necessarily mean that a contractor will be denied relief if adverse site problems arise. Several theories of recovery have been advanced in such cases. The most prominent of these theories are: (1) breach of warranty, (2) breach of a duty to disclose available information, (3) mutual mistake, (4) innocent misrepresentation, and (5) fraud. These theories are discussed briefly later in the chapter, after a more detailed review of the standard clauses and types of changed condition problems.

III. STANDARD DIFFERING SITE CONDITIONS CLAUSES

Today, virtually all construction industry "standard" form contracts between owners and contractors contain some type of differing site conditions clause. The first such standard clause appeared in 1927 in the federal government's standard fixed-price

[1] 435 F.2d 873, 887 (Ct. Cl. 1970). This seminal case has been cited in more recent decisions, including *Renda Marine, Inc. v. United States,* 71 Fed. Cl. 378 (2006) and *Asphalt Roads & Materials Co. v. Commonwealth,* 512 S.E.2d 804 (Va. 1999).

construction contract. Its purpose was, and is today, to place the risk of reasonably unexpected site conditions on the federal government by granting a price increase and time extension to contractors in the event that such conditions are encountered.[2]

A. Federal Government Contracts

The text of the current differing site conditions clause used in federal government contracts was adopted in 1984 and is set forth in the Federal Acquisition Regulation at FAR § 52.236–2.[3] The clause provides:

> **DIFFERING SITE CONDITIONS (APR 1984)**
>
> (a) The Contractor shall promptly, and before the conditions are disturbed, give a written notice to the Contracting Officer of (1) subsurface or latent physical conditions at the site which differ materially from those indicated in this contract, or (2) unknown physical conditions at the site, of an unusual nature, which differ materially from those ordinarily encountered and generally recognized as inhering in work of the character provided for in the contract.
>
> (b) The Contracting Officer shall investigate the site conditions promptly after receiving the notice. If the conditions do materially so differ and cause an increase or decrease in the Contractor's cost of, or the time required for, performing any part of the work under this contract, whether or not changed as a result of the conditions, an equitable adjustment shall be made under this clause and the contract modified in writing accordingly.
>
> (c) No request by the Contractor for an equitable adjustment to the contract under this clause shall be allowed, unless the Contractor has given the written notice required; provided, that the time prescribed in (a) above for giving written notice may be extended by the Contracting Officer.
>
> (d) No request by the Contractor for an equitable adjustment to the contract for differing site conditions shall be allowed if made after final payment under this contract.

B. ConsensusDOCS

ConsensusDOCS 200 Standard Form of Agreement and General Conditions Between Owner and Contractor (Where the Contract Price Is a Lump Sum) (2007 ed.), and

[2]*See generally* Thomas E. Abernathy IV & C. Michael Shull, Jr., Construction Business Handbook, Chapter 15 (Aspen Publishers, Inc. 2004); Overton A. Currie, Thomas E. Abernathy IV & Robert C. Chambers, Changed Conditions, Construction Briefings No. 84–12 (Federal Publications, Inc. 1984).

[3]The Federal Acquisition Regulation is found in 48 Code of Federal Regulations (C.F.R.). For a more detailed discussion of differing site conditions in the federal government contracts context, *see* Thomas J. Kelleher, Jr., Thomas E. Abernathy IV & Hubert J. Bell, Jr., Federal Construction Contracts—A Practical Guide for the Industry Professional, Chapter 7 (John Wiley & Sons 2008).

paragraph 3.16, entitled *Worksite Conditions,* contains a differing site conditions clause, which states:

> 3.16.1 WORKSITE VISIT The Contractor acknowledges that it has visited, or has had the opportunity to visit, the Worksite to visually inspect the general and local conditions which could affect the Work.
>
> 3.16.2 CONCEALED OR UNKNOWN SITE CONDITIONS If the conditions at the Worksite are (a) subsurface or other physical conditions which are materially different from those indicated in the Contract Documents, or (b) unusual or unknown physical conditions which are materially different from conditions ordinarily encountered and generally recognized as inherent in Work provided for in the Contract Documents, the Contractor shall stop Work and give immediate written notice of the condition to the Owner and the Architect/Engineer. The Contractor shall not be required to perform any work relating to the unknown condition without the written mutual agreement of the Parties. Any change in the Contract Price or the Contract Time as a result of the unknown condition shall be determined as provided in Article 8. The Contractor shall provide the Owner with written notice of any claim as a result of unknown conditions within the time period set forth in Paragraph 8.4.

The form also contains another clause under Article 4, *Owner's Responsibilities,* which discusses work site information and states:

> 4.3 WORKSITE INFORMATION Except to the extent that the Contractor knows of any inaccuracy, the Contractor is entitled to rely on Worksite information furnished by the Owner pursuant to this Paragraph 4.3. To the extent the Owner has obtained, or is required elsewhere in the Contract Documents to obtain, the following Worksite information, the Owner shall provide at the Owner's expense and with reasonable promptness:
>
> 4.3.1 Information describing the physical characteristics of the site, including surveys, site evaluations, legal descriptions, data or drawings depicting existing conditions, subsurface conditions and environmental studies, reports and investigations. …

The ConsensusDOCS contain similar differing site conditions clauses in Document No. 300, entitled Standard Form of Tri-Party Agreement for Collaborative Project Delivery (See Article 13, Section 14) and Document No. 500, entitled Standard Agreement and General Conditions Between Owner and Construction Manager (see Article 3, Section 13).

In ConsensusDOCS 410, entitled Standard Design-Build Agreement and General Conditions Between Owner and Design-Builder, the differing site conditions clause states:

> 9.4 UNKNOWN CONDITIONS If in the performance of the Work the Design-Builder finds latent, concealed or subsurface physical conditions which

materially differ from the conditions the Design-Builder reasonably anticipated, or if physical conditions are materially different from those normally encountered and generally recognized as inherent in the kind of work provided for in this Agreement, then the GMP, estimated Cost of the Work, the Design-Builder's Fee, the Date of Substantial Completion or the Date of Final Completion, and if appropriate the compensation for Design Phase services, shall be equitably adjusted by Change Order within a reasonable time after the conditions are first observed. The Design-Builder shall provide the Owner with written notice within the time period set forth in Paragraph 9.6.

C. Other Standard Forms

1. The American Institute of Architects

In 2007, the American Institute of Architects (AIA) released an update to its General Conditions contract document, the AIA A201 (2007 ed.), that contains the following differing site conditions clause:

> 3.7.4 Concealed or Unknown Conditions. If the Contractor encounters conditions at the site that are (1) subsurface or otherwise concealed physical conditions that differ materially from those indicated in the Contract Documents or (2) unknown physical conditions of an unusual nature that differ materially from those ordinarily found to exist and generally recognized as inherent in construction activities of the character provided for in the Contract Documents, the Contractor shall promptly provide notice to the Owner and the Architect before conditions are disturbed and in no event later than 21 days after first observance of the conditions. The Architect will promptly investigate such conditions and, if the Architect determines that they differ materially and cause an increase or decrease in the Contractor's cost of, or time required for, performance of any part of the Work, will recommend an equitable adjustment in the Contract Sum or Contract Time, or both. If the Architect determines that the conditions at the site are not materially different from those indicated in the Contract Documents and that no change in the terms of the Contract is justified, the Architect shall promptly notify the Owner and the Contractor in writing, stating the reasons. If either party disputes the Architect's determination or recommendation, that party may proceed as provided in Article 15.

In addition to moving the differing site conditions clause from § 4.3.4 to § 3.7.4, the AIA made these significant revisions to the A201 since the document was last issued in 1997:

- The burden of discovering differing site conditions and notifying the owner and architect has been shifted solely to the contractor.
- Clarified that the architect is the only party responsible for making the initial determination of whether a differing site condition differs materially from what

was expected on the site and whether the condition causes an increase or decrease in the contractor's costs or time.

- Removed from the differing site conditions clause the 21-day deadline for the owner or contractor to oppose the architect's determination regarding the effect or lack of effect of a differing site condition to the project.[4]
- Removed the provision regarding the automatic equitable adjustment to the contract sum and contract time in the event the architect determined that a differing site condition was materially different from what was expected.

Immediately following Section 3.7.4 is a new and separate differing site conditions clause that specifically addresses the responsibilities of parties to a construction contract when unanticipated human remains, archaeological sites, or wetlands are encountered on a project. Unlike the "prompt" notice requirement in the AIA's standard differing site conditions clause, Section 3.7.5 requires the contractor to "immediately suspend" any operations that would affect the unanticipated remains or features and to recommence operations only after receiving further instructions from the owner. Interestingly, the clause specifies that the contractor "shall continue with all other operations that do not affect those remains or features."

2. Engineers Joint Contract Documents Committee

Engineers Joint Contract Documents Committee (EJCDC), a coalition of stakeholders in the project delivery process consisting of the American Council of Engineering Companies (ACEC), the Associated General Contractors of America (AGC), the American Society of Civil Engineers (ASCE), and the National Society of Professional Engineers (NSPE), also issued updated standard contract forms in 2007 that contain differing site condition provisions. The EJCDC C-700 (2007 ed.) provides:

GC-4.02 Subsurface and Physical Conditions

A. *Reports and Drawings:* The Supplementary Conditions identify:
   (1) those reports known to Owner of explorations and tests of subsurface conditions at or contiguous to the Site; and
   (2) those drawings known to Owner of physical conditions relating to existing surface or subsurface structures at the Site (except Underground Facilities).

B. *Limited Reliance by Contractor on Technical Data Authorized*: Contractor may rely upon the accuracy of the "technical data" contained in such reports and drawings, but such reports and drawings are not Contract Documents. Such "technical data" is identified in the Supplementary Conditions. Except

[4]AIA A201, § 15.1.2 (2007 ed.) contains a requirement, however, that the claimant must initiate a claim "within 21 days after occurrence of the event giving rise to such Claim or within 21 days after the claimant first recognizes the condition giving rise to the Claim, whichever is later."

for such reliance on such "technical data," Contractor may not rely upon or make any claim against Owner or Engineer, or any of their officers, directors, members, partners, employees, agents, consultants, or subcontractors with respect to:

(1) the completeness of such reports and drawings for Contractor's purposes, including, but not limited to, any aspects of the means, methods, techniques, sequences, and procedures of construction to be employed by Contractor, and safety precautions and programs incident thereto; or

(2) other data, interpretations, opinions, and information contained in such reports or shown or indicated in such drawings; or

(3) any Contractor interpretation of or conclusion drawn from any "technical data" or any such other data, interpretations, opinions, or information.[5]

Additionally, the following "mandatory Supplementary Condition" as provided by EJCDC C-800 (2007 ed.) is to be inserted after Paragraph 4.02B and used for the purpose of identifying the known site condition documents.

C. The following reports of explorations and tests of subsurface conditions at or contiguous to the Site are known to Owner:

[LIST ALL SUCH REPORTS]

D. The following drawings of physical conditions relating to existing surface or subsurface structures at the Site (except Underground Facilities) are known to Owner:

[LIST ALL SUCH REPORTS]

[Use one of the following two subparagraphs:]

(a) All of the information in such drawings constitutes "technical data" on which Contractor may rely, except for ______________ appearing on Drawing No. ______________

[or]

(b) None of the contents of such drawings is "technical data" on which Contractor may rely.

E. The reports and drawings identified above are not part of the Contract Documents, but the "technical data" contained therein upon which Contractor may rely, as expressly identified and established above, are incorporated in the Contract Documents by reference. Contractor is not entitled to rely upon any other information and data known to or identified by Owner or Engineer.

F. Copies of reports and drawings identified in SC-4.02.C and SC-4.02.D that are not included with the Bidding Documents may be examined at [insert location] during regular business hours.

[5]If there are no known site-related reports or drawings, GC 4.02 A&B should be revised to read: No reports of explorations or tests of subsurface conditions at or contiguous to the Site, or drawings of physical conditions relating to existing surface or subsurface structures at the Site, are known to Owner.

GC-4.03 Differing Subsurface or Physical Conditions

A. *Notice:* If Contractor believes that any subsurface or physical condition that is uncovered or revealed either:

(1) is of such a nature as to establish that any "technical data" on which Contractor is entitled to rely as provided in Paragraph 4.02 is materially inaccurate; or

(2) is of such a nature as to require a change in the Contract Documents; or

(3) differs materially from that shown or indicated in the Contract Documents; or

(4) is of an unusual nature, and differs materially from conditions ordinarily encountered and generally recognized as inherent in work of the character provided for in the Contract Documents; then Contractor shall, promptly after becoming aware thereof and before further disturbing the subsurface or physical conditions or performing any Work in connection therewith (except in an emergency as required by Paragraph 6.16. A), notify Owner and Engineer in writing about such condition. Contractor shall not further disturb such condition or perform any Work in connection therewith (except as aforesaid) until receipt of written order to do so.

B. *Engineer's Review:* After receipt of written notice as required by Paragraph 4.03.A, Engineer will promptly review the pertinent condition, determine the necessity of Owner's obtaining additional exploration or tests with respect thereto, and advise Owner in writing (with a copy to Contractor) of Engineer's findings and conclusions.

C. *Possible Price and Times Adjustments:*

(1) The Contract Price or the Contract Times, or both, will be equitably adjusted to the extent that the existence of such differing subsurface or physical condition causes an increase or decrease in Contractor's cost of, or time required for, performance of the Work; subject, however, to the following:

(a) such condition must meet any one or more of the categories described in Paragraph 4.03.A; and

(b) with respect to Work that is paid for on a unit price basis, any adjustment in Contract Price will be subject to the provisions of Paragraphs 9.07 and 11.03.

(2) Contractor shall not be entitled to any adjustment in the Contract Price or Contract Times if:

(a) Contractor knew of the existence of such conditions at the time Contractor made a final commitment to Owner with respect to Contract Price and Contract Times by the submission of a Bid or becoming bound under a negotiated contract; or

(b) the existence of such condition could reasonably have been discovered or revealed as a result of any examination, investigation, exploration, test, or study of the Site and contiguous areas required by the Bidding Requirements or Contract Documents to be conducted by or for Contractor prior to Contractor's making such final commitment; or

(c) Contractor failed to give the written notice as required by Paragraph 4.03.A.

(3) If Owner and Contractor are unable to agree on entitlement to or on the amount or extent, if any, of any adjustment in the Contract Price or Contract Times, or both, a Claim may be made therefor as provided in Paragraph 10.05. However, neither Owner or Engineer, or any of their officers, directors, members, partners, employees, agents, consultants, or subcontractors shall be liable to Contractor for any claims, costs, losses, or damages (including but not limited to all fees and charges of engineers, architects, attorneys, and other professionals and all court or arbitration or other dispute resolution costs) sustained by Contractor on or in connection with any other project or anticipated project.

IV. TYPES OF CONDITIONS COVERED

An examination of the various clauses in the preceding form contracts reveal some basic similarities but also some important differences. The FAR and EJCDC clauses define differing site conditions as "subsurface or latent physical conditions" and "subsurface or physical conditions" respectively, while the ConsensusDOCS and AIA clauses refer to the site conditions as "concealed or unknown." Generally, all of the standard differing site conditions clauses cover similar situations, although circumstances arise where the wording in a particular clause can make a difference.

A. Type I and Type II Changed Conditions

The FAR, ConsensusDOCS, AIA, and EJCDC contract provisions identify two distinct types of unanticipated conditions that are compensable. These are usually designated as Type I and Type II changed conditions.

Type I changed conditions are, in the language of the FAR, conditions "differing materially from those indicated in the contract." ConsensusDOCS 200 speaks of such conditions as "materially different from those indicated in the Contract Documents."

The FAR and ConsensusDOCS clauses describe a Type II changed condition as unusual or unknown physical conditions at the site that differ materially from those ordinarily encountered and generally recognized as inhering in work of the character provided for in the contract. The language used to describe Type II changed conditions in the AIA and EJCDC provisions is similar to that used in the FAR and ConsensusDOCS.

B. Notice Requirements

Differences also exist among these standard clauses with respect to notice requirements. The FAR and ConsensusDOCS clauses require that the contractor stop work and give written notice upon encountering an unexpected condition before disturbing it so that the owner's representative will have an opportunity to inspect and evaluate the condition. The EJCDC clause, with its "promptly after becoming aware thereof and before further disturbing" language, appears to be closer in intent to the FAR than it is to AIA A201, which requires that notice of differing conditions be given to the owner and architect within 21 days after "first observance" by the contractor. Regardless of the exact language, it is always preferable to immediately notify the owner (or its agent, such as the project architect or engineer) when materially different conditions are encountered. By giving the owner the option of investigating the condition and, if appropriate, determining how best to proceed, the contractor greatly increases the likelihood of resolving any resulting claim in an expedient and mutually acceptable manner.

V. OPERATION OF THE DIFFERING SITE CONDITIONS CLAUSE

Although a differing site conditions clause provides a mechanism for dealing with an unanticipated site condition, a contract adjustment is not guaranteed. To obtain an adjustment under the clause, one must first establish that the changed condition falls within the scope of the clause. Before examining what typically must be proven, it is important to remember what one is not required to prove.

If you are a contractor, notification of a suspected differing site condition does not mean that you are attempting to establish fault, bad faith, or defective design by the owner or its representative. There are simply some situations where differing, unanticipated conditions are encountered. This is especially true when dealing with subsurface work or work on older structures where a limited construction history is available.

Because unexpected site conditions are common on subsurface construction projects, differing site conditions clauses are routinely found in the terms of agreements between soils engineers and owners. For example, the standard language incorporated in the ASFE's (formerly known as the Associated Soil and Foundation Engineers) suggested form contract between a geotechnical consultant and an owner (Client) provides:

> CLIENT recognizes that subsurface conditions may vary from those observed at locations where borings, surveys, or explorations are made, and that site conditions may change with time. Data, interpretations, and recommendations by the geotechnical engineer will be based solely on information available to geotechnical engineer. Geotechnical engineer is responsible for those data, interpretations, and recommendations, but will not be responsible for other parties' interpretations or use of the information developed.

The presence of a differing site conditions clause allows the contractor to be reimbursed for its reasonable additional costs, regardless of the owner's knowledge or ignorance of actual conditions. By agreeing to a differing site condition provision in the contract, the owner assumes a portion of the risk of such conditions in exchange for the contractor not feeling compelled to protect itself by including a contingency in its bid.

The converse is also true: The owner is entitled to a cost reduction if site conditions prove less onerous than expected. Although downward adjustments are not common, they do occur.[6] Such credits are consistent with the clauses' central purpose, which is to base the owner's cost and the contractor's compensation on the reasonable value of the work actually performed, thereby eliminating unnecessary risks to each party.

A. Recovery for a Type I Changed Condition

To recover for a situation where actual conditions are at variance with the conditions "indicated" by the contract documents—a Type I changed condition—the contractor must show:

(1) That certain conditions are indicated by the plans, specifications, and other contract documents;

(2) That it relied on the physical conditions indicated in the contract;

(3) The nature of the actual conditions encountered;

(4) The existence of a material variation between the conditions indicated and the conditions actually encountered;

(5) That notice, as required by the contract, was given; and

(6) That the changed condition resulted in additional performance costs, time or both, as demonstrated by satisfactory documentation or proof.

The initial emphasis in Type I changed condition situations is on those conditions that are "indicated" in the contract. The contract must contain some statement or representation as to the conditions to be expected, and the actual conditions must differ from that statement or representation.

What is meant by "indicated in the contract" has been considered in numerous court decisions. In some instances, it is not required that the indications (on which the contractor is reasonably entitled to rely) be affirmatively expressed on the plans or in specific contract provisions. Instead, such indications may be a reasonable inference based on reading the contract as a whole. Thus, the contractor may be able to compare actual conditions not only with the express representations in the contract documents but also with all reasonable inferences and implications that can be drawn from those documents.[7] As pointed out in *Metropolitan Sewerage Commission v. R. W. Construction, Inc.*,[8] it is not required that the contract indications be "explicit

[6] *AFGO Eng'g Corp. v. United States*, 227 Ct. Cl. 730 (1981).
[7] *Condon-Johnson & Assoc's, Inc. v. Sacramento Mun. Util. Dist.*, 57 Cal. Rptr. 3d 849 (Cal. Ct. App. 2007).
[8] 241 N.W.2d 371 (Wis. 1976).

or specific, but only enough to impress or lull a reasonable bidder not to expect the adverse conditions actually encountered."

In certain situations, a contract indication may be found from documents that are not a part of the contract. For example, one federal court of appeals held that soil borings were a "contract indication" even though the borings were contained in an appendix to the plans and specifications that was expressly excluded as a contract document. In this regard, the court stated:

> The test boring logs do not have to be strictly considered "a part of the contract documents" (which the Appendix states they are not) to be binding on the [owner] to the extent of their own accuracy. We can accept the [owner's] argument that the Appendix is not an item listed in the Table of Contents (but is in addition to the Table of Contents) and therefore the Appendix is not a part of the contract. However, the differing site conditions clause need not be interpreted to limit reimbursements to situations where the logs themselves are necessarily a part of the contract. The clause entitles the contractor to reimbursement when there are "conditions at the site differing materially from those indicated in this contract." Even though the logs may not be included in the contract, they are "indicated" in the contract. ...[9]

By contrast, a federal appeals court was not persuaded that "conditions indicated by the Contract Documents" could include soil reports that were not part of the actual contract documents.[10] Observing that "[e]ither the soil report is part of the contract documents or it is not," the court refused to consider any document that was not specifically incorporated into the contract and, therefore, strictly enforced the owner's written disclaimer of responsibility for soil conditions.

Examples of situations where express representations of conditions in the contract documents were found to have differed materially from the actual conditions encountered include:

(1) *Variance from actual field conditions.* During construction of a highway retaining wall, a contractor discovered that the actual interface point for two portions of the wall varied significantly from the point indicated in the specifications from the state department of transportation (DOT). In order to build the wall as required by the plans, the contractor had to remove a substantial amount of rock that was unforseen at the time of its bid. The contractor's justified reliance on the DOT's representations in the specifications entitled the contractor to additional compensation for the differing site condition.[11]

(2) *Muddy versus dry conditions.* The contract documents stated that when "test holes were drilled in the area, no water was noted in any of the test holes." During construction, the contractor encountered "subsurface mud covered by a cracked and deceptively dry looking surface. ..." The Supreme Court

[9] *City of Columbia v. Paul N. Howard Co.,* 707 F.2d 338, 340 (8th Cir. 1983).
[10] *Millgard Corp. v. McKeen/Mays,* 49 F.3d 1070, 1072 (5th Cir. 1995).
[11] *Thomas M. Durkin & Sons, Inc. v. Dep't of Transp.,* 742 A.2d 233 (Pa. Commw. Ct. 1999).

of Idaho found the subsurface mud to be materially different from the dry conditions indicated by the contract documents and affirmed the contractor's recovery for a Type I differing site condition.[12]

(3) *Variance from anticipated blow counts.* Soil conditions with actual blow counts that were one-third to one-half the strength indicated by the contract borings constituted a changed condition. The contractor that encountered this condition during the construction of two underground garages was entitled to additional compensation.[13]

(4) *Limitations on access.* A playground construction contract called for the contractor to furnish a certain brand of playground equipment, and included a drawing showing the placement and orientation of the equipment. The contractor, in reliance on the drawing, believed that it would be possible to use a dump truck and backhoe to bring in and spread sand after the equipment was in place. This proved impossible, however, and the sand had to be spread by hand. Since the contractor used the specified brand-name equipment, it was justified in relying on the government's drawing and was entitled to an equitable adjustment for a Type I changed condition.[14]

(5) *Hard clay versus soft mud.* The contract specifications required the contractor to remove soft mud, silt, and sand in a river-dredging project. When the contractor encountered hard, undisturbed clay instead of the soft materials specified, the contractor was entitled to an equitable adjustment for a Type I differing site condition.[15]

(6) *Excavated materials not suitable as fill.* In one case, the contract specifications required that soil materials located on-site be excavated and reused as fill, but the specified excavation and recompaction was prevented by the physical properties of the soil, which differed materially from the contract indications. Although the problem was further impacted by an abnormal amount of rainfall, the contractor was entitled to an equitable adjustment for a Type I changed condition.[16]

(7) *"Balanced" excavated materials.* Where a contract for airport service roads and taxiways contained defective specifications which incorrectly stated that the amount of dirt excavated from the project site was roughly equivalent to the amount needed for fill-in requirements, (a "balanced project"), the court held that the contractor encountered a Type I differing site condition and could recover its costs for the purchase of the additional soil necessary to comply with the contract's fill requirements.[17]

[12] *Beco Corp. v. Roberts & Sons Constr. Co.,* 760 P.2d 1120 (Idaho 1988), *rev'd on other grounds, Hougland Farms, Inc. v. Johnson,* 803 P.2d 978 (Idaho 1990).

[13] *Baltimore Contractors, Inc. v. United States,* 12 Cl. Ct. 328 (1987). *See also Granite-Groves v. Wash. Metro. Area Transit Auth.,* 845 F.2d 330 (D.C. Cir. 1988).

[14] *Torres Constr. Co.,* ASBCA No. 25697, 84–2 BCA ¶ 17,397.

[15] *C. J. Langenfelder & Son, Inc.,* Maryland Department of Transportation 1000 (Aug. 15, 1980).

[16] *S. Paving Corp.,* AGBCA No. 74–103, 77–2 BCA ¶ 12,813; *see also W.R. Henderson Constr.,* ASBCA No. 52938, 02–1 BCA ¶ 31,741.

[17] *Ace Constructors, Inc. v. United States,* 499 F.3d 1357, 1364 (Fed. Cir. 2007).

Some examples of nonexpress, or implied, contract indications include:

(1) *Owner's awareness of lead paint.* Although a state school system did not include a lead paint abatement provision in a renovation contract for one of its university buildings, the owner issued a separate notice to contractors that "[a]ll asbestos and lead containing materials affected by the project will be addressed." In reliance on the state's representation, the contractor performed the abatement after the discovery of lead paint stalled the renovation work. The state's awareness of a problem unknown to the contractor was sufficient to justify additional compensation for work not contemplated in the contract.[18]

(2) *Hidden roof system not disclosed.* An additional roof system not referenced in the contract specifications and drawings was determined to be a valid Type I differing site condition claim. Further, an inspection of the roof revealed no evidence that any additional roofing work had been performed after the as-built drawings had been prepared.[19]

(3) *Suitable equipment for work.* The Armed Services Board of Contract Appeals (ASBCA) upheld a differing site condition claim, holding that the "compaction, and clearing and grubbing" requirements were sufficient contract indications. The board held that, while the contract documents made no express representation regarding subsurface conditions, the compaction and clearing and grubbing requirements led the contractor to reasonably believe it could utilize heavy equipment to perform its work. The board stated that "where, as here, design requirements cannot be met and procedures and equipment reasonably anticipated cannot be used, the situation represents a classic example of a Type I differing site condition."[20] Similarly, courts have found an implied representation by requirements in the plans for a specific dredging technique,[21] where a grouting specification stated that groundwater could be controlled to a certain level during construction,[22] and where the contract documents indicated that "trench footings" could be used when in fact the soils on-site would not support them.[23]

(4) *Dry conditions implied by specified construction procedures.* When the construction procedures and design requirements set forth in the contract documents, read as a whole, indicated subsurface conditions permitting excavation "in the dry," but actual conditions made it impossible or impracticable to excavate in this manner, a changed condition was held to have been encountered.[24]

[18] *A.G. Cullen Constr., Inc. v. State Sys. of Higher Educ.,* 898 A.2d 1145, 1170 (Pa. Commw. Ct. 2006).
[19] *S. Cal. Roofing Co.,* PSBCA No. 1737 *et al.*, 88–2 BCA ¶ 20,803.
[20] *Kinetic Builders, Inc.*, ASBCA No. 32627, 88–2 BCA ¶ 20,657.
[21] *Midwest Dredging Co. v. McAninch Corp.,* 424 N.W.2d 216 (Iowa 1988).
[22] *S.A. Healy Co. v. Milwaukee Metro. Sewerage Dist.,* 50 F.3d 476 (7th Cir. 1995).
[23] *Sherman R. Smoot Co. v. Ohio Dept. of Admin. Serv.,* 736 N.E.2d 69 (Ohio Ct. App. 2000).
[24] *See Foster Constr., C.A. v. United States,* 193 Ct. Cl. 587 (1970). *But see Tricon-Triangle Contractors,* ENGBCA No. 5113, 88–1 BCA ¶ 20,317 (denying a Type I differing site condition claim where the presence of groundwater could be implied from the contract provision requiring the contractor to maintain a dewatering system).

(5) *Unanticipated sloughing of soils.* A tunneling contractor that encountered "running" ground conditions that were not disclosed by the contract soils information was granted relief under the differing site conditions clause for encountering a Type I condition. The contractor was required to grout in order to stop the sloughing.[25]

(6) *Implied thickness of concrete floor.* By comparing a 6-inch drain connection detail with the owner-provided cross-section drawings of the concrete floors, the contractor reasonably concluded that floors were about 6 inches thick. When the contractor encountered concrete floors between 18 and 24 inches thick, a changed condition claim was allowed based on the implied information contained in the contract drawings.[26]

B. Recovery for a Type II Changed Condition

Type II changed conditions differ significantly from Type I changed conditions. Under a Type II situation, it is possible to recover even where the contract is silent about the nature of the condition. To establish a Type II changed condition, one must show that the conditions encountered were unusual and differed materially from those reasonably anticipated, given the nature of the work and the locale. This can be a particularly heavy burden for excavating contractors, given the "wide variety of materials ordinarily encountered when excavating the earth's crust."[27]

To qualify as sufficiently "unknown and unusual," the condition encountered by the contractor does not have to be in the nature of a geological freak—for example, permafrost in the tropics.[28] Nevertheless, the standard under which a Type II case is evaluated can be somewhat vague. Whereas Type I cases use the contract as a basis of comparision, by definition, there is no clear point of reference for Type II changed conditions.[29] Generally the courts consider whether the site condition should have been anticipated based on the totality of the circumstances of each specific project.

The key to recovery for a Type II changed condition is the comparison of actual conditions with what was reasonably expected at the time of bidding. This inquiry into reasonable expectations will raise questions of the contractor's actual and constructive knowledge of working conditions in the particular area. For example, awareness of a condition at the site that is common knowledge to other contractors working in the area, and thus reasonably ascertainable by inquiry, may be attributed to the contractor. Moreover, a contractor's failure to visit the work site, particularly when alerted to potential problems by the plans and specifications, and the resulting failure to discover obvious physical conditions, may indicate that the bidder's judgment was simply a "guess ...premised in error," which forms no basis for recovery as a Type II changed condition.[30]

[25]*Shank-Artukovich v. United States,* 13 Cl. Ct. 346 (1987).

[26]*J.E. Robertson Co. v. United States,* 437 F.2d 1360 (Ct. Cl. 1971).

[27]*Manuel Bros., Inc. v. United States,* 55 Fed. Cl. 8 (2002).

[28]See *Ruff v. United States,* 96 Ct. Cl. 148 (1942); *W. Well Drilling Co. v. United States,* 96 F. Supp. 377 (D. Cal. 1951).

[29]*Servidone Constr. Corp. v. United States,* 19 Cl. Ct. 346 (1990).

[30]*See L.B. Samford, Inc.,* GSBCA No. 1233, 1964 BCA ¶ 4309.

The following are examples of Type II changed conditions:

(1) *Hard clay.* A site preparation contractor encountered an unusual amount of clay material during its attempt to install sand drains need for a bridge replacement project. The court held that the subsurface conditions encountered by the contractor differed substantially from those ordinarily encountered in the installation of sand drains using the methods specified in the contract.[31]

(2) *Subsurface water.* A water table found to be much higher than could have been anticipated has been held to be a changed condition, where dry and stable subsurface conditions were reasonably anticipated (but not indicated).[32]

(3) *Buried pipe and debris.* During the course of its installation of an underground electrical conduit, a contractor encountered asphalt, concrete, rebar, and other debris that damaged its directional drilling equipment on 19 separate occasions. Because the unanticipated subsurface materials differed considerably from the clay and occasional river rock common in the area, the contractor was compensated for the damages based on a Type II changed condition.[33]

    Similarly, submerged piling in a dredge-filled land area warranted Type II changed conditions relief.[34] Another case reached the opposite result where the presence of buried stumps should have been anticipated because the site was in a fill area that contained some protruding stumps, new sprouts, and new branches—indicating growth from buried stumps.[35]

(4) *Undersized floor joists/oversized walls.* A contractor entered into an agreement with the federal government to renovate certain family housing units. Rather than encountering typical 2-by-8 floor joists, the contractor found that over 80% of the joists were much closer to 7 inches in height. This required substantial shimming and other modifications, which resulted in extra costs. Because the actual joist dimensions differed significantly from the conditions an experienced contractor would reasonably expect to encounter on such a project, the contractor had a compensable Type II differing site condition.[36] Similarly, another contractor encountered a four-course-thick brick and masonry wall during a hospital renovation. Because the size of the wall was unusual for an interior partition, the contractor recovered the extra costs associated with the removal of the wall as a Type II differing site condition.[37]

(5) *Utilities.* When a contractor discovered that a third party had performed previous wiring in such a way that the phasing and wiring required by its contract could not be accomplished without extra work, this unknown condition

[31] *Sutton Corp. v, Metro. Dist. Comm'n,* 667 N.E.2d 838, 842 (Mass. 1996).
[32] *Loftis v. United States,* 110 Ct. Cl. 551 (1948).
[33] *Parker Excavating, Inc.,* ASBCA No. 54637, 06–1 BCA ¶ 33,217.
[34] *Caribbean Constr. Corp.,* IBCA No. 90, 57–1 BCA ¶ 1315.
[35] *Gilloz Constr. Co.,* W.D. BCA ¶ 826 (1944).
[36] *Kos Kam, Inc.,* ASBCA No. 24684, 88–1 BCA ¶ 20,246.
[37] *Hercules Constr. Co.,* VABCA No. 2508, 88–2 BCA ¶ 20,527.

warranted payment.[38] An undisclosed sewage line encountered in attempting to dig a manhole has been judged a Type II changed condition.[39] Similarly, a Type II differing site condition was found to exist when a contractor installing conduit pipe under an airfield perimeter road encountered a sewer line that was not indicated on the contract documents and was not a condition that would generally be expected.[40] In a different case, however, where a contractor encountered sewers, gas lines, water lines, and coaxial cables that were not shown on the plans, a changed conditions claim was denied because the site was in a heavily built-up area and manholes were shown on the plans.[41]

(6) *Peculiar structural conditions.* A dock-painting contractor was entitled to an equitable adjustment for extra work due to peculiar structural features that, in combination with the air pressure from incoming tides, caused a continuous water seepage or mist over the dock. Neither the contract documents, the prebid site inspection, nor the contractor's experience was sufficient to provide notice of this unusual condition.[42]

(7) *Thick paint.* Where an existing paint layer was much thicker than anticipated, a Type II changed condition was found.[43]

(8) *"Double-poured" roof.* Where an existing roof system was found to be "double-poured," and therefore much thicker and expensive to replace, a court found that a Type II changed condition existed.[44]

(9) *Hazardous materials or substances.* Where a contractor encountered creosote and other tarlike substances in an environment in which they were not anticipated, a Type II changed condition was found.[45] A differing site condition may even arise where the contractor encounters contamination of a different type from anticipated. For example, where the contractor encountered unanticipated free-flowing contaminant ("free product") as opposed to an anticipated layer of contaminated sand, the court found a Type II differing site condition.[46]

(10) *Miscellaneous items.* Where the contractor encountered beer cans, live ammunition, and ladies' underwear in cleaning a duct system in a military barracks, the contractor was granted relief for a Type II changed condition.[47]

A Type II differing site condition may result not only from a variance in the type or quantity of a material encountered, but also from the unusual performance of an expected material. Thus, even though clay was expected to be encountered, when, as

[38]*Dodson Elec. Co.,* ASBCA No. 5280, 59–2 BCA ¶ 2342.
[39]*Neale Constr. Co.,* ASBCA No. 2753, 58–1 BCA ¶ 1710.
[40]*Unitec, Inc.,* ASBCA No. 22025, 79–2 BCA ¶ 13,923.
[41]*H. Walter Schweigert,* ASBCA No. 4059, 57–2 BCA ¶ 1433.
[42] *Warren Painting Co.,* ASBCA No. 18456, 74–2 BCA ¶ 10,834.
[43]*R.J. Wildner Contracting Co. v. Ohio Turnpike Comm'n,* 913 F. Supp. 1031 (N.D. Ohio 1996).
[44]*Lathan Co., Inc. v. U.S.,* 20 Cl. Ct. 122 (1990).
[45]*Reliance Ins. Co. v. County of Monroe,* 604 N.Y.S.2d 439 (N.Y. App. Div. 1993).
[46]*All Power, Inc. v. United States,* 60 Fed. Cl. 675 (2004).
[47]*Cmty. Power Suction Furnace Cleaning Co.,* ASBCA No. 13803, 69–2 BCA ¶ 7963.

a result of percolating water, the clay behaved in an unusual, erratic fashion, with an unexpected tendency to slide, there was a changed condition.[48] Similarly, the unexpected shrinkage of soil, which materially increased the number of cubic yards of earth in a dam, was an unexpected property of the soil that constituted a changed condition.[49] Also, a contractor was allowed to recover for the additional cost of handling a subsurface water condition, although subsurface water was to be expected, when the place where it was encountered and the rate of its flow were unusual and unforeseeable.[50]

VI. STUMBLING BLOCKS TO RECOVERY

Although many contracts contain a differing site conditions clause, they often include other clauses that attempt to minimize or reduce claims under them. These include clauses relating to site inspection, notice, and various other clauses that seek to limit the ability of contractors to rely on information provided during the bidding process. Whether these additional contract clauses will bar or foreclose recovery under a differing site conditions clause usually depends on the specific circumstances of each case.

Even the most valid differing site condition claim is vulnerable if the contractor cannot prove how much the unanticipated condition increased the cost or time of performance under the contract. Good record-keeping and cost accounting procedures are important throughout any construction project, but they are essential for successfully substantiating damages incurred as a result of unforseen site conditions. (See **Chapter 13** Section entitled "Cost Accounting Records.")

A. Site Investigations

Bid invitations commonly require contractors to visit the site before submitting bids. Construction contracts routinely require the contractor to warrant that it has made a site inspection. One example of this type of clause reads:

> The Contractor shall be fully aware of all conditions that might affect successful completion of the work. Before submitting his proposal he shall examine the site and compare the actual conditions on site with those shown or represented by the plans and specifications, and shall determine the existence of all physical features, obstructions above or below the ground, ground elevations, etc., on or adjacent to the site, that might affect the work. No allowance will be made for the Contractor's failure to adequately familiarize himself with all conditions and no claim will be permitted for relief due to unforeseen conditions.

Such a requirement does not automatically nullify the effect of a differing site conditions clause if one is present and does not necessarily obligate the contractor to

[48]*Paccon, Inc.*, ASBCA No. 7643, 1962 BCA ¶ 3546.
[49]*Guy F. Atkinson*, IBCA No. 385, 65–1 BCA ¶ 4642.
[50]*Norair Eng'g Corp.*, ENGBCA No. 3568, 77–1 BCA ¶ 12,225.

discover hidden conditions at its peril.[51] A contractual requirement that the contractor make a site investigation does not obligate bidders to discover hidden subsurface conditions that would not be revealed by a reasonable preaward inspection.[52] The adequacy of the site investigation is measured by what a reasonable, intelligent contractor, experienced in the particular field of work involved, could be expected to discover—not what a highly trained expert might be able to find.[53]

The term "site investigation" is generally interpreted to mean, essentially, "*sight* investigation" and not to extend to making independent subsurface investigations.[54] This is not always the case. The contractor is deemed to have knowledge of all information reasonably made available to it as well as all information that could be gained by a "reasonable" site inspection under the circumstances. For example, in *Cook v. Oklahoma Board of Public Affairs,*[55] the Oklahoma Supreme Court overturned a contractor's recovery on a differing site conditions claim where the contractor had neglected to attend prebid conferences where site conditions were discussed and made only a cursory drive-through of the site.

Likewise, the ASBCA held in *Tri-Ad Constructors*[56] that a contractor was not entitled to an equitable adjustment for installing more electrical cable than anticipated because the contractor failed to conduct a prebid site inspection as required by the contract. From its reading of an electrical wiring diagram, the contractor believed that seven electrical substations were located immediately above the main bank of underground ducts running between two switching stations. If the contractor had inspected the site, it would have seen that the electrical substations, each the size of an automobile, were offset some 300 feet from the main line, requiring loops between this duct bank and each substation. The contractor was charged with the knowledge obtainable from a reasonable site inspection, when the ASBCA concluded that even the most cursory inspection would have revealed the need for additional cable between each substation and the main ductline.[57]

In addition to requiring that a contractor conduct a reasonable site investigation, some bid solicitations also require a contractor to review documents concerning the

[51]*Farnsworth & Chambers Co. v. United States,* 171 Ct. Cl. 30 (1965).

[52]*John G. Vann v. United States,* 420 F.2d 968, 983 (Ct. Cl. 1970); *Travelers Cas. & Sur. Co. v. United States,* 75 Fed. Cl. 696 (2007); *Atherton Constr., Inc.,* ASBCA No. 48527, 00–2 BCA ¶ 30,968; *Warren Painting Co.,* ASBCA No. 18456, 74–2 BCA ¶ 10,834; *Maint. Eng'rs,* ASBCA No. 17474, 74–2 BCA ¶ 10,760, *See also Warren Bros. Co. v. N.Y. State Thruway Auth.,* 309 N.Y.S.2d 450 (N.Y. App. Div 1970), *aff'd,* 314 N.E.2d 878 (1974); *L.J. McNulty, Inc. v. Vill. of Newport,* 187 N.W.2d 616 (Minn. 1971).

[53]*Stock & Grove, Inc. v. United States,* 493 F.2d 629, 631 (Ct. Cl. 1974); *Neal & Co. v. United States,* 36 Fed. Cl. 600, 620 (1996); *Commercial Mech. Contractors, Inc.,* ASBCA No. 25695, 83–2 BCA ¶ 16,768.

[54]*See, e.g., Martin K. Eby Constr. Co. v. Jacksonville Transp. Auth.,* 436 F. Supp. 2d 1276 (M.D. Fla. 2005); *Sherman R. Smoot Co. v. Ohio Dept. of Adm. Serv.,* 736 N.E.2d 69 (Ohio Ct. App. 2000); *Condon-Cunningham, Inc. v. Day,* 258 N.E.2d 264 (Ohio Misc. 1969).

[55]736 P.2d 140 (Okla. 1987). *See also Mega Constr. Co. v. United States,* 29 Fed. Cl. 396 (1993).

[56] ASBCA No. 34732, 89–1 BCA ¶ 21,250.

[57]*See also Interstate Contracting Corp. v. City of Dallas,* 407 F.3d 708 (5th Cir. 2005); *McCormick Constr. Co. v. United States,* 18 Cl. Ct. 259 (1989) (denying a contractor's differing site condition claim because a reasonable site investigation would have revealed the possible subsurface condition).

site conditions that are made available for inspection before bidding but are not provided to the contractor in the bid package. If the contractor fails to review the available documents before submitting its bid, it may later be precluded from recovering for conditions that are different from those expected but that could have been determined from a review of the documents made available. For example, a contractor's differing site conditions claim was denied on the ground that the contractor failed to review records of previous dredgings that contained information regarding the nature of the materials to be dredged and that were available to the contractor before bidding.[58] This obligation on the contractor will likely apply even where the referenced documents are remotely located.[59]

B. Exculpatory Clauses

Contracts frequently contain broad exculpatory clauses disclaiming liability for the accuracy of plans, specifications, borings, and other subsurface data. An example of such a clause follows.

> Information, data and representations contained in the contract documents pertaining to the conditions at the site, including subsurface conditions, are for information only and are not warranted or represented in any manner to accurately show the conditions at the site of the work. The CONTRACTOR agrees that he shall make no claims for damages, additional compensation or extension of time against the OWNER because of encountering actual conditions in the course of the work which vary or differ from conditions or information contained in the contract documents. All risks of differing subsurface conditions shall be borne solely by the CONTRACTOR.

Many courts have held that these clauses do not have the sweeping effect the drafter of the clause may have desired. Courts normally will not allow such clauses to eliminate the relief provided to the contractor by the differing site conditions clause.[60] For example, in *Woodcrest Construction Co. v. United States,*[61] the United States Court of Claims allowed a contractor to recover under the changed conditions clause despite the extremely broad exculpatory provisions in the contract. The court stated:

> The effect of an actual representation is to make the statements of the Government binding upon it, despite exculpatory clauses which do not guarantee the accuracy of a description. ...Here, although there is no (express) statement

[58]*Stuyvesant Dredging Co. v. United States,* 834 F.2d 1576, 1581 (Fed. Cir. 1987). *See also G&P Constr. Co., Inc.,* ASBCA No. 49524, 98–2 BCA ¶ 29,457.
[59]*See Billington Contracting, Inc.,* ASBCA Nos. 54147, 54149, 05–1 BCA ¶ 32,900.
[60]*URS Group, Inc. v. Tetra Tech FW, Inc.,* 181 P.3d 380 (Colo. Ct. App. 2008); *Roy Strom Excavating & Grading Co. v. Miller-Davis Co.,* 501 N.E.2d 717 (Ill. App. Ct. 1986), *opinion superseded* by 509 N.E.2d 105 (Ill. App. Ct. 1986); *Metro. Sewerage Comm'n v. R.W. Constr., Inc.,* 241 N.W.2d 371 (Wis. 1976); *Contra Cruz Constr. Co. v. Lancaster Area Sewer Auth.,* 439 F. Supp. 1202 (E.D. Pa. 1977).
[61]408 F.2d 406 (Ct. Cl. 1969).

> which can be made binding upon the Government, there was in effect a description of the site, upon which plaintiff had a right to rely, and by which it was misled. Nor does the exculpatory clause in the instant case absolve the Government, since broad exculpatory clauses...cannot be given their full literal reach, and "do not relieve the defendant of liability for changed conditions as the broad language thereof would seem to indicate."[62] [G]eneral portions of the specifications should not lightly be read to override the Changed Conditions Clause.[63]

Even when a contract lacks a differing site conditions clause and contains extensive exculpatory language, it may still be possible for the contractor to recover if it can show, for example, that an independent subsurface investigation was not feasible, and that it was thus forced to rely on information provided by the owner.[64] For example, in *Raymond International, Inc. v. Baltimore County, Maryland,*[65] the county solicited bids to repair bridge piers. Even though the contract required the contractor to verify all dimensions in the contract documents, the court held that requiring the contractor to make tests to verify the information supplied by the county was an unduly burdensome. The court allowed the contractor to recover the increased costs incurred due to misrepresented conditions in the contract documents.

Some courts choose instead to strictly interpret these broad exculpatory clauses. In *Interstate Contracting Corp. v. City of Dallas,* the Fifth Circuit Court of Appeals held that a disclaimer concerning subsurface conditions was enforceable even after the revelation that the owner withheld a key soils report.[66] Other courts have issued similar rulings.[67]

C. Notice Requirements

Ideally, the notice requirement in a changed conditions clause alerts the owner to the existence of the condition and provides the owner an opportunity to evaluate its potential impact on the project. Such an evaluation may cause the owner to change the design or alter the contractor's method of performance.

The contracting parties should fully and timely adhere to all notice provisions in their agreements. Failure to do so may preclude one from recovering damages caused by a changed condition. For example, a court decision arising out of a large Boston tunneling project held that a soil excavator's notice to the general contractor of unexpected levels of comtaminated clay four years after the condition was discovered was neither prompt

[62]*Fehlhaber Corp. v. United States,* 151 F. Supp. 817, 825 (Ct. Cl.), *cert. denied,* 355 U.S. 877 (1957).
[63]*See e.g., Appeal of Am. Wyott Corp.,* ASBCA No. 42024, 94–2 BCA ¶ 26,758; *United Contractors v. United States,* 368 F.2d 585, 598 (Ct. Cl. 1966).
[64]*See e.g., Morris, Inc. v. State ex rel. S.D. Dep't of Transp.,* 598 N.W.2d 520 (S.D. 1999); *Midwest Dredging Co. v. McAninch Corp.,* 424 N.W.2d 216 (Iowa 1988); *Robert E. McKee v. City of Atlanta,* 414 F. Supp. 957 (N.D. Ga. 1976).
[65]412 A.2d 1296 (Md. Ct. Spec. App. 1980).
[66]407 F.3d 708 (5th Cir. 2005).
[67]*Stabler Constr., Inc. v. Commonwealth, Dep't of Transp.,* 692 A.2d 1150 (Pa. Commw. Ct. 1997); *Brown Bros., Inc. v. Metro. Gov't of Nashville and Davidson County,* 877 S.W.2d 745 (Tenn. Ct. App. 1993).

nor before the condition was disturbed, as was specifically required by the contract.[68] The court held that the unexcused failure of the excavator to comply with the contractual notice provisions forever barred the otherwise valid differing site conditions claim.

In some instances, the lack of strict compliance may be excused by the courts. The underlying purposes of the contract requirements may be satisfied by substantial compliance with the terms of the notice requirement, actual knowledge of the condition by the owner or its agent, or if the owner suffers no prejudice from the contractor's failure to give written notice. In some instances, the owner's actions may operate as a waiver of its right to insist on strict compliance with certain contract requirements.

Examples of cases in which recovery for a differing site condition was permitted despite the lack of strict compliance with contractual notice requirements include:

(1) In *Ronald Adams Contractor, Inc. v. Mississippi Transportation Commission,*[69] the differing site condition clause in a highway improvement contract placed an equal burden on the contractor and the state transportation commission to discover and notify the other of unforeseen conditions. Due to the "mutuality requirement," the commission's investigation of the physical conditions of the roadway before commencement of the contractor's work excused the contractor from failing to give formal notice of the unsuitable soil conditions. The court reasoned that the commission's *actual knowledge* of the poor soil conditions made it pointless to require the contractor to give notice of those same conditions.

(2) In *Brinderson Corp. v. Hampton Road Sanitation District,*[70] the contractor maintained that extremely wet subsurface conditions on the project site differed from the soil conditions presented in the contract documents. The contractor, however, failed to give written notice in accordance with the contract until after the wet soils had been disturbed and at least partially removed. An owner's representative was present on-site when the unusual conditions were first encountered and inspected the conditions. The court stated that the owner had *actual knowledge* of the conditions and, therefore, the purpose of the notice requirement was satisfied. The court held that the contractor should be allowed to proceed to the merits of its differing site conditions claim, despite the lack of timely written notice.

(3) In *Weber Construction Inc. v. County of Spokane,* the court held that a county waived strict compliance with the contractual notice terms by failing to respond to a road contractor's repeated requests for guidance on the disposal of boulders that were unsuitable for use as fill.[71]

[68] *Earth Tech. Env't and Infrastructure, Inc. v. Perini/Kiewit/Cashman,* 2004 WL 2341397 (Mass. Supp. 2004). *See also Gratech Co., Ltd. v. N.D. DOT,* 676 N.W.2d 781 (N.D. 2004) (holding that subcontractor's failure to provide timely notice of unanticipated site condition waived its right to seek arbitration of claim for additional excavation work).

[69] 777 So. 2d 649 (Miss. 2000).

[70] 825 F.2d 41 (4th Cir. 1987).

[71] 98 P.3d 60 (Wash. Ct. App. 2004).

VII. RELIEF IN THE ABSENCE OF A CONTRACT PROVISION

The lack of a differing site conditions clause does not necessarily preclude recovery by the contractor. Instead, a contractor may base a claim on legal theories such as misrepresentation, breach of warranty, or mutual mistake. These theories sometimes are used even when the contract contains a differing site conditions clause.[72] For example, if the differing site conditions clause has been rendered inoperative by the contractor's failure to give notice or to adhere to the clause's express requirements, these legal theories may provide an alternative avenue for recovery.

A. Misrepresentation

Misrepresentation or fraud (i.e., intentional misrepresentation) by the owner may allow the contractor to recover extra costs incurred because the actual conditions encountered were not as they were represented. A contractor may also be able to recover, even when the owner has made no affirmative misrepresentation, on the basis that the owner has breached its duty to disclose available information, such as: (1) if the owner makes an accurate representation but does not disclose facts that materially qualify the facts disclosed; (2) if the facts are known or accessible only to the owner, and the owner knows they are not known to, or reasonably discoverable by, the contractor; or (3) if the owner actively conceals information from the contractor.[73]

Some courts have shown a willingness to allow the standard site investigation and disclaimer clauses to undercut a contractor's action for misrepresentation. Other courts are not nearly as harsh, and allow the contractor to utilize the owner's misrepresentations as a basis of liability in spite of site investigation and disclaimer clauses.[74]

In construction, misrepresentations relating to the amount or character of the work to be performed under the contract, or the cost of its performance, may give the contractor grounds to rescind the contract or to sue for damages under a fraud theory.[75] Frequently, however, statements of this nature are ruled to be mere estimates or approximations that would not support a fraud action. Each of these cases is controlled by its specific facts and circumstances.

In *Robert E. McKee Inc. v. City of Atlanta,*[76] the United States District Court for the Northern District of Georgia discussed the application of the theory of misrepresentation to changed conditions situations. The city had supplied the contractor with inaccurate information concerning the quantity of rock excavation required by the

[72]*Damon Pursell Constr. Co. v. Mo. Highway and Transp. Comm'n,* 192 S.W.3d 461 (Mo. Ct. App. 2006).
[73]*Warner Constr. Corp. v. City of L.A.,* 466 P.2d 996, 1001 (Cal. 1970). *See also Davis v. Comm'rs of Sewerage,* 13 F. Supp. 672 (W.D. Ky. 1936); *L.I. Waldman & Co. v. State,* 41 N.Y.S.2d 704 (Ct. Cl. 1943).
[74]*See, e.g., W. States Mech. Contractors v. Sandia Corp.,* 798 P.2d 1062 (N.M. Ct. App. 1990); *Metro. Sewerage Comm'n v. R.W. Constr., Inc.,* 241 N.W.2d 371 (Wis. 1976); *Fattore Co. v. Metro. Sewerage Comm'n,* 454 F.2d 537 (7th Cir. 1971).
[75]*Busch v. Wilcox,* 46 N.W.940 (Mich. 1890).
[76]414 F. Supp. 957 (N.D. Ga. 1976).

project. The court noted that a mere showing by the contractor of subsoil conditions that were not expected by either party would not automatically release the contractor from its contract obligations (and therefore the financial risk of the extra rock excavation). The court also recognized that the contract placed the burden of uncertainty on the contractor. Nevertheless, the court held that the city could be held liable if the contractor could show that (1) it was not reasonably able to discover the true facts through investigation and (2) the misrepresentation was material. Thus, the theory of misrepresentation may be available to a contractor even if it has "assumed" the risk of changed conditions in the contract.[77]

Misrepresentation may also be found notwithstanding generalized statements regarding subsurface data. In some cases, where the owner supplies specific information, such as sieve analysis or liquid-limit information, yet also provides general statements about boring logs that are in conflict with this information, misrepresentation still may be found.[78]

B. Duty to Disclose

Closely related to misrepresentation is another theory of recovery based on the failure of an owner (most often a public owner) to disclose all available information.[79] An example of recovery under this theory is found in the case of *P. T. & L. Constr. Co. v. New Jersey Dep't of Transportation,*[80] where the public owner neglected to disclose to bidders an engineering report that noted the presence of saturated soil conditions not otherwise known to exist or otherwise referenced in the contract.

In *Pinkerton & Laws Co. v. Roadway Express, Inc.,*[81] the court recognized the validity of a contractor's claim based on the owner's failure to disclose information concerning soil moisture and compaction criteria that the court found important to a contractor's ability to prepare a responsive bid. Interestingly, the *Pinkerton* case involved a private owner, not a public owner or governmental authority, unlike most cases where an owner has been held liable under a duty to disclose theory.

C. Breach of Implied Warranty

Just as recovery for a differing site condition may be possible in instances of misrepresentation or the breach of an express contractual warranty, it may also be possible when an owner breaches its implied warranty regarding the adequacy of its plans and specifications. This theory stems directly from the 1918 United States Supreme Court decision in *United States v. Spearin.*[82] In essence, this decision holds that the contractor should be able to construct the project in accordance with the plans and

[77]*See also Raymond Int'l, Inc. v. Baltimore County, Md.,* 412 A.2d 1296 (Md. Ct. Spec. App. 1980).
[78]*T. Brown Constructors, Inc. v. Pena,* 132 F.3d 724 (Fed. Cir. 1997).
[79]*Jacksonville Port Auth. v. Parkhill-Goodloe Co., Inc.,* 362 So. 2d 1009 (Fla. Dist. Ct. App. 1978); *Welch v. State of Cal.,* 188 Cal. Rptr. 726 (Cal. Ct. App. 1983).
[80]531 A.2d 1330 (N.J. 1987). *See also Alpert v. Commonwealth,* 258 N.E.2d 755 (Mass. 1970).
[81]650 F. Supp. 1138 (N.D. Ga. 1986).
[82]248 U.S. 132 (1918).

specifications furnished by the owner, and if so performed, the project should be acceptable.[83] This principle has become so widely associated with this case that the owner's implied warranty of the adequacy of the plans and specifications is sometimes referred to by courts as the *Spearin* doctrine.[84]

Under the *Spearin* doctrine, when an owner requires strict compliance with its plans and specifications, it is impliedly warranting to the contractor that the information is accurate and reliable. If the plans and specifications are, in fact, inaccurate, any defective construction resulting from the contractor's reliance on the plans will be the owner's responsibility.[85] Typically, an owner cannot avoid responsibility under the implied warranty theory simply by inserting into the contract a general disclaimer of responsibility for the accuracy of its plans and specifications.[86]

D. Mutual Mistake

"Mutual mistake" is another legal doctrine under which a contractor may obtain relief from a changed condition. Under the mutual mistake theory, a contractor may be successful in having the contract rescinded and in having the actual cost paid on a *quantum meruit*, that is, the "reasonable value" of the materials and services furnished, if the contractor can show the existence of a factual condition that both the contractor and owner were unaware of, and that goes to the "very essence" of the contract. In *Long v. Inhabitants of Athol,* the amount of work to be done under a sewer contract was stated "approximately" by the owner, and the contractor's bid was based on this "approximation."[87] When it became apparent that the project construction would require much greater quantities of work than estimated, the court found that a mutual material mistake existed that warranted a rescission of the contract. A court may look to the customs and practices of the local construction industry when determining whether a mutual mistake regarding responsibility for a known site condition gives rise to a valid differing site condition claim.[88] Be aware, however, that contractual relief under the mutual mistake theory may be precluded where the contract specifically allocates the risk of a mistake regarding the site conditions to one of the parties.[89]

[83]For a more detailed discussion of this landmark construction law decision, *see* **Chapter 5**, Section **VII.B,** "Warranty of Plans and Specifications."

[84]*Cent. Ohio Joint Vocational Sch. Dist. Bd. of Educ. v. Peterson Constr. Co.,* 716 N.E.2d 1210, 1215 (Ohio Ct. App. 1998); *W. States Mech. Contractors, Inc. v. Sandia Corp.,* 798 P.2d 1062 (N.M. Ct. App. 1990); *But see Willamette Crushing Co. v. State,* 932 P.2d 1350 (Ariz. Ct. App. 1997).

[85]*Green Constr. Co. v. Kan. Power & Light Co.,* 1 F.3d 1005, 1009 (10th Cir. 1993); *Ruby-Collins, Inc. v. City of Charlotte,* 740 F. Supp. 1159 (W.D.N.C. 1990); *Frank Briscoe Co., Inc. v. Clark County,* 857 F.2d 606 (9th Cir. 1988); *Midwest Dredging Co. v. McAninch Corp.,* 424 N.W.2d 216 (Iowa 1988). *But see DeAtley Constr., Inc. v. United States,* 71 Fed. Cl. 370 (2006) (finding that although contractor can assert separate differing site conditions and defective specifications claims, these separate claims will typically "collapse into one" where the facts supporting such costs are "so intertwined as to constitute a single claim.")

[86]*Hollerbach v. United States,* 233 U.S. 165 (1914).

[87]82 N.E.665 (Mass. 1907).

[88]*New England Rock Servs., Inc. v. Empire Paving, Inc.,* 731 A.2d 784, 788 (Conn. App. Ct. 1999).

[89]*Hall Contracting Corp. v. Entergy Servs., Inc.,* 309 F.3d 468 (8th Cir. 2002).

➢ POINTS TO REMEMBER

- Differing site conditions or changed conditions are a common source of construction problems, particularly early in a project.
- A differing site condition or changed condition is defined as a physical condition encountered in performing the work that was not visible and not known to exist at the time of bidding and that materially differs from the conditions believed to exist at the time of bidding.
- Under traditional contract law, the contractor generally assumes the risk of differing site conditions of which neither party is aware.
- If the contractor assumes the risk of differing site conditions, a contingency may be included in its bid to cover that risk. If there is no changed condition, the contractor receives a windfall.
- To balance the risk, the federal government and many public and private owners use a differing site conditions clause that places the risk of encountering a differing site condition on the owner and removes the contractor's need for a contingency.
- Several differing site conditions clauses used in construction industry contracts are quoted and discussed in this chapter. These include the differing site conditions clause contained in the standard federal government construction contract, the ConsensusDOCS contracts, the clause contained in the AIA 201 Standard General Conditions (2007 ed.), and the clause contained in the EJCDC C-700.
- Differing site conditions clauses generally recognize two types of changed conditions: (1) a Type I changed condition, which is a condition that is at variance with the conditions indicated in the contract documents; and (2) a Type II changed condition, which is a condition unusual in nature that differs materially from the conditions ordinarily encountered in performing the type of work called for in the geographic area where the project is located.
- Most clauses contain notice provisions, requiring the contractor to stop work and notify the owner before disturbing the condition, so that the owner will have an opportunity to inspect and evaluate it. Failure to give the required notice may jeopardize the contractor's ability to receive an adjustment for the additional costs, time, or both required to deal with the differing site condition.
- The contractor should follow the procedures outlined in the contract's differing site conditions clause in order to preserve its right to an adjustment.
- Good record keeping is critical for the recovery of the costs resulting from a differing site condition.
- To recover for a Type I changed condition, a contractor generally must show that: (1) the conditions were indicated in the contract documents; (2) the contractor relied on the conditions indicated in the contract documents; (3) the actual conditions encountered materially differed from those indicated; (4) proper notice was given; and (5) the changed condition resulted in additional performance costs, time, or both, as demonstrated by appropriate documentation.

- To recover for a Type II changed condition, a contractor generally must show that: (1) the conditions encountered were unusual and differed materially from those reasonably anticipated, given the nature of the work and the locale; (2) proper notice was given; and (3) the change resulted in additional performance costs or time, as demonstrated by appropriate documentation.
- Some contracts contain a site investigation clause, which requires the contractor to investigate and examine the existing conditions before submitting its bid.
- Where the contract contains both a site investigation clause and a differing site conditions clause, the contractor's ability to obtain an adjustment to the cost or time for performance may depend on whether the condition was one that a reasonable, intelligent contractor, experienced in the particular field of work involved, would discover based on a reasonable site investigation.
- In interpreting site investigation clauses, the term "site" often means "sight," and does not require an independent subsurface investigation.
- Many public and private owners use differing site conditions clauses; however, other exculpatory clauses purport to shift the risk of differing site conditions back to the contractor.
- Courts have held that such exculpatory clauses are generally not enforceable and have narrowly construed them and limited their effect.
- In the absence of a differing site conditions clause, a contractor may be able to recover additional costs resulting from a changed condition if the contractor can establish the facts necessary to support the legal theories of misrepresentation, breach of warranty, mutual mistake, or can establish superior knowledge and a duty to disclose on the part of the owner.

11

SCHEDULES, DELAYS, AND ACCELERATION

The contract between the parties is the first place to look to determine: (1) what the schedule is; (2) the extent to which the schedule can be changed; and (3) how the financial risks associated with schedule changes, delays, suspensions, and accelerations are allocated between the parties. A party entering into a construction contract in today's environment must carefully read and understand how the contract allocates these risks because delays and suspensions, as well as work accelerations required to overcome delays and suspensions, contribute significantly to job cost overruns.

Over the last two decades, there have been two important changes in construction scheduling and how delays, work suspensions, and accelerations are handled contractually: (1) contracts have become increasingly sophisticated, incorporate more elaborate scheduling requirements, and often subtly shift the risk of delay-related events from one party to another; and (2) the use of critical path method (CPM) scheduling software is customary on many projects. This chapter identifies these changes and provides important information about the typical causes of delay, contractual limitations to delay claims, and the delay claim process.

I. DEVELOPMENTS IN CONTRACTUAL RISK ALLOCATION

The legal rights and obligations of the parties associated with performance delays are either defined in the contract or arise from an implied obligation that each party will not delay, hinder, or interfere with the performance of the other party. A party that hinders or prevents performance by the other party, or that renders performance impossible, may not benefit from its wrong.[1] This rule of law prevents

[1] *See United States v. Killough,* 848 F.2d 1523, 1531 (11th Cir. 1988) (discussing within the context of quantification of damages the principle that a wrongdoer shall not profit by its wrongdoing at the expense of its victim).

a party from taking advantage of its own contract breach. The same rule also provides a basis for the recovery of costs generated by delays that are the fault or responsibility of one of the contracting parties.

Traditionally, the obligations associated with project scheduling and performance delays have been set forth as an express obligation to perform either by a specific date or under a specific schedule. The specific completion date becomes a material part of the contract when the contract also includes a "time is of the essence" clause. Those terms remain important today, but in recent years additional contract terms often appear in twenty-first century contracts that can strengthen or modify these traditional contract benchmarks for evaluating delay. For example, an owner might include a sentence in the "time is of the essence" clause that states: "Any additional or unanticipated cost or expense required to maintain the schedule shall be borne solely by the contractor." Similarly, a subcontract that lists a specific completion date also may include a clause that states: "The contractor shall have the right to decide the time, order, and priority in which the subcontractor's various work items shall be performed and all other matters relative to the timely and orderly conduct of the subcontractor's work." The next sections begin with the traditional clauses and progress to the more detailed contract provisions that affect how delays, suspensions, and accelerations are handled.

A. "Time Is of the Essence" Clause

Most contracts provide that "time is of the essence." A "time is of the essence" clause makes time a material requirement of the contractor's performance obligation and ensures that the owner can recover delay damages for missed milestone or completion dates. In the absence of such a clause or an expression in the contract as a whole that time is a material element of performance, delay damages may not be recoverable.[2]

A contractor generally is entitled to extra time for any excusable delay that occurs during the execution of the work. The contractor, however, should be aware of clauses that purport to preclude recovery of monetary damages for delays and limit the contractor's remedy for delays only to time extensions.

B. Contract Commencement and Completion Dates

The first factors to be analyzed in assessing a delay claim are the contract start and completion dates. Construction contracts usually specify performance periods either by setting forth start and completion dates or by establishing that the work shall be completed within a specified number of days after the notice to proceed or commencement of work.

Many contracts also include interim milestone dates, which indicate the dates certain portions of the work should be completed. The failure to meet milestone dates may provide the basis for claims that seek to recover actual or liquidated delay damages, termination of the contract, or an acceleration directive.

[2]*See Gunn v. Heggins,* 964 So. 2d 586 (Miss. Ct. App. 2007).

Where a contract specifies the date for the commencement of work, the owner may be deemed to have warranted the readiness of the work site on the specified date. If the work site is not prepared for the contractor to begin work on that date, the owner may be liable for delay damages.[3] To avoid liability for such delays, owners often include a statement in the contract that the specified commencement date is only a projection or an estimate. To avoid misunderstandings and disputes, all parties should clearly define contract start dates, interim completion milestones, and contract completion dates.

C. Substantial Completion and Final Completion

Most contract documents define "substantial completion" of the work as "the stage in the progress of the Work when the Work or a designated portion thereof is sufficiently complete . . . so the Owner can occupy or utilize the Work for its intended use."[4] Generally, an owner may not assess, and a contractor is not liable for, delay or liquidated damages after substantial completion.[5] Thus, even when a contractor has not fully completed the work specified by its contract, the owner may be prevented from collecting actual delay damages or liquidated damages if the contractor has advanced work sufficiently to have achieved substantial completion.

It is not unusual, however, for some contracts to contain separate liquidated damage provisions related to the specified date for substantial completion and to the date for final completion. The latter provision might provide a separate specified liquidated damages amount imposed if final completion of all punch-list work is not accomplished within a stipulated number of days following the substantial completion date.

D. Typical Contract Time and Scheduling Clauses

Contract clauses that address scheduling and time have become increasingly detailed. The parties often must look to, and in some cases reconcile, numerous provisions throughout the contract to understand what the schedule is and the nature, size, and scope of changes a party can make to the schedule before it must compensate the other party.

Parties must identify what clauses affect directly or indirectly how the project will be scheduled, how changes to the schedule are made, and which party bears the cost associated with such changes. After identifying the contract substantial and final completion dates, the next terms to find are clauses that clearly address delays and time extensions, such as § 8.3 (Delays and Extensions of Time) of the American Institute of Architects (AIA) A201 (2007 ed.) (General Conditions of the Contract for Construction) or Paragraph 6.3 (Delays and Extensions of Time) of the 2007 ConsensusDOCS 200 (Standard Agreement and General Conditions Between Owner and Contractor).

[3] *See, e.g., Renda Marine Inc. v. United States,* 66 Fed. Cl. 639, 718 (2005) (stating that the government would be in breach of contract for interfering with a contractor's access to a job site controlled by the government.).
[4] *See, e.g.,* AIA A201, § 9.8 (2007 ed.).
[5] *See U.S. Fid. & Guar. Co. v. W. Rock Dev. Corp.,* 50 F. Supp. 2d 127, 129 (D. Conn. 1999).

Section 8.3 of the AIA A201 (2007 ed.) provides:

8.3 DELAYS AND EXTENSIONS OF TIME

8.3.1 If the Contractor is delayed at any time in the commencement or progress of the Work by an act or neglect of the Owner or Architect, or of an employee of either, or of a separate contractor employed by the Owner; or by changes ordered in the Work; or by labor disputes, fire, unusual delay in deliveries, unavoidable casualties or other causes beyond the Contractor's control; or by delay authorized by the Owner pending mediation and arbitration; or by other causes that the Architect determines may justify delay, then the Contract Time shall be extended by Change Order for such reasonable time as the Architect may determine.

8.3.2 Claims relating to time shall be made in accordance with applicable provisions of Article 15.

8.3.3 This Section 8.3 does not preclude recovery of damages for delay by either party under other provisions of the Contract Documents.[6]

Paragraph 6.3 (Delays and Extensions of Time) of the ConsensusDOCS 200 provides:

6.3 DELAYS AND EXTENSIONS OF TIME

6.3.1 If the Contractor is delayed at any time in the commencement or progress of the Work by any cause beyond the control of the Contractor, the Contractor shall be entitled to an equitable extension of the Contract Time. Examples of causes beyond the control of the Contractor include, but are not limited to, the following: acts or omissions of the Owner, the Architect/Engineer or Others; changes in the Work or the sequencing of the Work ordered by the Owner, or arising from decisions of the Owner that impact the time of performance of the Work; transportation delays not reasonably foreseeable; labor disputes not involving the Contractor; general labor disputes impacting the Project but not specifically related to the Worksite; fire; terrorism, epidemics, adverse governmental actions; unavoidable accidents or circumstances; adverse weather conditions not reasonably anticipated; encountering Hazardous Materials; concealed or unknown conditions; delay authorized by the Owner pending dispute resolution; and suspension by the Owner under Paragraph 11.1. The Contractor shall submit any requests for equitable extensions of Contract Time in accordance with the provisions of Article 8.[7]

Division 1 (General Requirements) of the technical specifications is an easy-to-overlook contract document that could play a significant role in how the project is to be scheduled and how delays are to be treated. For example, an owner may include

[6] AIA A201, § 8.3 (2007 ed.).

[7] ConsensusDOCS 200, ¶ 6.3 (2007 ed.).

a clause within Division 1 that requires the contractor or construction manager to submit at the contractor's expense a recovery schedule in the event of a defined delay. A contract might include a provision that reads: "If the contractor's schedule is delayed on its critical path by fourteen (14) calendar days or longer, the contractor shall be required to, at no additional cost to the owner, prepare and submit to the owner and architect a supplementary recovery schedule to explain how those activities shall be rescheduled to regain compliance with the contract schedule."

Similarly, subcontracts may include provisions that identify the subcontractor's specific responsibilities in the event of a failure to perform according to a specific schedule. The following clause illustrates this point:

> In the event subcontractor fails to meet any interim completion date provided for in the subcontract for any reason, contractor shall be entitled to direct subcontractor to immediately implement whatever means are necessary to bring the Project back on schedule, including supplying additional manpower, working overtime at subcontractor's sole expense, or by re-sequencing and/or accelerating the Work, provided such re-sequencing and/or acceleration is possible.

II. USE OF SCHEDULES IN PROJECT MANAGEMENT

Today, most construction project schedules are generated with CPM personal computer–based scheduling software, such as Primavera Project Planner (P3), Primavera Suretrak, or Microsoft Project. Since the mid-1990s the Primavera products have utilized the Precedence Diagramming Method (PDM) of CPM scheduling, which connects activities based on logic connections. PDM allows the user to define relationship activities in more ways than just "start to finish," as the older arrow diagramming method did. Relationships can be defined in PDM as start to start (i.e., the start of one activity to the start of another) and finish to finish (i.e., the finish of one activity to the finish of another). For example, with the use of leads or lags, the user can define two activities, A and B, with a start-to-start relationship with a three-day lag where activity B can only start three days after activity A.

Due to the complex nature of most construction projects, most schedules call for critical path portions of the project to be performed simultaneously with noncritical portions of the project. The noncritical portions of the project consist of activities for which the schedule provides additional time beyond the predicted time of performance. This additional time is known as float. Float time keeps a schedule on track when noncritical portions of the project exceed the predicted time of performance since any delay of a noncritical portion that is not in excess of the float time will not delay the project. Only after all available float time is exhausted can the project be delayed by a noncritical activity.

With the wider availability of relatively inexpensive scheduling software comes other issues. For example, someone with relatively little experience in CPM scheduling can generate an impressive-looking, but potentially inaccurate, schedule. The party preparing the schedule still needs to have experience with, and understand, the specific work that is to be performed. Schedulers can overestimate the amount of time required for a particular task or deliberately build into a schedule extra time to allow greater flexibility to ensure avoidance of liquidated damages. Further, it is impossible to predict factors that may affect performance time, such as unusually good weather, a particularly efficient workforce, or especially productive subcontractors and prompt suppliers. Whatever the reason, contractors sometimes finish ahead of schedule, which allows them to save considerable sums of money in overhead and variable expenses.

Certainly no contract or court would penalize a contractor for completing a project early—unless, of course, the contractor achieves early completion by violating some implied duty of cooperation with other subcontractors still on the job. Nevertheless, a "constructive penalty" could be imposed on the contractor that is in a position to finish early but is prevented from doing so by an owner. If the delay does not extend performance beyond the originally scheduled completion date, the owner could argue that because the contractor finished "on time," either the contractor was not delayed or it suffered no damage as a result of the delay. Network analysis techniques (such as CPM) make it possible to overcome the first argument, and there is case law suggesting that a contractor is damaged when it is precluded from finishing early.[8]

Delay may not only impact the critical path, but it may independently consume float time as well. Loss of this valuable float time may reduce the contractor's flexibility in sequencing work activities and allocating resources, and may end up costing money in terms of increased costs of performance of certain activities, even if there is no overall project delay. To the extent that the contractor uses its own float time, it cannot complain, but owner-caused delays may consume float and ultimately have an impact on the critical path. When all of the float time is thus used, any further contractor-caused delays are critical; that is, they impact directly on the critical path and end up costing the contractor money. Thus, there can be a difference between owners and contractors as to who "owns" float.[9]

Two solutions to this issue have been suggested. One position is that the contractor owns float. In that context, the owner may use float without cost unless the contractor has need of it. To the extent the owner has used float time needed by the contractor, the owner must compensate the contractor. The other position is that the float belongs to the party that uses it first. Thus, once float is gone for one, it is gone for all. Some government agencies and private owners have begun to use clauses specifically dealing with ownership of float. Such clauses are becoming the rule rather than the exception, so contractors must be aware of the impact they can have on the project.

III. TYPES OF DELAYS

A. Excusable Delays versus Nonexcusable Delays

The occurrence of a construction delay raises the issue of which party should bear both the responsibility for and cost of the delay. In deciding this question, courts

[8]*See Weaver Bailey Contractors v. United States,* 24 Cl. Ct. 576 (1991); *Grow Constr. Co. v. State,* 391 N.Y.S.2d 726 (N.Y. App. Div. 1977); *Metro. Paving Co. v. United States,* 325 F.2d 241 (Ct. Cl. 1963).

[9]One of the more comprehensive treatments of float ownership issues is the paper by John C. Person, Who Owns the Float? Construction Briefings No. 91–7 (Federal Publications, Inc. May 1991).

look to the causes of the delay and to the express and implied obligations imposed by the contract. Determining the legal consequences that flow from a given delay, and identifying the party that will bear the legal consequences of the delay, depend on correctly identifying the type of delay that has occurred.

Construction delays fall into two major categories: excusable delays and nonexcusable delays. An *excusable delay* provides a basis under the contract for an extension of performance time. Excusable delays can also be further subdivided into either *compensable delays,* which permit the recovery of both time and money, or *noncompensable delays,* permitting solely the recovery of time.

If the cause of delay stems from the conduct of the owner, then the contractor's delay will be excused.[10] Accordingly, the owner cannot claim liquidated damages for the delays it causes by its own conduct.[11] In addition, the contractor may recover delay damages due to owner-caused delays to the project.[12]

In contrast to an excusable delay, a *nonexcusable delay* provides no basis for recovery of either the time or the monetary impact of the delay. Moreover, the party responsible for that delay may be liable for resulting delay damages or may be subject to other breach of contract claims.

B. Excusable Delays

Generally, the parties' contract dictates whether a delay is excusable. Typical examples of excusable delays to a contractor's work are differing site conditions, design problems, changes to the work, inclement weather, strikes, and acts of God.

Most contracts specifically enumerate the types of excusable delays for which a time extension is due.[13] These terms vary from contract to contract. Because contracts differently allocate the risk of both nonperformance and unanticipated occurrences beyond the control of the parties, the precise terms of the contract are critical.

Some contracts exhaustively list each type of excusable delay and seek to limit the granting of time extensions to the listed delays. For example, Section 52.249–10 of the Federal Acquisition Regulation (FAR) sets forth the standard default termination clause for all federal government fixed price construction contracts. Subparagraph (b) of that clause defines excusable delays in this way:

> (b) The Contractor's right to proceed shall not be terminated nor the Contractor charged with damages under this clause, if—
>
> (1) The delay in completing the work arises from unforeseeable causes beyond the control and without the fault or negligence of the Contractor. Examples of such causes include—
>
> (i) Acts of God or of the public enemy,
>
> (ii) Acts of the Government in either its sovereign or contractual capacity,

[10]*See RDP Royal Palm Hotel L.P. ex rel. PADC Hospitality Corp. I v. Clark Constr. Group, Inc.,* 168 Fed. Appx. 348, (11th Cir. 2006); *Hill v. City of Duluth,* 58 N.W. 992 (Minn. 1894).

[11]*See Higgins v. City of Fillmore,* 639 P.2d 192 (Utah 1981).

[12]*Id.*

[13]*See* AIA A201, § 8.3 (2007 ed.); FAR § 52.249–14.

(iii) Acts of another Contractor in the performance of a contract with the Government,
(iv) Fires,
(v) Floods,
(vi) Epidemics,
(vii) Quarantine restrictions,
(viii) Strikes,
(ix) Freight embargoes,
(x) Unusually severe weather, or
(xi) Delays of subcontractors or suppliers at any tier arising from unforeseeable causes beyond the control and without the fault or negligence of both the Contractor and the subcontractors or suppliers.[14]

Other contracts may contain somewhat less extensive lists but may conclude the enumeration of excusable delays with a catchall phrase such as "causes beyond the control, and without the fault or negligence, of the contractor." Each party to a construction contract must have a clear understanding of the intended scope and operation of such a clause when requesting time extensions or analyzing time extension requests.

C. Compensable Excusable Delays

Compensable excusable delays are delays for which the innocent party is entitled to both a time extension and additional compensation for the resulting costs. For example, when an owner causes a delay, the contractor may be entitled to both additional compensation and time if the contract does not include a provision exonerating the owner from financial responsibility for such delays.

D. Nonexcusable Delays

Nonexcusable delays are those for which the contractor is responsible. The reasons for the delays vary and can include lack of adequate manpower, slow progress due to unqualified workers, poor planning, defective work, failure to forward submittals timely, and so on. In such cases, the contractor is not going to receive additional compensation or time to complete the work and may be liable for damages to the owner caused by such delay.

IV. TYPICAL CAUSES OF EXCUSABLE COMPENSABLE DELAY

In most cases, faulty or negligent acts or omissions or other conduct by an owner in contravention of the contractor's rights will not only result in excusable delay but may entitle the contractor to recover its increased costs or damages for breach of contract. Some examples of acts or failures to act resulting in excusable compensable delay are discussed next.

[14]FAR § 52.249–10.

A. Defective Drawings or Specifications

Federal courts, as well as most states, have also adopted the *Spearin* doctrine, which reasons that an owner generally is held to impliedly warrant the plans and specifications it provides to the contractor.[15] If such plans are erroneous or are insufficient to allow the contractor to perform the work in accordance with the intended design, and the defects within the plans and specifications result in a delay, the owner may be liable for time extensions and delay damages. Conversely, defects and omissions in the plans and specifications will not entitle the contractor to an extension of time if the defects and omissions are not related to the delay.[16]

The extent to which the *Spearin* doctrine applies to delay cases is an issue that continues to generate debate. For example, the Ohio Supreme Court "declined the opportunity to extend the *Spearin* doctrine from job-site conditions cases to cases involving delay due to plan changes" because the contract in question contained an enforceable no-damages-for-delay clause.[17] The contract at issue in *Dugan & Meyers* was executed in 1997—one year before the Ohio General Assembly declared no-damages-for-delay clauses "void and unenforceable as against public policy 'when the cause is a proximate result of the owner's act or failure to act.'"[18]

Before 1998, such clauses were enforceable under Ohio law. *Dugan & Meyers* was decided under the grandfather clause in Section 4113.62(C)(1) of the Ohio Revised Code, which, as the court stated, provides: "'Nothing in [this section]. . .shall be construed as applying to any contracts, agreements, or understandings entered into before the effective date of this act.'"[19]

Because the no-damages-for-delay clause was, at that time, enforceable, the court explained: "In order to hold in favor of Dugan & Meyers, we would need, first, to find that the state had implicitly warranted that its plans were buildable, accurate, and complete, and second, to hold that the implied warranty prevails over express contractual provisions. To do so would contravene established precedent, which we will not do."[20]

B. Failure to Provide Access and Improper Site Preparation

The owner is generally required to provide the contractor/subcontractor access to the work site in a timely and properly sequenced fashion.[21] In order to limit its exposure, however, the owner or general contractor may insert an exculpatory clause into

[15] *See United States v. Spearin,* 248 U.S. 132 (1918); *A.G. Cullen Constr., Inc. v. State Sys. of Higher Educ.,* 898 A.2d 1145 (Pa. 2006); *Modern Cont'l. S. v. Fairfax County Water Auth.,* 2006 WL 3775938 (Va. Cir. Ct. 2006); *Cent. Ohio Joint Vocational Sch. Dist. Bd. of Educ. v. Peterson Constr. Co.,* 716 N.E.2d 1210 (Ohio Ct. App. 1998).

[16] *Blinderman Constr. Co., Inc.,* ASBCA No. 21966, 79–1 BCA ¶ 13,875; *The Blakely Corp. v. EFCO Corp.,* 853 N.E.2d 998 (Ind. Ct. App. 2006); *Martin K. Eby Constr. Co., Inc. v. Jacksonville Transp. Auth.,* 436 F. Supp. 2d (M.D. Fla. 2005).

[17] *Dugan & Meyers Constr. Co., Inc. v. Ohio Dept. of Admin. Servs.,* 864 N.E.2d 68, 73–4 (Ohio 2007).

[18] *Id.* at 74.

[19] *Id.*

[20] *Id.* at 75.

[21] *See, e.g., Blinderman Constr. Co. v. United States,* 695 F.2d 552 (Fed. Cir. 1983); *Howard Contracting Inc. v. G.A. McDonald Constr. Co., Inc.,* 83 Cal. Rptr. 2d 590 (Cal. Ct. App. 1998).

the construction contract or request that the contractor waive its rights or expressly assume the risk of restricted access.[22] By inserting an exculpatory clause into the contract, the owner shifts the risk of site-access delays, as well as the risk of many other types of owner-caused delays, to the contractor. Closely related to the owner's duty to provide the contractor with access to the work site is the owner's duty to prepare the work site properly.[23] In regard to this duty, an owner may require the contractor to inspect the work site before beginning work. Should the contractor fail to discover site preparation problems that a reasonable inspection would have revealed, or should it fail to raise any objections in a timely manner, the contractor may be precluded from recovering delay damages.[24]

C. Failure to Supply Materials or Labor

Many construction contracts make the owner responsible for supplying certain materials or equipment to the contractor. Should the owner breach this duty by failing timely to provide the materials or equipment, the owner may be liable for delay damages.[25]

D. Failure to Provide Plans/Approve Shop Drawings

The submittal of shop drawings, product technical data brochures, or "cut sheets" for owner review is an important part of the construction process. For contractors, the submittal and review process creates a potential source of delay because the contractor cannot proceed with the work until the owner or its agent has approved the submittal.[26] With the exception of design-build projects, the owner typically is responsible for providing plans and specifications to the contractor. Should the owner fail to timely provide plans and specifications, or to timely approve a contractor's shop drawings, the owner may be liable for the resulting project delays.[27] Where, however, approval

[22] *A. Kaplen & Son, Ltd. v. Hous. Auth.*, 126 A.2d 13 (N.J. Super. Ct. App. Div. 1956); *see also Burgess Constr. Co. v. M. Morrin & Son*, 526 F.2d 108 (10th Cir. 1975); *Broome Constr., Inc. v. United States*, 492 F.2d 829 (Ct. Cl. 1974); *Ind. Dept. of Transp. v. Shelley & Sands, Inc.*, 756 N.E.2d 1063 (Ind. Ct. App. 2001); *Weber Constr. Co. v. State*, 323 N.Y.S.2d 492 (N.Y. App. Div. 1971), *aff'd*, 282 N.E.2d 331 (N.Y. 1972).

[23] *Valentine Concrete, Inc. v. Ohio Dept. of Admin. Servs.*, 609 N.E.2d 623 (Ohio Misc. 1991); *Jennings v. Reale Constr. Co.*, 392 A.2d 962 (Conn. 1978) (deciding a suit brought by a subcontractor against a contractor); *E.C. Nolan Co. v. State*, 227 N.W.2d 323 (Mich. Ct. App. 1975); *Columbia Asphalt Corp. v. State*, 420 N.Y.S.2d 36 (N.Y. App. Div. 1979); *Fehlhaber Corp. v. State*, 419 N.Y.S.2d 773 (N.Y. App. Div. 1979); *Commonwealth State Highway & Bridge Auth. v. Gen. Asphalt Paving Co.*, 405 A.2d 1138 (Pa. Commw. Ct. 1979).

[24] *See Connor Bros. Constr. Co., Inc. v. United States*, 65 Fed. Cl. 657 (2005). *Morrison-Knudson Co. v. United States*, 84 F. Supp. 282 (Ct. Cl. 1949); *Public Constructors, Inc. v. State*, 390 N.Y.S.2d 481 (N.Y. App. Div. 1977); *Camarco Contractors v. State*, 253 N.Y.S.2d 827 (N.Y. App. Div. 1964); *A.E. Ottaviano, Inc. v. State*, 110 N.Y.S.2d 99 (N.Y. Ct. Cl. 1952).

[25] *See, e.g., Gymco Constr. Co. v. Architectural Glass & Windows, Inc.*, 884 F.2d 1362 (11th Cir. 1989).

[26] *See* Bruce Jervis, *Delay in Shop Drawing Review*, 25 Construction Claims Monthly 1 & 7 (July 2003).

[27] *See, e.g., Pathman Constr. Co.*, ASBCA No. 23392, 85–2 BCA ¶ 18,096; *see also Sydney Constr. Co., Inc.*, ASBCA No. 21377, 77–2 BCA ¶ 12,719; *TLT Constr. Corp. v. A. Anthony Tappe & Assoc.*, 716 N.E.2d 1044 (Mass. App. Ct. 1999); *Alamo Cmty. Coll. Dist. v. Browning Constr. Co.*, 131 S.W.3d 146 (Tex. App. 2004).

of plans is withheld because the contractor's submission is incomplete, no time extension may be due.[28]

Subparagraph 4.2.7 of AIA A201 (1997 ed.) provides that "[t]he Architect's action will be taken with such reasonable promptness as to cause no delay in the Work or in the activities of the Owner, Contractor or separate contractors, while allowing sufficient time in the Architect's professional judgment to permit adequate review."[29] Interestingly, the language "as to cause no delay in the Work or in the activities of the Owner, Contractor or separate contractors" was removed in the 2007 edition of AIA A201, and language was added to state that the architect's action would be undertaken under "the submittal schedule approved by the Architect or, in the absence of an approved submittal schedule," with reasonable promptness.[30] The "reasonable promptness" standard is not clear. One court found that a 10- to 12-week delay was normal for a highway project.[31]

E. Failure to Coordinate Prime Contractors

When an owner elects to execute a project with multiple prime contractors, many jurisdictions recognize a duty on the part of the owner to coordinate the work of the separate prime contractors.[32] Thus, the owner may be responsible to one prime contractor for delays caused by another. Even where the owner attempts to shift this duty to one of the prime contractors, the owner still may be liable for delays if that contractor is not also given the power to enforce its responsibilities.[33]

F. Failure to Give Timely Orders for Work

If the owner fails to issue the Notice to Proceed (NTP) within the time frame set forth in the contract, or within a reasonable time if the contract does not specify a time, the owner generally will be liable for the resulting delay. This rule also applies to delays in authorizing extra work, delays in responding to requests for information, and any unreasonable failure to approve materials.[34]

It has been held that, where the contract is silent as to time of issuance of the NTP, the owner or its designated representative must issue the NTP within a reasonable time after contract award. An unreasonable delay in issuing the NTP may amount to a breach of contract or a constructive suspension of work.[35]

[28]*See, e.g., Constr. Servs., Inc.*, GSBCA No. 2423, 68–2 BCA ¶ 7154; *Dept. of Transp. v. Brayman Constr. Corp.-Bracken Constr. Co.*, 513 A.2d 562 (Pa. Cmmw. Ct. 1986) (finding that 10 to 12 weeks was normal for a highway project and any time beyond that constitutes a compensable delay).

[29]AIA A201, § 4.2.7 (1997 ed.).

[30]*See* AIA A201, § 4.2.7 (2007 ed.).

[31]*See, e.g., Commonwealth of Pa. Dept. of Transp. v. Brayman Constr. Corp.-Bracken Constr. Co.*, 513 A.2d 562 (Pa. Commw. Ct. 1986).

[32]*See, e.g., United States ex rel. Va. Beach Mech. Servs., Inc. v. SAMCO Constr. Co.*, 39 F. Supp. 2d 661 (E.D. Va. 1999); *N. Harris County Junior Coll. Dist. v. Fleetwood Constr. Co.*, 604 S.W.2d 247 (Tex. App. 1988).

[33]*Shoffner Indus. v. W.B. Lloyd Constr. Co.*, 257 S.E.2d 50 (N.C. Ct. App. 1979), *rev. denied*, 259 S.E.2d 301 (N.C. 1979).

[34]*See, e.g., Rome Hous. Auth. v. Allied Bldg. Materials, Inc.*, 355 S.E.2d 747 (Ga. Ct. App. 1987).

[35]*Freeman Elec. Constr. Co.*, DOTCAB No. 74–23A, 77–1 BCA ¶ 12,258; *Nicon v. United States* 331 F.3d 878 (Fed. Cir. 2003); *GASA, Inc. v. United States*, 79 Fed. Cl. 325 (2007).

In determining what is a reasonable period of time within which to issue the NTP, the boards and the United States Court of Federal Claims will consider the time required to resolve any pending bid protest.[36] If the contractor wants the NTP to be issued expeditiously, it should make a specific request.[37]

G. Failure to Make Timely Payments to Contractors

Should the owner fail to make timely payment, the contractor may elect to terminate the contract as specified by the contract terms or may elect to continue with the contract work and seek damages. The contractor generally can recover interest on the late payments and in some jurisdictions may also recover consequential damages suffered due to late payment.[38]

H. Failure to Inspect

Under the typical contract, an owner may have the right or duty to inspect the contractor's work as it progresses. (See **Chapter 12**.) The owner may be liable to the contractor for inspections that are unreasonably intensive or repetitious, or for failure to timely and promptly inspect.[39]

I. Suspensions

A suspension is a form of delay that usually results from the owner's purposeful interruption of the work. The term is encountered primarily in federal government contracts; the standard Suspension of Work FAR clause provides, in part:

> SUSPENSION OF WORK (APR 1984)
>
> (a) The Contracting Officer may order the Contractor in writing to suspend, delay, or interrupt all or any part of the work of this contract for the period of time that the Contracting Officer determines appropriate for the convenience of the Government.[40]

This type of clause provides the owner the right to halt construction temporarily, if, for example, the owner experiences funding problems. The clause also provides equitable compensation for the resulting additional costs and extended performance time. Recovery is also allowed in federal government contracts where no formal suspension order is issued, but the effect of the government's action or inaction is to suspend the work.[41]

[36]*DeMatteo Constr. Co. v. United States,* 600 F.2d 1384 (Ct. Cl. 1979).

[37]*See Freeman Elec. Constr. Co.,* DOTCAB No. 74-23A, 77–1 BCA ¶ 12,258.

[38]*See, e.g., Anthony P. Miller, Inc. v. Willington Hous. Auth.,* 165 F. Supp. 275 (D. Del. 1958).

[39]Thomas E. Abernathy IV & Thomas J. Kelleher, Jr., Inspection Under Fixed-Price Construction Contracts, Briefing Papers, at 6–8, (Federal Publications, Inc. Dec. 1976).

[40]FAR § 52.242–14.

[41]*See, e.g., Blinderman Constr. Co. v. United States,* 695 F.2d 552 (Fed. Cir. 1983).

Some contract clauses recognize the concept of a constructive suspension of work. The Suspension of Work clause used in federal government contracting is an example of a contractual remedy for such delays or interruptions of the work. The clause becomes operative wherever there is an act or omission by the contracting officer resulting in delay to the contractor. It provides, in part:

> If the performance of all or any part of the work is, for an unreasonable period of time, suspended, delayed, or interrupted (1) by an act of the Contracting Officer in the administration of this contract, or (2) by the Contracting Officer's failure to act within the time specified in this contract (or within a reasonable time if not specified), an adjustment shall be made for any increase in the cost of performance of this contract (excluding profit), necessarily caused by the unreasonable suspension, delay, or interruption, and the contract modified in writing accordingly. However, no adjustment shall be made under this clause for any suspension, delay, or interruption to the extent that performance would have been so suspended, delayed, or interrupted by any other cause, including the fault or negligence of the Contractor, or for which an equitable adjustment is provided for or excluded under any other term or condition of this contract.[42]

Contractors performing on federal government contracts have been allowed to recover under either the express or the constructive suspension provisions for costs generated by: (1) delays in making the site available; (2) delays in issuing change orders; and (3) delays caused by defective plans and specifications.

J. Excessive Change Orders

When the owner orders an excessive or unreasonable amount of changed or extra work, the contractor may be allowed to recover its resulting delay damages.[43] These damages are commonly referred to as "impact costs."

K. Failure to Accept Completed Work

Should the owner unreasonably refuse to make final acceptance of the contractor's work, the owner may be liable for the contractor's resulting delay damages.

V. CONCURRENT DELAY

Concurrent delay, in addition to excusable delay and nonexcusable delay, is an analytical framework for identifying and evaluating construction delays. Concurrent delays are delays that occur, at least to some degree, during the same time period

[42]FAR § 52.242–14.

[43]*See Air-A-Plane Corp. v. United States,* 408 F.2d 1030 (Ct. Cl. 1969); *Coley Props. Corp.,* PSBCA No. 291, 75–2 BCA ¶ 11,514; *Linda Newman Constr. Co. v. United States,* 48 Fed. Cl. 231 (2000); *see generally* Reginald M. Jones, *Lost Productivity: Claims for the Cumulative Impact of Multiple Change Orders,* 31 Pub. Cont. L.J. 1 (Fall 2001).

and that impact the critical path of a planned sequence of events. The term "concurrent delay" refers to the situation that arises when an excusable compensable delay and an unexcusable delay occur during overlapping time periods. One example of a concurrent delay would be when a contractor cannot commence work on the second phase of a project because the owner has failed to obtain a necessary right-of-way, and simultaneously, the contractor is prevented from commencing the second phase by its own failure to timely complete antecedent first-phase work.

Concurrent delay creates complex legal issues regarding assessing responsibility for overall project delay. The analysis of concurrent delays may be further complicated if: (1) the delay periods are different lengths; (2) the delay periods are not totally concurrent; or (3) the delay periods have different impacts on the number and types of work activities they affect, and the severity of the impact on the affected work activities is different for each of the delays.

A. Traditional View: No Recovery by Either Party

Traditionally, when project performance was concurrently delayed, neither party was allowed an affirmative recovery from the other. The courts took the view that when each party proximately contributed to the delay, the law would not provide for the apportionment of damages between the parties.[44] In *J.A. Jones Constr. Co. v. Greenbriar Shopping Ctr.,* the prime contractor caused delay through the failure of its subcontractors and materialmen to timely perform, and the owner caused delay by changes to the design and late issuance of drawings. The court held that neither party was entitled to any affirmative recovery, meaning the owner and contractor each bore its own costs even though one may have been responsible for more delay than the other.[45]

B. Modern Trend: Apportionment of Delay Damages

In contrast to the analysis of concurrent delays found in *J.A. Jones Constr. Co.,* the current trend in concurrent delay analysis is toward apportionment of the delay between the delaying parties. This shift is due in part to the use of the CPM techniques for evaluating delay claims. These techniques make it possible to more accurately segregate and quantify the impact of concurrent delays.

Logically, if the impact of one delay exceeds that of the other, the party responsible for the lesser impact should be allowed to recover damages for the excess impact. Apportionment analysis, at least on its face, would seem to allow for more equitable results than nonapportionment analysis. In apportioning delays, if the effects of concurrent delay cannot accurately be segregated and quantified, the court will likely revert to nonapportionment type review and no damages will be awarded.[46]

[44] *Malta Constr. Co. v. Henninston, Durham & Richardson, Inc.,* 694 F. Supp. 902 (N.D. Ga. 1988); *J.A. Jones Constr. Co. v. Greenbriar Shopping Ctr.,* 332 F. Supp. 1336 (N.D. Ga. 1971), *aff'd,* 461 F.2d 1269 (5th Cir. 1972).

[45] *J.A. Jones Constr. Co. v. Greenbriar Shopping Ctr.*, 332 F. Supp. 1336 (N.D. Ga. 1971), *aff'd,* 461 F.2d 1269 (5th Cir. 1972).

[46] *See, e.g., SIPCO Servs. & Marine, Inc. v. United States,* 41 Fed. Cl. 196, 225–26 (1998); *Blinderman Constr. Co., Inc. v. United States,* 39 Fed. Cl. 529, 543–44 (1997).

In *Essex Electro Engrs., Inc. v. Danzig,* the United States Court of Appeals for the Federal Circuit reversed an Armed Services Board of Contract Appeals (ASBCA) decision denying a contractor's claim for delay damages. The court of appeals determined that each party's delay was apportionable and, thus, should have been allocated to each responsible party.[47]

In arriving at its decision, the Federal Circuit noted that:

> [T]he contractor generally cannot recover for concurrent delays for the simple reason that no causal link can be shown: a Government act that delays part of the contract performance does not delay "the general progress of the work" when the "prosecution of the work as a whole" would have been delayed regardless of the Government's act.[48]

Further, the court held that "in recent cases, the principle has been characterized as requiring the Government's act to have affected activities on the critical path."[49] Thus, the court held that "if 'there is in the proof a clear apportionment of the delay and the expense attributable to each party,' then the government will be liable for its delays."[50]

VI. NONCOMPENSABLE EXCUSABLE DELAYS

A. Weather

Under most contracts, unusually severe weather conditions can give rise to an excusable, but not compensable, delay.[51] For example, subparagraph 6.3.1 of Paragraph 6.3 (Delays and Extensions of Time) of the 2007 ConsensusDOCS 200 provides:

> If the Contractor is delayed at any time in the commencement or progress of the Work by any cause beyond the control of the Contractor, the Contractor shall be entitled to an equitable extension of the Contract Time. Examples of causes beyond the control of the Contractor include, but are not limited to, the following: . . . fire; terrorism, epidemics, adverse governmental actions, unavoidable accidents or circumstances; adverse weather conditions not reasonably anticipated.[52]

Unusually severe weather is weather that is unusual for the time of year and the place it occurred. This may be shown by comparing previous years' weather with the weather experienced by the contractor. The mere fact that the weather is harsh or

[47] *Essex Electro Eng'rs, Inc. v. Danzig,* 224 F.3d 1283 (Fed. Cir. 2000).
[48] *Id.* (*citing Coath & Goss, Inc. v. United States,* 101 Ct. Cl. 701, 714–15 (1944)).
[49] *Id.* (*citing Mega Constr. Co. v. United States,* 29 Fed. Cl. 396, 424 (1993)).
[50] *Id.* (*citing Coath & Goss, Inc. v. United States*, 101 Ct. Cl. 701, 714 (1944)); *see also United States v. Killough,* 848 F.2d 1523, 1531 (11th Cir. 1988).
[51] *See Fru-Con Constr. Corp. v. United States,* 43 Fed. Cl. 306, 328 (1999) (*citing Turnkey Enters., Inc. v. United States,* 597 F.2d 750, 754 (Ct. Cl. 1979)).
[52] ConsensusDOCS 200, ¶ 6.3.1 (2007 ed.).

destructive is not sufficient if the contractor reasonably should have anticipated that type of weather at the time and place it occurred. Some bad weather is always to be expected, and the contract period should take into account normal weather delays. If the contract period is 600 days, the contractor obviously does not have the right to expect 600 dry, sunny days with all of its subcontractors working at full force.

B. Acts of God

By definition, acts of God are beyond the control and without the fault or negligence of the contractor. The Comptroller General of the United States has defined an act of God as "some inevitable accident which cannot be prevented by human care, skill, or foresight, but results from natural causes such as lightning, tempest, floods and inundations."[53] Examples of acts of God are tornadoes, hurricanes, floods, earthquakes, and other natural disasters.

Unless the contract specifically provides otherwise, acts of natural forces entitle the contractor to an extension of contract performance time but not to a contract price adjustment. On this basis, a highway contractor was granted a time extension but denied a contract price increase for delays encountered when the government reasonably refused to alleviate flooding on the work site caused by heavy rains.[54]

C. Labor Problems

Similarly, delays resulting from most, but not all, strikes and labor disturbances generally constitute noncompensable excusable delays.[55] If a strike is in effect or anticipated at the time of contracting, it may be determined that because the labor problems were foreseeable, the contractor should have provided for them in the contract. Likewise, when a strike is provoked by an unfair labor practice on the part of the contractor, the delay might not be considered to be due to a "cause beyond the contractor's control." Finally, with few exceptions, delays that result from labor shortages unrelated to labor disputes or delays that are caused by nonperformance of subcontractors will not be excused.

VII. ACCELERATION

When the owner requires its contractor to complete the work by a date earlier than the contract completion date, an "acceleration" occurs. For the purpose of determining whether the contractor's work has been accelerated, the contract completion date should reflect time extensions due the contractor for excusable delays to its work.

[53] *To Guimond Farms*, 49 Comp. Gen. 733 (1970).

[54] *Sec. Nat'l Bank of Kan. City v. United States,* 397 F.2d 984 (Ct. Cl. 1968); *see also Vicari v. United States,* 47 Fed. Cl. 353 (2000).

[55] *See Murdock & Sons Constr., Inc. v. Goheen Gen. Constr., Inc.*, 461 F.3d 837 (7th Cir. 2006) (finding that an organized slowdown by laborers did not constitute a labor dispute although laborers were only laying 50 blocks per day as opposed to 150 as anticipated); *McNamara Constr. of Manitoba, Ltd. v. United States,* 509 F.2d 1166 (Ct. Cl. 1975).

An acceleration of the contractor's work occurs under two different circumstances: (1) actual (or directed) acceleration or (2) constructive acceleration. *Actual acceleration* occurs when the owner expressly directs the contractor to complete the project earlier than the contract completion date. *Constructive acceleration* occurs when the owner fails to grant its contractor time extensions to which it is entitled, and the contractor is required to achieve, or strive for, a completion date that is earlier than the properly extended contract completion date. A project can be considered accelerated even if the contractor is only striving to finish on the original contract date if the contractor is entitled to time extensions that have not been granted. Thus, acceleration may be a by-product of delay or other factors that justify a time extension, which the owner has not formally granted.

Acceleration damages usually include premium time pay in the form of overtime or shift work, the cost of added crews or increased crew sizes, the cost of additional tools and equipment required for added crews, the cost of additional supervision and job-site overhead, and the cost of labor inefficiency that may occur due to longer hours or increased crew sizes.

A. Directed Acceleration

In federal government contracts, the Changes clause for fixed-priced contracts found at FAR § 52.243–4 specifically allows the contracting officer to direct an acceleration in the performance of the work:

CHANGES (AUG 1987)

(a) The Contracting Officer may, at any time, without notice to the sureties, if any, by written order designated or indicated to be a change order, make changes in the work within the general scope of the contract, including changes—

(1) In the specifications (including drawings and designs);

(2) In the method or manner of performance of the work;

(3) In the Government-furnished facilities, equipment, materials, services or site; or

(4) Directing acceleration in the performance of the work.[56]

Thus, the government's order to a contractor to accelerate modification to an air-conditioning system in a medical facility to speed completion of operating rooms was a compensable change because the government was aware that excess costs for overtime would be incurred.[57]

[56]FAR § 52.243–4.

[57]*E.C. Morris and Son, Inc.*, ASBCA No. 20697, 77–2 BCA ¶ 12,622; *see also Gibbs Shipyard, Inc.*, ASBCA No. 9809, 67–2 BCA ¶ 6499; *Fru-Con Constr. Corp. v. United States,* 43 Fed. Cl. 306 (1999).

B. Constructive Acceleration

In the seminal case *Fermont Div., Dynamic Corp. of America,*[58] the five basic elements of a constructive acceleration claim based on a government failure to grant a time extension for excusable delay were summarized in this way:

(1) Existence of a given period of excusable delay; and

(2) Contractor notice to the government of the excusable delay, and request for extension of time together with supporting information sufficient to allow the Government to make a reasonable determination;

Exceptions:

(a) such notice, request, and information are not necessary if the government's order directs compliance with a given schedule expressly without regard to the existence of any excusable delay,

(b) the supporting information is unnecessary if it is already reasonably available to the government; and

(3) Failure or refusal to grant the requested extension within a reasonable time; and

(4) A government order, either express or implied from the circumstances, to (a) take steps to overcome the excusable delay, or (b) complete the work at the earliest possible date, or (c) complete the work by a given date earlier than that to which the contractor is entitled by reason of the excusable delay. Circumstances from which such an order may be implied include expressions of urgency by the government especially when coupled with (i) a threat of default or liquidated damages for not meeting a given accelerated schedule, or (ii) actual assessment of liquidated damages for not meeting a given accelerated schedule; and

(5) Reasonable efforts by the contractor to accelerate the work, resulting in added costs, even if the efforts are not actually successful.[59]

If these elements are proven, the contractor is entitled to recover the costs incurred in accelerating its performance.[60]

VIII. CONTRACTUAL LIMITATIONS TO RECOVERY

A. Requirement for Written Notice

Most construction contracts require the contractor to submit written notice to the owner or its designated representative, within a definite period of time after the delay-causing event, before submitting any claim for additional compensation or for

[58]*Fermont Div., Dynamic Corp. of Am.*, ASBCA No. 15006, 75–1 BCA ¶ 11,138; *see also Fraser Constr. Co. v. United States,* 384 F.3d 1354 (Fed. Cir. 2004).

[59]*Fermont Div., Dynamic Corp. of Am.*, ASBCA No. 15006, 75–1 BCA ¶ 11,138, at 52,999.

[60]*See Dept. of Transp. v. Anjo Constr. Co.,* 666 A.2d 753 (Pa. Commw. Ct. 1995); *Norair Eng'g Corp. v. United States,* 666 F.2d 546 (Ct. Cl. 1981).

an extension of time. Such notice requirements are imposed to protect the interest of the owner, who may be unaware of the causes of a particular delay and thereby precluded from taking immediate measures to rectify the situation and mitigate its cost. Failure to give prompt notice may result in a waiver of the contractor's rights or result in a time-consuming litigation effort, which may ultimately prove unsuccessful.[61]

Formal notice may be unnecessary when the owner has actual or constructive knowledge of the problem, or when the lack of notice does not prejudice a legitimate owner interest. The contractor, however, should never knowingly forego written notice on the assumption that one of those conditions is present. Some courts view notice as a condition precedent to recovery.[62] The contractor that gives the owner prompt written notice of delays and disruptions that are the owner's responsibility increases its opportunity to recover the costs generated by those problems.

B. No-Damages-for-Delay Clauses

Despite the widely recognized right of a contractor to recover damages stemming from delayed or out-of-sequence work caused by the owner, the owner (or the general contractor if the claimant is a subcontractor) may succeed in asserting one of several possible defenses to a delay claim. The most notable potential defense is the no-damages-for-delay clause.

A no-damages-for-delay clause typically provides that the owner will not be liable for monetary damages resulting from any delays or resulting from certain specified delays. Most clauses of this kind provide that a contractor's only relief for delays covered by the clause is a time extension. In the past, when the language of the clause was clear and unambiguous, a no-damages-for-delay clause was legally valid and enforceable. Traditionally, most jurisdictions handled exceptions to the enforceability of no-damages-for-delay clauses through their state's common law, and the majority view was clearly summarized by the court's opinion in *Corinno Civetta Constr. Corp. v. City of New York*:

> A clause which exculpates a contractee from liability to a contractor for damages resulting from delay in the performance in the latter's work is valid and enforceable and is not contrary to public policy if the clause and the contract of which it is a part satisfy the requirements for the validity of contracts generally. The rule is not without its exceptions, however, and even exculpatory language which purports to preclude damages from all delays resulting from any cause whatsoever are not read literally. Generally, even with such a clause, damages may be recovered for:
>
> (1) Delays caused by the contractee's bad faith or its willful, malicious or grossly negligent conduct,
>
> (2) Uncontemplated delays,

[61] *See, e.g., Allgood Elec. Co., Inc. v. Martin K. Eby Constr. Co., Inc.*, 959 F. Supp. 1573 (M.D. Ga. 1997), *aff'd*, 137 F.3d 1356 (11th Cir. 1998).

[62] *A.H.A. Gen. Constr., Inc. v. N.Y. City Hous. Auth.*, 699 N.E.2d 368 (N.Y. 1998).

(3) Delays so unreasonable that they constitute an intentional abandonment of the contract by the contractee, and
(4) Delays resulting from the contractee's breach of a fundamental obligation of the contract.[63]

Many states have begun to abandon the common law approach to no-damages-for-delay clauses in favor of a statutory restriction on the clauses. Statutes invalidating the enforcement of certain no-damages-for-delay clauses in public contracts have a relatively long history,[64] but for most states, legislation that bars enforcement of a no-damages-for-delay clause for private construction contracts has come to the forefront only recently.

Washington was one of the first states to enact a statute providing that any clause in a construction contract that "purports to waive, release, or extinguish the rights of a contractor, subcontractor or supplier to damages or an equitable adjustment arising out of unreasonable delay in performance which delay is caused by the acts or omissions of the contracted or persons acting for the contracted. . . is void and unenforceable."[65] This language is reflected in the more recent trend of statutory bars to the enforcement of no-damages-for-delay clauses.

For example, in 1998, Ohio enacted a statute that declares any provision of a construction contract or subcontract unenforceable as a matter of public policy where the provision waives liability for delay when the delay is caused by either the owner's or contractor's acts or failure to act.[66] A recent North Carolina statute also provides a blanket prohibition on no-damages-for-delay clauses.[67] Some states, however, have attempted to limit the no-damages-for-delay clause while leaving the clause itself intact by codifying well-recognized exceptions.[68] Virginia has also adopted a statute voiding any contractual attempt to "waive, release or extinguish" the rights of a contractor to recover delay damages in public projects.[69] Further, the Virginia Supreme Court recently held that the statute should be interpreted strictly and held: "Any provision to waive, release, or extinguish the rights of a contractor shall be void."[70]

C. Trade-to-Trade Clauses

Another recent development in contractual risk allocation for delays are the so-called trade-to-trade clauses that many owners, general contractors, and construction managers are inserting into their subcontract agreements. A trade-to-trade clause relieves the general contractor or construction manager of some of the risk of delay by requiring a delayed subcontractor to assert its delay claim against the subcontractor that caused the delay rather than immediately pursuing a claim against the general

[63] *Corinno Civetta Constr. Corp. v. City of N.Y.*, 493 N.E.2d 905, 909–10 (N.Y. 1986).
[64] *See, e.g.*, CAL. PUB. CONT. CODE § 7102 (2007); COLO. REV. STAT. § 24–91–103.5 (2008).
[65] WASH. REV. CODE § 4.24.360 (2008).
[66] *See* OHIO REV. CODE ANN. § 4113.62(C)(1) & (2) (2008).
[67] *See* N.C. GEN. STAT. § 143–134.3 (2007).
[68] *See, e.g.*, ARIZ. REV. STAT. § 41–2617 (2007).
[69] *See* VA. CODE ANN. § 2.2–4335(A).
[70] *Blake Constr. Co., Inc. v. Upper Occoquan Sewage Auth.*, 587 S.E.2d 711 (Va. 2003).

contractor or construction manager. These clauses establish accountability among the subcontractors for delay damages that they cause each other and relieve the general contractor or construction manager of some of the burden. A sample of that portion of a trade-to-trade clause shifting such risk is:

> The construction manager shall not be liable to subcontractor for any adjustments to the subcontract price, schedule of work, damages, costs, losses or expenses, including but not limited to attorney's fees, resulting from acts or omissions (whether or not negligent), failure to perform, delays in performance, or defaults of any other subcontractor or any supplier in connection with the performance of any of the work. Subcontractor agrees to file any claim for such directly against the other subcontractor or supplier which subcontractor contends is responsible, without making owner or construction manager a party to any such claim or action. Subcontractor agrees that other subcontractors or suppliers on the project shall have a direct right of action against subcontractor . . . for such claims.

IX. DELAY CLAIMS AND THE USE OF CPM SCHEDULES

To prove or refute delay claims, contractors and owners typically examine project records and project schedules in detail. Delay claims typically are based on contemporaneous CPM schedules that were generated during the project or on after-the-fact as-built CPM schedules reconstructed by a scheduling expert. The relative weight to be given to either approach has been vigorously contested. As in other areas of construction law, federal government contract decisions illustrate some of the key issues related to using CPM schedules to prove delay claims.

For example, the Veterans Affairs Board of Contract Appeals (VABCA) addressed the question of using after-the-fact CPM schedules in two decisions in 1984 and 1987, both bearing the same name, *Santa Fe, Inc.*[71] The *Santa Fe* decisions indicated the board's clear preference for contemporaneous schedules when assessing delay claims.

In 1984, the VABCA denied a contractor's claim for time extensions because "the delayed activities had not been on the critical path for completion of the project."[72] The VABCA held that the "critical path calculations had been made for each task involved, and these calculations were continuously updated during construction." The VABCA further elaborated that "it is the very existence of the contractually agreed upon CPM procedure which, when properly utilized, allows the contracting officer and subsequent bodies to determine with greater exactitude whether, and to what extent, a particular change order affects critical path and hence delays ultimate performance."

[71]*See Santa Fe, Inc.*, VABCA No. 1943 *et al.*, 84–2 BCA ¶ 17,340; *Santa Fe, Inc.*, VABCA No. 2168, 87–3 BCA ¶ 20,104.

[72]*Santa Fe, Inc.*, VABCA No. 1943 *et al.*, 84–2 BCA ¶ 17,340.

In the 1987 *Santa Fe, Inc.* decision, the VABCA determined that contract extensions "shall be based upon the computer-produced calendar dated schedule for the time period in question and all other relevant information."[73] In this decision, the VABCA held that "submission of proof based on revised activity logic durations and cost is obligatory to any approvals." The board relied on the most current CPM schedule developed when analyzing whether the contractor incurred delay on the project. According to this decision, "[t]here is a rebuttable presumption of correctness attached to CPM schedule upon which the parties have previously mutually agreed." "To put it another way," the VABCA stated, "in the absence of compelling evidence of actual errors in the CPMs, we will let the parties 'live or die' by the CPM applicable to the relevant time frames." The *Santa Fe* decisions illustrate the rationale and preference for contemporaneous schedules when calculating delay claims.

In *P.J. Dick,*[74] the VABCA reiterated its preference for a contemporaneous schedule as its baseline for evaluating delay claims. In this case, the VABCA was faced with a "rather unique situation." The CPM schedules used during the project were, according to the VABCA:

> properly constituted in [their] logic and assiduously and properly maintained throughout contract performance. This circumstance is in sharp contrast to the usual problems we encounter in dealing with CPMs where warring as-built schedules are constructed by the parties after the fact because the CPM was either never properly or timely prepared or was not updated in accordance with the contract scheduling requirements.[75]

The VABCA determined that "since there is no dispute concerning the validity of a CPM, these appeals present the circumstance where we have said in the past that we will let the parties 'live or die' by analysis of the CPM to determine the number of days of additional contract performance time."

The *P.J. Dick* decision underscores a recent theme discussed in several board of contract appeals decisions: the requirement of extensive documentary evidence that demonstrates the alleged claims before awarding the appropriate relief. Boards are becoming more demanding and are requiring more concrete evidence in their analysis of a contractor's delay claim.

These increased demands should have a direct, practical impact on those contractors that rely on government contracts as a source of business. Project schedules should be updated routinely, they should reflect all changes, and they should be approved and shared with all interested parties through the life of the project.

Does the *P.J. Dick* decision end the use of after-the-fact as-built schedules in the analysis of delay claims? Not necessarily. The strong preference expressed by some

[73] *Santa Fe, Inc.,* VABCA No. 2168, 87–3 BCA ¶ 20,104.

[74] *P.J. Dick, Inc.,* VABCA Nos. 5597, 5836, 01–2 BCA ¶ 31,647.

[75] *Id.*

tribunals for contemporaneous analysis notwithstanding, the quantification of delays also may be developed retrospectively. If contemporaneous schedules are flawed or do not exist, an accurate and realistic after-the-fact schedule can be useful in analyzing or presenting a delay claim.

Even where the contract contains detailed CPM scheduling requirements but the parties do not follow those requirements, delays still can be proven. For example, in *Whitesell-Green, Inc.,* the Armed Services Board of Contract Appeals (ASBCA) found "that the government abandoned the use of the [P]rimavera software network analysis as the sole method for determining time extensions."[76] In that case, the government did not use or require a P3 CPM analysis when the parties negotiated time extensions that resulted in three bilateral modifications, but the government did use a CPM analysis to assess liquidated damages at the end of the project. The ASBCA found that the government's attempt to require the contractor to use CPM scheduling in order to justify its claims for additional time came too late in the construction process. In the absence of a CPM analysis, the ASBCA relied on the contemporaneous project documentation to find that the government contributed to the project delays and was therefore not entitled to assess liquidated damages.

X. DELAY CLAIMS SUPPORTING DOCUMENTATION

CPM analysis is often used to assess or present a delay claim. Basic project documentation is often critical to establishing the validity of the analysis. The next checklist itemizes many of the sources of information to be evaluated when preparing, or attempting to rebut, a delay claim:

✔ SUPPORTING DOCUMENTATION CHECKLIST

(1) Estimates
(2) Original schedules
(3) Schedules used on the project, including look-ahead schedules, CPM logic diagrams, and tabular printouts
(4) As-built schedules
(5) Daily reports
(6) Diaries
(7) Manpower and manloading reports
(8) Cost accounting records
(9) Scheduling meeting minutes
(10) Material and equipment delivery tickets
(11) Job photographs and videotapes
(12) As-built drawings

[76] *Whitesell-Green, Inc.,* ASBCA Nos. 53938, 53939, 54135, 06–2 BCA ¶ 33,323.

(13) Shop drawing logs
(14) Project correspondence
(15) Change orders
(16) Contract documents
(17) Pay applications
(18) Internal memoranda

✔ CHECKLIST FOR EVALUATING POTENTIAL DELAY CLAIM

When evaluating a potential delay claim, the next checklist is a useful tool for reviewing the pertinent factual information.:

(1) *Accuracy.* Are the schedules used for the project accurate? Were they agreed on and used by the parties, or were they issued for "internal purposes" only? Courts may give more weight to schedules to which the parties have agreed previously.

(2) *Abandonment of schedule.* Was the selected scheduling technique abandoned during performance? If so, why? A contractor that committed by contract to a particular scheduling technique might be precluded from proving its claim with that technique if it did not meet its scheduling commitments.

(3) *Current schedule.* Was the schedule updated and kept current on a regular basis? Schedules often change dramatically during the project.

(4) *Changes.* Was the schedule revised to reflect the effect of change orders? The schedule should show whether the change impacted work along the critical path or consumed float. A change order impact analysis is a handy tool for negotiating the price of a change.

(5) *Change order compensation.* Was additional overhead included in the change order? Even if the changed work affects only float, it may result in less effective resource utilization or involve unforeseen overhead or job staffing.

(6) *Cross-references.* Are the project records tied into the project schedule by work activity code or designation? Doing so provides data for subsequent updates and for the preparation of an accurate as-built analysis.

(7) *Float ownership.* Does the contract bar the contractor from seeking compensation, time extensions, or both, for delays that consume only float time?

(8) *Coordination responsibility.* Which party is responsible for coordinating and scheduling the work?

(9) *Scheduling experts.* Solicit expert or in-house scheduling assistance early. This may aid in efficient record keeping and assist in minimizing the effect of a delay.

(10) *Choosing an expert*: An expert should be well versed in:

  (a) The theory and output of all scheduling techniques used.

  (b) The estimating process used and its relation to the contractor's resources.

  (c) The contractual relationships among owner, designers, contractors, subcontractors, and suppliers.

(d) Good project record keeping and cost accounting.

(e) The design and construction of the type of work involved.

➢ POINTS TO REMEMBER

- Many large-dollar construction disputes are based on changes to the construction schedule in the form of delays or acceleration.
- Most construction contracts contain an express obligation that the contractor will complete the work by a given date or within a specified time frame, which is accompanied by an implied obligation that neither party will do anything to delay, hinder, or interfere with the other's performance.
- Common causes of delay include inclement weather, labor disputes, untimely equipment delivery, defective specifications, changes, and differing site conditions. Which party bears responsibility for each of these delays depends on the language of the contract and the surrounding circumstances.
- The presence of a "time is of the essence" provision in a contract should not be viewed as mere boilerplate. Such a provision makes time a material element of the contract. The provision can be a predicate for the recovery of delay damages and is a signal that timely performance is required.
- An understanding of several key contract terms is important in determining the party that bears responsibility for construction delays:
  - An *excusable delay* entitles the contractor to a time extension under the contract terms. Moreover, an excusable delay may be either compensable or noncompensable.
  - A compensable excusable delay is not only excusable (entitling the contractor to a time extension) but also entitles the contractor to additional compensation for the resulting cost.
  - A noncompensable excusable delay entitles the contractor to a time extension but no additional compensation.
  - A *nonexcusable delay* does not entitle the contractor to a time extension and may subject the contractor to liability for delay damages arising out of the nonexcusable delay.
- To determine whether a given delay is nonexcusable or excusable and possibly a compensable delay, one must carefully study the contract, the nature of the delay, and all surrounding circumstances.
- Examples of delays that may constitute excusable noncompensable delays are weather, acts of God, and unforseeable labor problems.
- Most construction contracts require the contractor to provide the owner prompt written notice of any excusable delay. A contractor's failure to provide such notice may jeopardize the contractor's right to a time extension, additional compensation, or both.

- Examples of delays that may constitute excusable compensable delays include delays due to defective drawings or specifications, the owner's failure to provide access, improper site preparation by the owner or a parallel contractor, the owner's failure to timely supply owner-furnished materials or labor, the owner's failure to timely provide plans and approved shop drawings, the owner's failure to properly coordinate parallel prime contractors, the owner's failure to make timely payments, the owner's failure to perform timely inspections, the owner's suspension of the work, excessive change orders, and the owner's failure to timely accept complete work.
- Acceleration is another common source of construction claims and disputes. Acceleration may take two forms:
  - Directed acceleration
  - Constructive acceleration
- Directed acceleration occurs where the owner explicitly directs the contractor to complete the work earlier than the contractually required completion date.
- Constructive acceleration occurs where an owner fails to grant a contractor a time extension to which it is entitled, thereby requiring the contractor to complete, or attempt to complete, the work by a date earlier than contractually required. Thus, constructive acceleration is, in effect, a possible by-product of an excusable delay.
- The six essential elements of a constructive acceleration claim are:
  - An excusable delay
  - A timely request for a time extension
  - Failure or refusal by the owner to grant the request for time extension
  - Conduct by the owner that is reasonably construed as requiring the contractor to complete on a schedule that has not been properly extended
  - Effort by the contractor to accelerate performance
  - Additional costs incurred by the contractor as a result of the acceleration
- If the contractor can establish each of these six elements of a constructive acceleration claim, the contractor generally is entitled to additional compensation even though the contractor may have been unsuccessful in its attempt to complete the work by the nonextended completion date.
- Examples of acceleration damages include premium time, the cost of added crews or increased crew sizes, the cost of additional tools and equipment required for added or larger crews, the cost of additional supervision and jobsite overhead, and the cost of the labor inefficiency that may occur due to the longer hours or increased crew sizes or numbers.
- "Concurrent delay" refers to the situation where two different delays, caused by different parties, occur simultaneously or in overlapping time periods and one of the delays is a compensable delay, while the other is a nonexcusable delay.
- The early view stated by the courts and administrative boards regarding concurrent delay was that neither party was allowed any affirmative recovery from the other in the case of concurrent delay.

- The modern trend is to apportion responsibility for project delays between the parties whenever it is possible, using modern, sophisticated scheduling techniques such as critical path method (CPM) scheduling to segregate the impact of the concurrent delays.
- Some construction contracts contain a type of exculpatory clause referred to as a no-damages-for-delay clause. A no-damages-for-delay clause typically provides that the owner will not be liable to the contractor for monetary damages resulting from any delays or from certain specified types of delay.
- As a general rule, no-damages-for-delay clauses are legally valid and enforceable; however, in order to avoid harsh results, courts often narrowly construe such provisions and create various judicial exceptions to the general rule of enforceability.
- Some common exceptions to the enforcement of no-damages-for-delay clauses are:
    - Delays of a kind not contemplated by the parties
    - Delays that amount to an abandonment of the contract
    - Delays that were the result of fraud, bad faith, or arbitrary action
    - Delays that were the result of active interference
    - Delays that were unreasonable
- Some state legislatures have also passed statutes limiting the enforceability of no-damages-for-delay clauses.
- A good scheduling analysis and good project documentation are critical to the successful presentation of any delay claim.

12

INSPECTION, ACCEPTANCE, WARRANTIES, AND COMMISSIONING

From the owner's perspective, the primary objectives of any construction project generally fall into three categories: cost, schedule, and quality. The owner naturally desires high-quality construction, on schedule, and at a low cost. Unfortunately, these three objectives sometimes conflict with one another, and certain trade-offs are required. The natural give-and-take that occurs between these three project objectives is perhaps best illustrated by the remark often made by contractors to owners in jest: "Cost, schedule, and quality—pick any two; but you can't have all three." As the saying goes, many a truth is sometimes spoken in jest.

Other chapters of this book focus on what happens when construction costs escalate or schedule delays occur. This chapter focuses on the third prong of the cost/schedule/quality triumvirate by discussing issues that relate to construction quality: (1) inspections, (2) acceptance, (3) warranties, and (4) commissioning.

Inspections are the primary vehicle employed by an owner during the course of construction to ensure that appropriate quality standards are being met. Inspections typically are performed by the owner or the owner's authorized representative periodically during the course of construction and again upon project completion. Timely and appropriate inspections afford an informed owner and contractor an opportunity to address quality problems before the work is complete and allow any necessary corrective work to be implemented when it is less costly.

Acceptance is a power generally vested by contract in the owner or the owner's representative (e.g., the project architect or engineer). The owner's right to inspect and accept the contractor's work before payment can be a valuable tool if used properly. An owner should employ inspection and acceptance procedures that will identify and appropriately address detectable defects in the work and before they are "covered up." Most construction contracts state that the owner's "acceptance" of the work and payment for the work do not preclude the owner from later objecting to defective work. But an owner that fails to inspect the work and to reject nonconforming work may be doing itself a great disservice. Project *commissioning* is a comprehensive approach to quality assurance and acceptance, as well as a determination that building systems function as intended.

The word *warranties* has several different meanings in the construction context. One way is to refer to the various express and implied promises set out in every construction contract. Some, but not all, of these promises relate to quality issues. An example is the express warranty whereby the contractor promises to perform its work in a "good and workmanlike manner." Another use of the word "warranties" is to describe the obligations of the contractor or a subcontractor, supplier, or manufacturer to address any quality problems that may be discovered after construction is complete. A construction contract typically provides that the contractor "warrants" its work for a period of one year (or some other defined time period) after substantial completion. This is usually a "repair" warranty, which requires the contractor to correct defective work upon notice given within the one-year (or other contractually defined) period. In most contracts, for example, the AIA A201 General Conditions of the Contract for Construction (2007 ed.), the contractor also gives a warranty that its work is performed in a workmanlike manner and that all materials are new and conform to the contract requirements. The term of this warranty is limited only by the applicable statute of limitation for breach of contract claims.

It is also common that certain subcontractors and manufacturers of certain products and systems installed in a project will provide warranties. One purpose of such warranties is to allocate responsibility for defective work, equipment, and materials or for equipment and materials that cease to function properly after operating for a period of time.

I. INSPECTION

A. Introduction

Owners of both private and public construction projects generally employ representatives to inspect the quality of the contractor's work. In private construction, a third party specially retained by the owner often performs these inspections. In public construction, however, government employed inspectors often handle such inspections.

Inspection protects the owner, not the contractor. Therefore, the owner generally has no duty to inspect beyond its contract obligations. Even if the contract allows for owner inspection, such a provision generally will not obligate the owner to inspect.[1]

[1] *Blumenthal Kahn Elec. Ltd. P'ship v. Bethlehem Steel Corp.*, 708 A.2d 1 (Md. Ct. Spec. App. 1998).

If the owner assumes a contractual obligation to inspect, the owner can find itself liable for defective inspections or a failure to inspect. In *Continental Insurance Co. v. City of Virginia Beach,*[2] when the contractor filed bankruptcy during construction, the public owner called on the contractor's surety to complete the project. After discovering that the contractor's work had not been properly performed, the surety sued the city for payments made to the contractor, which the surety contended could have been withheld and available to the surety if the defective work had been discovered through the city's inspection. The court found that the city had assumed the duty of inspecting and testing the contractor's work. The court held that city had breached its inspection obligations, thereby prejudicing the surety when payments for defective work were made to the contractor. As a result, the court held that the surety was discharged from its bond obligations to the extent it was prejudiced by the city's conduct.[3]

Owners often place the burden of inspections and quality control on the contractor by requiring the contractor to adhere to stringent quality control specifications, which may include the use of comprehensive quality control procedures during construction. Failure to inspect effectively, however, may affect the owner's rights under applicable warranties once the project is accepted. Furthermore, the owner cannot, with impunity, perform inspections in such a manner as to delay or disrupt the contractor's work or to alter contract requirements.

B. Standard Inspection Clauses

The rights and responsibilities of the owner and contractor in a typical construction contract regarding inspections are illustrated by the standard provisions found in industry documents.. The standard clause used in federal construction contracting, entitled Inspection of Construction, is set forth in Federal Acquisition Regulation (FAR) § 52.246–12. Examples of standard clauses used in many private construction contracts are found in: ConsensusDOCS 200, Standard Agreement and General Conditions Between Owner and Contractor (2007 ed.) [hereinafter ConsensusDOCS 200], Paragraph 3.7, "Tests and Inspections"; AIA A201, Article 12, "Uncovering and Correction of Work"; and Engineers Joint Contract Documents Committee C-700, Standard General Conditions of the Construction Contract (2007 ed.) [hereinafter EJCDC C-700], Section 13.04, "Uncovering Work."

1. Federal Government Contracting Inspection Clauses

FAR § 52.246–12 specifies that the government can inspect "at all reasonable times before acceptance to ensure strict compliance with the terms of the contract." This clause provides that the inspection is solely for the government's benefit and does not constitute or imply acceptance of the contractor's work. The contractor, therefore, still must ensure compliance with contract requirements even though the government has conducted inspections.

[2]908 F. Supp. 341 (E.D. Va. 1995).
[3]*Id.* at 348.

Other standard federal government contract clauses relate to inspection as well. The "Material and Workmanship" clause, FAR § 52.236–5, provides that materials employed are to be "new and of the most suitable grade for the purposes intended" unless the contract specifically provides otherwise; that references to products by trade name are intended to set a standard of quality and not to limit competition; that anything installed without the required approval may be rejected; and that work must be performed in a "skillful and workmanlike manner."

The "Permits and Responsibilities" clause of the standard federal construction contract, FAR § 52.236–7, requires the contractor to take proper precautions to protect the work, the workers, the property of others, and third parties. The clause states that the contractor is responsible for damages to persons or property caused by the contractor's fault or negligence, and places responsibility on the contractor for all materials delivered and work performed up until completion and acceptance by the government.[4]

The standard federal "Use and Possession Prior to Completion" clause, FAR § 52.236–11, provides that the owner may take possession of or use a partially or totally completed part of a project without being deemed to have accepted the work. Before such possession or use, the contracting officer must give the contractor a list of work remaining to be done on the relevant portion of the project. Even if the owner fails to list a particular defect or item of work, however, the contractor still must comply with the contract terms.[5]

2. Industry Form Contract Inspection Clauses

For private contracts, forms such as the ConsensusDOCS 200 provide that the contractor must schedule all required tests, approvals, and inspections so as not to delay the project work and give proper notice to all required parties.[6] The ConsensusDOCS 200 also provides that the owner is responsible for retaining independent testing firms and paying for the inspections; but the contractor is responsible for obtaining the certificates of testing, approval, and inspections.[7] The ConsensusDOCS 200 contract further provides that the contractor will be responsible for the costs of correction and retesting.[8]

The ConsensusDOCS 200 also provides that the owner can direct the contractor to uncover work that the owner did not require to be inspected so that the work can be inspected.[9] If the work was properly performed or if the defective condition was caused by the owner or others, then the owner will pay for the costs of uncovering and inspecting the work and placing it back in its preinspection form. If the work does not conform to the contract requirements, the contractor must pay for the uncovering and correction of the work.[10]

[4]See FAR § 36.507.
[5]*See* FAR § 52.236–11.
[6]ConsensusDOCS 200, ¶ 3.7 (2007 ed.).
[7]*Id.* at ¶ 3.7.1.
[8]*Id.* at ¶ 3.7.3.
[9]*Id.* at ¶ 3.10.1.
[10]*Id.*

The AIA A201 provides that the contractor is responsible during construction for inspection of the work already performed to determine that the work conforms with the contract documents so additional work can be performed.[11] The contractor is also required to secure and pay for inspections necessary for the proper execution and completion of the contract work and to obtain any required certificates of testing, inspection, or approval.[12] The contractor bears the cost of correcting the failure if there is a defect in the contractor's work.[13]

The EJCDC C-700 generally provides that the contractor must "supervise, inspect, and direct the Work competently and efficiently. . . ."[14] Under this industry form document, the contractor also must inspect the work of others and report to the engineer any impacts of such other work on the contractor's work (except for latent defects and deficiencies in such other work).[15] The EJCDC C-700 provides that the owner must pay for an independent testing laboratory to perform all inspections, tests, and approvals required by the contract documents, provided, however, the contractor is responsible for testing, inspection and approvals: (1) required by any governing body having jurisdiction over the project; (2) necessary for the owner's and engineer's acceptance of materials, mix designs, or equipment incorporated into the work; (3) required as part of uncovering of defective work caused by the contractor; and (4) if expressly required by the contract documents.[16]

C. Safety-Related Inspection Obligations

Contracts may attempt to impose safety-related inspection obligations in specific circumstances as well. For example, Article 10 of AIA A201 also makes the contractor responsible for initiating, maintaining, and supervising all safety precautions and programs in connection with contract performance, and requires the contractor to take reasonable precautions for the safety and protection of employees and other persons, the work itself (and materials and equipment incorporated or to be incorporated therein), and other property at or adjacent to the site.[17] The ConsensusDOCS 200 also provides that the contractor is responsible for safety precautions and programs and that the contractor must provide the owner with notices required for safety purposes.[18] The EJCDC C-700 broadly states that the contractor "shall be solely responsible for initiating, maintaining and supervising all safety precautions and programs in connection with the Work."[19]

[11]AIA A201, §§ 3.3.3 and 13.5.1 (2007 ed.).
[12]*Id.* at §§ 3.7.1 and 13.5.1. AIA A201, § 4.2.2 (2007 ed.) limits the architect's responsibility for inspections. That provision states that "the Architect will not be required to make exhaustive or continuous on-site inspections to check the quality or quantity of the Work."
[13]*Id.* at § 13.5.3.
[14]EJCDC C-700, § 6.01A (2007 ed.).
[15]*Id.* at § 7.01C.
[16]*Id.* at § 13.03 B.
[17]AIA A201, §§ 10.1 and 10.2.1 (2007 ed.).
[18]*Id.* at § 3.11.
[19]EJCDC C-700, § 6.13 A (2007 ed.).

The contractor is not, however, an insurer and therefore is not responsible for all job-site injuries. For example, in Delaware, an employee of an independent contractor was denied recovery against the general contractor when the employee fell off the roof of a new home. The court stated that a general contractor that supervises job-site safety conditions by making checklists, reporting safety issues to the independent contractor, and even terminating the independent contractor if the safety issues persist, has not assumed a duty to protect the safety of the independent contractor's employees. The independent contractor was responsible for correcting any safety issues. Since the general contractor did not undertake responsibility for implementing safety measures, the employee of the independent contractor recovered nothing from the general contractor.[20]

D. Costs of Inspection

Although the owner may bear its own inspection costs, the contractor generally is required to bear the expense of providing the inspector with the facilities, labor, or material reasonably necessary to perform the test or inspection.[21] Circumstances may exist, however, that would entitle the contractor to be reimbursed for expenses incurred for inspection or testing. For example, if the owner increases the cost of conducting the inspection or test by changing the location or requiring special inspection devices, the contractor may recover additional costs.[22]

The owner generally may examine completed work and require the contractor to remove or tear out defective or nonconforming work. If the work is defective or does not conform to the specification, the contractor must pay the costs of both the inspection and correction of the work. If inspection reveals the work is satisfactory, the contractor is entitled to a price adjustment for the additional costs and a time extension if completion is delayed.[23]

The cost of reinspection generally is assigned to the party whose action or inaction resulted in the reinspection.[24] If, for example, the contractor's work was not sufficiently complete at the time of the original inspection, the contractor should pay the costs of reinspection. Similarly, if the reinspection is the result of an earlier rejection, the contractor is responsible for the additional costs.[25] Before any reinspection, however, the owner must provide a reasonable notification and a reasonable amount of time for the contractor to correct or complete the work.

[20] *Urena v. Capano Homes, Inc.*, 930 A.2d 877 (Del. 2007).

[21] *See Bauunternehmung GmbH & Co. KG*, ASBCA No. 48209, 99–2 BCA ¶ 30,547; *see also* ConsensusDOCS 200, ¶¶ 3.7 and 9.8.1 (2007 ed.) (owner generally responsible for inspection costs); *But see* AIA A201, § 13.5.1 (2007 ed.) (contractor is responsible for inspection and test costs, except inspections and tests that do not become requirements until after contract negotiations are concluded or are required by code or law to be paid by the owner).

[22] *See Gordon H. Ball, Inc.*, ASBCA No. 8316, 1963 BCA ¶ 3925; *Corbetta Constr. Co.*, ASBCA No. 5045, 60–1 BCA ¶ 2613.

[23] *See* ConsensusDOCS 200, ¶ 3.10.1 (2007 ed.).

[24] *Id.*; *see Pride Indus.*, ASBCA No. 55771, 2007 WL 4565891.

[25] *See* ConsensusDOCS 200, ¶ 3.10.1 (2007 ed.); *Bauunternehmung GmbH & Co. KG*, ASBCA No. 48209, 99–2 BCA ¶ 30,547.

The ConsensusDOCS 200 provides that the owner is responsible for inspection costs.[26] But the contractor will be responsible for the cost of correction and retesting if the contractor's work fails a test.[27] Section 12.2.4 of AIA A201 requires the contractor to "bear the cost of correcting destroyed or damaged construction, whether completed or partially completed, of the Owner or separate contractors caused by the Contractor's correction or removal of work that is not in accordance with the requirements of the Contract Documents." Section 12.3 of AIA A201 authorizes the owner to accept nonconforming work instead of having it removed and replaced, and to reduce the contract price to account for such defective work. EJCDC C-700 makes the contractor responsible for the costs to repair defective work, that is, work not in accordance with the contract documents. Section 13.08 of EJCDC C-700 allows the owner to accept defective work, but if the owner so chooses, the contractor is still responsible for:

> all claims, costs, losses, and damages (including but not limited to all fees and charges of engineers, architects, attorneys, and other professionals and all court or arbitration or other dispute resolution costs) attributable to Owner's evaluation of and determination to accept such defective Work . . . and for the diminished value of the Work to the extent not otherwise paid by Contractor pursuant to this sentence.

E. The Owner's Right to Inspect

1. The Right, Not the Duty

Thorough, but reasonable, contemporaneous inspections can be the contractor's best friend. Such inspections allow the owner or its representative to monitor the work periodically and inspect for deviations from the plans and specifications. If deficiencies do exist, and the owner or its representative reasonably objects, performance can be modified to make the work acceptable with minimal cost. In the event of an ambiguous requirement, the owner's acquiescence to the work, as performed by the contractor, may show that the owner agreed with the contractor's interpretation at the time of performance.[28]

Aware of the risks of overlooking defects during inspection, owners have sought to minimize contractors' ability to rely on owners' inspections. For example, AIA A201 provides in Section 9.4.2, with regard to the effect of issuing a Certificate for Payment, that "the issuance of a Certificate for Payment will not be a representation that the Architect has . . . made exhaustive or continuous on-site inspections to check the quality or quantity of the Work . . .."

The standard federal "Inspection of Construction" clause, FAR § 52.246–12, is more specific: "Government inspections and tests are for the sole benefit of the Government and do not [r]elieve the Contractor of responsibility for providing adequate quality control measures . . . [or] [c]onstitute or imply acceptance. . . ." This provision makes it clear that no inspection duty is imposed on the government; rather, the government has the right to inspect should it so desire.

[26]ConsensusDOCS 200, ¶ 3.7 (2007 ed.).
[27]*Id.* ¶ at 3.7.3.
[28]*Milaeger Well Drilling Co. v. Muskego Rendering Co.*, 85 N.W.2d 331 (Wis. 1957).

The owner has an affirmative duty to inspect the work when the contract specifically contemplates or requires that the owner perform certain tests during the work.[29] The owner may lose some of its specific rights and remedies if it fails to inspect or test in accordance with the contract terms, such as the right to reject items or have defects corrected if the contractor's work fails a test, when a reasonable inspection would have uncovered such defects.

2. Scope of Inspection

The scope of the owner's inspection rights often leads to disputes regarding the interpretation of specifications, quality of workmanship, and other "quality" determinations. The scope of an owner's inspection is usually set forth in the contract. Inspections must be reasonable in scope when no specific inspection requirements are set forth. In federal government work, the scope of the inspection requirements depends on an analysis of the type of work to be delivered.

The standard inspection clause generally controls construction contracts. The FAR, however, establishes four categories of contract quality requirements: (1) reliance on the contractor's existing quality assurance systems as a substitute for government inspection and testing for commercial items; (2) government reliance on the contractor to perform all inspections and testing; (3) "standard" inspection requirements contained in the standard clauses, calling for inspections to be performed by both the contractor and the government; and (4) "higher-level quality requirements" prescribing more stringent inspections to be performed by the government.[30]

In most construction projects, the government will perform either the standard inspection or the higher-level quality inspection. The requirements for the standard inspection are set forth in the inspection clause, which provides that: (1) the contractor must establish an inspection system; (2) the government may inspect during performance; and (3) the contractor must maintain inspection records.[31] The higher-level quality inspection requirements generally are set forth in special supplementary contract clauses implementing stricter quality control.[32]

Even if the scope of inspections is set forth in the contract, as a general rule, the federal government contract may impose an unspecified alternative test as a basis for determining contract compliance. The new test must reasonably measure contract compliance. If the "specified" test can be viewed as establishing a standard of performance, however, a different test increasing the level of performance cannot be substituted without a change to the contract price.[33]

Not only does the government have the right to inspect at all places and times; the government also has the right to reinspect the same performance. Generally,

[29]*See Cone Bros. Contracting Co.,* ASBCA No. 16078, 72–1 BCA ¶ 9444.
[30]FAR § 46.202, *et seq.*
[31]FAR § 46.202–3.
[32]FAR §§ 46.311 and 52.246–11.
[33]*Sw. Welding & Mfg. Co. v. United States,* 413 F.2d 1167 (Ct. Cl. 1969); *Roda Enters., Inc.,* ASBCA No. 22323, 81–2 BCA ¶ 15,419.

the government may conduct reasonable, continuing inspections at any time before acceptance.[34] There are exceptions regarding the government's right to reinspect. Multiple inspections cannot be wholly inconsistent. Subjecting the contractor to inconsistent inspections amounts to an unreasonable interference with the contractor's work and entitles the contractor to compensation.[35]

F. Rejection and Correction

After inspection, an owner has the right to accept the performance, reject the performance if it is nonconforming, require correction of nonconforming performance, or, in appropriate circumstances, terminate the contract for default.[36] To enforce its rejection/correction remedy, the federal government must provide the contractor with notice of the alleged discrepancy within a "reasonable time" after discovery of the defects. The notice must include the reasons for the rejection.[37] When the government fails to provide the reasons for the rejection in the initial notice and the contractor is prejudiced by such failure, the rejection can be overturned as ineffective. Furthermore, a failure to reject the performance in a reasonable time can be interpreted as an implied acceptance of the contractor's performance.[38]

If the federal government rejects performance, ordinarily it must give the contractor an opportunity to correct the defects if they can be cured within the contract schedule.[39] If the contracting officer orders correction instead of rejecting and requiring replacement of the work, the contractor is entitled to a reasonable time to make the correction, without regard to the original schedule.[40]

If the contractor fails to timely replace or correct rejected work, the federal government has three remedies. The government can: (1) terminate the contract for default and reprocure the supplies, services, or construction; (2) replace or correct the defective supplies, services, or construction by contract or by using government resources, at the contractor's expense, under the inspection clause; or (3) retain the nonconforming supplies, services, or construction and reduce the contract price based on the difference in value between the work as delivered and the work contemplated by the contract.[41]

G. Limitation on Owner's Inspections

Despite the owner's broad inspection rights, improper inspections can give rise to certain rights and remedies on the contractor's part—if, for example, "constructive

[34] *Elec. Contracting Corp. of Guam,* ASBCA No. 34337, 90–3 BCA ¶ 23,003.

[35] *H & S Mfg. v. United States,* 66 Fed. Cl. 301 (2005); *WRB Corp. v. United States,* 183 Ct. Cl. 409 (1968).

[36] FAR §§ 46.407(a) and 52.246–2(f).

[37] FAR § 46.407(g).

[38] *Id.*

[39] FAR § 46.407(b).

[40] *Baifield Indus., Div. A-T-O, Inc.,* ASBCA No. 14582, 72–2 BCA ¶ 9676.

[41] FAR §§ 52.249–8; 52.249–10; 52.246–2(h); 52.246–4(e); 52.246–12(g); 52.246–2(b); 52.246–4(d); 52.246–12(f).

changes" to the work or delays and disruptions result from the owner's inspections. Several issues must be addressed to determine whether an improper inspection might be a "constructive change."

1. Authority

Differences in opinion regarding the standards of performance required by the contract or the correct inspection test to be used often cause contractors to claim they are being required to perform extra work. Even if the contractor's interpretation was correct and the inspector was wrong, the contractor still may be confronted with the argument that the inspector lacked the authority to change the contract and bind the owner.

The issue of the inspector's authority can be complicated. Inspectors seldom have authority to change the contract requirements, but they do have authority to reject work. It is usually held, therefore, that an erroneous rejection is within the inspector's authority and can form the basis of a contract extra. This assumes, of course, proper notice by the contractor and "performance under protest."[42] To avoid disputes over authority, the best procedure is to routinely provide written notice to an authorized owner-representative whenever the actions of an inspector are causing performance delay and cost beyond that contemplated by the contract.

The federal government frequently argues that its inspectors lack the authority to effect a constructive change. In one case, the government's specifications for brick were strict, and the contractor's chief mason complied with the requirements by rejecting between 20 and 25 percent of the brick. The manufacturer, the government's on-site representative, and the architect agreed that the contractor was being overly critical and told the brick mason to stop rejecting brick. Thereafter, the government rejected the brickwork due to an undesirable basket-weave appearance and directed the contractor to remove and replace the brick.

The contractor demanded an equitable adjustment for its costs in removing and replacing the brick. The government argued that its on-site representative was not authorized to direct the contractor to stop rejecting brick. The board of contract appeals held that the inspector's authority depends on the facts and conduct of each case and that the contracting officer can authorize technical personnel (such as inspectors) to give guidance or instruction about specification problems. Therefore, the government was liable for the constructive change that caused the placement and removal of the defective brick.[43]

2. Higher Standards of Performance

The owner may perform any reasonable inspection. But if the owner requires a higher standard of performance through the use of inspection procedures or tests more stringent than those called for by the contract or inconsistent with industry practice, the

[42] C. P. Jhong, Annotation, *Effect of Stipulation, in Private Building or Construction Contract, That Alterations or Extras Must Be Ordered in Writing,* 2 A.L.R. 3d 620 (1965).
[43] *Jordan & Nobles Constr. Co.,* GSBCA No. 8349, 91–1 BCA ¶ 23,659.

contractor should be entitled to additional compensation.[44] Similarly, if the inspector requires the contractor to use materials or construction methods that the contract does not require and that cost more than the contractor's chosen materials or methods, a compensable change may result.[45]

Problems may occur where the contract does not clearly define either the standard of workmanship required of the contractor or the standard of inspection to be employed. In such cases, inspectors often will rely on industry standards and trade customs, or even on subjective standards such as "skillful and workmanlike" construction. Where the use of such criteria actually requires a level of performance in excess of that reasonably contemplated when the parties entered into the contract, the contractor may be entitled to extra compensation.

For example, an inspector's use of straightedges and other measuring tools to check stud alignment has been held to amount to a change when no such method was specified in the contract and the normal industry practice was to check such alignment by visual inspection.[46]

3. Rejecting Acceptable Work

An inspector's wrongful rejection of acceptable work involves issues similar to the imposition of increased standards of performance. If work that should have been accepted is "corrected" to a higher standard of quality and additional costs are incurred in the process, a compensable change has occurred.[47]

Where specifications are ambiguous, an inspector's silent acquiescence while the contractor performs in accordance with its own reasonable interpretation of the performance standards may establish that the contractor's approach was reasonable and the work acceptable.[48] Also, if the owner submits to the contractor what purports to be a complete list of defects in the work, the owner may later be prevented from rejecting work that had been corrected pursuant to such list on the grounds that its list amounted to a binding interpretation of ambiguous specifications.[49]

Generally, the owner can reject defective work at any time before acceptance of the work, and an inspector's observation of nonconforming work does not necessarily preclude later rejection.[50] If an owner's delay in rejecting nonconforming work substantially prejudiced the contractor, however, the owner may be estopped, or prevented, from later rejecting such work.[51] If the contractor has given clear notice of its interpretation of the standards and methods of performance that were used and that later became the subject of the dispute, then a finding of estoppel is more likely.

[44]*See Eris Painting & Gen. Corp.,* ASBCA No. 27803, 84–1 BCA ¶ 17,148.

[45]*See, e.g., Randolph & Co.,* ASBCA No. 52953, 03–1 BCA ¶ 32,080.

[46]*See Williams & Dunlap,* ASBCA No. 6145, 63–1 BCA ¶ 3834.

[47]*See Acme Missiles & Constr. Corp.,* ASBCA No. 13671, 69–1 BCA ¶ 7698; *Byson v. City of L.A.,* 308 P.2d 765 (Cal. Ct. App. 1957); *Denton Constr. Co. v. Mo. State Highway Comm'n,* 454 S.W.2d 44 (Mo. 1970).

[48]*See Dondlinger & Sons Constr. Co.,* ASBCA No. 13651, 70–2 BCA ¶ 8603.

[49]*See Frederick P. Warrick Co.,* ASBCA No. 9644, 65–2 BCA ¶ 5169.

[50]*Forsberg & Gregory, Inc.,* ASBCA No. 18457, 75–1 BCA ¶ 11,293.

[51]*See Baltimore Contractors, Inc.,* ASBCA No. 15852, 73–2 BCA ¶ 10,281.

4. Delay and Disruption

Each construction contract includes an implied obligation on the owner not to unduly delay or hinder the contractor's work. This duty extends to the owner's exercise of its inspection rights. The standard federal inspection clause for construction contracts, FAR § 52.246–12(e), includes this as an express obligation where it states: "[T]he Government shall perform all inspection and tests in a manner that will not unnecessarily delay the work."

This principle is similarly recognized in the AIA A201 Section 9.10.1, which requires that the architect, upon receipt of a final payment application and the contractor's written notice that the work is ready for final inspection and acceptance, "will promptly make such inspection." Paragraph 9.6.1 of ConsensusDOCS 200 contains a similar requirement for prompt inspections by the owner and its design professional (architect or engineer). Likewise, Section 14.06 of EJCDC C-700 requires the engineer to "promptly make a final inspection with Owner and Contractor. . . ." Where the owner has unreasonably delayed or interfered with the contractor in conducting the final inspection, the contractor may be entitled to a time extension and recovery of additional costs or breach of contract damages.

The surrounding facts and circumstances will determine whether a particular delay was unreasonable. The basic test is whether the inspector's actions were reasonably necessary to protect the owner's interests or whether the owner's legitimate objectives could have been accomplished by some other, less disruptive, means.[52]

Compensable delays also may be caused by multiple and inconsistent inspections.[53] Likewise, the owner's failure to make a timely inspection after a request by the contractor may result in owner liability.[54] What may be a timely inspection in one situation can amount to an unreasonable delay in another. For example, in one case, a government inspection three days after the contractor's request was held to be an unreasonable delay, but in another case a 10-day delay was not sufficient to make the inspection untimely.[55]

Unreasonable delays in reviewing and approving shop drawings, equipment submittals, material submittals, plans of operations, and the owner's determinations as to what corrective action is required when defects are discovered have all been held to entitle the contractor to relief. Likewise, when an inspector interferes with a contractor's employees, disrupts the performance sequence, or otherwise causes the work to be performed less efficiently, the contractor may be entitled to be reimbursed for the cost of resulting extra work. Such actions may also be deemed a breach of contract.[56]

[52]*See S. S. Silberblatt, Inc. v. United States,* 433 F.2d 1314 (Ct. Cl. 1970).
[53]*See WRB Corp. v. United States,* 183 Ct. Cl. 409 (1968); *see also H & S Mfg., Inc. v. United States,* 66 Fed. Cl. 301 (2005).
[54]*See Larco-Indus. Painting Corp.,* ASBCA No. 14647, 73–2 BCA ¶ 10,073.
[55]*Compare Kingston Bituminous Prod. Co.,* ASBCA No. 9964, 67–2 BCA ¶ 6638 *with Fullerton Constr. Co.,* ASBCA No. 11500, 67–2 BCA ¶ 6394.
[56]*See WRB Corp. v. United States,* 183 Ct. Cl. 409 (1968).

H. Inspection by the Design Professional or Inspector

The party inspecting the work must perform such inspections adequately and without negligence. Architects, engineers, construction managers, and government inspectors may be liable to the owner, contractor, or other third parties as a result of failing to fulfill their inspection duties.

The contract's inspection standards should be construed so as to reconcile inconsistencies. Where one contract provision called for inspections to be performed by sampling supplies by lot while another provision called for all supplies to meet specific standards, the court reconciled the alleged differences in the inspection standards, finding that the inspections of the lots did not override the rights of the owner and contractor to reject individual materials that did not meet the specific standards.[57]

Generally, a design professional is required to visit the site at regular intervals but is not required to perform exhaustive or continuous on-site inspections to check the quality or quantity of the work.[58] The design professional also generally must inform the owner of the work's progress and guard the owner against defects and deficiencies in the work.

If an inspecting party such as an architect or engineer fails to make adequate periodic inspections during the work, an owner may challenge whether that architect or engineer is immune from liability by virtue of a contract provision stating that the architect or engineer is not responsible for the contractor's acts or omissions. In one case, the court noted that the architect had to visit the site periodically to be familiar with the progress and quality of the work, keep the owner informed about the work's progress and quality, and guard the owner against defects in the work. Furthermore, the architect's obligation to issue certificates of payment required familiarity with both quantity and quality of work. Therefore, the exculpatory provision excusing the architect from responsibility for construction methods and for the acts or omissions of the contractor did not immunize the architect from liability flowing from a breach of its duty to the owner.[59]

In another case, a design professional was held liable to an owner and the contractor's surety for negligently inspecting a roof. Even after repeated warnings by a roofing expert that the roof was not being installed in accordance with the contract specifications, the design professional's resident inspector informed the owner that the roof was fine and that "you don't have to worry about it." In reliance on the inspector's assurances, the owner accepted the building and released all payments to the contractor. A few months later, the roof began to leak and the contractor's attempts to solve the problem were unsuccessful.

The court ruled that the design professional had a duty to inspect the roof construction and to protect the owner against poor work by the contractor. A design

[57]*Seabury Constr. Corp. v. Jeffrey Chain Corp.*, 289 F.3d 63 (2d Cir. 2002).

[58]AIA A201, § 4.2.2 (2007 ed.) states that "the Architect will not be required to make exhaustive or continuous on-site inspections to check the quality or quantity of the Work." *See also* EJCDC C-700, § 9.02 A (2007 ed.) (containing identical language).

[59]*Diocese of Rochester v. R. Monde Contractors,* Inc., 562 N.Y.S.2d 593 (N.Y. Sup. Ct. 1989), *aff'd*, 561 N.Y.S.2d 659 (N.Y. App. Div. 1990).

professional is required to exercise ordinary professional skill and diligence, and this duty is nondelegable. Since the design professional breached its obligation to the owner under the above circumstances, the design professional was liable.[60]

I. Inspection by the Contractor

The contractor's inspection duties in the routine performance of a construction contract typically include not only the inspection of the work in place, but an inspection of job conditions, including job cleanup, potential safety hazards, and monitoring work progress and schedule. In addition to inspecting its own work, the contractor must inspect the work of its subcontractors and material suppliers.

Many construction contracts impose specific duties on the contractor to perform such inspections. (See **Section II.B.2** of this chapter.) Even if no express contractual duty applies, prudence dictates that such inspections be carried out routinely. A regular process should be implemented for reporting and exchanging information in order for the contractor to promptly, expeditiously, and economically complete the project. Additionally, contractors generally cannot rely on inspection provisions allowing owners to perform inspections to relieve the contractor of its duty to perform its work properly if the owner's inspection fails to detect deficiencies in the contractor's work.[61]

The contractor also may have to obtain test results on work in place or materials to be used. Normally such tests are obtained through designated independent testing laboratories. For example, one usually must make test cylinders of structural concrete placed. Sometimes such tests are prescribed by the specifications, and in other cases they are imposed by industry standards incorporated in the contract documents. These inspections not only satisfy the contractor's obligations to the owner but also help the contractor monitor its own work. If a failure occurs, such test results can also serve as useful evidence in any corresponding dispute.

In federal government construction, the standard federal inspection clause places primary responsibility for contract compliance on the contractor. In addition, most federal agencies have included provisions in construction contracts that require the contractor to conduct inspections and ensure that the work complies with the plans and specifications. For example, one clause provides that "[t]he Contractor shall maintain an adequate inspection system and perform such inspections as will ensure that the work performed under the contract conforms to contract requirements. The Contractor shall maintain complete inspection records and make them available to the Government."[62] Similarly, the contract and applicable regulations also may include various contractor record-keeping and certification requirements. In one case, the board of contract appeals strictly interpretated such a provision.[63]

[60] *U.R.S. Co., Inc. v. Gulfport-Biloxi Reg'l Airport Auth.*, 544 So. 2d 824 (Miss. 1989). *But see Watson, Watson, Rutland/Architects, Inc. v. Montgomery County Bd. of Educ.*, 559 So. 2d 168 (Ala. 1990).

[61] *See Blumenthal Kahn Elec. Ltd. P'ship v. Bethlehem Steel Corp.*, 708 A.2d 1 (Md. Ct. Spec. App. 1998).

[62] FAR § 52.246–12(b).

[63] *See Acorn Specialty & Supply Co.*, GSBCA No. 7577, 85–2 BCA ¶ 17,995.

The contract required the contractor to designate an individual who would be responsible to specifically test each unit before delivery and to issue a certification. Instead, the contractor relied on the supplier's testing procedure and certification, which did not comply with the contract's requirements. The contracting officer terminated the contractor for default because of the contractor's failure to provide the required inspection. The default termination was upheld on the grounds that the government was entitled to the specific type of inspection set forth in the contract.

II. ACCEPTANCE

A. Overview

Acceptance of a construction project has great significance. Acceptance generally limits the owner's ability to complain of defects and reject work. Acceptance also may commence the running of warranties. Contractors and owners often dispute when the project is complete, which frequently results in the owner withholding its formal acceptance as well as some amount of money. The theory of constructive acceptance, however, has evolved to help contractors avoid the harsh consequences of the unreasonable withholding of formal acceptance. A theory closely related to constructive acceptance is substantial completion. This theory recognizes the point at which the owner has received the benefit of the bargain—usually in the form of the owner's ability to occupy and use the project for its intended purpose—even though every detail may not have been completed or corrected.[64]

It is a well-recognized rule of contract law that a party entitled to performance may waive strict performance.[65] In the context of a construction contract, this means that the owner may acquiesce to the contractor's failure to perform according to the strict terms of the agreement. Such waiver or acquiescence often is established through acceptance. Courts often are reluctant, however, to bar an owner's right to recover for defective construction "merely" because of "acceptance."[66]

The standard Inspection of Construction clause in federal government contracts indicates that "[a]cceptance shall be final and conclusive except as regards latent defects, fraud, gross mistakes amounting to fraud, or the Government's rights under any warranty or guarantee."[67]

In private construction, final payment typically constitutes a waiver of claims, except those reserved in writing at the time of final payment and claims relating to liens or similar encumbrances, warranties, defective work, and latent defects.[68]

[64]*See Sletto v. Wesley Constr. Inc.*, 733 N.W.2d 838 (Minn. Ct. App. 2007).

[65]*See Aguiar v. Segal,* 167 S.W.3d 443 (Tex. App. 2005).

[66]*Davidge v. H. H. Constr. Co.,* 432 So. 2d 393 (La. Ct. App. 1988).

[67]FAR § 52.246–12(i).

[68]*See, e.g.,* ConsensusDOCS 200, ¶ 9.8.6 (2007 ed.); AIA A201, § 9.10.4 (2007 ed.); EJCDC C-700, § 14.09 (2007 ed.).

B. Types of Acceptance: Formal versus Constructive

Under private contracts, acceptance typically occurs when there has been final completion and final payment.[69] In federal contracting, execution of the proper form by an authorized government representative may constitute acceptance. Further, where the owner has taken no positive action to accept or reject, a failure to reject noncomplying work within a reasonable time may constitute a "constructive" acceptance.[70]

In *Tranco Industrial Tires, Inc.,*[71] constructive acceptance was found after the government failed to inspect the painting of fuel tanks for three months. The board ruled that the proper standard for timely acceptance or rejection is "a reasonable time for prompt action under the circumstances." In this case, the board held that a three-month delay from paint sample approval to final inspection was unreasonable and that, while a change in contracting officers justified a two-week delay, it did not justify the balance of the government's tardiness.

In contrast, it has long been recognized that mere occupation and use of a structure by the owner does not constitute acceptance or waiver of defects therein.[72] This rule applies as well when the owner has taken possession of the project with the express understanding that defects will be remedied at a later date.[73]

The ConsensusDOCS 200 provides that occupancy or use of a completed or partially completed portion of the project constitutes substantial completion of the portion of the work used or occupied.[74] The contractor is still responsible to complete uncompleted items.[75]

The AIA A201 contains a section regarding partial occupancy or use of the project by the owner, which provides that "[u]nless otherwise agreed upon, partial occupancy or use of a portion or portions of the Work shall not constitute acceptance of Work not complying with the requirements of the Contract Documents."[76]

The EJCDC C-700 allows for the owner's partial utilization of "any substantially completed part of the Work which has specifically been identified in the Contract Documents, or which Owner, Engineer, and Contractor agree constitutes a separately functioning and usable part of the Work[,]" so long as such portion of work does not significantly interfere with the contractor's remaining work.[77] Further, the owner is prohibited from partially utilizing the work until it has put in place contractually required property insurance.[78]

[69] *See* ConsensusDOCS 200, ¶ 9.8 (2007 ed.); AIA A201, § 9.10 (2007 ed.); EJCDC C-700, § 14.07 (2007 ed.).
[70] *See Havens Steel Co. v. Randolph Eng'g Co.,* 613 F. Supp. 514 (W.D. Mo. 1985), *aff'd,* 813 F.2d 186 (8th Cir. 1987).
[71] ASBCA No. 26305, 83–2 BCA ¶ 16,679.
[72] *See Granite Constr. Co.,* ENGBCA No. 4642, 89–3 BCA ¶ 21,946.
[73] *See Brouillette v. Consol. Constr. Co. of Fla., Inc.,* 422 So. 2d 176 (La. Ct. App. 1982).
[74] ConsensusDOCS 200, ¶ 9.7 (2007 ed.).
[75] *Id.* at ¶ 9.6.4.
[76] AIA A201, § 9.9.3 (2007 ed.).
[77] EJCDC C-700, § 14.05 (2007 ed.).
[78] *Id.; see also* EJCDC C-700, §§ 5.06 & 5.10 (detailing owner property insurance insurance requirements).

In federal construction, the concept of constructive acceptance is better developed than in private construction and is recognized in the procurement regulations, which define acceptance as "the act of an authorized representative of the Government by which the Government, for itself or as agent of another, assumes ownership of existing identified supplies tendered or approves specific services rendered as partial or complete performance of the contract."[79]

Similarly, the "Use and Possession Prior to Completion" clause prescribed by FAR §§ 36.511 and 52.236–11, provides that "the Government shall have the right to take possession of or use of any completed or partially completed part of the work. . . ." but that "such possession or use shall not be deemed an acceptance of any work under the contract." In certain cases where that clause was not employed, however, government use, possession, or control of the project, coupled with a failure to indicate that the work was not complete, has been deemed to amount to an acceptance.[80] In federal procurement, constructive acceptance, as with formal acceptance, commences the contract's warranty period.[81]

One should also realize that individual actions under the contract, standing alone, may not constitute an acceptance. For example, the boards have held that:

- Payment, by itself, does not constitute an implied acceptance.[82]
- Visits by government representatives during fabrication of equipment do not constitute inspection and acceptance.[83]
- The government's failure to inspect is not an implied acceptance waiving strict compliance.[84]
- Mere acceptance of the delivery of supplies is not an implied acceptance.[85]

The specific facts and circumstances must be viewed in light of the applicable contract language and the surrounding circumstances to determine if the owner's actions amount to an acceptance.

C. Authority as an Element of Constructive Acceptance

The actual authority of the individual whose action or inaction is being relied on may determine whether there has been a formal or constructive acceptance. Acceptance in federal government work will be binding only if made by a person authorized to accept on behalf of the government.[86] In private agency law, an employer may be bound by an employee under the legal theory of apparent authority if the employer

[79]FAR § 46.101.
[80]*See Bell & Flynn, Inc.*, ASBCA No. 11038, 66–2 BCA ¶ 5855.
[81]*See Paul Tishman Co.*, GSBCA No. 1099, 1964 BCA ¶ 4256.
[82]*See Abney Constr. Co.*, ASBCA No. 26358, 83–1 BCA ¶ 16,246; *G. M. Co. Mfg.*, ASBCA No. 5345, 60–2 BCA ¶ 2759.
[83]*See J. W. Bateson Co.*, GSBCA No. 3157, 71–1 BCA ¶ 8820.
[84]*See Waterbury Co.*, ASBCA No. 6634, 61–2 BCA ¶ 3158.
[85]*See Lox Equip. Co.*, ASBCA No. 8518, 1964 BCA ¶ 4469.
[86]*See Inter-Tribal Council of Nev., Inc.*, IBCA No. 1234–12–78, 83–1 BCA ¶ 16,433.

has permitted the employee to assume authority or has held the employee out as possessing the requisite authority.[87]

Recognizing the importance of effective government control over the conduct of its agents, the boards and courts generally have rejected the apparent authority rule, holding that actual authority is required to bind the government.[88] Even government personnel with official-sounding titles such as contract specialists, negotiators, and administrators, who handle the government's daily contracting activities, generally do not have authority to order additional work or otherwise obligate the government.[89]

D. Limitations on the Finality of Acceptance

In order for the owner to waive strict performance or to acquiesce in the deviation from the contract documents through acceptance, the owner or its authorized representative generally must know of the defect or the deviation.[90] Courts have ruled that an owner waives deviations from the contract where the owner should have known of the deviations. Whether particular acts or conduct amount to an acceptance, and thus a waiver of strict performance, is a fact question that depends on the circumstances of each case.

On virtually all construction projects, defects that are not apparent and that cannot be discovered until a later date (i.e., latent defects) are not deemed to have been accepted. Also, on federal construction projects, where a contractor knowingly misrepresents the condition or quality of its work with the intent to deceive, the government is considered to have been induced to accept defective work as a result and may recover the costs of repairing such defects from the contractor.[91]

When the contractor makes gross mistakes or misrepresents a material fact without the intent to deceive the government, the contractor still may be liable for such defects despite government acceptance of the work. One such case involved contractor changes to a drawing that had been previously approved by the government, and the contractor's subsequent failure to alert government officials to the change before the drawing was used in checking the contractor's production run.[92]

Closely related to waiver of defects by acceptance is the issue of whether progress payments constitute a waiver. Courts appear evenly divided. Some state courts have ruled that partial payments constitute a waiver of defects while other states have ruled the opposite way.[93] The argument for waiver of defects by partial payment is much more persuasive when the defects were known at the time of payment, when the owner or its representative failed to protest, and when there was no express agreement

[87]Restatement (Second) of Agency § 8 (1958).

[88]*See Winter v. Cath-dr/Balti Joint Venture,* 497 F.3d 1339 (Fed. Cir. 2007), *reh'g denied, reh'g en banc denied.*

[89]*See, e.g., Gen. Elec. Co. v. United States,* 412 F.2d 1215 (Ct. Cl. 1969), *reconsideration denied,* 416 F.2d 1320 (Ct. Cl. 1969).

[90]*See Upchurch Plumbing, Inc. v. Greenwood Utils. Comm'n,* 964 So. 2d 1100 (Miss. 2007).

[91]*See Nasatka & Sons, Inc.,* IBCA No. 1157–6–77, 79–2 BCA ¶ 14,064.

[92]*See Catalytic Eng'g & Mfg. Corp.,* ASBCA No. 15257, 72–1 BCA ¶ 9342.

[93]*Compare Coats v. Or. Dep't of Transp.,* 927 P.2d 108 (Or. Ct. App. 1996), *with Guschl v. Schmidt,* 63 N.W.2d 759 (Wis. 1954).

to remedy them. Generally, however, the contract will provide that the making of a progress payment does not constitute acceptance of noncompliant work.[94]

The owner's waiver or acceptance of defective performance may preclude its refusing to pay the contractor the reasonable value of the work or its price according to the contract terms.[95] Similarly, where the owner accepts the work as fully compliant with the contract, it is generally not possible for the owner later to maintain an action against the contractor to recover for deviations from the contract documents or to recoup such damages in an action brought by the contractor for compensation.[96]

E. Contract Provisions Related to the Finality of Acceptance

Contracts often include clauses aimed at qualifying the significance of acceptance. For example, ConsensusDOCS 200 provides that final payment does not constitute a waiver of the owner's claims reserved in writing at the time of final payment and claims relating to liens or similar encumbrances, warranties, defective work, and latent defects.[97] AIA A201 Section 9.10.4, quoted above, provides that final acceptance and the making of final payment do not constitute waiver by the owner of any claims work that does not conform to contract requirements. Finally, EJCDC C-700 states that the making of final payment shall constitute "a waiver of all Claims by Owner against Contractor, except Claims arising from unsettled Liens, from defective Work appearing after final inspection. . ., from failure to comply with the Contract Documents or the terms of any special guarantees specified therein, or from Contractor's continuing obligations under the Contract Documents."[98]

In the absence of a nonwaiver contract provision, a waiver may be implied if the owner or its representative had an opportunity during the progress of the work to inspect and reject work or materials that obviously did not comply with the contract requirements but failed to do so.[99] In such a case, a failure to object during the progress of the work may amount to a waiver of the defect. The mere presence of the owner or the owner's representative at the site, however, does not necessarily constitute a waiver, such as when a defect is not readily discoverable.[100]

Similarly, under the standard "Inspection of Construction" clause in federal government contracts, also quoted above, there are several exceptions to the finality of acceptance, such as latent defects, fraud, and gross mistakes amounting to fraud. In addition, the government has its rights under the warranty or guarantee provisions of the contract. (See **Section III.A** of this chapter.)

[94] *See, e.g.*, AIA A201, § 9.6.6 (2007 ed.).

[95] *See Milaeger Well Drilling Co., Inc. v. Muskego Rendering Co.*, 85 N.W.2d 331 (Wis. 1957).

[96] *City of Gering v. Patricia G. Smith Co.*, 337 N.W.2d 747 (Neb. 1983); *John Price Assocs., Inc. v. Davis*, 588 P.2d 713 (Utah 1978).

[97] ConsensusDOCS 200, ¶ 9.8.6 (2007 ed.).

[98] AIA A201, § 9.10.4 (2007 ed.); EJCDC C-700, § 14.09A.1 (2007 ed.); *see also* **Section III.A** of this chapter.

[99] *See Fla. Ice Mach. Corp. v. Branton Insulation, Inc.*, 290 So. 2d 415 (La. Ct. App. 1974); *see also Brand S. Roofing*, ASBCA No. 24688, 82–1 BCA ¶ 15,513.

[100] *See Shaw v. Bridges-Gallagher, Inc.*, 528 N.E.2d 1349 (Ill. App. Ct. 1988).

F. Substantial Completion

Substantial completion has been defined in numerous ways and has been the subject of extensive litigation. Generally, when the owner has the use and benefit of the contractor's work and the project is capable of being used for its intended purpose, substantial completion has occurred.[101]

Section 9.8.1 of the AIA A201 defines substantial completion as "the stage in the progress of the Work when the Work or a designated portion thereof is sufficiently complete in accordance with the Contract Documents so that the Owner can occupy or utilize the Work for its intended use." Similarly, paragraph 2.4.17 of ConsensusDOCS 200 provides that substantial completion occurs when the contractor's work is "sufficiently complete in accordance with the Contract Documents so that the Owner may occupy or utilize the Project, or a designated portion, for the use for which it is intended, without unscheduled disruption." Likewise, EJCDC C-700, Section 1.01 A.44 defines substantial completion as:

> The time at which the Work (or a specified part thereof) has progressed to the point where, in the opinion of Engineer, the Work (or a specified part thereof) is sufficiently complete, in accordance with the Contract Documents, so that the Work (or a specified part thereof) can be utilized for the purposes for which it is intended.

The substantial completion date is a crucial date in construction claims and disputes for three reasons:

(1) Once substantial completion has been attained, the owner has received essentially what it bargained for and the contractor has substantially performed its obligations; thus the contractor usually is entitled to the balance of the contract price less the cost of remedying minor defects.

(2) Liquidated damages generally stop after this date.[102]

(3) Substantial completion may trigger the warranty period under the contract.[103]

In many states, substantial completion marks the commencement of a special limitation period, sometimes called a "statute of repose," for actions against persons performing or furnishing design, planning, supervision, observation of construction, or construction of any improvement to real property. Having substantially completed its contractual obligations, the contractor also may recover on the contract and need not rely on the equitable theory of *quantum meruit* to recover for the value of the work performed and materials supplied, or some other equitable theory.[104]

[101]*See Sletto v. Wesley Constr., Inc.*, 733 N.W.2d 838 (Minn. Ct. App. 2007).

[102]*See Seacoast Builders Corp. v. Rutgers*, 818 A.2d 455 (N.J. Super. Ct. App. Div. 2003); *Phillips v. Ben M. Hogan Co.*, 594 S.W.2d 39 (Ark. Ct. App. 1990).

[103]*See, e.g.*, ConsensusDOCS 200, ¶ 9.6.3 (2007 ed.).

[104]*See RAJ Partners, Ltd. v. Darco Constr. Corp*, 217 S.W.3d 638 (Tex. App. 2006).

G. Final Completion

Upon final completion (generally defined as when punch-list work has been completed after substantial completion and the project is ready for final inspection and acceptance), the contractor's work generally should be finally accepted.[105] At final completion, in some states, the contractor will be relieved of liability to third parties with which it did not have a contract.[106] This doctrine, often referred to as the doctrine of completion and acceptance or the accepted work doctrine, is based on the concept of avoiding excessive litigation.[107] Even in those states that still accept the doctrine of completion and acceptance, courts have found exceptions, including where the defect is imminently dangerous, the defect is hidden such that it cannot be discovered through a reasonable inspection, the contractor knows of the defect but has failed to reveal it, or the contractor's negligence was not part of the contract work.[108]

Other states have rejected the completion and acceptance doctrine, finding that third parties are not barred from suing contractors for injuries after final completion.[109] These courts generally have found that it is unreasonable for an owner's acceptance to prevent third parties from pursuing claims against contractors whose conduct leads to an injury to a third party.

H. Revocation of Acceptance

Revocation of acceptance may be available under special circumstances. For federal government contracts, the government may revoke acceptance if it discovers latent defects, fraud, or gross mistakes amounting to fraud. "Latent defects" are flaws in the contractor's work that existed when the government accepted the work but were not discovered and could not have been discovered by a reasonable inspection.[110] Latent defects must be distinguished from "patent defects," or defects that had been discovered by the time of acceptance or that should have been discovered through a reasonable and competent inspection. Although latent defects give the government the right to revoke acceptance, the government only has warranty rights against the contractor after acceptance of patent defects.

[105]*See, e.g.,* ConsensusDOCS 200, ¶ 9.8 (2007 ed.); AIA A201, § 9.10.1 (2007 ed.); EJCDC C-700, § 14.07 (2007 ed.).

[106]*See, e.g., First Church of Christ Scientist v. City of Seattle,* 964 P.2d 374 (Wash. Ct. App. 1998); *Fisher v. State Highway Comm'n of Mo.,* 948 S.W.2d 607 (Mo. 1997); *David Allen Co., Inc. v. Benton,* 398 S.E.2d 191 (Ga. 1990).

[107]*See, e.g., First Church of Christ Scientist v. City of Seattle,* 964 P.2d 374 (Wash. Ct. App. 1998).

[108]*See, e.g., Irwin v. Hoover Treated Wood Prods., Inc.,* 906 F. Supp. 530 (E.D. Mo. 1995) (imminently dangerous defect, hidden defect, and superior contractor knowledge exceptions); *First Church of Christ Scientist v. City of Seattle,* 964 P.2d 374 (Wash. Ct. App. 1998) (negligence in noncontract work). *See also David Allen Co., Inc. v. Benton,* 398 S.E.2d 191 (Ga. 1990) (citing *Shetter v. Davis Bros., Inc.,* 293 S.E.2d 397 (Ga. Ct. App. 1982)) (as to contractor that follows plans without negligence and whose work is accepted, there is only one exception: when contractor is an expert in the design of the type of work being done).

[109]*See, e.g., Suneson v. Holloway Constr. Co.,* 992 S.W.2d 79 (Ark. 1999); *Pierce v. ALSC Architects,* P.S., 890 P.2d 1254 (Mont. 1995).

[110]*See Moreland Corp. v. United States,* 76 Fed. Cl. 268 (2007).

Upon discovery of a latent defect, the government may revoke its previous acceptance and demand the contractor take corrective action. Alternatively, the government may correct the problem and charge the costs to the contractor. The question therefore typically turns on whether the defect should have been discovered through a reasonable inspection performed with ordinary care. What is "reasonable" varies with the facts and circumstances of each case. For example, welding defects in a heat distribution system that later caused leaks were held to be latent in one case.[111] Yet in another contract where the government was responsible for conducting X-ray inspections of structural steel, welding defects were held not to be latent.[112]

Contractor quality control inspection requirements may help determine whether a defect is latent. A defect, even if discoverable through a reasonable government inspection, may be found to be latent if the contractor has the primary responsibility to inspect.[113]

The burden of proof is on the government to prove that the defect exists, was latent, and was unknown at the time of acceptance. Finally, to have a cause of action against the contractor, the government must show injury and a connection between the defect and the injury. The burden of proving each of these elements can be difficult. The passage of time, the departure of witnesses, intervening events, subsequent construction work, improper maintenance or repair, and normal wear and tear are only some of the problems that the government may have to sort out.

Once the government discovers a latent defect, it must act promptly to demand correction from the contractor. In one case, a two-year delay between the time the government discovered the defect and the time it demanded correction from the contractor was constructive acceptance of the defect.[114]

Fraud also may justify revocation of acceptance. Where a contractor knowingly misrepresents the condition or quality of its work with intent to deceive, and the government is induced to accept defective work as a result, the government may revoke its acceptance and recover the costs of repairing such defects.[115] (In addition to recovering the costs of repair, the government also may pursue the contractor under both civil and criminal fraud statutes.)

The government also may revoke acceptance of work where the contractor makes a "gross mistake amounting to fraud." This type of situation involves a major mistake by the contractor "so serious or uncalled for it was not to be reasonably expected, or justifiable, in the case of a responsible contractor" or a mistake that "cannot be reconciled in good faith."[116]

[111]*See Cottman Mech. Contractors, Inc.*, ASBCA No. 11387, 67–2 BCA ¶ 6566.
[112]*See Gordon H. Ball, Inc.*, ASBCA No. 8316, 1963 BCA ¶ 3925.
[113]*See Tricon-Triangle Contractors,* ENGBCA No. 5553, 92–1 BCA ¶ 24,667.
[114]*See Utley-James, Inc.*, GSBCA No. 6831, 88–1 BCA ¶ 20,518.
[115]*See Henry Angelo & Co.*, ASBCA No. 30502, 87–1 BCA ¶ 19,619.
[116]*See Catalytic Eng'g & Mfg. Corp.*, ASBCA No. 15257, 72–1 BCA ¶ 9342.

III. CONTRACTUAL WARRANTIES

It is generally recognized in both commercial law and government contract law that there are two kinds of warranties accompanying virtually any construction contract. These are: (1) express warranties, which are express promises, either oral or written, and (2) implied warranties, which commercial law implies from the nature of the transaction between the parties unless the contract expressly provides that such warranties are inapplicable.

A. Express Warranties

Express warranties in construction contracts can be complex and do not have to be labeled as a warranty or guarantee in order to have the effect of an express warranty. An express warranty has been defined as "[a]ny affirmation of fact or any promise by the seller relating to the goods . . . if the natural tendency of such affirmation or promise is to induce the owner to purchase the goods, and if the buyer purchases the goods relying thereon."[117] An example of an express warranty provision in an industry form construction contract follows:

> A. Contractor warrants and guarantees to Owner that all Work will be in accordance with the Contract Documents and will not be defective. Engineer and its officers, directors, members, partners, employees, agents, consultants, and subcontractors shall be entitled to rely on representation of Contractor's warranty and guarantee.
>
> B. Contractor's warranty and guarantee hereunder excludes defects or damage caused by:
>
> (1) abuse, modification, or improper maintenance or operation by persons other than Contractor, Subcontractors, Suppliers, or any other individual or entity for whom Contractor is responsible; or
>
> (2) normal wear and tear under normal usage.[118]

ConsensusDOCS 200, Paragraph 3.8, requires the contractor to warrant that all materials and equipment are new, of good quality, in conformance with the contract documents, and free from defective workmanship and materials. That contract provision also states that, upon the owner's request, the contractor will give evidence of the quality and type of materials furnished and that the contractor's work shall be free from material defects not intrinsic to the design or materials required by the contract. The contractor also must provide any special or extended warranties required by the contract. ConsensusDOCS 200, Paragraph 3.9, further requires that the contractor will correct any defective work within one year of substantial completion.

[117] *Rite Aid Corp. v. Levy-Gray*, 876 A.2d 115 (Md. Ct. Spec. App. 2005) (citing the Uniform Sales Act).
[118] EJCDC C-700, § 6.19 A, B (2007 ed.).

Similarly, AIA A201 contains a number of express warranties. One such warranty is a general warranty that guarantees all equipment and materials are new, in conformance with the contract documents, and that all work is of good quality. Section 3.5 of AIA A201 provides:

> The Contractor warrants to the Owner and Architect that materials and equipment furnished under the Contract will be of good quality and new unless the Contract Documents, require or permit otherwise. The Contractor further warrants that the Work will conform to the requirements of the Contract Documents and will be free from defects, except for those inherent in the quality of the Work the Contract Documents require or permit.

Both parties to a construction contract must review all the contract documents carefully to determine whether requirements and language that may create an express warranty are intended and are desirable under all the surrounding facts and circumstances. Without this type of careful review and analysis, the owner may find that it is buying protection that it did not desire. Conversely, the contractor may find that it warranted a certain result or performance and that the risk attending such a warranty was not considered in the preparation of the bid or proposal for the work.

The contractor may "expressly warrant" that the material and workmanship furnished for the project are free from defects for a specified time period after completion and acceptance of the work. Liability under these express warranties expands the scope of the contractor's responsibility for defective work beyond the final acceptance date. At least one court has held that this express warranty may be limited only by the applicable statute of limitations. Thus the provision could extend the contractor's obligations for a substantial period.[119] This period may be extended in some states even further by the "discovery rule." This rule provides that a cause of action does not accrue, and that the statute of limitations clock does not even begin to "tick" until the defect is, or should have been, discovered.[120]

Section 12.2.2.1 of the AIA A201 also requires the contractor to correct any defective work within one year of substantial completion. Warranties that expressly state a specific duration, such as 12 months from substantial completion, generally do not limit other avenues of recovery and do not reduce applicable statutes of limitation.[121] So long as the defect occurred within the warranty period, an action to enforce the warranty can be brought any time within the applicable statute of limitation, although the equitable doctrine of "laches" may limit that time. Laches provides that when a claimant delays asserting its claim for so long that the defendant is materially prejudiced (such

[119]*See, e.g., Corp. of Mercer Univ. v. Nat'l Gypsum Co.*, 368 S.E.2d 732 (Ga. 1988) (statute of limitation in construction-related cases begins to run upon substantial completion and runs for four years in Georgia); *Colormatch Exteriors, Inc. v. Hickey*, 569 S.E.2d 495 (Ga. 2002) (reiterating *Mercer* decision but further holding that applicable statutes of limitation do not begin to run as to claims against a contractor/owner making improvements to its own property for the express purpose of sale of the property until the initial sale of the improved property, regardless of the date of substantial completion).

[120]*See* FLA. STAT. § 95.11(3)(c).

[121]*See Turner v. Westhampton Court, LLC*, 903 So. 2d 82 (Ala. 2004).

as defense witnesses have died or can no longer be located), it would be inequitable to allow the claimant to proceed.

Where an express warranty covers the same subject of an implied warranty, some courts will enforce only the express warranty and not the implied warranty.[122] Additional express warranties may be included in a contract in connection with equipment supplied by the contractor. Such specific warranties usually are spelled out under the provisions of the specifications to which they apply rather than in the general conditions. They often appear as performance guarantees or an agreement to repair defects for a specified period of time.

A warranty generally is not waived by final payment or completion.[123] The express warranties required by the contract documents generally begin to run from the date of substantial completion.[124] Other special warranties may commence at delivery of the machinery or commencement of operations.

Contractors performing projects with phased completion requirements must ensure that the warranties from equipment suppliers are consistent with the contractor's warranty to owner. It is not unusual for a contract with phased completion dates to provide that the contractor's warranty to the owner for all systems begins at the date of substantial completion or acceptance of the final phase, even though systems in the earlier phases have been previously put in operation. Equipment suppliers' warranties often are triggered by either delivery or first operation. In this situation, the contractor may face a gap in the warranty coverage. Although a supplier may be willing to agree to an extended warranty during the bid or proposal phase at no extra cost, an effort to extend the supplier's warranty at project completion may be very costly.

B. Implied Warranties

Both private and government contracts have been held to contain warranties that are implied by law for the benefit of one of the contracting parties. These implied warranties typically can be excluded (or disclaimed) by express contract language and may not exist unless there is privity of contract between the parties.[125] In construction, one common implied warranty arises when there are no express contractual warranties. This is the warranty of good and workmanlike construction in accordance with customary trade standards.[126] This implied warranty differs from state to state.

[122]*See, e.g., Graham Constr. Co., Inc. v. Earl,* 208 S.W.3d 106 (Ark. 2005).

[123]*See* AIA A201, § 9.10.4 (2007 ed.).

[124]*See, e.g.,* AIA A201, § 12.2.2.1 (2007 ed.).

[125]*Compare Hayden Bus. Ctr. Condos, Ass'n v. Pegasus Dev. Corp.,* 105 P.3d 157 (Ariz. Ct. App. 2005) (privity required) *with Lofts at Fillmore Condo. Ass'n v. Reliance Commercial Constr. Inc.,* 190 P.3d 733 (Ariz. 2008) (privity not required for home buyer's action).

[126]*See, e.g., AGF Marine Aviation Transp. v. LaForce Shipyard, Inc.,* 2006 WL 2402345 (S.D. Ala. 2006) (unpublished opinion) (finding that the implied obligation to perform construction contracts in a good and workmanlike manner is breached when the contractor does not exercise "a reasonable degree of care, skill, and ability under similar conditions and like surrounding circumstances as is ordinarily employed by others in the same profession"); *see also Allstate Enters., Inc. v. Brown,* 907 So. 2d 904, 912 (La. Ct. App. 2005). *Accord Heath v. Palmer,* 915 A.2d 1290 (Vt. 2006) (holding that absent clear and unambiguous exclusionary language the duration of the implied warranty good workmanship is determined by a reasonableness standard).

Courts have stated that contractors impliedly warrant that they will perform in a "workmanlike manner and without negligence"[127] or that the work will be done in a "fit and workmanlike manner."[128]

Another implied warranty arises in home or condominium construction. This is the implied warranty of habitability, which imposes a duty on the homebuilder to construct the home so that it is fit for habitation.[129] For example, in *Roland v. Heritage Litchfield, Inc.,*[130] the plaintiff condomium owners sued the developer and builder after discovering mold in the firewall area of the condominium buildings. The plaintiffs claimed damages based on, among other things, breach of the implied warranty of habitability.[131] The trial court found, as a matter of law, that the developer/builder was liable to the condo owners for its breach of this implied warranty where the condo owners had presented undisputed facts that the toxic mold rendered the condo units unsafe.[132]

This implied warranty may exist even where the homebuilder attempts to exclude it in the contract documents. For example, in *McGuire v. Ryland Group, Inc.,*[133] the federal district court held that, under Florida law, a homebuilder's general exclusion of implied warranties is insufficient to preclude as a matter of law a claim for breach of the implied warranty of habitability. On reconsideration, the court held, however, that where the contract contained express performance specifications, such specific standards sufficiently disclaimed the implied warranty.[134]

Probably the best-known implied warranty in construction is the warranty of design information furnished by one of the parties to the contract. This is commonly referred to as the *Spearin* doctrine, which provides that "where a contractor must build according to plans and specifications of an owner, the contractor will not be responsible for the consequences of defects in the plans and specifications, even though the contractor is required to check the plans and inform itself of the requirements."[135] In *AAB Joint Venture v. United States,*[136] the plaintiff contractor claimed increased costs due to defective specifications for the construction of a military storage base in Israel. The court held under the *Spearin* doctrine, the implied warranty that is imposed on the government owner is "that the specifications will result in a satisfactory, acceptable or adequate result; [and,] short of that, the specifications are

[127]*See Larchmont Nurseries, Inc. v. Daly,* 827 N.Y.S.2d 56 (N.Y. App. Div. 2006); *see also Farmer v. Rickard,* 150 P.3d 1185 (Wyo. 2007).

[128]*See Young v. Oak Leaf Builders, Inc.*, 626 S.E.2d 240 (Ga. App. 2006).

[129]*See Elovic v. Nagar Constr. Co., Inc.*, 2007 WL 1149205 (N.D. Ill. 2007) (finding that plaintiffs had stated a claim for breach of implied warranty of habitability and defining the implied warranty of habitability as a public policy doctrine that "forces homebuilders to bear the cost when it is proven that a home has a latent defect caused by improper design, material, or workmanship that renders the property unsuitable for use as a home").

[130]641 S.E.2d 465 (S.C. Ct. App. 2007).

[131]*Id.* at 466–67.

[132]*Id.* at 468.

[133]497 F. Supp. 2d 1347 (M.D. Fla. 2007).

[134]*McGuire v. Ryland Group, Inc.*, 497 F. Supp. 2d 1356, 1360 (M.D. Fla. 2007).

[135]*United States v. Spearin,* 248 U.S. 132, 136 (1918).

[136]Fed. Cl. 414 (2007).

defective and the contractor is entitled to an equitable adjustment."[137] This implied warranty is not, however, absolute.

For example, the owner will likely not be held to have breached the implied warranty of adequate design plans and specifications where the alleged defect is only minor and not a fundamental flaw or series of flaws requiring the contractor to make major revisions.[138] Likewise, if the alleged defect is a patent (or obvious) ambiguity about which the contractor failed to inquire, it should not give rise to a claim for breach of the implied warranty of design plans and specifications.[139] For a more detailed discussion on the *Spearin* doctrine, see **Chapter 5 and Chapter 9**.

C. Statutory Warranties

Many states have enacted statutes providing warranties for the benefit of purchasers of new homes or condominium units. The items covered by the warranty, the length of time the warranty lasts, and the ability to waive the warranty vary from state to state. These statutory warranties protect purchasers who failed to protect themselves through contract.

Several of these statutory warranties allow purchasers of new homes or condominium units to seek relief for defective construction against the parties that sold the homes or condominium units as well as the contractor responsible for the defective construction. For example, in Florida, the developer of a condominium, the contractor, and all subcontractors and suppliers grant to the purchaser of each condominium unit a three-year warranty as to certain key elements of the condominium, such as the roof, structural components, mechanical elements, and plumbing elements. Additionally, the developer and various contractors grant the purchaser one-year warranties as to all other improvements and materials.[140] These warranties assure that the purchaser can seek relief against all potentially responsible parties.

As an added level of protection for home purchasers, several statutory warranties cannot be waived by the owner or reduced by the builder.[141] In Mississippi, for example, a builder cannot waive its warranty to new home purchasers that the home will be free from defects due to noncompliance with the building standards.[142]

IV. PROJECT COMMISSIONING AND POST-ACCEPTANCE FACILITY OPERATIONS

One emerging trend with some owners is to combine the traditional construction quality control procedures with a requirement that the contractor be responsible for the operation and maintenance of the facility for a fixed period of time after the final

[137] *Id.* at 428–29.
[138] *Caddell Constr. Co., Inc. v. United States,* 78 Fed. Cl. 406 (2007).
[139] *Travelers Cas. & Sur. Co. of Am. v. United States,* 74 Fed. Cl. 75 (2006).
[140] FLA. STAT. § 718.203; *see also* CONN. GEN. STAT. § 47–74e.
[141] MISS. CODE ANN. § 83–58–5.
[142] *Id.*

completion and acceptance. The price for such work typically is fixed at the time of award. The rationale for this added requirement is that this operational responsibility will provide a practical incentive to the contractor to ensure that the equipment, systems, and workmanship meet all of the quality control standards.

Building commissioning is not a new process in the sense that owners, designers, and contractors always have tried to ensure that a construction project results in a building that is functional and meets the owner's performance objectives. There is increased emphasis on commissioning as an integrated *process* beginning before the design is prepared and extending after the building is complete and in use. The ConsensusDOCS, AIA, and EJCDC contract documents do not deal with commissioning as a process so it is difficult to predict how the proponents of those forms will modify them as commissioning becomes the norm in future construction project delivery systems. Owners incorporating commissioning in contract documents probably may choose to follow the federal government's guides to specifiers, discussed below, in drafting the project specifications.

A. Commissioning Programs

The federal government recently has begun to include operation, maintenance, and repair provisions in some contracts, requiring contractors to provide maintenance and inspection services during the course of the project, and after project acceptance. In one recent contract for construction of a research facility, the Department of the Navy Facilities Engineering Command included provisions for service calls, preventive maintenance inspections, repair work, and other services that would begin during the construction of the project and last for five years after beneficial occupancy. These services and work activities would be provided at a fixed price established during the bidding or proposal phase for the contract. The specifications stated that the contractor would provide labor, transportation, equipment, materials, tools, supplies, management, and supervision for preventive and corrective maintenance and repairs to maintain building systems, including structural components, exterior and interior finishes, plumbing, HVAC (heating, ventilation, and air conditioning), and electrical systems. The contract obligates the contractor to maintain the facility, including specialty equipment and associated equipment within the facility, "to a standard that prevents deterioration and corrects deficiencies in a timely manner to insure full life expectancy of the facility, structures, and associated equipment."

As part of the requirements, the contractor is to develop and implement a preventive maintenance, inspection, and service program (PMIS). The PMIS is to be submitted to the government's designated representative 15 days prior to the beneficial occupancy date. An annual PMIS schedule is to be submitted with the initial program plan and every 12 months thereafter during the contract period. The program plan is to include all inspection checks and maintenance activities called for in the operation and maintenance manuals. Minor deficiencies requiring no more than three hours of labor and a material cost of less than $250 are considered to be part of the contractor's contract price. Deficiencies exceeding those limits are to be corrected as service calls, repair work, or indefinite quantity work; but excessive or repeated system breakdowns or deficiencies are considered an indication of unsatisfactory performance of the PMIS by the contractor. Changes to correct deficiencies are to be performed at no cost to the government.

Although the systems will be commissioned initially on the beneficial occupancy date, the contractor is also required to "recommission" the HVAC systems at two years after the beneficial occupancy date and again at the end of the contract maintenance period (five years after the beneficial occupancy date). Recommissioning includes performance tests, the results of which are to be compared with the initial commissioning test results. Any areas showing degraded performance are to be addressed with required maintenance to repair or arrest and correct any reductions in performance identified in the recommissioning process. In effect, the contractor is required to maintain the systems in the project for five years at the level they were performing at beneficial occupancy. This places a very high burden on the contractor because it requires the contractor to ensure that five years after beneficial occupancy the systems operate as effectively as they were required to operate at beneficial occupancy.

A contractor taking on these responsibilities, in effect, becomes a maintenance contractor and not just a construction contractor. The contractor is assuming the responsibility of providing inspection services and maintenance services at its expense for five years after beneficial occupancy. Construction contractors that are not used to providing maintenance services probably will find it necessary to retain inspection and maintenance subcontractors to perform the work. Contractors will need to select carefully the inspection and maintenance subcontractors to be sure they can perform the work in a cost-effective manner and that they have the financial ability to do so.

The federal government is actively promoting commissioning as a part of constructing sustainable and efficient buildings. In January 2006 the "White House Summit on Federal Sustainable Buildings" was held. The President later signed Executive Order (EO) 13423, which addressed principles developed at the Summit for environmental, energy, and transportation performance and accountability.[143] EO13423 requires federal agencies to employ commissioning practices for all new federal construction and major renovations and for 15% of existing federal capital asset building inventory by 2015.[144]

A Federal Green Construction Guide for Specifiers (Green Guide) has been prepared to implement EO 13423 by a partnership of federal agencies and private sector industry organizations, led by the Environmental Protection Agency and including the Federal Environmental Executive and the National Institute of Building Sciences (which publishes the Whole Building Design Guide and other sustainable building documents).[145] The Green Guide is in an open format, and comments and suggestions for improvement are solicited. This document is not protected by a copyright and may be used and modified by any user to custom-fit any project.

[143]EO 13423, Jan. 24, 2007.
[144]*Id.*
[145]*www.wbdg.org/design/greenspec.php.*

The Green Guide section on commissioning specifies the commissioning activities and the general order in which they occur. These include:[146]

- Design review and documentation
- Commissioning scoping meeting
- Commissioning plan
- Submittals review
- Start-up/prefunctional checklists
- Functional performance testing
- Short-term diagnostic testing
- Deficiency report and resolution record
- Operations and maintenance training
- Record documents review
- Final commissioning report and documentation
- Deferred testing

Additional information on commissioning and best practices may be found in the General Services Administration's "Building Commissioning Guide."[147] The GSA Guide assigns responsibilities for holding commissioning review meetings during the construction phase, reviewing equipment warranties, developing training programs for personnel, and delivering the turnover package and commissioning record.

During the postconstruction stage, the commissioning team assigns responsibilities to correct deficiencies, conducts deferred and seasonal testing, reviews building operation 10 months into a 12-month warranty period, addresses operating facility concerns, performs satisfaction review ten months after occupancy, and recommissions the facility three to five years after turnover to reset optimal performance.[148]

The GSA provides an expanded definition of commissioning quoting the National Conference on Building Commissioning in its official definition of "Total Building Commissioning":

> Systematic process of assuring by verification and documentation, from the design state to a minimum of one year after construction, that all facility systems perform interactively in accordance with the design documentation and intent, and in accordance with the owner's operational needs, including preparation of operation personnel.[149]

The GSA Guide provides a comprehensive list of commissioning resources and contact information in Appendix C.

[146]Green Guide, 3.1 A.
[147]*www.wbdg/ccb/GSAMAN/buildingcommissioningguide.pdf.*
[148]GSA, "The Building Commissioning Guide," 11.
[149]*Id.,* Appendix D, 80.

Commissioning is now required for all new federal building construction projects, and these principles will become increasingly important in state and local government construction and in the private sector. Commissioning is an essential element in "green building" construction and achieving LEED (Leadership in Environmental and Energy Design) certification. (See **Chapter 17**.)

B. Bonding Considerations

A contractor subjected to postacceptance inspection and maintenance obligations that extend for months or years may be faced with extended bonding requirements, not just the extensive inspection and maintenance obligations. Where a contract requires ongoing inspection and maintenance by the contractor after beneficial occupancy, the contractor's performance bond (and possibly its payment bond) may remain at risk for the entire inspection and maintenance period, plus for some period of time after the maintenance period expires. This may create additional problems for contractors that find their bonding capacity affected for years by the ongoing inspection and maintenance obligations, in addition to extra premiums for bonds at risk for an extended period of time. Contractors facing the extended inspection and maintenance requirements need to clarify during the bidding or proposal stage the extent to which their bonds are to remain in effect if the contract requires extended inspection and maintenance.

➢ POINTS TO REMEMBER

INSPECTION

- Good quality control and quality assurance on construction projects is a must, and contractors are often responsible for performing this activity.
- Construction project delivery team members should consider costs of inspection no later than the estimating stage of proposal preparation.
- Although both the contractor and owner will be responsible for the costs of their own inspections, the contractor usually must assist the owner in inspecting work.
- The owner has the right to conduct reasonable and timely inspections and to reinspect and reject work previously inspected.
- The owner, its design professional, or its designated inspector can be liable to the contractor for delays in inspecting or for improper, incorrect, or unnecessary inspection.
- The contractor should request required inspections of the owner or the owner's designated representative in writing, and timely advise the owner in writing of any impacts (time or money) caused by the owner's failure to inspect or by the owner's improper or unnecessary inspection.

ACCEPTANCE

- Determine who has the authority, actual or apparent, to accept the work. Typically, in public contracting work, authority is limited. In private sector work, such authority likely will extend beyond the owner to its designated representatives.
- There can be "constructive" acceptance of the contractor's work, which typically occurs where the owner uses portions of the project and does not object to the contractor's performance or where the owner unreasonably delays formally accepting the work.
- Substantial completion of the work is a commonly accepted construction practice that usually means the owner has received what it has bargained for and can occupy and use the project for its intended purpose. Upon substantial completion, the contractor typically is entitled to the remaining contract balance less the cost of completing all punch-list work, and most contract liquidated damages provisions would cease to apply.
- The owner or its designated inspector can waive known or reasonably foreseeable defects in the work, unless the contractor has misrepresented or intentionally concealed such defects.
- Final acceptance of the contractor's work requires final completion by the contractor and final payment by the owner.
- The contractor generally is relieved of its performance obligations once the work is finally accepted; however, the contractor still will have postcompletion warranty obligations and also will be responsible for latent (or unknown hidden) defects in its work.

WARRANTIES

- Read the contract to better understand the scope and extent of express warranties.
- Contracting to perform construction work typically gives rise to certain implied warranties by both the owner and the contractor.
- Warranties related to performance of the work are, in general, only limited by the applicable statute of limitation.
- Make sure that postcompletion warranties are limited to a defined time period after final completion and acceptance of the work (typically one year).

COMMISSIONING

- Project commissioning typically extends the contractor's performance obligations beyond final completion by requiring the contractor to provide performance guarantees, operations and maintenance programs, or specified repair programs.
- The federal government strongly encourages project commissioning, an example of which can be found in the Federal Green Construction Guide for Specifiers.
- For contracts requiring performance and payment bonds, contractors that perform project commissioning work or that guaranty work well beyond project completion should confirm with their sureties whether such bonds cover this work or whether it would be prudent to purchase some type of warranty bond.

13

MANAGEMENT TECHNIQUES TO LIMIT RISKS AND AVOID DISPUTES

I. CONSTRUCTION: A RISK-PRONE BUSINESS

Every business enterprise involves risk identification and management. Construction is especially risky, for nearly every project is unique. Unlike a manufacturing operation in which standard products are assembled in a controlled environment by a relatively stable workforce, construction projects are typically unique structures built on unique sites by a team that has been assembled for that specific purpose. To that extent, each construction project can be viewed as involving the development of a prototype. Even if the actual structure or building is essentially repetitive, the possible variation in the site conditions and locale, the historical low-profit margins, the vagaries of weather, and the changing team of trade contractors and vendors create the potential for significant risks and losses. The allocation of risk and the economic consequences of risk are significant contributors to the fact that construction is a dispute-prone industry. Therefore, prudent participants in the construction process strive to identify and manage risks and seek to avoid disputes by implementing proven management techniques.

Even successful projects can have claims. Claims are a natural outgrowth of a complex and highly competitive process during which the unexpected often happens. Careful organization and coordination of numerous parties are required, and outside parties may control many of the circumstances and events that generate claims. The possibility that a claim might develop should not be ignored. The responsible owner, contractor, trade contractor, or design professional must recognize the need to anticipate the circumstances that foster claims and develop effective strategies for dealing with them.

The best way to handle claims is to anticipate them and avoid them to the extent possible. Despite the uniqueness of each project and its participants, certain recurring problems generate disputes. History repeats itself. Some of those recurring problems can be avoided, or their impact mitigated. At a minimum, some preparation can be made to more effectively address a dispute if one should occur. Of course, too intense a focus on eliminating all risks and anticipating claims and disputes can also create problems or paralysis, which impairs one's ability to conduct business effectively. A certain element of risk must be recognized and accepted. Risk can only be mitigated, not eliminated.

This chapter focuses on those measures that can be taken at the outset of a project to identify risks, avoid them, or effectively prepare for and successfully deal with disputes when they cannot be avoided. Common sense, planning, skill, and experience can help identify those strategies for avoiding and effectively dealing with disputes.

II. QUALIFYING THE PROJECT AND THE PARTICIPANTS

Construction is a cooperative enterprise involving numerous entities and disciplines. These include the owner or developer, the design professional, the lender, the prime contractor, subcontractors and suppliers, sureties, and others. Each has an essential function. A failure by any participant to perform its obligations properly can mean disaster for the entire project and the rest of the participants. Equally important to a successful project is the nature of the work, its location, and any other special requirements or conditions affecting the proposed work. Consequently, it is essential to risk management and dispute avoidance to qualify the project and the key players well before making any commitment to participate in the project.

A. Qualifying the Project

Qualifying the project requires that management evaluate all aspects of the work and the probable conditions affecting performance. This evaluation should be done systematically to provide management with an effective tool to qualify the project and to evaluate the company's ability to complete the work successfully. A practical first step in this process is to answer a series of yes/no questions about the potential project, such as those set forth in Table 13.1.

The term "relevant" is especially important and is intended to ascertain that the experience reflects, in large part, contemporaneous experience of the company's current employees. Depending on the nature of the question and the response, senior management should carefully review the decision to compete for the work. That review should include an analysis of the project risks, the actions to mitigate those risks, as well as the potential benefit or profit from performing the work.

Every company should develop an overall business plan, which addresses the company's goals in terms of volume, project size and type, and geographic market. There is no right plan for all companies. One firm might decide to seek only work

Table 13.1 Qualifying the Project

Topic/Issue	Yes	No
Does your firm have sufficient relevant experience with this type of work?		
Does your firm have sufficient relevant experience with projects of this size and complexity?		
Does your firm have sufficient relevant experience with the performance of similar work in terms of weather conditions, location, and does it have a reliable source of trained workers?		
Has your prior experience with this type of work included meeting or exceeding the anticipated project margin?		
Do you have sufficient, relevant experience with the proposed project delivery method?		
Do you have available a project team and job-site staff with successful experience in managing a project of this type, size, and complexity?		
Do you have the home office support required for this project?		
Does the realistic profit potential offset the project risks?		
Can your company bond the project?		
Does the bonding requirement substantially impact the company's overall bonding capacity?		
Does the project involve bonded, postcompletion obligations, such as long-term warranty or commissioning etc.?		

within a defined geographic area or radius of the home office. Another might focus on particular types of work. The key is to develop a realistic plan and evaluate potential projects against that plan.

B. Qualifying the Project Participants

Project participants often dictate the project's destiny, hence qualifying the project participants is one of the first steps in avoiding claims. Many headaches and actual losses can be avoided simply by investigating the past performance record of the other parties rather than looking solely at the lowest price or the opportunity to obtain new work. By dealing with reputable companies and individuals with a proven ability to perform, by running credit checks, and by inquiring about the experience of others with that particular company, major risks and big mistakes may be avoided.

Traditionally, the owner is in the best position to control the selection of the project participants, because under many project delivery systems, the owner separately selects the designer and the contractor.[1] The owner also can have a significant

[1]This process reflects the traditional design-bid-build delivery method. On a design-build project, an owner often evaluates the contractor and designer as a team. (See **Chapter 2**.)

impact on subcontractor selection through the use of a subcontract award approval clause. The prime contractor, standing in an analogous position to the owner, has greater control over the selection of subcontractors, as does the architect or lead designer over the selection of its subconsultants. When the choice of a specific party would create a risk of performance failure, claims, or disputes, to the extent that any reasonable return would be jeopardized, that party should not be used on the project, regardless of price. Although the prime contractor cannot select the project's owner or developer, it can decide whether it will do business with a particular party. Likewise, subcontractors can, and should, carefully select the contractors with whom they do business. Sometimes the risks of a project or of doing business with a particular owner, design professional, contractor, or subcontractor are simply too great, and prudence dictates that certain opportunities be forgone.

Higher-volume and backlog figures are meaningless if they engender unnecessary risks and do not translate to profit. Despite the time crunch and euphoria often associated with the beginning of a project, everyone involved should consider certain key factors when evaluating or selecting project participants: the financial condition of the parties, their qualification for bonds, evidence of their technical skills, and their reputation in the industry. Even a cursory investigation of potential project participants may yield clues to future problems.

Similar to the routine use of a checklist to qualify the project, the development of a checklist approach to qualifying the potential participants in a project may help identify situations with an unacceptable level of potential risk or a potential for claims. Table 13.2 contains a series of topics and questions that are intended to facilitate the qualification of the project participants.

Depending on the nature of the answers to these questions, the potential project participants may be able to "qualify" the other participants on the project and to identify circumstances or conditions that indicate unusual risks or a potential for disputes. Prudent application of a tailored prequalification checklist should be as essential to risk management as the customary site visit.

Development and consistent use of a standard form to obtain a statement of qualifications on a contractor or subcontractor can facilitate the receipt of information consistent with the goal of qualifying prospective participants in the project. By providing prospective contractors or trade contractors with a document outlining the categories to be addressed, it is possible to increase the likelihood of making apples-to-apples comparisons and to elicit more relevant information. Attached to this chapter at **Appendix 13.1** is a copy of ConsensusDOCS 221 Contractor's Statement of Qualifications for a Specific Project. This document is contained in ConsensusDOCS's *DocuBuilder®* contract document software as part of its 200 Series of documents and provides an excellent tool to solicit prequalification information in its current form or modified for specific trade contractors. See **Section IV.A** in this chapter for a discussion of standard contract forms, including the ConsensusDOCS.

Obviously, money is the source of many disputes and claims that arise on a construction project. The owner's financial resources are of paramount concern. An underfinanced owner virtually dooms any project. Although the possibility of lien rights might provide comfort, if the owner goes under, the probability of full and

Table 13.2
Qualifying the Participants

Topic/Issue	Yes	No
Have you previously constructed a project for this owner?		
If yes, are the key personnel different?		
Has the owner provided adequate assurance of sufficient financing?		
Does the owner have experience with the project delivery system selected for the project?		
Has the owner previously built a project of similar size and complexity to this project?		
Does the owner have an experienced staff to facilitate the resolution of problems?		
Does the owner have a history of disputes and litigation?		
Does the proposed contract allocate risks consistent with customary practices in the industry?		
If not, is the owner willing to modify the proposed contract documents?		
Does the proposed contract contain a requirement to exchange documents electronically? Is there a protocol established for that purpose?		
Will the project use a project Web site or a third party provider to exchange information electronically?		
Have the owner and design professional clearly defined those portions of the design that will be delegated to the contractor?		

Contractor/Subcontractor

Topic/Issue	Yes	No
Have you previously worked with this concern?		
If yes, are the key personnel different?		
Has your prior experience with this firm been satisfactory?		
Do you have information on the contractor's/subcontractor's current backlog?		
Do you have adequate information on the contractor's/subcontractor's finances?		
Does the contractor/subcontractor have a reputation for promptly paying its subcontractors/vendors?		
Can the contractor/subcontractor provide bonds?		
If yes, have you investigated the financial capacity of the proposed surety?		
Has the contractor/subcontractor identified key project personnel?		

continued

Table 13.2 ***continued***

Topic/Issue	Yes	No
If so, do they have experience with projects of this type, size, and complexity?		
Does the contractor/subcontractor have a history of disputes and litigation?		
Has the contractor/subcontractor been assessed liquidated damages in the past five years?		
Has the contractor/subcontractor ever failed to complete a project for any reason?		
Has the contractor/subcontractor ever been terminated for default?		

Design Team (Design Professional)

Topic/Issue	Yes	No
Have you previously built a project with this design professional?		
If so, are the same people assigned to this project?		
If so, was the prior experience satisfactory?		
Is this project similar to projects in size, type, and complexity to prior projects designed by this design professional?		
Is there any indication that the design professional was not provided adequate fee or time to complete the design?		
Does the design professional have a good reputation for fairness?		
Does the design professional have a good reputation for prompt decision making?		
Do the design documents appear to be complete and coordinated?		
Are there missing details, incorrect cross references, dimensions that do not add up, discrepancies between large-scale drawings and details?		
Do the contract documents contain language purporting to shift the responsibility for the design or the adequacy of the design to the contractor?		
Do the documents contain unusual requirements for coordination of the various disciplines?		
Does the design appear to have sufficient space for MEP system installation?		

complete payment to the contractor is very low. Considering the owner's preeminent influence on the success or failure of a project, contractors and design professionals are wise to subject the owner's background to an informal "prequalification" process, like that used on other project participants, to confirm that the owner has the capacity to meet its commitments. In addition to other independent sources that may be available to obtain information about the owner's finances, some industry standard form contracts expressly obligate the owner to provide evidence of project financing at the project's inception. Other form contracts require the owner to provide reasonable evidence when requested in writing by the contractor.

For example, Paragraph 4.2 of the Standard Agreement and General Conditions Between Owner and Contractor, ConsensusDOCS 200 (2007 ed.), obligates the owner to provide to the contractor evidence of financing of the project, and states that evidence of financing is a condition precedent to the contractor's obligation to proceed with the work. The standard General Conditions published by the American Institute of Architects (AIA), AIA Document A201–2007, contains a provision requiring the owner to provide reasonable evidence of project financing but only if requested in writing by the contractor.

Both the ConsensusDOCS and the AIA form documents address the owner's obligation to furnish information pertaining to the property title. ConsensusDOCS mandate that the owner disclose such information and other relevant information regarding the site.[2] The comparable section of AIA A201 requires the owner to furnish information regarding title only when requested by the contractor.[3] Regardless of the form of the contract, prospective contractors should routinely inquire about and seek from the owner any relevant information regarding the site and its characteristics.[4]

The financial condition of contractors and subcontractors is also extremely important. A subcontractor that has insufficient working capital may bring myriad problems, such as slow deliveries of materials as suppliers grow concerned about the subcontractor's ability to pay. This can have a ripple effect on other work. Similarly, a contractor needing cash flow may front-load its proposal and pay requests. The early overpayment caused by front-loading may result in the contractor suffering cash flow problems or, worse, defaulting during the latter part of the project as contract funds run out. Unfortunately, there are few, if any, effective remedies against an unbonded, insolvent contractor. The typical action to recover completion costs is often pointless when the default resulted directly from the contractor's financial problems.

An obvious source of financial protection for owners is to require payment and performance bonds. Bonding serves two purposes. First, the contractor's competence and financial well-being are endorsed by the surety's underwriting department, which is also trying to avoid bad risks. If a contractor is incapable of obtaining bonding, it means sureties are not convinced that the contractor has the resources to complete a project. That warning is probably best heeded. Second, and more directly, the bonds represent a financial guarantee. A performance bond usually means that if the contractor defaults and fails to complete, the surety will complete performance or pay damages up to the limit of the bond's penal sum. In contrast, a labor and material payment bond helps assure the owner that labor and materials will be paid for, and creates

[2]*See* ConsensusDOCS 200, ¶ 4.3 (2007 ed.).

[3]*See* AIA A201, § 2.1.2 (2007 ed.).

[4]Paragraph 4.3 of ConsensusDOCS 200 (2007 ed.) can provide a useful checklist of possible types of information to be requested and reviewed.

alternatives to the filing of liens on the project. (Many general contractors require that subcontractors furnish payment and performance bonds for similar reasons.) (See **Chapter 14 and Chapter 15** for discussions of payment and performance bonds, respectively.)

Even if provided, payment and performance bonds are not a cure-all. It is also necessary to carefully consider the financial stability of the surety itself. Sureties can suffer bankruptcy. Moreover, even solvent, well-financed sureties are far from an automatic source of relief. Claims under the bond can themselves be the subject of lengthy disputes and litigation. Trade associations such as the Associated General Contractors of America (AGC) can provide useful materials on topics such as bonds and bond claims. For, example, the AGC's Surety Bonding Committee has released an educational publication on bonds and bond claims, which is available from that organization.

Of course, there are concerns about technical qualifications that go beyond money. For example, licensing requirements provide some protection from incompetent and inexperienced contractors, particularly in the skilled areas such as electrical and mechanical work. Licensing should be deemed a bare minimum requirement, however, not an endorsement of qualifications for any type of work authorized by a particular license. Inquiries into the contractor's experience on particular types and sizes of projects should also be made. There is a significant difference between installing plumbing in a low-rise apartment building and installing the mechanical systems for a major healthcare facility. The owner's technical capabilities and qualifications to handle a particular type or size of project are also relevant. The owner's shortcomings may, however, be offset by the association of capable consultants.

More subjective reports about other parties should also be considered, but perhaps be given lesser weight. For example, engaging a subcontractor with a reputation for shoddy or defective work may result in the general contractor being required to remedy unsatisfactory work at its own expense. A particularly litigious owner may refuse to negotiate a settlement in the event of a dispute in this type of situation, forcing the contractor into more expensive arbitration or court battles.

Even if an owner, design professional, contractor, or subcontractor appears to have good qualifications and an established track record, it is important to consider the personnel it will assign to the particular project. Companies can be too successful, causing them to be stretched too thin, with all their capable and experienced personnel assigned to and consumed by other projects. The company is certainly important, but the individuals representing those companies, executing responsibilities, and working in the field are no less important.

C. Qualifying the Site and Locale

Similar to qualifying the type of project and the project participants, experience demonstrates that the site for the project also requires qualification. This inquiry extends beyond the physical characteristics at the location of the project and includes questions about state and local license requirements, lien laws, taxes, availability of labor, and so on. Table 13.3 contains examples of topics that should be addressed when

Table 13.3 Qualifying the Project Site/Locale

Project Site

Topic/Issue	Yes	No
Has there been an adequate prebid site and subsurface investigation?		
Is there a changed conditions (differing site conditions) clause in the contract?		
Do the design documents indicate that environmental permits have been obtained and environmental restrictions observed?		
Are existing utilities clearly located on design documents?		
Are there site access restrictions?		
Have you requested that the owner furnish all available site, subsurface, and environmental information?		
Is there any prior history of environmental hazards on or near the site?		
Have you evaluated the likely impact of conditions affecting the site (e.g., weather, traffic, work restrictions, storage problems, etc.)?		

Project Locale

Topic/Issue	Yes	No
Does your company have sufficient prior experience in the project jurisdiction?		
Was the prior experience in this jurisdiction profitable?		
Does your company possess the required licenses to work in this jurisdiction?		
Are local licenses/permits required?		
Are the insurance policies/coverages adequate for work in this locale?		
Does your company understand bond/lien claim rights and requirements in this jurisdiction?		
Do the laws in this jurisdiction create special risks or require changes to subcontract forms or documentation systems?		
Is the available local labor force adequate and sufficiently skilled?		
Is the labor market in site's locale predominantly union or nonunion?		
Is there a required project/union agreement for this work?		
Are there state/local taxes to be considered?		
Are materials and equipment available locally?		
Has your company identified any local or required labor rates and fringes?		

continued

Table 13.3 ***continued***

Project Site

Topic/Issue	Yes	No
Are there project/local Minority Business Enterprise/Women Owned Business Enterprise (MBE/WBE) requirements?		
Are there unusual business or political circumstances that may affect the project?		
Are there federal/state/local immigration laws or regulations to consider?		

contemplating the locale of any project and, particularly, a project location that is unfamiliar.

Answering these questions is critical to qualifying the project and evaluating the risks (potential or actual) in the performance of the work. Many of the questions on each of the foregoing tables flag a risk depending on whether the answer is yes or no. Possibly a greater potential risk is signaled if the answer to any question is "do not know." If the answer is unknown, it is very likely that the evaluation of the potential risk and reward is a guess. That only compounds the inherent risks of working in the construction industry.

Environmental risks and potential issues associated with performance in a different state from a firm's traditional or home base of operations require careful review to avoid unforeseen risks or obligations. Attached to this chapter at **Appendix 13.2** and **Appendix 13.3,** respectively, are checklists addressing Prebid/Proposal Environmental Considerations and Contracts in Foreign States. These checklists reflect examples of questions or concerns that should be addressed before submitting a bid or proposal in order that the firm more completely evaluates the risks and obligations associated with the work. (See **Chapter 17** for a detailed discussion on environmental risks typically found on construction projects.)

III. DEFINING RIGHTS, RESPONSIBILITIES, AND RISKS: PARTIES AND THEIR CONTRACTS

A written contract generally provides the foundation for each of the numerous relationships and binds the disparate project participants into a cohesive force to get the job built. Keeping those participants together requires anticipating issues and events that might generate disputes and detract from the goal of prompt and cost-effective project completion. This is done both by allocating risks among the parties, so it is clear who will have to bear the burden if the risk becomes reality, and providing mechanisms for resolving disputes when the risk allocation is not clear or there is disagreement. A well-drafted contract is another important element in effectively managing a construction project and avoiding or efficiently dealing with claims.

Clarity, common sense, and precision are essential when drafting contract language. Such efforts will, it is hoped, limit later uncertainty and misunderstanding among the parties and the need to refer to some third-party decision maker, court, or arbitration, to determine how the contract will be interpreted. Unreasonable and overly burdensome terms should be avoided, as they can unnecessarily drive up the cost of the work through inflated contingencies and may be difficult to enforce. But a contractor or subcontractor should not ignore such terms in an unrealistically optimistic belief that they will not be enforced. The parties must grapple with the tough issues raised by their conflicting interests in the contract-preparation stage or face the prospect of much more serious disagreements and disputes during contract performance.

IV. CONTRACT FRAMEWORK

Establishing the contract framework for the project is a threshold decision that must be made by the owner. The selection depends on a variety of factors, including the owner's needs and its expertise and capabilities. Construction projects traditionally have been designed, bid, built, and paid for within a framework of strictly defined roles, relationships, and procedures. This has proven satisfactory for many construction projects, but perceived weaknesses in the traditional method have led to consideration and use of new, alternative methods, such as the various forms of construction management, multiprime contracting, and design-build.[5] The new methods have provided many advantages, but their divergence from clearly defined practices and roles requires careful attention in the contract drafting phase to be certain that the advantages are not lost through unanticipated problems and disputes.

A. Standard Contract Forms

There are a number of available standard contract forms that establish the various relationships on a construction project. A new family of construction contract documents endorsed by the Associated General Contractors of America are fast gaining favor. Also, the series, or families of documents published by the American Institute of Architects, and the Engineers Joint Contract Documents Committee (EJCDC) have been widely used in the construction industry. These common forms permit all parties to focus on critical variables when negotiating construction transactions and obviate the need to start from scratch with each new construction transaction.

Reference to or use of a family or series of documents can be advantageous, as those documents have been screened by committees composed of industry participants and counsel in an effort to provide a more integrated and comprehensive approach to the contracting process.

[5]See **Chapter 2**.

1. ConsensusDOCS

In the fall of 2007, a new family of standard construction contracts was published: ConsensusDOCS. These contracts and forms are a result of the collaboration of 20 leading construction associations representing owners, contractors, subcontractors, and sureties.[6] The group's goal was to draft a series of contract documents that would represent the best interests of the construction project, rather than a single association. The ConsensusDOCS purport to more fairly allocate project risks to the party in the best position to control the risk. The ConsensusDOCS library of forms and documents, which is included as part of its *DocuBuilder*® software package, contains six series of documents:

GENERAL CONTRACTING DOCUMENTS

ConsensusDOCS 200: Standard Agreement and General Conditions Between Owner and Contractor [Lump Sum]

ConsensusDOCS 200.1: Amendment No. 1 Potentially Time and Price-Impacted Materials

ConsensusDOCS 200.2: Electronic Communication Protocol Addendum

ConsensusDOCS 202: Change Order

ConsensusDOCS 203: Interim Directed Change

ConsensusDOCS 205: Standard Short Form Agreement Between Owner and Contractor [Lump Sum]

ConsensusDOCS 220: Construction Contractor's Qualification Statement for Engineered Construction

ConsensusDOCS 221: Contractor Statement of Qualifications for a Specific Project

ConsensusDOCS 222: Key Personnel

ConsensusDOCS 240: Standard Form of Agreement Between Owner and Architect/Engineer

ConsensusDOCS 245: Standard Short Form Agreement Between Owner and Architect/Engineer

[6]The organizations behind ConsensusDOCS: National Association of State Facilities Administrators (NASFA); The Construction Users Roundtable (CURT); Construction Owners Association of America (COAA); Associated General Contractors of America (AGC); Associated Specialty Contractors, Inc. (ASC); Construction Industry Round Table (CIRT); American Subcontractors Association, Inc. (ASA); Associated Builders and Contractors, Inc. (ABC); Lean Construction Institute (LCI); Finishing Contractors Association (FSA); Mechanical Contractors Association of America (MCAA); National Electrical Contractors Association (NECA); National Insulation Association (NIA); National Roofing Contractors Association (NRCA); Painting and Decorating Contractors of America (PDCA); Plumbing Heating Cooling Contractors Association (PHCC); National Subcontractors Alliance (NSA); Sheet Metal and Air Conditioning Contractors' National Association (SMACNA); National Association of Surety Bond Producers (NASBP); The Surety & Fidelity Association of America (SFAA); Association of the Wall and Ceiling Industry (AWCI). See *www.consensusdocs.org/about_member-organizations.html.*

ConsensusDOCS 260: Performance Bond

ConsensusDOCS 261: Payment Bond

ConsensusDOCS 262: Bid Bond

ConsensusDOCS 263: Warranty Bond (For Correction of Work Period)

ConsensusDOCS 270: Instructions to Bidders on Private Work

ConsensusDOCS 280: Certificate of Substantial Completion

ConsensusDOCS 281: Certificate of Final Completion

ConsensusDOCS 291: Application for Payment [GMP]

ConsensusDOCS 292: Application for Payment [Lump Sum]

ConsensusDOCS 293: Schedule of Values

ConsensusDOCS 907: Equipment Lease

COLLABORATIVE DOCUMENT

ConsensusDOCS 300: Standard Form of Tri-Party Agreement for Collaborative Project Delivery

CONSTRUCTION MANAGEMENT DOCUMENTS

ConsensusDOCS 500: Standard Agreement and General Conditions Between Owner and Construction Manager [GMP]

ConsensusDOCS 500: Amendment No. 1 Standard Agreement and General Conditions Between Owner and Construction Manager [GMP with Preconstruction Services Option]

ConsensusDOCS 510: Agreement and General Conditions Between Owner and Construction Manager [Cost of Work]

ConsensusDOCS 525: Change Order [Construction Manager Fee Adjustment]

DESIGN-BUILD DOCUMENTS

ConsensusDOCS 400: Preliminary Design-Build Agreement Between Owner and Design-Builder

ConsensusDOCS 410: Standard Design-Build Agreement and General Conditions Between Owner and Design-Builder [Cost Plus with GMP]

ConsensusDOCS 415: Standard Design-Build Agreement and General Conditions Between Owner and Design-Builder [Lump Sum]

ConsensusDOCS 420: Standard Agreement Between Design-Builder and Architect/Engineer for Design-Build Projects

ConsensusDOCS 421: Design-Builder's Statement of Qualifications for a Specific Project

ConsensusDOCS 450: Standard Form of Agreement Between Owner and Design-Builder and Subcontractor

ConsensusDOCS 460: Standard Agreement Between Design-Builder and Design-Build Subcontractor[GMP]

ConsensusDOCS 470: Design-Build Performance Bond [Surety Liable for Design Costs]

ConsensusDOCS 471: Design-Build Performance Bond [Surety NOT Liable for Design Services]

ConsensusDOCS 472: Design-Build Payment Bond [Surety Liable for Design Costs]

ConsensusDOCS 473: Design-Build Payment Bond [Surety NOT Liable for Design]

ConsensusDOCS 491: Design-Builder's Application for Payment [Cost Plus, with GMP]

ConsensusDOCS 492: Design-Builder Application for Payment [Lump-Sum Contract]

ConsensusDOCS 495: Design-Build Change Order [Cost Plus, with GMP]

ConsensusDOCS 496: Design-Build Change Order [Lump Sum]

SUBCONTRACTING DOCUMENTS

ConsensusDOCS 702: Standard-Form Purchase Order

ConsensusDOCS 705: Invitation to Bid/Subbid

ConsensusDOCS 706: Subcontract Performance Bond

ConsensusDOCS 707: Subcontract Payment Bond

ConsensusDOCS 710: Subcontractor's Application for Payment

ConsensusDOCS 721: Subcontractor's Statement of Qualifications for a Specific Project

ConsensusDOCS 750: Standard Form of Contract Agreement Between Contractor and Subcontractor

ConsensusDOCS 750.1: Standard Form Rider Between Contractor and Subcontractor for Storage of Materials at Subcontractor's Yard

ConsensusDOCS 751: Standard Short Form Agreement Between Contractor and Subcontractor [Contractor Assumes Risk of Owner Payment]

ConsensusDOCS 760: Subcontract Bid Bond

ConsensusDOCS 795: Subcontract Change Order

ConsensusDOCS 796: Subcontract Interim Direct Change

PROGRAM MANAGEMENT DOCUMENTS

ConsensusDOCS 800: Standard Form of Program Management Agreement and General Conditions Between Owner and Program Manager

ConsensusDOCS 801: Standard Owner and Construction Manager Agreement [CM Agency]

ConsensusDOCS 802: Owner/Trade Contractor Agreement (CM is Owner's Agent)

ConsensusDOCS 803: Owner/Architect-Engineer Agreement [CM is Owner's Agent]

ConsensusDOCS 810: Standard Agreement Between Owner and Owner's Representative

2. AIA Documents

The AIA revised many of its key contract docments in 2007, including its A201 General Conditions of the Contract for Construction. The AIA documents are also fairly well integrated, with the terms of the various contract forms coordinated with and complementing each other.[7] Consistency potentially enhances usefulness of these documents.

Historically, AIA documents have enjoyed acceptance by owners, particularly in the traditional design-bid-build delivery system, where the owner first engages a design professional. The AIA documents may not, however, meet the needs of each and every project, and some modification may be required for each specific situation. To the extent that documents such as the AIA documents may protect the interests of the design professional, at the expense of the owner and contractor, alternate clauses can be introduced into standard AIA contracts to address any perceived bias.[8] Similarly, documents in the other families of construction contract forms may warrant review and modification to fit particular needs or requirements.

3. EJCDC Documents

Completing the trifecta for private construction industry document revisions, the EJCDC updated its Construction Series Documents in 2007.[9] EJCDC Documents, which first became available in the 1970s, are sponsored by the National Society for Professional Engineers (NSPE) Private Practice Group as well as other owner, legal, and risk management organizations.[10]

EJCDC's Construction Series documents include:

C-700: Standard General Conditions of the Construction Contract

C-520: Suggested Form of Agreement Between Owner and Contractor for Construction Contract (Stipulated Price)

C-525: Suggested Form of Agreement Between Owner and Contractor for Construction Contract (Cost-Plus)

C-620: Contractor's Application for Payment

[7]*See generally* J. Sweet, *Sweet on Construction Industry Contracts* (John Wiley & Sons, Inc. 3d ed. 1996).

[8]*See* Glower W. Jones, *Alternative Clauses to Standard Construction Contracts,* (Aspen Law & Business 2d ed. 1990).

[9]EJCDC also issues what it terms "Engineer/Subconsultant" and "Owner/Engineer" documents. EJCDC updated these series of documents in 2007 as well.

[10]"Fair and Objective Engineering Documents Get an Update—EJCDC Releases New Construction Series Documents," *NSPE*, Aug. 23, 2007.

EJCDC makes its documents available electronically through licensing arrangements. Once a license is obtained, the documents can be tailored to meet the specific project requirements. Like the AIA contract documents or the ConsensusDOCS, use of EJCDC documents will likely include modifications to meet specific project needs.

B. Critical Contract Provisions

Whether reliance is placed on a standard-form contract, a custom-drafted contract, or some combination of the two, certain contract provisions are of critical importance in anticipating, avoiding, and resolving claims. They are:

- Payment
- Time for completion and time extensions
- Damages for delay
- Changes in the work
- Termination for default and for convenience
- Changed conditions
- Dispute resolution (negotiation, mediation, dispute review boards [DRBs], arbitration, litigation)
- Insurance
- Indemnity
- Warranty

Careful attention should also be paid to the use of liquidated damages or no-damage-for-delay clauses as well as other exculpatory clauses and attorneys' fees provisions, which can weigh heavily in the resolution of claims. It is also worthwhile to consider whether the parties intend for Article 2 of the Uniform Commercial Code, which governs the sale of goods, to apply to the transaction. See **Chapter 6** for a discussion of these provisions and their application to the construction process. Arbitration as a means of resolving construction claims is discussed separately in **Chapter 21**.

V. AVOIDING AND PREPARING FOR DISPUTES THROUGH PROPER MANAGEMENT AND DOCUMENTATION

The prudent and realistic contractor designs and uses systems and procedures to manage, monitor, and document the work and progress on the project. This serves two important functions. First, these systems and procedures ensure an adequate flow of information to facilitate proper project control and coordination, including adjustments needed to respond to unexpected circumstances. Second, they aid in the compilation of an accurate and complete record of job conditions and problems and their impact on the project. The contractor certainly bears the bulk of responsibility during construction as it installs the work and generally controls the means and methods employed. The design professional and owner should not, however, abdicate all responsibility and oversight and totally remove themselves from the construction process to insulate themselves from liability. They cannot avoid all liability. Moreover, some interaction and monitoring of the construction is always required of the owner and design professional, and is in their interests. If owners or design professionals become too removed from the construction, they can neither anticipate nor promptly address problems requiring their assistance. The level of activity and monitoring will vary depending on the type and terms of the contract involved, but should not be so active or intrusive as to constitute interference and disruption of the contractor's work. Although it may be somewhat unpleasant to begin a project with an eye to possible future claims, a failure to adopt such prudent management procedures almost ensures that disputes will develop.

VI. PRUDENT AND RESPONSIBLE ESTIMATING

Efforts to effectively manage work on the project and avoid claims should begin for the general contractor or trade contractor before it even mobilizes or reaches the site. Many risks and claims arise not in the field but in the estimating department. Prudent estimating and bidding can avoid a host of performance problems and claims. A project that starts out in the hole because of bad estimating generally cannot climb out. Instead, the hole gets bigger and deeper, expanding the problem and drawing more parties into it.

Failure at any level to accurately perceive and then price the scope of the work or the associated risks results in unnecessary losses and difficulties which tend to ripple throughout the project. Estimates and bids should be supported by worksheets and backup documentation of sufficient detail under the circumstances. Such backup, and the entire estimating process, should be subject to standard forms and procedures and management review to ensure their accuracy.

Overly optimistic estimates based on vague or incomplete designs should be avoided or at least clearly identified and qualified as such. Performance specifications often entail more responsibility and cost more than may be initially apparent and are often another soft spot in the estimating effort. The zeal applied to selling the project or submitting an early guaranteed maximum price to satisfy the owner must be balanced with caution against establishing an unrealistic budget or inflated expectations that, regardless of any contractual significance, are bound to cause disappointment, distrust, and disagreement when they are not met.

This scrutiny must be applied to bids received from subcontractors as well as those generated in-house. It is not always the case that the contractor may have recourse against the subcontractor. More important, the contractor has its own obligations to the owner and other subcontractors and may not evade that responsibility or liability simply by pointing to the subcontractor that cannot perform because of an estimating error. Owners should likewise be wary of a bid that seems too good to be true; it probably is! Success in initially enforcing a mistaken or reckless bid can reap bitter returns later in the project when the contractor's financial problems adversely affect the project.

Much construction estimating is now done with the help of computers. The yellow estimate sheet no longer dominates the industry. Computer software for estimating varies from the simple electronic spreadhseet to programs that perform take-offs from digital plans. The information developed in many estimating software applications can be exported for use in purchasing, scheduling, and job-cost reports. The original electronic estimate also can be used later in the project for estimating change orders. The value of powerful yet easy-to-use estimating software is hard to overstate. The project's success or failure often depends on an accurate estimate. Prudent owners, construction managers, contractors, and subcontractors should carefully evaluate the many software applications available to determine the most cost-effective solution for their business. Some of the more prevalent brands of scheduling software are: RS Means Costworks, Sage Timerline, Microsoft® Excel, MII-PIC, Hard Dollar, and Autodesk Constructuare.

VII. ESTABLISHING STANDARD OPERATING PROCEDURES

Construction projects run by the seat of their participants' pants are accidents waiting to happen. Every project should have formalized, standard operating procedures with which all project personnel are completely familiar. These procedures should identify the specific authority and areas of responsibilities for each project staff position. Ideally, these should be standardized within a company and be consistent from project to project. Standard job descriptions then can be used to define the roles of the individuals on a particular project. The standard procedures should address the duties and responsibilities for processing change orders and extra work, purchasing and receiving, submittal review, project documentation, and costs and accounting.

As the project team is being assembled and mobilized and the standard procedures are adjusted, defined, and implemented for the particular project, it is a good idea to reassess the project in terms of estimating, scheduling, procurement, cost accounting, and the like before construction begins. This reassessment can serve as an additional safeguard for the early identification and correction of problems that might otherwise have a serious impact on the project at some later date if left undetected.

VIII. ESTABLISHING LINES OF COMMUNICATION

The ability of the parties on the project to establish and maintain constructive lines of communication is essential to a successful project. Prosecution of the work must be recognized as a cooperative effort that demands a team approach rather than adversarial conflict. The owner, design professional, contractor, and subcontractors must establish some method to both discover what is going on at the job site (e.g., what problems the subcontractors are having and what problems they anticipate) and relay suggestions, recommendations, and requirements regarding how these problems can be avoided or solved. Satisfactory communications can be achieved only if the parties have personnel who can sustain professional and reasonable working relationships with one another. The subcontractor's workers should have sufficient confidence in the prime contractor's on-site personnel so that they will not hesitate to report difficulties and seek the prime contractor's recommendations as to how those difficulties can be avoided or resolved. Efforts to develop this confidence should begin at the preconstruction conference and continue throughout the contract period.

One activity that could improve communications is partnering. Partnering can be seen as an organized effort at team building. Typically, the process begins contemporaneously with the preconstruction meeting. One of the primary purposes of a partnering program is to help the project participants recognize and affirm that they have many interests in common and that a collaborative approach to problem solving is in the general interest. Partnering meetings can be conducted by the project participants, with or without the involvement of a third-party facilitator. Regardless of the label, or the use of a facilitator, the key is to promote a project atmosphere that is not adversarial.

An important procedural aid to establishing and maintaining the required lines of communications is to hold regular job meetings. Weekly, biweekly, or at least monthly meetings should be scheduled *and* held. The participants and the frequency of the meetings will depend on their purpose and the status or level of activity on the job. Field coordination meetings should involve the project superintendent, subcontractor superintendents, and key foremen. Brief but regular meetings can aid the process of coordinating and scheduling the work. They also can help identify problem areas and information needed for progress before a situation becomes critical.

Regular meetings between the designer's staff and the contractor's personnel are helpful for keeping up on the status of submittals, shop drawings, and areas requiring clarification. Meetings involving the contractor, design professional, and owner also should occur, but probably less frequently. These meetings can be used to apprise the owner of important developments and to work out contractual issues such as changes. Further, the parties can discuss problems that are not being worked out on a more operational level and require the owner's intervention. The contractor should be wary of allowing the owner to get too far removed from the construction effort.

IX. PROJECT DOCUMENTATION

Consistent and complete project documentation is the key to successfully asserting or defending against construction claims. The process of project documentation should not be reserved for "problem" jobs. If adequate documentation is not maintained from start to finish, the circumstances giving rise to a dispute often will go unrecorded and perhaps even unnoticed. The "paperwork" on a construction project may seem to be overwhelming, but it is essential. The contractor typically generates and maintains the bulk of the documentation on a construction project, but all participants have an interest in it.

Contracts often require that the contractor maintain certain documentation, with access available to the owner and others. Project documentation creates an accessible history of the project that serves two roles: (1) planning and managing the project and (2) aiding in resolving claims and disputes. It must be organized and maintained in such a manner that it is a help, not a hindrance, to effective project management and prosecution or defense of claims. Routine and uniform documentation is essential to an effective project documentation system. The procedures should be standardized not only for the project but for the entire company. Only with that level of emphasis and indoctrination can all the benefits of such a system be reaped.

Evaluation of the adequacy of the project's documentation system is an essential pre-performance step. Even though an analysis of risk allocation is part of the prebid/proposal project evaluation, a company's established project documentation system should be reevaluated in light of the specific requirements of each project. Table 13.4 illustrates the type of questions to be addressed at the inception of a project.

Senior management should review the answers to these questions. Based on that review, the firm's documentation system can be adapted to the particular project's requirements. Notice checklists and standard form letters should be prepared or modified, as needed, for a particular project. **Appendix 13.4** to this chapter contains examples of notice checklists and various forms and logs, which can be modified for use. In particular, it is essential to identify all notice requirements using a checklist

Table 13.4 Qualifying the Project Documentation System

Topic/Issue	Yes	No
Have we identified and reviewed *every* document that will be part of our contract?		
Has the contract been reviewed to determine if modifications to our "standard" project documentation system are required?		
Does the contract impose specific cost accounting requirements?		
Does the contact define recoverable direct or indirect costs?		
Does the contract contain special notice requirements?		
Have we identified notice and claim deadlines?		
Are special procedures or cost limitations imposed on change order work?		
Does the contract contain unusual risk-shifting clauses that our project documentation system should consider?		
Do we need the input of legal counsel on documentation practices in light of contract risk-shifting provisions?		

approach and then to develop a specific project checklist addressing the contract provision, subject matter of notice, time requirements, form of notice, and the stated consequences of a lack of notice. Based on that study, standard forms can be adapted as necessary.

These system and procedures must be in writing. The length and level of detail of the written description will vary with the size and complexity of the project. Some description may be obviated by the use of electronic tracking systems, such as job cost accounting programs, as well as those for tracking submittals, modifications, requests for information (RFIs), and so on. Regardless of how extensive the procedures are, they must be clear and specific. If they are vague and general or allow for personal interpretation and selective application, there will be no system at all. Instead, a hodgepodge of personal record keeping and filing systems will result.

Simply writing out procedures is not enough. The procedures must be reviewed with all levels of personnel who will be responsible for implementation so they are understood, used, and enforced. The critical importance of project documentation must be emphasized, and that emphasis must be maintained throughout the project and from project to project.

Certain basic information should be maintained and organized in separate files include:

- The contract, including all its components, and all change orders or amendments, including a bid or original set of project plans and specifications
- All documents, worksheets, and forms associated with the original bid estimate and subsequent revisions
- Subcontractor or vendor files, including bids, quotes, subcontracts, or purchase orders, together with changes and correspondence
- Project schedules, including the original ("baseline") schedule and all updates and look-aheads
- Insurance requirements and information for all parties

The standard procedures relating to documentation also should address the creation, maintenance, and orientation of certain specific types of documentation:

(1) *Correspondence.* Procedures for date-stamping, copying, routing, filing, and indexing incoming and outgoing correspondence should be the responsibility of secretarial or clerical support staff to perform in accordance with standard procedures. Copies of all correspondence should go in a master correspondence file. The party responsible for responding to or acting on incoming correspondence should be identified. Even if there is a decision to use a browser or network-based project management system, a disciplined approach must be used. In addition, each project participant should carefully consider the extent to which hard copies need to be maintained as backup.

As a matter of routine, project management personnel should be drilled on the importance of complying with the contract's technical notice requirements. Discussion with other parties likewise should be confirmed in writing with the involved parties, with copies to the file. Such confirmation should help resolve any misunderstanding that might exist and preserve the substance of the discussion if a dispute arises.

(2) *E-mail.* Many contracts acknowledge and anticipate that the parties will communicate electronically with one another during a construction project (i.e., by e-mail). The time and cost savings of electronically transmitting project correspondence, pay applications, submittals, shop drawings, RFIs, notice, minutes, and other project correspondence can be significant. Because e-mail is so easy and quick, however, there may be a tendency to use it less carefully than other, more formal means of communication. Remind the project team that e-mail is a part of the project documentation system. As such, it constitues a *discoverable* collection of project communications in a construction dispute.

Just as with written forms of communication, the project team should be trained to avoid the following practices in connection with e-mail messages:

- Avoid attempts at humor or sarcasm. E-mail messages intended to be funny may not seem humorous to the juror viewing that e-mail message much later. Instead, the message sender's sarcasm may be interpreted as a lack of professionalism, or worse.
- Do not say anything in an e-mail that a person would not write in a letter. The convenience and expediency of the e-mail process encourages a more informal, or less professional, level of communication. Avoid the temptation.
- Document facts, not feelings.
- Do not be misled into thinking that a deletion of an e-mail message will prevent its later recovery and use as a project record.

Unfortunately, e-mail can be "signed" by someone who did not actually create the message. It can be forwarded or routed without the originator's knowledge or consent. It can be modified, saved, printed, and distributed very easily. Establish clear written policies on the use/misuse of e-mail. Frequently police the firm's e-mail use.

E-mail messages often are stored as part of computer backup operations. Deleting a copy of the e-mail message from a hard drive on one computer may not prevent recovery of the e-mail message. Even if the e-mail message is an internal comunication, and not printed out in hard-copy form, it still may be part of a company's computerized backup files. Destruction of the project e-mail *after* a dispute arises only creates problems and harmful inferences. If a company's documentation retention system specifically provides for the "dumping" of e-mail files after a certain period of time, and if those practices are routinely followed, a company may eliminate some of the potential harm resulting from embarrassing internal e-mail. See **Section XIV** in this chapter for a discussion on preserving electronically stored information.

(3) *Meeting notes.* Regular job-coordination meetings between the various parties on the project, on a cumulative basis, probably cover more issues and can

contribute more to the exchange of information necessary to complete the work than all the project correspondence. Therefore, what occurs at such meetings is of great importance. Someone should be designated to maintain the minutes or notes for each meeting, preferably the same person at each meeting. That person should record the subjects covered, the nature of the discussion, the future actions to be taken, who has responsibility for the future action, and the applicable deadline for action. The name, title, and affiliation of each participant should be listed. Notes should be concise but accurate and informative. The items discussed should be indexed or designated so that they can be located for future reference. The notes should then be distributed to all participants and those affected on a regular basis.

A computer can be valuable for updating regular meeting notes, as certain items likely will remain open to discussion through several meetings. At the opening of each regular meeting, the notes from the previous meeting can be reviewed to confirm their accuracy and the mutual understanding of the participants. By identifying those items that remain outstanding, the previous week's minutes also can serve as the current meeting's agenda.

(4) *Job-site logs or daily reports.* Job-site logs or daily reports generally are maintained by the project superintendent and can provide the best record of what happens in the field. They help keep management and office personnel informed of progress and problems. In the event of a claim, they are often among the most helpful documents in re-creating the job progress and as-built schedules. They can be essential documents in the evaluation of claims for disruption and loss efficiencies.

The daily log or report must be a part of the superintendent's daily routine. If it is too burdensome, it either will be ignored or will detract from the superintendent's primary function of getting the job built. Key information should be elicited, requiring as little narrative as possible to describe the conditions. The information covered at a minimum should include:

- Manpower, preferably broken down by subcontractors
- Equipment used and idle
- Major work activities
- Any delays or problems and the activities affected
- Areas of work not available
- Safety and accidents
- Oral instructions and informal meetings
- Weather summary
- Job-site visitors
- Key inspections

The burden on the superintendent can be eased and the information maintained in a more organized manner by using a standard form. The process can be expedited further simply by allowing the superintendent to dictate entries and having the report typed up by office staff. An e-mail format also can be used for ease of transmission of the information to the home office.

All key project personnel, such as foremen, project engineers, and project managers, should also be encouraged to maintain personal daily logs and follow the procedures established to facilitate this effort. The information they should record should be similar to that maintained in the job log or daily report, but it need not be as extensive or detailed.

These types of routine, contemporaneous descriptions of work progress, site conditions, labor and equipment usage, and the contractor's ability (or inability) to perform its work can provide valuable information necessary to accurately reconstruct the events of the project in preparation of a claim. In maintaining these reports or logs, project personnel must be consistent in recording the events and activities on the job, particularly those relating to claims or potential claims. Failure to record an event, once the responsibility of a daily report or log is undertaken, carries with it the implication that the event did not occur or was insignificant and threatens the credibility of the entire log.[11]

(5) *Standard forms and status logs.* There is a constant flow of information among the project participants by means of a variety of media. Drawings are revised; shop drawings are submitted, reviewed, and returned; field orders and change orders are issued; questions are asked; and clarifications are provided. Cumulatively and individually, these bits and pieces of information are essential for building the job and for reconstructing the progress of events on paper in the event of a claim. The standard procedures must include the means for providing, eliciting, recording, and tracking this mass of data so that it can be used during the course of the job and efficiently retrieved in an after-the-fact claim setting.

Routine transmittal forms should be customized to address specific, routine types of communications in order to expedite the process but also to ensure that required information is provided. For example, separate specialized forms can be prepared for transmittal of shop drawings and submittals, requests for clarification, drawing revisions, and, of course, field orders and change orders. When possible, the forms should provide space for responses, including certain standard responses that simply can be checked off or filled in. At a minimum, the forms should identify the individual sender, the date issued, and specific and self-descriptive references to the affected or enclosed drawings, submittal, or specification. If a response is requested by a certain date, that date should be identified on the form. Again, although a computer network- or browser-based system can significantly expedite the prompt exchange of information, the key to good project management is the development of a systematic and consistently followed routine. If a computer network system is employed, it does not mean backups or hard copies can be eliminated.

Ideally, each discrete type of communication or specialized form should be numbered or somehow identified in a chronologically sequential manner based on the date it is initiated. Shop drawings and submittals, however, are best identified by

[11] *See* Fed. R. Evid. 803(7).

specification section, with a suffix added to indicate resubmittals. This provides a basis for easy reference and orientation. Copies of the completed forms should be maintained in binders in reverse chronological/numerical order. Although various project staff members may require working copies, a complete master file should be maintained as a complete reference source and historical document.

In order to maintain the status of and track these numerous and varied communications, which can number many thousands, logs should be maintained. These logs need only address key information, such as number assigned, date, and a self-descriptive reference. Proposed change orders and change order logs also should identify any increase or decrease in contract amount as well as time extensions. Such logs can be kept on personal computers using inexpensive, commercially available software to expedite updating. Logs should be maintained for internal record keeping and for distribution to other parties on the project. The logs serve as a reminder of outstanding items and can highlight action required to keep the work progressing.

The contractor should use standard forms and procedures for communications with subcontractors as well as with the owner and design professional. Ideally, subcontractors should be encouraged to standardize their communications so there will be a more integrated approach for the entire project. The adoption of a network-based project management system can facilitate the use of standardized approach to project communications.

(6) *Photographs and videotapes.* Photographs and videotapes are helpful, easy, and inexpensive means to monitor, depict, and preserve conditions of the work as those conditions change and the work progresses. They are particularly helpful in claims situations. One approach, incorporated in many contracts, is to accumulate a periodic pictorial diary of the job through a series of weekly or monthly photographs of significant milestones in the construction. This encourages personnel to take photographs of site conditions on a routine basis, perhaps concentrating on problem areas and those areas associated with crucial construction procedures and scheduling. Photographs are also the best evidence of defective work or problem conditions that are "cured" or covered up and cannot be viewed later.

Cameras capable of producing quality photographs and negatives should be used. Digital cameras offer an excellent method for taking, storing, and transmitting project images. A digital camera also allows the party responsible for taking pictures to check the content and clarity of the photos while he or she is still at the site and before conditions are altered.

Pictures always should be marked with notations as to time, date, location, conditions depicted, personnel present, and photographer. Some of this information can be imprinted on the negative or stored digital image. This should be done when the photograph is taken if a self-developing camera is used. Otherwise, a log should be kept as the photos are taken, the log should be checked immediately when the photos are developed, and the appropriate entries should be made on the back of prints. Without this information correlated to specific photographs, the utility of the entire effort can be substantially undermined. If film is used, negatives also should be retained in an organized, retrievable manner.

In some situations, a video recording can be considerably more informative than a still photograph, such as when attempting to depict an activity or the overall status of the project. Static conditions, however, are best photographed. The availability of a contemporaneous narrative as part of the video can give the after-the-fact viewer a much better idea of what is being depicted and why. A monthly video recording is an excellent way of preserving and presenting evidence. Again, properly trained job-site personnel can operate the video recorder and later testify in conjunction with the showing of the video.

X. ELECTRONIC COMMUNICATIONS ON CONSTRUCTION PROJECTS

Paper (drawings, letters, sketches, RFIs, etc.) has been the traditional or historical medium for the exchange of information on construction projects, but electronic communications using computer technology is supplementing that traditional medium for project documentation. Ideally, electronic communications increase productivity by providing an accurate record of data exchanges and greater consistency in the content of the information provided to the project participants. Electronic communications also reduce the time and cost associated with the transmission of hard copies (paper) documents by the mails or commercial carriers.

Although the medium of electronic document exchange provides a potential for substantial benefits and efficiency to the project, the effective use of electronic communications requires the commitment of the project participants to its use in a systematic and coordinated manner. In addition to obtaining the commitment of each of the parties, implementation of electronic communications for the exchange of project information can be more effective with careful preplanning. To a large extent, attention to the details at the inception of the project can help minimize later issues or problems.

A. Industry Forms Addressing Electronic Communications

Industry documents, such as the ConsensusDOCS and the AIA contract documents, have standard forms that allow the parties to address the acceptable formats and technology for sharing electronic information. A comparison of the AIA's current approach to sharing electronic information to the ConsensusDOCS family of documents illustrates the varied approaches to the use of electronic communications on construction projects.

First, the approach adopted by the AIA is relatively simplistic. Section 1.6 of AIA A201 states:

> § 1.6 TRANSMISSION OF DATA IN DIGITAL FORM
>
> If the parties intend to transmit Instruments of Service or any other information or documentation in digital form, they shall endeavor to establish necessary protocols governing such transmissions, unless otherwise already provided in the Agreement or the Contract Documents.

In addition to this general statement, the AIA has developed two specific documents that can be incorporated into its form contract documents to address specifically how data on a particular project is transmitted in a digital working environment and who has control over its future use. AIA C106, Digital Data Licensing Agreement (2007 ed.), is a licensing agreement between two parties covering the use and transmission of Digital Data[12] on the project.[13] Pursuant to this form, the party receiving the digital data is granted a "nonexclusive limited license" to use such data "to perform services or construction for the Project" in accordance with agreed upon licensing conditions that can be specified in the document.[14]

AIA E201, Digital Data Protocol Exhibit (2007 ed.), allows the contracting parties to determine and agree on: (1) the method for transmitting project data; (2) the data format; and (3) permitted uses for the data throughout the life of the project. These Digital Data protocols are set forth in a table, which includes such data as:

- Project agreements and modifications
- Project communications (such as meeting notices, minutes, and RFIs)
- Architect's preconstruction submittals
- Architect's drawings and specifications
- Contractor's and subcontractor's submittals (such as product data and shop drawings)
- Modifications
- Project payment documents
- Notices and claims
- Close-out documents[15]

The term "data format" as used by AIA E201 means the medium in which the document was generated. For example, a document prepared using Microsoft® Word would appear as being in a *.doc (W) data format.[16] The transmitting parties are simply the different project team members, such as the owner (O), architect (A), and contractor (C).[17] AIA E201 discusses various means by which the parties can agree to transmit digital data, including: (1) e-mail (EM); (2) an attachment to an e-mail transmissiong (EMA); (3) via compact disc (CD); (4) posted to the project Web site (if a Web-based project management system is utilized) (PS); and (5) file transfer protocol (FTP) transfer to receiving FTP server.[18] Permitted uses for such digital data

[12]AIA broadly defines "digital data" as "information, communications, drawings, or designs created or stored for the Project in digital form." AIA E201, § 1.2.1 (2007 ed.)

[13]AIA C106, § 1.1 (2007 ed.).

[14]*Id.* at §§ 2.1 and 3.1.

[15]*See* AIA E201, § 3.1 (2007 ed.) (containing Project Protocol Table).

[16]*Id.* at § 3.2.

[17]*Id.*

[18]*Id.*

would include: (1) store and view only (S); (2) reproduce and distribute (R); (3) integrate (incorporate additional digital data without modifying data received) (I); and (4) modify as required to fulfill project obligations (M).[19]

According to the AIA, if the parties want to agree on the form, method, and use of digital data on a construction project, they can do so by incorporating AIA E201 into the contract documents. This form allows the parties to do so in a relatively simplistic format. Although this approach has the possible advantage of providing flexibility to the project participants, it essentially leaves the participants to their own resources and collective experience.

The ConsensusDOCS collaborators elected to prepare a detailed thirteen page document entitled ConsensusDOCS 200.2 Electronic Communications Protocol Addendum (2007). This addendum seeks to provide the parties a list of specific issues to be considered and addressed when adopting electronic communications. The topics addressed include:

- Integration of protocol addendum with the other contract documents.
- Parties including subcontractors and material suppliers participating in the agreed protocol.
- Information technology (IT) administrators, IT management team and Webmaster for the project
- System parameters, such as file formats, hardware, operating system, hardware, and software[20]
- Transmission/access and security/encryption requirements
- Allocation of risk of translation errors caused by software utilities
- Test protocols if a new party is added
- System upgrades
- Proprietary and copyrighted information
- Types of documents to be exchanged electronically with or without hard-copy backup
- Data compilation and archiving responsibility
- Version/revision control of documents
- Third-party service provider agreements

The detail provided in ConsensusDOCS 200.2 illustrates the need for and value of careful planning if a project is to adopt electronic communications as a primary method to exchange information and documents. Without that careful planning, glitches may negate all of the potential savings of time and money and create disputes.

[19]*Id.*

[20]Primavera Contract Manager, discussed further herein, is one software package intended to meet project communication needs.

B. Web-Based Project Management Systems

The Internet has dramatically impacted construction project management. Web-based project management is now quite common on large and complex construction projects. As previously mentioned, industry form documents now recognize that project delivery team members can prepare, disseminate, and share project documentation and information in real time.[21] In Web-based project management, sometimes called online collaboration and project management (OCPM), project team members use the Internet to send, share, update, track, review, and store project documents.

Project team members should select a Web-based project management vendor early in the planning process. The preferred form of service in construction is through an application service provider (ASP). The ASP provides the servers and software, and project participants access the service through the Internet. Typically, a project-specific Web site is set up that authorized team members access through a standard Web browser. Industry leaders in Web-based project project management services include Primavera, Meridian, Autodesk, Bentley, and Sage Timberline.

Primavera's Contract Manager is an illustrative example of a Web-based project management system used in the construction industry. This network-based system is specifically designed to generate and track construction project documentation. Primavera is also a well-recognized provider of construction project scheduling software, and its Contract Manager system can be bundled with its scheduling sytems, such as P6™.

Primavera Contract Manager enables the construction industry professional to manage documents, job costs, and project controls. Contract Manager allows the user to identify impacts to potential cost and schedule delays and distribute such information to the parties impacted. Further, use of this Web-based project management system can improve decision making by allowing the tracking of RFIs, shop drawings, drawing logs, turnaround graphs, alerts, issue tracking, daily reports, submittals, change orders, meeting minutes, and electronic communications from any location at any time. Good information management can reduce the risks of delays, disputes, miscommunication, and rework.

Although the Internet will most likely never completely replace paper documents and filing cabinets, it can be a critical tool for construction project management.

XI. BUILDING INFORMATION MODELING

Another computer-based electronic communications trend sweeping the construction industry affects the planning and design and construction of a project from the outset, throughout the project's duration, and beyond. Through such a process, the owner and other members of the project delivery team, such as the design professional, the contractor, major trade subcontractors, and suppliers, utilize virtual design and construction concepts, or building information modeling (BIM), to interactively

[21] *See, e.g.*, AIA E201 (2007 ed.); ConsensusDOCS 200.2 (2007 ed.).

model project design and construction. Multiple software programs currently on the market offer BIM capabilities. Industry leaders include Autodesk's Revit Building, Graphisoft's Virtual Construction, Bentley's Bentley Architecture, and VectorWorks Architect.[22]

Various models or processes can broadly be categorized as BIM. The Associated General Contractors of America offers this definition:

> *Building Information Modeling* is the development and use of a computer software model to simulate the construction and operation of a facility. The resulting model, a *Building Information Model,* is a data-rich, object-oriented, intelligent and parametric digital representation of the facility, from which views and data appropriate to various users' needs can be extracted and analyzed to generate information that can be used to make decisions and improve the process of delivering the facility.[23]

Another industry organization categorizes BIM "three ways, as [an IT] product . . . , a collaborative process, and a facility lifecycle management requirement."[24] The remainder of this section focuses on BIM's collaborative process and the related issues that construction project delivery team members should understand.

A. Collaborative Uses of BIM

In its simplest form, BIM represents to the construction industry a means of interactively sharing project design and construction data to more economically and timely achieve the desired finished project that meets the owner's objectives. More specifically, BIM allows the project delivery team to achieve this goal in a number of ways, including:

- Design visualization and comprehension
- Space utilization
- Structural analysis
- Energy analysis
- Preparation of design drawings
- Systems (for example mechanical, electrical, and plumbing) coordination
- Interference or conflict identification and resolution
- Constructability reviews
- Communication
- Integration of various players' models
- Multi-dimensional modeling ("XD")[25]

[22]For a detailed listing of available BIM software products, training, and interoperability with other construction industry software as of 2008, *see* AGC, Contractor's Guide to BIM, App. B (Associated General Contractors of America, Inc., 2006).

[23]AGC, Contractor's Guide to BIM, 3 App. B (Associated General Contractors of America, Inc. 2006).

[24]National Institute of Building Sciences, National Building Information Modeling Standard 18 (2007).

[25]From a modeling standpoint, BIM offers users the ability to view facility design and construction in three-dimensions ("3D"). Most modeling software also allows the user to integrate other "dimensions" into the modeling, such as time ("4D"), costs ("5D"), and beyond ("XD").

- Work scheduling and sequencing
- Site planning and utilization
- Cost estimating
- Layout and fieldwork
- Prefabrication of materials and equipment
- Safety analysis and management
- Operations and maintenance[26]

Members of the project delivery team interactively collaborate to provide the information necessary to achieve these goals. Ideally, this collaboration serves to achieve a desired finished construction project more timely and economically than more conventional methods, such as paper RFIs, by which project members would collaborate and communicate.

B. BIM's Legal Implications

Given the recent development of BIM and the contract documents implementing its use, it is difficult to predict all of the legal issues that might be presented by this data-sharing technique. Needless to say, such sharing of information would give rise to intellectual property issues such as copyright and ownership issues—that is, who owns the collaborative work product. The most effective way to resolve such an issue is to address it in the parties' contract. Beyond that, any further discussion of intellectual property issues exceeds the scope of this chapter or this book. Other more "traditional" construction law questions or issues that might arise in the use of this collaborative method would include such items as:

- If there is a contract addendum or specification addressing BIM, does it clearly define key concepts or terms?
- If the owner is seeking competitive proposals addressing the use of BIM in the design and construction of the project, what level of detail is required in the proposal? How costly will that effort be?
- Does the BIM model become a contract document once it is completed? If there are differences between the final BIM model and the owner's original statement of requirements or specifications, which controls?
- Who is responsible for project design, and, by collaborating in such design, does one assume added risk or liability for such design?
- Who is responsible for project submittals, such as shop drawings and material samples, and, by collaborating in such submittals, does one assume added risk or liability?

[26]Howard W. Ashcraft, Jr., *Building Information Modeling: Electronic Collaboration in Conflict with Traditional Project Delivery,* 27 Construction Litig. Rep. 335, 338–39 (2006). *See also* Dwight A. Larson and Kate A. Golden, *Entering the Brave, New World: An Introduction to Contracting for Building Information Modeling*, 34 WM. MITCHELL L. REV. 75, 79 (2007).

- Who is responsible for delays in design or submittals?
- What impact, if any, does participation in this collaborative process have on the recognized standards of care for design and construction?
- If there is a serious construction failure involving property damage or physical injury, can each party's responsibility be clearly traced? If the responsibility is mixed, does the contract address the allocation or sharing of that risk? Does the contract address responsibility for economic losses?
- If multiple contracts contemplate collaboration by multiple parties in the BIM process, does a party's responsibility for losses or damages extend to those with whom it does not have a direct contractual relationship?
- Do potential insurance coverage gaps exist between traditional coverage under comprehensive general liability policies and errors and omissions policies?
- Are there any statutory limitations of risk transfer applicable where the project is located?
- Who is responsible if there are any software errors?

Ultimately, the goal of such a collaborative project design, construction, and management system should be to reduce risks and liabilities. The legal issues just mentioned are not unique to BIM. They arise in other contexts, as discussed throughout this book. As in those cases, to avoid these risks, the parties need to understand how they manifest themselves during the project and attempt to address and mitigate them from the outset. This can be accomplished in careful drafting of the contract documents.

As previously discussed, currently available industry form documents do attempt to address this. Very generally, AIA E201, which covers digital data, could be applied to cover BIM and its uses on a project. Likewise, although in more detail, ConsensusDOCS 200.2 could also be utilized to address BIM issues.[27] Regardless of whether industry contract documents or specially drafted documents are used, in the context of BIM, the contract documents should clearly identify, at a minimum, the scope and extent to which BIM is intended to be used on a project, the key concepts, and the parties' respective duties and responsibilities in this collaborative process.

XII. COST ACCOUNTING RECORDS

The use of effective cost accounting methods and the maintenance of appropriate cost records can minimize many of the proof problems inherently associated with construction claims. Unfortunately, even though a claimant may be able to prove that an event has occurred that entitles it to additional compensation, it will be able

[27] Various organizations, such as the Army Corps of Engineers, the federal government's General Services Administration, the AIA, and ConsensusDOCS, have published specifications or addenda addressing the use of BIM. There are significant differences in these organizations' use of BIM.

to recover only the amount of damages that it can prove with reasonable certainty. Proving the actual dollars lost is crucial to the claim.

Cost accounting systems used by contractors vary dramatically in their level of sophistication. Often the accounting function suffers from a lack of priority by senior management until a dispute arises and the claim development process begins. The procedures described earlier to effectively capture and document events or occurrences to prove liability are only half the battle. Without effective job cost accounting systems, the development of a clear, concise, and winning claim is haphazard at best. In addition, accounting rules for the construction industry are sometimes subjective, and in many cases accounting professionals and businesspeople can reach different conclusions on a single set of circumstances. Hence it is important that accounting policies and procedures be documented and that management appropriately monitor their application for compliance. This approach tends to improve the consistency of the manner in which items pertaining to all contracts are accounted for and therefore improves the credibility of the way in which matters are treated in developing a claim. Further, and perhaps of primary importance, a good job cost accounting system is invaluable in providing timely information to management for decision-making purposes and for monitoring the construction project's financial performance.

XIII. MONITORING THE WORK THROUGH SCHEDULING

The general contractor or construction manager should continuously monitor the work of all subcontractors or trade contractors to determine that each is meeting its deadline so that the work of other trades can proceed as originally scheduled. The owner must perform the same task when multiple prime contractors are involved. Even when the contractor has primary scheduling responsibility, which is most often the case, the owner should nonetheless monitor the progress of the work and the scheduling effort. A project schedule, required on most construction projects, provides the easiest means of monitoring the work. The critical path method (CPM) schedule required by the prime contract on many large projects can be even more valuable as a scheduling tool if properly developed, updated, and used.

The input of subcontractors and all project participants in the development and updating of any project schedule is critical to its usefulness. As a practical matter, a schedule that is developed without the input of the parties actually performing the work may result in an unworkable product; hence the schedule as an instrument of coordination will be wasted. If all involved parties participate in the preparation of the schedule, it becomes a much more meaningful and productive project management device. In addition, through its involvement, each party has in effect admitted or acknowledged what was reasonable and expected of it. If a party later fails to perform or follow the schedule, its ability to dispute the relevance of the project schedule and what was required of it can be substantially reduced.

A project schedule can be a double-edged sword for the prime contractor, particularly if it is a CPM that shows the interrelationship of all activities and trades.

A properly developed schedule can be used to demonstrate how a subcontractor is behind schedule and how its delayed performance is impacting the entire project.[28] Conversely, a subcontractor also may use a project schedule against the prime contractor to show how the subcontractor reasonably expected and planned to proceed with the work and how that plan was disrupted by the prime contractor, another subcontractor, or the owner, for which the affected subcontractor may be entitled to additional compensation.[29] If the schedule is not properly maintained, updated, and enforced such that it bears little relationship to the actual progress of the work or the parties' contractual obligations, it may be dismissed by a court or arbitrators as merely representing "theoretical aspirations rather than practical contract requirements."[30] But if a contractor regularly meets with the subcontractors to verify the timing and duration of construction activities, it will have a better chance of enforcing the schedule.[31] The heavy use of scheduling information and analysis in resolving claims underscores the importance of preparing, and maintaining through updates, a realistic schedule that secures subcontractor involvement and agreement. Computer software applications are available, such as Primavera P6, which allow collaborative online scheduling. Proper use of these tools can help management reduce the risks of costly project delays.

XIV. PRESERVING ELECTRONICALLY STORED INFORMATION

In 2006, the Federal Rules of Civil Procedure were revised to specifically address and allow for the discovery (pretrial investigation and review) of electronically stored information (ESI) in the federal court system. Most states will likely develop similar if not identical rules. ESI is defined broadly to include "writings, drawings, graphs, charts, photographs, sound recordings, images, and other data or data compilations stored in any medium from which information can be obtained."[32] ESI sources include: forensic copies of personal computers, company file servers, e-mail servers, e-mail, electronic documents and spreadsheets, scanned documents and images, backup tapes, floppy disks, CDs, Web-based e-mail, instant messaging (IM), voicemail, calendars, internal database systems, sales management systems, network access information, personal digital assistants, iPods, BlackBerrys, and other portable storage devices.

If a party reasonably anticipates that a construction claim or dispute will end up in federal court litigation (or in a state that has adopted electronic-discovery rules),

[28] *See, e.g., Williams Enters., Inc. v. Strait Mfg. & Welding, Inc.*, 728 F. Supp. 12 (D.D.C. 1990), *aff'd in part, remanded in part on other grounds,* 938 F.2d 230 (D.C. Cir. 1991); *Carrington v. W.A. Soefker & Son, Inc.* (Tenn. Ct. App. 1981); *Ill. Structural Steel Corp. v. Pathman Constr. Co.,* 318 N.E.2d 232 (Ill. App. Ct. 1974); *Santa Fe, Inc.,* VABCA No. 2168, 87–3 BCA ¶ 20,104; *Santa Fe Eng'rs, Inc.,* ASBCA No. 24578, 94–2 BCA ¶ 26,872; *Kaco Contracting Co.,* ASBCA No. 44937, 01–2 BCA ¶ 31,584.

[29] *See United States ex. rel. R.W. Vaughn Co. v. F.D. Rich Co.,* 439 F.2d 895 (8th Cir. 1971).

[30] *Id.* at 900.

[31] *Williams Enters., Inc. v. Strait Mfg. & Welding, Inc.,* 728 F. Supp. 12 (D.D.C. 1990).

[32] Fed. R. Civ. P. 34(a).

it has an affirmative duty to ensure preservation of all relevant documents, including ESI. The party must suspend its routine document retention/destruction policy and put in place a "litigation hold." The party also must take steps to ensure that all sources of ESI are identified and made subject to the hold. Electronic evidence can play a key role in a construction dispute. Prudent participants in the construction process may wish to consult litigation counsel regarding the policies and procedures for the storage, preservation, and retrieval of ESI.

XV. CONCLUSION

Effectively dealing with construction disputes begins with the recognition that disputes are best avoided. The identifiable recurring causes of claims permits planning and preparation to try to steer clear of major risks or handle claims responsibly when they cannot be avoided. The same policies and procedures that aid in limiting claims also contribute to comprehensive and effective preparation of claims. The potential for claims cannot be ignored. Skillful and determined management are required both before and during construction to handle the threats and challenges they present.

➢ POINTS TO REMEMBER

- When considering a new project, evaluate it against the firm's business plan, experience, and available resources. Find out if the project locale presents unusual risks or requirements.
- Parties can reduce the risk of construction disputes by dealing with reputable companies that have demonstrated their ability to perform. Carefully consider the prospective project participants' financial strength, bonding capacity, licensing, available project management personnel, and experience on similar projects.
- Allocating risks and providing for effective dispute resolution can be accomplished by a well-drafted written contract, particularly when nontraditional construction contract methods are employed.
- Use of generally accepted standard contract forms and an integrated family of documents provides a starting point for parties to negotiate critical variables such as clauses addressing payment terms, time, damages, changes, termination, insurance, and dispute resolution.
- Prudent estimating and bidding, as well as proper project management and documentation, will help reduce the risk that disputes will develop. Moreover, proper project documentation can provide valuable information necessary to accurately reconstruct the events of the project in preparation of a claim.
- Proper project documentation involves procedures to systematically maintain bidding documents, vendor files, correspondence, meeting notes, job-site logs

or daily reports, schedules, standard forms and status logs, photographs, and video recordings.

- Company operating procedures should be standardized, consistent, and followed from project to project.
- Project participants should establish and maintain open lines of communication by engaging in regular job meetings.
- Because problems of proof are inherently associated with construction claims, implementation and maintenance of a good cost accounting system is crucial.
- A realistic project schedule that secures the involvement and agreement of all project participants should be prepared, routinely updated, and maintained.
- The efficient use of collaborative processes, such as building information modeling (BIM) and Web-based project management, can help reduce the risks of delays and disputes.
- If a project participant reasonably anticipates that a construction dispute could involve litigation, it should immediately determine if it must suspend its routine document retention/destruction policy and consult counsel.

APPENDIX 13.1 CONSENSUSDOCS 221—CONTRACTOR'S STATEMENT OF QUALIFICATIONS FOR A SPECIFIC PROJECT

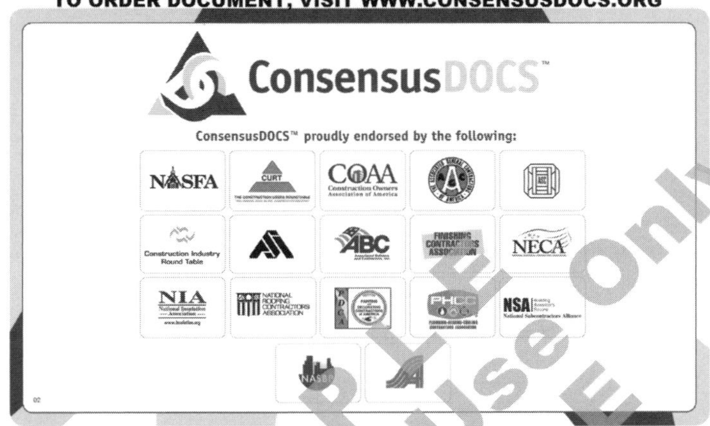

CONSENSUSDOCS 221

CONTRACTOR'S STATEMENT OF QUALIFICATIONS FOR A SPECIFIC PROJECT

This docume t was developed through a collaborative effort of e tities represe ti g a wide cross-sectio of the co structio i dustry. The orga izatio s e dorsi g this docume t believe it represe ts a fair a d reaso able co se sus amo g the collaborati g parties of allocatio of risk a d respo sibilities i a effort to appropriately bala ce the critical i terests a d co cer s of all project participa ts.

These e dorsi g orga izatio s recog ize a d u dersta d that users of this docume t must review a d adapt this docume t to meet their particular eeds, the specific requireme ts of the project, a d applicable laws. Users are e couraged to co sult legal, i sura ce a d surety advisors before modifyi g or completi g this docume t. Further i formatio o this docume t a d the perspectives of e dorsi g orga izatio s is available i the Co se susDOCS Guidebook.

The CONTRACTOR

_______________________________________ (Name)

submits this Stateme t of Qualificatio s dated _____________ to

the OWNER

IMPORTANT: A vertical li e i the margi i dicates a cha ge has bee made to the origi al text. Prior to sig i g, recipie ts may wish to request from the party produci g the docume t a "redli ed" versio i dicati g cha ges to the origi al text. Co sultatio with legal a d i sura ce cou sel a d careful review of the e tire docume t are stro gly e couraged.

in connection with the following PROJECT

The contents of this Contractor's Statement of Qualifications shall be treated as confidential by the Owner.

1 CONTRACTOR'S ORGANIZATION

1.1 General Information

Address:

Telephone and Facsimile: ______________

E-mail address: ____________

Web site: _______________

If address given above is a branch office address, provide principal home office address:

Attach brochure or promotional information.

1.2 Type of Organization

The Contractor's Organization is a:

__ Corporation

Date and State of Incorporation: ______________

Executive Officers: (Names and Addresses)

__ Partnership

Date and State of Organization: ______________

Type of Partnership: __ General __ Limited __ Limited Liability __ Other: _____________

Current General Partners: (Names and Addresses)

__ Joint Venture

Date and State of Organization: ______________

Joint Venturers: (For each indicate the name, address and form and state of organization, as well as the managing or controlling Joint Venturer if applicable.)

__ Limited Liability Company

Date and State of Organization: ________________

Members: (Names and Addresses)

__ Sole Proprietorship

Date and State of Organization: _______________

Owner or Owners: (Names and Addresses)

__ Other

Type of Organization: ______________________

State of Organization: ____________________

Owners and/or Principals: (Names and Addresses)

In addition to the above categories of business entities, indicate whether Contractor's organization is certified as a:

__ Disadvantaged Business Enterprise Certified by: ___________________

__ Minority Business Enterprise Certified by: ___________________

__ Women's Business Enterprise Certified by: ___________________

__ Historically Underutilized Business Zone Small Business Concern Certified by: _________

2 LICENSING AND REGISTRATION

2.1 Jurisdictions in which Contractor is legally qualified to practice: (Indicate license or registration numbers for each jurisdiction, if applicable, and type of license or registration. Attach separate sheet as necessary.)

2.2 In the past five (5) years, has Contractor had any business or professional license suspended or revoked?

__ Yes __ No

If yes, describe circumstances on separate attachment, including jurisdiction and bases for suspension or revocation.

3 CONTRACTOR'S PERSONNEL AND APPROACH

3.1 Key Construction Personnel List on Schedule A, attached, the construction experiences of Contractor's Key Construction Personnel who are directly involved in construction operations and the percentage of time that is anticipated to be devoted to the Project.

3.2 List types of work generally performed by Contractor's own work force:

3.3 Subcontractors

a. Indicate criteria used in the selection of subcontractors (Indicate if Not Applicable).

__ Price

__ Financial strength

__ Bonding capacity

__ Previous experience with Contractor

__ Previous experience in industry

__ Subcontractor's reputation in industry

__ Availability of sufficient personnel

__ Safety record

__ Other: ____________________

b. State Contractor's policy on the bonding of its subcontractors:

3.4 Describe Contractor's proposed technical and management approach to the Project, including approaches to quality, time and cost control: (Attach additional sheets as necessary.)

4 CONTRACTOR'S RELEVANT EXPERIENCE

4.1 Past Projects List on Schedule B, attached, at least five construction projects Contractor has worked on in the past five (5) years with project delivery systems similar to the one to be employed for this Project. (For Joint Ventures, list each joint venturer's projects separately). The Owner may contact the owners of the projects listed on Schedule B.

4.2 Current Projects List on Schedule C, attached, all current projects of the Contractor, including projects not yet underway, approximate dollar value or each and the percentage of completion of each project. (For Joint Ventures, list each joint venturer's projects separately).

4.3 Annual Construction Volume Indicate the annual volume of work completed for the past three years:

Year ____ ______________________

Year ____ ______________________

Year ____ ______________________

4.4 In the past five (5) years, has Contractor defaulted, been terminated for cause or failed to complete a construction contract awarded to it?

__ Yes __ No

If yes, describe circumstances on separate attachment, including dates and owner, and if applicable, Contractor's surety.

4.5 In the past five (5) years has any officer, partner, joint venturer or proprietor of the Contractor ever failed to complete a construction contract awarded to that person or entity in their name or on behalf of another organization?

__ Yes __ No

If yes, describe circumstances on separate attachment, including dates and Owner, and if applicable, surety.

4.6 Describe all litigation arising from Contractor's active projects or projects worked on within the last five (5) years. (Attach additional sheets as necessary.)

5 CONTRACTOR'S SAFETY PROGRAM

5.1 If Contractor has a written safety program, attach a copy.

5.2 Does the Contractor's safety program include instructions on the following: (If yes, attach.)

a. Safety work practices

__ Yes __ No

b. Safety supervision

__ Yes __ No

c. Toolbox safety meetings

__ Yes __ No

d. Emergency procedures

__ Yes __ No

e. First aid procedures

__ Yes __ No

f. Accident investigation

__ Yes __ No

g. Fire protection

__ Yes __ No

h. New workers' orientation

__ Yes __ No

5.3 Do you have a safety officer/department in your company?

__ Yes __ No

If yes,

Name: ____________________

Title: ____________________

Phone: ____________________

5.4 Do you conduct project safety inspections?

__ Yes __ No

If yes, how often? ______________

Who conducts this inspection?

Name: ____________________

Title: ____________________

5.5 Do you hold project safety meetings for field supervisors?

__ Yes __ No

If yes, how often? __ Weekly __ Bi-weekly __ Monthly __ Less often as needed

5.6 Do you have in place an instruction program on safety for newly hired or promoted supervisors?

__ Yes __ No

If yes, please attach a copy of program format.

5.7 If craft "toolbox" safety meetings are held, what is their frequency?

__ Weekly __ Bi-weekly __ Monthly __ Less often as needed

5.8 Do you have a drug and alcohol testing policy?

__ Yes __ No

If Yes, attach a copy of the policy.

5.9 Provide Contractor's OSHA No. 300 Log and Summary of Occupational Injuries and Illnesses for the past five (5) years.

5.10 List all OSHA Citations and Notifications of Penalty, monetary or other, received within the last five (5) years: (Indicate final disposition as applicable. Attach additional sheets as necessary.)

5.11 List all safety citations of violations under state law received within the last five (5) years: (Indicate final disposition as applicable. Attach additional sheets as necessary.)

5.12 List Contractor's Workers' Compensation Experience Modification Rate (EMR) for the past five (5) years. (The EMR may be obtained from your insurance agent. Attach a copy of the insurance agent's EMR verification letter.)

Year _____ EMR: _______________________

Year _____ EMR: _______________________

Year _____ EMR: _______________________

Year _____ EMR: _______________________

Year _____ EMR: _______________________

5.13 List Contractor's Total Recordable Frequency Rate (TRFR) for the past five (5) years. (The TRFR may be obtained from your insurance agent. Attach a copy of the insurance agent's TRFR verification letter.)

Year _____ TRFR: _______________________

Year _____ TRFR: _______________________

Year _____ TRFR: _______________________

Year _____ TRFR: _______________________

Year _____ TRFR: _______________________

5.14 List Contractor's total number of man hours worked for the past five (5) years:

Year: _____ Total Number of Man Hours Worked: ________

Year: _____ Total Number of Man Hours Worked: ________

Year: _____ Total Number of Man Hours Worked: ________

Year: _____ Total Number of Man Hours Worked: ________

Year: _____ Total Number of Man Hours Worked: ________

5.15 Comment on any additional areas of your company's safety program and policies that you feel will be appropriate in our evaluation:

6 SURETY AND INSURANCE

6.1 Surety Company: (Name and Address)

6.2 Agent: (Name, Address and Telephone Number)

6.3 Total bonding capacity:$ ________________

Limit per project: $ ______________________

6.4 Available bonding capacity as of this date: $ ________________

6.5 Can Contractor provide a bid bond for this project, if applicable?

__ Yes __ No

6.6 Commercial General Liability Carrier and summary of liability coverage:

7 CONTRACTOR FINANCIAL INFORMATION

7.1 List principal banks used, the approximate value of outstanding loans and general repayment history, as well as the Name, Address and Telephone Number of a contact person:

7.2 Attach audited financial statements for the past three (3) years, including latest balance sheet, containing but not limited to the following information:

a. Current Assets

b. Net Fixed Assets

c. Other Assets

d. Current Liabilities (i.e. accounts payable, notes payable, accrued expenses, provision for income taxes, advances, accrued salaries and accrued payroll taxes)

e. Other Liabilities (i.e. capital, capital stock, authorized and outstanding shares par values, earned surplus)

f. Retained Earnings and Net Worth

g. Date of Statement

h. Name of firm preparing statement

7.3 State whether Contractor, or any of the individuals identified in Article 1, has/have been the subject of any bankruptcy proceeding within the last five (5) years.

__ Yes __ No

If yes, describe circumstances on separate attachment.

8 INDUSTRY AGREEMENTS, AFFILIATIONS, MEMBERSHIPS, AWARDS AND HONORS

8.1 List Trade Unions or Associations with which Contractor has an Agreement:

Trade ________________

National Agreement ________________

Local Agreement ________________

Expires __________

Trade ________________

National Agreement ________________

Local Agreement ________________

Expires __________

Trade ________________

National Agreement ________________

Local Agreement ________________

Expires __________

8.2 Industry Affiliations and Memberships:

8.3 Industry Awards and Honors and dates:

9 STATEMENT OF POTENTIAL CONFLICTS OF INTEREST

Provide information about any business associations, financial interests or other circumstances that may create a conflict of interest with the Owner or any other Party known to be involved in the Project.

10 OTHER INFORMATION

10.1 Within the past five (5) years, has Contractor, or any of the individuals identified in Paragraph 1.2 and/or Schedule A been the subject of any criminal indictment or judgment of conviction for any business-related conduct constituting a crime under state or federal law?

__ Yes __ No

If yes, describe circumstances on separate attachment.

10.2 Within the past five (5) years, has Contractor or any of the individuals identified in Paragraph 1.2 and/or Schedule A been the subject of any federal or state suspension or disbarment?

__ Yes __ No

If yes, describe circumstances on separate attachment.

10.3 Within the past five (5) years, has Contractor, or any of the individuals identified in Paragraph 1.2 and/or Schedule A been the subject of any formal proceeding or consent order with a state or federal environmental agency involving a violation of state or federal environmental laws?

__ Yes __ No

If yes, describe circumstances. (Attach additional sheets as necessary.)

11 REFERENCES

Provide references for each of the following categories. (Additional references may be provided on a separate attached sheet.)

Owner

Name: ______________________

Address: ______________________________

Telephone No.: ____________________

Contact Person: _________________________

Architect/Engineer

Name: ______________________

Address: ______________________________

Telephone No.: ____________________

Contact Person: _________________________

Subcontractor

Name: ______________________

Address: ______________________________

Telephone No.: ____________________

Contact Person: _________________________

The Undersigned, on behalf of the Contractor, certifies under oath that the information provided herein, including any schedule or attachment, is true and sufficiently complete so as not to be misleading.

CONTRACTOR

By: ....

Title: ______________________________________

Date: _________________________

APPENDIX 13.2 CHECKLIST: PREBID/PROPOSAL ENVIRONMENTAL CONSIDERATIONS

RISK ASSESSED BY:	a	* *
Contract Documents		
Do the plans or specs include any remediation work, asbestos abatement, or lead-based paint removal?		
Do the plans or specs refer to the performance by others of remediation work, asbestos abatement, or lead based paint removal?		
Do the Contract Documents attempt to allocate the risk or responsibility for hazardous materials or asbestos?		
Is there a contract requirement for the indemnification of the Owner or others for losses associated with hazardous materials or asbestos?		
Is the Contractor/Subcontractor indemnified against losses associated with hazardous waste or asbestos encountered on the Project?		
If unanticipated asbestos, lead-based paint, tanks, drums, contaminated soil, PCBs or other hazardous material or waste are encountered, does the Contract specify the Contractor's rights and duties?		
If the Contract completion is delayed by unexpected environmental problems, does the Contract provide for a time extension, an equitable Contract price adjustment, or termination rights?		
Are any environmental permits necessary?		
Who is contractually responsible for any necessary environmental permits?		
What is the extent of the site investigation required by the Contract Documents?		
Site Observation		
Did the site visit indicate any distressed vegetation, hydrocarbon or chemical contamination, underground storage tanks, transformers, drums, suspect lead-based paint or suspect asbestos-containing materials?		
Has any prior cleanup taken place at the site?		
Have tanks been removed or filled?		
Is there any Closure Report?		
Are there structures on the site which must be moved or demolished?		
What is the extent of the required excavation and grading operation?		
Will fill material be brought to the site?		
Will waste material be hauled off to another site?		

APPENDIX 13.2 (CONTINUED)

RISK ASSESSED BY:	a	* *
Owner-Furnished Information		
Has the Owner been asked to disclose any available information regarding any hazardous materials, asbestos, lead-based paint, waste materials, contaminated soil, or water that might affect the Contractor's work?		
Has a request been made for the results of any Environmental Site Assessment?		
Has a request been made for any building surveys or inspection reports on asbestos, lead-based paint, underground storage tanks, soil or water quality, or hazardous waste present at the site?		
Has a request been made for all information regarding any prior clean-up on or near the site?		
Environmental Risk Allocation		
If environmental problems are encountered, is insurance coverage available?		
Are there insurance policy exclusions or limits which affect potential insurance coverage?		
Are environmental risks shared with appropriate Subcontractors?		
If Subcontractors will be or may be involved in work which could encounter environmental problems, do the Subcontractors have adequate insurance coverage or proper bond protection?		
Personnel Readiness		
Are responsible project personnel familiar with and able to recognize likely environmental hazards?		
Is there an in-house response plan for dealing with environmental hazards?		
Other Environmental Considerations		

APPENDIX 13.3 CHECKLIST: CONTRACTS IN FOREIGN STATES

RISK ASSESSED BY:	✓	COMMENTS
Foreign Corporation Registration Requirements		
Determine Required Fees		
Obtain Certificate of Authority		
Establish Registered Agent/Office		
State Revenue Department Requirements		
Tax Bond Requirements		
Consequences of Violation		
Sales and Use Tax Liability		
Ad Valorem Taxes		
State Licensing Requirements		
Application Process		
Determine Testing Requirements		
Obtain Before Bid?		
Consequences of Violation		
Environmental Permits/Filings		
Vehicle/Equipment Licenses		
Local Licensing/Permit Requirements		
City/County License		
Permit Requirements		
Building Code Requirements		
Other Requirements		
Special Insurance Requirements		
Nature of Insurance Required		
Additional Rider Required?		
Filing/Reporting Requirements		
Preservation Lien/Bond Rights		
Public or Private Project?		
Verify Existence of Lien Rights		
Property Legal Description/Owner Information		
Verify Payment Bond(s)		
Pre-Contract Filing Requirements		
Preliminary/Early Notice Requirements		

APPENDIX 13.3 (CONTINUED)

RISK ASSESSED BY:	a	COMMENTS
Post-performance Notice Requirements		
Limitations on Lien/Bond Recovery		
Request for Bond/Lien Information		
Mandatory Lien Waiver Forms		
Payment Security Verification		
Verify Adequacy of Financing		
Notice to Lender Requirements		
Lender Approvals Required		
Extra Work Payment Limitations		
Change Order Authority Limitations		
Existence of State Prompt Payment Statute		
Trust Fund/Retainage Requirements		
Existence/Validity of Pay-If-Paid Clause		
Existence/Validity of No-Damages-For-Delay Clause		
Impact of Foreign Jurisdiction's Laws		
Identification of Controlling Law		
Public Policy Limits on Exculpatory Clauses		
Impact on Subcontract/Purchase Orders		
Unusual State/Local Requirements		
Labor, Equipment, and Material Impacts		
Available Local Labor		
Union/Non-Union Market		
Materials/Equipment Availability		
Applicable Sales Taxes		
Applicable Labor Rates/Fringes		
Project/Union Agreements		
Political Climate Assessment		
MBE/WBE Requirements		
Identify Special Political Ties/Circumstances		
Business Climate Considerations		
Federal, State, Local Immigration Laws and Regulations		

APPENDIX 13.3 (CONTINUED)

RISK ASSESSED BY:	a	COMMENTS
Other Special Risks		

APPENDIX 13.4 LOGS AND FORMS

13.4A	Format for Notice Checklist
13.4B	Sample Partial Notice Checklists
13.4C	Forms [Five sample notice letters]
13.4D	Request for Information
13.4E	Telephone Conversation Memorandum
13.4F	Sample Daily Report
13.4G	Notice of Backcharge Work to Be Performed
13.4H	Field Order Status Chart
13.4I	Correspondence Log—Incoming
13.4J	Correspondence Log—Outgoing

APPENDIX 13.4A FORMAT FOR NOTICE CHECKLIST

Clause Reference	Subject Matter Of Notice	Time Requirement for Notice	Form of Notice	Stated Consequences of Lack of Notice
Changes Paragraph # ___	Proposal for adjustment	(Sent)(Rec'd) in ___________ days Triggering Event: Other Action Required: ________________ ________________	__ Written __ Certified __ Registered Sent to: ___________	
Constructive Changes Paragraph # ___	Date, circumstances, and source of the order and that the contractor regards the order as a contract change	(Sent)(Rec'd) in __________ days Triggering Event: Other Action Required: ________________ ________________	___ Written ___ Certified ___ Registered Sent to: ___________	
Differing Site Conditions Paragraph # ___	Existence of unknown or materially different conditions affecting the contractor's cost	(Sent)(Rec'd) in ___ days Triggering Event: Other Action Required: ________________ ________________	___ Written ___ Certified ___ Registered Sent to: ___________	
Suspension of Work Paragraph # ___	The act or failure to act involved and the amount claimed	(Sent)(Rec'd) in ___ days Triggering Event: Other Action Required: ________________ ________________	___ Written ___ Certified ___ Registered Sent to: ___________	
Time Extensions Paragraph # ___	Causes of delay beyond contractor's control	(Sent)(Rec'd) in ___ days Triggering Event: Other Action Required: ________________ ________________	___ Written ___ Certified ___ Registered Sent to: ___________	

Clause Reference	Subject Matter Of Notice	Time Requirement for Notice	Form of Notice	Stated Consequences of Lack of Notice
Claims Paragraph # ___	Notice of event or condition giving rise to a claim	(Sent)(Rec'd) in _____ days Triggering Event: Other Action Required: ________________ ________________	___ Written ___ Certified ___ Registered Sent to: ____________	
Termination for Default Paragraph # ___	Notice of intent to terminate for default	(Sent)(Rec'd) in ___ days Triggering Event: Other Action Required: ________________	___ Written ___ Certified ___ Registered Sent to: ____________	
Termination for Convenience Paragraph # ___	Notice of intent to invoke right to terminate for convenience	(Sent)(Rec'd) in ___ days Triggering Event: Other Action Required: ________________ ________________	___ Written ___ Certified ___ Registered Sent to: ____________	
Injury or Damage to Person or Property Paragraph # ___	Claim of injury or damage to property caused by act or omission of other party or agent	(Sent)(Rec'd) in ___ days Triggering Event: Other Action Required: ________________ ________________	___ Written ___ Certified ___ Registered Sent to: ____________	
Arbitration Notices and Demands Paragraph # ___	Demand for arbitration	(Sent)(Rec'd) in ___ days Triggering Event: Other Action Required: __________________ __________________	___ Written ___ Certified ___ Registered Sent to: ____________	
Disputes Paragraph # ___	Appeal of A/E or C.O. Final Decision	(Sent)(Rec'd) in ___ days Triggering Event: Other Action Required: __________________ __________________	___ Written ___ Certified ___ Registered Sent to: ____________	
Mechanic's Lien Paragraph # ___	Notices to be sent or filed to preserve lien rights	1st Notice Required: ___________ Notice Deadline: ______ Other Action Required: __________________ __________________ Foreclosure Deadline:	___ Written ___ Certified ___ Registered Sent to: ____________	

APPENDIX 13.4B SAMPLE PARTIAL NOTICE CHECKLISTS

Sample Notice Checklist Federal Government Contracts

Clause Reference	Subject Matter of Notice	Time Requirements For Notice	Writing Required	Stated Consequences Of a Lack of Notice
Changes FAR 52.243–4	Proposal for adjustment.	**30 DAYS** from receipt of a written change order from the Gov't or written notification of a constructive change by the contractor.	Yes	Claim may not be allowed. Notice requirement may be waived until final payment.
Constructive Changes FAR 52.243–4	Date, circumstances, and source of the order & that the contractor regards the Gov't's order as a contract change.	No starting point stated, but notice within **20 DAYS** of incurring any additional costs due to the constructive change fully protects the contractor's rights.	Yes	Costs incurred more than **20 DAYS** prior to giving notice cannot be recovered, except in the case of defective specifications.
Differing Site Conditions FAR 52.236–2	Existence of unknown or materially different conditions affecting the contractor's cost.	From the time such conditions are identified, notice must be furnished "promptly" and before such conditions are disturbed.	Yes	Claim not allowed. Lack of notice may be waived until final payment.
Suspension of Work FAR 52.242–14	(1) Of "the act or failure to act involved," and	(1) Within **20 DAYS** from the act or failure to act by the C.O. (not including a suspension order.) (2) "As soon as practicable" after termination of the suspension, delay or interruption.	(1) Yes (2) Yes	(1) Costs incurred more than **20 DAYS** prior to notification cannot be recovered. (2) Claim not allowed, but claim may be considered until final payment.

Clause Reference	Subject Matter of Notice	Time Requirements For Notice	Writing Required	Stated Consequences Of a Lack of Notice
Termination for Default Damages for Delay—Time Extensions FAR 52.249–10	Causes of delay beyond contractor's control.	**10 DAYS** from the beginning of any delay.	Yes	Contractor's right to proceed may be terminated and the Government may sue for damages.
Disputes FAR 52.233–1	Appeal of any final decision by the Contracting Officer (C.O.).	(1) Boards of Contract Appeals—**90 DAYS** from receipt of C.O.'s final decision. (2) U.S. Court of Fed. Claims—**1 YEAR** from receipt of C.O.'s final decision.	(1) Yes—Notice of Appeal (2) Yes—Filing of Complaint	C.O.'s decision becomes final and conclusive. C.O.'s decision becomes final and conclusive.

Sample Notice Checklist: AIA A201 General Conditions (2007)

Clause Reference	Subject Matter of Notice	Time Requirements For Notice	Writing Required	Stated Consequences Of a Lack of Notice
Time Limits on Claims: Articles 15.1.2	Claims by either party. Contractor claims for time shall include cost estimate; Claims for increase in price due before starting work, except for emergency.	Within **21 DAYS** after claimant first recognizes condition giving rise to claim or **21 DAYS** after occurrence of event giving rise to claim, whichever is later. (Note: Claim must be submitted to other party and Initial Decision Maker (IDM)).	Yes	?????????
Concealed or Unknown Site Conditions: Article 3.7.4	Materially different, subsurface or otherwise concealed physical site conditions.	Promptly, before disturbing the conditions but no later than **21 DAYS** after observing such conditions. If the Architect does not agree, a claim must be filed within **21 DAYS**.	Yes	?????????

APPENDIX 13.4B (CONTINUED)

Clause Reference	**Subject Matter of Notice**	**Time Requirements For Notice**	**Writing Required**	**Stated Consequences Of a Lack of Notice**
Injury or Damage to Person or Property Article 10.2.8	Claim of injury or damage to property caused by act or omission of other party or agent.	Within a reasonable time not exceeding **21 DAYS** after discovery. Notice must provide enough detail to permit the other party to investigate the matter.	Yes	?????????
Mediation Notices and Demands Article 15.3	Demand for Mediation. (Note: Mediation is a condition precedent to binding dispute resolution.)	The IDM's decision on claim is subject to immediate Mediation. Party may file for Mediation at any time, subject to the following: Either party may, within **30 DAYS** from the date of IDM's decision, demand in writing that the other party file for Mediation within 60 days of initial decision.	Yes	If 15.2.6.1 is invoked, failure to meet the **60 DAY** period for filing Mediation demand will result in a waiver of Mediation and binding dispute resolution rights. IDM's decision will be final. If 15.2.6.1 is not invoked, party may file for Mediation at any time following initial decision.

APPENDIX 13.4B (CONTINUED)

Clause Reference	Subject Matter of Notice	Time Requirements For Notice	Writing Required	Stated Consequences Of a Lack of Notice
Arbitration Notices and Demands Article 15.4	Demand for Arbitration. (Note: Mediation is not mandatory default, but must be selected by the parties. See A101 6.2. Mediation must precede Arbitration.)	If the parties selected Arbitration, once the IDM has rendered a decision on Claim and Mediation has not resulted in settlement, that decision is subject to immediate Arbitration. Demand for Arbitration shall be made no earlier than concurrently with Mediation request, but in no event after applicable statute of limitations period.	Yes. See 15.4.1 regarding need to assert other Claims when filing the Arbitration demand.	If 15.2.6.1 is invoked, failure to meet the **60 DAY** period for filing Mediation demand will result in a waiver of Arbitration rights. IDM's decision will be final. If 15.2.6.1 is not invoked, party may file for Arbitration concurrently with Mediation, or at any time within the statute of limitations period following Mediation.
General Notes on Preparation of a Checklist				
The **Tables** above are sample formats for **Notice Checklists**. Regardless of your familiarity with the contract, each contract should be carefully reviewed as **special notice requirements are often in *"standard"* contracts!** The checklist should identify the clause, time requirements for notice, the subject of the notice, whether notice must be in writing and the stated consequences for failing to give notice. **The checklist should not be provided to the project staff**. Rather, those responsible for giving timely notice should prepare the checklist for every contract. The checklist can be contained on a single sheet of paper, three hole punched, and retained in the project manual.				

APPENDIX 13.4C FORMS

SAMPLE NOTICE LETTER: EXTENSION OF TIME FOR DELAYS (AND EXTRA COSTS IF APPROPRIATE)

ECC # _____

SENT VIA: [U.S. Mail, Electronic Mail, Overnight Delivery, Facsimile, etc.]

Addressee:

(To Prime Contractor) or

(Owner and Design Professional)

Dear:

We are continuing to pursue the completion of our work as rapidly as is reasonably possible under the current circumstances. We have, however, recently encountered certain delays to our performance through no fault of our own and which are beyond our control. We have continued to keep your job representatives informed of these delays and of their effect on overall job completion. You may be assured that we will diligently seek to reasonably minimize the effects of these delays on our work.

Specifically, we have been delayed in the following particulars:

Accordingly, we hereby request an extension of [_______ days]* to our contract completion to take into consideration the above delays under Clause _______ of the contract provisions.

**[The foregoing delays have also impacted our costs of performance, where it has taken additional time to perform the work. Such increased costs include, but may not be limited to: additional labor and premium time labor; additional costs for supervision; overhead; equipment; rentals; and loss of efficiency for direct labor. Accordingly, this is to place you on notice that we are entitled to additional compensation for all costs flowing from these delays and interference that have been imposed on us through no fault of our own. We will provide you with the specific amount of additional compensation covered by this notice as soon as we research this matter and have computed it.]

Sincerely yours,

Eager Construction Company, Inc.

By ______________________
(Title)

* To be inserted where specific time of delay is known.

** To be used where extra money is claimed for delay.

APPENDIX 13.4C (CONTINUED)

SAMPLE NOTICE LETTER: CLAIM FOR EXTRAS

ECC # _____

SENT VIA: [U.S. Mail, Electronic Mail, Overnight Delivery, Facsimile, etc.]

Addressee:

(To Prime Contractor) or

(Owner and Design Professional)

Re: (Describe Extra Work)

Dear:

This is to notify you that (on _______________________ we will begin) (we are about to begin) this extra work and are expecting to be compensated for it. If you do not want us to perform this work as an extra to the contract, please immediately notify us before we incur additional costs in the preparation for performance of this extra work. If we do not hear from you right away, we will proceed on the basis that you agree with our plan to perform this work.

OR

This work was performed pursuant to your representative's requirement and entitles us to additional compensation. We have proceeded to complete this work so as to minimize the cost of the work and any delay to (our work) (the job). We will be pleased to review this matter with you at your convenience.

We will provide you with a detailed cost breakdown for this added work as soon as we are able to compute it.

Sincerely yours,

Eager Construction Company, Inc.

By _______________________
(Title)

APPENDIX 13.4C (CONTINUED)

SAMPLE NOTICE LETTER: CONFIRMING CHANGE DIRECTIVE

ECC # _____

SENT VIA: [U.S. Mail, Electronic Mail, Overnight Delivery, Facsimile, etc.]

DATE

SUBJECT (Contract Name)

Dear:

We were given instructions by (insert name) on (date) (put in time also if pertinent) to (describe work added or changed).

This change order is for work not within the scope of our present contract, and we therefore request a written modification to cover the added (material, labor, equipment, etc.) required to perform the work as ordered. (Give notice of other factors involved such as delay, acceleration, diversion of men or equipment from contract work, material shortages, etc.)

Our proposal for the added cost resulting from this change order is being prepared and will be submitted for your approval as soon as possible. We cannot determine at this time the effect on contract completion date or other work under the contract, and will advise when a full analysis has been made.

As ordered, we (are proceeding) (have proceeded) at once to (procure materials) (perform the work) in order to complete this change order at the earliest possible time. In the event you do not approve of such action, please advise immediately in order that we may stop this effort and minimize the cost involved.

Your signature at the bottom of this letter will satisfactorily confirm the oral instructions.

Very truly yours,

Eager Construction Company, Inc.

By ________________________
(Title)

Confirmation:

The above-stated report of our instruction is confirmed.

COMPANY: ________________________

BY: ________________________

Title: ________________________

File No.: ________________________

APPENDIX 13.4C (CONTINUED)

SAMPLE NOTICE LETTER: ORAL DIRECTIONS OF EXTRA WORK

ECC # _____

SENT VIA: [U.S. Mail, Electronic Mail, Overnight Delivery, Facsimile, etc.]

DATE

SUBJECT (Contract Name)

Dear:

On the _______ day of _______________, 20_______, we received certain oral instruction (or orders, approvals, changes, as the case may be) from (insert name). These instructions were confirmed by our _______________, 20_______, letter and should have been given to us in writing under the terms of our agreement. Your (insert name) has refused to confirm the oral instructions (or orders, approvals, changes, as the case may be) that we have recited in our referenced letter. Accordingly, we must advise that we will not (proceed with) (continue to follow) these verbal instructions unless we receive your immediate written confirmation. In any event, we will expect reimbursement for all costs reasonably incurred in reliance upon your direction.

We understand that it may take time to go though all the steps necessary to bring about a written authorization for extra work, and that sometimes it is more practical to do the work before that written authorization can be obtained. It has been our past practice to try to recognize your need to follow this method of operation. However, in this case, and in order to avoid any misunderstanding, we think it appropriate that you first provide us with a formal written authorization for changed work.

Very truly yours,

Eager Construction Company, Inc.

By _________________________
(Title)

NOTE: Where the work already has been performed, it may be important to establish a prior history of reliance by the parties on oral directives. If the work has been fully performed, then the second paragraph should be deleted and the last sentence of the first paragraph replaced with the following:

As you know, we proceeded immediately as directed to perform this additional work. We did so in order to minimize your extra cost, and in the same manner in which we have handled other verbal directives in the past. Consistent with that past practice, we will provide you with our costs as soon as they are fully known and expect your prompt reimbursement.

APPENDIX 13.4C (CONTINUED)

SAMPLE NOTICE LETTER: CONFIRMING EXTRA COST DIRECTIVE

ECC # _____

SENT VIA: [U.S. Mail, Electronic Mail, Overnight Delivery, Facsimile, etc.]

DATE

SUBJECT (Contract Name)

Dear:

We were given direction by (name or letter dated __________) on (date) (put in time also if pertinent) to (describe work and specific location). This directive stipulates and orders that we are to complete this work by (date and time).

This directive necessarily (accelerates, delays, diverts men and equipment from contract work, involves inefficiencies, interrupts contract work, involves excessive working hours, shortages, causes manpower shortage for contract work, creates inefficient working conditions, involves work under hazardous conditions, etc.) and thereby will result in increased cost to __________ on this contract.

Our proposal for the added costs resulting from this directive is being prepared and will be submitted for your approval as soon as possible. We cannot determine at this time the effect on contract completion and will advise after a full analysis has been made.

Very truly yours,

Eager Construction Company, Inc.

By ________________________
(Title)

File No.________________________

APPENDIX 13.4D REQUEST FOR INFORMATION

REQUEST FOR INFORMATION

TO: ____________________________ **DATE:** ____________________

____________________________ **PROJECT:** ____________________

____

ATTENTION:____

We are this date requesting the following information, clarification, or direction:

____

____

____

____

____

____

____

____

The above information is needed:

- As soon as possible, to avoid *disrupting* the work
- Immediately, to minimize *disruption* and added costs already being incurred
- Not later than ____________, or the work may experience *disruptions* and added costs

Thank you for your prompt attention to this matter.

COMPANY: ________________________

BY: ____________________________

(Signature and Title)

Response:

____

____

____

____

____

____

COMPANY: ________________________

BY: ____________________________

(Signature and Title)

DATE: ________________________

APPENDIX 13.4E TELEPHONE CONVERSATION MEMORANDUM

TO: ______________________

DATE: ______________________

PROJECT: ______________________

I talked to ______________________
at telephone number ______________________
regarding ______________________
____
____
____

To: ______________________

For your files and use, I am providing you with this confirmation of our telephone conversation described above.

____
____

Signature

Copies to:

- Construction Manager
- Project Manager
- General Superintendent
- Job Superintendent
- Construction Accounting

A Confirmation Letter or E-Mail:

- Has been sent
- Will be sent
- Need not be sent

APPENDIX 13.4F SAMPLE DAILY REPORT

DAILY REPORT

Job or Area: ______________________ **Date:** ______________

Weather Conditions: ______________________

____

____

____

Temperature: ______________________

____

High: __________ Low: __________

Rainfall Amount: ______________________

Critical Activities Affected Duration of:

By Adverse Weather ______ By Adverse Weather ______

Personnel	G. Foreman	Foreman	Journeyman	Apprentice	Labor	New Hires	Laid Off
Carpenters							
Laborers							
Operators							
Finishers							
Teamsters							
Pipe Fitters							
Pipe Laborers							

Work Performed Today: ______________________

Remarks: (Such as testing, conflicts, verbal instructions, delays, safety problems, visitors)

Major Equipment and Materials Received:

Item *Carrier* *Description*

Subcontractors *Work Performed and Number of People*

Superintendent:

APPENDIX 13.4G NOTICE OF BACKCHARGE

WORK TO BE PERFORMED

Location: ____________________________ Backcharge No.: ________________________
Charge To: ___________________________ Subcontractor No. ______________________
Date: ________________________________ P.O. Reference: ________________________

Description: ____
____
____
____

Notes:

☐ Labor shall be charged at actual costs plus ______ % to cover payroll additives.
☐ Material shall be charged at actual delivered cost.
☐ Equipment rental shall be charged at prevailing job site rates of the area.
☐ _____ % shall be added for indirect costs, overhead, supervision, and administration.

APPROVALS

____

Representative **Date** **Supplier/Subcontractor**

APPENDIX 13.4H FIELD ORDER STATUS CHART

Project Name: ______________________ **Page No.** ______________________

Project Number: ______________________ **Status Date:** ______________________

Owner: ______________________ **Contractor:** ______________________

Job NO.	Owner's Field Order #	Authorized By	Date of Order	Description of Work	Work Performed By	Date Work Started	Date Work Completed	Comments on Work	Invoice No.	Date of Bill	Amount of Bill	Amount Paid
									1.			
									2.			
									1.			
									2.			
									1.			
									2.			
									1.			
									2.			
									1.			
									2.			
									1.			
									2.			
									1.			
									2.			
									1.			
									2.			
									1.			
									2.			
									1.			
									2.			
									1.			
									2.			
									1.			
									2			

To: ____________ To keep you aware of the status of added work directed in the field, we will submit updates of this status chart as frequently as is practical. With this form you should be able to track our planned start dates for field work and the status of the work. In the absence of a field order number from you, we will assign our own field order number to each field directive.

APPENDIX 13.4I CORRESPONDENCE LOG—INCOMING

Incoming Correspondence From ____________________ **(Contractor/Supplier)**							
				Reply Required		Reply	
Ref. No.	Date	Subject	Author	Yes	No	Ref. No.	Date
						Page No. ______	

APPENDIX 13.4J CORRESPONDENCE LOG: OUTGOING

Incoming Correspondence To ______________________________

(Contractor/Supplier)

				Reply Required		Reply	
Ref. No.	Date	Subject	Author	Yes	No	Ref. No.	Date
						Page No. _______	

14

PAYMENT BONDS

When subcontractors or suppliers are not paid for the labor and services they have provided to a construction project, they may be able to recover payment by making claims on a payment bond, if the contractor provided one for the project. A *payment bond* is an agreement between a contractor and a surety by which the surety guarantees payment for the labor and materials contracted for and used by the contractor on a particular project. Payment bonds have long been required by statute for public construction projects and frequently are mandated in private construction contracts.[1] When the owner—whether a governmental or private entity—requires a payment bond, the contractor obtains a payment bond from a surety and furnishes the bond to the owner.

Payment bonds required on federal, state, or local government construction projects are governed by statute.[2] The most influential such statute is the federal Miller Act, 40 U.S.C. §§ 3131–3134 (the "Miller Act"), which governs payment bonds on federal government construction projects. Payment bonds on private projects are essentially private contractual undertakings governed by the terms of the bond itself. Whether statutory or private, payment bonds are subject to notice and timing requirements relating to claims and litigation. Potential claimants must comply with these requirements or risk waiving otherwise valid claims.

In exchange for providing a payment bond, the surety typically requires contractors or their principals to indemnify the surety for any expenses incurred in evaluating and paying claims on that payment bond. The contractor may be required to compensate the surety for its expenses even if the surety neither pays nor litigates payment bond claims arising on the project.

[1]*See* ConsensusDOCS 200, ¶ 10.7 (2007 ed.); AIA A201, § 11.4.1 (2007 ed.) *and* AIA A201, § 11.5.1 (1997 ed.).

[2]*See, e.g., U.S. Filter Distribution Group Inc. v. Katspan, Inc.,* 72 P.3d 1103 (Wash. Ct. App. 2003) (surety's liability to subcontractors and suppliers for a public works project governed by statute and surety agreement, not by the contract between a supplier and a primary contractor).

This chapter examines the scope of protection offered by statutory and private payment bonds: which project participants may recover; which types of work are covered; and whether and to what extent delay damages, extra work, and other costs qualify for payment bond coverage. The chapter differentiates between performance bond claims and payment bond claims and addresses the relationship between payment bond claims and lien rights. Finally, the chapter explores the defenses to payment available to payment bond sureties: time of notice, time of lawsuit; misrepresented status of payments, and claim or lien waivers.

I. PAYMENT BONDS REQUIRED BY STATUTE

A. The Miller Act

The most influential statute governing payment bonds is the federal Miller Act, which requires:

> [b]efore any contract of more than $100,000 is awarded for the construction, alteration, or repair of any public building or public work of the Federal Government, a person must furnish to the Government … [a] payment bond … for the protection of all persons supplying labor and material in carrying out the work provided for in the contract.
>
> * * *
>
> Accordingly, a general contractor entering into a public building or public works contract with the federal government must furnish a payment bond in an amount equal to the contract price, unless the contracting officer determines that it is impractical to obtain a bond in that amount and specifies an alternative bond amount.[3]

Miller Act payment bonds guarantee payment to parties supplying labor and materials to contractors or subcontractors engaged in the construction, alteration, or repair of any public building or public work of the United States. A payment bond may provide the only remedy—a right of recovery against the surety issuing the payment bond—available to unpaid subcontractors and material suppliers on public projects. If the contractor failed to furnish a payment bond where one should have been required under the Miller Act, the would-be bond claimants have no recourse against the government.[4] The Miller Act does not grant jurisdiction to federal courts

[3]On projects in excess of $30,000 up to $100,000, the contracting officer is required to obtain either a payment bond or alternative payment protection equal to 100% of the contract price. 40 U.S.C. § 3132; FAR § 28.101–1(b)(1); FAR § 28.102–2(c).

[4]*Sloan Constr. Co., Inc. v. Southco Grassing, Inc.,* 629 S.E.2d 372 (S.C. Ct. App. 2006) (citing *Active Fire Sprinkler Corp. v. U.S. Postal Serv.,* 811 F.2d 747, 752 (2d Cir. 1987)) ("The Miller Act does not provide subcontractors with a right of recovery against the United States."); *Devlin Lumber & Supply Corp. v. United States,* 488 F.2d 88 (4th Cir. 1973); *Acousti Eng'g Co. of Fla. v. United States,* 15 Cl. Ct. 698 (1988).

or give subcontractors or suppliers a right of action against the federal government if the contractor failed to furnish Miller Act bonds.[5]

Statutory payment bonds generally provide subcontractors and suppliers on public projects with the same type of protection available to them on private construction projects under applicable state lien laws.[6] When subcontractors or material suppliers on a private construction project are not paid for the work or materials they provided, they have the right under state law to file a mechanics' (or materialmen's) lien on the property. By filing a lien on the property, they reserve their rights to collect unpaid sums from the property owner. They also may have protection under a payment bond. Public property, however, is not subject to legal or equitable liens, and the Miller Act was designed to provide an alternative remedy to the mechanics' liens ordinarily available on private construction projects.[7]

1. Factors Used to Determine Whether the Miller Act Applies

When a contractor has provided a payment bond for a project, but the parties disagree about whether the Miller Act applies, courts must determine whether the bond is a Miller Act payment bond.[8] The Miller Act provides no rights of action on common law bonds (i.e., those furnished pursuant to private contractual arrangements) or state law bonds.[9] Whether a payment bond is governed by the Miller Act is important to lower-level subcontractors and suppliers, who may have rights under the Miller Act but not under the terms of the bond itself.[10]

Whether the Miller Act applies to a given payment bond is not determined by its title.[11] Rather, courts look to the underlying contract and circumstances related to the bond to determine whether it should be treated as a Miller Act bond. The Miller Act applies to payment bonds:

- Related to contracts for the "construction, alteration, or repair" of federal buildings or federal public works, and not for services
- Where the solicitation for bids characterized the contract as a "construction, alteration, or repair" contract or mentioned the Miller Act

[5] *See Dep't of Army v. Blue Fox, Inc.*, 525 U.S. 255, 262 (1999) and *United States ex rel. Owens v. Olympic Marine Servs., Inc.*, 827 F. Supp. 1232 (E.D. Va. 1993).

[6] See **Chapter 3** (generally discussing subcontractor/supplier lien rights); 17 Am. Jur. 2d *Contractors' Bonds* § 26 (2008); *Am. Home Assurance Co. v. Plaza Materials Corp.*, 908 So. 2d 360 (Fla. 2005); *United States ex rel. E. Gulf, Inc. v. Metzger Towing, Inc.*, 910 F.2d 775, 780 (11th Cir. 1990); *F.D. Rich Co. v. Indus. Lumber Co.*, 417 U.S. 116, 122 (1974).

[7] *See, e.g., Dep't of Army v. Blue Fox, Inc.*, 525 U.S. 255, 262 (1999); *United States ex rel. Johnson Pugh Mech., Inc. v. Landmark Constr. Corp.*, 318 F. Supp. 2d 1057 (D. Colo. 2004).

[8] *See, e.g., United States ex rel. Polied Envtl. Servs., Inc. v. Incor Group, Inc.*, 2003 WL 1797846 (D. Conn. 2003).

[9] *United States ex rel. Polied Envtl. Servs., Inc. v. Incor Group, Inc.*, 2003 WL 1797846 (D. Conn. 2003)

[10] *See, e.g., United States ex rel. Polied Envtl. Servs., Inc. v. Incor Group, Inc.*, 2003 WL 1797846 (D. Conn. 2003); *Acro-Tek Commc'ns v. Comnet, LLC*, 2007 WL 4162873 (E.D. La. 2007) (where payment bond was a common law bond, it was governed by its express terms rather than by statute).

[11] *Regal Indus. Corp. v. Crum and Forster, Inc.*, 890 A.2d 395 (Pa. Super. Ct. 2005).

- Where the bond was furnished to the United States, or one of its agencies, and not a higher-level subcontractor
- Where the project is owned by the United States, or one of its agencies, not a private, state, or local government entity

The Miller Act applies to contracts for "construction, alteration, or repair" of public buildings or works but not to "service contracts."[12] Distinguishing between service contracts and Miller Act contracts on the basis of the work being contracted for is not always straightforward; the language the contracting officer used in the solicitation for bids—rather than the work itself—is often determinative. For example, a solicitation for bids on a contract for the installation of a telephone switching system at an Army depot initially required Miller Act bonds; the contracting officer later amended the solicitation to treat the subject as a "services contract." The contracting officer's determination that the contract was for "services" stood.[13]

The terms of the contract between the contractor and the federal agency may include determinative language for the payment bond. If the contract mentions the Miller Act specifically, the Miller Act probably applies. An agreement to comply with applicable federal law "in connection with the performance of the contract," however, is not specific enough to implicate the Miller Act.[14]

Where a contract for debris removal after Hurricane Katrina did not mention the Miller Act but did incorporate provisions of the Service Contract Act, the court found that it was a services contract and the Miller Act did not apply to the payment bonds on the project.[15] This ruling was consistent with the language of the Miller Act, in that "construction, alteration, or removal" of a public building or public work does not seem to implicate "debris removal."[16] The court analogized hurricane debris removal contracts to contracts solely for the demolition of federal buildings, which do not fall within the ambit of the Miller Act.[17]

2. *Is the Contract One for the Construction, Alteration, or Repair of "Public Buildings" and "Public Works"?*

When deciding whether the Miller Act applies to a given contract, a key question is whether the contract constitutes construction or alteration of a "public building or public

[12]*See, e.g., Dep't of Army v. Blue Fox, Inc.*, 525 U.S. 255 (1999) (contracting officer treated contract for installation of a telephone switching system at an army depot as a services contract, so payment bond was not required).

[13]*Dep't of Army v. Blue Fox, Inc.*, 525 U.S. 255 (1999).

[14]*See, e.g., Tradesmen Int'l, Inc. v. U.S. Postal Serv.*, 234 F. Supp. 2d 1191 (D. Kan. 2002) (contract for the supply and installation of automated parcel sorting equipment at several post offices was not governed by the Miller Act).

[15]*Acro-Tek Commc'ns v. Comnet, LLC*, 2007 WL 4162873 (E.D. La. 2007).

[16]*See, e.g., Acro-Tek Commc'ns v. Comnet, LLC*, 2007 WL 4162873 (E.D. La. 2007).

[17]*Id.* (*citing Chi. Rigging Co. v. Uniroyal Chem. Co.*, 718 F. Supp. 696, 700 (N.D. Ill. 1989)) *See also United States ex rel. Warren v. Kimrey*, 489 F.2d 339 (8th Cir. 1974).

work of the United States."[18] Courts have been left to define and apply the terms "construction, alteration, or repair," "public buildings," and "public works," because the statute is silent on these terms. In contrast to the decisions that debris removal and demolition contracts are not subject to the Miller Act, courts have established that contracts for a broad variety of other undertakings that may not seem to involve "any public building or public work" were covered under the statute. Contracts for the design, construction, and repair of vessels,[19] building a highway,[20] and raising a sunken towboat from a canal[21] all have been judged subject to the Miller Act. Considering how broadly courts have defined "public works," cases rarely turn on that definition.

Instead, whether the Miller Act applies often turns on whether the project is sufficiently "federal." Most courts have concluded that federal funding alone is not enough to make a project a federal public work, which would bring a project into the purview of the Miller Act.[22] For the Miller Act to apply, the United States, or one of its agents or agencies, must have contracted for the work in question,[23] and the payment bond must have been furnished directly to the United States or one of its agents or agencies, and not a contractor or nonfederal agency.[24]

A court tasked with deciding whether a contract for the construction of a pedestrian tunnel at the Washington Reagan National Airport was subject to the Miller Act focused on whether the contracting agency was part of the federal government. Although the contract was federally funded, the Miller Act did not apply because the contracting agency was a political subdivision created by state statute, independent of the federal government, and the project was not a "public work of the United States."[25]

When the Miller Act requires a prime contractor to furnish a payment bond to the contracting officer, that prime contractor may require its subcontractors to furnish payment bonds in turn. Depending on many factors, the subcontractors may or may not provide payment bonds. If so, the subcontractor submits its payment bond to the prime contractor, not to the contracting officer. Accordingly, the payment bond

[18] *United States ex rel. Blumenthal-Kahn Elec. Ltd. P'ship v. Am. Home Assurance Co.,* 219 F. Supp. 2d 710 (E.D. Va. 2002).

[19] *Regal Indus. Corp. v. Crum and Forster, Inc.,* 890 A.2d 395 (Pa. Super. Ct. 2005); *United States ex rel. Owens v. Olympic Marine Servs., Inc.,* 827 F. Supp. 1232 (E.D. Va. 1993).

[20] *United States ex rel. Motta v. Able Bituminous Contractors,* 640 F. Supp. 69, 71–2 (D. Mass. 1986).

[21] *United States ex rel. Shlager v. MacNeil Bros. Co.,* 27 F. Supp. 180, 181 (D. Mass. 1939).

[22] *See, e.g., U.S. for Use of Gen. Elec. Supply Co., a Div. of Gen. Elec. Co. v. U.S. Fid. & Guar. Co.,* 11 F.3d 577 (6th Cir. 1993); *United States ex rel. of Miss. Rd. Supply v. H. R. Morgan, Inc.,* 542 F.2d 262, 266 (5th Cir. 1976) ("existence of government funding alone is not enough"); *TIJ Materials Corp. v. Green Island Constr. Co., Inc.,* 131 F.R.D. 31, 33 (D. R.I. 1990).

[23] *Diversified Carting, Inc. v. City of N.Y.,* 423 F. Supp. 2d 85 (S.D.N.Y. 2005); *United States ex rel. Tri-State Rd. Boring, Inc. v. U.S. Fid. & Guar. Co.,* 959 F. Supp. 345 (E.D. La. 1996).

[24] *See Socony-Vacuum Oil Co. v. Cont'l Cas. Co.,* 219 F.2d 645, 647 (2d Cir. 1955) (distinguishing the rights of a subcontractor's supplier against the subcontractor's surety on a private bond from its rights against the prime contractor's surety on a bond furnished under the Miller Act).

[25] *United States ex rel. Blumenthal-Kahn Elec. Ltd. P'ship v. Am. Home Assurance Co.,* 219 F. Supp. 2d 710 (E.D. Va. 2002).

the subcontractor furnishes to the prime contractor is not a Miller Act payment bond, even though the project is a "public work."[26]

B. Little Miller Acts

Many states have enacted statutes requiring and governing payment bonds for certain state and local government construction projects. These statutes, known as Little Miller Acts, usually follow the policies and procedures of the federal Miller Act.[27] Where such a state statute is patterned after the Miller Act, state courts look to federal case law interpreting the Miller Act to aid in interpretation of the state statute.[28] Little Miller Acts, like the federal version, are viewed as remedial statutes, and bonds provided pursuant to these statutes are construed liberally in favor of claimants.[29]

Although decisions interpreting the federal act are not binding on a court's interpretation of a state's "Little Miller Act," they do provide persuasive authority.[30] Keep in mind, however, that decisions of federal courts and the sundry state courts can be inconsistent. Many Little Miller Acts define more terms than the federal Miller Act does; when a statutory definition conflicts with a definition from federal case law, the statutory definition prevails. Furthermore, since statutory terms are incorporated by law into public works bonds, familiarity with a state's specific requirements regarding payment bond recovery is essential to ensure that a claimant does not lose any of its bond rights.

In some states, when a contractor has failed to furnish a payment bond in violation of a statute requiring one, a subcontractor or supplier may obtain some relief, unlike a party to a federal government contract in the same situation. Under Georgia law, for example, a government agency is statutorily liable to the extent of the payment bond had it been furnished.[31] Absent such a statutory provision, a subcontractor or supplier still may be able to recover from the governmental agency on a theory that the agency negligently failed to enforce the statutory requirements, which proximately caused the subcontractor's inability to recover from the payment bond surety.[32]

[26]*See, e.g., United States ex rel. Polied Envtl. Servs., Inc. v. Incor Group, Inc.,* 2003 WL 1797846 (D. Conn. 2003); *Socony-Vacuum Oil Co. v. Cont'l Cas. Co.,* 219 F.2d 645 (2d Cir. 1955); *United States ex rel. DeGeorge Glass Co. v. R.M. Walker Constr. Co., Inc.,* 1992 WL 178682 (E.D. La. 1992).

[27]*See, e.g.,* ALA. CODE § 39–1–1; ARIZ. REV. STAT. ANN. § 34–222; CAL. CIV. CODE §§ 3096, 3098, 3225–27, 3235–42, 3247–52; FLA. STAT. § 255.05; GA. CODE ANN. § 36–91–90; LA. REV. STAT. ANN. §§ 38:2216, 38:2241(A)(2); NEV. REV. STAT. §§ 339.025, 408.357; N.C. GEN. STAT. § 44A-26; S.C. CODE ANN. § 11–35–3030; WIS. STAT. § 779.14.

[28]*See, e.g., Younge Mech., Inc. v. Max Foote Constr. Co., Inc.,* 869 So. 2d 1079 (Miss. Ct. App. 2007); *Imperial Mfg. Ice Cold Coolers, Inc. v. Shannon,* 101 P.3d 627 (Alaska 2004); *Gulf Ins. Co. v. GFA Group, Inc.,* 554 S.E.2d 746 (Ga. Ct. App. 2001).

[29]*See, e.g., Imperial Mfg. Ice Cold Coolers, Inc. v. Shannon,* 101 P.3d 627 (Alaska 2004).

[30]*See, e.g., Gen. Fed. Constr., Inc. v. D.R. Thomas, Inc.,* 451 A.2d 1250 (Md. 1982); *Syro Steel Co. v. Eagel Constr. Co.,* 460 S.E.2d 371 (S.C. 1995); *Rish v. Theo Bros. Constr. Co.,* 237 S.E.2d 61 (S.C. 1977).

[31]*See* GA. CODE ANN. § 13–10–61; *Hall County Sch. Dist. v. C. Robert Beals & Assocs., Inc.,* 498 S.E.2d 72 (Ga. Ct. App. 1998); *Atlanta Mech., Inc. v. DeKalb County,* 443 S.E.2d 856 (Ga. Ct. App. 1994) *See al s o* TEX. GOV'T. CODE ANN. § 2253.027.

[32]*See, e.g., Hous. Auth. of Prattville v. Headley,* 360 So. 2d 1025 (Ala. Civ. App. 1978) (Housing Authority was not immune from suit).

In *Sloan Constr. Corp. v. Southco Grassing, Inc.*, the South Carolina Supreme Court specifically held that a statute requiring payment bonds on public projects "gives rise to a private right of action against a government entity for failure to ensure that a contractor is properly bonded."[33] Furthermore, "a government agency's failure to secure and maintain statutory bonding as required by the [relevant statute] gives rise to a third-party beneficiary breach of contract action by a subcontractor."[34] Thus, a subcontractor may bring an action both in tort—for negligence—and in contract against the government based on its failure to comply with, or ensure the prime contractor complied with, the statutory payment bond requirements.[35] The South Carolina Supreme Court chided the lower court—which denied redress for the subcontractor on either theory—for analyzing the case under federal Miller Act rubric. The court found that the statute in queston had not been characterized as a Little Miller Act, did not appear to be patterned after the Miller Act, and indeed provided stronger payment protection than the Miller Act.

Other jurisdictions, however, will deny recovery in spite of the public body's failure to require the mandated bond.[36] The outcome may depend on whether the contracting body is shielded by sovereign immunity.

II. PAYMENT BONDS ON PRIVATE PROJECTS

Private owners may require contractors to provide payment bonds.[37] Some states have enacted legislation governing these private payment bonds. Massachusetts, for example, enacted a statute stating that any person who furnishes labor or materials to a project is entitled to recover against the private payment bond on the project, if any, without proving reliance on the bond.[38] Other states have enacted statutes outlining the requirements a private payment bond must meet to shield the project owner from mechanics' liens.[39] In Texas, if a private payment bond meets the statutory requirements, a claimant must look to the surety as the presence of the bond negates any lien rights against the property owner.[40] States that have not enacted such statutes consider payment bonds on private projects "common law" bonds, which are governed by standard contract law principles.[41]

[33]*Sloan Constr. Co. v. Southco Grassing, Inc.*, 659 S.E.2d 158, 164 (S.C. 2008) (where original payment bond surety became insolvent and prime contractor failed to respond to SCDOT's request for replacement bond, SCDOT was liable to subcontractor up to the amount remaining on the contract at time of subcontractor's notice).
[34]*Id.* at 165.
[35]*Id.* at 165-166.
[36]*See Haskell Lemon Constr. Co. v. Indep. Sch. Dist. No. 12,* 589 P.2d 677 (Okla. 1979).
[37]*See* A201 A201 § 11.4.1 (2007 ed.); A201 A201, § 11.5.1 (2007 ed.)
[38]Mass. Gen. Laws ch. 149, § 29A .
[39]*See. e.g.,* Tex. Prop. Code Ann. § 53.202; *A. F. Blair Co. v. Mason,* 406 So. 2d 6 (La. Ct. App. 1981).
[40]*Laughlin Envtl., Inc. v. Premier Towers, L.P.,* 126 S.W.3d 668 (Tex. App. 2004).
[41]*See, e.g., Triboro Hardware & Supply Corp. v. Fed. Ins. Co.,* 841 N.Y.S.2d 600 (N.Y. App. Div. 2007); *Mai Steel Serv., Inc. v. Blake Constr. Co.,* 981 F.2d 414 (9th Cir. 1992) (common law surety bonds are construed in the same manner as other contracts).

III. QUALIFYING FOR PAYMENT BOND COVERAGE

To qualify for coverage—that is, the right to recover under a payment bond—a potential claimant must show that it is sufficiently close to the owner or prime contractor in the contracting chain to be considered a first- or second-tier subcontractor or supplier. The potential claimant also must show that it supplied labor or material to the project with a good-faith belief that the labor or materials were intended for the work as provided in the contract and that it has not been paid. [42]

A. "Subcontractors" and "Suppliers"

The language of the Miller Act—"all persons supplying work and materials in the prosecution of the work"—is misleading in its broadness. The Miller Act does not cover *any* party that supplies labor or services remotely connected to the project or *all* materials that eventually end up in the project. Many parties contribute to the progress of a given project, but not all of them are sufficiently close to the contractor that obtained and furnished the payment bond to recover under that bond.

The United States Supreme Court identified two tiers of claimants entitled to protection under the federal Miller Act.[43] The tier system assumes that the prime contractor contracted with and furnished a payment bond to the contracting officer. The first tier comprises the subcontractors and suppliers (also known as materialmen) that contracted directly with the prime contractor. The second tier comprises the subcontractors and suppliers that contracted with a first-tier *subcontractor* but not those who contracted with a first-tier supplier. Suppliers to suppliers are not covered under the Miller Act.[44] In other words, only those parties that have direct relationships with the prime or a first-tier subcontractor may recover on Miller Act payment bonds.[45]

Whether a first-tier entity is considered a subcontractor or a supplier is critical to determining whether the Miller Act covers suppliers to that first-tier entity. For example, in *United States ex rel. E & H Steel Corp. v. C. Pyramid Enters., Inc.* to determine whether a supplier was covered under a Miller Act bond, the court first focused on whether the first-tier entity with whom the supplier had contracted was a supplier or a subcontractor.[46] The court decided that the first-tier entity was a subcontractor, meaning its supplier was covered by the payment bond.

[42] *United States ex rel. Andrews Marine Servs., Inc. v. United Sur. and Indem. Co.,* 2005 WL 1308919 (D.P.R. 2005).
[43] *Clifford F. MacEvoy Co. v. United States ex rel. Tomkins Co.,* 322 U.S. 102 (1944).
[44] *See United States v. G & C Enters., Inc.,* 29 F. Supp. 2d 49 (D. P.R. 1998); *J. W. Bateson Co. v. United States ex rel. Bd. of Trs.,* 434 U.S. 586 (1978).
[45] *United States v. G & C Enters., Inc.,* 29 F. Supp. 2d 49 (D. P.R. 1998) (*citing J.W. Bateson Co. v. United States ex rel. Bd. of Trs. of the Nat'l Automatic Sprinkler Indus. Pension Fund,* 434 U.S. 586 (1978)) *See also U.S., Dep't of the Navy v. Norden Enters., LLC,* 2004 WL 42318 (N.D. Ill. 2004); *United States ex rel. Polied Envtl. Servs., Inc. v. Incor Group, Inc.,* 238 F. Supp. 2d 456 (D. Conn. 2002).
[46] 509 F.3d 184 (3d Cir. 2007).

When determining whether a party should be considered a supplier or a subcontractor under the Miller Act, many federal courts apply a balancing test and favor finding a "subcontractor" relationship when the party in question has assumed a "significant and definable part of the construction project."[47] Other factors include: whether a payment or performance bond was required, whether the price included sales tax, whether progress payments and retainage were withheld, and whether shop drawings and certified payrolls were submitted.[48]

The definition and scope of the term "subcontractor" varies among the Little Miller Acts, and many of them include their own statutory definitions of terms left undefined in the Miller Act. Coverage under the Little Miller Acts may be broader than under the federal Miller Act.[49] When distinguishing between suppliers and subcontractors, state courts may consider whether the claimant is considered a subcontractor or a supplier by the custom of the trade, or whether the claimant's performance satisfied a substantial and definite portion of the prime contract.[50] Other states require a claimant's work to have been completed at the construction site for that claimant to qualify as a subcontractor.[51] But where no statutory definition applies and no state law exists on the subject, state courts will apply the definition of "subcontractor" developed by federal courts under the Miller Act.[52]

B. Suppliers of Customized Materials

Although a party that merely supplies materials probably will fail to qualify as a subcontractor, specialty firms that provide customized materials may be considered subcontractors for federal Miller Act purposes. Courts consider whether the materials were designed or fabricated specially for the project and whether they have commercial value outside the particular project.[53] A supplier of customized materials may qualify as a "subcontractor" even if those materials were not incorporated into

[47] *United States ex rel. Conveyor Rental & Sales Co. v. Aetna Cas. & Sur. Co.*, 981 F.2d 448 (9th Cir. 1992); *Aetna Cas. & Sur. Co. v. United States ex rel. Gibson Steel Co.*, 382 F.2d 615 (5th Cir. 1967).

[48] *United States ex rel. Conveyor Rental & Sales Co. v. Aetna Cas. & Sur. Co.*, 981 F.2d 448 (9th Cir. 1992); *United States ex rel. Consol. Pipe & Supply Co. v. Morrison-Knudsen Co.*, 687 F.2d 129 (11th Cir. 1982); *United States ex rel. Pioneer Steel Co. v. Ellis Constr. Co.*, 398 F. Supp. 719 (E.D. Tenn. 1975).

[49] *See Tom Barrow Co. v. St. Paul Fire & Marine Ins. Co.*, 421 S.E.2d 85 (Ga. Ct. App. 1992) (supplier to second-tier subcontractor could recover on Little Miller Act payment bond, although it would not qualify under Federal Miller Act); *D&L Bldg., Inc. v. State ex rel. Maltby Tank & Barge, Inc.*, 747 P.2d 517 (Wyo. 1987) (Wyoming Little Miller Act coverage not limited to first two tiers).

[50] *See, e.g., Preussag Int'l Steel Corp. v. March-Westin Co.*, 655 S.E.2d 494 (W. Va. 2007); *B. J. Cecil Trucking, Inc. v. Tiffany Constr. Co.*, 597 P.2d 184 (Ariz. Ct. App. 1979); *Tiffany Constr. Co. v. Hancock & Kelley Constr. Co.*, 539 P.2d 978 (Ariz. 1975).

[51] *See, e.g., Preussag Int'l Steel Corp. v. March-Westin Co.*, 655 S.E.2d 494 (W.Va. 2007) (*citing Leonard B. Hebert, Jr. & Co. v. Kinler,* 336 So. 2d 922 (La. Ct. App. 1976)).

[52] *See, e.g., Imperial Mfg. Ice Cold Coolers, Inc. v. Shannon,* 101 P.3d 627 (Alaska 2004).

[53] *See, e.g., ISSC, Inc. v. Baugh Skanska Inc.*, 160 Fed. Appx. 628 (9th Cir. 2005) (supplier of structural steel that was responsible for customizing every piece of fabricated steel was a "subcontractor" for purposes of the Miller Act).

the project, so long as they were "specially fabricated" for the project.[54] In a case where a supplier had design responsibility, prepared shop drawings, and alleged that it manufactured custom conduit for the project, the court found that it was not a "subcontractor."[55] The court held "that the coating, cutting and threading operations" of the conduit did "not constitute custom manufacturing." [56]

C. "Substantiality and Importance" of Relationship with Prime Contractor

Before lower courts developed the preceding factors, the United States Supreme Court held that whether an entity is a supplier or a subcontractor depends on "the substantiality and importance of his relationship with the prime contractor."[57] In *F. D. Rich Co., Inc. v. United States ex rel. Industrial Lumber Co.,*, the Supreme Court applied the "substantiality and importance" test and held that a firm providing plywood sheets—that were not unique or customized—was a subcontractor for the purposes of the Miller Act.[58] The decision was based on the strength of the firm's relationship with the prime. Under this approach, almost any first-tier supplier would qualify as a "subcontractor" for the purposes of the Miller Act. The outcome would have been different if the Court had applied the factors that lower courts have developed in the years since *F. D. Rich* was decided. Noting this disparity, some courts recently have eschewed the factors in favor of returning to the Supreme Court's broad "substantiality and importance" test.[59]

In one such case, *United States ex rel. E & H Steel Corp. v. C. Pyramid Enters., Inc.*, the court of appeals overruled the lower court's determination that a steel fabricator was a "supplier," holding instead that it was a "subcontractor" because it had a contract with the prime contractor.[60] The lower court's determination was based on a number of factors, including the nonspecialized nature of the material the steel fabricator supplied; the determinaton that the steel fabrcator was a supplier foreclosed recovery on the payment bond for its suppliers (because a supplier's suppliers are not covered). The court of appeals, overturning the lower court and holding that the steel fabricator was a subcontractor, warned against overly emphasizing the nature of the materials and the "laundry list" of factors listed earlier.[61] In the words of the court, "[a]lthough furnishing customized or complex material may in some cases be a helpful

[54]*See Aquatic Plant Mgmt., Inc. v. Paramount Eng'g, Inc.*, 977 So. 2d 600 (Fla. Dist. Ct. App. 2007) (firm providing specially fabricated plants was a "subcontractor" even though the plants were never incorporated into a wetland project); *United States ex rel. Parker-Hannifin Corp. v. Lane Constr. Corp.*, 477 F. Supp. 400 (M.D. Pa. 1979) (manufacturer of specialized dam gates was a "subcontractor" since the sole purpose and usefulness of the gates would be their functioning as an integral portion of the dam).
[55]*E. Indus. Mktg. Inc. v. Desco Elec. Supply*, 651 F. Supp. 140 (W.D. Pa. 1986)
[56]*Id.*
[57]*United States ex rel. E & H Steel Corp. v. C. Pyramid Enters., Inc.*, 509 F.3d 184, 187 (3d Cir. 2007) (quoting *F.D. Rich Co. v. United States ex rel. Indus. Lumber Co.*, 417 U.S. 116, 123 (1974)).
[58]*F.D. Rich Co. v. United States ex rel. Indus. Lumber Co.*, 417 U.S. 116, 122 (1974).
[59]*See. e.g., United States ex rel. E & H Steel Corp. v. C. Pyramid Enters., Inc.*, 509 F.3d 184 (3d Cir. 2007).
[60]*Id.* at 184.
[61]*Id.* at 188.

indication of the strength of the supplier's relationship with the prime contractor, it does not follow that the absence of such characteristics in the material supplied establishes a lack of 'subcontractor' status."[62]

D. "Dummy" Subcontractors, Alter Egos, and Joint Ventures

The Miller Act does not permit contractors to manipulate recovery under a payment bond by telescoping or creating alter egos or joint ventures to enter subcontracts. A prime contractor will not be allowed to insert a dummy subcontractor between itself and actual performing subcontractors simply to avoid Miller Act liability.[63] When subcontractors or suppliers go unpaid by a dummy subcontractor, they still may sue on the payment bond, as if they were in contract directly with the dummy's parent company.[64]

Moreover, a prime contractor will not be allowed to create Miller Act liability by failing to pay itself, that is, by failing to pay a subcontractor that is actually its own alter ego or joint venturer. The Miller Act was not designed to protect the contractors furnishing payment bonds, and thus, a prime contractor's partner or joint venturer is not protected by the Miller Act.[65] As one court has said, "[i]n sum, payment and performance bonds do not cover an entity controlled by the same person that controls the principal named in the bond or [that controls an entity that] has otherwise agreed to indemnify the surety."[66]

E. Claimants on Private Payment Bonds

The payment bond form used frequently on private projects, as well as on many state and local public works, is the American Institute of Architects' (AIA) Document A312 (1984 ed.),[67] which defines a claimant as:

> An individual or entity having a direct contract with the Contractor or with a subcontractor of the Contractor to furnish labor, materials or equipment for use

[62] *Id.* at 189. In addition to the existence of the subcontract, the court cited basic elements of the subcontract as evidence of a "strong relationship": The material supplied, while not customized, had to be "carefully manufactured" so the prime contractor could work efficiently and delivery terms had to comply with prime contractor's schedule. Again, the mere existence of a contract with the prime contractor is enough to qualify any supplier to the prime as a "subcontractor" for Miller Act purposes.

[63] *See, e.g., Ragan v. Tri-County Excavating, Inc.,* 62 F.3d 501 (3d Cir. 1995) (court "pierced the corporate veil" and found that subcontractor was a "dummy" of the prime contractor, allowing claimant to recover on payment bond); *Cont'l Cas. Co. v. United States ex rel. Conroe Creosoting Co.,* 308 F.2d 846 (5th Cir. 1962) *But see Gateco, Inc. v. Safeco Ins. Co. of Am.,* 2006 WL 2077011 (E.D. Pa. 2006) (court did not find a dummy or alter ego even where surety had treated two entities as one in the past and the two entities shared an address, some employees, and equipment).

[64] *Ragan v. Tri-County Excavating, Inc.*, 62 F.3d 501 (3d Cir. 1995); *Cont'l Cas. Co. v. United States ex rel. Conroe Creosoting Co.*, 308 F.2d 846 (5th Cir. 1962).

[65] *See, e.g., United States ex rel. Johnson Pugh Mech., Inc. v. Landmark Constr. Corp.,* 318 F. Supp. 2d 1057 (D. Colo. 2004) (where subcontractor was controlled by same stockholder as the contractor that obtained the bond, the subcontractor could not recover on the bond).

[66] *Id.* at 1073 (also setting out 10 factors used to determine whether a subcontractor should be considered an alter ego or joint venturer of the prime contractor).

[67] Although the AIA published updated forms in 1997 and 2007, the A312 Payment Bond form has not changed and the 1984 version is still in use.

in the performance of the Contract. The intent of this Bond shall be to include without limitation in the terms "labor, materials or equipment" that part of water, gas, power, light, heat, oil, gasoline, telephone service or rental equipment used in the Construction Contract, architectural and engineering services required for performance for the work of the Contractor and the Contractor's subcontractors, and all other items for which a mechanic's lien may be asserted in the jurisdiction where the labor, materials or equipment were furnished.

This definition is much more specific than that contained in its predecessor, AIA A311(1970 ed.). Although this newer definition maintains the limitation of claimants down to the second tier, it also relies on local lien laws for a description of the type of work covered by the bond.

ConsensusDOCS 261 (2007 ed.) replaces the previous standard payment bond form published by the Associated General Contractors of America (AGC) Document 261 (2000 ed.) Like the earlier AGC payment bond form, ConsensusDOCS 261 defines a claimant as "an individual or entity having a direct contract with the Contractor or having a contract with a subcontractor having a direct contract with the Contractor." Likewse, the Engineer's Joint Committee on Construction and Design Documents (EJCDC) payment bond form C-615 (2002 ed.) defines a claimant as "[a]n individual or entity having a direct contract with Contractor, or with a first-tier subcontractor of Contractor to furnish labor, materials, or equipment for use in the performance of the Contract." Essentially, the classes of protected parties are the same under these industry standard forms.

IV. WORK QUALIFYING FOR PAYMENT BOND COVERAGE

A. Labor and Materials

After determining whether a given entity is protected by a payment bond, the next step is to determine if that entity's work qualifies for protection. The Miller Act affords payment protection for "labor and materials" provided "in the prosecution of the work," but it does not define those terms.[68] Courts have defined "labor," for purposes of the Miller Act, to mean physical or manual labor.[69] That means that clerical and administrative tasks are not "labor" for the purposes of the act, even if they are performed on the job site.[70] Note that payment bond protection extends only to

[68]40 U.S.C. § 3131(b)(2).

[69]*See United States v. Fed. Ins. Co.,* 251 Fed. Appx. 269 (5th Cir. 2007) (citing *United States ex rel. Constructors, Inc. v. Gulf Ins. Co.,* 313 F. Supp. 2d 593, 597 (E.D. Va. 2004) (paying invoices, reviewing proposals, and supervising hiring were clerical or administrative tasks that do not involve the physical toil or manual work necessary to bring them within the scope of the Miller Act, even if performed on the job site); *United States ex rel. Barber-Colman Co. v. U.S. Fid. & Guar. Co.,* 19 F.3d 1431 3 (4th Cir. 1994) ("labor" includes "physical toil, but not work by a professional, such as an architect or engineer." (citations omitted); *United States ex rel. Olson v. W.H. Cates Constr. Co.,* 972 F.2d 987, 990 (8th Cir. 1992); and *Glassell-Taylor Co. v. Magnolia Petroleum Co.,* 153 F.2d 527, 529–30 (5th Cir.1946).

[70]*United States ex rel. Constructors, Inc. v. Gulf Ins. Co.,* 313 F. Supp. 2d 593, 597 (E.D. Va. 2004).

contracts for providing labor and materials, and not to lenders that provided funds to purchase them. For example, a bank that made loans to a subcontractor, which the subcontractor used to pay for materials and labor, could not recover on the subcontractor's payment bond.[71] As to "materials," the Miller Act generally covers the costs of materials that are substantially consumed in the prosecution of the work.[72] This includes parts and equipment necessary to and wholly consumed by the project and material used in construction but not incorporated into the project.[73]

Material or labor supplied "in the prosecution of the work" means material or labor incorporated in the project or material or labor supplied for the benefit of the project.[74] Even materials that were damaged in transit and were not incorporated into the project have been considered covered under a Miller Act payment bond; the court held that the material was furnished "in the prosecution of the work" because the subcontractor had assumed the risk of loss or damage during shipment.[75] Similarly, a supplier recovered for delivery of equipment to a subcontractor despite the subcontractor's subsequent removal of the material from the job site and use on another project.[76] Following this general line of federal cases, state courts have ruled that a supplier need not show that the materials delivered to the site actually were incorporated into the project.[77] When construing these terms for the purposes of the Little Miller Acts, state courts tend to follow the definitions deveoped by federal courts interpreting the Miller Act.[78]

B. Equipment Repairs and Rental

The cost of incidental repairs necessary to maintain equipment during its use on the project also may be recoverable under a payment bond.[79] Substantial "replacement" repairs, however, are not covered, on the theory that they add value to the construction

[71] *Wasatch Bank of Pleasant Grove v. Sur. Ins. Co. of Cal.*, 703 P.2d 298 (Utah 1985).

[72] *See United States ex rel. Sunbelt Pipe,* 785 F.2d 468 (4th Cir. 1986); *United States ex rel. Skip Kirchdorfer, Inc. v. Aegis/Zublin Joint Venture,* 869 F. Supp. 387 (E.D. Va. 1994); and *United States ex rel. Tom P. McDermott, Inc. v. Woods Constr. Co.,* 224 F. Supp. 406 (N.D. Okla. 1963).

[73] *Id. See also United States ex rel. Chemetron Corp. v. George A. Fuller Co.,* 250 F. Supp. 649 (Mont. 1966).

[74] *See Sunbelt Pipe Corp. v. U.S. Fid. & Guar. Co.,* 785 F.2d 468 (4th Cir. 1986); *United States ex rel. Westinghouse Elec. Supply Co. v. Endebrock-White Co.,* 275 F.2d 57 (4th Cir. 1960).

[75] *United States ex rel. Nat'l U.S. Radiator Corp. v. D. C. Loveys Co.,* 174 F. Supp. 44 (D. Mass. 1959). Although nearly 50 years have passed since this decision, it still represents the current state of the law.

[76] *See Glassell-Taylor Co. v. Magnolia Petroleum Co.,* 153 F.2d 527 (5th Cir. 1946).

[77] *See, e.g., Solite Masonry Units v. Piland Constr. Co.,* 232 S.E.2d 759 (Va. 1977); *see also Key Constructors, Inc. v. H & M Gas Co.,* 537 So. 2d 1318 (Miss. 1989); *Mid-Continent Cas. Co. v. P&H Supply, Inc.,* 490 P.2d 1358 (Okla. 1971) (evidence that materials were delivered to project site creates rebuttable presumption that materials were actually consumed in construction).

[78] *See, e.g., Dixie Bldg. Material Co. v. Liberty Somerset, Inc.,* 656 So. 2d 1041 (La. Ct. App. 1995) (concrete used in the ordinary course of the performing the contract work was covered by the Little Miller Act payment bond), *and Quality Equip. Co. v. Transamerica Ins. Co.,* 502 N.W.2d 488 (Neb. 1993).

[79] *See Finch Equip. Corp. v. Frieden,* 901 F.2d 665 (8th Cir. 1990); *McGee Steel Co. v. State ex rel. McDonald Indus. Alaska, Inc.,* 723 P.2d 611 (Alaska 1986); *United States ex rel. Miss. Rd. Supply Co. v. H.R. Morgan, Inc.,* 542 F.2d 262 (5th Cir. 1976), *cert. denied,* 434 U.S. 828 (1977); *But see Transamerica Premier Ins. Co. v. Ober,* 894 F. Supp. 471 (D. Me. 1995).

equipment by extending its useful life beyond the project in question.[80] The cost of the fair rental value of equipment leased for use in the prosecution of the contract work also is covered by a Miller Act bond.[81] As with equipment repairs, however, the payment bond covers only that portion of equipment rental costs actually associated with use on the project, not any portion of rental costs for any other uses of that equipment.[82]

In keeping with the trend of broader, more explicit coverage under state statutes than under the Miller Act, many Little Miller Acts expressly cover the cost of leased equipment under statutory payment bonds.[83] In one Little Miller Act case, the surety was found liable to an equipment rental firm for the costs of repairing leased equipment that the contractor had returned damaged.[84] Similarly, the forms for private payment bonds such as AIA A312, quoted earlier, expressly contemplate coverage for leased equipment.[85]

Miller Act payment bonds also have been construed to cover transportation and delivery costs.[86] Food and lodging have been found to be covered when they are a necessary and integral part of performance.[87] Furthermore, as the Miller Act does not limit recovery for services rendered to wages, union dues and contributions to welfare and health funds may be covered.[88] In contrast to the broad definitions of "labor" and "materials," for these employee benefits expenses, courts usually require some direct involvement with and benefit to the project.[89]

V. RECOVERY UNDER PAYMENT BONDS FOR EXTRA WORK, DELAY DAMAGES, OR LOST PROFITS, AND OTHER COSTS

Generally, courts require a payment bond claimant to state its claim with enough certainty so that a court can determine its damages "with reasonable certainty and accuracy, without resort to conjecture, guess or speculation."[90] This rule applies to

[80]*Houston Gen. Ins. Co. v. Maples,* 375 So. 2d 1012 (Miss. 1979).

[81]*See United States ex rel. Skip Kirchdorfer, Inc. v. Aegis/Zublin Joint Venture,* 869 F. Supp. 387 (E.D. Va. 1994); *United States ex rel. Miss. Rd. Supply Co. v. H.R. Morgan, Inc.,* 542 F.2d 262 (5th Cir. 1976), *cert. denied,* 434 U.S. 828 (1977); *Friebel & Hartman, Inc. v. United States ex rel. Codell Constr. Co.,* 238 F.2d 394 (6th Cir. 1956); *United States ex rel. D&P Corp. v. Transamerica Ins. Co.,* 881 Supp. 1505 (D. Kan. 1995) (may recover rental value of owned equipment).

[82]*See, e.g., Trestle & Tower Eng'g, Inc. v. Star Ins. Co.,* 13 F. Supp. 2d 1166 (D. Kan. 1998).

[83]*See, e.g., Quality Equip. Co. v. Transamerica Ins. Co.,* 502 N.W.2d 488 (Neb. 1993) (*citing* NEB. REV. ST. § 52–118); 8 PA. CONS. STAT. § 193(a)(2) (2008); R.I. GEN. LAWS § 37–12–1.

[84]*Equip. World, Inc. v. Int'l Fid. Ins. Co.,* 90 P.3d 590 (Okla. Civ. App. 2004).

[85]AIA A312 Payment Bond (2007 ed.).

[86]*See United States ex rel. Benkurt Co. v. John A. Johnson & Sons, Inc.,* 236 F.2d 864 (3d Cir. 1956); *United States ex rel. Carlisle Constr. Co. v. Coastal Structures, Inc.,* 689 F. Supp. 1092 (M.D. Fla. 1988); *Javeler Constr. Co. v. Fed. Ins. Co.,* 472 So. 2d 258 (La. Ct. App. 1985).

[87]*Brogran v. Natl'l Sur. Co.,* 246 U.S. 257 (1918); *United States ex rel. T.M.S. Mech. Contractors, Inc. v. Millers Mut. Fire Ins. Co. of Tex.,* 942 F.2d 946 (5th Cir. 1991) (*citing Brogran v. Nat'l Sur. Co.,* 246 U.S. 257 (1918)).

[88]*S. Elec. Health Fund v. Kelley,* 308 F. Supp. 2d 847 (M.D. Tenn. 2003).

[89]*United States ex rel. Carlisle Constr. Co. v. Coastal Structures, Inc.,* 689 F. Supp. 1092 (M.D. Fla. 1988).

[90]*S. Elec. Health Fund v. Kelley,* 308 F. Supp. 2d 847, 866 (M.D. Tenn. 2003) (*citing United States ex rel. Moody v. Am. Ins. Co.,* 835 F.2d 745, 748 (10th Cir. 1987)).

claims for extra work, delay damages, and all other costs the claimant is seeking to recover on the payment bond.

A. Extra Work

Work done by a qualifying claimant under a change order is generally within the payment bond's protection. The rationale for covering authorized or approved "extra work" is that it has been incorporated into and has benefited the project; the change order ratifies the fact that the extra work went into the project and that the contractor approved or authorized it. Although sureties have attempted to limit their liability strictly to the contract amount when a payment bond is issued, courts have held that Miller Act payment bonds necessarily involve "some amount of uncertainty."[91]

Generally, if a contractor fails to compensate a subcontractor for extra work or fails to approve its change order requests, and the subcontractor can show the extra work comprised labor and materials provided to the project, it may be able recover the costs of the extra work from the surety.[92] For example, in a case where the prime contractor and a subcontractor had agreed on a $175,000 increase in the subcontract amount, the surety was found liable for the contract price plus a portion of the increased amount. The contractor had agreed to pay the subcontractor the additional $175,000 in exchange for the subcontractor agreeing to both an extended payment schedule and a release of any claims the subcontractor had asserted against the contractor (i.e., change order requests). The court found that the surety was liable for the portion of the $175,000 attributable to the claims for extra work waived by the subcontractor.

Courts must determine whether a subcontractor's claim for extra work is covered by the payment bond when the parties disagree about the nature of the extra work and the surety denies the subcontractor's claim for that work. To determine how changes should be treated and whether extra work is recoverable under the payment bond, courts look to the terms of the payment bond and the terms of the underlying contract between the owner/government and the prime contractor. For example, in a case involving a Little Miller Act payment bond, the court rejected the surety's arguments that certain change order work was not "fairly within the contemplation of the parties [to the orginal contract]" (between the state and prime contractor), and thus was not covered by the payment bond.[93] The court found that a subcontractor's claim was covered under the bond because it involved a relatively small amount of additional work that was similar to the original contract work and was necessary to achieve the contract's essential purpose.[94]

[91] *United States ex rel. Cortez III Serv. Corp. v. PMR Constr. Servs., Inc.,* 117 Fed. Appx. 661, 666 (10th Cir. 2004) (*citing Am. Auto Ins. Co. v. United States ex rel. Luce,* 269 F.2d 406 (1st Cir. 1959)); *United States ex rel. IBM v. Hartford Fire Ins. Co.,* 112 F. Supp. 2d 1023 (D. Haw. 2000).
[92] *United States ex rel. I.B.M. v. Hartford Fire Ins. Co.,* 112 F. Supp. 2d 1023 (D. Haw. 2000).
[93] *Beninati Roofing & Sheet Metal Co. v. Gelco Builders, Inc.,* 720 N.Y.S.2d 37 (N.Y. App. Div. 2001).
[94] *Id.*

B. Damages for Delay and Lost Profits

Although delay damages were once viewed as outside the scope of Miller Act coverage, the traditional prohibition of delay damage claims against payment bonds has been reversed.[95] Now it appears that the Miller Act favors allowing full recovery of a subcontractor's delay damages from a general contractor (or its surety) " *regardless of the general contractor's fault.* "[96] This broad coverage of subcontractors' delay damages is justified because "general contractors have privity of contract with the government and can thus recover delay damages directly from the government, while subcontractors cannot."[97] Even where state law barred damages for delay and the underlying contract included a no-damages-for-delay clause, a subcontractor was permitted to recover its delay damages on a Miller Act payment bond because the delay had been caused by the prime contractor's interference.[98] (Note: A restriction on a claim for delay damages under a payment bond does not affect the claimant's general right to collect delay damages from the contractor or subcontractor causing the damage.[99])

In another case, *United States ex rel. Pertun Construction Co. v. Harvester's Group, Inc.,* although the subcontract contained a no-damages-for-delay clause, the court read the clause as conditioned on the subcontractor being granted reasonable time extensions for delays.[100] The court found that the prime contractor wrongfully and prematurely terminated the subcontractor, and as a result, neither the contractor nor its surety could claim protection under the no-damages-for-delay clause.[101]

As to the amount of damages for delay recoverable on a Miller Act payment bond, courts look to the labor and materials that the subcontractor actually furnished to the project as a result of the delay.[102] Generally, a subcontractor may recover its actual out-of-pocket increases as delay costs, but not profit.[103]

[95]*See Lighting & Power Servs., Inc. v. Roberts,* 354 F.3d 817, 821 (8th Cir. 2004); *United States ex rel. Metric Elec., Inc. v. Enviroserve, Inc.,* 301 F. Supp. 2d 56 (D. Mass. 2003); *Mai Steel Serv. Inc. v. Blake Constr. Co.,* 981 F.2d 414, 419 (9th Cir. 1992); *United States ex rel T.M.S. Mech. Contractors, Inc. v. Millers Mut. Fire Ins. Co. of Tex.,* 942 F.2d 946 (5th Cir. 1991); *McDaniel v. Ashton-Median Co.,* 357 F.2d 511 (9th Cir. 1966); *United States ex rel. Pittsburgh-Des Moines Steel Co. v. MacDonald Constr. Co.,* 281 F. Supp. 1010 (E.D. Mo. 1968).

[96]*Lighting & Power Servs., Inc. v. Roberts,* 354 F.3d 817, 821 (8th Cir. 2004) (emphasis added).

[97]*Lighting & Power Servs., Inc.,* 354 F.3d 817, 821 (8th Cir. 2004) (emphasis in original) (citing *Mai Steel Serv. Inc. v. Blake Constr. Co.*, 981 F.2d 414, 419 (9th Cir. 1992). "The Miller Act does not limit a subcontractor's recovery to situations where the general contractor is at fault. Indeed, we have allowed a subcontractor to recover against a Miller Act surety for labor and materials furnished to a subcontractor where the general contractor was blameless."); and *United States ex rel T.M.S. Mech. Contractors, Inc. v. Millers Mut. Fire Ins. Co. of Tex.,* 942 F.2d 946, 951 (5th Cir. 1991) ("subcontractor can recover increased out-of-pocket costs for labor and materials furnished in the course of performing its subcontract caused by contractor or government delay.") (citation omitted).

[98]*United States ex rel. Wallace v. Flintco Inc.,* 143 F.3d 955 (5th Cir. 1998).

[99]*See U.S. Fid. & Guar. Co. v. Ernest Constr. Co.,* 854 F. Supp. 1545 (M.D. Fla. 1994).

[100]918 F.2d 915 (11th Cir. 1990).

[101]*Id.*

[102]*United States ex rel. Mandel Bros. Contracting Corp. v. P. J. Carlin Constr. Co.,* 254 F. Supp. 637 (E.D.N.Y. 1966).

[103]*United States ex rel T.M.S. Mech. Contractors, Inc. v. Millers Mut. Fire Ins. Co. of Tex.,* 942 F.2d 946 (5th Cir. 1991); *see also Mai Steel Serv. Inc. v. Blake Constr. Co.,* 981 F.2d 414, 419 (9th Cir. 1992).

The Miller Act does not provide for lost profits as a remedy, but it does not prevent project participants from bringing state law breach of contract claims for lost profits either.[104] Thus, a contractor must specifically assert a state law claim for breach of contract to recover claimed lost profits, in addition to raising a Miller Act claim to recover out of pocket expenses.[105]

C. Attorneys' Fees and Other Costs

Attorneys' fees are not recoverable under the Miller Act, even if the public policy of the state where the project is located might allow attorneys' fees in a similar situation. This exclusion is based on the fact that the Miller Act provides a federal cause of action, and the courts cannot refer to state law to determine the substance of the rights created by that act.[106] The Miller Act treats attorneys' fees like lost profits: It neither provides them as a remedy nor prevents a party from seeking them in a separate count or claim. Where the underlying contract provides for the recovery of attorneys' fees, the court hearing a Miller Act dispute arising from that contract also can hear accompanying breach of contract claims; thus the court can enforce all the provisions of the contract, including the attorneys' fee provision.[107] The questions of a claimant's right to interest and amount of interest allowed are determined by reference to the law of the state where the contract was performed.[108]

VI. DISTINGUISHING BETWEEN PAYMENT BOND CLAIMS AND PERFORMANCE BOND CLAIMS

The surety's specific obligations are governed by the terms of the payment bond, and the surety's obligations to payment bond claimants are separate from any performance bond obligations. (See **Chapter 15.**)As such, the surety must respond to payment bond claims even though the surety may have a valid defense to performance bond obligations. Furthermore, when a replacement subcontractor has come in to fulfill the surety's performance bond obligations, that replacement sub is covered under the payment bond, as a "subcontractor" furnishing labor or materials, or both, to the project.[109]

VII. PROCEDURAL REQUIREMENTS

A. Time of Notice

Under the Miller Act and most Little Miller Acts, notice must be received by the prime contractor within a certain number of days from the date the claimant last

[104]*Lighting and Power Servs., Inc. v. Roberts,* 354 F.3d 817 (8th Cir. 2003).

[105]*Consol. Elec. and Mechs., Inc. v. Biggs Gen. Contracting, Inc.,* 167 F.3d 432 (8th Cir. 1999).

[106]*United States ex rel. Metric Elec., Inc. v. Enviroserve, Inc.,* 301 F. Supp. 2d 56 (D. Mass. 2003).

[107]*Contractors Equip. Maint. Co. v. Bechtel Hanford, Inc.,* 150 Fed Appx. 585 (9th Cir. 2005); *United States ex rel. Varco Pruden Bldgs. v. Reid and Gary Strickland Co.,* 161 F.3d 915 (5th Cir. 1998).

[108]*Jani-King of Memphis Inc. v. Yates,* 965 S.W.2d 665, 668 (Tex. App. 1998).

[109]*See, e.g., United States v. Pickus Constr. and Equip. Co.,* 2000 WL 190574 (N.D. Ill. 2000); *United States ex rel. CTI Ltd. v. Mellon Stuart Co.,* 860 F. Supp. 556, 559 (N.D. Ill. 1994).

performed work or supplied materials for which the claim is made. The Miller Act notice period is 90 days for second-tier claimants.[110] The notice period for most Little Miller Act statutes is 90 days as well.[111]

Although notice under the Miller Act must be sent to the prime contractor, the surety does not have to receive notice.[112] The Miller Act requires notice to be sent "by any means which provides written third party verification of delivery."[113] Oral notice by itself, however, generally will be insufficient.[114] Since notice is intended to protect the contractor providing the payment bond, written notice must expressly or impliedly inform the contractor that the claimant is looking to it or the surety for payment.[115] As set out in 40 U.S.C. § 3133(b)(2), the notice also must state with substantial accuracy the amount claimed and the name of the party to which the materials or services were provided. The Miller Act notice requirements, however, do not apply to subcontractors and suppliers in direct privity with the prime contractor.

B. Time of Lawsuit

A lawsuit to enforce the provisions of a payment bond under the Miller Act generally must be brought within one year of "the day on which the last of the labor was performed or material was supplied by [claimant]."[116] Federal courts consider this one-year rule jurisdictional in nature, and, as such, it cannot be waived by the surety.[117]

A substantial body of law has developed defining "the day on which the last of the labor was performed or material was supplied by [claimant]." For example, in *General Insurance Co. of America v. United States ex rel. Audley Moore & Son,*[118] the court refused to regard the act of "inspecting" within the definition of "labor" as used in the Miller Act. The correction of prior work, however, has been held to constitute "labor" where the government refused to accept the project until such work has been completed. The correction of defects or warranty work done after completion of the

[110] 40 U.S.C. § 3133 (b)(2); *see also United States ex rel. B&R, Inc. v. Donald Lane Constr.,* 19 F. Supp. 2d 217 (D. Del. 1998).

[111] *See, e.g.,* GA. CODE ANN. § 36–82–104.

[112] *See Cont'l Cas. Co. v. United States ex rel. Robertson Lumber Co.,* 305 F.2d 794 (8th Cir. 1962), *cert. denied,* 371 U.S. 922 (1962).

[113] 40 U.S.C. § 3133(b)(2)(A).

[114] *See United States ex rel. Bros. Builders Supply Co., v. Old World Artisans, Inc.,* 702 F. Supp. 1561 (N.D. Ala. 1988); *Fleischer Eng'g & Constr. Co. v. United States ex rel. Hallenback,* 311 U.S. 15 (1940).

[115] *See MacCaferri Gabions, Inc. v. Dynateria, Inc.,* 91 F.3d 1431 (11th Cir. 1996), *reh'g denied,* 102 F.3d 557, *cert. denied,* 520 U.S. 1167; *Bowden v. United States ex rel. Malloy,* 239 F.2d 592 (9th Cir. 1956), *cert. denied,* 353 U.S. 957 (1957).

[116] 40 U.S.C. § 3133(b)(4).

[117] *S ee United States ex rel. Celanese Coatings Co. v. Gullard,* 504 F.2d 466 (9th Cir. 1974); *United States ex rel. Soda Montgomery,* 253 F.2d 509 (3d Cir. 1958); *But see United States ex rel. Am. Bank v. C.I.T. Constr., Inc. of Tex.,* 944 F.2d 253 (5th Cir. 1991).

[118] 406 F.2d 442 (5th Cir. 1969), *cert. denied,* 396 U.S. 902 (1969).

original subcontract work most likely will not constitute the furnishing of labor or materials for purposes of the Miller Act's time limitation.[119]

These Miller Act cases distinguish "guarantee work" from "punch list work." In other words, work that the government demands to be finished in accordance with the contract plans and specifications by a punch list or other similar device is considered to be contract work. Performance of this work normally will toll the Miller Act's notice and limitations provisions. Work performed under a warranty or to repair latent defects, however, is regarded by the courts as being non-contract work and, as such, outside of the Miller Act's recognized defintion of the term "labor."[120]

Despite the profusion of federal case law fixing the date from which the one-year limitation period runs, the interpretation of the same issue under a Little Miller Act may differ. Ultimately, the issue is often a fact question for the court to decide. In *Johnson Service Co. v. Transamerica Insurance Co.,*[121] the court observed: "Common to all of these decisions . . . is the notion that each case must be judged on its own facts and that sweeping rules about 'repairs' offer little help in the necessary analysis."[122]

C. Surety Response to Notice of Claim

When a payment bond surety receives notice of a claim on a payment bond, it must deal with the claimant in good faith. Payment bonds often contain terms governing the surety's response to claims; these terms are binding and enforceable against the surety. For example, in *J.C. Gibson Plastering Co., Inc. v. XL Specialty Ins. Co.,*[123] the AIA A312 bond form set forth the surety's obligations subsequent to the receipt of a notice of claim in this way:

> § 6 When the Claimant has satisfied the conditions of Section 4, the Surety shall promptly and at the Surety's expense take the following actions:
>
> § 6.1 Send an answer to the Claimant, with a copy to the Owner, within 45 days after receipt of the claim, stating the amounts that are undisputed and the basis for challenging any amounts that are disputed.
>
> § 6.2 Pay or arrange for payment of any undisputed amounts.

[119]*See United States ex rel. Light & Power Utils. Corp. v. Liles Constr. Co.,* 440 F.2d 474 (5th Cir. 1971); *United States ex rel. Automatic Elevator Co. v. Lori Constr.,* 912 F. Supp. 398 (N.D. Ill. 1996) *See also S. Steel Co. v. Union Pac. Ins. Co.,* 935 F.2d 1201 (11th Cir. 1991) (involving a private bond but using Miller Act decisions as precedent); *Johnson Serv. Co. v. Transamerican Ins. Co.,* 485 F.2d 164 (5th Cir. 1973).

[120]*See, e.g., United States ex rel. State Elec. Supply Co. v. Hesselden Constr. Co.,* 404 F.2d 774 (10th Cir. 1968); *United States ex rel. Hussman Corp. v. Fid. & Deposit Co. of Md.,* 999 F. Supp. 734 (D. N.J. 1998).

[121]485 F.2d 164 (5th Cir. 1973).

[122]*Id.* at 173.

[123]521 F. Supp. 2d 1326 (M.D. Fla. 2007).

The surety failed to satisfy this requirement when it merely acknowledged receiving the payment bond claimant's evidentiary support for the claim and notified the claimant that it was referring the claim to the principal (contractor) for review. As a result, the court held that the surety waived its rights to defend against the payment bond claim.[124]

VIII. EFFECT OF PAYMENT BONDS ON LIEN RIGHTS

In some states, the right to claim under a payment bond supplements rather than replaces a subcontractor's lien rights on a private construction project. The existence of a payment bond simply provides a subcontractor or other qualified claimant with a separate right of recovery, in addition to any lien rights.[125] In other states, however, the existence of a right to claim under a payment bond abrogates the claimant's lien rights.[126] In those states, a private owner can shield itself from mechanics' and materialmen's liens by requiring the general contractor to provide a payment bond; that way, any unpaid subcontractors and suppliers can seek payment by filing a claim on the payment bond rather than by filing a lien on the owner's property.[127]

IX. THE SURETY'S DEFENSES TO PAYMENT BOND LIABILITY

In any claim against a payment bond, the surety is entitled to assert all defenses of the principal (contractor), including the defense of offset or recoupment. The surety may have additional, independent defenses according to the applicable bond statute, the terms of the bond, or both. The most common surety defenses are a claimant's failure to comply with notice requirements and time limitations outlined earlier. When asserting defenses relating to the timing and sufficiency of the required notice, however, the surety must deal with bond claimants in good faith.[128] Note also that at least one court, applying the Miller Act, held that a surety defendant may be equitably estopped from using the statute of limitations as a defense if the facts of the case so require.[129]

It is imperative that potential claimants review the terms of the payment bond, as well as all applicable statutes and case law, to determine the exact timing, nature, recipient of notice, or any other requirements necessary to secure their rights under

[124]*Id.* at 1332.

[125]57 C.J.S. Mechanics Liens § 258.

[126]*See* Fla. Stat. Ann. § 713.23; *Scheifer v. All-Shores Constr. & Supply Co.*, 260 So. 2d 270 (Fla. Dist. Ct. App. 1972); *Globe Indem. Co. v. W. Tex. Lumber Co.*, 34 S.W.2d 896 (Tex. Civ. App. 1930).

[127]*See* Fla. Stat. Ann. § 713.23; *Resnick Developers S., Inc. v. Clerici, Inc.*, 340 So. 2d 1194, 1197 (Fla. Dist. Ct. App. 1976); *Scheifer v. All-Shores Constr. & Supply Co.*, 260 So. 2d 270 (Fla. Dist. Ct. App. 1972); *Globe Indem. Co. v. W. Tex. Lumber Co.*, 34 S.W.2d 896 (Tex. Civ. App. 1930).

[128]*See Szarkwoski v. Reliance Ins. Co.*, 404 N.W.2d 502 (N.D. 1987) (Supreme Court of North Dakota recognized a cause of action by an unpaid subcontractor against a surety for its alleged bad-faith refusal to pay on the bond); *United States ex rel., Ehmcke Sheet Metal Works v. Wausau Ins. Cos.*, 755 F. Supp. 906, 909 (E.D. Cal. 1991) (state law may provide a cause of action against Miller Act surety for breach of covenant of good faith and fair dealing); *see also K-W Indus. v. Nat'l Sur. Corp.*, 855 F.2d 640 (9th Cir. 1988).

[129]*United States ex rel. United Rentals, Inc. v. Hartford Fire Ins. Co.*, 339 F. Supp. 2d 799 (W.D. Tex. 2004).

the bond. Any deviation may defeat an otherwise valid claim.[130] This careful review is equally important when responding to a payment bond claim.

A. Pay-If-Paid Clauses

Pay-if-paid clauses are found in many subcontracts and are the subject of considerable litigation regarding their interpretation and enforcement. (See **Chapter 8**.) If an effective pay-if-paid clause is included in the subcontract, a question remains as to whether that defense to the contractor's payment obligation also applies to the surety's obligation under the payment bond. Some courts have held that the payment bond is a separate agreement and that the inability to bring an action on the subcontract does not necessarily preclude recovery on the payment bond.[131] Some states have enacted statutes allowing a payment bond surety to include conditional payment language in the payment bond.[132] Typically, the required language is specified in the statute.[133] Contractors and their sureties need to carefully evaluate such language and its application. For example, in *Everett Painting Co., Inc. v. Padula & Wadsworth Const., Inc.*,[134] the court held that the subcontract's conditional language did not bar an action on the payment bond on a public school project. Moreover, since the payment bond was issued pursuant to Florida Statute § 255.05, the inclusion of conditional payment language would have been ineffective as that public bond statute does not authorize the use of conditional payment bonds on public projects.

B. Misrepresented Status of Payments

Often a subcontractor may require a supplier (or materialman) to sign lien waivers in order for the subcontractor to receive final payment from the prime contractor. To induce the signing of these lien waivers, the subcontractor promises to pay the supplier as soon as the subcontractor receives payment from the prime contractor. If a subcontractor does not make good on this promise, because of the intervention of bankruptcy or judgments, the supplier then sues on the payment bond for any outstanding balance. When this occurs, a surety may raise as a defense the supplier's misrepresentation to the contractor of the status of the subcontractor's account and, by so doing, waiver of the supplier's own right to payment. In denying relief to the supplier, courts likely would rely on the theory of estoppel to prevent the supplier from recovering under the payment bond because the prime contractor relied on the

[130] *See United States ex rel. B&R, Inc. v. Donald Lane Constr.*, 19 F. Supp. 2d 217 (D. Del. 1998).

[131] *OBS Co. v. Pace Constr. Corp.*, 558 So. 2d 404 (Fla. 1990).

[132] *See, e.g.*, FLA. STAT. § 713.245.

[133] *Id.* Florida law requires the bond to contain the following language, on the front page, in at least 10-point type: "THIS BOND ONLY COVERS CLAIMS OF SUBCONTRACTORS, SUB-SUBCONTRACTORS, SUPPLIERS, AND LABORERS TO THE EXTENT THE CONTRACTOR HAS BEEN PAID FOR THE LABOR, SERVICES, OR MATERIALS PROVIDED BY SUCH PERSONS. THIS BOND DOES NOT PRECLUDE YOU FROM SERVING A NOTICE TO OWNER OR FILING A CLAIM OF LIEN ON THIS PROJECT." FLA. STAT. § 713.245(c).

[134] 856 So. 2d 1059 (Fla. Dist. Ct. App. 2003)

supplier's misrepresentation to the contractor's and the surety's detriment. By signing the waivers, the materialman may discharge the prime contractor and its surety from any obligation on the bond.[135]

C. Claim or Lien Waivers

In addition to the defense of estoppel based on misrepresentation by a claimant, a claimant may simply waive its bond rights by signing a claim or lien waiver. Subcontracts used by general contractors may provide that a subcontractor waives its lien rights and bond rights. On private projects, a subcontractor may be asked to waive any lien rights and instead look only to the payment bond. As discussed previously, in most states, payment bonds are mandated for public construction works.

On federal government projects, a waiver of Miller Act rights by a party otherwise protected must be specific and otherwise comply with the act. One court has said that to be an effective waiver of Miller Act rights, the Miller Act must be mentioned. Courts do not favor a finding that a subcontractor has contractually waived its rights under the Miller Act.[136]

The Miller Act also addresses the timing of any waiver of the right to sue on the payment bond. In this regard, 40 U.S.C. § 3133(c) provides:

> A waiver of the right to bring a civil action on a payment bond required under this subchapter is void unless the waiver is—
>
> (1) in writing;
>
> (2) signed by the person whose right is waived; and
>
> (3) executed after the person whose right is waived has furnished labor or material for use in the performance of the contract.

This text reflects the substance of a 1999 amendment to the Miller Act, and any comparable state Little Miller Act legislation should be considered carefully when drafting subcontract or purchase order terms and conditions, especially those addressing subcontractor/supplier recovery on claims or the application of the disputes process in the prime contract to the subcontractor or supplier.

Many general contractors typically require waiver forms to be submitted with each pay application. Such waiver forms vary considerably in form and content. Often the form will include a waiver of lien rights, payment bond rights, or both, through a certain date or for work performed up to a certain date. In defending a payment bond suit, a surety will examine the underlying contract, pay applications, and monthly waiver forms to determine if a waiver has occurred.

In some states, a subcontractor still may be able to recover even if it has waived all of its rights against payment bonds posted by a prime contractor on a public works

[135]*See United States ex rel. Krupp Steel Prods. Inc. v. Aetna Ins. Co.*, 923 F.2d 1521 (11th Cir. 1991); *United States ex rel. Gulfport Piping Co. v. Monaco & Son, Inc.*, 336 F.2d 636 (4th Cir. 1964) *See also United States ex rel. Westinghouse Elec. v. James Stewart Co.*, 336 F.2d 777 (9th Cir. 1964).

[136]*See H.W. Caldwell & Son, Inc. v. United States ex rel. John H. Moon & Sons, Inc.*, 406 F.2d 21, 23 (5th Cir. 1969); *United States ex rel. DDC Interiors, Inc. v. Dawson Constr. Co.*, 895 F. Supp. 270 (1995), *aff'd*, 82 F.3d 427 (1996).

contract. In *Coastal Caisson Drill Co. v. American Casualty Co. of Redding Pa.,*[137] a prime contractor on a state bridge project required its subcontractors to sign a subcontract stating that the subcontractor "waived all rights under any bond . . . executed by Contractor and its surety."[138] When the owner failed to pay, the subcontractor sued the prime's surety on the statutory public works payment bond. The court held that the subcontract agreement provision waiving rights against the surety was void as against public policy. The court found that the public works statute, requiring bonds on such projects, expressed a strong public policy, and that such waivers undermine public policy, making subcontractors reluctant to bid on public projects.

➢ POINTS TO REMEMBER

- A payment bond provides a financial guarantee of payment from the surety to certain suppliers and subcontractors, but subject only to the terms of that bond and applicable statutes and case law.
- Payment bonds are most prevalent on public works projects on which they are required by statute to replace the protection provided by lien laws, which generally do not apply to public projects.
- A payment bond required by statute generally will be construed according to the terms of the statute. Private payment bonds (those not required by statute) are essentially private contractual undertakings where the terms of the bond will be determinative. In any event, potential claimants should review and be familiar with the terms of the payment bond relevant to a particular project.
- The federal Miller Act governs payment bond requirements for federal construction projects. Most states have enacted Little Miller Acts, modeled after the federal Miller Act, which govern payment bonds on state and local construction projects. Regardless of similarities, claimants should familiarize themselves with the requirements applicable to the jurisdiction governing the project.
- In most jurisdictions, payment bond coverage does not extend beyond second-tier subcontractors and suppliers—subcontractors to subcontractors and suppliers to subcontractors. Under the federal Miller Act and many Little Miller Acts, coverage does not extend to suppliers to suppliers.
- Payment bonds, whether statutory or private, involve various notice and timing requirements relating to claims and litigation. Such requirements must be identified and followed. Failure to comply may bar otherwise valid claims.

[137] 523 So. 2d 791 (Fla. Dist. Ct. App. 1988); *Ruyon Enter., Inc. v. S.T. Wicole Constr. Corp. of Fla.,* 677 So. 2d 909 (Fla. Dist. Ct. App. 1996).
[138] *Coastal Caisson Drill Co. v. Am. Cas. Co. of Redding, Pa.,* 523 So. 2d 791 (Fla. Dist. Ct. App. 1988).

15

PERFORMANCE BONDS AND TERMINATION

Although most construction projects end with the satisfactory completion of the contracting parties' respective obligations, there is some potential risk on any project that the construction contract might not be completed to the satisfaction of one or more parties. Project owners or developers can face significant financial disruption and harm when a general contractor fails to perform its contractual obligations. These risks can be mitigated by obtaining a third-party surety's pledge (or guarantee) to arrange for completion of the contract if the contractor fails to do so, and by including express terms in the written contract that allow for termination of the construction contract in the event of such a failure by the contractor. A general contractor also may benefit from the inclusion of express contract terms permitting termination of the construction contract, where the owner becomes unable or unwilling to perform as agreed. Terminating a construction contract before completion of the work is a drastic step that should not be considered lightly. Even so, most construction contracts anticipate that a contract can be terminated for nonperformance. Some contracts also require a third-party to promise it will help complete the contract, if there is a termination.

This chapter analyzes two distinct, but related, topics concerning management of the risk of nonperformance under a construction contract. First, it discusses surety performance bonds, instruments by which a third party gives assurances that a construction contract will be completed, even if the contractor runs into trouble performing. Second, this chapter discusses contract termination provisions, which typically set out steps that will be followed if the owner or the contractor terminates the construction contract.

I. SURETY PERFORMANCE BONDS

A. Fundamentals of Suretyship Law

A claim usually is made on a performance bond upon the occurrence of a default or alleged default by the bond principal, the contractor. The claim is directed to the bond surety, calling on the surety to perform as it pledged to do under the bond. A basic explanation of the legal principles applicable to the default aspects of the performance bond is set out in the next several sections.

Using a surety to ensure or guarantee that one party to a contract will perform as promised is an ancient practice, mentioned as long ago as biblical times. The surety provides a second contract, usually referred to as the surety's "bond." Under the bond, the surety gives its additional promise to guarantee the performance of the contract agreed to by the contractor, if the contractor does not perform the contract. Historically, the surety was often a relative, business associate, or friend of the contractor, who gave the additional pledge that the surety would use its own wealth or resources, if necessary, to support the promise of the contractor who agreed to perform the contract. In modern times, the surety is typically in the business of providing bonds, pledging the surety's credit and financial resources to ensure completion of a contract as promised by the party who agreed to complete it, in exchange for the payment of a bond premium by that party.

The typical surety relationship involves three parties: (1) the principal that is primarily responsible for performing the contract (usually the prime contractor); (2) the party to whom the principal is obligated to perform the contract, known often as the obligee or the owner; and (3) the surety, known also as the obligor, that provides a bond pledging financial support to complete the contract, if that becomes necessary.

The surety for a construction contractor also often writes a separate bond, known as a payment bond, which ensures that the bond principal/contractor will pay its subcontractors and suppliers on the project. This chapter, however, discusses only surety performance bonds assuring that the contractor will perform the work agreed to with the obligee or owner. **Chapter 14** addresses payment bonds.

This chapter frequently refers to the "obligee" as the "owner" of the project. Performance bonds also may be posted by a surety to ensure the completion of a subcontractor's subcontract with a general contractor. In such case, the general contractor is the obligee and the subcontractor is the bond principal.[1] This three-party bonding relationship can exist at any tier of the contractual chain. Thus, a surety also can post a bond to assure the completion of a sub-subcontractor's work to a subcontractor, and so on. Bonding a lower-tier subcontractor's work is more common on larger projects, where the subcontract value would justify the effort and expense of such a bond. Nevertheless, this situation could in theory arise on any size construction project.

Although the form of the performance bond as a three-party contract is constant, sureties, contractors, and owners regularly struggle over the rights and liabilities each

[1]In such a case the owner might also be named as a co-obligee on the subcontractor's performance bond.

has under the specific terms of their particular bond, because bond terms vary widely.[2] Contract terminations, and the performance bond obligations that often follow the termination, may be encountered less frequently than other construction law issues discussed in this book, but extraordinary risks often are involved for those affected by a contract termination, and any resulting demands on the performance bond. Therefore, every construction industry participant should have some basic understanding of the rights and potential liabilities associated with performance bonds.

Performance bonds also affect indirectly the selection of contractors on private construction and on government building or public works projects. For example, performance bonds ensuring completion of a contractor's duties have long been required by law on public works projects, and standard-form contracts used for private construction projects may require performance bonds as well.[3] Thus, many jobs require performance bonds. Of course, sureties do not want to write bonds for projects on which there is an unusual (i.e., heightened) risk of default. Therefore, sureties tend to decline bonds for contractors on projects for which the contractors lack expertise or adequate financing. A contractor that cannot get a performance bond may be unqualified to bid for a given contract. Such inability to qualify for bonding will be, at the least, a fact the owner or prime contractor may take into consideration in selecting their contractor.

1. Performance Bonds Are Not Insurance Policies

Under the law, a surety bond is not treated as an insurance policy, despite the apparent similarities between the two.[4] "Insurance has been defined as a contract whereby one undertakes to [indemnify] another against loss, damage or liability arising from an unknown or contingent event; whereas a contract of suretyship is one to answer for the debt, default or miscarriage of another."[5] The two parties to an insurance contract (the insurer and the insured) agree on terms for protecting the insured from the occurrence of a future, unknown accident ("coverable risk").

A performance bond involves three parties: the "principal," the "obligee," and the "surety", who are not protecting against an unknown future accident that largely is beyond the control of any party but rather are guarding against the possible failure of the contractor to perform its existing obligations under the bonded contract. In this tripartite relationship, the surety does not agree to indemnify any of the parties but rather to protect the owner from loss or damage resulting from the failure or default

[2]See **Section I.F** of this chapter (discussing industry bond forms).

[3]*See* 40 U.S.C. §1331 (requiring performance bonds on federal projects); ConsensusDOCS 200, ¶ 10.7 (2007 ed.) (making performance bond optional); AIA A201, § 11.4 (2007 ed.) (owner may require performance bond).

[4]*See Travelers Cas. & Sur. Co. of Am. v. J.K. Merz Constr., Inc.*, 2007 WL 4468680 (N.D. Cal. 2007); *Schmitt v. Ins. Co. of N. Am.*, 281 Cal. Rptr. 261 (Cal. Ct. App. 1991); *Farmers Ins. Exch. v. Midwest Emery Freight Sys., Inc.*, 215 N.W.2d 623 (Minn. 1974).

[5]*Meyer v. Bldg. & Realty Servs. Co.*, 196 N.E. 250, 253–54 (Ind. 1935). *See also FIDC v. Ins. Co. of N. Am.*, 105 F.3d 778, 785–86 (1st Cir. 1997); *Cates Constr. Inc. v. Talbot Partners,* 980 P.2d 407, 419 (Cal. 1999).

of the contractor to perform.[6] In other words, the surety must perform under the performance bond only when the principal fails to perform as promised. This differs from insurance, where the covered risk is an accident over which the insured has little or no control. Further, because the surety must perform under the performance bond only when the principal fails to perform its promise, the surety may recover its losses on the bond from the principal responsible for the default. This is a basic difference from insurance, where the insurer typically cannot recover its losses from the insured. Further, different legal duties and standards apply to the relationship between an insurer, where fiduciary standard applies, and a surety, where a fiduciary standard is not applicable.[7]

Because the peformance bond surety must act when the principal is "in default" of its contract obligations, the obligee must give a clear declaration of default. A surety has no duty to act in the face of "mere obligee complaints" or threats to declare the principal in default.[8] Further, the default must constitute a material breach of the contract. A surety is not liable to the obligee for partial or minor breaches that do not rise to a level of a material contract breach.[9]

2. *Indemnification—The Bottom Line*

Because the surety promises to perform the obligations for which the principal is liable, the law implies a promise by the principal to repay the surety the amount the surety had to pay owing to the principal's failure to perform. The law gives the surety a right to sue its principal for such repayment.[10] This legally implied promise by the principal to indemnify its surety may be expressly stated and strengthened in a written indemnification agreement (commonly called a "general agreement of indemnity" or "agreement of indemnity"). A surety will regularly require its principal to execute an agreement of indemnity before the surety will write a performance bond for the principal. The surety often requires that the contractor's individual owners, shareholders, or officers (and often their spouses) add their respective indemnity promises by personally signing such indemnity agreements as well.

Surety indemnity agreements typically include many provisions to protect the surety if it incurs a liability or loss under the performance bond (or payment bond—see **Chapter 14**). A few of such provisions are: (1) a promise to assign to the surety,

[6] *See Airlines Reporting Corp. v. U.S. Fid. & Guar. Co.,* 37 Cal. Rptr. 2d 563 (Cal. Ct. App. 1995).
[7] *See, e.g., Ins. Co. of the W. v. Gibson Tile Co., Inc.,* 134 P.3d 698 (Nev. 2006).
[8] *Seaboard Sur. Co. v. Town of Greenfield,* 370 F.2d 215 (1st Cir. 2004); *Elm Haven Constr. Ltd. P'ship v. Neri Constr., LLC,* 376 F.3d 96 (2d Cir. 2004); *L&A Contracting Co. v. S. Concrete Servs.,* 17 F.3d 106 (5th Cir. 1994); *Bank of Brewton, Inc. v. Int'l Fid. Ins. Co.,* 827 So. 2d 747 (Ala. 2002). The difference between a proper declaration of default and an improper one, or whether a declaration is required at all, may pose questions that require careful analysis.
[9] *See L&A Contracting Co. v. S. Concrete Servs., Inc.,* 17 F.3d 106, 110 (5th Cir. 1994); *Vill. of Fox Lake v. Action Cas. & Sur. Co.,* 534 N.E.2d 133 (Ill. App. Ct. 1989).
[10] *See, e.g., Union Switch & Signal, Inc. v. St. Paul Fire & Marine Ins. Co.,* 226 F.R.D. 485 (S.D.N.Y. 2005); *Kimberly-Clark Corp. v. Alpha Bldg. Co.,* 591 F. Supp. 198 (N.D. Miss. 1984); *Frank Lerner & Assocs., Inc. v. Vassy,* 599 N.E.2d 734 (Ohio Ct. App. 1991).

in the event the principal defaults, all tools, materials, and equipment on the job site and the principal's interest in future contract payments; (2) a promise to indemnify and legally protect the surety from any loss, expense, or claims under the bond; and (3) a promise to deposit funds with the surety, as additional security, if the surety so demands, or if it receives or becomes liable for a bond claim.

Although indemnity agreements are drafted for the purpose of making indemnitors broadly responsible to the surety for any loss incurred under the bond, an indemnitor is not without defenses against its surety. The obligation of the surety comes into play only if the principal is properly terminated.[11] If the principal is not properly defaulted, the surety has no duty to perform under the bond.[12] A surety may have no right to recover costs from its principal incurred in complying with demands by an owner that the surety does not in good faith believe to be proper.[13] Conversely, if the surety unreasonably or in bad faith rejects a favorable settlement offer that would have reduced the indemnitors' losses, the indemnitors may be able to claim a discharge from liability to the extent that the surety's damages exceeded the amount for which it could have settled.[14]

Although indemnitors may defend themselves if the surety acts unreasonably, indemnity agreements typically give the surety the right to settle bond claims, and that right is generally enforceable.[15] The surety must deal in good faith and fairly with all parties interested in the bond, including the principal, the obligee, and the indemnitor, but the surety likely will be able to enforce its indemnity agreement, unless it has violated its duty of good faith and fair dealing in taking the actions for which it seeks enforcement.[16] It may not be a simple matter to establish a claim for breach of the surety's good-faith duty. The duty of good faith is not a fiduciary duty, or a special duty the surety owes to the obligee in particular.[17] Where an indemnitor sought to avoid its obligation to the surety based on the surety's "bad-faith" settlement of the obligee's claims, a Connecticut court stated that the indemnitor must "establish something more than mere negligence," but something less than fraud.[18]

[11]*Airline Reporting Corp. v. U.S. Fid. & Guar. Co.,* 37 Cal. Rptr. 563 (Cal. Ct. App. 1995).

[12]*See, e.g., Seaboard Sur. Co. v. Town of Greenfield,* 370 F.3d 215 (1st Cir. 2004).

[13]*See Seaboard Sur. Co. v. Dale Constr. Co.,* 230 F.2d 625 (1st Cir. 1965); *Arntz Contracting Co. v. St. Paul Fire & Marine Ins. Co.,* 54 Cal. Rptr. 2d 888, 898–99 (Cal. Ct. App. 1996); *Angle v. Banker's Sur. Co.,* 210 F. 289 (N.D.N.Y. 1913).

[14]*See, Briggs v. Travelers Indem. Co.,* 289 So. 2d 762 (Fla. Dist. Ct. App. 1974); *see generally Dadeland Depot, Inc. v. St. Paul Fire & Marine Ins. Co.,* 945 So. 2d 1216, 1228 (Fla. 2006) (discussing good-faith duty in surety contracts).

[15]*See, e.g., First Nat'l Ins. Co. v. Joseph R. Wunderlich, Inc.,* 358 F. Supp. 2d 44 (N.D.N.Y. 2004); *U.S. Fid. & Guar. v. Napier Elec. & Constr. Co.,* 571 S.W.2d 644 (Ky. Ct. App. 1978).

[16]*See, e.g., W. Sur. Co. v. Bradford Elec. Co., Inc.,* 483 F. Supp. 2d 1114 (N.D. Ala., 2007); *U.S. Fid. & Guar. Co. v. Stanley Contracting, Inc.,* 303 F. Supp. 2d 1169 (D. Or. 2004); *Artnz Contracting Co. v. St. Paul Fire & Marine Ins. Co.,* 54 Cal. Rptr. 2d 888 (Cal. Ct. App. 1996).

[17]*See, e.g., Ins. Co. of the W. v. Gibson Tile Co., Inc.,* 134 P.3d at 698 (Nev. 2006); *Travelers Cas. & Sur. Co. of Am. v. Amoroso,* 2004 WL 1918890 (N.D. Cal. 2004).

[18]*PSE Consulting, Inc. v. Frank Mercede & Sons, Inc.,* 838 A.2d 135 (Conn. 2004); *see also Atl. Contracting & Material Co., Inc., v. Ulico Cas. Co.,* 844 A.2d 460 (Md. 2004) (similar analysis of good-faith settlement under payment bond).

The court explained that the claimant must show that the surety acted with an "improper motive" or a "dishonest purpose."[19]

3. "Common Law" versus "Statutory" Bonds

Performance bonds often are required by statutes in connection with public construction projects at the federal, state, and even municipal level. For example, with certain limited exceptions, the federal Miller Act[20] requires performance bonds on federal government construction and public works projects. The statutes requiring performance bonds frequently list specific terms and provisions that the bonds must contain. Generally, courts read such a bond in light of the purpose of the statute requiring it, and some standard bond forms expressly provide that the parties intend the bond to be read in that way.[21] This may result in reading statutory provisions into a bond, even if the bond *does not* expressly state those requirements, or reading provisions out of the bond that are contrary to the applicable statute.[22] If the bond exceeds the statute's requirements, thus creating more rights than the statute requires, however, the bond may be interpreted as giving the claimant all the statutory rights and additional rights under the provisions that exceed the statutory bond requirements.[23]

Statutory provisions may be critical in determining whether a bond claim is allowed, and the fine points of the type of bond given in a case (i.e., statutory, non-statutory, or a "statutory plus" common law bond) can be both decisive and difficult to determine. The primary test involves examining the bond, concentrating on the obligations imposed on the principal and its surety, and comparing the language contained within the bond with the minimal requirements enunciated in any applicable statute.[24] If the surety's obligations have not been extended beyond the statutory minimum requirements, the bond is statutory.[25] Even a bond furnished incident to a public works project, however, may not necessarily be construed as a statutory bond.[26] Furthermore, merely because a bond fails to reference a particular statute does not automatically render it a common law bond.[27] Given these nuances, interpreting a bond together with applicable statute may require care.

[19]*PSE Consulting, Inc. v. Frank Mercede & Sons, Inc.*, 838 A.2d 135 (Conn. 2004); *Cf. Arntz Contracting Co. v. St. Paul Fire & Marine Ins. Co.*, 54 Cal. Rptr. 2d 888, 898-99 (Cal. Ct. App. 1996) (surety that failed to show its action was "desirable or necessary" acted in bad faith).
[20]40 U.S.C. §§ 3131–34.
[21]*See, e.g.,* AIA A312, § 11 (1984 ed.) (performance bond form). *See also May v. Cont'l Cas. Co.,* 936 A.2d 747, 751 (D.C. 2007); *S. Ins. Co. v. ADESA Austin,* 239 S.W.3d 423, 426 (Tex. App. 2007).
[22]*See generally Gloucester City Bd. of Educ. v. Am. Arbitration Ass'n,* 755 A.2d 1256 (N.J. Super. Ct. App. Div. 2000); Restatement (Third) of Suretyship & Guaranty § 71(2) (1996).
[23]*Am. Bldg. Contractors Ass'n, Inc. v. Mica & Wood Creations, LLC,* 804 N.Y.S.2d 109, (N.Y. App. Div. 2005); *A.C. Legnetto Constr., Inc. v. Hartford Fire Ins. Co.,* 702 N.E.2d 830 (N.Y. 1998).
[24]*See Am. Home Assurance Co. v. Plaza Materials Corp.,* 908 So. 2d 360 (Fla. 2005); *Fla. Keys Cmty. Coll. v. Ins. Co. of N. Am.,* 456 So. 2d 1250 (Fla. Dist. Ct. App. 1984). Statutory bonds are those that meet the minimal requirements of the statute; common law bonds are those that provide coverage in excess of the minimum statutory requirements.
[25]*Id.*
[26]*Id.*
[27]*Id.*

4. *Surety's Obligations*

The nature of the surety's obligation is financial, but the terms of the bond itself are the bedrock of that obligation and must be analyzed case by case. The performance bond surety generally binds itself with its principal, to pay to the obligee a sum equal to the contract amount. Performance of the bond obligation becomes due if the principal wrongfully and without justification defaults on the primary obligation (the bonded contract), and only to the extent of the loss covered by the bond. Generally, the surety's liability to perform is strictly limited by both the terms of the bond and by the terms of the underlying contract, which typically is incorporated into and made a part of the bond.[28]

Because the bond is intended to secure the performance of the underlying contract, a performance bond and the bonded contract are read together.[29] As a result, the surety's obligations usually are defined by the scope of the work as set out in the incorporated contract documents.[30]

Since the surety guarantees only the principal's promise to perform the underlying contract, the surety can assert all contractual defenses that the principal would have been able to assert.[31] "Since a surety's liability is commensurate with that of the principal, where the principal is not liable on the obligation, neither is the guarantor."[32]

B. Extension of Performance Bond Surety's Liability to Third Parties

Although a single obligee typically is named in a performance bond, sureties increasingly are being held liable to other project participants on a "third-party beneficiary" theory. Under this theory, a claimant that has a claim against the contractor, although not named in the performance bond (such as a subcontractor or supplier), is nonetheless deemed as intended to have a claim against the surety. For example, a federal court has held that the federal Miller Act's dual requirement of performance and payment bonds would allow suppliers whose payment bond had been exhausted by

[28]*N. Am. Specialty Ins. Co. v. Chichester Sch. Dist.*, 158 F. Supp. 2d 468 (E.D. Pa. 2001); *Dawson Corp. v. Nat'l Union Fire Ins. Co.*, 666 A.2d 604 (N.J. Super. Ct. App. Div. 1995).

[29]*N. Am. Specialty Ins. Co. v. Chichester Sch. Dist.*, 158 F. Supp. 2d 468 (E.D. Pa. 2001); *Pac. Employers Ins. Co. v. City of Berkeley*, 204 Cal. Rptr. 387 (Cal. Ct. App. 1984).

[30]J. Milana, *The Performance Bond and the Underlying Contract: The Bond Obligations Do Not Include All of the Contract Obligations*, 12 Forum 187, 188 (1976); *See Beard Family P'ship v. Commercial Indem. Ins. Co.*, 116 S.W.3d 839 (Tex. App. 2003); *Am. Home Assurance Co. v. Larkin Gen. Hosp. Ltd.*, 593 So. 2d 195 (Fla. 1992); *St. Paul Fire & Marine Ins. Co. v. Woolley/Sweeney Hotel,* 545 So. 2d 958 (Fla. Dist. Ct. App. 1989) (arbitration agreement in contract bound surety); *U.S. Fid. & Guar. Co. v. Gulf Fla. Dev. Corp.*, 365 So. 2d 748 (Fla. Dist. Ct. App. 1978).

[31]*See* Restatement (Third) of Suretyship §§ 34–36 (1996); *State Athletic Comm'n v. Mass. Bonding & Ins. Co.*, 117 P.2d 80 (Cal. Ct. App. 1941).

[32]*U.S. Leasing Corp. v. DuPont,* 70 Cal. Rptr. 393, 403 (Cal. 1968); *see also State v. Fed. Ins. Co.*, 2005 WL 3495001 (Ohio Ct. App. 2005); *Riley Constr. Co. v. Schillmoeller & Kroff Co.*, 236 N.W.2d 195, 198 (Wis. 1975). This is not true, however, for defenses that are "personal" to the contractor, such as bankruptcy.

claims exceeding the penal sum to claim against the performance bond.[33] In reaching such results, courts reason that, if the contractor had a duty to pay for the work of subcontractors or suppliers under the construction contract guaranteed by the performance bond, then those subcontractors or suppliers may make their claims for payment directly under the contractor's performance bond.[34]

In addition, where there are several primes working together on a multiprime (or parallel prime) construction project, some of them may attempt to assert claims against the performance bond of the other parallel primes. Performance bond sureties have attempted to defend against claims by other prime contractors on the basis that the applicable performance bond gives rights only to the named obligee, the owner, not to third parties. This defense has been successful in some cases.[35] In other instances, the performance bond surety has been held liable to a multiple-prime contractor under a third-party beneficiary theory. Usually the surety's liability depends on the degree to which the principal's duty to coordinate and cooperate with other prime contractors is incorporated in or referred to by the bonded contract and the terms of the bond itself. A surety may be held liable for damages sustained by the principal's failure to perform its duty.[36]

C. Increase of the Surety's Liability under a Performance Bond

A surety's obligation is limited to the terms stated in its bond. As such, courts historically have declined to hold a surety liable for an amount greater than the limit, known as the penalty or penal sum, that is usually stated in each bond.[37] Increasingly, however, performance bond sureties face claims and damages beyond the express penal sum of the bond.

Continental Realty Corp. v. Andrew J. Crevolin Co.[38] was an important early case holding that a surety could be liable, in an amount beyond the bond penal sum, for

[33]*See United States ex rel. Blount Fabricators, Inc. v. Pitt Gen. Contractors, Inc.*, 769 F. Supp. 1016 (E.D. Tenn. 1991); *see also United States ex rel. Edward Hines Lumber Co. v. Kalady Constr. Co.*, 227 F. Supp. 1017 (N.D. Ill. 1964). *But see Titan Stone, Tile & Masonry, Inc. v. Hunt Constr. Group, Inc.*, 2007 WL 869556 at *16 (D. N.J. 2007) (subcontractor not a beneficiary under performance bond); *Transamerica Premier Ins. Co. v. Ober*, 894 F. Supp. 471 (D. Me. 1995) (no subcontractor claim for payment existed under Miller Act performance bond).

[34]*See C&M Warehouse, Inc. v. E. Trucking*, 1994 WL 551238 (Conn. Super. Ct. 1994); *Dealers Elec. Supply v. U.S. Fid. & Guar. Co.*, 258 N.W.2d 131 (Neb. 1977); *Royal Indem. Co. v. Alexander Indus., Inc.*, 211 A.2d 919 (Del. 1965); *Amelco Window Corp. v. Fed. Ins. Co.*, 317 A.2d 398 (N.J. 1974); *Edward Hines Lumber Co. v. Kalady Constr. Co.*, 227 F. Supp. 1017 (N.D. Ill. 1964).

[35]*Angelo Lafrate Constr., LLC v. Potoshmich Constr., Inc.*, 370 F.3d 715 (8th. Cir. 2004); *J. Louis Crum Corp. v. Alfred Lindgren, Inc.*, 564 S.W.2d 544 (Mo. Ct. App. 1978); *M.G.M. Constr. Corp. v. N.J. Educ. Facilities Auth.*, 532 A.2d 764 (N.J. Super. Ct. Law Div. 1987).

[36]*See, e.g., M. T. Reed Constr. v. Va. Metal Prods. Corp.*, 213 F.2d 337 (5th Cir. 1954); *see also Aetna Cas. & Sur. Co. v. Doleac Elec. Co.*, 471 So. 2d 325 (Miss. 1985).

[37]A surety, however, voluntarily waives the penal sum of its bond if the surety agrees to perform all remaining work itself, thereby consenting to pay whatever costs are necessary to perform that promise. *See, e.g., Employers Mut. Cas. Co. v. United Fire & Cas. Co.*, 682 N.W.2d 452 (Iowa 2004).

[38]380 F. Supp. 246 (S.D. W. Va. 1974).

breaching the surety's duty—independent of the principal's default—to: (1) complete the surety's duties in a timely manner; or (2) to pay the cost of completion. In *Crevolin,* the obligee recovered its actual damages, which exceeded the penal sum of the bond. In essence, the surety was held liable in the *Crevolin* case for costs that could have been avoided by the surety, had it performed promptly and properly. For instance, water damage occurred after the contractor abandoned the project, which the surety could have avoided by performing its performance bond obligations timely.

Statutes and regulations increasingly provide independent claims against a surety for reasonable attorneys' fees and other damages, in addition to the surety's liability stated under its bond. For example, in *Boland v. Trans Coastal Roofing,*[39] the Florida Supreme Court held that a Florida statute required a surety to pay the attorneys' fees of a prevailing claimant, even where those fees exceeded the penal sum and the surety had not engaged in any independent misconduct.[40] Similarly, in *Town of Clarkstown v. North River Insurance Co.,*[41] a federal court awarded interest in excess of the penal sum under a New York statute permitting the award of interest to put a claimant in the same financial position that it would have been in had the surety paid promptly.[42]

Performance bond sureties may be liable for attorneys' fees or expenses (such as interest) even in the absence of an express bond provision or a specific statutory right. If the bond incorporates a construction contract that provides for recovery of attorneys' fees, the surety may be required to pay those fees.[43]

Although punitive damages generally are not allowed in breach of contract actions,[44] some statutes and regulations permit claimants to recover penalties, punitive damages, or both, against sureties as a remedy for wrongful conduct.[45] For example, in *Fisher v. Fidelity & Deposit Co. of Maryland,*[46] the owner recovered attorneys' fees and costs in addition to its contract damages for a surety's vexatious and unreasonable delay in settling the owner's performance bond claim.

[39]851 So. 2d 724 (Fla. 2003); *see also Dadeland Depot v. St. Paul Fire & Marine Ins. Co.,* 945 So. 2d 1216 (Fla. 2006).

[40]*See David Boland, Inc. v. Trans Coastal Roofing Co.*, 851 So. 2d 724 (Fla. 2003).

[41]803 F. Supp. 827 (S.D.N.Y. 1992).

[42]*Id.*

[43]*Austin v. Parker,* 672 F.2d 508 (5th Cir. 1982); *HOH Co. v. Travelers Indem. Co.,* 903 F.2d 8 (D.C. Cir. 1989) (pass-through agreement under general contractor/subcontractor contract obligated surety to reimburse attorneys' fees). *But see N. Am. Specialty Ins. Co. v. Chichester Sch. Dist.,* 158 F. Supp. 2d 468 (E.D. Pa. 2001) (under Pennsylvania law, terms of contract allowing attorneys' fees could not extend the surety's obligation beyond that stated in the bond).

[44]*Oliver B. Cannon & Son, Inc. v. Fid. & Cas. Co. of N.Y.,* 484 F. Supp. 1375 (D. Del. 1980).

[45]*See, e.g.,* GA. CODE ANN. §10–7–31; FLA. STAT. ANN. §§ 627.428, 627.756 (1996) (Florida has enlarged an insurer's liability for attorneys' fees for bad-faith failure to settle to cover sureties issuing payment bonds). *See also Danis Indus. Corp. v. Grand Implement Techniques, Inc.,* 629 So. 2d 985 (Fla. Dist. Ct. App. 1993); *Gen. Elec. Supply Co. v. Downtown Church of Christ,* 746 S.W.2d 386 (Ark. Ct. App. 1988); *Ray Ross Constr. Co., Inc. v. Raney,* 587 S.W.2d 46 (Ark. 1979). *See generally* 44 AM. JUR. 2D, *Insurance,* §§ 1798–99; A. S. Klein, Annotation, *Insurer's Liability for Consequential or Punitive Damages for Wrongful Delay or Refusal to Make Payments Due under Contracts*, 47 A.L.R. 3d 314 (1973).

[46]466 N.E.2d 332 (Ill. App. Ct. 1984).

Improper conduct by a surety also may give rise to complaints of bad faith and, consequently, claims for negligence and punitive damages.[47] As discussed in **Section I.A.2** of this chapter, a surety's acts of bad faith may breach the requirement that the parties to a contract exercise good faith and engage in fair dealing with one another. When surety misconduct allegedly results from actual malice, fraud, or willful and wanton disregard for the obligee's rights, the surety may have committed a wrong independent of its bond obligations, for which punitive damages may be recovered.[48] For example, a Colorado court, in finding that a performance bond surety breached its duty of good faith and fair dealing by unreasonably refusing to settle a claim,[49] concluded that the surety should be held liable for punitive damages, in order to discourage such conduct when it was motivated by a desire to save interest by deferring payment or to force compromise claim settlements under duress. As previously noted, the burden of proving "bad faith" can be significant.[50]

In summary, sureties are increasingly held liable for improper conduct in handling claims. This may give the obligee or principal some remedy for such conduct by sureties. Further, because the liability arises from the surety's wrongful act, the surety may be denied a claim to recover from its indemnitors the consequent loss it incurs.[51] Some states, however, are reluctant to recognize bad-faith claims against performance bond sureties. In these states, courts are likely to focus on the differences between insurance contracts and contracts of suretyship, and to conclude that the state can deter bad-faith acts by sureties through exercise of regulatory powers.[52]

D. Surety's Defenses to Performance Bond Liability

The foregoing discussion of the surety's liability for failure to perform its obligations properly in no way diminishes the many defenses on which sureties may rely to deny bond claims. A surety generally may have a number of independent defenses arising out of the language of the performance bond itself (also known as technical defenses) or out of the nature of the surety relationship. As stated in **Section I.A.4** of this chapter, a surety's liability for the bonded obligations cannot be greater than the principal's liability.[53] Conversely, a defense normally available to the principal,

[47]*City of Westminster v. Centric-Jones Constructors,* 100 P.3d 472 (Colo. Ct. App. 2003) (punitive damages not available where bad-faith damages are not proven).

[48]*See, e.g., Hoskins v. Aetna Life Ins. Co.,* 452 N.E.2d 1315 (Ohio 1983); *Riva Ridge Apartments v. Robert G. Fisher Co. Inc.,* 745 P.2d 1034 (Colo. Ct. App. 1987).

[49]*Brighton Sch. Dist. 27J v. Transamerica Premier Ins. Co.,* 923 P.2d 328 (Colo. Ct. App. 1996). *See also Horst Masonry Constr., Inc. v. ProControls Corp.,* 208 F.3d 218 (8th Cir. 2000) (awarding attorneys' fees for bad-faith conduct).

[50]*PSE Consulting, Inc. v. Frank Mercede & Sons, Inc.,* 838 A.2d 135 (Conn. 2004).

[51]*See id.*

[52]*See, e.g., Tudor Dev. Group, Inc. v. U.S. Fid. & Guar. Co.,* 692 F. Supp. 461 (M.D. Pa. 1988).

[53]*See, generally*, P. Alces, The Law of Suretyship & Guaranty § 1:1 (2003). *See also* Cal. Civ. Code § 2809; *State Athletic Comm'n v. Mass. Bonding & Ins. Co.,* 117 P.2d 80 (Cal. 1941) (the liability of the contractor is the measure of the surety's liability); *Aetna Cas. & Sur. Co. v. Warren Bros. Co.,* 355 So. 2d 785 (Fla. 1978); *Vill. of Rosemont v. Lentin Lumber Co.,* 494 N.E.2d 592 (Ill. App. Ct. 1986).

but which the principal is barred from asserting, is likewise unavailable to the surety.[54]

A surety may waive its defenses if it fails to raise them in the time and manner required by its bond. For example, a performance bond surety may waive its right to require the performance of conditions precedent under the bond, if the surety declares that it will not complete the project because it determined that its principal had performed properly.[55] More generally, courts have held that a surety which fails to respond to a claim within the time required by its bond waives the right to dispute that claim.[56]

A surety generally may defend based on the language of the performance bond requiring the obligee to comply with conditions precedent under a performance bond.[57] If the obligee fails to perform a necessary condition for making a claim under the bond, the surety may be discharged from a liability it would otherwise have had under the bond. The standard form American Institute of Architects' (AIA) A312 Performance Bond (1984 ed.) (AIA A312) contains several conditions the obligee must observe to make a valid bond claim.[58] These conditions include: (1) notifying the contractor and surety that the owner may declare a contractor default; (2) trying to meet and discuss with the surety how to perform the remaining contract; (3) properly declaring the contractor in default and terminating the contractor's right to complete the contract;[59] and (4) agreeing to pay the remaining contract funds to the surety, or to the surety's designee for completing the contract, in accordance with the terms of the construction contract.[60] If the surety fails to perform its duties under the AIA A312 Performance Bond, the owner must give additional notice to the surety demanding that it perform and wait 15 days before the owner can deem the surety to be in default of its duty to perform.[61] Courts have held that the surety is entitled to insist on the obligee's strict compliance with these conditions of notice, declaration of default, termination of the contractor, and right to investigate and cure the default.[62] Other courts have reached the opposite results, however, by carefully

[54]*Indem. Ins. Co. v. United States,* 74 F.2d 22 (5th Cir. 1934). *See also R.I. Hosp. Trust Nat'l Bank v. Ohio Cas. Ins.* Co., 789 F.2d 74 (1st Cir. 1986).

[55]*See, e.g., J.C. Gibson Plastering Co., Inc. v. XL Specialty Ins. Co.*, 521 F. Supp. 2d 1326, 1336 (M.D. Fla. 2007).

[56]*See, e.g., J.C. Gibson Plastering Co., Inc. v. XL Specialty Ins. Co.,* 521 F. Supp. 2d 1326 (M.D. Fla. 2007)*; Nat'l Union Fire Ins. Co. of Pittsburgh v. David A. Bramble, Inc.*, 879 A.2d 101, 110 (Md. 2005).

[57]*See Easton v. Boston Inv. Co.,* 196 P. 796 (Cal. Ct. App. 1921).

[58]*See* AIA A312—Performance Bond Between Owner, Contractor, and Surety (1984 ed.). *See also Bank of Brewton, Inc. v. Int'l Fid. Ins. Co.,* 827 So. 2d 747 (Ala. 2002); *Enter. Capital, Inc. v. San-Gra Corp.,* 284 F. Supp. 2d 166 (D. Mass. 2003); 120 *Greenwich Dev. Assocs. v. Reliance Ins. Co.,* 2004 WL 1277998 (S.D.N.Y. 2004).

[59]AIA A312, §3.1 (1984 ed.).

[60]*Id.* at § 3.3.

[61]*Id.* at § 5.

[62]*See, e.g., Seaboard Sur. Co. v. Town of Greenfield,* 370 F.3d 215 (1st Cir. 2004); *Elm Haven Constr. Ltd. P'ship v. Neri Constr., LLC,* 376 F.3d 96 (2d Cir. 2004); *L&A Contracting Co. v. S. Concrete Servs.,* 17 F.3d 106 (5th Cir. 1994); *Bank of Brewton, Inc. v. Int'l Fid. Ins. Co.,* 827 So. 2d 747 (Ala. 2002). *See also 150 Nassau Assocs., LLC v. Liberty Mut. Ins. Co.,* 826 N.Y.S.2d 567 (N.Y. App. Div. 2007).

reading the bond language, including language from the bonded contract incorporated into the bond, or by finding that the failure to promptly declare a default was a nonmaterial breach of the owner's duties under the bond.[63]

In determining the extent of any waiver, discharge, or conditions precedent to liability under the bond, the terms of the bond as well as the terms of the principal's construction contract must be reviewed. For example, statutory performance bonds may incorporate conditions precedent from the governing statutes, thus creating another set of conditions that may discharge the surety.[64]

An additional defense, arising in part from the contract, may be available where the surety is not notified of, and does not consent to, a material change to the contract it has bonded. The surety may be discharged either in whole or to the extent of injury caused by such a material alteration.[65] In recognition of this rule, many standard performance bond forms provide that the surety consents in advance to permit the owner and the principal to modify the work scope in the normal course of the construction contract without further consent of the surety.[66] Further, a surety normally must show it suffered some injury before it can claim a discharge from the work scope alteration.[67]

The surety has another defense based in part on the bond language when the owner makes unauthorized or premature payments or fails to withhold contractually required retainage.[68] For example, in *Southwood Builders, Inc. v. Peerless Insurance Co.*,[69] a prime contractor that paid for workers and materials to help bring a subcontractor back on schedule, without terminating or notifying the subcontractor's surety, could not recover its cost overruns in completing the work after the subcontractor failed to get back on schedule. The prime contractor's advance payment for the

[63]*See, e.g., Colo. Structures, Inc. v. Ins. Co. of the W.*, 167 P.3d 1125 (Wash. 2007) (bond closely read did not disclose a breach); *Dooley & Mack Constructors, Inc. v. Developers Sur. & Indem. Co.*, 972 So. 2d 893 (Fla. Dist. Ct. App. 2007) (bond incorporated contract, which permitted owner to complete, and hold surety liable, without declaration of breach); *Kilpatrick Bros. Painting v. Chippewa Hills Sch. Dist.*, 2006 WL 664210 (Mich. Ct. App. 2006) (owner's beginning corrective work before declaring default was not a material breach of bond conditions).

[64]*See, e.g., United States v. Am. Mfrs. Mut. Cas. Co.*, 901 F.2d 370 (4th Cir. 1990), *cert. denied*, 498 U.S. 851 (explaining statutory notice requirements for government claim for taxes under Miller Act).

[65]*See, e.g., Ramada Dev. Co. v. U.S. Fid. & Guar. Co.*, 626 F.2d 517 (6th Cir. 1980); *City of Peekskill v. Cont'l Ins. Co.*, 999 F. Supp. 584 (S.D.N.Y. 1998), *aff'd*, 166 F.3d 1199 (2d Cir. 1998); *Cf. Cont'l Bank & Trust Co. v. Am. Bonding Co.*, 605 F.2d 1049 (8th Cir. 1979) (finding no change increasing surety's risk).

[66]*See, e.g.*, ConsenusDOCS 200, ¶ 10.7.1 (2007 ed.); *see also Am. Ins. Co. v. United States*, 62 Fed. Cl. 151 (2004); *Data Sales Co., Inc. v. Diamond Z Mfg.*, 74 P.3d 268 (Ariz. Ct. App. 2003); *Trinity Universal Ins. Co. v. Gould*, 258 F.2d 883 (10th Cir. 1958); *Mass. Bonding & Ins. Co. v. John R. Thompson Co.*, 88 F.2d 825 (8th Cir.), *cert. denied*, 301 U.S. 707 (1937).

[67]*See* Restatement (Third) of Suretyship & Guarantee § 37 (1996); *Peerless Ins. Co. v. Cerny & Assocs., Inc.*, 199 F. Supp. 951 (D. Minn. 1961).

[68]*See Pa. Nat'l Mut. Cas. Ins. Co. v. City of Pine Bluff*, 354 F.3d 945 (8th Cir. 2004); *Nat'l Sur. Corp. v. United States*, 118 F.3d 1542 (Fed. Cir. 1997); *Prairie State Nat'l Bank v. United States*, 164 U.S. 227 (1896); *Fireman's Fund Ins. Co. v. United States*, 15 Cl. Ct. 225 (1988), *rev'd on other grounds*, 909 F.2d 495 (Fed. Cir. 1990).

[69]366 S.E.2d 104 (Va. 1988).

subcontractor's workers and materials was a "material variation" in the subcontract terms, which released the surety from its bond obligations.

The rationale for this rule is that a surety relies on provisions of the bonded contract in executing the bond. Contract terms, such as payment according to a set schedule and retainage requirements, benefit the surety by controlling the expenditure of contract funds. Overpayment by the owner to the contractor may discharge the surety's obligations, because such voluntary action by the owner in violation of the contract terms reduces the contract balance the surety would expect to have available for completion of the project in the event of a default.[70] Thus, if the owner or its designated representative fraudulently certify that progress payments are due beyond the defaulting principal's actual progress, the surety may defend based on overpayment.[71]

Overpayment may not discharge the surety, however, if the obligee makes payments in good faith.[72] Unauthorized prepayments made by an owner/obligee may not discharge a compensated surety so long as (1) the funds were used in actual construction of the project and (2) the surety was not prejudiced by the prepayments.[73] When an architect has negligently certified progress payments to a contractor, thereby damaging the surety by reducing the contract balance available to the surety upon completion of the work, however, the surety may have a cause of action against the architect for the resulting damage.[74]

The principal's fraud in obtaining a bond from the surety will not alone discharge the surety, although the surety may have a claim against the principal for its fraud,[75] because such a discharge would injure the obligee without its having done any wrong. If the obligee has perpetrated a fraud on the surety, or even participated in it, however, the surety will be discharged from its obligation.[76] A closer question arises when the obligee has practiced no active fraud but has knowledge of facts that might

[70] *Pa. Nat'l Mut. Cas. Ins. Co. v. City of Pine Bluff,* 354 F.3d 945, 952–53 (8th Cir. 2004).

[71] *See, e.g., U.S. Fid. & Guar. Co. v. United States,* 16 Cl. Ct. 541 (1989) (discussing burden that must be met in making such a claim on a federal project).

[72] *Argonaut Ins. Co. v. Town of Cloverdale,* 699 F.2d 417 (7th Cir. 1983); *see Balboa Ins. Co. v. Fulton County,* 251 S.E.2d 123 (Ga. Ct. App. 1978) (overpayments made in reliance on progress payment requests certified by architect).

[73] *See Basic Asphalt & Constr. Corp. v. Parliament Ins. Co.,* 531 F.2d 702 (5th Cir. 1976); *Fireman's Fund Ins. Co. v. United States,* 15 Cl. Ct. 225 (1988), *rev'd on other grounds,* 909 F.2d 495 (1990).

[74] *See generally Mid-State Sur. Corp. v. Thrasher Eng'g, Inc.,* 2006 WL 1390430 (S.D. W. Va. 2006) (claim by surety against engineer for professional negligence allowed to proceed); *Carolina Cas. Ins. Co. v. R.L. Brown Assocs., Inc.,* 2006 WL 3625891 (N.D. Ga. 2006) (discussing limits on a claim by surety against owner's architect for improper payments).

[75] *Commercial Money Ctr., Inc. v. Ill. Union Ins. Co.,* 508 F.3d 327, 342–43 (6th Cir. 2007); *Kvaerner Constr., Inc. v. Am. Safety Cas. Ins. Co.,* 847 So. 2d 534 (Fla. Dist. Ct. App. 2003); *Chrysler Corp. v. Hanover Ins. Co.,* 350 F.2d 652 (7th Cir. 1965), *cert. denied,* 383 U.S. 906 (1966).

[76] *Pinkerton & Laws, Inc. v. Macro Constr., Inc.,* 485 S.E.2d 797 (Ga. App. 1997); *Filippi v. McMartin,* 10 Cal. Rptr. 180 (Cal. Ct. App. 1961); *see also St. Paul Fire & Marine Ins. Co. v. Commodity Credit Corp.,* 646 F.2d 1064 (5th Cir. 1981).

affect the surety's willingness to provide a bond.[77] In such a situation, the duty of the obligee to disclose those facts may depend on whether the obligee: (1) had reason to believe that the facts materially increase the risk beyond that which the surety intended to assume; (2) had reason to believe such facts are unknown to the surety; and (3) had a reasonable opportunity to reveal such facts to the surety.[78] Given these considerations, the nature and extent of the dealings between the obligee and surety may become relevant, if, for example, the obligee knows from past dealings that the surety relies on the obligee for certain types of information. The law will not favor an unseeing surety, however, that fails to exercise reasonable diligence to learn relevant facts for itself.[79]

E. Surety Entitlement to Contract Funds

In addition to its defenses, the surety has the important right upon a default termination of the contractor-principal to receive the contract funds then due to the principal and to receive all remaining contract funds as they become due.[80] The surety "earns" the right to receive these funds by performing its bond obligation. These contract funds must be applied to reduce or satisfy the loss incurred by the surety in performing its bond obligations toward the defaulted principal's contract.[81]

The surety's right to these funds is solid to the extent the funds are needed to pay losses incurred under the performance bond, being superior to a claim by a financial institution with a security interest in those funds[82] and even to a bankruptcy trustee's rights when the principal has filed for bankruptcy.[83] In some cases, a performance bond surety has been given priority to the contract funds before permitting setoffs by the obligee.[84] To guard against the potential setoff right of the owner, a surety often will require the owner to agree, in exchange for the surety's performing its obligations, that the owner will pay remaining contract balances to the surety as the project is completed. There are limits, however, to a surety's right to the contract proceeds. For example, a surety that had entered into a completion agreement to perform its defaulted principal's work, in exchange for the contract balance, could not claim

[77]*See Ransom v. United States,* 17 Cl. Ct. 263 (1989), *aff'd,* 900 F.2d 242 (Fed. Cir. 1990).

[78]*See Sumitomo Bank v. Iwasaki,* 447 P.2d 956 (Ca. 1968); *Rachman Bag Co. v. Liberty Mut. Ins. Co.,* 46 F.3d 230 (2d Cir. 1995). *See generally* Restatement (First) of Security § 124 (1941).

[79]*See Ransom v. United States*, 17 Cl. Ct. 263 (1989).

[80]*See Nelson Constr. Co. v. United States,* 79 Fed. Cl. 81 (2007).

[81]*Aetna Cas. & Sur. Co. v. United States,* 845 F.2d 971 (Fed. Cir. 1988); *U.S. Fid. & Guar. Co. v. Mo. Highway & Transp. Comm'n,* 783 S.W.2d 516 (Mo. Ct. App. 1990). *But see Aetna Cas. & Sur. Co. v. United States,* 12 Cl. Ct. 271 (1987), *rev'd on other grounds,* 845 F.2d 971 (Fed. Cir. 1988) (funds advanced or paid out under the payment bond do not give the surety the same right).

[82]*See Prairie State Nat'l Bank v. United States,* 164 U.S. 227 (1896); *Nat'l Am. Ins. Co. v. United States,* 498 F.3d 1301, (Fed. Cir. 2007); *Kan. City v. Tri-City Constr. Co.,* 666 F. Supp. 170 (W.D. Mo. 1987); *Mid-Continent Cas. Co. v. First Nat'l Bank & Trust Co.,* 531 P.2d 1370 (Okla. 1975); *Interfirst Bank Dallas, N.A. v. U.S. Fid. & Guar. Co.,* 774 S.W.2d 391 (Tex. App. 1989).

[83]*Pearlman v. Reliance Ins. Co.,* 371 U.S. 132 (1962).

[84]*Nat'l Fire Ins. Co. v. Fortune Constr. Co.,* 320 F.3d 1260 (11th Cir. 2003); *Trinity Universal Ins. Co. v. United States,* 382 F.2d 317 (5th Cir. 1967), *cert. denied,* 390 U.S. 906 (1968); *Covenant Mut. Ins. Co. v. Able Concrete Pump,* 609 F. Supp. 27 (D.C. Cal. 1984).

entitlement to the contract balance as a priority over a claim by a state's Department of Labor for unpaid back wages or by the Internal Revenue Service for taxes. In that case, a state law deemed the contract balance to be funds held in trust, and, as such, the surety's claim to funds did not take priority to the extent such funds were needed to pay wages or satisfy taxes owing for work done on a public works project.[85]

The surety may be able to claim only the funds from the specific project that it bonded and on which it completed performance. For example, a performance bond surety for a contractor on two federal government contracts, having incurred losses when it took over and completed performance of a second contract, could not recover funds payable by the government to the contractor on the first contract.[86]

F. Industry Performance Bond Forms: A Comparison

The most widely used "industry" form for performance bonds is the AIA A312 Performance Bond. Although the AIA has revised and reissued many of its standard construction industry forms since 1984, most recently in 2007, the performance bond form remains unchanged. Other leading industry professional organizations have developed performance bond forms that compete with the AIA A312. These include bond forms issued as part of the ConsensusDOCS series of contract documents[87] and the Engineers Joint Contract Documents Committee (EJCDC) construction series documents. These other bond forms may be more advantageous to the users depending on the type of project or project delivery method.

ConsensusDOCS offers four different performance bond forms to choose from, for different sets of parties or project delivery methods. ConsensusDOCS 260 (2007 ed.) is the standard performance bond, which may be used with a variety of contracts. ConsensusDOCS 470 (2007 ed.) is a performance bond form designed for use on a design-build project (one party responsible to owner for both project design and construction), where the surety agrees to be liable for performance of both design and construction services of the bond principal. ConsensusDOCS 471 (2007 ed.) is also a performance bond for a design-build contract, but under this form, the surety expressly excludes liability for performance of design services from the scope of the bond obligations. Finally, ConsensusDOCS 706 (2007 ed.) is a performance bond specifically tailored for the contractor-subcontractor relationship.

EJCDC Form C-610 Performance Bond (2007 ed.) is the only performance bond form issued as part of this industry group's construction documents. This performance bond is very similar to the AIA A312 form. Both forms require, and impose time limits for, a pre-default meeting among owner, surety, and contractor as well as time constraints on how long the owner must wait before it can terminate the contract.[88] Also,

[85] *Titan Indem. Co. v. Triborough Bridge & Tunnel Auth., Inc.,* 135 F.3d 831 (2d Cir. 1998).

[86] *Transamerica Ins. Co. v. United States,* 989 F.2d 1188 (Fed. Cir. 1993).

[87] See **Chapter 13** for a complete list of the ConsensusDOCS.

[88] *Compare* AIA A312, § 3 (1984 ed.) *with* EJCDC C-610, §§ 2.1–2.2 (2007 ed.).

both forms specify the same four options available to the surety upon declaration of contractor default.[89]

These standard industry performance bond forms enjoy many similarities despite employing different terminology. For example, ConsensusDOCS 260 refers to the maximum liability under the bond as the "Bond Sum"; AIA A312 and EJCDC C-670 use the term "Penal Sum." Yet these forms do contain distinct differences, touching on such important issues as when the surety's bond obligation is triggered, the surety's options for performing when there is a contractor default, and when the limitation of action for bringing an action against the surety starts to run. An important difference in AIA A312 is that it does not specifically address a surety's potential design liability, were that bond form to be issued to a design-build contractor. Liability for performing design services differs significantly from liability for construction services, and the unwary surety, owner, or contractor might face exposure to unintended risks if the performance bond form does not adequately address such liability and risk where the contractor takes on design liability.

G. Effect on the Surety of Arbitration of Construction Disputes

Arbitration is a common method of dispute resolution in the construction industry, especially in the private sector. Many standard construction contract forms provide that disputes between the owner and contractor may or will be resolved by arbitration.[90] Typically, no specific reference to the surety is made in the construction contract's arbitration provisions, and the basic rule is that arbitration can be enforced only when there is an agreement between the parties to arbitrate. Thus, the mere fact that the principal has entered into an arbitration agreement may not be a basis to compel the surety to arbitrate any disputes involving the performance bond.[91] If the obligee has no agreement to arbitrate with the performance bond surety, a potential problem is created in that the obligee might be forced to pursue relief in two separate forums: in court against the surety and through arbitration against the principal. Such a piecemeal dispute resolution process creates additional expense and the risk of inconsistent results. As a partial solution to this problem, and in an effort to avoid multiple proceedings in different forums, the standard construction contract forms have become more lenient toward allowing the contractor and the owner to join related parties (e.g., sureties) in arbitration.[92]

A more effective solution to the multiple proceedings problem has been built on the fact that performance bonds regularly refer to and incorporate the underlying

[89]*Compare* EJCDC C-610, § 2 (2007 ed.), *with* AIA A312, § 4 (1984 ed.).

[90]The AIA contract forms historically required binding arbitration of unresolved disputes. *See, e.g.,* AIA A201, § 4.6 (1997 ed.). But the most recent revisions to the AIA contract documents allow the parties to choose between arbitration and other means of dispute resolution. *See, e.g.,* AIA A101, § 6.2 (2007 ed.) and AIA A201, § 15.4 (2007 ed.). Similarly, the ConsensusDOCS 200 (2007 ed.) standard form permits the parties to choose between arbitration and litigation. *See* ConsensusDOCS 200, ¶ 12.5 (2007 ed.).

[91]*See generally Employers Ins. of Wausau v. Bright Metal Specialties, Inc.,* 251 F.3d 1315, 1322 (11th Cir. 2001); *Transamerica Ins. Co. v. Yonkers Contracting Co.,* 267 N.Y.S.2d 669 (N.Y. App. Div. 1966); *Windowmaster Corp. v. B.G. Danis Co.,* 511 F. Supp. 157 (S.D. Ohio 1980).

[92]*Compare* AIA A201, § 15.4.4 (2007 ed.) *with* AIA A201, § 4.6.4 (1997 ed.).

construction contract. Currently, a strong federal policy and federal statutes favor arbitration. Recognizing this policy and law, courts have held that a surety that has not directly signed an arbitration agreement may nonetheless be held to have agreed to arbitrate claims, if the surety's bond incorporates a bonded contract requiring arbitration.

For example, in *Cianbro Corp. v. Empresa Nacionale de Ingenieria,*[93] a general contractor was allowed to compel its subcontractor's surety to arbitrate a dispute arising out of a subcontract containing an arbitration clause that was incorporated by reference into the surety's performance bond. Although the surety did not sign the subcontract agreement containing the arbitration clause, it was bound by the incorporation of that agreement into the bond.

Rules governing which parties may be forced to arbitrate in a given case are complex, and outcomes may be hard to predict, as they may require the simultaneous interpretation of multiple contracts. This poses a problem, because a party may suffer serious consequences if it fails to participate in the correct proceeding. For example, the surety in *Fidelity & Deposit Co. of Maryland v. Parsons & Whitemore Contractors Corp.,*[94] which bonded a subcontract containing an arbitration clause, filed multiple appeals asserting that the surety was not obligated to arbitrate claims and issues under its bond for that subcontract. Even after several appeals, however, a court decided that the issue of whether the surety had to participate in the subcontractor's arbitration with the contractor was not ready for determination. Still, the court held that the arbitration would go forward on the claims between the contractor and the bonded subcontractor. The court further held that the arbitrator's decision as to those claims would be binding on the surety, regardless of whether it participated in the arbitration.[95] This result was based on a rule, applied in lititgation, that a surety that knows of and could participate in a suit alleging its principal was liable for breach of a bonded contract may be bound by the determination in that suit, whether the surety participated or not.[96]

Sureties have, in fact, been bound by an arbitration award against the contractor (the principal) when the surety knew of the arbitration and declined to take the opportunity to participate.[97] If the mere fact that a surety knows of an arbitration may bind the surety on matters determined in the arbitration, obviously the surety may be forced as a practical matter to defend its interests in the arbitration, even if the surety feels it is not legally bound to participate and does not wish to do so. In such

[93]697 F. Supp. 15 (D. Me. 1988). *See also Employers Ins. of Wausau v. Bright Metal Specialists, Inc.,* 251 F.3d 1315, 1323 (11th Cir. 2001) (surety incorporated arbitration clause of contract by signing ratification agreement); *U.S. Fid. & Guar. v. W. Point Constr. Co.,* 837 F.2d 1507 (11th Cir. 1988); *But see AgGrow Oils, LLC v. Nat'l Union Fire Ins. Co.,* 242 F.3d 777 (8th Cir. 2001) (incorporation of contract arbitration clause by bond did not permit the surety to require arbitration of every dispute arising under the bond, but requiring arbitration of some claims might be proper).

[94]397 N.E.2d 380 (N.Y. 1979).

[95]*Id. See also Kearsarge Metallurgical Corp. v. Peerless Ins. Co.,* 418 N.E.2d 580 (Mass. 1981).

[96]*See, e.g., Drill S., Inc. v. Int'l Fid. Ins. Co.,* 234 F.3d 1232 (11th Cir. 2001); *Am. Cas. Ins. Co. v. C.G. Mitchell Constr., Inc.,* 601 S.E.2d 633 (Va. 2004).

[97]*Von Eng'g Co. v. R. W. Roberts Constr. Co.,* 457 So. 2d 1080 (Fla. Dist. Ct. App. 1984).

an arbitration, the surety should have a right to plead any of its own defenses, such as material alteration of the obligation or failures of conditions precedent under the bond. The surety likewise could raise the defense that the arbitration included matters outside of the liability of the surety or matters not covered by the performance bond, and in some jurisdictions, the surety may raise defenses that could have been raised by the principal but were not.[98]

H. Alternatives to Bonds: "Subguard" Programs

"Subguard" is a subcontractor default insurance policy developed as an alternative method of ensuring subcontractor performance without requiring subcontractor performance bonds.[99] Subguard policies are designed for use by general contractors on projects involving multiple subcontractors. Although the product is available to owners, its primary users have been large general contractors.

There are several significant advantages to the use of Subguard insurance. First, it allows the general contractor direct control over evaluating the financial stability of its selected subcontractors. Further, upon the failure of a subcontractor to perform, the subcontractor default insurance is more flexible than a bond in allowing the general contractor to fashion an appropriate remedy. This insurance policy allows the general contractor to control insurance and financial concerns at all levels of subcontractors on the project through one master policy issued to the general contractor, as opposed to dealing potentially with multiple surety bonds and multiple surety companies. Use of a Subguard policy allows the general contractor to prequalify its subcontractors, thereby avoiding subcontractors that cannot provide a bond after being awarded a subcontract for part of the work. Finally, subcontractor default insurance may allow for a negotiated contract of insurance in which the terms can be modified, added, or deleted to customize the policy for a specific project.

Subcontractor default insurance is a useful tool in some circumstances, but it is not appropriate in all cases. Like surety bonds, the cost of coverage varies with the type of project, the parties involved, and the risks covered. The subcontractor insurance policy typically requires the insured general contractor to fund any default and make a claim only after the covered loss has occurred. Further, unlike the surety bond, the Subguard insurance policy includes a significant deductible per event, which ranges from $500,000 to much higher, depending on the value and type of project, and includes an aggregate deductible often in excess of $2 million before recovery on the policy is realized. Thus, the general contractor takes on a significant financial risk in relying on the Subguard insurance policy.

A subcontractor default policy may be canceled for nonpayment, material change, or postissuance violation of policy terms or applicable qualifications. This represents another disadvantage when compared to a surety bond, which generally is not cancelable absent fraud or misrepresentation of the obligee. Further, the subcontractor insurance policy may not satisfy any statutory payment and performance surety

[98] *P. R. Post Corp. v. Md. Cas. Co.*, 271 N.W.2d 521 (Mich. 1978).

[99] *See, e.g., SMI-Owen Steel Co., Inc. v. St. Paul Fire & Marine Ins. Co.*, 199 F.R.D. 209 (S.D. Tex. 2001).

bond requirements on public projects. Thus, the subcontractor insurance policy may have to be modified in order to meet the statutory bond requirements in many jurisdictions.

The Subguard insurer will require the general contractor to first qualify itself and the project under the policy requirements before a subcontractor insurance policy is issued. At that point, the insurer will assist the general contractor in establishing a subcontractor control program by which each eligible subcontractor must also qualify for the policy.

II. TERMINATION

A. Overview

The performance bond surety's involvement on a project is generally the result of the owner's (or general contractor's) termination or threatened termination of the bonded contract due to the contractor's (or subcontractor's) default. Virtually all construction contracts expressly recognize the right to terminate the contract for the default of a breaching party in certain specific circumstances. Such provisions usually afford the nonbreaching party a right to recover damages resulting from the termination.

Even in the absence of a contractual termination provision, the right to terminate a contract generally is implied against a party that has materially breached the contract. Many public contracts, and an increasing number of private construction contracts, contain "termination for convenience" clauses authorizing the owner (or prime contractor) to terminate the prime contractor (or subcontractors) even without cause. Practically, a termination for convenience clause acts as a very specific changes clause in a contract—the change being that the owner no longer wants to proceed with the work through no fault of the contractor. When the termination is for convenience, payments to the terminated contractor usually are limited to actual costs incurred by that contractor, and ordinarily do not include overhead or other standard breach of contract damages or lost profits.

B. Termination for Default

The right to terminate a construction contract for default generally arises only when a material or substantial provision of the contract has been breached or a party has failed to perform a material obligation. The rights and liabilities of the parties in such a termination are determined by the common law (made up of prior court decisions applying the law in similar situations), by regulations or statutes applicable to the contract, and by the contractual terms governing the termination. At common law, minor deviations in performance or failures to meet nonmaterial provisions of the contract were not viewed as grounds to terminate for default.[100] Although such deviations or

[100] *See L&A Contracting Co. v. S. Concrete Servs., Inc.*, 17 F.3d 106, 110 (5th Cir. 1994); *Vill. of Fox Lake v. Action Cas. & Sur. Co.*, 534 N.E.2d 133 (Ill. App. Ct. 1989).

failures in performance may give the nonbreaching party a right to recover damages caused by the failure, they are not viewed as the basis for a complete termination of the contract. The common law has measured the materiality of a breach by the importance of the event or act to the purpose of the contract as a whole, or to the stated basis of the agreement set out in the contract between the parties. By the terms set out in their contract, therefore, the parties can expand, limit, or redefine the grounds for termination for default, compared to the grounds that would otherwise exist at common law. They also can provide for the payment of damages and for the imposition of other obligations after the contract is terminated for default.

C. Standard-Form Contracts: Grounds for Default Termination

Federal construction procurement contracts provide an example of how rights and duties can be allocated by the parties' contract when there is a default. The default clause used by the federal government in fixed-price construction contracts is set forth at Federal Acquisition Regulation (FAR) § 52.249–10. This standard federal government construction contract termination clause affords the government several remedies in addition to the relief generally available at common law. If a contractor fails to perform by the date specified, produces defective or nonconforming work, or refuses or fails to prosecute the work so as to ensure its timely completion, the government may pursue its administrative remedy under the terms of the default clause. That is, under FAR § 52.249–10(a) the government may terminate the contract and take over and complete the work at the contractor's expense. The clause makes the contractor liable for any liquidated or actual damages caused by unexcused delays in the completion of the work.

At the same time, FAR § 52.249–10 (b) gives the contractor the right to contest a default termination based on excusable delays that extend the time for the contractor's performance duty. Under FAR § 52.249–10(c), however, if the contractor successfully challenges a wrongful termination for default, the termination for default is converted automatically into a termination for the government's convenience, thereby limiting the terminated contractor's remedies. The remedy available under a termination for convenience is less than what a terminated contractor could seek in a suit for wrongful contract termination under common law.

Like the federal default termination clause, industry contract forms address, and supplement, common law remedies for the owner in the event of a default termination. Section 14.2.1 of AIA A201 (2007 ed.), for example, describes the instances in which an owner may terminate a contract for cause. In addition to circumstances relating to the contractor's financial situation, the owner may terminate a contract for cause if the contractor:

(1) Repeatedly refuses or fails to supply enough properly skilled workers or proper materials;

(2) Fails to make payment to Subcontractors for materials or labor in accordance with the respective agreements between the Contractor and the Subcontractors;

(3) Repeatedly disregards applicable laws, statutes, ordinances, codes, rules and regulations, or lawful orders of a public authority; or

(4) Otherwise is guilty of substantial breach of a provision of the Contract Documents.[101]

The AIA A201 form also requires that the "Initial Decision Maker" (often an architect) certify that sufficient cause exists for the termination.[102] The owner must also give notice to the contractor and its surety seven days before terminating the contract.

Article 11.2 of ConsensusDOCS 200—Standard Agreement and General Conditions Between Owner and Contractor (2007 ed.), sets forth grounds for a default termination similar to the AIA A201 terms, provided that the contractor fails to remedy the grounds for termination after receiving notices to cure required in Article 11.2. Unlike the AIA form, however, the ConsensusDOCS form does not require an architect's certification that cause exists to justify a termination for default.

For the most part, the federal FAR termination clause, as well as the ConsensusDOCS and AIA termination clauses, provides for termination based on facts that would likely be deemed to constitute material breaches of contract at common law. Specific situations in which sufficient grounds for termination of a construction contract have been found to exist are discussed next. Remember also that the parties may define in their construction contracts those breaches of contract which they deem sufficiently "material" to justify a default termination.

1. Refusal or Failure to Prosecute the Work

A typical default termination clause will include a provision stating, in substance, that: "If the contractor refuses or fails to prosecute the work, or separable parts of the work, with the diligence that will insure completion within the time specified in [the] contract," the contract may be terminated for default.[103] In the case of such a termination provision, the terminating party has the burden of showing both that a significant lag in performance has occurred and that the contractor, at the time of the termination, could not have completed performance on time.[104]

The fact that the contractor's progress lags during one stage of the work may not, alone, support termination. For example, a termination for default has been overturned as premature where sufficient time remained in the contract schedule for the contractor to complete performance.[105] Although factual determinations vary from case to case, courts have interpreted the standard default clause to require a showing that timely completion was clearly in jeopardy.[106]

[101]AIA A201, § 14.2.1 (2007 ed.).

[102]*Id.* at § 14.2.2.

[103]*See, e.g.*, FAR §52.249.10(a).

[104]*Litcom Div., Litton Sys.*, ASBCA No. 13413, 78–1 BCA ¶ 13,022.

[105]*Strickland Co.*, ASBCA No. 9840, 67–1 BCA ¶ 6193.

[106]*See, e.g., McDonald Douglas Corp. v. United States,* 76 Fed. Cl. 385 (2007); *Hannon Elec. Co. v. United States,* 31 Fed. Cl. 135 (1994); *Discount Co. v. United States,* 554 F.2d 435 (Ct. Cl. 1977), *cert. denied,* 434 U.S. 938 (1977).

2. Failure to Complete on Time

As a general rule, a contract stating that time is "of the essence" and containing fixed or specific performance dates is interpreted to mean that the time for performance is a material element of contract's performance. In a federal government construction contract containing such a time clause, if timely performance does not occur, the government can terminate immediately without notice and without affording the contractor an opportunity to cure.[107] When time is not described as being of the essence, however, late performance is only one of the factors to be considered in determining the adequacy of performance and the justification for a default termination. For example, a delay in final completion may not justify a default termination if the work is sufficiently complete that it can be used for its intended purpose, and the contract does not require full and final, timely completion.[108]

3. Repudiation by the Contractor

The termination of a contractor for repudiating or declaring it will not perform its contract generally arises from the contractor's refusal or complete failure to prosecute the work in a timely manner. To terminate a contract on the basis of repudiation by the contractor, the owner must show that the contractor manifestly cannot or will not perform the contract. The cases speak in terms of a "positive, definite, unconditional and unequivocal manifestation of intent, by words or conduct, on the part of a contractor" not to perform.[109] It must be apparent that the contractor is unable to, or will not, perform the contract. A failure to make progress alone does not constitute anticipatory repudiation absent some objective manifestation of an inability to, or intent not to, perform.

When a contractor repudiates the existence of the contract it has entered into, the contractor is subject to default termination regardless of its rate of progress in performance.[110] In federal government contracts, a contractor's refusal to perform during a dispute about contract interpretation, after the contracting officer properly directs the contractor to proceed with the work, also constitutes grounds for finding that the contractor has repudiated the contract.[111] Similarly, private contracts may obligate the contractor to continue the work during a contract dispute.[112] In some circumstances, however, a contractor's refusal to perform may be justified and will not be regarded as an anticipatory breach or repudiation of the contract. For example, when the contract specifications are impossible to perform, the contractor has a

[107]*Nat'l Farm Equip. Co.,* GSBCA No. 4921, 78–1 BCA ¶ 13,195; *see Dallas–Fort Worth Reg'l Airport Bd. v. Combustion Equip. Assocs., Inc.,* 623 F.2d 1032 (5th Cir. 1980).
[108]*Franklin E. Penny Co. v. United States,* 524 F.2d 668 (Ct. Cl. 1975).
[109]*Mountain State Constr. Co.,* ENGBCA No. 3549, 76–2 BCA ¶ 12,197.
[110]*See First Nat'l Bank of Aberdeen v. Indian Indus.,* 600 F.2d 702 (8th Cir. 1979).
[111]*Tester Corp.,* ASBCA No. 21312, 78–2 BCA ¶ 13,373.
[112]AIA A201, § 15.1.3 (2007 ed.) requires continued performance by the contractor during a dispute, as have past AIA standard contract forms.

right to stop work on the basis that no other practical alternative exists.[113] Further, a contractor should not be terminated for default where it is reasonably awaiting clarification from the government of drawings, specifications, or other central requirements. If, for example, the contractor cannot proceed until it obtains guidance from the government, the delay in work is justified.[114]

Many contracts allow the parties to make necessary changes to the contract as the work progresses, but most standard changes clauses include only changes "within the general scope of the contract." Changes beyond that scope are termed "cardinal changes."[115] Contractors cannot validly be terminated for refusing to perform a contract that has been altered by an actual or constructive cardinal change.[116] Stopping performance on the grounds of an alleged cardinal change, however, may be risky. Whether the particular change to the contract amounted to a cardinal change may be contested strongly in the course of a suit or arbitration over the validity of the termination for default.

If an owner materially breaches the contract, the contractor's subsequent repudiation may be justified and provide no basis for a default termination. For example, when the owner withholds or fails to make payment that is due under a contract certificate of payment, the contractor may stop performance.[117]

4. *Failure to Comply with Other Material Provisions of the Contract*

The failure of a contractor to comply with other terms of the contract may be a material breach for which the owner can terminate the contract for default, but the owner will have to prove that the contract provision not complied with is material to the contract and that the failure to perform is a material breach of the contract. In *Antonio Santisteban & Co.,*[118] a contractor's failure to furnish a performance bond constituted a material breach. Likewise, some federal boards of contract appeals have held that a contractor's failure to furnish Miller Act bonds pursuant to 40 U.S.C. § 3131 *et seq.* justifies default termination.[119]

D. Defenses of the Contractor

1. *Excusable Delay*

Virtually all construction contracts excuse certain types of delays from counting against the time allowed for performance. As an example, paragraph (b) of the default

[113]*Chugach Elec. Ass'n v. N. Corp.,* 562 P.2d 1053 (Alaska 1977); *L. J. Casey Co.,* AGBCA No. 75–148, 76–2 BCA ¶ 12,196.
[114]*Electromagnetic Indus., Inc.,* ASBCA No. 11485, 67–2 BCA ¶ 6545.
[115]See **Chapter 9** for a more detailed discussion on cardinal changes.
[116]*Allied Materials & Equip. v. United States,* 569 F.2d 562 (Ct. Cl. 1978); *P. L. Saddler v. United States,* 287 F.2d 411 (Ct. Cl. 1961); *Cray Research Inc. v. Dep't of the Navy,* 556 F. Supp. 201 (D. D.C. 1982).
[117]*United States ex rel. E. C. Ernst, Inc. v. Curtis T. Bedwell & Sons, Inc.,* 506 F. Supp. 1324 (E.D. Pa. 1981); *Contract Maint., Inc.,* ASBCA No. 19603, 75–1 BCA ¶ 11,097. *See* AIA A201, § 14.1.1.3 (2007 ed.).
[118]ASBCA No. 5586, et al., 60–1 BCA ¶ 2497; *see also In re: Cole's Constr. Co., Inc.,* ENGBCA No. 607494–3, BCA ¶ 2,497; *Marvin J. Hargrove,* GSBCA No. 5117 78–2, BCA ¶ 13,386.
[119]*H.L. & S. Contractors, Inc.,* IBCA No. 1085–11–75, 76–1 BCA ¶ 11,878.

clause typically used by the federal government excuses delays and therefore will not allow such delays to be used as a cause for default termination, where a contractor's work is delayed by "unforeseeable causes beyond the control and without the fault or negligence of the contractor."[120] Although the contract may excuse certain kinds of delays, it generally will do so only if the contractor follows the contract's procedure for obtaining a time extension (e.g., by giving notice of the excusable delay and requesting a time extension request within a short period of time after the delay occurs).

2. *Waiver by the Owner*

A contractor may defend against a default termination by showing that the owner waived strict compliance with the contract completion date or other contract performance provisions, and thereby gave up its right to terminate for that reason. Even if the contract specifically states that its terms may be waived by the parties only if they follow required steps, the parties may nonetheless waive the contract requirements by expressly consenting to, or by agreeing to disregard, a failure of performance. A waiver can be effective without an exchange of value or a new contract agreement.[121] There will be no waiver, however, if the owner's forbearance in terminating was based on a misrepresentation by the contractor.[122]

The right to terminate a contract for default may be waived if not timely exercised, and it also can be expressly reserved. In *Indemnity Insurance Co. of North America v. United States,*[123] the government expressly reserved its right to terminate a takeover surety for default by informing the surety that the owner was continuing to assess liquidated damages and by demanding that the surety cure when the completion date had passed. The government's failure to terminate the contract immediately after expiration of the completion date did not constitute a waiver under those circumstances.

An owner may not be deemed to have waived available bases for a default termination simply because it relied on a different basis when it declared a default termination. In *Joseph Morton Co. v. United States,*[124] the government's decision to terminate for default was upheld based on the contractor's fraud, even though the government was unaware of the fraud when it terminated the contract. In *FJW Optical Systems, Inc.,*[125] the Armed Services Board of Contract Appeals upheld a termination for default even though the supplier claimed that the government had decided that the items being purchased were too expensive and therefore was really

[120]*R.P. Wallace, Inc. v. United States,* 63 Fed. Cl. 402 (2004).

[121]*See United States ex rel. EPC Corp. v. Travelers Cas. & Sur. Co. of Am.,* 423 F. Supp. 2d 1016 (D. Ariz. 2006).

[122]*W.M.Z. Mfg. Co.,* ASBCA No. 28347, 85–3 BCA ¶ 18,169.

[123]14 Cl. Ct. 219 (1988).

[124]757 F.2d 1273 (Fed. Cir. 1985); *see also Glazer Constr. Co., Inc. v. United States,* 52 Fed. Cl. 513 (2002).

[125]ASBCA No. 29780, 85–2 BCA ¶ 18,049.

terminating the contract for the government's convenience. The United States Claims Court has held, however, that a termination for default may be reversed if the contracting officer's decision to terminate was motivated by hostility toward the contractor and was not in the best interests of the government, even when a contractor was technically in default.[126]

✔ CONTRACTOR RESPONSE TO TERMINATION NOTICES

When faced with a termination or threat of termination, the contractor first should seek advice of counsel familiar with construction contract termination issues. Decisions must be made quickly and competently to protect the contractor's position. Additionally, the following list suggests some general considerations for a contractor or subcontractor to formulate a plan of action when faced with a termination threat:

- Assess and document the factual accuracy of the notice. If the complaints are inaccurate, document their inaccuracy—in writing if possible, using photographs or video, and any other appropriate means. Rebut, in writing, any inaccuracies in the termination threat.
- Assess the legal adequacy of the termination threat. Check contract documents to determine whether the termination notice meets the contract requirements and was issued when and as the contract requires. Does the contract provide a right of termination for the contract breach set out in the notice? Does the termination notice follow the procedural requirements (adequate notice, right to cure, given at the correct time, and to the correct parties, etc.) specified in the contract?
- Invoke the right to cure. If there is any legitimacy to the complaints being made, address those complaints in a curative plan of action developed and communicated to the complaining party within the contractual cure period.
- If the contract provides no specific cure period, develop and communicate a curative plan as soon as possible.
- Implement a cure. Take action immediately to show an effort to implement the plan to cure any legitimate complaints.
- If the termination notice is improper, because it really seeks to accelerate work without paying for acceleration, act accordingly. When the termination threat is unwarranted, determine whether to treat the threat as an order to accelerate under the contract change clause. If so, advise the complaining party of an intention to claim a change based on acceleration.
- Document the status of the work at the time of the termination threat (e.g., by marking up plans to show as-built conditions) and any obstacles to performing the work.
- Inventory material and equipment on-site. If the contract gives the owner or surety a right to take possession of materials and equipment, adequately record the materials and equipment on site.

[126]*Quality Envtl. Sys., Inc. v. United States,* 7 Cl. Ct. 428 (1985).

- Assess subcontractor's and supplier's termination rights. Review various subcontract and purchase order provisions addressing such termination rights and liabilities. Consider notice obligations to the subcontractors and their entitlement to termination costs under their subcontracts.
- Consider the possible loss of project staff. Termination of a construction contract frequently is accompanied by the loss of valuable members of a contractor's project staff. Memorialize the knowledge and advice of project staff about issues likely to arise in a dispute over the termination decision. Memories will fade and employees will leave, or will lose interest in a former employer's job problems.
- Ensure the protection of project records. After termination, the terminating party may request project records to assist in the completion of the contractor's work. Protect confidential or sensitive information.
- Reassure the performance bond surety. A threat to terminate a contract likely will draw the scrutiny of a contractor's performance bond surety. Assure your surety of a proper response to the threat.
- Anticipate potential subcontractor and supplier claims. Assess and document the status of work by subcontractors and suppliers, in anticipation of potential termination claims.

E. Remedies of the Owner

When a contractor is terminated for default, the owner typically may (1) complete the contract work itself, (2) contract with another contractor to complete the work, or (3) allow the defaulted contractor's surety (if it is a bonded contract) to complete, or arrange to complete, the contract.[127] Usually an owner will complete the work with its own employees or forces only if the project is substantially complete and little work remains. Otherwise, the owner commonly will enter into another contract or will rely on the surety to complete, assuming the project is bonded and the surety agrees.

An owner completing a contract either on its own or through a completion contractor must take reasonable steps to minimize its completion costs. This duty to mitigate damages limits the recovery the owner can obtain, due to breach of contract, to the cost of reasonable efforts to complete the contract economically and efficiently.[128] The reasonableness of these actions will be determined in light of the facts and circumstances of each case.

Section 6 of the AIA A312 Performance Bond obligates the defaulting contractor's surety to cover the cost of completion up to the amount of the bond. Although the surety may not be obligated to complete the contract work, at least where the default is not contested by the contractor, most sureties will cooperate with the owner

[127]*See, e.g.*, FAR § 52.249–10(a); AIA A312, §§ 4.1–4.3 (1984 ed.).

[128]*Mega Constr., Inc. v. United States*, 29 Fed. Cl. 396, 484 (1993); *Marley v. United States*, 423 F.2d 324, 333 (Ct. Cl. 1970); *Bell BCI Co. v. HRGM Corp.*, 2004 WL 3222885 (D. Md. 2004).

in arranging for completion of the contract. In most cases, the surety will be allowed to take over the work unless there is reason to believe that the party proposed by the surety to complete the work is incompetent or unqualified to such an extent that the owner's interests would be substantially prejudiced by such efforts.[129]

Where the surety tenders another contractor to complete the work, the bonds of the surety under the original contract remain effective.[130] In addition, the owner often may receive the benefit of additional surety bonds provided by the completion contractor.

Where a surety elects to have the work completed by another firm, without either a valid assignment or a takeover agreement, it may seek to assert the defaulted contractor's claims against the owner. The surety's ability to do this may be affected by other laws. For example, on federal government contracts, a surety cannot prosecute a claim against the federal government under the contract "disputes" clause in the surety's own name.[131] Under such circumstances, the surety must be assigned the defaulted contractor's claim, and thus submit the claim as a representative of the contractor.[132]

1. Reprocurement Costs

Under the default clause, the federal government can recover its "increased costs"—that is, those additional costs in excess of the original contract price that are necessarily incurred in completion of the work. The government cannot recover the administrative costs of reprocurement.[133]

If the government procures more work than the terminated contract required, the government may recover from the defaulted contractor only those costs attributable to the work in the original contract.[134] Also, a court or a board of contract appeals generally will not award excess reprocurement costs the government could have avoided by mitigating damages.[135] If the reprocurement contract contains significant deviations from the original contract, no excess costs may be assessed against the defaulting contractor.[136]

In the absence of a default clause in the contract, or perhaps supplementing the rights granted by such a clause, the common law rights for breach of contract apply. Paragraph 14.2.4 of the AIA A201 (2007 ed.) states that the owner is entitled to the payment of damages that would put it in the same position as if the contract had not

[129]FAR § 49.404(c).

[130]*See, e.g., Transamerica Ins. v. United States,* 31 Fed. Cl. 532 (1994).

[131]*Fireman's Fund Ins. Co.,* ASBCA No. 50657, 00–1 BCA ¶ 30,802, *reconsideration denied,* 00-1 BCA ¶ 30905, *aff'd,* 313 F.3d 1344 (Fed. Cir. 2002). *Cf. In re Maharaj Constr., Inc.,* BCA No. 2001-BCA-3, 2005 WL 166315 (L.B.C.A.)

[132]*Sentry Ins.,* ASBCA No. 21918, 77–2 BCA ¶ 12,721.

[133]*Evans,* ASBCA No. 10951, 66–1 BCA ¶ 5316.

[134]*M.S.I. Corp.,* VACAB No. 599, 67–2 BCA ¶ 6643; *see also Transamerica Ins. v. United States,* 31 Fed. Cl. 532 (1994).

[135]*See, e.g., A & W Gen. Cleaning Contractors,* ASBCA No. 14809, 71–2 BCA ¶ 8994.

[136]*Blake Constr. Co.,* GSBCA No. 4013 *et al.,* 75–2 BCA ¶ 11,487.

been breached. Under FAR § 52.249–10(d), the government default clause damages are "in addition to any other remedies provided by law."

2. *Delay Damages*

The federal government construction liquidated damages clause, FAR § 52.211–12, provides that regardless of whether the contractor is terminated for default or not, the contractor is liable for damages resulting from the contractor's refusal or failure to complete the work within the specified time. Absent a liquidated damages provision, the owner is entitled to actual damages caused by the contractor's delay. Actual damages under the FAR include, for example, the cost of keeping a government inspector on the job after the specified completion date.[137]

When the contract contains a liquidated damages provision, the federal government is entitled to damages for the period from the specified contract completion date to the actual date of completion, whether the contract is completed by the contractor, the surety, or a reprocurement contractor. The default clause provisions attempt to secure for the federal government the same measure of recovery it could obtain under common law, that is, to put it in as good a position as it would have enjoyed had the delay not occurred.

F. Termination for Convenience

The parties to a construction contract may include a clause allowing one party (usually the owner) the right to terminate another party without cause—that is, for convenience. The federal government's right to terminate a contract for convenience generally is deemed to be implied as a matter of law.[138] Although the federal government's right to terminate for its convenience has been implied at law, the FAR includes a specific clause, incorporated into most federal government construction contracts, that expressly provides for this right.[139] Likewise, state or local governments often include express contract provisions affording the government entity the right to terminate for convenience.[140]

1. *Standard Industry Forms—Private Contracts*

In order for one party to a private commercial contract to terminate for convenience, an express provision in the contract must give it the right to do so. If one party terminates another for convenience, the terminated party's right to compensation upon such

[137]*B&E Constructors, Inc.*, 1967 WL 224 (I.B.C.A. 1967).

[138]*See, e.g., G.L. Christian & Assocs. v. United States,* 312 F.2d 418 (Ct. Cl. 1963), *cert. denied,* 375 U.S. 954 (1963); *Cf. Advanced Team Concepts, Inc., v. United States,* 68 Fed. Cl. 147 (2005) (termination for convenience clause not read into a contract when government terminated in bad faith).

[139]*See* FAR § 49.101 (6).

[140]*See, e.g.,* Standard Specification for Roads and Structures North Carolina Department of Transportation § 108–13(c) (2006).

termination may be set out in, and limited by, the contract's terms. The private owner's right to terminate for convenience is as varied as the specific convenience termination provisions that parties may choose to include in their contracts. There are several fairly typical convenience termination clauses. For example, ConsensusDOCS 200 provides, in Paragraph 11.4.1, for termination of the contract at the convenience of the owner in this way:

> Upon written notice to the Contractor, the Owner may, without cause, terminate this Agreement. The Contractor shall immediately stop the Work, follow the Owner's instructions regarding shutdown and termination procedures, and strive to minimize any further costs.

The ConsensusDOCS contract then describes the payment that will be made for the work performed before the convenience termination of the contract, and identifies the contractor's obligations to the owner upon termination of the contract for either convenience or cause.[141]

The AIA A201 also contain a termination for convenience clause. Section 14.4 allows the owner to terminate the contractor at any time and without cause.[142] This contract provision also sets out three specific steps the contractor must take upon receiving written notice of a termination for convenience from the owner.[143]

Another construction industry contract form for private work also addresses termination for the owner's convenience. EJCDC C-700, at Article 15.03, allows the owner to terminate the contract for conveinence upon seven days' written notice to the contractor and the engineer.[144]

2. *"Bad-Faith" Convenience Terminations: Theory and Reality*

Does a good-faith requirement exist for termination without cause in the private contract setting? The question is not resolved in many jurisdictions.[145] At least one court has

[141]ConsensusDOCS 200, ¶¶ 11.4.2, 11.4.3 (2007 ed.).

[142]AIA A201, § 14.4.1 (2007 ed.).

[143]*Id.* at § 14.4.2 (these steps include: cease operations; take precaution to preserve the work; and terminate all existing subcontracts and purchase orders except to the extent necessary to perform work before the effective date of the termination).

[144]EJCDC C-700, § 15.03 (2007 ed.).

[145]*See, e.g.,* Restatement (Second) of Contracts § 205 (1981); *Davis Tatera, Inc. v. Gray-Syracuse, Inc.,* 716 F. Supp. 1078, 1088 (S.D. Ohio 1992); *Randolph v. New England Mut. Life Ins. Co.,* 526 F.2d 1383, 1386 (6th Cir. 1975); *Interboro Packaging Corp. v. Fulton County Schs.,* 2006 WL 2850433 (N.D. Ga. 2006) (no Georgia cases apply federal good-faith standard set out in *Torncello*); *RSG Caulking & Waterproofing, Inc. v. J.P. Morgan Chase & Co.,* 831 N.Y.S.2d 350 (N.Y. Sup. Ct. 2006) (suggesting law on inquiry into good faith in convenience termination is not really settled); *Linan-Faye Constr. Co., Inc. v. Hous. Auth. of Camden,* 49 F.3d 915 (3d Cir. 1995) (predicting N.J. law, as no state cases exist on this issue); *Cf. Jan Rubin Assocs., Inc. v. Hous. Auth. of Newport,* 2007 WL 1035016 (E.D. Ky. 2007) (good-faith duty is implied under Kentucky law); *A.J. Temple Marble & Tile, Inc. v. Long Island R.R.,* 682 N.Y.S.2d 422 (N.Y. App. Div. 1998) (no court inquiry into good faith where termination for convenience clause is unconditional).

held that there are circumstances in which a termination, even under a seemingly unrestricted convenience termination clause, may not be exercised in bad faith.[146] Other courts have hesitated to hold that there is a good-faith requirement in terminating for convenience. A federal court applying New York law found "on the basis of existing precedents, that such a [good-faith] restriction would probably not be imposed by New York courts."[147]

In the federal procurement context, the government's right to terminate for convenience, although extremely broad, cannot be exercised in bad faith. Termination is permitted "whenever the Contracting Officer shall determine that such termination is in the best interest of the Government."[148] That provision expresses a right that has long been recognized, even in the absence of a termination for convenience clause in the contract.[149] Yet a terminated contractor alleging bad faith by the contracting officer in issuing a convenience termination may make a claim by presenting evidence to demonstrate bad faith.[150] "The termination of a contract for the convenience of the government is only valid in the absence of bad faith or a clear abuse of discretion."[151] To prove bad faith in a termination for convenience, however, the contractor must show clear and convincing evidence (a heightened evidentiary standard) that the government acted with malice and the specific intent to harm the contractor.[152]

In *Municipal Leasing Corp. v. United States,*[153] the U.S. Claims Court ruled that a finding of bad faith was not necessary to invalidate a termination for convenience. The court held that:

> The termination for convenience clause can appropriately be invoked only in the event of some kind of change from the circumstances of the bargain or in the expectations of the parties.
>
> The termination for convenience clause will not act as a constructive shield to protect defendant from the consequences of its decision to follow an option considered but rejected before contracting with plaintiff.[154]

Thus, while the government has broad discretion to terminate for its convenience, it may not do so in bad faith or without a change in circumstance warranting the action.

[146]*See Randolph v. New England Mut. Life Ins. Co.*, 526 F.2d 1383, 1386 (6th Cir. 1975).

[147]*Niagara Mohawk Power Corp. v. Graver Tank & Mfg.*, 470 F. Supp. 1308 (N.D.N.Y. 1979); *see also Triangle Mining Co. v. Stauffer Chem. Co.*, 753 F.2d 734 (9th Cir. 1985) (no good-faith requirement under Idaho law); *Masarjian v. Mark Lighting Fixtures Co.,* 595 F. Supp. 869 (D. Conn. 1984) (no good-faith requirement under Connecticut law).

[148]FAR § 52.249–2(a).

[149]*United States v. Corliss Steam Engine Co.,* 91 U.S. 321 (1875).

[150]*See Nat'l Factors, Inc. v. United States,* 492 F.2d 1383 (Ct. Cl. 1974).

[151]*Id.* at 1385. *See also Fields v. United States,* 53 Fed. Cl. 412 (2002).

[152]*Am-Pro Protective Agency, Inc. v. United States,* 281 F.3d 1234 (Fed. Cir. 2002).

[153]7 Cl. Ct. 43 (1984).

[154]*Id.* at 47. *See Torncello v. United States,* 681 F.2d 756 (1982); *Cf. Pacificorp Capital, Inc. v. United States,* 25 Cl. Ct. 707, 720 (1992) (discussing limits of *Torncello* and *Mun. Leasing*).

H. Convenience Termination Costs

In federal procurement law, the contractor whose contract is wholly or partially terminated for convenience must be made financially whole by compensation for the direct consequences of the government's termination. Negotiated termination settlements are limited only by the requirements of reasonableness of cost and the original contract price. Individual cost items need not be negotiated. Costs incurred may even be recovered where the contract is terminated for convenience before any actual work was performed.[155]

In addition, a federal contractor is entitled to a "fair and reasonable" profit on the work it performed, unless it can be demonstrated that the contractor would have lost money on the contract.[156] Expected profits on future work are not recoverable in the absence of a wrongful termination. Profit is allowable only on the work performed.[157] If it can be shown that the contractor would have incurred a loss in completing the contract, the contractor is not entitled to any profit markup, and the termination costs may be reduced by the anticipated loss the contractor would have suffered.

The standard federal termination for convenience clause under FAR § 52.249.2 entitles the contracting officer to terminate all or part of a contract. Where only a portion of the contract is terminated for convenience, the contractor's measure of cost recovery is governed by the same principles applicable to termination of the entire contract. The contractor may recover costs incurred until the date of the termination, including a fair and reasonable profit, and the cost of settlement with subcontractors, suppliers, and the government. The partial termination settlement proposal must be submitted within 90 days after the effective date of the partial termination. In addition, the contractor is entitled to an equitable adjustment in the price of the work that is not terminated, if it is shown that eliminating a portion of the work caused an increase in the cost of the remaining work.

The AIA A201 spells out the payment that the terminated contractor is entitled to receive upon termination for the owner's convenience, which includes "payment for Work executed, and costs incurred by reason of such termination, along with reasonable overhead and profit on the Work not executed."[158] The EJCDC C-700 form provides more broadly for the contractor to recover the costs of termination, including the costs of: (1) completed and acceptable work, done before the effective date of termination, including the reasonable overhead and profit on such work; (2) expenses incurred before the effective termination date for services performed in accordance with the contract, relating to the work that will not be completed, plus reasonable overhead and profit, on these expenses; (3) all claims costs, losses and damages (including attorney's and other professional fees, and court or arbitatration costs)

[155] *Nicon v. United States,* 329 F.3d 848 (Fed. Cir. 2003).

[156] *C. W. McGrath, Inc.,* GSBCA No. 4586, 77–1 BCA ¶ 12,379.

[157] *See, e.g., Maxima Corp. v. United States,* 847 F.2d 1549, 1552 (Fed. Cir. 1988) (contractor may receive "profits on work done" when terminated for convenience).

[158] AIA A201, § 14.4.3 (2007 ed.).

incurred by the contractor in settling terminated subcontractor or suppliers contracts; and (4) reasonable expenses directly caused by the termination.[159] Use of these or similar provisions in private construction contracts should afford the parties a better understanding of their resepctive rights if the owner decides to terminate the contract for its convenience.

➢ POINTS TO REMEMBER

- Performance bonds are not insurance. Sureties may not react to claims in the same way that an insurance company might respond to, for example, a general liability insurance policy claim.
- The language of the bond is critical; read it carefully. It may contain claim requirements or refer to a statute that contains such requirements.
- Determine if the bond is a statutory or common law bond—this distinction may affect claim requirements.
- The penal sum of the bond may not always be the surety's limit of liability.
- A surety has all of the defenses of its principal plus additional technical (strictly bond-related) defenses.
- A surety may be bound by arbitration requirements in the bonded contract. Obligees should insist that surety bonds specifically incorporate by reference the principal's contract documents.
- Subcontractor performance (or Subguard) insurance is available and should be considered as an alternative to performance bonds for each subcontractor, although these programs may be most suitable on large, complex construction projects.
- Termination for default is a remedy of last resort. Consult with counsel to consider carefully your rights, duties, obligations, and options if this situation arises on your project.
- Any attempt to terminate usually requires an opportunity to cure performance deficiencies, even if the contract does not specifically provide a cure period.
- The primary defenses to termination are excusable delay and the failure to pay for work as required by the contract.
- Converting a termination for default into a termination for convenience generally limits the contractor to recovery of the costs of performance on the work completed, plus reasonable profit thereon, as well as reasonable termination costs.
- A contract clause that provides for the automatic conversion of a wrongful termination for default to a termination for convenience may reduce substantially the risks associated with a contract termination decision.

[159] EJCDC C-700, § 15.03A (2007 ed.).

16

PROVING COSTS AND DAMAGES

The issue in construction disputes that generally receives the most attention is liability. Does a differing site condition exist? Who caused the delay, and is it compensable? But the issue of damages (or cost flowing from the events giving rise to liability) is no less important. Too often calculating costs and proving damages takes a backseat, with little precision or scrutiny applied until late in the dispute resolution process. That approach can result in an entirely misguided claim effort, missed opportunities for settlement, and loss at trial or in arbitration. The inability to prove damages with a reasonable degree of certainty may prevent the claimant from recovering the full amount, or even a substantial portion, of the damages to which it may be entitled. An early and realistic analysis of damages can help determine whether a claim really exists and the best means of preparing and positioning the claim for the affirmative recovery sought.

I. BASIC DAMAGE PRINCIPLES

A. The Compensatory Nature of Damages

Several basic premises underlie the theory of damages. For example, when a claimant seeks to recover damages resulting from another party's breach of contract, the court generally will attempt to put the claimant in the same position it would have been in had the contract been performed by all parties according to its terms.[1] This theory

[1] *Merrill Lynch & Co. v. Allegheny Energy, Inc.*, 500 F.3d 171 (2d Cir. 2007); *United States ex rel. Morgan & Son Earthmoving, Inc. v. Timberland Paving & Constr. Co.*, 745 F.2d 595 (9th Cir. 1984); *Bennett v. Associated Food Stores, Inc.*, 165 S.E.2d 581 (Ga. Ct. App. 1968); *E. B. Ludwig Steel Corp. v. C. J. Waddell Contractors, Inc.*, 534 So. 2d 1364 (La. Ct. App. 1988).

of the measure of damages applies to all breach of contract actions, not just those arising from construction contracts.[2] The law of contract damages is compensatory in nature and, as such, is designed to reimburse the complaining party for all "losses caused and gains prevented" by the other party's breach.[3]

In contrast, the goal in a tort (noncontractual wrong) case is to put the injured party in the same position it would have been in had the tort not been committed.[4] In the construction area, tort claims generally are asserted for negligence, misrepresentation, and, in rare cases, on the basis of strict liability.[5] Computation and proof of tort damages are often complex, requiring an evaluation of the foreseeability of the injury and the possible contributory or comparative negligence of other parties, or assumption of risk by a party. Although tort damages may be broader in scope than contract damages, many courts limit tort damages to cases involving either personal injury or property damage, and deny recovery for purely economic loss.[6] For these reasons, and because most construction claims are based primarily on a breach of contract, this section focuses on the computation of contract damages.

B. Categories of Damages

Damages resulting from breach of a construction contract are generally of two basic types: direct and consequential. Although it may be difficult to determine the category within which a claimant's damages fit, the increased use of damage waivers in contracts has made this analysis more important than ever. Specifically, both the 1997 and 2007 versions of American Institute of Architects (AIA) A-201 General Conditions and the 2007 version of ConsensusDOCS 200 contain a waiver by both the owner and the contractor of consequential damages.[7] Therefore, before deciding whether to execute a contract containing such a clause, one will certainly need to be familiar with the type of damages included within the waiver.

1. *Direct Damages*

Direct damages, sometimes referred to as general damages, are those that result from the direct, natural, and immediate impact of the breach, and are recoverable in all cases where proven.[8] A contractor's direct damages may include idle labor

[2]*See, e.g., Merrill Lynch and Co. v. Allegheny Energy, Inc.*, 500 F.3d 171 (2d Cir. 2007).

[3]John D. Calamari & Joseph M. Perillo, The Law of Contracts § 327 (1st ed. 1970).

[4]Restatement (Second) of Torts § 901 (1997).

[5]For a discussion of the application of strict liability principles in construction litigation, *see* Edie Lindsay, *Strict Liability and the Building Industry,* 33 Emory L.J. 175 (1984).

[6]*Test Drilling Serv. Co. v. Hanor Co., Inc.*, 322 F. Supp. 2d 965 (C.D. Ill. 2004); *See, e.g., Casa Clara v. Toppino,* 588 So. 2d 631 (Fla. Dist. Ct. App. 1991); *State v. Mitchell Constr. Co.*, 699 P.2d 1349 (Idaho 1984); *Bates & Rogers Constr. Corp. v. N. Shore Sanitary Dist.*, 471 N.E.2d 915 (Ill. App. Ct. 1984). *See also* **Chapter 7**.

[7]*But see* EJCDC C-700, § 95.07 B.2 (2007 ed.) (only the owner waives its claims to consequential damages).

[8]*Spang Indus., Inc. v. Aetna Cas. & Sur. Co.*, 512 F.2d 365 (2d Cir. 1975); *EBWS, LLC v. Brilly Corp.*, 928 F.2d 495 (Vt. 2007).

and machinery, material and labor escalations, labor inefficiency, extended job-site general conditions, and home office overhead. The owner's direct damages generally are those costs incurred in completing or correcting the contractor's work and the cost of delay, which is either its actual cost in terms of lost rent or loss of use, or liquidated damages.

2. Consequential Damages

The second category of contract damages is consequential damages, sometimes referred to as special damages. Consequential damages do not flow directly from the alleged breach but are an indirect source of loss. In order to be included within the claimant's recovery, consequential damages must have been within the contemplation of the parties, or flow from special circumstances attending the contract known to both parties, when the contract was executed. These losses, which are related only indirectly to the breach, may include loss of profits or a loss of bonding capacity. These are more difficult to prove, because the causal link between such damages and the act constituting the breach may be tenuous and uncertain.

The most frequently sought types of consequential damages are lost profits, interest on tied-up capital, and damage to business reputation. Both owner and contractor will most often seek these types of consequential damages in connection with delay claims.

Recovery of consequential damages for breach of contract requires proof of several things: (1) the consequence was foreseeable in the normal course of events; (2) the breach is a substantial causal factor in the damages; and (3) the amount of the loss can be reasonably ascertained.[9] The first element is satisfied by showing the particular type of injury was reasonably foreseeable to the other party at the time of contracting. The "reasonably foreseeable" test was originally enunciated in the English case of *Hadley v. Baxendale*[10] and has since been widely adopted by American courts.[11]

The second element the claimant must prove is the damages flowed "naturally" or "proximately" from the breach.[12] In lay terms, this means the injury must be the result of the breach rather than some other cause.

The third limitation on the recovery of consequential damages is that the damages sought must not be too remote or speculative.[13] This general requirement is frequently codified under state law. Questions and issues of the "remote and speculative" nature of claimed damages frequently arise when a claimant seeks to recover profits that have allegedly been lost as a result of the breach—for example, as a result of tied-up capital or reduced bonding capacity. Although statutes covering consequential

[9] *Die Casters Int'l, Inc. v. United States,* 73 Fed. Cl. 174 (2000).

[10] (1854) 156 Eng. Rep. 145.

[11] *Archdale v. Am. Int'l Specialty Lines Ins. Co.,* 64 Cal. Rptr. 3d (Cal. Ct. App. 2007); *Tousley v. Atl. City Ambassador Hotel Corp.,* 50 A.2d 472 (N.J. 1947); *Bumann v. Maurer,* 203 N.W.2d 434 (N.D. 1972).

[12] *Die Casters: Int'l Inc. v. United States,* 73 Fed. Cl. 174 (2000).

[13] *Dileo v. Nugent,* 592 A.2d 1126 (Md. Ct. Spec. App. 1991); *Baker v. Riverside Church of God,* 453 S.W.2d 801 (Tenn. Ct. App. 1970).

damages may require "exact computation," most courts have taken a somewhat less stringent approach. The court in one such case explained:

> [T]he pecuniary amount of consequential damages need only be proven with reasonable certainty and not absolute precision. Once a defendant has been shown to have caused a loss, he should not be allowed to escape liability because the amount of the loss cannot be proved with precision. Consequently, the reasonable level of certainty required to establish the amount of a loss is generally lower than that required to establish the fact or cause of a loss. The certainty requirement is met as to the amount of lost profits if there is sufficient evidence to enable the trier of fact to make a reasonable approximation. What constitutes such an approximation will vary with the circumstances. Greater accuracy is required in cases where highly probative evidence is easy to obtain than in cases where such evidence is unavailable.[14]

In seeking consequential damages, the claimant assumes a much heavier burden of proof as compared to direct damages. Moreover, many contracts, and particularly public construction contracts, by their terms exclude claims for consequential damages.[15]

3. Punitive Damages

Punitive damages are awarded where there is evidence of oppression, malice, fraud, or wanton and willful conduct on the part of the defendant, and are above what would ordinarily compensate the complaining party for its losses. These damages are not compensatory in nature but rather are intended to punish the defendant for its wrongful behavior or to make an example in order to deter others from similar conduct.[16] Many jurisdictions refer to punitive damages as exemplary damages and define them similarly. California defines punitive or exemplary damages as "damages other than compensatory damages which may be awarded against a person to punish him for outrageous conduct."[17]

Traditionally it has been held punitive damages are not recoverable in an action for breach of contract, absent proof of fraud or malicious intent.[18] This is generally so even if the breach is intentional.[19] In some states, however, the courts may permit the award of punitive damages where there is sufficient evidence of "malice" or

[14] *Kraatz v. Heritage Imports,* 71 P.3d 188, 201 (Utah Ct. App. 2003).

[15] ConsensusDOCS 200, ¶ 6.6 (2007 ed.), AIA A201, § 4.3.10 (1997 ed.), and AIA A201, § 15.1.6 (2007 ed.) are all examples of an express waiver of consequential damages.

[16] *See, e.g., Banks v. Mario Indus. of Va., Inc.,* 650 S.E.2d 687 (Va. 2007).

[17] *Wetherbee v. United Ins. Co. of Am.,* 95 Cal. Rptr. 678, 680 (Cal. Ct. App. 1971).

[18] *Smith v. Nationwide Prop. & Cas. Ins. Co.,* 505 F.3d 401 (6th Cir. 2007); *Roger Lee, Inc. v. Trend Mills, Inc.,* 410 F.2d 928 (5th Cir. 1969); *Otto v. Imperial Cas. & Indem. Co.,* 277 F.2d 889 (8th Cir. 1960); *Smith v. Johnston,* 591 P.2d 1260 (Okla. 1979).

[19] *See Storm v. Farm Bureau Life Ins. Co.,* 601 N.W.2d 339 (Iowa 1999); *Pogge v. Fullerton Lumber Co.,* 277 N.W.2d 916 (Iowa 1979).

utter disregard for the rights of others so as to constitute a willful and wanton course of action, or other tortious conduct amounting to fraud.[20] Additionally, some states allow for recovery of punitive or exemplary damages through their unfair trade practice statutes, many times referring to them as "treble damages."[21] Finally, expenses of litigation may be allowed as damages in an action for breach of contract, if specifically allowed for under the contract or if provided by statute.

C. Causation

In order to prosecute a claim successfully, a claimant must establish the liability of the other party and the amount of its own damages, and prove the damages were caused by the acts giving rise to liability. It is essential that the claimant demonstrate causation, meaning the damages presented flow directly or indirectly from the liability issues presented. Without making this link, even the most thoroughly prepared and well-documented construction claim will not be able to withstand competent attack. It is not essential to establish the extent of the damage with absolute certainty if there is no question as to the fact damage did occur.[22] Again, although speculative damages are not recoverable, the courts generally recognize that there is a difference in the measure of proof needed to show the claimant sustained damage and the measure of proof needed to fix those damages.

D. Cost Accounting Records

The availability of proper cost accounting techniques when the claim is identified can substantially reduce the problem of calculating and proving damages. In fact, utilizing such techniques consistently may even help in the early identification of a potential claim if the actual recorded costs vary from the anticipated costs. Accounting measures can be established to segregate and carefully maintain separate records. If such a procedure is followed, proof of damages may be reduced to little more than the presentation of evidence of separate accounts. Unfortunately, this ideal situation seldom exists; either the problem is not recognized in time to set up separate accounting procedures, the maintenance of separate accounts is simply not possible because of an inability to isolate costs, or no attempt is made to establish the requisite procedures. These circumstances necessitate the development of some formula sufficiently reliable to permit the court or arbitrators to allow its use as proof of damage.

E. Mitigation of Damages

In a breach of contract action, the amount of recovery generally is limited to those losses and damages caused by the breach that are considered unavoidable.[23]

[20] *Walker v. Signal Cos.*, 149 Cal. Rptr. 119 (Cal. Ct. App. 1978).
[21] *See, e.g.*, N.C. Gen Stat. Ann § 75–161; Ga. Code Ann. § 10–1–399(c).
[22] *C. L. Maddox, Inc. v. Benham Group, Inc.*, 88 F.3d 592 (8th Cir. 1990); *Kaatz v. Heritage Imports*, 71 P.3d 188, 201 (Utah Ct. App. 2003).
[23] 22 Am. Jur. 2d *Damages* § 33.

The complaining party may not stand idly by and allow the losses to accumulate and increase, when reasonable effort or cost could have reduced the losses. This requirement is known as the duty to mitigate damages. In construction cases, it usually arises where, upon a breach by one of the parties, there is a need for protection of partially completed work, timely reprocurement, assignment of equipment or work crews, or reduction of delay costs. In particular, an owner that makes no effort to obtain a reasonable contract price upon reprocurement on a defaulted project cannot expect to recover the full difference between the original contract and the increased cost to complete.[24]

The duty to mitigate calls for reasonable diligence and ordinary care. The party's actions need only be reasonable under the circumstances.[25] The law does not require that the defaulted party undertake extraordinary expense or effort to avoid losses flowing from a breached contract.[26]

F. Betterment

A related concept holds a party cannot expect compensation for more than the loss arising from a breach of contract. For example, necessary repairs or replacement of a structure may, in fact, provide the owner with a "better" building than provided for in the plans and specifications.[27] In such instances, where the owner obtains a "betterment" from the efforts of the contractor, any award of damages for breach must be reduced by the value of the betterment the owner receives.[28]

II. METHODS OF PRICING CLAIMS

There are several basic methods for pricing construction claims. The simplest method is the total cost method. The oversimplicity of the total cost method causes it to be frowned on and accepted only in extreme cases. The modified total cost method attempts to address those weaknesses but still faces similar criticisms. Another more complicated, but more widely accepted, method is the discrete or segregated cost method. Finally, there is the *quantum meruit* approach to pricing a claim, which ignores costs and focuses instead on the value of the material and services provided.

[24]*See e.g., Metal Bldg. Prods. Co. v. Fid. & Deposit Co.,* 144 So.2d 751 (La. Ct. App. 1962). *See also In Re: New River Shipyard, Inc.,* 355 B.R. 894 (S.D. Fla. 2006).
[25]*Yankee Atomic Elec. Co. v. United States,* 73 Fed. Cl. 249 (2006); *Brewster Wallcovering Co. v. Blue Mountain Wallcovering, Inc.,* 864 N.E.2d 518 (Mass. App. Ct. 2007).
[26]*Blake Homes, Ltd. v. First Energy Corp.,* 877 N.E.2d 1041 (Ohio Ct. App. 2007).
[27]*Main St. Corp. v. Eagle Roofing Co.,* 168 A.2d 33 (N.J. 1961).
[28]*Nicholson & Laupe, Inc. v. Carl E. Woodward, Inc.,* 596 So. 2d 374 (La. Ct. App. 1992); *Davidge v. H&H Constr. Co.,* 432 So. 2d 393 (La. Ct. App. 1983); *Correlli Roofing Co. v. Nat'l Instrument Co.,* 214 A.2d 919 (Md. 1965).

A. The Total Cost Method

A total cost claim is simply what the name implies. It essentially seeks to convert a standard fixed-price construction contract into a cost-reimbursement arrangement. The contractor's total out-of-pocket costs of performance are tallied and marked up for overhead and profit. Payments already made to that contractor are deducted from that amount, and the difference is the contractor's damages. This approach can be refined or adjusted to meet particular needs and circumstances, but the basic components and approach remain: Costs associated with the basis for the claim are not segregated. The total cost method often is used for impact or disruption claims when the segregation of costs may be more difficult.

The total cost approach, although preferred by claimants because of the ease of computation, is generally discouraged by courts. This method assumes that the contractor was virtually fault-free. The total cost method is also fraught with uncertainties such as whether the contractor's bid was reasonable, and the manner in which job costs are accounted. For these reasons, numerous court decisions have established fairly rigorous requirements for the presentation of total cost claims:

(1) Other methods of calculating damages are impossible or impractical.
(2) Recorded costs must be reasonable.
(3) The contractor's bid or estimate must have been accurate (i.e., contained no underbidding).
(4) The actions of the plaintiff must not have caused any of the cost overruns.[29]

The second requirement, the reasonableness of recorded costs, is typically not a difficult assumption to prove. The claimant must demonstrate the appropriateness of costs, the reliability of the contractor's accounting methods and systems, and a relationship to industry practices and standards. Ironically, the more detailed and well documented the claimant's costs are, the more vulnerable the claimant is to an argument that it can use another method for calculating damages and, therefore, fails to satisfy the first requirement for use of the total cost method.

The last two requirements are the most difficult to meet. Proving the contractor's bid was strictly accurate is challenging. That proof might require a comparison of other bids and supplier and subcontractor quotes to bid amounts as well as the comparison of material quantity estimates to contract drawings. Presenting such an analysis is expensive, often difficult to follow, and easily refutable because most bids rely on assumptions. Due to the nature of the bidding process, many of these assumptions are accumulated in the absence of accurate information.

[29]*Hi-Shear Tech. Corp. v. U.S.*, 356 F.3d 1372 (Fed. Cir. 2004); *MCI Constructors, Inc.*, DCCAB No. D-924 (June 4, 1996), *citing Servidone Constr. Corp. v. U.S.*, 931 F.2d 860 (Fed. Cir. 1991). *See also Batteast Constr. Co., Inc.*, ASBCA Nos. 35818 & 36609, 92–1 BCA ¶ 24,697; *Chi. Coll. of Osteopathic Med. v. George A. Fuller Co.*, 719 F.2d 1335 (7th Cir. 1983); *John F. Harkins Co. v. Sch. Dist. of Phila.*, 460 A.2d 260 (Pa. Super. Ct. 1983).

Establishing the claimant was blameless for any overruns is perhaps the most difficult aspect of a total cost claim and why the method so rarely succeeds. The claimant essentially attempts to prove causation by showing the damages were not its own fault and therefore must be due to the acts of the other party. The premise is easily attacked by demonstrating only a single area of potential blame attributable to the contractor, which could erode the credibility of the entire total cost claim. That is why this method often has been defeated in practice and why parties may instead pursue a "modified" total cost method, as discussed in **Section II.C** of this chapter.

Owners probably view contractor total cost claims with even greater suspicion and distrust than do the courts, so much so that the credibility of the entire claim and the claimant can be undermined. The difficulties of establishing the prerequisites for use of a total cost calculation in court, combined with the skepticism it can generate, counsel against use of the total cost method whenever possible.

B. Segregated Cost Method

The segregated cost method of pricing claims is more difficult to use than the total cost method or the modified total cost method, but it is usually a more accurate, reliable, and persuasive way of presenting damages. Under this approach, the additional costs associated with the events or occurrences giving rise to the claim are segregated from those incurred in the normal course of performance of the contract. For example, on an extra work claim, the pricing would reflect an allocation (actual or estimated) for the additional labor, materials, and equipment used in performing the extra work. If the project was delayed, costs of added (or extended) field overhead and home office overhead would also be calculated.

The use of this cause-and-effect methodology most often yields an accurate, well-defined, and defensible presentation of damages. It may, however, be extremely difficult to accomplish in the absence of detailed and contemporaneous job cost record keeping and sophisticated cost-control systems that segregate changed or impacted work. This method tends to have added credibility when the person presenting the damage shows the sum of all specifically identified damages does not equal the total difference between the bid cost and total cost (i.e., the total cost method). The difference remaining represents the costs related to contractor-caused events that have been excluded from the claim.

C. Modified Total Cost Method

Another general approach to calculating and presenting damages borrows from the concepts of both the specific identification (segregated costs) and total cost methods. The modified total cost method employs the inherent simplicity of the total cost approach but modifies the calculation to demonstrate more direct cause-and-effect relationships that exist between the costs and acts giving rise to liability. The success of the approach often depends on the extent of the modifications that demonstrate the cause-and-effect dynamics.

The initial step in calculating damages using the modified total cost method involves adjusting the contractor's bid for any weaknesses uncovered during job performance, whether they were judgment or simple calculation errors. A reasonable bid (an as-adjusted bid) is thus established. The recorded project costs are then similarly examined for reasonableness, and reductions are made for costs that cannot be attributed to the owner, such as unanticipated labor material cost escalations that are not tied to the alleged basis for liability.

Focusing on specific areas of work can further refine the modified total cost calculation or cost categories related tothe claim issues. For example, if the claim relates to a differing site condition that affected only site work and foundations and not the balance of the project, the claimant should focus only on those areas in the calculation and eliminate extraneous costs and issues that complicate and dilute the credibility of the pricing.

Although the modified total cost approach often is viewed as a method of avoiding the unfavorable scrutiny generally givento a total cost analysis, courts and boards of contract appeals may apply the same standards of admissibility to modified total cost claims as they have in the past to traditional total cost claims.[30] If the claimant has modified the cost calculation properly and not relied on a simplistic approach, the modified total cost method is far more likely to withstand scrutiny and offer a credible means of quantifying a claim.

D. *Quantum Meruit* Claims

A final type of damage theory involves an analysis of the "reasonable value" of the work performed. *Quantum meruit* is an equitable doctrine meaning "as much as deserved." This method of recovery typically measures damages under an implied contract theory and is based on the concept that no one should unfairly benefit from another's labor or materials. In these circumstances, the law creates a promise to pay a reasonable amount for the services furnished, even without a specific contract. For instance, mechanic's liens allow contractors and subcontractors to recover the reasonable value of their labor and materials used in improving property.

Quantum meruit generally is used in damage claims when the contractor does not have a written agreement to perform the work and the other party has been "unjustly enriched" by work performed.[31] *Quantum meruit* also may be available if there has been a material breach of the contract and the contractor elects to rescind the contract and seek the reasonable value of the benefits.[32] In most states, the contractor is not limited to the price specified in the initial agreement. The proper preparation and presentation of such a claim often can render proof of damages much easier, avoid the problems inherent in seeking a "total cost" recovery on a pure contract breach theory, and bring about recovery in excess of actual contract prices.[33]

[30] *Servidone Constr. Corp. v. United States,* 931 F.2d 860 (Fed. Cir. 1991).

[31] *River's Bend Red-E-Mix, Inc. v. Parade Park Homes, Inc.,* 919 S.W.2d 1 (Mo. Ct. App. 1996).

[32] *Ken's Carpets Unlimited, Inc. v. Interstate Landscaping Co., Inc.,* 37 F.3d 1500 (6th Cir. 1994).

[33] *See, e.g., United States ex rel. Susi Contracting Co. v. Zara Contracting Co.,* 146 F.2d 606 (2d Cir. 1944); *Murray v. Marbro Builders, Inc.,* 371 N.E.2d 218 (Ohio 1977).

III. CONTRACTOR DAMAGES

The material in this section is not intended to provide an all-inclusive listing of categories of potential claims or cost elements. Instead, cost elements will vary depending on the circumstances of each case. The elements that follow are examples of the more common types of claim items in construction disputes. Many items of damages a contractor will suffer may fall into more than one of these categories.

A. Contract Changes and Extras

In most instances, a contractor presenting an affirmative claim to the owner will be seeking damages arising from changes in the anticipated quality, quantity, or method of work. Obviously, quality and quantity changes are relatively easy to discern. One clear, colorful, and effective way of describing such changes is to say that "The owner had the contractor build a Mercedes rather than the Ford required by the contract."

The owner that requires a contractor to perform an additional quantity of work is susceptible to a claim if the contractor has kept accurate records of the amount of work performed and can prove the difference between actual performance and what a reasonable interpretation of the contract would require.[34] The contractor seeking to recover additional costs associated with changes in the anticipated method or sequence of construction—a category that includes delay, disruption, and acceleration damages—faces potential problems. These damages are more difficult to understand and prove with reasonable certainty. In addition, the contractor's records may furnish little support for the total claimed impact of a change in the anticipated method or sequence of work.

In recognition of the frequency of disagreements over pricing, contracts frequently include terms dictating the methodology for determining the price of extra or changed work. For example, the AIA A201 (2007 ed.) provides several methods of determining costs for changes, including: (1) mutual acceptance of a substantiated lump-sum price, (2) contract or subsequently agreed-on unit prices, (3) cost plus a mutually acceptable percentage or fixed fee, or (4) a method determined by the architect when the contractor fails to respond promptly or disagrees with the chosen method for adjustment in the contract sum.[35] When a method determined by the architect is employed, the contractor must present an itemized accounting with supporting data, and generally is limited to recovery of the costs of labor, materials, supplies, equipment, rentals, bond and insurance premiums, permit fees, taxes, direct supervision, and other related cost items.[36]

[34]*Sornsin Constr. Co. v. State,* 590 P.2d 125 (Mont. 1978).
[35]AIA A201, § 7.3.3 (2007 ed.)
[36]*Id.* at § 7.3.7

Paragraph 8.3, "Determination of Cost," in ConsensusDOCS 200, (2007 ed.), contains the following provisions regarding the pricing of a change in the work:

8.3 DETERMINATION OF COST

8.3.1 An increase or decrease in the Contract Price or the Contract Time resulting from a change in the Work shall be determined by one or more of the following methods:

8.3.1.1 unit prices set forth in this Agreement or as subsequently agreed;

8.3.1.2 a mutually accepted, itemized lump sum;

8.3.1.3 costs calculated on a basis agreed upon by the *Owner* and *Contractor* _______ % Overhead and _______ % profits; or

8.3.1.4 if an increase or decrease cannot be agreed to as set forth in Clauses .1 through .3 above, and the *Owner* issues an Interim Directed Change, the cost of the change in the Work shall be determined by the reasonable actual expense and savings of the performance of the Work resulting from the change. If there is a net increase in the Contract Price, the *Contractor's* Overhead and profit shall be adjusted accordingly. In case of a net decrease in the Contract Price, the *Contractor's* Overhead and profit shall not be adjusted unless ten percent (10%) or more of the Project is deleted. The *Contractor* shall maintain a documented, itemized accounting evidencing the expenses and savings.

8.3.2 If unit prices are set forth in the Contract Documents or are subsequently agreed to by the Parties, but the character or quality of such unit items as originally contemplated is so different in a proposed Change Order that the original unit prices will cause substantial inequity to the *Owner* or the *Contractor,* such unit prices shall be equitably adjusted.

* * *

Based either on common sense or a contract clause, the reasonableness of any claimed cost will be a critical factor in any pricing dispute. In federal government construction contracting, the controlling guidance on what is a reasonable cost is provided by the following provision of the Federal Acquisition Regulation (FAR):

FAR § 31.201–3 Determining Reasonableness

(a) A cost is reasonable if, in its nature and amount, it does not exceed that which would be incurred by a prudent person in the conduct of competitive business. Reasonableness of specific costs must be examined with particular care in connection with the firms or their separate divisions that may not be subject to effective competitive restraints. No presumption of reasonableness shall be attached to the incurrence of costs by a contractor. If an initial review of the facts results in a challenge of a specific cost by the contracting officer or the contracting officer's representative, the burden of proof shall be upon the contractor to establish that such cost is reasonable.

(b) What is reasonable depends upon a variety of considerations and circumstances, including

(1) Whether it is the type of cost generally recognized as ordinary and necessary for the conduct of the contractor's business or the contract performance;

(2) Generally accepted sound business practices, arm's length bargaining, and Federal and State laws and regulations;

(3) The contractor's responsibilities to the Government, other customers, the owners of the business, employees, and the public at large; and

(4) Any significant deviations from the contractor's established practices.[37]

If the contract dictates a method for pricing extra or changed work, such as the private industry form contract terms just quoted, the contractor should understand this methodology and implement it with respect to the construction project accounting system being utilized. If the contract does not provide a formula or basis for pricing such work, the prudent contractor should still implement accounting methods that contemporaneously account for these added costs.

B. Wrongful Termination or Abandonment

A contractor may be damaged if there is a breach of contract by the owner that prevents the contractor from performing the contract. In such a situation, the contractor is entitled to be placed in as good a position as it would have been in had the contract been performed. This generally means the contractor can seek lost profits so long as they can be proven with reasonable certainty.[38] This rule also applies to subcontractors that may be prevented from performing, as in the case of wrongful termination of the subcontract by the general contractor.[39]

Once the contractor begins performance, it may find the owner (or, if it is a subcontractor, the general contractor) has committed a material breach of some obligation, either express or implied. In these circumstances, and if the breach is a major one, the contractor could treat the breach as terminating its contract, suspend construction activities, and seek to recover damages.

If such an election were made, the contractor could pursue a number of means of establishing damages. First, it might seek to recover out-of-pocket expenses, less the value of any materials on hand, plus lost profits.[40] Alternatively, the contractor might

[37]FAR § 31.201–3.

[38]*Innkeepers Int'l, Inc. v. McCoy Motels, Ltd.*, 324 So. 2d 676 (Fla. Dist. Ct. App. 1975); *Cetrone v. Paul Livoli, Inc.*, 150 N.E.2d 732 (Mass. 1958).

[39]*Tolar Const., LLC v. Kean Elec. Co.*, 944 So. 2d 138 (Ala. 2006).

[40]5 Arthur L. Corbin, Corbin On Contracts § 1094 (1964). *See also Autrey v. Williams & Dunlop*, 343 F.2d 730 (5th Cir. 1965).

seek recovery based on the total contract price, less the total cost of performing the contract, plus any expenses incurred in performing until the material breach.

In addition, if there is some reasonable basis for pursuing this method, the contractor might seek the contract price of work performed through the time of material breach, plus a profit on all unperformed work.[41] Obviously the election made by the contractor will depend on the applicable rule of damages in the particular jurisdiction, the extent of the contractor's records, whether the contractor would have made a profit, and provisions of the construction contract, which may limit the remedies available.

More typically, however, the contractor will continue to perform after the "breach" by the owner (or the general contractor) and will seek to recover damages after completion of the contract.[42]

C. Owner-Caused Delay and Disruption

In the "typical" delay-disruption case, one of the contractor's tasks is to establish and isolate the period of delay caused by the adverse party. Once this is done, proof of damages involves itemizing those fixed (ongoing) costs incurred during that period of delay. If, however, the period of delay itself cannot be isolated in this manner, the ensuing problems can be difficult. For example, courts generally will not make any effort to apportion damages in a situation where both parties are found to have contributed to the delays in completion of the contract.[43] In those courts, where each party proximately contributes to the delay, "the law does not provide for the recovery or apportionment of damages occasioned thereby to either party."[44] Some courts, however, depart from this general rule and allow each party to recovery for delays caused by the other.[45]

One indirect cost attributable to a delay is extended overhead costs incurred by the contractor. Overhead "refers to indirect costs that a contractor must expend for the benefit of the business as a whole, as opposed to direct costs, which are costs specifically identified with a final cost object such as a contract."[46] There are two specific types of overhead costs that are allocated to every project that a construction contractor performs.

1. Field Office Overhead

The first type of overhead is known as *field office overhead*, which is also commonly referred to as extended general conditions.[47] Extended field office overhead consists

[41] *See M & R Contractors & Builders, Inc. v. Michael,* 138 A.2d 350 (Md. 1958).

[42] *Underground Constr. Co. v. Sanitary Dist. of Chi.,* 11 N.E.2d 361 (Ill. 1937).

[43] *See United States v. United Eng'g Contracting Co.,* 234 U.S. 236 (1914).

[44] *Malta Constr. v. Henningson, Durham & Richardson,* 694 F. Supp. 902 (N.D. Ga. 1988); *J.A. Jones Constr. Co. v. Greenbrier Shopping Ctr.,* 332 F. Supp. 1336 (N.D. Ga. 1971), *aff'd,* 461 F.2d 1269 (5th Cir. 1972).

[45] *See United States ex rel. Heller Elec. Co. v. William F. Klingensmith, Inc.,* 670 F.2d 1227 (D.C. Cir. 1982). *See also Wilner v. United States,* 23 Cl. Ct. 241 (1991); *Inversiones Arunsu S.A.,* ENGBCA No. PCC-77, 91–2 BCA ¶ 24,584; *JEM Dev. Corp.,* VABCA No. 3272, 91–2 BCA ¶ 24,010.

[46] *See MCI Constructors, Inc.,* DCCAB No. D-924 (June 4, 1996).

[47] *See generally Clark Concrete Contractors, Inc.,* GSBCA No. 14340, 99–1 BCA ¶ 30,280, at 49,755–56.

of "[c]osts incurred at the job site incident to performing the work, such as the cost of superintendence, timekeeping and clerical work, engineering, utility costs, supplies, material handling, restoration and cleanup, etc."[48] Unlike home office overhead, this category of overhead is attributable to a particular contract or project.

A contractor generally is entitled to recover field overhead costs on changes that extend the time of contract performance. These changes include both change orders that add to the scope of the contractor's work as well as changes or events that impact contract performance and that cause a compensable delay on the project. Such costs are attributable to the increase in costs for continuing to run and staff a project for a time period not contemplated at the time of contracting.[49] As with any other type of delay or disruption damage, these costs are recoverable "in order to make the contractor whole."[50]

A contractor may allocate these costs as direct costs or indirect costs, depending on the contractor's generally used accounting method. For example, if project supervision and administration are stated as direct costs, then "an equitable adjustment for extended field supervision and administration is calculated as a direct cost item."[51] If field overhead is charged to a general expense pool as an indirect cost, it should not be part of any direct cost claim; instead, it should be listed as a separate item in any equitable adjustment request.[52] One recognized means of calculating direct costs for extended field overhead is "to compute a daily rate by dividing total labor supervisionand administration costs on the project by the total days of contract performance and then multiplying the result by the number of days of compensable delay."[53] This same accounting method also may be used during the project by dividing expected field overhead costs by expected total days of contract performance to obtain the per-day rate,then multiplying that number by the number of days performance is extended.

As with other claims, it is necessary that a contractor show first that it is entitled to damages for delay. Once this is done, the contractor can recover costs for field overhead, even those not included in the original bid. An illustrative case on this comes from the District of Columbia Contract Appeals Board in *MCI Constructors.*[54] In *MCI,* a default termination was converted to a termination for convenience by a prior board decision. The board determined MCI's proper amount of resulting compensation, including a claim for extended field overhead costs. The District of Columbia argued that MCI significantly underbid its supervisory costs, failing to include bid items for project engineering, clerical work, main office engineering, and main office expediting—items MCI had intended for the project manager and superintendent to handle at the project outset.

In awarding recovery for all these items as field overhead costs, the board stated that it could find "no reason to deny MCI a pro rata share of such costs actually

[48]FAR § 31.105(d)(3).

[49]*M.A. Mortenson,* ASBCA Nos. 40750 *et al.,* 98–1 BCA ¶ 29,658, *aff'g on recon. on other grounds,* 97–1 BCA ¶ 28,623.

[50]*See MCI Constructors, Inc.,* DCCAB No. D-924 (June 4, 1996).

[51]*Id.*

[52]*Id.*

[53]*Id.*

[54]*Id.*

incurred during a compensable delay period...[e]ven though MCI commenced charging directly to the job some supervision and administrative costs it had not bid."[55] Such a decision was reached despite the alleged underbidding, because the board was not giving "MCI any undue recovery[,] because this computation applies only to the delay period and covers labor supervision and administration that MCI would not have incurred but for the District's constructive changes."[56] Due to the inability to identify what portion of the amount the equation provided had already been recovered through change orders signed during the project and what portion had been allocated to a claim item for pending change orders, the board then awarded an amount based in part on a "jury verdict" finding.[57]

The lesson of *MCI Constructors* is apparent: Even if a contractor incurs field overhead costs that were not bid originally, it is important to segregate all these costs and apply them to change orders and delay claims. To add additional specificity to the claim, further record keeping should identify the costs allocated to each change order item.

2. *Home Office Overhead*

The second type of overhead cost is referred to as *home office overhead* costs, which include such expenditures as administrative staff salaries, accounting and payroll services, general insurance, office supplies, telephone charges, depreciation, taxes, and utility costs. These costs are "expended for the benefit of the whole business, which by their nature cannot be attributed to any particular contract."[58]

Contractors typically rely on project revenue to support their home office operations, and, accordingly, contractors include a markup in their bids for new work to absorb such costs. Allocated over the project's planned duration, this markup provides monthly revenue that is used to pay, or absorb, rent, home office staff salaries, accounting and payroll services, general insurance, office supplies, telephone charges, depreciation, taxes, and utility costs. If the project time is extended without increasing the amount of the markup, the dollars available to absorb monthly overhead are reduced.

Several recent U.S. Court of Appeals for the Federal Circuit decisions appear to make it more difficult for a federal government contractor to recover unabsorbed home office overhead using the *Eichleay* formula for government-caused delays.[59]

a. *The Original* Eichleay *Decision*

The *Eichleay* formula originated with *Eichleay Corp.,* an ASBCA decision, in which the board specifically approved the contractor's three-step formula for determining

[55] *MCI Constructors, Inc.,* DCCAB No. D-924 (June 4, 1996).
[56] *Id.*
[57] *Id.*
[58] *Id.*
[59] *See P.J. Dick, Inc. v. Principi,* 324 F.3d 1364 (Fed. Cir. 2003); *Nicon, Inc. v. United States,* 331 F.3d 878 (Fed. Cir. 2003); *Charles G. Williams Constr., Inc. v. White,* 326 F.3d 1376 (Fed. Cir. 2003).

the amount of recoverable home office overhead.[60] Since the original decision, this formula has become the mandatory method for calculating claims for unabsorbed home office overhead on federal government construction contracts.[61] The three steps are:

(1) Determine the overhead allocable to the contract by multiplying the total overhead by a ratio of the contract's billings to the total billings of the contractor for the contract period.
(2) Calculate a daily overhead rate for the contract by dividing the overhead allocable to the contract by the number of days of contract performance.
(3) Determine the total overhead recoverable by multiplying the daily overhead rate for the contract by the number of days of delay.

This formula may be schematically summarized as:

$$\left(\frac{\text{Total Contract Price)}}{\text{Total Company Billings for the Contract Period}}\right) \times \begin{array}{l}\text{Total Home Office Overhead}\\ \text{for the Contract Period}\end{array}$$

= Allowable Home Office Overhead, or Home Office Overhead allocated to the Contract

$$\left(\frac{\text{Allowable Home Office Overhead}}{\text{Total Days of Performance}}\right) \times \text{Number of Days of Delay}$$

= Amount Recoverable

b. *Post* Eichleay *Decisions*

Since *Eichleay* was first decided, courts and boards have imposed two major requirements as a prerequisite to the recovery of *Eichleay* damages: (1) standby requirement, that is, the government required the contractor to stand by during a government-caused delay of indefinite duration; and (2) replacement work requirement, that is, while and as a result of standing by, the contractor was unable to take on other work.[62]

In *P.J. Dick, Inc. v. Principi,* the Federal Circuit reversed a decision by the Department of Veterans Affairs (VA) Board of Contract Appeals (VABCA) denying the contractor's claim for unabsorbed home office overhead.[63] In rendering its decision, the Federal Circuit summarized its view of the standby requirement in great detail.

In *P.J. Dick,* the VA awarded a contract to construct an addition to a medical center in Ann Arbor, Michigan.[64] During contract performance, the VA issued more than 400 change orders, increasing the contract price by more than 5% and granting time extensions totaling 107 days. The contractor signed each of the changes and reserved

[60] *See Eichleay Corp.,* ASBCA No. 5183, 60–2 BCA ¶ 2688, *reheard and reaffirmed,* 61–1 BCA ¶ 2894.
[61] *See E.R. Mitchell Constr. Co. v. Danzig,* 175 F.3d 1369, 1372 (Fed. Cir. 1999).
[62] *See, e.g., Interstate Gen. Gov't Contractors v. West,* 12 F.3d 1053 (Fed. Cir. 1993).
[63] *P.J. Dick, Inc. v. Principi,* 324 F.3d 1364 (Fed. Cir. 2003).
[64] *Id.* at 1368.

its right to seek delay damages under the Suspension of Work clause. Ultimately, the contractor completed the contract 260 days after the original contract completion date and 153 days after the revised date. The VA and the contractor stipulated that, to the extent the contractor could prove entitlement of delay under the Suspension of Work clause, the contractor's recovery for field and home office overhead would be calculated at stated daily rates "without the need for future proof of costs or damages."[65] The VABCA held that the contractor was not entitled to recover unabsorbed home office overhead under the *Eichleay* formula because it had been able to continue work on other unaffected portions of the project and had never been placed in a standby position.[66] On appeal, the Federal Circuit rejected the contractor's argument that a contractor is automatically on standby any time there is a government-caused delay of uncertain duration extending the performance of the contract, at the end of which the contractor can be required to immediately resume work.

The Federal Circuit stated that in the absence of a written order suspending work for an indefinite duration and requiring the contractor to remain ready to resume work immediately or on short notice, the contractor must prove standby through indirect evidence.[67] Accordingly, in order to satisfy the standby requirement, a contractor must show three things:

(1) The government-caused delay was not only substantial but was of indefinite duration.

(2) During the delay, the contractor was required to be ready to resume work on the contract at full speed, as well as immediately.

(3) Work on much, if not all, of the contract was suspended due to the delay.

In *P.J. Dick,* although the contractor characterized its direct billings as "minor," it billed 47% of the contractually permitted progress payment amounts before the delay. Based on this, the court concluded that the contractor was able to perform substantial amounts of work during the suspension periods.

The Federal Circuit went on to summarize the current status of recovery for unabsorbed home office overhead as follows:

A court evaluating a contractor's claim for Eichleay damages should ask the following questions:

(1) Was there a government-caused delay that was not concurrent with another delay caused by some other source;

(2) Did the contractor demonstrate that it incurred additional overhead (i.e., was the original time frame for completion extended);

(3) Did the government contracting officer issue a suspension or other order expressly putting the contractor on standby;

[65] *Id.* at 1374.

[66] *P.J. Dick, Inc.,* VABCA Nos. 5597 *et al.,* 01–2 BCA ¶ 31,647.

[67] *P.J. Dick, Inc. v. Principi,* 324 F.3d at 1364, 1371 (Fed. Cir. 2003).

(4) If not, can the contractor prove there was a delay of indefinite duration during which it could not bill substantial amounts of work on the contract and at the end of which it was required to be able to return to work on the contract at full speed and immediately;

(5) Can the government satisfy its burden of production showing that it was not impractical for the contractor to take on replacement work (i.e., a new contract) and thereby mitigate its damages; and

(6) If the government meets its burden of production, can the contractor satisfy its burden of persuasion that it was impractical for it to obtain sufficient replacement work?[68]

The Federal Circuit elaborated that entitlement under the Suspension of Work clause "requires proof entirely different, and less demanding, than that required to show entitlement to Eichleay damages."[69] Accordingly, the court set forth a four-part test to recover under the Suspension of Work clause:

(1) There must be a delay of unreasonable length extending the contract completion time.

(2) The delay must have been proximately caused by the government's action or inaction.

(3) The delay resulted in some injury.

(4) There is no delay concurrent with the suspension that is the fault of the contractor.[70]

In addition to extended overhead, other delay and disruption types of damages include:

(1) Increased or protracted equipment rentals[71]

(2) Increased labor costs, including wage or benefit increments, such as when an owner-caused delay forces contract performance into a new labor contract period[72]

(3) Increased material costs[73] and

(4) Costs of an idle workforce and equipment[74]

As noted earlier, in certain circumstances, a contractor may be able to recover future profits lost as a result of owner-caused delays. A delay on one project can cause the contractor's bonding capacity to be reduced, which in turn prevents the

[68]*Id.* at 1373.
[69]*Id.* at 1374.
[70]*Id.* at 1375; *Nicon, Inc. v. United States,* 331 F.3d 878 (Fed. Cir. 2003).
[71]*See Shore Bridge Corp. v. State,* 61 N.Y.S.2d 32 (Ct. Cl. 1946).
[72]*See Weaver Constr. Co.,* ASBCA No. 12577, 69–1 BCA ¶ 7455.
[73]*See Samuel N. Zarpas, Inc.,* ASBCA No. 4722, 59–1 BCA ¶ 2170.
[74]*See State v. Feigel,* 178 N.E. 435 (Ind. 1931).

contractor from bidding on new projects that would generate future profits. Courts generally view such damages as too speculative to allow recovery.[75]

Contractors can, however, recover future lost profits arising from a loss of bonding capacity in rare and severe circumstances. In the recent case of *Begl Construction Co. v. Los Angeles Unified School Dist.*,[76] a California Court of Appeals court allowed the contractor to recover lost future profits as special damages.[77] The court reasoned such damages were recoverable because the owner in that case was well aware at the time of contracting that its termination of the contractor and subsequent suit against the contractor's surety would "break" the contractor by crippling bonding capacity and precluding future profits.[78]

D. Owner-Caused Acceleration

Acceleration arises when the contractor must complete performance of the contract, or a portion thereof, on a date earlier than originally specified by the contract or as required by a properly adjusted schedule. There are two types of acceleration that may entitle the contractor to damages: (1) directed acceleration occurs when the owner directs the contractor to complete the contract (or a portion thereof) before the scheduled completion date; and (2) constructive acceleration occurs when the owner refuses to grant the contractor a time extension in the event of owner-caused or other excusable delays. When the contractor must accelerate the pace of performance, the contractor's increased costs are generally compensable.[79] These costs may include, among other things, overtime and shift premiums, supervision costs, extra equipment costs, loss of efficiency, overhead, and profit.[80] If prolonged overtime is required, the effect may be an overall decrease in worker productivity for both overtime and regular (straight) time activities.[81]

To recover, the contractor must demonstrate it has incurred extra costs due to the accelerated efforts. For example, one court denied a contractor's claim for acceleration where the owner clearly directed the contractor to accelerate its work, but the contractor could produce no evidence it did anything different or suffered any damage as a result of the acceleration directive.[82]

[75]*See, e.g., H.H.O., Inc. v. United States,* 7 Cl. Ct. 703 (1985); *Lewis Jorge Constr. Mgmt., Inc. v. Pomona Unified Sch. Dist.,* 102 P.3d 257 (Cal. 2004); *But see Laas v. Mont. State Highway Comm'n,* 483 P.2d 699 (Mont. 1971).

[76]66 Cal. Rptr. 3d 100 (Cal. Ct. App. 2007).

[77]*But see Lewis Jorge Constr. Mgmt., Inc. v. Pomona Unified Sch. Dist.,* 102 P.3d 257 (Cal. 2004) (finding that loss of potential profits due to impaired bonding capacity were not recoverable as "general" damages).

[78]*See Laas v. Mont. State Highway Comm'n,* 483 P.2d 699 (Mont. 1971).

[79]*J. W. Bateson Co.,* ASBCA No. 6069, 1962 BCA ¶ 3529; *Tyee Constr. Co.,* IBCA No. 692–1–68, 69–1 BCA ¶ 7748.

[80]*See* Ralph C. Nash, Government Contract Changes (Federal Publications, Inc. 1975).

[81]*See* Business Roundtable Construction Industry, Scheduled Overtime Effect on Construction Projects (Nov. 1980).

[82]*Utley-James, Inc. v. United States,* 14 Cl. Ct. 804 (1988).

E. Defective Drawings or Specifications

Where the owner supplies the plans and specifications to be used in construction, it is usually held to impliedly warrant the plans and specifications will be adequate to achieve the purposes contemplated. This principle is known as the *Spearin* doctrine, after the famous 1918 case, *United States v. Spearin*[83], in which the principle was articulated by the United States Supreme Court. If the plans and specifications are defective or contain omissions, the contractor may incur substantially increased costs of performance. These costs, including the costs of identifying and correcting defects in the drawings or specifications, along with any delay costs arising therefrom, may be recovered if the contractor properly relied on such drawings in attempting to perform its contractual obligations.[84] In addition, where defective specifications create "wasted effort" and hinder the contractor's performance, the resulting delay may be excusable, and usually no damages can be recovered by the owner in the event of delayed completion.[85]

F. Inefficiency Claims

An inefficiency (or lost productivity) claim arises when productivity is impacted by events for which the contractor is not responsible and is, therefore, entitled to additional compensation. As one board of contract appeals put it, inefficiency costs typically are increased labor costs that "stem from the disruption to labor productivity resulting from a change in working conditions. Productivity is inversely proportional to the man-hours necessary to produce a given unit of product...if productivity declines the number of man-hours of labor to produce a given task will increase."[86] A contractor may assert a claim for lost productivity when the anticipated means, methods, techniques, scheduling, or work sequence are altered by events or circumstances outside the contractor's control and the contractor is entitled to relief for the loss. Productivity losses can be attributed to several things, such as: multiple owner changes, stacking of trades, out-of-sequence work, adverse weather, start-stop work, overtime, dilution of supervision, acceleration, change in local labor market conditions, availability of tools and equipment, and overmanning. Not all of these conditions, however, will necessarily entitle the contractor to additional compensation or time. To recover additional compensation for project inefficiencies, the contractor must prove (1) liability (i.e., the owner was contractually responsible for the impact), (2) causation (i.e., the impact caused the labor overruns), and (3) resultant cost increase (i.e., the impact actually caused a compensable loss).[87]

[83]*United States v. Spearin,* 248 U.S. 132 (1918). *But see Dugan and Meyers Constr. Co., Inc. v. Ohio Dep't of Admin. Serv.,* 864 N.E.2d 68 (Ohio 2007) (declining to expand *Spearin* doctrine from job-site condition cases to cases involving delays due to plan changes in the face of a "no damages for delay" provision).

[84]*La Crosse Garment Mfg. Co. v. United States,* 432 F.2d 1377 (Ct. Cl. 1970); *Hol-Gar Mfg. Corp. v. United States,* 360 F.2d 634 (Ct. Cl. 1966); *Celesco Indus., Inc.,* ASBCA No. 18370, 76–1 BCA ¶ 11,766.

[85]*Warner Constr. Corp. v. City of L.A.,* 466 P.2d 996 (Cal. 1970); *Souza & McCue Constr. Co. v. Superior Court of San Benito County,* 20 Cal. Rptr. 634 (Cal. 1962).

[86]*Centex-Bateson Constr. Co.,* VABCA Nos. 4613 *et al.,* 99–1 BCA ¶ 30,153.

[87]*Hoffman Constr. Co. v. United States,* 40 Fed Cl. 184 (1998); *aff'd in part, rev'd in part,* 178 F.3d 1313.

In simple terms, to recover for lost productivity, the contractor must prove that a particular work activity was impaired by an action taken by the owner or for which the owner bears contractual responsibility. Typically, this means showing the normal or expected level of performance for the type of work at issue and the extent to which the owner's action impacted that performance.[88]

As noted, lost productivity costs are not limited to labor inefficiencies. They also can result from equipment inefficiencies and even material inefficiencies, although the latter is uncommon. Consequently, where the contractor demonstrates owner-caused disruptions impacted its use of equipment, the contractor may be able to recover the resulting inefficiency costs.[89]

1. Lost Productivity Entitlement

In *Centex-Bateson,* the VABCA addressed a lost productivity claim caused by the cumulative impact of approximately 1,500 events, including change orders and Requests for Information (RFI), taking place during the course of a hospital construction project.[90] The board recognized the contractor when proving cumulative impact is subject to the "fundamental triad of proof" necessary to recover lost productivity costs: (1) liability, (2) causation, and (3) resultant injury. The VABCA found the contractor satisfied the first element because it showed the government was liable for the multiple changes. To satisfy the second element, a contractor must demonstrate the "government exceeded the permissible limits of its discretion under the changes clause and ordered changes that materially alter the nature of the bargain agreed upon."[91] In other words, the contractor must show a causal link between the changes and the labor overruns by proving the undifferentiated group of changes affecting the changed and unchanged work resulted in the loss of productivity on that work.[92]

In order to do this, the contractor should maintain and base its claim on "contemporaneous, detailed, objective logs and other records relating to the performance of the contract and impact of the events on labor productivity."[93] Although proving an inefficiency claim based on cumulative impact may be more difficult than proving inefficiencies from other, more discrete conditions, the board's decision in *Centex-Bateson* is instructive for those making inefficiency claims because it delineates the proof standards to which the contractor may be held.[94]

In 2000, the VABCA again provided a detailed analysis for evaluating inefficiency claims in *Clark Construction Group, Inc.*[95] There the general contractor

[88] *Stroh Corp.,* GSBCA No. 11029, 96–1 BCA ¶ 28,265.
[89] *Servidone Constr. Corp.,* ENGBCA No. 4736, 88–1 BCA ¶ 20,390.
[90] *Centex-Bateson Constr. Co.,* VABCA Nos. 4613 *et al.,* 99–1 BCA ¶ 30,153.
[91] *Id.*
[92] *See, e.g., Homer J. Olsen, Inc. v. Santa Clara Valley Transp. Auth.,* 2005 WL 3462739 (Cal. Ct. App. 2005)
[93] *Id.*
[94] *See Coates Indus. Piping,* VABCA No. 5412, 99–2 BCA ¶ 30,479, *See also J.A. Jones Constr. Co.,* ENGBCA Nos. 6348 *et al.,* 00–2 BCA ¶ 31,000.
[95] *Clark Constr. Group,* VABCA No. 5674, 00–1 BCA ¶ 30,870.

sponsored a claim on behalf of its mechanical subcontractor for lost productivity on a large, multiyear hospital project. The subcontractor contended several government impacts—including a stop-work order, late RFI responses, and poor site conditions. Citing *Centex,* the board applied the fundamental "triad" of proof—liability, causation, and resulting injury—to determine each aspect of the subcontractor's claim. The board first noted the testimony of the subcontractor's project managers and experts was "forthright." Nonetheless, it stated that, to prove a labor inefficiency claim, the claimant must present contemporaneous project records demonstrating the impact and its effect on the claimant's work.[96] The board found the subcontractor met this burden on some occasions but failed to meet it on others.

For example, with respect to the out-of-sequence work experienced by the subcontractor, the board found the subcontractor's only reasonable response to the stop-work order was to resequence its activities. The subcontractor's evidence showed the new sequence of work resulted in more difficult material handling, crew supervision, and work area access problems. This hurt labor productivity, which entitled the subcontractor to compensation. The board concluded, however, that the resequencing did not cause all of the subcontractor's inefficiencies. For instance, it found the general contractor was responsible for some of them. In addition, it found the subcontractor failed to prove the resequencing of work was the sole cause of the labor overruns rather than the subcontractor's own failure to adjust parts of its work to accommodate the stop-work order.

The key for the contractor is to establish a system whereby productivity losses and impact events are identified and documented *when they occur*. For example, if out-of-sequence mechanical work must be performed in order to accommodate the delayed arrival of owner-controlled equipment, then the contractor will be well served to isolate the impacted work from the unimpacted work. This is done by documenting: (1) increases or decreases in the number of workers being used to perform the impacted work; (2) additional expenditures for the impacted work (discretely identifying the impacted work through use of specific cost accounting codes); (3) work stoppages or slowdowns; (4) the amount of material being installed each day (e.g., amount of piping runs installed); (5) adjustments in the schedule due to the impacts; and (6) the amount of delay, usually through use of the CPM schedule.

2. Calculating Lost Productivity Costs

Once liability is established, the contractor then must demonstrate the costs associated with the lost productivity. Productivity claims can be difficult to prove from a cost standpoint.[97] As noted by one board, "it is rare when the loss of productivity can be proven by books and records," and the contractor often must resort to expert testimony and other forms of calculating the inefficiency.[98]

[96]*Id.*
[97]*See generally Luria Bros. v. United States,* 369 F.2d 701 (Ct. Cl. 1966).
[98]*Stroh Corp.,* GSBCA No. 11029, 96–1 BCA ¶ 28,265.

Project documents are critical in demonstrating costs as well as causation. When proving lost productivity, the contractor may need to evaluate its original estimate and supporting worksheets, daily reports showing on-site labor forces and equipment, job cost reports, material quantity sheets, time logs or sheets showing labor hours expended per day, weekly progress reports, and the like. This documentation is not always available, and, without such data, a contractor will rarely be able to absolutely quantify lost inefficiency costs caused by the owner.

Lost productivity costs can be demonstrated in a variety of ways, including (1) the total cost/modified total cost approach (previously discussed in **Sections II. A and C** of this chapter), (2) measured mile (or differential studies), and (3) industry studies. The latter two approaches are discussed in the sections immediately following.

3. *Measured Mile (Differential Studies)*

A common method for proving lost productivity costs is the "measured mile." The measured mile takes a designated period on a project where a particular activity is unimpacted and compares it with the period where the activity is impacted. In some instances, the unimpacted (or representative) period also may be taken from a project different from the one at issue if the activities and other relevant conditions are similar.

The measured mile approach is exemplified in the General Services Administration Board decision, *Clark Concrete Contractors, Inc. v. General Services Admin.*[99] In *Clark Concrete,* the contractor undertook to construct a multistory concrete building for the Federal Bureau of Investigation. During construction, the government elected to implement significant design changes to make the building capable of withstanding a bomb blast. Because the changes occurred after construction began, the impact from the changes was significant. Specifically, the changes resulted in resequencing of work, rescheduling work crews, site congestion, overtime, and delays. The impact from the changes was then compounded by the government's failure to respond timely to contractor questions relating to the changed work. Despite the government's attempt to shift blame for the inefficiencies to the contractor, the board easily found the government responsible for those inefficiencies.[100] The board then focused on calculating these costs.

The board adopted the contractor's use of the measured mile as the appropriate method for determining the costs to which the contractor was entitled. There, the measured mile analysis permitted a comparison of the labor costs of performing work during different time periods. The contractor's expert was able to isolate a period of work before the design changes and then compared that work with work during both severely and moderately impacted periods. The government objected to the use of the measured mile as the work during the unimpacted period was not the

[99]*Clark Concrete Contractors, Inc.*, GSBCA No. 14340, 99–1 BCA ¶ 30,280.

[100]Although the board's analysis of liability and causation was not especially detailed, the design changes by the government were so extensive and pervasive that any such type of analysis would not likely have been possible.

same as the work during the impacted period, and therefore a comparison between the two did not yield an accurate lost productivity rate. The board recognized the work performed during an unimpacted period might not always be identical to the work performed during the impacted period; but it is generally accepted that labor inefficiency costs are not susceptible to "absolute exactness."[101] Consequently, the board concluded a comparison between kinds of work that are reasonably alike is acceptable. As such, the board provided a significant recovery to the contractor, including a mark-up for field overhead and general and administrative expenses.

In *Lamb Eng'g. & Constr. Co.,* the Army Corps of Engineers Board of Contract Appeals also allowed an expert's use of the "measured mile" in order to calculate the contractor's inefficiency claim for its site work on a project.[102] The board noted the probative value of the measured mile "depends upon the comparability of the circumstances surrounding the sample to the circumstances which would have prevailed for the work which could not be directly measured. Its use is limited to circumstances where...proof of actual costs of the work as contracted is not feasible."[103] The board recognized that in using the measured mile, an expert must establish the basis for its conclusions, including demonstrating the comparability between the measured, unimpacted work and the impacted work

In *P.W. Constr., Inc. v. United States,* the Court of Appeals for the Federal Circuit reached a different conclusion, noting that the expert's use of the measured mile was not appropriate because the type of work was not equivalent for the pre- and post-disruption periods.[104] Again, this decision does not necessarily mean the unimpacted work (measured mile) must be the exact type of work as the impacted work. Rather, the unimpacted work must be similar enough to the impacted work to serve as a reliable benchmark to measure against the impacted work.[105]

4. *Industry Studies and Guidelines*

On occasion, a contractor's work is impacted from the outset of the project. Consequently, a measured mile approach may not be feasible. In such cases, recent board decisions reflect contractors' frequent use of certain industry studies and guidelines to calculate lost productivity costs. Such studies include those performed by the Mechanical Contractors Association of America (MCAA), the National Electrical Contractors Association, the Construction Industry Institute, and other professional groups.[106]

In addition to endorsing the measured mile method, the General Services Administration Board of Contract Appeals in *Clark Concrete Contractors, Inc.* also allowed

[101] *Clark Concrete Contractors, Inc.*, GSBCA No. 14340, 99–1 BCA ¶ 30,280.
[102] *Lamb Eng'g & Constr. Co.,* ENGBCA No. C-9304172, 97–2 BCA ¶ 29,207.
[103] *Id.*
[104] *P.W. Constr., Inc. v. United States,* 53 Fed. Appx. 555 (Fed. Cir. 2002).
[105] *P.J. Dick, Inc.,* VABCA Nos. 6020 *et al.,* 01–2 BCA ¶ 31,647; *rev'd and vacated on other grounds,* 324 F.3d 1364 (Fed Cir. 2003).
[106] Factors Affecting Productivity, Mechanical Contractors Association of American, Bulletin, No. 58 (1976).

for the use of industry guidelines in calculating lost productivity costs.[107] In *Clark Concrete,* one of the contractor's subcontractors based its claim on the labor productivity rates established by the MCAA.

The MCAA manual lists several types of impacts that may occur on a project and then, for each impact, assigns a percentage representing loss of labor productivity for minor, average, and severe impact events. Based on the MCAA manual, the subcontractor's president in *Clark Concrete* concluded the company sustained a 60% loss of productivity in its work. The impacts considered included: stacking of trades (20% loss), concurrent operations (15% loss), dilution of supervision (5% loss), site access (5% loss), out-of-sequence work (10% loss), and competition for labor (not in the MCAA manual but recognized by the board as an impact—5% loss). The subcontractor then multiplied the lost productivity percentage (60%) by the work-hours estimated to perform the work in order to determine the additional work-hours spent due to the impactful events. This figure was in turn multiplied by the blended hourly rate the subcontractor paid its workers to perform the impacted work, resulting in the lost productivity costs incurred by the subcontractor. The board accepted this approach but reduced the subcontractor's claim because it found some of the additional lost productivity costs were the result of contractor delays rather than agency disruptions.[108]

In 2002, the VABCA again approved the contractor's use of the MCAA factors to prove its loss of efficiency claim in *Fire Security Systems, Inc.*[109] There, the contractor claimed it suffered lost productivity when it encountered unforeseen asbestos on a renovation project. To support its claim, the contractor relied primarily on its daily logs, which identified disruptions in work due to asbestos. The contractor then used the labor productivity factors in the MCAA manual to calculate its lost productivity claim. The board concluded because the contractor was never able to perform efficiently, the proper method for determining the efficiency loss was to use productivity factors from the MCAA manual. The case, however, presented a unique twist. Rather than adopt the contractor's MCAA analysis, the board significantly reduced the efficiency loss factors applied by the contractor's expert and selected its own factors to fit its evaluation of the facts.[110]

IV. OWNER DAMAGES

Liability and damages can flow both ways in a construction dispute. The owner's costs can be the same type of costs a contractor incurs if the owner is required to

[107] *Clark Concrete Contractors, Inc.*, GSBCA No. 14340, 99–1 BCA ¶ 30,280. *See also Stroh Corp.*, GSBCA No. 11029, 96–1 BCA ¶ 28,265

[108] *See also Hensel Phelps Constr. Co. v. Gen. Servs. Admin.* GSBCA Nos. 14744 *et al.*, 01–1 BCA ¶ 31,249; *Normant Sec. Groups, Inc. v. Ohio Dept. of Rehab. and Correction,* 2003 WL 22890088 (Ohio Ct. Cl. 2003) (criticizing use of MCAA factors by expert).

[109] *Fire Sec. Sys. Inc.*, VABCA Nos. 5559 *et al.*, 02–2 BCA ¶ 31,977.

[110] *Id.*

complete the project or remedy defective work. The owner's consequential damages, however, can be very different from those of a contractor, and potentially of a much larger magnitude. Generally speaking, the law provides that the owner is entitled to receive the benefit for which it contracted.

A. Direct Damages

A contractor can breach a construction contract in a variety of ways: It can fail to commence work at all; it can commence work but fail to complete it as required by the contract (i.e., abandon the project or fail to complete it within the specified time period); or it can substantially complete the work but have deviated from the plans in some major or minor respect. In each of these cases, the owner may be entitled to recover damages for the contractor's breach.

Where the contractor has breached the contract by its failure to commence work at all, no consistent method of computing damages has been recognized and applied by the courts.[111] One approach utilized by certain courts is to allow recovery by the owner of the difference between the price at which the original contractor agreed to complete the construction and the price at which the owner was able to obtain a replacement contractor[112] or the fair market price of erecting the building.[113]

Where a contractor abandons the work before completion, the generally recognized measure of direct damages is the amount it costs the owner to complete the work in accordance with the terms of the original agreement. The contract is the first place an owner should look to determine what damages it is entitled to seek from the contractor in the event of default. For example, subparagraph 11.3.1 of ConsensusDOCS 200 provides that, after termination, the owner is entitled to the difference between the contract price and the owner's costs arising out of the contractor's failure to cure, including the cost of completing the work and any reasonable attorneys' fees.

In the absence of a contractual provision to the contrary, courts generally allow similar methods of direct damage calculation. For instance, courts usually allow an owner, after a contractor's failure to perform, to take possession of the project and complete the work itself. Those courts then generally hold the failing contractor liable for any excess in the reasonable cost of completion over and above the contract balance.[114]

When the contractor fails to complete the contract within the time period specified, the owner may be entitled to damages for delayed completion.[115] The amount of such damages may be stipulated in the contract by a liquidated damages provision. Liquidated damages are intended to represent a fair approximation of the actual damages that an owner will incur due to the contractor's delayed completion.

[111]*See generally* John P. Ludington, Annotation, *Modern Status of Rule as to Whether Cost of Correction or Difference in Value of Structure Is Proper Measure of Damages for Breach of Construction Contract,* 41 A.L.R. 4th 131, § 15 (1985).

[112]*Ross v. Danter Associates, Inc.,* 242 N.E.2d 330 (Ill. App. Ct. 1968).

[113]*Bell v. McCann,* 535 P.2d 233 (Colo. Ct. App. 1975).

[114]*See, e.g., In Re Ferguson,* 183 B.R. 122 (Bankr. N.D. Tex 1995).

[115]*Potter v. Anderson,* 178 N.W.2d 743 (S.D. 1970).

(See **Section IV.C** of this chapter.) Where liquidated damages are not specified as the only recovery for delay damages, the proper measure of owner damages for delayed completion in construction contracts generally is held to be the loss of reasonable income (e.g. rentals) from the property for the period of the contractor's unexcused delay.[116]

Where the contractor has substantially completed work under the contract but has deviated to some extent from the plans and specifications, the measure of direct damages to the owner generally will be based on some variation of one of these two theories: (1) the "cost" rule or (2) the "value" rule.[117] The cost rule measures either the cost of completing or the cost of repairing the contractor's work. The value rule measures the difference between the actual value of the building as delivered and the value promised.

Under either of these theories of recovery, the owner, like the contractor, must fully document and prove its damages. Where the damage or injury is so slight as to be insignificant, no recovery is allowed.[118] These two methods of computing damages are widely accepted. The cost-to-complete measure appears to be the more commonly used, particularly in those cases where the specific defect or omission of which the owner is complaining can be remedied at reasonable cost and without destroying work that has already been done.[119]

When as a practical matter defects and omissions cannot be remedied except at unwarranted expense and with excessive economic waste, the courts usually apply the value rule. This diminution-in-value rule measures damages by calculating the difference between the value of the work if it had been performed in accordance with the contract and the value of the work as it was actually done, or, alternatively, the difference between the value of the defective structure and the structure if properly completed.[120] In some situations, however, when it is necessary to recover the benefit of the bargain, an owner may be able to recover the costs of repair, even when it would result in economic waste.[121]

Finally, these two methods of calculating damages can, under some circumstances, be used together to arrive at an acceptable, overall damage figure. Some of the items of damage may properly be calculated under the cost-to-complete approach, while others may be more suited to a diminution-in-value calculation.

B. Consequential Damages

The previous discussion focused on direct damages recoverable by an owner. Owners, like contractors, also can recover consequential damages attributable to a breach

[116]*Kaltoft, Inc. v. Nielson,* 106 N.W.2d 597 (Iowa 1960); *Hemenway Co. v. Bartex, Inc.,* 373 So. 2d 1356 (La. Ct. App. 1979).

[117]Dan B. Dobbs, Law Remedies § 12.21, at 897 (1973).

[118]*Gen. Refrigeration Co. of Lake Charles, Inc. v. Style Home Builders, Inc.,* 379 So. 2d 1211 (La. Ct. App. 1980).

[119]*R.I. Turnpike & Bridge Auth. v. Bethlehem Steel Corp.,* 379 A.2d 344 (R.I. 1977).

[120]*Robbins v. C.W. Myers Trading Post, Inc.,* 111 S.E.2d 884 (N.C. 1960); *City of Charlotte v. Skidmore, Owings & Merrill,* 407 S.E.2d 571 (N.C. Ct. App. 1991).

[121]*Lapierre v. Samco Dev. Corp.,* 406 S.E.2d 646 (N.C. Ct. App. 1991).

under certain circumstances (if not excluded by contract, such as in the 1997 and 2007 versions of AIA Document A201 or the 2007 version of ConsensusDOCS 200) and upon adequate proof. An excellent example of such a situation arose in *Northern Petrochemical Co. v. Thorsen & Thorshov, Inc.*[122] There the owner brought an action against the general contractor, the architect, and the structural engineer for defective construction and design. The presence of numerous major defects in design and construction required reconstruction of a major portion of the facility, with a resulting eight-month delay in its availability for use. In addition to the costs of reconstruction and the overall diminution in value of the building, the owner was also allowed to recover for the eight-month loss of use of the facility because the court determined the loss was a direct and proximate result of the injury and the parties could have reasonably contemplated such damages when entering into the contract.

The court also permitted the owner to recover its lost profits, explaining as follows:

> Loss of profits has been recognized as an appropriate measure of the damages resulting from the loss of use when the anticipated profits can be proved to a reasonable, although not necessarily absolute, certainty. In the instant case, the evidence on loss of profits was superbly marshaled and, in part because of the industry-wide inability to produce sufficient extruded plastic film to meet the demand, easily met the burden of reasonable certainty required for recovery of lost profits .... [W]e have no trouble finding the proof sufficiently certain to affirm the court's award in the instant case.[123]

The contractor must therefore recognize a real risk of substantial consequential damages may exist, depending on the specific details and circumstances of the particular project and contract.

C. Liquidated Damages

Owners often include in the request for proposal and contract documents a provision that states the contractor will pay the owner a certain stipulated sum of money for each day of delay in contract completion beyond a specified milestone date. In theory, the owner and the prospective contractor have agreed at the time of contract formation on the method of calculating the owner's damages in the event of the contractor's breach at a future date. In practice, however, very little negotiation or "agreement" among the parties is experienced. Due to the difficulty inherent in determining actual damages on a construction project, the owner, or its representative, will determine a dollar figure that appears to be sufficient to cover any anticipated excess costs arising out of delayed completion and insert that figure into the documents. Often this figure is stated as a per diem dollar amount of the actual damages an owner will suffer. If a dispute subsequently arises as to the application of the liquidated damages provision,

[122] 211 N.W.2d 159 (Minn. 1973).
[123] *Id.* at 163.

the courts will try to determine the reasonableness of the figure by examining the condition and positions of the parties at the time of contract execution rather than at the time of breach.

A liquidated damages provision is not necessarily invalid merely because it is at variance with the usual legal rules for computing damages or because it allows an innocent party to recover despite the fact that it is unable to prove exactly the amount of its actual damages. The crucial factors to consider are:

(1) Whether the liquidated damage amount bears some reasonable relationship to actual or anticipated damages contemplated by the parties when the contract was made so that it does not constitute a penalty;[124]

(2) Whether actual damages were difficult to compute at the time the contract was signed;[125]

(3) Whether the breach in question is one covered by the liquidated damage provision[126]

Many states now presume liquidated damage clauses are valid and place the burden on the party challenging the provision to prove the clause amounts to a penalty.[127] The rationale behind upholding a valid liquidated damages clause has been explained as follows:

> The modern trend is to look with candor, if not with favor, upon a contract provision for liquidated damages when entered into deliberately between parties who have equality of opportunity for understanding and insisting upon their rights, since an amicable adjustment in advance of difficult issues saves the time of courts, juries, parties, and witnesses and reduces the delay, uncertainty, and expense of litigation.[128]

Liquidated damages should be reasonably proportionate to actual damages. This requirement stems from the fact that courts traditionally have refused to enforce what amounts to a penalty for breach of contract.[129] One primary objection to penalties is that while the law favors reimbursement for loss, it does not approve of granting

[124]*Ga. Income Property Corp. v. Murphy,* 354 S.E.2d 859 (Ga. Ct. App. 1987); *Mattingly Bridge Co. v. Holloway & Son Constr.,* 694 S.W.2d 702 (Ky. 1985); *Psaty & Fuhrman Inc. v. Hous. Auth. of Providence,* 68 A.2d 32 (R.I. 1949); *Ray v. Elec. Prods. Consol.,* 390 P.2d 607 (Wyo. 1964).

[125]*Osceola County v. Bumble Bee Constr., Inc.,* 479 So. 2d 310 (Fla. Dist. Ct. App. 1985); *J.R Stevenson Corp. v. Westchester County,* 493 N.Y.S.2d 819 (N.Y. App. Div. 1985).

[126]*Grant Constr. Co. v. Burns,* 443 P.2d 1005 (Idaho 1968).

[127]*Farmers Export Co. v. M/V Georgis Prois, Etc.,* 799 F.2d 159 (5th Cir. 1986); *Hubbard Bus. Plaza v. Lincoln Liberty Life Ins. Co.,* 649 F. Supp. 1310 (D. Nev. 1986); *Coe v. Thermasol Ltd.,* 615 F. Supp. 316 (W.D.N.C. 1985), *aff'd,* 785 F.2d 511 (4th Cir. 1986).

[128]*Gorco Constr. Co. v. Stein,* 99 N.W.2d 69 (Minn. 1959).

[129]*See, e.g., Dahlstrom Corp. v. State Highway Comm'n of State of Miss.,* 590 F.2d 614 (5th Cir. 1979); *Precon, Inc. v. JRS Realty Trust,* 47 B.R. 432 (D.C. Me. 1985); *Unis v. JTS Constructors/Managers, Inc.,* 541 So. 2d 278 (La. Ct. App. 1989).

windfalls or unearned profits, even to an innocent party. To allow an injured party to recover an amount in excess of the actual damages it has suffered would, in effect, put that party in a better position than it would have been in had the contract been performed as originally anticipated. Such a result would be inconsistent with the basic theory of contract damages.

For these reasons, if the liquidated damage amount set by the parties is found to be a penalty, the party seeking to enforce the provision will not be allowed to rely on it but instead will be required to prove actual damages caused by the breach of contract. Whether a stipulated sum should be upheld as allowable liquidated damages or struck down as a penalty is a question of law the court must decide.[130]

Another condition most courts impose in connection with liquidated damages is that the actual damages resulting from the breach must be difficult or impossible to prove. Substitution of damages agreed on in advance would not be justified if actual damages could be measured readily. For this reason, a liquidated damages provision generally will not be upheld where actual damages are readily ascertainable by some adequate and approved legal standard. That said, the reasonableness of the liquidated damages must be viewed at the time of contracting, not at the time of the breach.

The rules governing liquidated damages have wide application in construction contracts, particularly in those situations involving late completion of the contract. The typical liquidated damages provision provides that the contractor will be assessed a certain dollar amount for each day the project remains uncompleted beyond a specified milestone or completion date. Such a provision avoids the difficulties of calculating lost revenues the owner would otherwise be receiving. Where a contractor has been granted time extensions for whatever reason, liquidated damages do not begin to accrue until the period of the extensions has passed.

Three specific factors should be noted in connection with the assessment of liquidated damages:

(1) Liquidated damages cannot be imposed where the owner is the sole cause of a delay suffered by the contractor.[131]

(2) Because of the difficulties inherent in apportioning responsibility for delays caused by the owner and the contractor, many courts will not assess liquidated damages where a concurrent delay has occurred.[132] Numerous cases have held the owner-caused portion of the delays so confuses the issue of responsibility for time overruns, liquidated damages cannot fairly be assessed against the contractor.

(3) An owner generally has no right to assess liquidated damages for delay after the contractor has achieved substantial completion of the project. The justification

[130] *Farmers Export Co. v. M/V Georgis Prois, Etc.*,799 F.2d 159 (5th Cir. 1986); *United States v. Swanson*, 618 F. Supp. 1231 (E.D. Mich. 1985); *Allied Informatics, Inc. v. Yeruva*, 554 S.E.2d 550 (Ga. Ct. App. 2001).

[131] *Dep't. of Transp. v. W. P. Dickerson & Son, Inc.*, 400 A.2d 930 (Pa. Commw. Ct. 1979).

[132] *Gillioz v. State Highway Comm'n*, 153 S.W.2d 18 (Mo. 1941); *Cf. Sw. Eng'g Co. v. United States*, 341 F.2d 988 (8th Cir. 1965); *J.A. Jones Constr. Co. v. Greenbriar Shopping Ctr.*, 332 F. Supp. 1336 (N.D. Ga 1971).

for this rule is that the owner has received essentially the benefit for which it contracted, and therefore the assessment of liquidated damages would constitute a penalty.[133]

Similarly, final payment from the owner to the contractor may waive the owner's right to later seek payment of liquidated damages for the contractor's late completion. In *Centerre Trust Co. v. Continental Insurance Co.*,[134] the owner was found to have waived its right to liquidated damages by making final payment rather than retaining funds to offset accruing liquidated damages. The owner would have been within its rights to withhold all remaining contract balances to recover the accruing liquidated damages if it had not made final payment to the contractor.

Additionally, an owner may waive its right to recover liquidated damages from a contractor by failing to place the contractor in default. In *Sun-Cal, Inc. v. United States*,[135] the General Services Administration (owner) terminated the contractor for default. The contractor disputed the termination, claiming the default was not authorized because the original contract completion date had been waived and a new completion date had never been established. The contract between the parties contained a liquidated damages clause that assessed $1,000 per day as damages against the contractor for each calendar day completion was delayed beyond the contract completion date.

Due to owner delay, the parties had agreed to delay the contract completion date for one month. Although the contractor did not complete the construction by the newly established contract completion date, the owner encouraged the contractor to continue with its work on the project. A new and final contract completion date was never established by the owner.

The court in *Sun-Cal* determined that the owner had surrendered its right to terminate the contractor by failing to enforce the newly established contract completion date within a reasonable time period. The court also determined that the parties failed to agree on a new contract completion date. Consequently, since the owner chose not to place the contractor in default when it did not meet the original contract completion date or the newly established contract completion date, and since a new contract completion date was never reached, the owner could not assess liquidated damages against the contractor.

Occasionally a liquidated damages provision may operate to the contractor's benefit. If such a provision is the exclusive contractual method of calculating damages and is held enforceable, the clause cannot be attacked by the owner merely because the fixed amount turns out to be less than the owner's actual damages. In *Brower Co. v. Garrison*,[136] a liquidated damages provision fixed damages at $50 per day and was upheld despite the fact the owner's actual damages were $230 per day. Likewise, in

[133] *See Cont'l Ill. Nat'l Bank v. United States*, 101 F. Supp. 755 (Ct. Cl. 1952); *Ed. L. Powers Contracting Co.*, ASBCA No. 1430 (Aug. 31, 1953).
[134] 521 N.E.2d 219 (Ill. App. Ct. 1988).
[135] 21 Cl. Ct. 31 (1990).
[136] 468 P.2d 469 (Wash. Ct. App. 1970).

Mars Associates, Inc. v. Facilities Development Corp.,[137] the court precluded the owner from recovering its actual damages, and instead held the owner to liquidated damages, even when the contractor admitted those liquidated damages were only a fraction of the actual damages. Thus, a liquidated damages provision, if upheld, will preclude the owner's right to bring an action for actual damages in most situations.

Contractors may include liquidated damages in their subcontracts. Depending on the wording of the subcontract clause, collection by the prime contractor of liquidated damages from a subcontractor may be limited to the amount the prime contractor has paid the owner.[138]

Finally, there is no apparent reason why a liquidated damages provision cannot be inserted in a contract to cover anticipated damages incurred by the contractor as a result of owner-caused delay, disruption, and so on. Such clauses are being discussed and utilized more frequently, under the proper circumstances.

➢ POINTS TO REMEMBER

- The general purpose of damages is to compensate the injured party for its losses, not to create a windfall.
- Direct damages are those costs flowing directly from the wrongful acts of the other party, such as increased costs of project completion.
- Consequential damages are the reasonably foreseeable, but indirect, injury incurred from contract breaches, such as lost profits on other projects due to lost bonding capacity.
- Punitive damages are not compensatory. Their purpose is to punish the wrongdoer. Punitive damages are generally not recoverable in contract actions or construction disputes.
- Liquidated damages are stipulated amounts the parties agree will be assessed for a failure of performance, typically late completion. Liquidated damages are generally in lieu of, not in addition to, actual (direct and consequential) damages.
- In order to recover for a claim, the claimant must establish liability, the amount of the additional costs incurred, and the existence of a causal relationship between the facts giving rise to liability and the damages claimed.
- Every party to a contract must mitigate or minimize their damages, even if the damages are the result of another's wrongful conduct. A claimant cannot unreasonably incur extra costs.

[137] 508 N.Y.S.2d 87 (N.Y. App. Div. 1986).

[138] *See Indus. Indem. Co. v. Wick Constr. Co.*, 680 P.2d 1100 (Alaska 1984); *Hall Constr. Co. v. Beynon;* 507 So.2d 1225 (Fla. Dist. Ct. App. 1987); *Mattingly Bridge Co. v. Holloway & Son Constr. Co.*, 694 S.W.2d 702 (Ky. 1985).

- Ideally, contractor's claims for direct damages are priced using the discrete cost method, in which specific costs are tied to specific acts of the other party. When it is not practical to do so, other less desirable methods, such as the modified total cost method or the total cost method, are available.
- The total cost method is frowned on by the courts and should be used only if absolutely necessary.
- Good cost accounting is necessary to credibly price and pursue a claim.
- Even with good documentation, the assistance of experts is often necessary to calculate and prove damages, particularly when the claim is based on inefficiency costs rather than pure extra work or delay.

17

CONSTRUCTION INDUSTRY ENVIRONMENTAL AND SAFETY CONCERNS

Construction projects must be approached with the knowledge that both environmental and job-site safety laws and regulations will impact many aspects of the work and may, at times, completely dictate the pace of job progress. Although the risk of encountering environmentally hazardous substances or "materials" and the associated risk of potentially limitless liability may be small on many projects, the risk of encountering other environmental problems always exists. For example, the improper discharge of storm water runoff could result in civil and criminal liabilities. Similarly, every project carries the risk of workplace injury, with complications such as lost worker time, resulting investigations, job-site shutdowns, as well as fines and penalties.

A maze of federal, state, and local environmental and safety laws and regulations can affect a construction project in both good and bad ways. It is important to appreciate and generally understand the scope of these statutes, and how the courts are interpreting them, in order to better plan for the project. The time to address safety and environmental considerations on construction projects begins at the proposal stage. Contractors must plan on satisfying applicable licensing, insurance, or qualification requirements before even considering submitting a bid or proposal. Contingency plans and management policies must be in place and ready for execution. Contract terms and insurance coverage must be scrutinized in light of the possibility, regardless of how remote, that hazardous materials will be encountered or created or that a worker will be injured on the job. This chapter should further sensitize, educate, and alert the reader to the urgent need to be prepared and to expect the unexpected when it comes to liability for environmental and safety issues.

I. SOURCES OF ENVIRONMENTAL REGULATION AND LIABILITY

The federal government's regulation and imposition of liability for hazardous materials is far-reaching and grabs significant media attention. The Comprehensive Environmental Response, Compensation and Liability Act of 1980,[1] commonly known as CERCLA or "Superfund," is the flagship of environmental liability laws. But it is certainly not the only environmental statute that may apply to the construction industry professional. In addition to federal laws, states have enacted their own statutory schemes to regulate and assess liability for environmental issues. Regardless of the existence of any specific statutes, liability also may arise under traditional common law theories of tort, nuisance, and trespass. The consequence of running afoul of this web of environmental liability is not limited to job delays and civil liability payable by a company, but can involve fines and criminal penalties assessed by the federal or state governments against individuals. This section focuses on the civil liability aspects of CERCLA and other key federal environmental statutes, which, despite their importance, are by no means the only legal requirements that may apply when hazardous materials affect a construction project.

A. Encountering Hazardous Materials on a Construction Site—CERCLA Liability

Broadly speaking, CERCLA addresses liability for cleanup costs associated with existing hazardous waste contamination at specific (or listed) sites. Other federal statutes, such as the Resource Conservation and Recovery Act of 1976 (RCRA),[2] deal with the current handling, storage, treatment, and disposal of listed hazardous materials. Because contamination may occur during handling, storing, treating, or disposing hazardous materials, there can be considerable overlap between CERCLA and RCRA.

CERCLA imposes liability for cleanup costs on those parties responsible for the contamination at the listed site. The parties subject to CERCLA liability are referred to as "potentially responsible parties," or "PRPs." Courts have interpreted this statutory term much more broadly than the conventional definition of a "responsible party" might allow. The breadth of CERCLA liability and the extraordinarily low level of involvement at which this liability is triggered makes CERCLA liability a major threat to many construction projects. In addition, CERCLA enforcement is not left to the government. Private individuals who are forced to incur cleanup costs can pursue other PRPs to recover those costs.

1. *Strict Liability under CERCLA*

If a party meets the statutory definition of PRP, it is liable. Taking reasonable precautions and operating in a legally and generally acceptable manner does not avoid liability. CERCLA imposes "strict liability," which means automatic liability without

[1] 42 U.S.C. §§ 9601–75 (2000).
[2] 42 U.S.C. §§ 6901–92k (2000).

regard to fault or negligence. Only a few and very limited defenses are allowed under CERCLA. CERCLA was created specifically to deal with the serious need to address environmental damage from hazardous wastes. The tough measures were deemed necessary to fulfill the statute's important remedial purpose.

There is an exception to strict liability under CERCLA for the contractor specifically engaged to clean up hazardous materials—called a response action contractor (RAC)—that affords immunity from strict liability. This somewhat anomalous response is due to the recognition that without some insulation from strict liability, it would be impossible to get any responsible contractor to perform cleanup work. The RAC is not totally relieved from CERCLA liability; it remains liable for negligence, gross negligence, and intentional wrongdoing in performing cleanup work. The CERCLA exception to strict liability for RACs applies only to federal statutes and does not in any way limit the RAC's potential liability, including strict liability, that may be imposed under state law.

2. *PRPs*

CERCLA lists four PRP categories. If a party falls into one of these categories, that party is liable:

(1) The facility's current owner or operator

(2) The owner or operator of the facility when the hazardous substance was released

(3) Any party who, by contract or otherwise, arranged for the disposal of the hazardous material owned by them or by another party

(4) Any person who accepted any hazardous material for transport or disposal that results in a release of hazardous materials

On their face, these categories are broad, and over time, courts have interpreted them even more broadly. These categories would appear inapplicable to the contractor that encounters an unexpected hazardous material since such waste is not something for which the contractor assumed responsibility. However, liability is imposed without regard to fault or negligence; that is termed strict liability. The contractor that unwittingly stumbles on and innocently transports or disturbs hazardous materials resulting in contamination of a site can be just as liable for cleanup costs under CERCLA as the party that knowingly handles and purposely disposes of hazardous materials in an illegal fashion. That may sound like a theoretical worst-case scenario with no relevance in the real world, but it is not. Consider the contractor's treatment in *Kaiser Aluminum & Chemical Corp. v. Catellus Development Corp.*[3]

In *Kaiser,* the court held that a contractor could be liable under CERCLA simply by cutting and filling soil on a site that, unknown to the contractor, had been contaminated by hazardous materials decades earlier. The property owner engaged

[3] 976 F.2d 1338 (9th Cir. 1992).

the contractor to grade and prepare the site for a housing development. This work required the contractor to "excavate" and "disperse" some of the soils on-site. The contractor did not remove soil from or import fill material onto the site. After the contractor started work, it was discovered that, decades earlier, the site and some of the soil handled by the contractor had been contaminated by hazardous chemicals.

Once hazardous material contamination was discovered, the owner had to clean up and restore the site per CERCLA and the United States Environmental Protection Agency's (EPA) corresponding regulations. The owner then sued the developer, which had sold the property, to recover the site cleanup costs and restoration. The developer responded by also suing the grading contractor under CERCLA. CERCLA allows a responsible party to seek contribution from other responsible parties for cleanup costs. In its claim, the developer claimed that the contractor had worsened the situation by excavating contaminated soil and then spreading it onto uncontaminated areas at the site. The claim against the contractor was initially dismissed, but that decision was reversed on appeal. The court of appeals found that, under the facts alleged, the contractor could be an "operator" of the property and also a "transporter" of hazardous materials and therefore liable for the cleanup costs under CERCLA.

a. Liability as an Operator The *Kaiser* court ruled that the "yardstick" for determining whether a party was an "operator" of a facility is the degree of control the party exerts over the activity causing the contamination when the contamination occurs. In *Kaiser*, the court found that the activity producing the contamination was site excavation and grading, which occurred during construction and while the site was under the contractor's control. Therefore, the court concluded that the contractor had sufficient control over this phase of the development to be an "operator" under CERCLA.[4]

Although site contamination originally occurred in the 1940s, decades before the contractor ever moved a shovel of dirt, this did not matter to the court. Regrading, or simply moving contaminated soils to uncontaminated areas of the same site, was sufficient to constitute disposal under CERCLA. In reaching this conclusion, the court relied on CERCLA's broad definition of "disposal," which includes "the discharge, deposit, injection, dumping, spilling, or placing of any . . . hazardous waste into or on the land." The court further reasoned that "disposal" should not be limited solely to initial contamination of the site. Instead, consistent with the remedial purpose of CERCLA, the term "disposal" can be read broadly to include subsequent movement, dispersal, or release of hazardous materials during landfill excavations and fillings. In the court's view, to limit liability for "disposal" to the initial contamination would result in "a crabbed interpretation [which] would subvert Congress' goal that parties who are responsible for contaminating property be held accountable for the costs of cleaning it up."

b. Liability as a Transporter The *Kaiser* court also found the contractor potentially liable as a "transporter" of hazardous materials. CERCLA defines "transportation"

[4]*But see United States v. Qwest Corp.*, 353 F. Supp. 2d 1048 (D. Minn. 2005) (finding that CERCLA operator liability will not attach to the contractor where there is not a direct relationship between the contractor and the property owner).

as "the movement of a hazardous substance by any mode." In the court's view, the contractor's movement of contaminated soil by excavation and grading was well within that definition. The court did note reference in the statute that transportation applied "to . . . sites selected by" the contractor, but ruled that transporting hazardous materials to an uncontaminated area of the same site was no different from transporting them to another site.

3. *Joint and Several Liability*

In addition to strict liability, CERCLA imposes joint and several liability. Under CERCLA, joint and several liability means that if a party is liable for any part of the contamination (e.g., 20% of the contamination), it is liable for 100% of the cleanup costs. Thus, the party seeking to recoup the cleanup costs at a Superfund site, whether the federal government or a private party, can pursue any responsible party for the entire cost, regardless of the respective contributions of the various responsible parties to the contamination. Not surprisingly, this joint and several liability provision makes solvent companies with deep pockets even bigger targets for prosecution by the government and private litigants.

4. *Contribution*

The impact of joint and several liability on a responsible party under CERCLA can be diluted by the ability of one responsible party to seek contribution for cleanup costs from other responsible parties. Congress added the contribution provision to CERCLA by the CERCLA Amendments and Reauthorization Act of 1986, known as SARA.

As previously mentioned, "contribution" means that if just one responsible party is sued, it can bring a separate action against another responsible party which it contends is partially or completely responsible for CERCLA cleanup costs actually assessed.[5]

Frequently, initial PRPs will claim against other potentially responsible parties, and those responsible parties then can directly assert contribution claims against each other in the same lawsuit. If the claimant has not named all responsible parties, the responsible party named might bring additional responsible parties into the original lawsuit. Otherwise, a separate contribution action may be initiated by one responsible party against another. If a responsible party has previously resolved its liability for cleanup costs with the federal or state government (usually by executing a written settlement agreement commonly referred to as a consent order), it is protected from contribution claims by other responsible parties.

The costs and burdens of CERCLA litigation are so great that they provide considerable financial incentive for a named responsible party to contribute to a settlement with the federal or state government, regardless of the strength of any defense.

[5]*See, e.g., United States v. Honeywell Int'l, Inc.*, 542 F. Supp. 2d 1188 (E.D. Cal. 2008) (finding defendant developer liable for contribution associated with EPA's response costs to remediate arsenic contaminated soils that developer and developer's subcontractors excavated and spread at residential development site).

Consequently, once a CERCLA action starts, it is generally in the interest of the claimant and each individual PRP to promptly involve most, if not all, other responsible parties in the litigation. Even if a right of contribution from other responsible parties exists, however, it does not diminish the basic liability to the party prosecuting the underlying claim for cleanup costs. Consequently, if the other responsible parties no longer exist or lack financial resources to contribute meaningfully to a settlement or judgment, the deep-pocket responsible party can be left to pay the complete tab, regardless of its relative fault. Because of these dynamics, many entities want to revamp CERCLA so that it focuses less on protracted litigation and attorneys' fees and more on dollars actually expended on hazardous waste cleanup.

B. Water Quality

The federal Clean Water Act[6] makes it unlawful for any person to discharge any pollutant into a defined body of water unless a permit for such a discharge is issued under the act. This applies to discharge of storm water from construction sites larger than one acre.[7] The permitting process is expedited by EPA's adoption of a general permit for storm water discharges from construction sites. EPA's general permit does not apply in states that have opted to administer the permit process on their own, but the federal requirements constitute the minimum requirements for such state-administered programs.[8] Construction industry professionals charged with responsibility for storm water discharge must consult with state agencies and local EPA offices to confirm the applicable standards and permitting procedures.

In order to come under EPA's general construction permit, the "operator" of the site must establish a site-specific storm water pollution control plan and then file a "notice of intent" with the EPA. In most cases, the general contractor will be the operator, along with the owner and developer and sometimes a grading subcontractor. The site-specific control plan must meet EPA or state requirements, include storm water control and maintenance measures, and identify contractors or subcontractors that will implement the plan.[9] The key to proper job-site management and the reduction of potential Clean Water Act liability is the continual implementation of best management practices.

When storm water discharges associated with construction are eliminated, the operator may file a "notice of termination" of the permit. With such notice, the operator must execute a certificate stating that storm water discharges have ended or that the permit holder no longer "operates" at the site. Violators of the Clean Water Act are subject to civil penalties of up to $25,000 per day and criminal penalties of up to $25,000 per day and prison up to one year.

[6]33 U.S.C. §§ 1251–1376 (2000).
[7]NPDES Storm Water Phase II Final Rule, 64 Fed. Reg. 68,722–770 (1999).
[8]*See, e.g.,* GA. CODE ANN. § 12–7–1, *et seq.* (requiring sedimentation and erosion control training and certification for persons covered by the act, including developers, contractors, builders, utility contractors, grading contractors, and others involved in land-disturbing activities on covered sites).
[9]Office of Water, U.S. Environmental Protection Agency, EPA No. 832 R-92-005 Storm Water Management For Construction Activities: Developing Pollution Prevention Plans and Best Management Practices (Sept. 1992).

C. Air Quality

The federal Clean Air Act[10] enables the EPA to establish outdoor air-quality standards. The EPA has set standards for the concentrations of a number of air pollutants, including sulfur oxides, hydrocarbons, carbon monoxide, nitrogen oxides, lead, ozone, and total suspended particulate matter. Regulation of air pollutants at both the federal and state level can impact a construction project. Moreover, as concerns about climate change, global warming, and increased carbon footprints continue to grow, air-quality regulations will continue to impact construction, and industry professionals will increasingly be called on to take steps to refine construction processes to mitigate such threats.

1. Greenhouse Gas Emissions

In 2007, the United States Supreme Court held that the EPA must regulate so-called greenhouse gas emissions (GHGs) from motor vehicle emissions.[11] The chemical compounds that make up GHGs consist predominantly of carbon dioxide, methane, nitrous oxide, and hydrofluorocarbons. GHGs are considered to be a significant contributor to the relatively recent global warming or climate change phenomenon.

Regulation of GHGs by EPA, and possibly by state and local governments, will likely impact construction projects in several ways. First, governments may attempt to regulate emissions from off-road vehicles, including heavy equipment used on construction projects. For example, California has enacted new Low-Carbon Fuel Standards (LCFS).[12] The LCFS apply to all gasoline and diesel, including freight and off-road applications. Construction industry professionals should be concerned that these rules will render obsolete older off-road equipment and force equipment owners to spend significant additional dollars on new or retrofitted equipment that can operate using the lower carbon fuels. The Associated General Contractors of America has estimated that these regulations could have a $13 billion impact in the California construction industry alone.[13]

Second, regulation of GHGs likely will impact construction of heavy industrial facilities (e.g., power plants). States likely will also require mandatory GHG reporting by industrial facility operators. States and local jurisdictions might rely on GHG restrictions to deny or delay permits for construction of facilities that produce GHGs. New construction of heavy industrial facilities likely will have to factor in further controls on GHGs; the incumbent risks for controlling these regulated airborne substances will fall on one or more members of the construction project delivery team.

[10]42 U.S.C. §§ 7401–7671 (2000).

[11]*Mass. v. Env't Prot. Agency,* _____ U.S. _____, 127 S.Ct. 1438 (2007).

[12]Assembly Bill 32—Global Warming Solutions Act (2006).

[13]Other states are expected to follow California's lead in the efforts to require more environmentally friendly fuels. Florida, Washington, Oregon, Minnesota, Arizona, and New Mexico are currently considering similar standards. *See* Press Release, *Five Western Governors Announce Regional Greenhouse Gas Reduction Agreement* (Feb. 26, 2007). California and New Mexico have enacted rules mandating GHG reporting to become effective in 2009.

2. Asbestos and Lead

Renovation and demolition projects, particularly projects involving buildings constructed before 1980,[14] can present more significant risks of exposure to airborne asbestos and lead as well as special risks to workers. These risks are so common and serious that the U.S. Occupational Safety and Health Administration (OSHA) has promulgated regulations specifying safety and health measures to be followed on construction projects involving asbestos and lead.[15] Most of these procedures and requirements are similar for both of these airborne substances. But there are important differences that cannot be overlooked. State and local regulations also may apply.

Regulations covering asbestos and lead apply to all construction work, including demolition or salvage, renovation, remodeling, repair, installation of products containing these materials, transportation, disposal, and emergency cleanup situations. Each regulation contains an action level that represents the maximum concentration of the substance allowable over a specified time period before the contractor must comply with most aspects of the regulation. These regulations also list a permissible exposure level (PEL), which is the maximum level of exposure to the substance that the contractor's employees can endure without respiratory protection.

Each contractor must initially determine if any employee on the job site may be exposed to one of the substances at or above the action level. This initial determination can be waived, however, when the contractor can prove through objective data that its employees cannot be exposed to concentrations of the substance above the action level even in worst-case scenarios, or when the contractor has monitored for substance exposures within the last 12 months during work operations conducted under workplace conditions closely resembling the current operation.

When the initial determination shows the possibility of substance exposure at or above the action level, the contractor must conduct monitoring that is representative of the exposure for each employee in the workplace who is exposed to the substance at least every six months and must notify affected employees of the monitoring results. If exposure is above the PEL, periodic monitoring must be performed every three months. If the initial monitoring shows no possibility of substance exposure above the action level, no further monitoring is necessary. Regardless of previous measurements, however, the contractor must repeat the initial monitoring process whenever there has been a change in process, control equipment, personnel, or work practices that may result in new or additional exposures above the action level limit.

The contractor must employ engineering controls and work practices to reduce and maintain employee exposure to the substances below the PEL to the extent such controls are feasible. Wherever the feasible engineering controls and work practices are not sufficient to reduce the substance concentration below the PEL, the contractor must supplement its controls with respiratory protective devices, protective clothing, changing and shower facilities, special signage, and training programs, among other

[14]*See* 29 C.F.R. § 1926.1101 (defining "presumed asbestos containing material" as "thermal system insulation and surfacing material found in buildings constructed no later than 1980").

[15]*See generally* 29 C.F.R. § 1926.1101 (2006) (asbestos); 29 C.F.R. § 1926.62 (2006) (lead).

measures. The contractor also is required to institute a rigorous medical surveillance program of exposed employees, including blood tests and physical examinations.

a. Inadvertent Asbestos Abatement The general construction or renovation project may be transformed quickly into an asbestos abatement project if the contractor unknowingly demolishes or disturbs asbestos-containing building materials. Under the EPA's National Emission Standards for Hazardous Air Pollutants (NESHAP) for Asbestos-Containing Materials,[16] demolition and renovation work become an asbestos-abatement project whenever regulated asbestos-containing materials in quantities of greater than 260 linear feet on pipes, such as pipe insulation, or 160 square feet of "other" facility components, are disturbed. The EPA's Asbestos NESHAP standards also are triggered if the total asbestos-containing waste product generated from facility components exceeds 35 cubic feet. These quantities are all relatively small amounts of material. If a contractor disturbs asbestos-containing materials exceeding these thresholds without notification, it is conducting an illegal asbestos abatement operation and is subject to penalties under each subsection of the asbestos regulations that it is violating.

To avoid possible violations, generally it is necessary to notify the regulators 10 working days before beginning any asbestos abatement project and update the notice if the quantity of asbestos-containing materials changes by 20% or more. The only exceptions to this notification requirement would be for emergency operations and unsafe buildings. Further, a new notification is required if the commencement date of the asbestos abatement changes. If the contractor inadvertently renovates or demolishes regulated asbestos-containing building materials, then, in addition to being subject to citation for violation of the notification requirement, it also may be fined for failing to properly control airborne asbestos emissions.

If a contractor unknowingly disturbs asbestos-containing materials, it could commit these violations:

- Failure to properly contain the work area
- Failure to provide workers with proper respiratory protection and protective clothing, hygiene facilities, and training
- Failure to employ proper and necessary medical surveillance of workers
- Failure to wet asbestos-containing building materials adequately during demolition
- Failure to seal and label, transport, or dispose of asbestos waste properly
- Failure to maintain proper records of an asbestos-abatement project

In addition to fines or penalties levied on contractors by regulatory agencies, the contractor that wrongly or unlawfully disturbs regulated asbestos-containing materials also faces liability to third parties. Employees, building owners, tenants, and occupants exposed to such materials will likely assert claims against the contractor

[16] 40 C.F.R. Pt. 61, Subpt. M (2008).

resulting from such exposure. Given these very real risks, the prudent contractor is wise to determine whether the building materials that will be disturbed or demolished contain asbestos well before it starts work.

b. Inadvertent Lead-Based Paint Abatement Similarly, if lead is present in building materials, typically in the form of lead-based paint, and the contractor accidentally disturbs it or does not perform an initial exposure assessment, the contractor must assume exposure at a level of 10 times the maximum permissible exposure level. With that assumption, the contractor must provide its workers with respiratory protection and protective clothing, provide hygiene facilities, conduct medical surveillance, conduct employee training, maintain extensive records, and post warning signs at all entrances and exits to the work areas. Failure to comply with these requirements makes the contractor liable for violating each subsection of the standard.

Under the standard, if lead is present, the contractor must presume that its construction activities will cause airborne emissions of lead in excess of permissible exposure limits. Therefore, the contractor should obtain all sampling and test data from the owner before commencing its demolition and renovation activities. Further, the contractor should ask the owner to perform appropriate testing in all areas that will be affected by the construction activities and should not start work until it receives the requested test results.

II. MINIMIZING ENVIRONMENTAL RISKS

Before bidding on or negotiating a contract for a conventional construction project, the contractor should assess the work's environmental aspects, so as to minimize potential environmental risks. This detailed process involves much more than a routine site investigation.

A. Conduct a Prebid Environmental Review of the Contract Documents

The prudent contractor will adopt and implement procedures that ensure a prebid environmental review of each proposed contract. Although reviewing the plans and specifications for specifically delineated hazardous material work is essential, it is not enough. One common trap that has snared more than its share of contractors are plans and specifications that do not identify any particular work involving hazardous materials but instead contain generic provisions that address asbestos, lead-based paint, or hazardous materials. In this contractual context, owners, including public owners, have argued successfully that the parties to the contract obviously contemplated the contractor performing hazardous material work by virtue of such generic contract clauses. At best, this puts the contractor in the position of subcontracting out the work and hoping that no claims arise.

1. Exclude Hazardous Materials from the Scope of Work

If the contractor wants to minimize its exposure on a private job that, according to the contract documents, contains no work involving hazardous substances or materials, the generic hazardous materials specification sections and related clauses should be stricken and the contractor's bid submission should take exception to the performance of any work involving hazardous materials. If it is awarded the work based on this submission, the contractor should make sure that the hazardous material condition (or exception) in its bid is expressly incorporated into the contract. On a similar public project, however, such a qualification exception may render the bid nonresponsive. Accordingly, the public works contractor must decide whether it wants to take the risk of possibly being directed by the government to perform hazardous material work.

2. Determine What Materials Will Be Encountered

Any renovation or demolition work on a project necessarily requires the contractor to address existing building materials. Before signing the contract, the contractor must know what materials it is dealing with, or run a very significant risk of environmental violations and lawsuits by workers or others exposed to hazardous materials in the workplace. Therefore, the contractor should be able to answer these questions before submitting a bid or proposal for such work:

(1) Do the contract documents include any hazardous material remediation, or asbestos or lead-based paint abatement?

(2) Do the contract documents refer to standards for performing hazardous material remediation, or asbestos or lead-based paint abatement?

(3) Is any part of the work affected by hazardous materials, asbestos, lead-based paint, underground storage tanks, wetlands, or protected natural resources?

(4) If the contract documents describe environmental remediation, a contractor must determine if it has insurance coverage for such environmental risks.

(5) Does the contract allocate risk or responsibility for hazardous materials (including asbestos) or lead-based paint? Is the contractor required to indemnify the owner or others for losses associated with such materials?

(6) Is the owner aware of any hazardous materials (including asbestos), lead-based paint, wastes, or contaminated soil or water that might affect the contractor's scope of work?

(7) Have any environmental site assessments been conducted? Are there any building surveys or inspection reports on asbestos, lead-based paint, underground storage tanks, soil or water quality, or other hazardous materials on the site?

(8) Did the site visit and inspection indicate distressed vegetation, hydrocarbon or chemical contamination, underground storage tanks, transformers, drums, suspect lead-based paint, or suspect asbestos-containing materials?

(9) Has the site ever been subject to or regulated by any environmental permits issued by regulatory agencies, or has the site ever been the subject of a regulatory enforcement action?

(10) Has any environmental cleanup occurred at the site? Have tanks been removed or filled? Is there a closure report? Is the owner aware of any residual soil or water contamination?

(11) If unanticipated asbestos, lead-based paint, tanks, drums, contaminated soil, poly-chlorinated biphenyls (PCBs), or other hazardous materials are encountered, what are the contractor's obligations to address these?

In order to obtain answers to some of these initial questions, a prudent contractor should submit a written environmental prebid inquiry to the owner or others on the project delivery team.

Given the risk of unlimited liability, the prudent contractor should conduct a comprehensive site visit before submitting its bid or proposal. Record what was observed and observable with photographs, video recordings, and trip reports. Look for distressed vegetation, unusual depressions, storage tanks, drums, wells (or pipes protruding from the ground), suspect lead-based paint, and suspect asbestos-containing materials. Obtain and review all as-builts for additional information as to building components and materials. Follow up the site visit with written questions to the owner asking about specific questionable areas or conditions and building materials that are suspected to contain asbestos or be layered with lead-based paint. Ask for copies of any bulk sample results and paint analyses.

This approach shifts at least some of the initial burden of identifying hazardous materials onto, and forces disclosure by, the owner as to what it knows or should know about the building or property. Although this is a good start, the prudent contractor should be aware that courts assume contractors are fully knowledgeable about building materials. As such, courts will not allow a contractor to escape liability from environmental law violations based on its blind reliance on plans and specifications. The contractor should thoroughly review each phase of the proposed project with an eye toward possible environmental risks and concerns. The contractor should not submit its proposal or bid until after it has assessed and, if necessary, allocated or removed such risks and concerns from its scope of work.

B. Contract Provisions and Indemnification

The risk of unanticipated hazardous materials should be the subject of specific contract terms and conditions appropriately and equitably allocating such risk among the proper parties. Routine contract clauses imposing considerable risk on the contractor for the unknown can apply to unanticipated hazardous materials. Consider a standard site investigation clause requiring the contractor to examine the site and shifting responsibility for concealed or unexpected conditions. What if the unexpected condition is a concealed hazardous material contamination rather than some additional rock or muck removal? Likewise, broad indemnity clauses that benefit the owner generally are considered burdensome under routine circumstances. If applied to a hazardous material situation, they can be devastating. A standard differing site conditions clause may afford the contractor some protection from the costs associated with unanticipated hazardous material on construction sites, but far greater specificity is preferred to ensure protection.

The need for specific contract provisions to address the unfortunate possibility of contamination by hazardous materials has been recognized, but is by no means fully addressed nor universally applied. The American Institute of Architects' (AIA) Document A201, General Conditions of the Contract for Construction (2007 ed.), addresses the discovery of unanticipated hazardous materials. AIA A201, however, does not protect the contractor from liability for all hazardous materials that may be encountered on the construction project. Instead, this version of AIA A201 puts an affirmative obligation on the contractor to comply "with any requirements included in the Contract Documents regarding hazardous materials."[17] This means that the contractor performing work under the AIA A201 General Conditions must review all contract documents to determine whether there are "any" hazardous material requirements.

AIA A201 2007 includes an indemnification provision in favor of the owner, which provides:

> The Contractor shall indemnify the Owner for the cost and expense the Owner incurs (1) for remediation of a material or substance the Contractor brings to the site and negligently handles, or (2) where the Contractor fails to perform its obligations under Section 10.3.1, except to the extent that the cost and expense are due to the Owner's fault or negligence.[18]

This indemnification provision in favor of the owner further underscores the need for the contractor to fully understand whether hazardous material work is in any way referenced in the contract documents of which AIA A201 (2007 ed.) is a part. AIA A201 does require the owner to indemnify the contractor from third-party claims if the unanticipated hazardous material causes personal injury or property damage (other than to the work) and only to the extent the third party is not at fault.[19]

Under AIA A201, a contractor must stop work in the affected area and notify the owner and architect: (1) if it encounters a hazardous material or substances not addressed in the contract documents; and, (2) if, through reasonable precautions, the contractor cannot prevent foreseeable bodily injury or death resulting from such unanticipated hazardous materials. Just what constitutes "unanticipated" hazardous materials and "reasonable precautions" to prevent injury or death under AIA A201 is unclear, however, so the prudent course for the contractor when encountering hazardous materials is to stop work. Under AIA A201, work cannot resume until the owner and contractor enter into a written agreement.

The C-700 Standard General Conditions of the Construction Contract published by the Engineers Joint Contract Documents Committee (EJCDC) (2007 ed.) afford protection by expressly placing the responsibility for virtually all unanticipated "Hazardous

[17] AIA A201, § 10.3.1 (2007 ed.).
[18] AIA A201, § 10.3.5 (2007 ed.).
[19] AIA A201, § 10.3.4 (2007 ed.).

Environmental Conditions" on the owner. The EJCDC C-700 General Conditions set forth specific procedures for notice by the contractor to the owner about the hazardous environmental condition encountered and the required response by the owner that must be implemented before the contractor can be directed or required to resume work in the affected area.

Whether the project is the traditional design-bid-build with a competitive bid lump sum or a design-build project, the potential issues and risks related to hazardous materials warrant consideration and treatment in the contract documents. One example of the treatment of hazardous materials in the context of a design-build project is found in Paragraph 3.6 in the Standard Design-Build Agreement and General Conditions Between Owner and Design-Builder, ConsensusDOCS 410 (2007 ed.), endorsed by the Associated General Contractors of America. This provision reads:

3.6 HAZARDOUS MATERIALS

3.6.1 A Hazardous Material is any substance or material identified now or in the future as hazardous under any federal, state or local law or regulation, or any other substance or material which may be considered hazardous or otherwise subject to statutory or regulatory requirements governing handling, disposal or cleanup. The Design-Builder shall not be obligated to commence or continue work until all Hazardous Material discovered at the Worksite has been removed, rendered or determined to be harmless by the Owner as certified by an independent testing laboratory approved by the appropriate government agency.

3.6.2 If after the commencement of the Work, Hazardous Material is discovered at the Project, the Design-Builder shall be entitled to immediately stop Work in the affected area. The Design-Builder shall report the condition to the Owner and, if required, the government agency with jurisdiction.

3.6.3 The Design-Builder shall not be required to perform any Work relating to or in the area of Hazardous Material without written mutual agreement.

3.6.4 The Owner shall be responsible for retaining an independent testing laboratory to determine the nature of the material encountered and whether it is a Hazardous Material requiring corrective measures or remedial action. Such measures shall be the sole responsibility of the Owner, and shall be performed in a manner minimizing any adverse effects upon the Work of the Design-Builder. The Design-Builder shall resume Work in the area affected by any Hazardous Material only upon written agreement between the Parties after the Hazardous Material has been removed or rendered harmless and only after approval, if necessary, of the governmental agency or agencies with jurisdiction.

3.6.5 If the Design-Builder incurs additional costs or is delayed due to the presence or remediation of Hazardous Material, the Design-Builder shall be entitled to an equitable adjustment in the GMP, compensation for Design Phase services, the Design-Builder's Fee or the Date of Substantial Completion or the Date of Final Completion.

3.6.6 To the extent not caused by the negligent acts or omissions of the Design-Builder, its Subcontractors, Material Suppliers and Sub-subcontractors, and the agents, officers, directors and employees of each of them, the Owner shall defend, indemnify and hold harmless the Design-Builder, its Subcontractors and Sub-subcontractors, and the agents, officers, directors and employees of each of them, from and against all claims, damages, losses, costs and expenses, including but not limited to reasonable attorneys' fees, costs and expenses incurred in connection with any dispute resolution process, to the extent permitted pursuant to Paragraph 6.5, arising out of or relating to the performance of the Work in any area affected by Hazardous Material.

3.6.7 Material Safety Data (MSD) sheets as required by law and pertaining to materials or substances used or consumed in the performance of the Work, whether obtained by the Design-Builder, Subcontractors, the Owner or Others, shall be maintained at the Project by the Design-Builder and made available to the Owner and Subcontractors.

3.6.8 During the Design-Builder's performance of the Work, the Design-Builder shall be responsible for the proper handling of all materials brought to the Worksite by the Design-Builder. Upon the issuance of the Certificate of Substantial Completion, the Owner shall be responsible under this Paragraph for materials and substances brought to the site by the Design-Builder if such materials or substances are required by the Contract Documents.

3.6.9 The terms of this Paragraph 3.6 shall survive the completion of the Work under this Agreement or any termination of this Agreement.[20]

This is a thorough, comprehensive, and balanced contract provision dealing with the possibility of unexpected hazardous materials and the associated risks and liabilities.[21]

Regardless of whether a form contract or a custom-drafted contract is utilized, that contract must specifically allocate the risk of unexpected hazardous materials. Definitive procedures also should be set forth so that the parties know precisely how to respond and react to the situation if encountered.

Obtaining indemnification from the consequences of hazardous materials is critical, but by no means foolproof. Many state laws limit the enforceability of indemnity clauses. These limitations reflect a public policy against allowing anyone to obtain indemnification from its own negligent acts. As liability under CERCLA is strict and without regard to negligence, this limitation may not apply in many situations. Nonetheless, the prudent contractor should consider the law applicable to its contract and the extent of the enforceability of any indemnity terms associated with hazardous materials.[22]

[20]A similar provision is found at ConsensusDOCS 200, ¶ 3.13 (2007 ed.). Standard Agreement and General Conditions Between Owner and Contractor (Where the Contract Price Is a Lump Sum).

[21]*See also* ConsensusDOCS 200 (2007 ed.), Standard Agreement Between Owner and Contractor (Where the Contract Price Is a Lump Sum) and ConsensusDOCS 300 (2007 ed.) Tri-Party Collaborative Agreement (containing substantially similar Hazardous Materials clauses).

[22]*See, e.g., Richmond Am. Homes of Colo., Inc. v. United States,* 75 Fed. Cl. 376 (2007) (finding that statutory indemnification provision of National Defense Authorization Act extended to contractors disturbing hazardous materials during demolition work performed at Lowry Air Force Base redevelopment).

Many contractors assume tremendous potential liability by naively believing that they can subcontract away the project's environmental risk. Although it is true that specialty environmental remediation contractors are available, it is also true that a contractor still is exposed to regulatory enforcement actions and personal injury actions despite favorable subcontract language and indemnification clauses. Given the potential liability, the prudent contractor should subcontract environmental remediation work to only the most experienced, and financially healthy environmental remediation subcontractors.

From the general contractor's perspective, it is much better, if possible, to delete environmental remediation work from the proposed work scope and rely on the owner to contract directly with an environmental remediation contractor. By doing so, the owner maintains responsibility for the adequacy and timing of the environmental remediation work as well as the proper disposal of the resulting waste stream. This approach can significantly reduce the environmental risk for the general construction contractor.

If the contractor begins work with no knowledge of the job's environmental hazards and then encounters such hazards during project performance, many owners will attempt to get the general contractor to simply subcontract out the environmental remediation work. This approach is extremely risky for the general construction contractor. It also likely places the contractor into an uninsured position while possibly compromising any potential claim it might have for suspension of its work due to the undisclosed environmental hazards.

C. Insurance

Once the magnitude of environmental liability in the construction industry was recognized, insurance coverage rapidly receded. Now new forms of coverage are becoming available as the insurance market readjusts. Still, coverage is far from complete and often comes at substantial cost. Generally, standard forms of insurance, such as commercial general liability (CGL) insurance, will not afford protection from liability associated with hazardous materials because CGL policies contain so-called pollution exclusion clauses. Therefore, specific pollution coverage must be procured. (See **Chapter 18**.)

Complete insurance protection for unanticipated hazardous materials is either impossible or financially impractical to obtain. Evaluating specific environmental risk coverage is nonetheless an appropriate step. Insurance planning for construction projects, particularly as it relates to insuring against environmental risks, is too complex and specialized to be left to the inexperienced layperson. Even contractors that employ full-time insurance specialists or risk managers should rely on insurance professionals to assist in procuring proper coverage at a competitive price. This is particularly true in the environmental liability insurance market, which continues to experience rapid change and evolution as new environmental hazards continue to surface.

Before the contractor decides to assume responsibility, for example, for asbestos abatement, it is important to scrutinize the contractor's own insurance to verify that there will be coverage or a defense if any claims are made against the contractor. As previously mentioned, a standard CGL policy contains a pollution exclusion that affords the contractor no coverage for any of the risks inherent in dealing with asbestos. Unless a specialized policy is purchased, there is no insurance coverage available, yet the potential liability is virtually unlimited even where the environmental remediation work has been subcontracted.

III. PROPER MANAGEMENT TECHNIQUES

The increasing use and intensity of the environmental site assessment by owners and lenders should reduce the likelihood of encountering unexpected hazardous materials. But the risk still exists. Constant recognition and review of this risk must be maintained at the management level, and vigilance and preparedness enforced on the job site, in order to mitigate the risk of unanticipated hazardous materials.

A. Management Review of Environmental Risks

The prudent contractor should delegate overall responsibility for environmental matters to a senior company official. This individual would be a resource for the entire company aiding in the evaluation of and protection against unanticipated hazardous materials on company projects. This individual should become thoroughly familiar with applicable laws and regulations affecting the company's work. This person must evaluate environmental risks not only in light of existing laws but also in the specific context of applicable contract provisions, the nature of the construction, the peculiarities of any given job site, and the ability to procure insurance. The company official responsible for environmental matters must be authorized to address unacceptable environmental risks directly or through direct reporting to other company officials. Environmental risk management is far too important to the future of a company to delegate it to an individual without enforcement authority. The contractor also should have a specific and mandatory written procedure to be implemented immediately upon encountering unexpected hazardous materials. The contents of the procedure and the nature of the response are discussed in greater detail in the next section.

One individual, however, cannot completely manage all environmental risks that a contractor may encounter in its ongoing operations. Therefore, all personnel, especially project management, must be educated and sensitized to such risks. If they do not know how to respond immediately to a situation involving hazardous materials, they will at least appreciate the severity of the situation and recognize the need to seek advice and direction from someone who does know how to respond.

All planning, evaluation, and actions to address environmental risks at the management level will be wasted if such efforts do not extend to each job site and the individuals working in the field.

B. Have a Response Plan

When environmental risks are encountered, the response of project personnel should not be left to chance. The appropriate response should be contained in a written and mandatory procedure reinforced through in-house training and indoctrination. As each project begins and before construction starts, this standard procedure should be reviewed and supplemented with the areas of responsibility for specific individuals. Contact information should be listed for all federal, state, or local agencies requiring notification or from whom an emergency response may be required. The procedure also should be coordinated with representatives of the owner, design professional, and construction manager as well as subcontractors. Although the response will be the primary responsibility of construction project management, the response cannot be triggered or properly implemented without the aid of all project personnel, including subcontractors and their employees. Once the response plan is developed, it should be reviewed in project meetings, subcontractor coordination meetings, and toolbox safety meetings at the level of detail appropriate for the gathering to ensure that it is understood.

C. Immediately Stop Work in the Affected Area

Generally, unanticipated hazardous materials are not the contractor's responsibility. If they are exposed or discovered but not disturbed or released, the hazardous materials certainly present a problem, but, from the contractor's perspective, a problem of limited scope. If, however, the contractor goes beyond merely exposing the hazardous materials and disturbs them in any manner, the contractor may well have unwittingly thrown itself into one or more of the categories of "potentially responsible parties" under CERCLA or stepped into other statutory or common law liability. This is what happened in the *Kaiser* case discussed earlier. That is why the first response to encountering hazardous materials or any unknown substance or material that might be hazardous is to STOP! As previously mentioned, this is a requirement in the ConsensusDOCS 200, 300, and 410, the AIA A201 General Conditions, and the EJCDC C-700 General Conditions. A prudent contractor should avoid performing any further work in the affected area or any other areas where the same condition might exist. Stopping work should be a reflexive response to encountering anything suspicious, whether it is an unusual color in the soil, a faint but unfamiliar odor, or anything out of the ordinary.

D. Provide Immediate Notice

The contractor should immediately notify the owner, the design professional, and any construction manager that a suspected hazardous material has been uncovered, even if the contractor cannot identify the substance. Initial notice should be verbal, with written confirmation provided as soon as practical thereafter. This notice should make it clear that the unanticipated hazardous materials are the responsibility of the owner and that the contractor is awaiting direction from the owner or the owner's representative as to how to proceed. The contractor also should expressly reserve its rights to seek indemnification from all claims due to encountering hazardous materials and to seek adjustments in contract time and price based on any work stoppage. In this notice, the contractor also should consider addressing any other concerns created by uncovering these materials, such as notifying regulatory agencies, insurance carriers, and other project delivery team members.

The owner or its representative should report the release or spill of a hazardous material or substance to the appropriate government officials. Still, the contractor may also be deemed a "person in charge" of the site and therefore required to promptly report the release or spill. The prudent contractor will exercise caution and also report the incident.

E. Do Not Resume Work without Proper Authorization

Once hazardous materials are uncovered, the contractor should not resume work in the affected areas without written authorization from the owner and the government agency with authority over the situation. All such authorizations should be explicit and written. If the owner directs the contractor to clean up the hazardous material, the contractor should insist on specific written cleanup instructions and confirmation that such instructions have been approved by the responsible government agency. The contractor also should require that the owner specifically indemnify the contractor from the risks of undertaking the cleanup. Even if the owner obliges on these points, the contractor should consider whether it wants to undertake the additional risks of environmental cleanup work. Remember, the owner's indemnification is only as good as its ability to pay and does not insulate the contractor from third-party claims. Undertaking the cleanup of hazardous materials also may cause some problems with the contractor's insurance carrier and surety if cleanup work is not within the contractor's typical line of work on which its insurance and bonding are based.

IV. MOLD: DEVELOPING A PROGRAM TO LIMIT LIABILITY

"Toxic mold" has become another in the modern lexicon of environmental risk terms. The fact that the adverse health effects from exposure to mold are not fully known[23] has not stopped a steady procession of construction defect claims and multimillion-dollar lawsuits. The frequent occurrence of mold during or as a result of the construction process, combined with the perceived health risks and a heightened public awareness of such risks, has created a breeding ground for litigation and the corresponding costs and uncertainty of such litigation.

Preventing the rampant growth and spread of mold requires controlling temperature, humidity, and moisture in the construction environment. To prevent, limit, and control the presence of moisture requires a cooperative and coordinated effort by the project delivery team, which includes: (1) a good design, (2) consideration of mold risks in the selection and specification of building materials as well as construction means and methods, (3) a thorough construction quality assurance/quality control program (including, but not limited to, documented inspections), and (4) proper project documentation.

One of the first questions a construction professional should ask in assessing and addressing the risk of mold is: "Do I have insurance for this?" As is often the case, the answer may be unclear. The standard CGL policy carried by most contractors contains a number of exclusions that may allow the contractor's insurance carrier

[23] *See, e.g., Fraser v. 301-52 Townhouse Corp.*, 831 N.Y.S.2d 347 (N.Y. Sup. Ct. 2006).

to assert that the risk of mold is not covered. These include certain "business risk" exclusions and the "pollution exclusion." Courts that have addressed the issue have reached conflicting conclusions as to whether mold-related claims are covered by a contractor's CGL policy or whether the insurance carrier is relieved of responsibility for such claims under the standard pollution exclusion.

The primary legal theory under which mold contamination litigation arises is the common law doctrine of negligence.[24] In such cases, the design professional, contractor, subcontractor, or developer owes a duty of care to the plaintiff. To determine whether a party has breached its duty of care, the plaintiff must determine the "duty" by examining relevant construction standards. By demonstrating a specialist's failure to abide by the practices of other similarly situated specialists, claims of negligence involving a design professional often are accompanied by a professional malpractice claim.

Fraud and misrepresentation claims arise from a false statement that is made by a defendant if the defendant knew or reasonably should have known that the statement was false when made, it intended the plaintiff to rely upon the statement, and the plaintiff relied on the statement to its detriment. Under such cause of action, an owner may be liable for fraud and misrepresentation if the owner represents that the structure was inspected and found to be "clean" when such an inspection was never conducted and mold was found to be present. If such a fraudulent statement or false impression causes another party to enter into a contract in reliance on such a statement or impression, the innocent individual may have a right to have the contract rescinded due to the fraudulent inducement to enter into the contract.

V. ENVIRONMENTALLY FRIENDLY CONSTRUCTION: GREEN BUILDINGS

In addition to environmental risks in construction due to hazardous materials, water quality, and air quality, industry professionals should understand the ever increasing movement toward "environmentally friendly" construction. The so-called Green Building concept has become an industry trend, the purpose of which is to promote design and construction utilizing means, methods, and materials that present less risk to the environment and less risk of exposure to environmental contaminants.

A. LEED Certification

The United States Green Building Council (USGBC)—a nonprofit organization comprised of owners, developers, facility managers, design professionals, general contractors, subcontractors, suppliers, and government agencies—has tasked itself with identifying Green Building practices. This effort resulted in the Leadership in Energy and Environmental Design (LEED) Green Building Rating System.™

[24]*See Stanley Martin Cos., Inc. v. Universal Forest Prods. Shoffner LLC,* 396 F. Supp. 2d 606 (D. Md. 2005); *Sabella v. Wisler,* 377 P.2d 889 (Cal. 1963).

LEED is the emerging shorthand for what makes a construction project green, environmentally friendly, or energy efficient. LEED is actually a series of checklists targeted at specialized projects. Currently, the USGBC has developed nine LEED classifications:

(1) New Construction and Major Renovations (NC)
(2) Existing Buildings: Operations & Maintenance (EB)
(3) Commercial Interiors (CI)
(4) Core and Shell (CS)
(5) Schools
(6) Retail
(7) Healthcare
(8) Homes
(9) Neighborhood Development

Projects accumulate points by meeting criteria and, based on those points, are designated LEED certified (the baseline), LEED Silver, LEED Gold, or LEED Platinum.

Familiarity with the LEED certification system is required for government contractors and certainly is recommended for contractors working in the private sector. Many public entities at the federal, state, and local level now require that their public projects meet LEED certification standards and offer incentives to private LEED projects. Although the public sector has led the way with LEED projects, Green Building construction has gone mainstream, including private construction too.

Various federal government agencies require LEED Certification for their projects.[25] For example, the General Services Administration (GSA) requires that all building projects meet the LEED Certified level and target the LEED Silver level.[26] Likewise, the U.S. Navy requires "all applicable projects" to meet the requirements for LEED Certification but does not require submission to the USGBC for actual certification.[27] The EPA requires that all new facility construction and acquisition projects 20,000 square feet or larger not just meet but actually achieve LEED Gold certification.[28] The U.S. Department of Agriculture requires all new or major renovation construction of covered facilities to achieve LEED Silver certification.[29] As time goes on, more and more government agencies may add such requirements, and current requirements, as noted above, may become more stringent.

[25]*See* Exec. Order No. 13423 (Jan. 24, 2007) (requiring federal agencies to "lead by example" in advancing national policy for enhanced environmental performance of buildings).
[26]United States General Services Administration Facilities Standards, § 1.6 Environmental Policies & Practices (2003).
[27]*www.usgbc.org/publicpolicy*.
[28]*www.epa.gov/greeningepa* (Green Building Vision and Policy Statement (Aug. 1995)).
[29]U.S.D.A. Departmental Regulation 5500–001 (June 19, 2006).

B. Design and Construction Issues Affected by Green Building Construction

The earlier in the planning phase that the owner decides to seek LEED certification, the lower the construction costs. As a rule, design costs for LEED projects exceed design costs for non-LEED projects. Arguably, however, the increased emphasis on front-end design reduces construction costs by potentially reducing the number of changes required. If an owner wants or requires LEED certification, the project participants need to understand the LEED process when estimating, bidding, and building the job. LEED touches many aspects of the construction project including but not limited to site selection, selection of materials, handling of construction waste, storm water, MEP design, and building finishes.

When an owner decides to obtain LEED certification for a project, the first step is designating a LEED Accredited Professional (AP) to oversee and shepherd the process. Often the LEED AP is a member of the design team, although it could be a construction manager or contractor on a design-build project. No matter who serves as the LEED AP, for a LEED project to be cost effective, all project participants—not just the designers—need to understand the requirements for LEED certification from the beginning. Contractors may be surprised that they are likely working in ways that would earn points for LEED Certification. Projects accumulate points in six core areas:

(1) Sustainable sites (e.g., by conforming to the required storm water permit requirements)

(2) Water efficiency (e.g., by capturing rainwater, using native landscaping, and using innovative water-efficient fixtures)

(3) Energy and atmosphere (e.g., by conforming with the American Society of Heating, Refrigeration, and Air Conditioning Engineers[30] standards reducing energy demand, using HVAC and fire suppression systems without Hydro Chloro Fluoro Carbons (HCFCs) or Halons, and using on-site renewable energy (e.g., photo-voltaics) or other green energy

(4) Materials and resources (e.g., by storing and collecting recyclables; reusing shell and nonshell components in renovations; diverting construction waste from the landfill; using recycled or rapidly renewable materials; and using local or regional materials)

(5) Pollutants and indoor environmental air quality (e.g., by choosing low- or no-emitting materials adhesives, sealants, paints, carpets, and composite wood products; flushing or filtering out any pollutants)

(6) Innovation and the design process

The LEED AP must submit documentation to support every point the project is claiming, so all project participants involved with a particular point must document it.[31]

[30]For more information, see *www.ashrae.org.*

[31]*See www.greenbuild365.org or www.buildinggreen.com.*

AIA B214 Standard Form of Architect's Services: LEED Certification (2007 ed.) specifically addresses an architect's duties and responsibilities when the owner seeks LEED certification. Under AIA B214, an architect's services include, among other things: (1) conducting a predesign workshop where the LEED rating system will be reviewed and LEED points will be targeted; (2) preparing a LEED Certification Plan; (3) monitoring the LEED Certification process; (4) providing LEED specifications for inclusion in the contract documents; and (5) preparing a LEED Certification Report detailing the LEED rating the project achieved. LEED Certification is obtained after submitting an application documenting compliance with the requirements of the LEED rating system.

C. Bearing the Risk: Legal Issues Raised by LEED

Although LEED certification is an emerging trend, the common law has not yet developed as to the nature or extent of liability related to such certification. Potential issues and risks will likely include:

(1) Who warrants the content of materials for LEED certification purposes?

(2) Does LEED certification operate as an affirmative defense to building defect claims after construction?

(3) Is a contractor liable for moisture (or mold) issues related to LEED-mandated mechanical or moisture-control systems?

(4) Who bears the risk of the project not meeting LEED criteria?

(5) Is it enough that the project meet LEED criteria or that it actually be certified by the USGBC?

(6) Who has the authority to decide whether a project meets LEED requirements if not the USGBC?

(7) Who bears the risk of a project not meeting the level of certification required by the owner? What is the remedy?

As with all projects, the participants should evaluate the contract documents carefully to determine the potential risks associated with environmentally friendly Green Building projects. The prudent construction industry professional should pay close attention to any mention of LEED and know its responsibilities, if any, with regard to the LEED certification process.

VI. CONSTRUCTION SAFETY

Despite ever-increasing safety standards and improving safety records, construction remains a dangerous business. Accidents still happen on construction projects, and when one does, everyone on site can be a source of blame. Responsibility typically rests with the injured worker's employer and those in direct contract with the employer, such as subcontractors, general contractors, owners, or others that exercise some degree of control over the employer's activities. Of course, government regulators with enforcement authority over job-site safety also can play an integral role in determining the contractor's liability and risks for construction workplace safety.

A. Sources of Safety Requirements: OSHA-Specific Project Procedures

In the United States, the predominant regulatory scheme governing job-site safety on construction projects is the federal Occupational Safety & Health Act[32] and the corresponding regulations written and enforced by the Occupational Safety and Health Administration.[33] Violation of these regulations may lead to substantial civil and criminal penalties. Although such violations do not constitute negligence as a matter of law, or "negligence per se," they are admissible as evidence of the applicable standard of care in third-party civil suits.[34]

Some states also have enacted safety statutes and other regulations, such as training and licensing requirements, that can affect liability. This often occurs in higher-risk construction operations such as electrical work, underground construction, and environmental remediation. Violations of state regulations also give rise to civil and criminal penalties and can constitute evidence of negligence or establish negligence as a matter of law.[35] Violations of such state regulations even can give rise to private causes of action against the violator.[36]

B. Successfully Working with OSHA

The best way to avoid liability for job-site personal injuries is to prevent accidents from ever happening. Preventing accidents will: (1) reduce exposure to fines and liability under federal and state safety laws, such as those enforced by OSHA; (2) reduce premiums for workers' compensation and liability insurance; and (3) reduce the odds that the company may be liable for injuries that are not covered by indemnification or insurance. Four basic steps to preventing accidents follow.

(1) Every company working a construction site should develop and implement a written safety program, commonly referred to as a safety manual.
(2) Management and employees must be committed to safety and complying with the company's safety manual.
(3) Companies should appoint a person trained or qualified in safety to coordinate the program by conducting inspections and employee training, establishing minimum safety standards, and keeping proper records.
(4) Job foremen should continuously perform project safety inspections and conduct necessary safety training or toolbox safety meetings.

[32] 29 U.S.C. §§ 651–78 (2000).
[33] 29 C.F.R. Parts 70–71, 1900–06, 1908, 1910–13, 1915, 1917–22, 1924–28, 1949, 1952–56, 1977–2200–05, 2400 (2006).
[34] *See, e.g., Charter Oak Fire Ins. Co. v. Nat'l Wholesale Liquidators,* 279 F. Supp. 2d 358, 361 (S.D.N.Y. 2003).
[35] *See, e.g., Larabee v. Triangle Steel, Inc.,* 451 N.Y.S.2d 258 (N.Y. App. Div. 1982).
[36] *See, e.g., R.J. Gaydos Ins. Agency, Inc. v. Nat'l Consumer Ins. Co.,* 773 A.2d 1132 (N.J. 2001).

C. Who Is Responsible for Project Safety?

Because of the numerous different entities involved in virtually every construction project, a complete answer to this question is extremely difficult. Laws such as OSHA regulations often provide part of the answer. The contractual relationships among owners, general contractors, subcontractors, suppliers, construction managers, and other design professionals can shuffle, shift, and ultimately allocate primary responsibility for many recurring safety issues and therefore help to further answer the question. Common law principles, such as tort negligence, also help resolve the issue.

1. Contractors

A contractor, just like anyone else, is liable for any act that directly causes an injury. A contractor also can be liable, even though it does not directly cause an injury. Although one contractor is generally not responsible for acts of an independent contractor,[37] there are many exceptions to this traditional rule of nonliability. For example, under certain circumstances, a general contractor can be held liable for the injuries to a subcontractor's employee even though the employee's injuries were directly caused by the subcontractor's negligence.

Some jurisdictions impose responsibilities on contractors to prevent subcontractor employees from being exposed to hazards not ordinarily found on a construction project. Liability even can be imposed for "ordinary" hazards if the contractor knows, or should know, that the subcontractor employee is failing to protect itself against a known hazard.[38] Other jurisdictions impose liability on contractors for injuries to subcontractor employees where the contractor retains "control" over common work areas or the subcontractor's means and methods.[39] This is especially common when a contractor assumes some type of safety-related supervisory duty, such as providing a safety supervisor, initiating a safety program, or ensuring compliance with safety rules and regulations.[40]

Other courts have predicated liability on the general contractor's safety responsibilities in its contract with the owner.[41] This is troublesome for general contractors, since many owner contracts impose extensive safety obligations on the general contractor. For example, AIA A201 (2007 ed.) Subsections 10.1.1, 10.2.1, 10.2.3, and 10.2.6 address the contractor's safety obligations. Similarly, ConsensusDOCS 200 (2007 ed.) provides this safety provision:

> 3.11 SAFETY OF PERSONS AND PROPERTY
>
> 3.11.1 SAFETY PRECAUTIONS AND PROGRAMS The Contractor shall have overall responsibility for safety precautions and programs in the performance of the Work. While this Paragraph 3.11 establishes the responsibility for safety between the Owner and Contractor, it does not relieve Subcontractors of their responsibility for the safety of

[37]*See, e.g., Cunnington v. Gaub,* 153 P.3d 1 (Mont. 2007).

[38]*See Vega v. Griffiths Constr., Inc.,* 833 P.2d 717 (Ariz. Ct. App. 1992).

[39]*See Cunnington v. Gaub,* 153 P.3d 1 (Mont. 2007).

[40]*See Kamla v. Space Needle Corp.,* 52 P.3d 472 (Wash. 2002).

[41]*See Cochran v. Gehrke, Inc.,* 305 F. Supp. 2d 1045 (N.D. Iowa 2004).

persons or property in the performance of their work, nor for compliance with the provisions of applicable laws and regulations.

3.11.2 The Contractor shall seek to avoid injury, loss or damage to persons or property by taking reasonable steps to protect:

3.11.2.1 its employees and other persons at the Worksite;

3.11.2.2 materials and equipment stored at onsite or offsite locations for use in the Work; and

3.11.2.3 property located at the site and adjacent to Work areas, whether or not the property is part of the Work.

3.11.3 CONTRACTOR'S SAFETY REPRESENTATIVE The Contractor's Worksite Safety Representative shall act as the Contractor's authorized safety representative with a duty to prevent accidents in accordance with Subparagraph 3.11.2. If no individual is identified in this paragraph 3.11, the authorized safety representative shall be the Contractor's Representative. The Contractor shall report immediately in writing to the Owner all recordable accidents and injuries occurring at the Worksite. When the Contractor is required to file an accident report with a public authority, the Contractor shall furnish a copy of the report to the Owner.

3.11.4 The Contractor shall provide the Owner with copies of all notices required of the Contractor by law or regulation. The Contractor's safety program shall comply with the requirements of governmental and quasi-governmental authorities having jurisdiction.

3.11.5 Damage or loss not insured under property insurance which may arise from the Work, to the extent caused by the negligent acts or omissions of the Contractor, or anyone for whose acts the Contractor may be liable, shall be promptly remedied by the Contractor.

3.11.6 If the Owner deems any part of the Work or Worksite unsafe, the Owner, without assuming responsibility for the Contractor's safety program, may require the Contractor to stop performance of the Work or take corrective measures satisfactory to the Owner, or both. If the Contractor does not adopt corrective measures, the Owner may perform them and deduct their cost from the Contract Price. The Contractor agrees to make no claim for damages, for an increase in the Contract Price or for a change in the Contract Time based on the Contractor's compliance with the Owner's reasonable request.

2. Design Professionals

A design professional, such as an architect or engineer, also can be liable for the injuries of someone else's employee even though the design professional is not directly responsible for the employee's injuries. Such liability can be based on either the contractual or voluntary assumption of certain responsibilities.[42]

[42] *See, e.g., Mid-W. Elec., Inc. v. DeWild Grant Reckert & Assocs. Co.,* 500 N.W.2d 250 (S.D. 1993).

Although most construction contracts place primary responsibility for project safety on the general contractor, they also may name the design professional as the owner's representative during construction and require the design professional to inspect or observe the work to ensure compliance with plans and specifications. Often an injured worker will try to establish that the design professional failed adequately to fulfill these duties, resulting in an unsafe work condition that caused the resultant injury.

Most courts, however, hold that design professionals have no safety duties unless such duties are clearly written into their contracts. These courts have found that the design professional's supervisory duties on a project only run to the owner and are just to ensure that work is completed in accordance with the construction documents.[43] Other courts have found a duty to provide safety supervision even though a general contractor may have contractually assumed primary responsibility for project safety. These courts hold that the design professional has a duty to ensure that the contractor complies with its contractual duties.[44] In response to these decisions, design professionals often place provisions in their contracts clarifying their safety responsibilities on a project. An example of such a provision is AIA A201 § 4.2.3 (2007 ed.).

Even though a design professional may not be contractually responsible for project safety, the design professional can be liable for personal injuries on the job by voluntarily assuming extracontractual responsibilities given these factors: (1) the scope of the design professional's actual supervision or control over the work; (2) the extent of the design professional's participation in the project; and (3) whether the design professional assumed responsibility for safety precautions.[45] A design professional also can be held liable for failing to take appropriate action where it actually knows of a safety violation even though it has not contractually or voluntarily assumed safety responsibilities.[46]

AIA A201 § 11.1.4 (2007 ed.) attempts to provide the architect with an additional measure of protection, even where the architect might be liable for project safety. In this regard, that section requires the general contract to name not only the owner but also the architect as an additional insured on the contractor's comprehensive general liability insurance policy. Thus, even where the architect might incur liability for project safety, it would seek to transfer this risk to the contractor and the contractor's insurance provider, under the AIA contract.

3. Construction Managers

Generally, a construction manager is responsible for reviewing safety programs developed by each contractor and coordinating a safety program for the entire project. A construction manager with such responsibilities may be responsible for personal injuries on the job even though the construction manager did not directly cause

[43] *See Nat'l Found. Co. v. Post, Buckley, Schuh & Jernigan, Inc.*, 465 S.E.2d 726 (Ga. Ct. App. 1995); *Frampton v. Dauphin Distrib. Servs. Co.*, 648 A.2d 326 (Pa. Super. Ct. 1994).
[44] *See Herczeg v. Hampton Transp. Mun. Auth.*, 766 A.2d 866 (Pa. Super. Ct. 2001).
[45] *See id.;* Restatement (Second) of Torts § 324A (1965).
[46] *See Frampton v. Dauphin Distrib. Servs. Co.*, 648 A.2d 326 (Pa. Super. Ct. 1994).

the injuries. On at least one occasion, a construction manager contractually obligated to supervise overall project safety was liable for an injury on the project where an employee of the construction manager negligently failed to prevent or observe a subcontractor laying an unsecured plank on a project that caused an injury. The court found the construction manager liable even though an indemnification clause was set forth in the construction manager's contract.[47]

4. Owners

An owner usually delegates responsibility for project safety to an independent contractor, such as a general contractor, design professional, or construction manager. As a general rule, and absent some form of control over an independent contractor's method of operation, the construction project owner is not liable for injuries to the contractor's employees.[48]

There are three recognized exceptions to this general rule: (1) where there is a nondelegable duty based on a contract; (2) where the activity is "inherently or intrinsically dangerous"; and (3) where the owner negligently exercises or retains control over an independent contractor's work.[49] For this last exception to apply, the owner generally must retain some degree of control over the manner in which the independent contractor performs its work. It usually is not enough merely to retain the right to stop the work, to inspect the contractor's progress, or to make suggestions or recommendations that need not be followed. Instead, there must be a retention of supervision or control such that the contractor is not entirely free to do its work in the way it sees fit.[50]

D. Indemnification

One way to avoid exposure to job-site safety liability is to seek contractual indemnification for any losses. Although it may not always be possible to obtain favorable contract language, at the very least, a contract should specifically state the parties' responsibilities for project safety. One will always be responsible for the safety of its own employees. The contract, however, should clearly exclude responsibility for the safety of others on the project.

Agreeing to be contractually responsible for project safety likely will include responsibility for nonemployees injured on the project even though the company did not directly cause their injuries. If a contractor assumed such responsibilities, it may be possible to demand some sort of indemnification agreement from other companies working on the project.

E. Workers' Compensation Statutes

The most typical claim a construction company is likely to encounter will be from a worker injured during construction. The applicable state workers' compensation

[47] *Wausau Bus. Ins. Co. v. Turner Constr. Co.,* 143 F. Supp. 2d 336, 344 (S.D.N.Y. 2001).
[48] *See, e.g., Cunnington v. Gaub,* 153 P.3d 1 (Mont. 2007).
[49] *See, e.g., id.*; Restatement (Second) of Torts § 414 (1965).
[50] *See* Restatement (Second) of Torts § 414 cmt. (1965).

statute will in large part determine liability for such a claim. Every state has some type of workers' compensation statute that requires employers to obtain insurance usually covering medical expenses, disability income, and death benefits. Standard workers' compensation insurance policies usually cover third-party claims against the employer as well. For example, an employee may sue a jackhammer supplier for his or her injuries, claiming that the injury was caused by a defect in the jackhammer. The jackhammer supplier then may try to bring the employer into the lawsuit, alleging that the employee's injuries were not caused by a defect in the jackhammer but by the employer's failure to maintain the jackhammer properly. Such a third-party claim usually is covered by the workers' compensation policy.

Under the applicable workers' compensation statute, the employer will be liable for its employee's injuries regardless of fault. In return, the amount of compensation benefits for which the employer is liable to the employee are limited by the applicable statute. Recovery of statutorily mandated benefits is the employee's exclusive remedy against the employer. The employer can lose statutory immunity from civil suit under two circumstances: (1) the employer's immunity may be stripped if the employer fails to procure the required workers' compensation insurance;[51] and (2) an employer generally will not have any immunity for "willful" or "intentional" conduct. What constitutes willful or intentional conduct varies from state to state. Courts in many states have interpreted this exception as applying only if the employer specifically intended to cause the employee's injuries.[52] Some courts and legislatures, however, have required less than intentional harm, and apply this exception to situations where an employer is aware of a high degree of risk created by the violation of a safety statute.[53]

Finally, as previously mentioned, workers' compensation laws are intended to address payment to employees injured, disabled, or killed in the performance of their job, without regard to liability or fault of the employer. This strict liability scheme may not insulate the construction professional who is in privity of contract with the injured worker's employer. Reliance on workers' compensation laws should not be a substitute for the inclusion of contractual provisions limiting responsibility for project safety and containing reasonable indemnification provisions.

[51]*See, e.g., Griffin v. Baker Petrolite Corp.*, 99 P.3d 262 (Okla. Civ. App. 2004).

[52]*See, e.g., Banes v. Am. Mut. Liab. Ins. Co.*, 544 So. 2d 700 (La. Ct. App. 1989). *See also* David B. Harrison, Annotation, *What Conduct is Willful, Intentional, or Deliberate Within Workmen's Compensation Act Provision Authorizing Tort Action For Such Conduct*, 96 A.L.R. 3d 1064 (1979).

[53]*See, e.g.*, W. VA. CODE § 23–4–2 (2004).

➢ POINTS TO REMEMBER

- Many overlapping federal and state laws create a complex web of rules relating to hazardous materials and substances, water and air quality, environmental risk issues, and workplace safety affecting construction. These laws impose broad liability as well as regulatory and reporting requirements.
- Conduct a prebid environmental review of the contract documents. Assess potential safety or environmental risks for each particular project before entering into a contract.
- Make an environmental prebid inquiry on every project.
- During contract negotiations, be wary of standardized contract clauses that may impose burdensome responsibility and liability for workplace safety or environmental risks. Allocate the risk of encountering unexpected hazardous materials in the contract through specifically drafted clauses.
- Appoint a senior manager or risk assessment team to become thoroughly familiar with federal, state, and local environmental and safety laws and to oversee compliance with these laws.
- Educate the job-site workforce regarding environmental and job-site safety laws and what must be done to comply with them. Make everyone part of the environmental and safety risk management team.
- Develop a comprehensive safety plan and environmental response plan. Employees should be safety conscious and sensitive to environmental risks.
- Evaluate insurance coverage with respect to environmental and job-site safety risks and liabilities.
- Carefully review and understand the rights, obligations, and risks under any applicable indemnity provision.
- Understand the enhanced design and construction issues and potential risks associated with environmentally friendly Green Building construction.
- Understand the workers' compensation laws in each state and ensure that subcontractors and sub-subcontractors actually are insured as required by state laws and regulations.

18

CONSTRUCTION INSURANCE

I. IMPORTANCE OF INSURANCE PLANNING

Insurance is a crucial consideration in construction project planning as a means of covering losses and minimizing disputes among the parties on a project. No party involved in any aspect of the construction industry can afford to neglect or overlook the substantial impact and effects that insurance considerations may have on potential costs, fees, rights, and damages. Such neglect is a mistake that can leave a party exposed to liability for which it was never prepared.

Construction insurance is a highly specialized and complex field, which is often confusing and mysterious to the uninitiated. In addition, in an effort to control and predict insurers' relative liabilities, the insurance industry is constantly changing and revising policy language in reaction to construction industry changes as well as to court decisions interpreting and applying the language of insurance policies that determine the rights and obligations of both the insurers and insured. There is no all-encompassing policy or coverage for any construction project or party to such a project. Instead, a web of policies, insuring agreements, coverages, exclusions, limitations, endorsements, deductibles, policy limits, and excess and umbrella coverages are stitched together among the parties to cover those relatively limited types of risks that are insurable on a construction project. Insurance policies generally begin by providing broad coverage for a variety of risks and potential damages in the insuring agreement found at the beginning of the policies. Upon more careful review, however, policies dramatically scale back purported coverage through exclusions, deductibles, implied exceptions, definitions, endorsements, policy limits, and other tools to limit the insurers' liability.

Insurance planning for construction projects is extremely complex and specialized; therefore, it would be ill-advised to leave such planning to an inexperienced layperson. Even companies that employ full-time insurance specialists or risk managers must further rely on professional insurance brokers to assist in the procurement of the proper coverage at a competitive price.

This chapter presents an overview of (1) the types of insurance that are available and frequently at issue on construction projects, (2) the basic scope and framework of different insurance coverages and exclusions, (3) practical considerations in responding to potential insurance claims, and (4) some recurring coverage issues. The unique terms of individual policies and the peculiarities of individual state laws limit the generalizations that can be made about insurance issues. The broad principles described in this chapter, therefore, cannot substitute for specific reference to the contracts, the insurance policies actually at issue, and applicable state law.

II. TYPES OF INSURANCE

Insurance policies that typically are purchased for construction projects, and therefore are discussed in this chapter, are commercial general liability, builder's risk, and errors and omissions. Many other forms of insurance also may come into play on a construction project, which can range from everyday reliance on workers' compensation insurance to a more unusual claim against a corporate officers' and directors' liability policy. Although some of the general principles described in this chapter may apply to those other forms of insurance, this chapter focuses on the aforementioned three most common forms of insurance relied on in resolving construction-related claims.

A. Commercial General Liability

Commercial general liability (CGL) insurance protects the insured party from injuries it may cause to other parties or property. Coverage for contactors under the CGL policy is limited by exclusions, which embody the well-established business risk doctrine. These business risk exclusions are designed to exclude coverage for defective workmanship by the contractor causing damage to the work itself.[1] The principle behind the business risk exclusion is based on the distinction made between the two kinds of risk incurred by a contractor.[2] First, the contractor bears the business risk of replacing or repairing defective work to make the building or project conform to the agreed contractual requirements. In the presence of the business risk exclusions, this type of risk is not covered by the CGL policy. Second, the contractor bears the business risk that defective or faulty workmanship may cause injury to people or damage to other property. As the liability associated with this risk is potentially limitless, it is the type for which the CGL coverage is contemplated.[3] Accordingly, a CGL policy is "third-party insurance" in that it protects the insured from claims of loss by third parties, but does not protect the insured from losses it directly incurs.

[1]*Bituminous Cas. Corp. v. N. Ins. Co. of N.Y.*, 548 S.E.2d 495 (Ga. Ct. App. 2001).
[2]*Glens Falls Ins. Co. v. Donmac Golf Shaping Co., Inc.*, 417 S.E.2d 197 (Ga. Ct. App. 1992).
[3]*Gary L. Shaw Builders, Inc. v. State Auto Mut. Ins. Co.*, 355 S.E.2d 130 (Ga. Ct. App. 1987).

The insurance industry's rationale for this limited coverage is that business risks that are under the control of the contractor (e.g., additional costs incurred because of inefficient or substandard work), for which the contractor seeks a profit, should not be the subject of insurance. In the insurance industry's view, a different result would essentially make the CGL policy into something that is similar to a performance bond issued by a surety. (See **Chapter 15**.)

1. CGL Insurance: "Occurrence" versus "Claims Made"

A CGL policy can be either an "occurrence policy" or a "claims-made policy"; the difference between the two relates to the timing of coverage. An occurrence policy provides coverage regardless of when the claim actually is made, provided the insured event occurred during the policy period. A claims-made policy is basically the opposite, extending coverage to claims made during the policy period, regardless of when the insured risk occurred.[4]

2. CGL Insurance: Costs of Defense and Deductibles

Another important distinction between types of CGL policies is whether the cost of defense is excluded from the policy coverage limits. If the cost of defense is excluded or is in addition to the policy limits, the costs the insurance company incurs in defending a claim do not erode the policy limits. For example, if there is a $100,000 claim on which the insurance company expends $25,000 to defend, the insured's coverage is still the full $100,000 policy limit. If, however, the cost of defense is included in the policy limits, the liability protection afforded is considerably reduced. In the prior example, the $25,000 in defense costs would reduce the insured's liability coverage to $75,000.

Policy limits define the ceiling of financial protection afforded by insurance, but there is also a floor to that protection, the deductible, which is the loss or expense the insured must absorb before the insurance company has any obligation to pay. The combined effects of a high deductible, policy limits, and included costs of defense can dramatically limit the real financial protection afforded by a CGL policy on many claims.

3. CGL Insurance: Layers of Insurance

CGL coverage often is purchased in multiple layers. The basic coverage is the "primary" policy, with the additional increments of coverage being "excess" or "umbrella" policies. The excess carrier typically will have no duty to defend or indemnify until the primary or other lower coverage is collected, reduced, or exhausted, similar to the deductible requirement. The specific language used in an excess policy to determine where the excess coverage will begin, the so-called attachment point, will control

[4]*See Nat'l Am. Ins. Co. v. Am. Re-Ins. Co.*, 358 F.3d 736, 741 (10th Cir. 2004); *Suter v. Gen. Accident Ins. Co. of Am.*, 424 F. Supp. 2d 781, 785 (D.N.J. 2006).

the events that will trigger that excess coverage. An insured must pay careful attention to the specific terms of those policies.[5] Regardless of when the excess coverage attaches, prompt notice by the insurance to the excess carrier(s) is good practice.

4. CGL Insurance: Additional Insured

The overall impact of CGL coverage on a construction project can be expanded substantially by adding other parties as "additional insureds," which then also enjoy coverage.[6] Historically, it was not uncommon for owners to require the general contractor and subcontractors to include the owner as an additional insured on their CGL policies. Many insurance clauses in contracts now seek to extend additional insured coverage on CGL policies to other members of the construction project delivery team. For example, the AIA A201 (2007 ed.) requires the contractor to add the owner, the architect, *and* the architect's consultants as additional insureds under the contractor's CGL policy.[7]

General contractors, likewise, typically require subcontractors to name the general contractor as an additional insured on the subcontractor's CGL policy. Inclusion as an additional insured not only provides protection to the named party against claims by third parties, it also protects the additional insured from subrogation claims by the insurance carrier, as insurance carriers are not legally permitted to subrogate against their own insured. Subrogation is the theory by which the insurance company, after paying on a covered claim, steps into the shoes of its insured and asserts the insured's rights against a third party that allegedly is responsible for the loss, and is thereby reimbursed for the payment.[8]

Whenever an owner or contractor requires that it be named as an "additional insured," it should confirm its additional insured status by obtaining a copy of each policy rather than by accepting a broker's binder or even a certificate of insurance. The terms of that policy should be reviewed to determine which policy is primary in the event of a claim and to understand the coverage exclusions in that policy.

5. CGL Insurance: Environmental Liability

The potentially far-reaching liability for environmental hazards is an ever-increasing concern in construction. Unfortunately, the protection afforded under a typical CGL policy is significantly limited when compared to the risk. Most CGL policies exclude coverage for all pollution damages except for that which is "sudden and accidental." Policies are available to separately address broader environmental risks, but such

[5] *Axis Surplus Ins. Co. v. Innisfree Hotels, Inc.*, 2006 WL 2882373 (S.D. Ala. 2006); *La. Ins. v. Interstate,* 630 So. 2d 759, 761–62 (La. 1994); *Kelly v. Weil,* 563 So. 2d 221, 222 (La. 1990).

[6] Douglas R. Richmond & Darren S. Black, *Expanding Liability Coverage: Insured Contracts and Additional Insureds*, 44 Drake L. Rev. 781, 790 (1996).

[7] General Conditions of the Contract for Construction, AIA A201, § 11.1.4 (2007 ed.).

[8] *RLI Ins. Co. v. CNA Cas. of Cal.*, 45 Cal. Rptr. 3d 667, 670-71 (Cal. Ct. App. 2006). Applicable state laws should be checked to determine if there are any limitations on the types of coverage which may be provided under "additional insured" clauses.

coverage is costly and far from all-inclusive. For a broader discussion of the issue, especially in the context of mold claims, see **Section VIII** of this chapter.

B. Builder's Risk Insurance

Builder's risk insurance provides coverage for damages to the project during construction. Accordingly, it protects against the insured's loss, as distinguished from compensating the losses of others as with a CGL policy.[9] Builder's risk insurance typically is purchased for large private projects by the owner but often includes coverage extending to the general contractor and subcontractors. The general contractor and subcontractors also have an insurable interest and can procure their own builder's risk coverage.

Builder's risk coverage comes in two similar, but different, types of policies: "all-risk" insurance and "specified" perils. Builder's risk insurance is typically written on an all-risk basis, which covers damage due to any cause not otherwise expressly excluded.[10] The label "all-risk" is essentially a misnomer. All-risk policies are not all-loss policies. All-risk policies contain express written exclusions and implied exceptions developed by the courts over the years.

A specified perils policy, in contrast, provides coverage only for listed risks and damage. In reality, although these two types of policies approach coverage from two extremes, the scope of coverage actually afforded typically is not that different.

As to the timing of an "occurrence" under a builder's risk policy, courts look to the same "occurrence trigger" options as they have with CGL policies. See **Section VI (A)** of this chapter for a more detailed description of the possible "occurrence trigger" tests used in various jurisdictions.

C. Errors and Omissions Insurance

Design professionals are covered for professional liability or malpractice claims under professional liability insurance policies. Such policies frequently are referred to as errors and omissions (E&O) policies.[11] For example, an architect furnishes defective plans to the owner. The building is constructed, and, as a result of the plans, the building contains a defect. Ordinarily, the architect's professional liability policy will pay the owner for its expenses in correcting the building defect. Or the defect may result in a catastrophic loss. Again, the architect's professional liability policy should cover any damages incurred as the result of the loss up to the E&O policy's limit of liability.

[9] *W.N. McMurry Constr. Co. v. Cmty. First Ins., Inc. Wyo.*, 160 P.3d 71, 74 (Wyo. 2007). *See also* Annotation, *Coverage Under Builder's Risk Insurance Policy*, 97 A.L.R. 3d 1270 (1980).

[10] *See Victory Peach Group, Inc. v. Greater N.Y. Mut. Ins. Co.*, 707 A.2d 1383 (N.J. Super. Ct. App. Div. 1998).

[11] *Brown Daltas & Assocs. v. Gen. Accident Ins. Co. of Am.*, 48 F.3d 30 (1st Cir. 1995); *1325 N. Van Buren, LLC v. T-3 Group, Ltd.*, 716 N.W.2d 822 (Wis. 2006).

Professional liability insuring design professionals and CGL policies insuring contractors both cover claims arising from bodily injury and physical damage to property. CGL policies, however, ordinarily do not usually cover economic loss claims. Such claims would include the cost of repairing defective construction or delay damages. Such claims are normally not covered because CGL policies generally only cover bodily injury and property damage claims, and because of the business risk exclusions of the standard CGL policy. Professional liability policies insuring design professionals, however, do cover claims for economic loss as well as claims for bodily injury and property damage.

As in all insurance policies, E&O policies contain a number of exclusions. All E&O policies exclude coverage for dishonest, fraudulent, or criminal acts; employment liability; automobile liability; workers' compensation claims; and liability assumed under contract. Many E&O policies typically exclude coverage for express warranties and guaranties. Most of these exclusions will, however, provide coverage for warranties or guaranties that the work will meet generally accepted architectural or engineering standards or for work that the design professional would otherwise be liable for regardless of the written warranty or guaranty.

Many E&O policies will not cover claims that arise from the activities of related controlled entities. By way of example, a design firm may be affiliated with a contractor. Typically, E&O policies exclude coverage for any claims made against the design professional by a company that the insured operates, manages, or controls, or any entity that is owned more than 50% by the insured. This could be significant to firms doing design-build work that set up a design-build entity which contracts with the design firm.

The cost of E&O insurance is substantial and, like most insurance, ever-increasing. Moreover, many design professionals believe that investing in E&O coverage actually encourages claims rather than protects against them.

Owners are increasingly requiring construction managers to carry professional liability coverage similar to that maintained by design professionals. Similarly, design-build contracts often require contractors to obtain the coverage of an E&O policy or to require the design member of the team to carry such insurance.

D. Alternatives to Traditional Insurance Programs

Although notoriously cyclical, the insurance industry, like any business, is dynamic and adaptive. Policies evolve as underwriters adjust the premiums, terms, and conditions consistent with the perceived risk. Further, underwriting is susceptible to competitive pressures. In addition, new insurance products are regularly rolled out, especially at the upper ranges of premium risk. Insurance is available for contractor and subcontractor default. An owner can procure an insurance policy that will pay liquidated damages. Loss mitigation policies are available in the extreme situation when a risk must be completely transferred from a company's balance sheet. Owner-Controlled Insurance Programs (OCIPs) are constantly evolving in their comprehensive nature of coverage for a particular construction project.

Even when an owner is providing builder's risk coverage, whether through its regular property insurance or through a separate policy purchased just for that specific project, most contractors also have a master builder's risk policy. Many contractors, especially those doing design-build work, also have a professional liability policy, which can cover errors and omissions on the part of the design professional.

The evolution of insurance recently has been most apparent in the reinsurance market, which provides a secondary layer of coverage for primary insurers. Recent catastrophic losses, consolidation, and bankruptcies in the primary insurance market have resulted in significantly tightened underwriting standards in the reinsurance market.

As the commercial insurance market in general has grown more restrictive in its underwriting, so-called insurance captives are being utilized to provide direct insurance and/or reinsurance. These captives allow more direct management of the risk as well as potential savings and rebates. Captives also can provide more expansive underwriting.

III. CONTRACT REQUIREMENTS FOR INSURANCE

Insurance coverage must be planned for each individual project. Although there are certain industry standards regarding which party provides what kind of coverage, insurance is too important and too specialized to assume anything. Each party must review its respective contract on each project to confirm how insurance requirements are addressed. This review is necessary to confirm compliance with the contract and to verify that others are providing the protection required by the terms of the contract. It is important to confirm that those contractually defined requirements are satisfied by all involved. Unexpected gaps in coverage are unpleasant surprises.

A. Standard Contract Clause

The typical insurance arrangement is reflected in the 2007 versions of the AIA A201 and the ConsensusDOCS, both of which are recognized industry form contracts.

Article 11 of the AIA A201 (2007 ed.) makes several changes in the insurance required by the contractor from its 1997 edition, the most notable relating to additional insureds and completed operations coverage. As previously mentioned, the AIA A201 now requires the contractor to add the owner, the architect, and the architect's consultants as additional insureds under the contractor's CGL policy for claims arising from the contractor's negligent acts or omissions during the contractor's operations.[12] The contractor also must obtain completed operations coverage, or coverage for claims arising out of the contractor's negligent acts or omissions after the contractor has completed its work, and name the owner as an additional insured.[13] The endorsement, adding the owner as an additional insured, does not require

[12]General Conditions of the Contract for Construction, AIA A201, § 11.1.4 (2007 ed.).
[13]*Id.*

the contractor's insurance carrier to cover claims arising solely out of the acts or omissions of the owner or architect.

Because AIA A201 now requires the contractor to add the owner, the architect, and the architect's consultants as additional insureds in the contractor's CGL policy, the AIA removed the requirement to provide project management protective liability insurance (PMPLI) to cover the owner's, contractor's, and architect's vicarious liability for construction operations. Since the coverage terms vary between the AIA A201 (2007 ed.) and earlier versions of AIA A201, it becomes ever more important to understand exactly what coverage is required in order to avoid being faced with a contractual requirement for insurance coverage, such as PMPLI, which may no longer be available.

Under AIA A201, the owner also can obtain liability coverage but is not required to do so.[14] The owner is, however, required to obtain builder's risk insurance on an all-risk basis,[15] and this builder's risk protection extends to project subcontractors.

Under Standard Form of Agreement and General Conditions Between Owner and Contractor, ConsensusDOCS 200 (2007 ed.), contractors are not required to purchase and to maintain liability coverage for the owner that is primary to the owner's own liability insurance, but it is an option.[16] If such coverage is mandated, then owners and contractors are given a choice of two options regarding additional liability coverage. Contractors either can purchase an Owners' and Contractors' Protective Liability Insurance (OCP) policy or can name the owner as an additional insured on the contractor's CGL policy.[17] The additional insured coverage is limited to liability caused by the negligent acts or omissions of the contractor or those acting on behalf of the contractor.[18]

Ideally, the general contractor would like the subcontractor's insurance to mirror what the general contractor is required to procure.[19] Such broad coverage for subcontractors, however, often is not practical, necessary, or cost effective. After certain minimums are applied, the specific nature and scope of a subcontractor's work should bear on the extent of the coverage required. If particularly hazardous work is required of the subcontractor, even greater levels of insurance may be necessary.

B. Waiver of Subrogation

Waiver of subrogation rights is an important component in planning insurance coverage on a project. As noted earlier, subrogation is the theory by which the insurance

[14]General Conditions of the Contract for Construction, AIA A201, § 11.2.1 (2007 ed.).

[15]General Conditions of the Contract for Construction, AIA A201, §11.3.1 (2007 ed.); Standard Form of Agreement and General Conditions Between Owner and Contractor, ConsensusDOCS 200, ¶ 10.3.1 (2007 ed.).

[16]Standard Form of Agreement and General Conditions Between Owner and Contractor, ConsensusDOCS 200, ¶ 10.5.1 (2007 ed.).

[17]Standard Form of Agreement and General Conditions Between Owner and Contractor, ConsensusDOCS 200, ¶ 10.5.2 (2007 ed.).

[18]Standard Form of Agreement and General Conditions Between Owner and Contractor, ConsensusDOCS 200, ¶ 10.5.2.1 (2007 ed.).

[19]*See* Standard Form Construction Subcontract, AGC Document No. 650 (1999 ed.).

company, after paying on a claim, can pursue the rights of the insured against the party responsible for the damages. By including a waiver of subrogation clause in a contract, however, the parties can insulate themselves from such subrogation claims and have any potential claims primarily resolved through insurance. Because the waiver of subrogation can substantially affect the insurance company's risk by cutting off its potential rights to recoup losses on claims, the insurance company should be notified of the inclusion of such waiver in a contract to avoid any jeopardizing of coverage.[20]

Both the ConsensusDOCS 200 (2007 ed.) and the AIA A201 (2007 ed.) require the owner and contractor to waive their subrogation rights and require them to obtain similar waivers with other contracts, which, it is hoped, protect all parties on the project from subrogation claims.[21] This waiver of subrogation is also often incorporated into standard subcontracts. In addition to the contractual provisions waiving subrogation, parties effectively waive the right of subrogation through the procurement of insurance that requires all parties to be named as additional insureds. An insurance carrier often is legally barred from subrogating against its own insured.[22]

C. Proof of Insurance

It is not sufficient merely to identify what each party is required to do with respect to insurance coverage. The greatest insurance planning is wasted if there is no execution and follow-through in accordance with the plan and contract requirements. Compliance with insurance requirements should be confirmed by written certificates of insurance describing the coverage and identifying the insurance company. Copies of the actual policies should be obtained, whenever possible. Proof of insurance by all parties should be a contract requirement and, further, should be confirmed as soon as possible before starting work.[23] If a loss occurs before insurance is procured, the ability to get coverage for that event is extremely limited.

Failing to promptly require proof of insurance may do more than leave gaps in coverage. If work has commenced without proof of required insurance coverage, the owner may try to terminate a contractor or subcontractor for default. The contractor, however, might argue that failure to procure insurance is not a material breach justifying termination for default or that the insurance requirement was waived by allowing work to proceed without insurance. In the face of these uncertainties, the better course is to obtain satisfactory proof of coverage before allowing the contractor to start work. These same guidelines should be followed regarding proof of insurance in the contractor/subcontractor relationship. In many contracts the contractor must submit proof of specified insurance as a condition of starting work.

[20]*See Liberty Mut. Ins. Co. v. Altfillisch Constr. Co.*, 139 Cal. Rptr. 91 (Cal. Ct. App. 1977).

[21]Standard Form of Agreement and General Conditions Between Owner and Contractor, ConsensusDOCS 200, ¶ 10.3.3 (2007 ed.); General Conditions of the Contract for Construction, AIA A201,, § 11.3.7 (2007 ed.).

[22]*Reeder v. Reeder*, 348 N.W.2d 832 (Neb. 1984); *Richards v. Allstate Ins. Co.*, 455 S.E.2d 803 (W. Va. 1995).

[23]Standard Form of Agreement and General Conditions Between Owner and Contractor, ConsensusDOCS 200, ¶ 10.3.1 (2007 ed.); General Conditions of the Contract for Construction, AIA A201, § 11.1.3 (2007 ed.).

IV. PROMPT ACTION TO PROTECT POTENTIAL COVERAGE

A. Sensitivity to Insurance Issues

Effective insurance planning not only involves the acquisition of coverage, it also must involve sensitizing project personnel to the potential for coverage of losses and claims as well as the need to react and preserve such potential relief. Project management personnel should be sufficiently trained and experienced to deal affirmatively with insurance notice requirements. At a minimum, project management personnel must know when to alert others with expertise or responsibility for insurance matters when potential coverage issues arise on a project. Ideally, there should be at least one resource person in every construction organization who, through special training and experience, can deal with specific situations and coverage questions beyond the broad concepts introduced in this chapter.

A starting point for dealing effectively with potential coverage issues is obtaining familiarity with all the insurance policies that may apply to the organization as well as the specific project. Reading the policies is essential but, unfortunately, not necessarily enough. Without some background in insurance, much of the policy language affirmatively describing coverage and the litany of exclusions may not make any sense.

B. Immediate Notice

When confronted with a potential loss or claim, the possibility of insurance coverage must be considered first, not last, after all other possible avenues of relief or defense have been exhausted. Prompt notice to the insurance company of a potential claim is critically important to satisfy policy requirements and protect coverage. Early consideration of coverage in dealing with claims or losses is also important in developing the documentation or posture that will enhance an argument in favor of coverage on potentially disputed claims.

If a claim is asserted, the defense of that claim should be affirmatively and immediately tendered to the insurer. Even if the defense is tendered to the insurer, however, the insured should take all immediate actions necessary to preserve any defense or rights available to either the insured or the insurer.

Lack of timely notice to the insurer can be the basis for loss of coverage in some states, even if the insurer experienced no actual prejudice as a result of the timing of the notice.[24] In other states, late tender of the defense will bar recovery of attorneys' fees and cost of defense incurred prior to the tender.

The tender should be as complete and as informative as possible. The insured party should provide copies of demand letters or pleadings as well as other documents that describe the nature and circumstances of the claim. If the basis for coverage is not entirely clear, it is also a good idea to provide documents or an explanation that

[24]*See* Allan D. Windt, Insurance Claims and Disputes: Representation of Insured and Insurers 1–23 (5th ed. 2007); *Wolverine Ins. Co. v. Sorrough,* 177 S.E.2d 819 (Ga. Ct. App. 1970).

highlights or depicts the circumstances that may establish coverage. For example, in tendering a claim under a CGL policy, the property damage or bodily injury of the claimant should be pointed out if it is not obvious.

In preparing any explanation or description that is provided to the insurer, the insured should bear in mind that the coverage may be disputed or that communication may be subject to discovery in subsequent litigation with the claimant. This concern should not preclude an insured party from taking appropriate action or keeping the insurer informed, but it should be a consideration.

V. INSURER'S RESPONSE TO CLAIMS

Upon receipt of any claim, the insurance company must answer two questions: (1) Is the contractor entitled to coverage for the claim? and (2) Does the claim have merit? In deciding whether the contractor is entitled to coverage for the claim, the insurer turns to the policy, to make a coverage determination. Based on its assessment of coverage for the claim, an insurance company has three basic choices in responding to a tender of the defense of a claim: (1) The tender can be accepted without reservation; (2) the tender can be accepted with a reservation of rights disputing coverage; and (3) the tender of defense can be flatly refused.

Acceptance of the defense without reservation by the insurer is obviously the best result. With the exception of deductibles and policy limits, the claim becomes the financial responsibility of the insurance company.[25] The insured still has a duty to cooperate with the insurer in defense of the claim, however, and this can have its own financial and administrative burdens in protracted litigation.

A. Reservation of Rights

If coverage of a claim is subject to some dispute, the insurance company frequently will err in favor of caution by accepting the defense with a "reservation of rights." In making this reservation of rights, the insurer describes why the claim may not be covered and confirms its right to subsequently refuse coverage if the facts or policy language demonstrate no coverage exists. In the meantime, the insurer provides the defense.[26]

Unfortunately, the insurer's reservation of rights can create a conflict of interest with the insured. In defending the claim under a reservation of rights, an insurance company could seek to protect its interest by defeating or settling the claim on the merits or try to push the claim or pursue the defense so that the outcome supports the insurance company's position that there is no coverage; the insured could end up with liability but without coverage.[27] Fortunately, it is generally recognized that counsel

[25]44 Am. Jur. 2d *Insurance* § 1413 (2003); *Henning v. Cont'l Cas. Co.*, 254 F.3d 1291 (11th Cir. 2001).

[26]*W.E. O'Neil Constr. Co. v. Gen. Cas. Co. of Ill.*, 748 N.E.2d 667 (Ill. App. Ct. 2001).

[27]*See also All Am. Ins. Co. v. Broeren Russo Constr., Inc.*, 112 F. Supp. 2d 723 (C.D. Ill. 2000); *Cassey v. Stewart*, 727 So. 2d 655 (La. Ct. App. 1999).

for an insured, even though appointed and paid by an insurance company, cannot put that insurance company's interests ahead of the insured's interests. Nevertheless, if the insured cannot pay the cost for legal counsel, it is a good idea to have the insurance company pay counsel, which the insured should have the right to select under the applicable policy.

B. Litigation with the Insurer

Coverage disputes themselves can be the subject of separate litigation while the underlying claim is also in dispute. Either the insured or the insurance company may initiate an action for declaratory relief against the other to resolve the issue of whether the underlying claim is covered under the policy.[28] Such a lawsuit asks the court to declare the rights of the parties under the applicable policy so they know how to deal with the defense of the underlying claim. If the insurance company denies coverage, however, neither the insurance company nor the contractor may initiate a declaratory action. By the issuance of a denial, there is no longer a controversy as to whether there is coverage. In effect, the insurance company has made a decision and must now live with that decision. At this point, the insured must file suit against the insurer for breach of contract (the contract being the underlying policy) to enforce the policy coverage.

A declaratory relief action can, however, be brought while the insurance company is defending the claim under a reservation of rights. If so, the insurance company may seek to stay any litigation of the underlying claim until the declaratory relief action is concluded and the coverage issue is resolved. The insured typically will want the litigation of the underlying claim to proceed first. As a practical matter, the farther the insurance company gets into the defense, the harder it is to get out, or the more likely the insurance company will participate in some settlement of the underlying claim, which will also invariably include a settlement of the coverage dispute.

Usually the insured has the upper hand on the question of which litigation should go first. In construction cases, there typically are similar or identical factual issues in the declaratory relief action and the underlying claim litigation. As the resolution of the underlying claim will be more likely to resolve those issues, the insured is in a better position to succeed in getting the declaratory relief action stayed. Of course, if the insurance company has not accepted the defense and the insured faces substantial costs of defending the underlying claim, the insured may want the litigation of the underlying claim stayed while coverage is resolved.

VI. ROUTINE COVERAGE ISSUES

The language of specific policies, changes to the standard policy language over time, and the divergent views of courts around the country make it difficult and dangerous to

[28] *See generally State Farm Lloyds v. C.M.W.*, 53 S.W.3d 877 (Tex. App. 2001); *Gallant Ins. Co. v. Wilkerson*, 720 N.E.2d 1223 (Ind. Ct. App. 1999).

presume anything regarding insurance coverage. The application of general principles to specific situations is risky. There are, however, a number of insurance coverage issues that routinely arise, the results of which can be somewhat predicted. These routine coverage issues generally involve the type of occurrences and damages that do not enjoy coverage.

For example, construction defects generally are not covered under either CGL or builder's risk policies. In addition to the specific exclusions that bar coverage for construction defects, public policy favors barring protection for one's own defective work. The contractor has incentive to exercise care in performing its work, and limiting coverage in this manner should reduce insurance costs.[29]

A. CGL Coverage Issues

1. *Continuing Damages*

Construction-related claims often involve a continuing problem rather than one catastrophic instance, such as a leak or other defect that becomes worse or creates more damage over time. Because several policies may have been in effect over the extended duration of the problem and the resulting damages, the question arises of which or how many policies may afford coverage. On claims-made policies, the coverage will be limited to the policy in effect when the first claim was made against the insured, regardless of whether the damages continue to increase after that policy's period. If, however, occurrence policies are involved, the coverage is less clear.

In a CGL "occurrence policy," there still remains the question of when the "occurrence" actually occurs for purposes of triggering coverage. This question generally arises when an insured has several CGL policies spanning a period of time and certain damages began to be incurred during one policy period but were not actually discovered until a later and different policy period. In cases with injuries that evolve over time, there are generally five separate theories for the "trigger" of coverage that courts in different jurisdictions, with differing results, will use to determine which insurers will be liable for the loss. These theories are the: (1) exposure theory; (2) manifestation theory; (3) continuous-trigger theory; (4) injury-in-fact approach; and (5) double-trigger theory.[30]

In the context of property damage claims, with an exposure trigger, coverage is triggered when the injury-producing agent first makes contact with the property.[31] With a manifestation trigger, coverage is triggered only when damage occurs and is discovered, that is, the damage manifests itself as readily obvious, within the policy period.[32] With a continuous trigger, all liability policies in effect from the exposure to manifestation provide coverage and are responsible for the loss.[33] With an injury-in-fact trigger,

[29]*See W. Employers Ins. Co. v. Arciero & Sons, Inc.*, 194 Cal. Rptr. 688 (Cal. Ct. App. 1983).
[30]*Atchison, Topeka & Santa Fe Ry. Co. v. Stonewall Ins. Co.*, 71 P.3d 1097, 1126 (Kan. 2003); *Owens-Ill., Inc. v. United Ins. Co.*, 650 A.2d 974 (N.J. 1994).
[31]*Auto Owners Ins. Co. v. Travelers Cas. & Sur. Co.*, 227 F. Supp. 2d 1248, 1266 (M.D. Fla. 2002).
[32]*James Pest Control, Inc. v. Scottsdale Ins. Co.*, 765 So. 2d 485, 491 (La. Ct. App. 2000).
[33]*Am. Family Mut. Ins. Co. v. Am. Girl, Inc.*, 673 N.W.2d 65, 85 (Wis. 2004).

coverage is triggered when actual injury first occurs.[34] With a double trigger, coverage is triggered both at the time of exposure and at the time of manifestation, but coverage for intervening policies in place between those two times is not triggered.[35]

A further theory provides for "stacking," or multiple, coverage by all policies in effect from the time the injury began until it was discovered, similar to the "continuous-trigger theory." Stacking provides the insured the protection of multiple policy limits.[36] This is an area of frequent dispute and can be addressed only by reference to the terms of the policies and the applicable court decisions.[37]

2. Diminution in Value

Under most CGL policies, physical damage, destruction, or loss of use of tangible property is required for coverage. Tangible property is property that is capable of being handled, touched or physically possessed.[38] A recurring issue in construction disputes is whether installation of defective or substandard work or materials constitutes "physical damage," which is covered, or merely "diminution in value," which is not covered. This is an issue on which apparently very similar facts and operative policy language have resulted in divergent treatment by courts in different jurisdictions.[39]

In one case, coverage was found for the diminished value of a building caused by defective plaster, even though the policy required "injury to or destruction of property."[40] But in another case, in which the policy required "physical injury to or destruction of tangible property," there was no coverage for the costs of replacing defective studs.[41]

3. Completed Operations

"Completed operations" coverage or "products-completed operations hazard" coverage, defined as liability arising for damage or injury incurred after the contractor has completed or abandoned its work, is an important provision that is too often overlooked by the contractor or subcontractor.[42] Failing to obtain or confirm this coverage can prove to

[34]*In re Silicone Implant Ins. Coverage Litigation,* 667 N.W.2d 405, 420 (Minn. 2003).

[35]*Zurich Ins. Co. v. Raymark Indus., Inc.*, 514 N.E.2d 150, 161 (Ill. 1987).

[36]*Cole v. Celotex Corp.*, 599 So. 2d 1058 (La. 1992).

[37]*See generally* Barry L. Bunshoft & Robert L. Seabolt, Contractor's Insurance Coverage Under Its Liability and Builder's Risk Policies, Construction Litigation: Representing The Contractor, 196–97 (2d ed. 1997)).

[38]*State Farm Fire & Cas. Ins. Co. v. White*, 777 F. Supp. 952 (N.D. Ga. 1991).

[39]*See, e.g., Mut. Ben. Group v. Wise M. Bolt Co., Inc.*, 227 F. Supp. 2d 469 (D. Md. 2002); *Structural Bldg. Prods. Corp. v. Bus. Ins. Agency, Inc.*, 722 N.Y.S.2d 559 (N.Y. App. Div. 2001).

[40]*Hauenstein v. St. Paul-Mercury Indem. Co.*, 65 N.W.2d 122 (Minn. 1954); *But see Minneapolis Soc'y. of Fine Arts v. Parker-Klein Assocs. Architects, Inc.*, 354 N.W.2d 816 (Minn. 1984).

[41]*Wyo. Saw Mills v. Transp. Ins. Co.,* 578 P.2d 1253 (Or. 1978); *But see Isspro Inc. v. Globe Indem. Co.*, 106 F.3d 407 (9th Cir. 1997).

[42]*Am. Family Mut. Ins. Co. v. Am. Girl, Inc.*, 673 N.W.2d 65 (Wis. 2004); General Conditions of the Contract for Construction, AIA A201, § 11.1.2 (2007 ed.); Standard Form of Agreement and General Conditions Between Owner and Contractor, ConsensusDOCS 200, ¶ 10.2.4 (2007 ed.).

be expensive by barring all such coverage for damage occurring after the contractor has completed or abandoned its work.

4. *Care, Custody, and Control*

Most CGL policies exclude coverage for damages to the property in the "care, custody, and control" of the insured. Policies vary, however, on whether this exclusion applies to both real and personal property or simply to personal property.[43] Regardless of the precise scope of the exclusion, its purpose is to coordinate coverage between CGL policies, which address liability to third parties, and builder's risk policies, which address the insured's own loss.[44] Because the general contractor generally will be deemed to maintain "care, custody, and control" of the entire project, this exclusion is usually broadly effective against general contractors. Subcontractors, however, typically are performing a discrete scope of work rather than the entire project, so this exclusion may only narrowly apply to subcontractors.

5. *Contractual Liability Exclusion*

Virtually all CGL policies contain an exclusion against contractually assumed liability. This exclusion is intended to avoid coverage for the insured's contractual obligations, including indemnity agreements between the insured and a third party (such as a bond surety), and thereby limit the insurer's coverage to the insured's direct liability.[45]

6. *Real Property Exclusion*

The typical CGL coverage form provides that the insurance does not apply to property damage to that particular part of real property on which the contractor, or subcontractors working directly or indirectly on its behalf, are performing operations, if the property damage arises out of such operations. As this exclusion is intended to preclude coverage for defective workmanship causing injury to the work itself, the cost of repairing these problems is excluded from coverage. Many courts interpreting this exclusion have found it to apply only to ongoing operations and not to completed operations.[46] The exclusion therefore has a temporal element such that the exclusion applies only to damage that was caused when the contractor was working on the property.

7. *Incorrect Performance of Work Exclusion*

Typical CGL policies exclude damage to property that must be restored, repaired, or replaced where the insured incorrectly performed its work. CGL policies typically

[43]Donald M. Zupaned, Annotation, *Scope of Clause Excluding from Contractor's or Similar Liability Policy Damage to Property in Care, Custody, or Control of Insureds*, 8 A.L.R. 4th 563 (1981).

[44]*See Estrin Constr. Co. v. Aetna Cas. & Sur. Co.,* 612 S.W.2d 413 (Mo. Ct. App. 1981). *But see Peters v. Employers Mut. Cas. Co.,* 853 S.W.2d 300 (Mo. 1993).

[45]*Thermo Terratech v. GDC Enviro-Solutions, Inc*., 265 F.3d 329, 336 (5th Cir. 2001).

[46]*Action Auto Stores, Inc. v. United Capitol Ins. Co*., 845 F. Supp. 428, 434 (W.D. Mich. 1993).

state, however, that this exclusion does not apply to property damage included in the "products-completed operations hazard." The products-completed operations hazard provision applies if the work is completed when the damage is incurred.[47] Thus, if any consequential damages arise after the contractor has completed its work, the resulting property damage is included in the products-completed operations hazard.

8. "Your Product" Exclusion

The typical CGL policy provides that the insurance does not apply to property damage to "your product" arising out of it or any part of it. The "damage to your product" exclusion makes clear that the CGL policy provides protection for personal injury or for property damage caused by the completed product, but not for the replacement and repair of that product.[48]

As with the "care, custody, and control" exclusion, the "your product" exclusion can be an effective bar to claims by the general contractor, as the entire project may be deemed the general contractor's product. Courts are split on whether the construction of a building constitutes "your product." Some courts have held that an entire home or building constructed by a contractor was its "product" for purposes of the exclusion.[49] Other courts have decided that the "your product" exclusion does not apply to the construction of a provisional building or site preparations.[50]

9. "Your Work" Exclusion

Perhaps the most hotly contested coverage exclusion is that for damage to "your work." This exclusion eliminates coverage for losses arising from a contractor's poor workmanship in a completed operations context. The exclusion for damage to "your work" in a CGL policy excludes coverage for property damage to the insured's work arising out of it or any part of it and included in the products-completed operations hazard. Unlike the "your product" exclusion, the question of whether the structure qualifies as "your product" is a nonissue. Here the exclusion focuses on the broad application of "your work" without any consideration of the involvement of a "product." Many CGL policies, however, contain what has come to be known as the "subcontractor exception" to the "your work" exclusion. This exception typically provides that the "your work" exclusion does not apply if the damaged work or the work from which the damage arises was performed on the insured's behalf by a

[47]*See Glen Falls Ins. Co. v. Donmac Golf Shaping Co., Inc.*, 417 S.E.2d 197 (Ga. Ct. App. 1992)

[48]*Reliance Ins. Co. v. Povia-Ballantine Corp.*, 738 F. Supp. 523, 526 (S.D. Ga. 1990); *Gary L. Shaw Builders, Inc. v. State Auto. Mut. Ins. Co.*, 355 S.E.2d 130 (Ga. Ct. App. 1987).

[49]E.g., *Taylor-Morely-Simon, Inc. v. Mich. Mut. Ins. Co.*, 645 F. Supp. 596 (E.D. Mo. 1986), *aff'd*, 964 F.2d 789 (8th Cir. 1992).

[50]*See McKellar Dev. of Nevada v. N. Ins. Co. of N.Y.*, 837 P.2d 858 (Nev. 1992); *Md. Cas. Co. v. Reeder*, 270 Cal. Rptr. 719, 727 (Cal. Ct. App. 1990); *Mid-United Contractors, Inc. v. Providence Lloyds Ins. Co.*, 754 S.W.2d 824 (Tex. App. 1988).

subcontractor. Given that the definition of "your work" typically includes not only work performed by the insured but "on your behalf," the subcontractor exception becomes very significant for certain classes of contractors, such as general contractors.

Most courts interpreting the "subcontractor exception" have determined that the provision requires coverage for property damage to the insured's own work when performed on the insured's behalf by a subcontractor.[51] Some courts, however, have reached the opposite conclusion and give almost no force and effect to the subcontractor exclusion.[52]

As of December 2001, the Insurance Services Organization (ISO) offered endorsements that eliminate the subcontractor exception. Specifically, endorsements CG 22 94 and CG 22 95 eliminate the subcontractor exception, one on a blanket basis and one on a project-specific basis. It is crucial for the insured to know whether these endorsements are included in its CGL policy.

B. Builder's Risk Coverage Issues

1. Policy Periods

Builder's risk coverage generally covers damage to the project during construction. The precise time period of coverage, however, is often the subject of dispute. Some builder's risk policies describe the policy period in terms of fixed calendar dates.[53] This level of precision, however, can create problems due to the uncertainties of the construction process and the possibility of delays. For example, if the policy expires on a date certain that is geared to the original contract completion date, delays to the project can result in lack of coverage. This problem can be avoided by extending the policy period before it runs out.

Many builder's risk policies simply describe the policy period as "during construction."[54] In that case, it is well settled that for coverage to commence, some actual, physical construction must have started. Defining completion of "construction," however, is frequently a more difficult task if it has not been defined in the policy or the construction contract. Completion frequently will be found if the project is sufficiently complete to permit the owner to utilize the structure for its intended purpose (commonly referred to as "substantial completion"). Many builder's risk policies expressly provide for termination of coverage when the building becomes occupied.

The AIA A201 (2007 ed.), addresses the potential impact of owner occupancy on builders' risk coverage by precluding the owner's use of the structure without advising and obtaining the consent of the builder's risk insurer.[55]

[51]*Lamar Homes, Inc. v. Mid Continent Cas. Co.,* 501 F.3d 435 (5th Cir. 2007); *U.S. Fire Ins. Co. v. J.S.U.B., Inc.*, 979 So. 2d 871 (Fla. 2007).

[52]*Lassiter Constr. Co. v. Am. States Ins. Co.,* 699 So. 2d 768 (Fla. Dist. Ct. App. 1997).

[53]Annotations, *Coverage Under Builder's Risk Policy*, 97 A.L.R. 3d 1270, § 4 (1980).

[54]*Id.*

[55]General Conditions of the Contract for Construction, AIA A201, § 11.3.1.5 (2007 ed.).

2. *Fortuitous Loss*

A fundamental aspect of builder's risk coverage is that it protects only against "fortuitous" losses. This means the loss must be a matter of chance and not something that is certain to occur, such as depreciation or ordinary wear and tear.[56] Losses arising from the insured's deliberate conduct are likewise excluded.[57] Losses from the negligence of the insured or its employees or subcontractors, however, are considered fortuitous, as they are not intended by the insured.[58]

Some builder's risk policies require that losses result from an external cause in order to establish coverage.[59] In general, the so-called external cause requirement excludes damages arising from an inherent defect in the property. This limitation can be viewed as another version of the requirement for a fortuitous loss.[60]

3. *Exclusions*

In addition to the coverage issues addressed, many builder's risk policies also contain express exclusions that bar claims for defective design, faulty workmanship and material, inherent defects, equipment breakdown, and earth movement.[61] The prudent contractor should compare its contract requirements for builders' risk coverage. If there are such gaps, then the contractor needs to address these in its underlying contract for the work or attempt to secure additional insurance, if available, to cover these risks.

VII. CONCURRENT CAUSES

Losses on construction projects often arise from a combination of forces. Some policies specifically exclude coverage in cases of multiple or concurrent causation.[62] If the policies do not address the issue of multiple causes, often the courts will be asked to decide this issue. Courts are divided on how to resolve the question of coverage in light of concurrent causation. At least one court has held that if at least one cause is not excluded, coverage exists.[63]

Most courts, however, use the "efficient proximate cause rule" to resolve coverage issues involving the concurrence of covered and excluded perils. The efficient proximate

[56] *City of Burlington v. Indem. Ins. Co. of N. Am.*, 332 F.3d 38 (2d Cir. 2003); *Johnson Press of Am., Inc. v. N. Ins. Co. of N.Y.*, 791 N.E.2d 1291 (Ill. App. Ct. 2003).

[57] CAL. INS. CODE § 533 (1935).

[58] *J. Aron & Co. v. Chown*, 647 N.Y.S.2d 8 (N.Y. App. Div. 1996). *See also C.H. Leavell & Co. v. Fireman's Fund Ins. Co.*, 372 F.2d 784 (9th Cir. 1967), *But see Trinity Indus., Inc. v. Ins. Co. of N. Am.*, 916 F.2d 267 (5th Cir. 1990).

[59] *Wolstein v. Yorkshire Ins. Co., Ltd.*, 985 P.2d 400 (Wash. Ct. App. 1999).

[60] *See Standard Structural Steel v. Bethlehem Steel Corp.*, 597 F. Supp. 164, 191–93 (D. Conn. 1984).

[61] *Swire Pac. Holdings, Inc. v. Zurich Ins. Co.*, 845 So. 2d 161 (Fla. 2003).

[62] *See* 43 Am. Jur. 2d *Insurance* § 480 (2003).

[63] *Phillips Home Builders, Inc. v. Travelers Ins. Co.*, 700 A.2d 127 (Del. 1997).

cause rule applies when two or more identifiable causes contribute to a single property loss, at least one of them covered under the policy and at least one of them excluded under the policy.[64] Under this approach, which is the prevailing rule in a majority of jurisdictions today, if multiple concurrent causes exist, and if the dominant or predominant cause is a covered peril, then coverage would exist for the entire loss, even though other concurrent causes are not covered under the policy. If the dominant or predominant cause is an excluded peril, however, coverage would not exist for the entire loss, even though other concurrent causes are covered by the policy.

VIII. CONSTRUCTION INSURANCE AND MOLD CLAIMS

The construction industry continues to experience a dramatic and substantial rise in both the volume of and costs associated with litigation relating to the presence of what is commonly and collectively referred to as mold. Consequently, construction insurers, including CGL (third-party) insurers, Builder's Risk (first-party) insurers, and Errors and Omissions (third-party) insurers, have sought interpretation of certain policy exclusions in a manner that would insulate them from liability for the vast potential damages associated with mold, encountering varying results depending on the specific language of the policy and the jurisdiction interpreting the same.

A. The Insurance Industry Responds

Before the mid-1980s, most pollution exclusion language contained in construction insurance policies provided an exception to the exclusion for "sudden and accidental" releases of pollutants. Many courts at that time accepted the insurance industry's arguments that, since environmental injury had to be both "sudden" and "accidental" in order to be covered, gradual release or development of mold causing damage was not covered. Many courts, however, began to hold the language of "sudden" to be ambiguous and thereby broadened coverage for environmental injury in favor of the insureds. The insurance industry responded to such unfavorable rulings with a new and, presumably, stronger exclusion for mold liability.

In the mid-1980s, the so-called absolute pollution exclusion, containing no exception for "sudden and accidental" injury, was introduced and quickly became standard policy language throughout the insurance industry.[65] Generally, such provisions purported to broadly exclude coverage for injury related to the "discharge, dispersal, seepage, migration, release, or escape" of a "pollutant." Since that time, jurisdictions have split as to

[64]*See Burgess v. Allstate Ins. Co.*, 334 F. Supp. 2d 1351 (N.D. Ga. 2003); *Kelly v. Farmers Ins. Co.*, 281 F. Supp. 2d 1290 (W.D. Okla. 2003).

[65]William H. Howard, *2003 Claim Trends & Litigation, Asbestos, Environmental and Mold Claims, Litigation and Administrative Practice Course Handbook Series: Insurance Coverage,* Practicing Law Inst., 166 (2003).

whether this exclusionary language is broadly or narrowly construed and, thus, whether it will or will not exclude damages arising out of the presence environmental agents, including mold. According to one court,

> [t]he scope of the [absolute] pollution exclusion is an evolving area of law, subject to differing interpretations. The [absolute] pollution exclusion is one of the most frequently litigated exceptions found in a staple insurance industry product. . . . Cases may be found for and against every issue any litigant has ever raised, and often the cases reaching the same conclusion as to a particular issue do so on the basis of differing, and sometimes inconsistent, rationales.[66]

Overall, it is difficult to make a general statement regarding the viability of the absolute pollution exclusions in mold cases, as the states vary in their application of the exclusions. Ultimately, it is essential to look closely at the law of the state controlling the interpretation of the policy before any coverage opinion may be rendered.

B. Mold Claims: Establishing whether Coverage is Available

1. Pollutant

The first general question in considering coverage for mold under a policy containing an absolute pollution exclusion is whether mold is considered a "pollutant" as provided in such exclusions. As a practical matter, mold is a natural substance and is not similar to the typical industrial chemicals traditionally considered and recognized as pollutants. Insurance companies often counter with the argument that the question of whether mold is a pollutant should be resolved by considering whether the injurious agent at issue (mold) is an irritant or a contaminant, is toxic, or poses a health hazard due to exposure.[67] In the end, courts that have interpreted the pollution exclusion in this context have reached varying conclusions.[68]

2. Discharge, Dispersal, Seepage, Migration, Release, or Escape

The second general question in considering coverage for mold under a policy containing an absolute pollution exclusion is whether the environmental injury at issue would qualify as a "discharge, dispersal, seepage, migration, release, or escape" as provided in such exclusions. Some courts have held that this language requires "movement" or an "active and clearly perceived physical event."[69] In such instances,

[66]*Porterfield v. Audubon,* 856 So. 2d 789, 800 (Ala. 2003), *quoting Kemper Nat'l Ins. Cos. v. Heaven Hills Distilleries, Inc.,* 82 S.W.3d 869 (Ky. 2002).

[67]*Porterfield v. Audubon Indem. Co.,* 856 So. 2d 789, 801 (Ala. 2003); *Anderson v. Highland House Co.,* 757 N.E.2d 329, 332 (Ohio 2001).

[68]*See, e.g., Cooper v. Am. Family Mut. Ins. Co.,* 184 F. Supp. 2d 960 (D. Ariz. 2002).

[69]*Sphere Drake Ins. Co. v. Y.L. Realty Co.,* 990 F. Supp. 240, 243 (S.D.N.Y. 1997); *Byrd ex rel. Byrd. v. Blumenreich*, 722 A.2d 598 (N.J. Super. Ct. App. Div. 1999).

courts have found certain environmental injuries that are "continual, imperceptible, and inevitable" as covered.[70] Further, some courts will find in favor of an insured if the court deems such language in an absolute exclusion to be "ambiguous" as to what damage is excluded.[71]

In other jurisdictions, however, courts have found the "transmission of . . . mold spores" constitutes a "release" or other action provided in the absolute pollution exclusion and, therefore, is not covered under the applicable policy.[72]

3. *Efficient Proximate Cause*

Some jurisdictions have applied the so-called efficient proximate cause rule to conclude that coverage exists when an insured can "identify an insured peril as the proximate cause of the loss even if subsequent or concurrent events are specifically excluded from coverage."[73] Therefore, in such jurisdictions, if mold is caused by water leakage that, in turn, was originally caused by a covered injury, the resulting mold will be deemed covered, regardless of an excluded intervening injury.

Other jurisdictions have ignored the "efficient proximate cause" rule and held "there is no coverage for losses caused by mold, even though a covered water event may have [caused] the loss."[74]

4. *Reasonable Expectations*

Some jurisdictions, in reading any ambiguity in liability exclusions against the insurer and in favor of the insured, look to the "reasonable expectations" of the insured, instead of the technical definitions of the language contained in the policy. If, based on the purpose of the policy and the language at issue, the insured could have "reasonably expected" that its policy would cover mold damages, such courts will find coverage.[75]

Other jurisdictions, however, have found the absolute pollution exclusion is not ambiguous, and, therefore, the insured is not entitled to such a favorable interpretation.[76] Such jurisdictions have held that the absolute pollution exclusion "exclude[s] coverage for all claims alleging damage cause[d] by pollutants."[77] Further, some courts will narrowly evaluate whether the insured's expectations were "reasonable," and "a subjective belief developed after the loss [will be deemed] insufficient to create coverage where none exists under the policy."[78]

[70]*Lititz Mut. v. Steely*, 785 A.2d 975, 981–82 (Pa. 2001).

[71]*Cincinnati Ins. Co. v. Becker Warehouse, Inc.*, 635 N.W.2d 112, 118–19 (Neb. 2001).

[72]*Lexington Ins. Co. v. Unity/Waterford-Fair Oaks, Ltd.*, 2002 WL 356756 (N.D. Tex. 2002); *W. Am. Ins. Co. v. Band & Desenberg*, 925 F. Supp. 758, 761 (M.D. Fla. 1996).

[73]*Bowers v. Farmers Ins. Exch.*, 991 P.2d 734, 738 (Wash. Ct. App. 2000).

[74]*Cooper v. Am. Family Mut. Ins. Co.*, 184 F. Supp. 2d 960, 962 (D. Ariz. 2002).

[75]*Anderson v. Highland House Co.*, 757 N.E.2d 329, 332 (Ohio 2001); *Reg'l Bank of Colo. v. St. Paul Fire and Marine Ins. Co.*, 35 F.3d 494, 498 (10th Cir. 1994).

[76]*Cincinnati Ins. Co. v. Becker Warehouse, Inc.*, 635 N.W.2d 112, 118–19 (Neb. 2001).

[77]*Id.*

[78]*Darner Motor Sales v. Universal Underwriters Ins. Co.*, 682 P.2d 388, 395 (Ariz. 1984).

5. Specific Mold Exclusions

As mentioned earlier, the insurance industry takes notice of varying courts' interpretations of policy language and responds to protect its interests. Accordingly, due to the rise in volume and cost of mold claims, many insurers have begun to include additional policy exclusions that expressly and specifically address mold and mold-related damage. The mold exclusion typically bars claims arising from "rust, rot, mold or other fungi." Although an insured can obtain additional coverage in such circumstances that would provide mold coverage, such additional coverage is often cost prohibitive.

As with all insurance-related issues, it is important to read and know the specific language of the applicable insurance policy as well as the law in the relevant jurisdiction. Slight variations in either of these circumstances can have drastic and costly repercussions when determining who must pay for damage.

➢ POINTS TO REMEMBER

INSURANCE PLANNING

- Planning is crucial for avoiding loss and limiting disputes.
- It requires special training and experience.
- The assistance of an insurance professional who is familiar with construction is necessary.

TYPES OF INSURANCE

Commercial General Liability Insurance

- CGL provides protection against claims by third parties and not the insured's own losses.
- A CGL policy can be a "claims-made" or "occurrence" policy; which type it is determines the timing of coverage.
- If costs of defense are not excluded from the policy limit, the financial protection afforded can be greatly reduced.
- Liability insurance can be purchased in multiple layers. Prompt notice of potential claims is required at all layers.
- Inclusion of others on a CGL policy as additional insureds extends protection and eliminates subrogation claims.
- CGL coverage affords extremely limited protection against environmental liability.

Builder's Risk Insurance

- Builder's risk coverage is first-party insurance; it protects against the insured's loss rather than against the claim of third parties.

- Builder's risk coverage typically is procured by the owner, with protection extended to the contractor and subcontractors.
- Contractors and subcontractors can obtain their own builder's risk coverage.

Errors and Omissions Insurance

- E&O insurance is professional liability insurance for design professionals.
- Similar insurance is increasingly required for professional construction managers.
- E&O coverage by the design professional is not always a contract requirement.

Contract Requirements for Insurance

- Never assume insurance coverage; always review the project's contract documents to confirm how it is set up, and what coverage is specified.
- Typically, the contractor procures CGL coverage; the owner procures builder's risk insurance.
- Most construction contracts include "waiver of subrogation" clauses that preclude the insurer from trying to recoup its losses from others on the project.
- It is imperative to require proof of insurance and to confirm that the parties have complied with the contract's insurance requirements before they start work.

Prompt Action to Protect Potential Coverage

- Project management personnel should be educated to deal with potential insurance issues and refer them to an appropriate individual with special training in insurance.
- Prompt notice of potential claims should be made to the insurance company to avoid the risk of losing coverage.
- The notice or tender of defense should be as complete and informative as possible.

The Insurance Company's Response to Claims

- The insurance company often will accept the defense of a claim with a reservation of rights to later deny coverage.
- In that situation, a potential conflict of interest may exist that requires the insured to retain independent counsel, hopefully at the insurance company's expense.
- Disputes over coverage may have to be resolved through separate litigation, a declaratory judgment action, between the insured and the insurance company.

Routine Coverage Issues

- Construction defects generally are not covered by CGL or builder's risk insurance.
- CGL coverage issues:
  - A claim resulting in continuing damages can raise the question of which of several policies apply.

  - CGL policies generally require physical damage or destruction or loss of use of tangible property for coverage so that diminished value is not covered; however, the cases conflict as to what constitutes diminished value.
  - Completed operations cover liability for claims arising after the contractor completes its work; it is often, but not always, part of a CGL policy.
  - The exclusion for property under the "care, custody, and control" of the insured in effect extends to the entire project for the general contractor but may be more limited for subcontractors.
  - The "real property" exclusion to the CGL policy precludes coverage for the costs of repairing or replacing real property on which a contractor or subcontractor is performing operations.
  - The "incorrect performance" exclusion to the CGL policy precludes coverage for damage to property that must be restored, repaired, or replaced because the insured incorrectly performed work on it. This exclusion does not apply to property damage included in the "products-completed operations hazard."
  - The "damage to your product" exclusion makes clear that the CGL policy provides protection for personal injury or for property damage caused by the completed product but not for the replacement and repair of that product.
  - The exclusion for damage to "your work" in a CGL policy excludes coverage for property damage to the insured's work arising out of it or any part of it and included in the products-completed operations hazard. Many CGL policies, however, contain a "subcontractor exception" to the "your work" exclusion.
  - CGL policies exclude contractually assumed liability that does not involve the insured's negligence.

- Builder's risk coverage issues:
  - The policy period extends to a specific date or to when the project is completed, which is often disputed.
  - Builder's risk coverage extends only to "fortuitous" losses, not that which can be expected, such as ordinary wear and tear.
  - The loss also must be the result of some external cause, not an inherent defect in the property.
  - The existence of concurrent causes contributing to a loss may impact coverage.

Mold-Related Issues

- Courts have widely split as to whether certain language in construction insurance policies will exclude damages arising out of the presence environmental agents, including mold.
- Courts have been fairly consistent in finding that mold is a "pollutant" in the context of the absolute pollution exclusion.

- Courts have split as to whether mold constitutes the type of "discharge, dispersal, seepage, migration, release, or escape" that will be excluded under the absolute pollution exclusion.
- Courts have split as to whether the "efficient proximate cause" rule will establish that if a covered damage eventually leads to mold, the resulting damage will be covered, regardless of an excluded intervening injury.
- Courts have split as to whether ambiguity exists in the absolute pollution exclusion, and, therefore, a court can look to the parties' "reasonable expectations" as to coverage of mold instead of the technical definitions of the language contained in the policy.
- Many insurers have begun to include additional policy exclusions that expressly and specifically address mold and mold-related damage.

19

LABOR AND EMPLOYMENT ISSUES AFFECTING THE CONSTRUCTION INDUSTRY

In operating within today's labor market, construction industry employers are faced with the government's enforcement of employment laws in various arenas and the looming possibility that their employees may institute legal proceedings to address perceived concerns about their working conditions. Workplace laws and regulations govern all aspects of the employment relationship including immigration, wage payment, hiring, safety, conduct in the workplace, and termination. Accordingly, government agencies and employees have myriad federal, state, and local laws to rely on when seeking redress to concerns about employment decisions or working environments. To protect themselves from the risks associated with defending against the variety of infractions and claims that can be asserted, employers must be aware of the duties imposed on them by employment laws, and implement workplace policies that satisfy these obligations.

This chapter provides an overview of and discusses the most common issues facing construction industry employers. Due to the variety of state and local laws that construction employers may face as a result of the location of a specific project, this chapter focuses primarily on applicable federal statutes and regulations. Keep in mind, however, that many states have enacted legislation that may provide coverage similar to the federal laws described herein. Moreover, in some instances, state laws provide employees with greater protection than that afforded by the federal statutes.

I. IMMIGRATION ISSUES

The public debate surrounding both legal and illegal immigration and employment has reached a fever pitch. Some estimates calculate the number of undocumented aliens in the United States at above 12 million. Given this large number of undocumented workers, groups on both sides of the debate currently are pushing both state and federal legislatures to take action on the immigration issue or to enforce existing laws more vigorously. Regardless of where one stands on this issue politically, the construction industry employer must be sensitive to this hot-button issue and the potential pitfalls of employing non-U.S. citizen workers.

Congress passed the Immigration Reform and Control Act of 1986 (IRCA) in an effort to curb the flow of illegal immigrants into the United States.[1] The IRCA imposes a dual burden on employers: (1) Employers must ensure the lawful right to work of all applicants; and (2) in confirming the lawful right to work, employers must not discriminate on the basis of the applicant's race or national origin.[2] Violations of the IRCA can result in fines against employers and even jail time for the violators.[3]

To verify an applicant's right to work, employers must check documents specified by the Immigration and Naturalization Service (INS) and the Employment Eligibility Verification Form (I-9). The I-9 requires employers to check both the identity of the applicant and the right to work. Generally, identity may be proven by presenting certain types of government-issued identification, and the right to work may be proven by presenting documents, including a Social Security card and a birth certificate. Some documents, such as a passport, establish both identity and eligibility under the IRCA.[4] When an employer complies with the I-9 verification requirements, it may assert as a defense to any unlawful employment action that it did not know that the applicant was not, in fact, eligible for employment.[5]

A. Immigration Documentation and Control

1. Form I-9

As noted, the central tool used by the government to cause employers to verify the citizenship or work status of employees the is I-9. The Department of Homeland Security, U.S. Citizenship and Immigration Services (USCIS) issues the I-9. This form, however, should not be filed with U.S. Immigrations and Customs Enforcement (ICE) or USCIS. Instead, the employer must keep Form I-9 on each employee either for three years after the date of hire or for one year after employment is terminated, whichever is later. The form must be available for inspection by authorized U.S. Government officials (e.g., Department of Homeland Security, Department of Labor, Office of Special Counsel).

Currently, an employer cannot affirmatively verify the employee's status with USCIS without joining the Department of Homeland Security's "Basic Pilot" program, which utilizes automated employee verification systems to confirm the work authorization of a newly hired employee. An employer who finds out that its

[1]8 U.S.C. § 1324.
[2]*Id.*
[3]8 U.S.C. § 1324a.
[4]8 U.S.C. § 1324a(b)(1)(B).
[5]8 U.S.C. § 1324a (a)(3).

employee has been working without the proper authorization (or whose authorization documents have expired), however, should immediately take steps to reverify the employee's work authorization by demanding the requisite documents to complete a new Form I-9. Absent such a reverification, an employer that knows or should know that its employee is working without the appropriate authorization may be subject to serious penalties for "knowingly continuing to employ" an unauthorized worker.[6]

In November 2007, the USCIS released a revised Form I-9, which must now be used for all new employees. Every U.S. employer is required to complete a Form I-9 for each employee hired within the United States. The new I-9 form has been applauded by many commentators as adding clarity to the immigration documentation process; however, the improved clarity and efficiency of the new form are directed at enhancing enforcment of the immigration laws and not necessarily on easing the burden on employers.[7]

Essentially, the revised Form I-9 is similar to the earlier versions of the document. The new form, however, includes key revisions to the documentation process. These changes mainly focus on acceptable forms of identification for proof of lawful work status. For instance, in the new Form I-9, five types of immigration documents are no longer considered "Acceptable Documents" to establish proof of lawful work status. These are:

(1) Alien Registration Receipt Card (I-151)
(2) Unexpired Re-entry Permit (Form I-327)
(3) Unexpired Refugee Travel Document (Form I-571)
(4) Certificate of U.S. Citizenship (Form N-560 or N-561)
(5) Certificate of Naturalization (Form N-550 or N-570)

Other changes to Form I-9 include the addition of one type of immigration document to List A of the List of Acceptable Documents: Unexpired Employment Authorization Document (I-766). Further, all Employment Authorization Documents with photographs are now included as one item on List A: Forms I-688, I-688A, I-688B, and I-766. The new Form I-9 became effective on December 7, 2007.

2. *E-Verify*

On June 6, 2008 President Bush issued an executive order entitled "Economy and Efficiency in Government Procurement through Compliance with Certain Immigration and Nationality Act Provisions and Use of an Electronic Employment Eligibility Verification System". In essence, this executive order directed the adoption of a new provision in the Federal Acquisition Regulation (FAR) to require certain contractors

[6]*See New El Rey Sausage Co., Inc. v. U.S. Immigration & Naturalization Servx.*, 925 F.2d 1153 (9th Cir. 1991).
[7]*See* Michael Newman and Shane Crase, *Challenges to the 'No-Match Rule,'* 55 Fed. Law 16, 17–18 (2008).

and subcontractors involved in the performance of federal government contracts to utilize the E-Verify System to verify the employment eligibility of certain categories of employees.

Following that executive order, the FAR Councils issued a proposed regulation on June 12, 2008. (See 73 FR 33374, June 12, 2008). The proposed FAR provision contains the following key sections:

- Requires the insertion of a new clause in all federal government prime contracts except those that do not exceed the micro-purchase threshold (usually $3,000.00) or those for commercially available off-the-shelf items.
- Requires the contractor to enroll in the United States Citizenship and Immigration Service's E-Verify System within 30 days of contract award.
- Requires the use of the E-Verify System to verify employment eligibility of **all new** employees that are hired after enrollment in the system.
- Requires the use of the E-Verify System to confirm the employment eligibility of **all existing** employees who are directly engaged in the performance of the covered federal government contract.
- Requires that prime contractors flow down a clause to all subcontracts for construction in excess of $3,000.00 that include work performed in the United States to impose comparable requirements on the subcontractors.

The proposed FAR clause will not require a contractor to perform an additional employment verification on employees who were previously verified through the E-Verify program. Nothing in the proposed FAR clause would appear to eliminate or preclude base or installation specific background checks on employees before such employee is allowed on the installation or federal government facility.

B. Responses to Immigration Inquiries

One of the most common inquiries as to immigration documentation that most construction industry employers will face is the "no-match letter" generated by the Social Securty Administration (SSA). The Department of Homeland Security's (DHS) final "no-match" regulation tool is designed to make it harder for illegal immigrants to use false Social Security numbers to get a job.[8] Letters are to be sent from the SSA to notify businesses that an employee's reported name and Social Security number do not match official government records. Similar "Notice of Suspect Document" may also be sent out by ICE to address other inconsistencies with the immigration status or employment-authorization documents of an employee.

Upon receiving a no-match letter, an employer must follow "reasonable steps" outlined by DHS in order to qualify for "safe harbor" protection from fines and penalties. Within 30 days of receiving the no-match letter, the employer must check its

[8]*See* 8 C.F.R. § 274a; *see also* Safe Harbor Procedures for Employers Who Receive a No-Match Letter, 72 Fed. Reg. 45611–45624. (proposed Aug. 15, 2007).

records to ensure there are no transcription errors in the records sent to the SSA or DHS. Whether there is an error or not, such information must then be verified or reverified with the relevant agency. A "reasonable employer" should keep records of the manner, date, and time of verification.[9] If such actions do not resolve the discrepancy, the employer must reverify the information with the employee. If the information originally submitted is not correct, the employer must then take steps to correct it. If, according to the employee, the information is correct, the employer should ask the employee to pursue the matter personally with the relevant agency."[10] As such, when the employer has verified all the information it can, the burden is then on the employee to resolve the issue; however, an employer should not continue to employ individuals with false or suspect documents that the employee cannot or will not verify.

In the event of a violation of the immigrant hiring rules, ICE is authorized to investigate whether an employer has violated the prohibitions against knowingly employing unauthorized alien(s) and failing to properly complete, present, or retain the Form I-9 for newly hired workers. If ICE believes that a violation has occurred, it may issue a warning notice or a technical or procedural failures letter notifying the employer of technical or procedural failures in need of correction, or a notice of intent to fine (NIF).[11] If the ICE issues a NIF, the employer may request a hearing within 30 days of service of the NIF to contest the notice before an administrative law judge of the Office of the Chief Administrative Hearing Officer (OCAHO), Executive Office for Immigration Review, U.S. Department of Justice.[12] Hearing requests must be in writing and filed with the ICE office designated in the NIF. If a hearing is not requested within the 30-day period, ICE will issue a final order to cease and desist and to pay a civil money penalty. Once a final order is issued, the penalty is unappealable. If the employer requests a hearing, ICE will file a complaint with OCAHO to begin the administrative hearing process, which may end in settlement, dismissal, or a final order for civil money penalties.

C. Immigration Issues in Public Contracting

Any company performing federal government contracting work should review its immigration documentation procedures to ensure they will meet current laws and regulations governing the employment of immigrant labor—what will likely be an increasingly high standard. In addition to federal government contracting laws and regulations affecting the employment of immigrants, many states and political subdivisions of those states are enacting such laws applicable to contracts with state or local governments.[13] Essentially, most of these state laws serve to bar from public contracts contractors that knowingly employ or contract with an illegal alien to

[9]*See* 8 C.F.R. § 274a.2.
[10]*Id.*
[11]*See* 8 C.F.R. § 274a.9.
[12]*See id.*
[13]*See, e.g.*, COLO. REV. STAT. ANN. § 8–17.5–101-02 (West 2008).

perform work under the contract, or that knowingly contract with a subcontractor that knowingly employs or contracts with an illegal alien to perform work under the contract. Further, these laws generally require the contractor to certify compliance with the relevant regulations and take action to verify the employee's worker status that may go beyond the federal requirements.

II. EMPLOYEE SAFETY AND HEALTH

Congress enacted the Occupational Safety and Health Act[14] (OSHA) in 1970 to "regulate commerce among the several States and with foreign nations and to provide for the general welfare, to assure so far as possible every working man and woman in the Nation safe and healthful working conditions and to preserve our human resources."[15] In light of the potential hazards that are often present on construction sites, OSHA has particular significance to the construction industry. Accordingly, to avoid civil and criminal liability, construction industry employers must know of their responsibilities under the applicable OSHA provisions.

OSHA requires employers to comply with certain safety standards and furnish a work environment for employees that is "free from recognized hazards that are causing or are likely to cause death or serious physical harm to his employees." Employer liability for OSHA violations could potentially include injunctions as well as civil and criminal penalties ranging from $5,000 to $70,000 per violation, depending on the severity of each violation, and imprisonment.[16]

Employers have dual responsibilities under OSHA. Principally, employers must follow codified regulations regarding unique aspects of their respective work environments.[17] In addition to the regulatory guidelines, however, OSHA also imposes a "general duty" on employers to maintain a safe and healthful work environment by eliminating otherwise unregulated working conditions that may be hazardous to the health or safety of employees.[18] OSHA's coverage extends to all persons engaged in a business affecting commerce who have employees but excludes the United States or any state or political subdivision.[19] The definition of a business affecting commerce is extremely broad, and employers engaged in the construction industry fall under OSHA's coverage.

OSHA empowers the secretary of labor (Secretary) with the responsibility of implementing safety standards through rule-making proceedings (or regulations).[20] The Secretary is also responsible for conducting on-site inspections to ensure employer compliance with OSHA requirements.[21] Compliance checks can

[14]29 U.S.C. § 651 *et seq.*
[15]29 U.S.C. § 651(b).
[16]29 U.S.C. § 666.
[17]29 U.S.C. § 654(2).
[18]29 U.S.C. § 654(1).
[19]29 U.S.C. § 652(5).
[20]29 U.S.C. § 655.
[21]29 U.S.C. § 657.

be initiated as a result of routine inspections or employee complaints. The Secretary has the authority to obtain a warrant for inspection if the employer refuses to allow inspectors access to the facility. If the employer receives a citation for alleged workplace hazards, it can challenge the citation by seeking review before the Occupational Safety and Health Review Commission.[22] If the employer is unsuccessful in its challenge before that commission, it can seek redress in the federal court system.[23]

III. WAGE AND HOUR REQUIREMENTS

The Fair Labor Standards Act (FLSA) is a federal law requiring certain covered employers to pay their employees a minimum wage. This statutue also imposes requirements on overtime rates of pay and regulates the occupational job duties and hours of labor for persons under the age of 18. In addition to the provisions contained in the FLSA, several other federal and state laws affect the employer's wage scheme. Many states have their own wage and hour laws that apply to individual employers. Because penalties for noncompliance with federal and state wage and hour laws can be significant, it is particularly important to be aware of these laws and how they affect each particular enterprise.

The wage and hour provisions of the FLSA apply to all employers that are engaged in "[interstate] commerce or in the production of goods for [interstate] commerce." Employers are engaged in interstate commerce if their work involves any movement of goods, services, people, or communications across state lines. Construction work that has a close tie with interstate production (producing or receiving goods shipped or providing services across state lines) is also covered.[24] For example, even if an employer's enterprise consists only of unloading goods that came from another state, FLSA applies. Coverage applies to all employers that expect or have reason to believe that any goods used or moved by their employees will cross or have crossed state lines, even if the goods are sold locally.

The activities of the employee, not the activities of the employer, determine whether the FLSA covers the employees in question. For example, most construction trade workers perform duties covered by the minimum wage and overtime provisions of the FLSA. But work performed by executive, administrative, or professional employees as defined in the FLSA are exempt from the statute's minimum wage and overtime requirements, even if performed in conjunction with the work of other employees on a construction site. At times, an employee may be subject to FLSA coverage for some work and not covered for other work performed during the same week. In that case, the employee is entitled to coverage for the entire week, as long as the covered work was not isolated and sporadic.[25]

[22] 29 U.S.C. § 661.
[23] 29 U.S.C. § 660.
[24] 29 C.F.R. § 776.24(b).
[25] 29 C.F.R. § 776.24(c).

Any work (including maintenance, repair, reconstruction, redesigning, improvement, replacement, enlargement, or extension) performed on a covered facility is subject to the FLSA. If the project is covered, then everyone who works on the project is covered. Even on FLSA-exempt projects, some employees still may be subject to federal wage provisions if their individual activities involve interstate activity. For example, employees who regularly order or procure materials and equipment from outside the state, or receive, unload, check, or watch such goods while they are still in transit, are covered by the FLSA.[26]

Coverage also depends on the existence of an employer-employee relationship. Although the FLSA covers individual "employees," it does not cover independent contractors hired to perform a service for the employer. This distinction is misleading, however, because FLSA broadly defines the employer-employee relationship. FLSA defines an employee as "any individual employed by an employer."[27] FLSA defines an employer as "any person acting directly or indirectly in the interest of an employer in relation to an employee."[28]

A. Basic Calculation

The regular rate of pay is based on the number of hours worked during a standard workweek. The workweek is defined as seven consecutive days, or 168 hours. The employer must calculate the wage by considering each workweek separately and may not average weeks in which less than the statutory minimum hourly wage was earned. The employee is entitled to straight-time pay for the first 40 hours worked during a workweek and one and a half times the regular rate of pay for each hour worked in excess of 40 hours per week.

The FLSA does not require that employees be paid hourly. Compensation systems involving weekly, monthly, or yearly salaries or piecework rates are perfectly acceptable as long as the total straight-time compensation divided by the straight-time hours worked equals the minimum hourly wage. Dividing straight-time compensation by the number of straight-time hours results in the "regular rate of pay." The regular rate of pay determines the amount of overtime due to a particular employee.

Currently, there is a several-tiered schedule of minimum wages at the federal level, with the the federal minimum wage of not less than $6.55 per hour effective July 24, 2008; and $7.25 per hour effective July 24, 2009, all for the first 40 hours worked. On the state level, the required minimum wage and exemptions vary from jurisdiction to jurisdiction, with some states having no minimum wage, other states having a low minimum wage (such as Kansas's $2.65-per-hour minimum wage), and still other states having a higher minimum wage (such as California's $8.00-per-hour minimum wage).[29] In most cases, employees must receive payment free and clear in

[26] *See, e.g., Mitchell v. Lublin, McGaughy & Assocs.*, 358 U.S. 207 (1959).
[27] 29 U.S.C. § 203(e)(1).
[28] 29 U.S.C. § 203(d).
[29] *See* Kan. Stat. Ann. § 44-1203(a); *see also* Cal. Minimum Wage Order (MW-2007).

cash or negotiable instrument. The only exception occurs in certain situations when employers may credit against the minimum wage the reasonable costs incurred in paying for the employee's room, board, or other facilities customarily provided to employees.[30]

Other expenses incurred may not be credited against the minimum wage. For example, although the FLSA does not prevent employers from requiring employees to wear uniforms, it does prevent employers from forcing employees to pay for the uniforms or the cleaning of the uniforms if doing so would push the standard rate of pay below the required minimum wage. Additionally, employers may not deduct expenses for tools of the trade, breakage, or suspected theft if the deduction will send the weekly wage below the statutory minimum. Deductions for theft resulting in a weekly wage below the minimum standard may be applied only after the guilt of an employee has been determined in a criminal proceeding.

Of course, the minimum wage for nonexempt employees is subject to all applicable federal, state, and local taxes. This would include federal income tax withholding and Social Security and Medicare tax withholdings up to the statutory maximum.

B. Overtime Calculation

In most instances, calculating overtime is not difficult. The employer simply multiplies the nonexempt employee's regular rate of pay by 1.5 for each hour worked in excess of 40 hours per week. There are, however, certain situations where it is difficult for the employer to discern whether time spent by the employee constitutes compensable or noncompensable work time. Any time considered "work time" will affect the regular rate of pay as well as the overtime calculation for each employee. Determining what constitutes compensable work time is vitally important to complying with FLSA.

1. Preliminary and Postliminary Time

One problem occurs when the employer must determine whether time spent by the employee preparing for the day's work should be compensated. Courts have recognized that small amounts of time beyond scheduled work hours may be disregarded when determining compensation. Time spent in excess of an extremely small preliminary time span is usually considered compensable. The key question is whether the time spent by the employee outside of scheduled work time predominantly benefits the employer.[31] If so, then the employee should be compensated.

Congress enacted the Portal-to-Portal Act[32] to clarify this particular situation. This statute allows employers to exclude activities that occur either before the time on any given workday at which an employee begins working ("preliminary time") or after the time on any given workday at which he or she stops working ("postliminary time").[33]

[30] 29 U.S.C. § 203(m).
[31] *See Adams v. United States*, 471 F.3d 1321, 1327 (Fed. Cir. 2006).
[32] 61 Stat. 84; 29 U.S.C. § 251.
[33] 29 U.S.C. § 254(a).

The Portal-to-Portal Act eliminates from compensable time activities such as travel and walking time before and after work. Preliminary and postliminary time are compensable, however, if considered an integral part of the principal job. If considered integral, then the activity is characterized as "preparatory" and is compensable work time. For example, although time spent washing hands and changing clothes at the end of a workday is usually not compensable, time spent filling up the fuel tanks of delivery vehicles is most likely compensable.

The distinction between preparatory and preliminary is difficult to determine in many situations. As in most instances of wage and hour law, the key inquiry is to determine who is the main beneficiary of the time in question. If the questioned activity primarily benefits the employee, then the time is most likely not considered work time. If, however, the employer is the prime beneficiary, then wages for the time spent during the activity must be paid.

2. Waiting and On-Call Time

Employers must pay nonexempt employees for all time spent "on duty." In many situations, it is fairly simple to determine when a particular employee is on or off duty. Problems develop, however, when employers attempt to determine whether to compensate employees who are "on call," or waiting to be called to work. When considering whether to compensate such employees, employers must pay close attention to all of the factors in order to ensure FLSA compliance.

Employees who are waiting for materials to arrive or waiting to work while on duty generally must be compensated.[34] It is particularly important to compensate employees for all time in which they are under the control of the employer and are unable to use that time for their own benefit. For example, time spent waiting because of machinery breakdown and delivery delays usually is compensable. Although there are no specific guidelines for each individual circumstance, it is fairly clear that if the employee is completely under the control of the employer and cannot pursue his or her own interests, then compensation is appropriate.

Employees who are completely relieved of duty are not entitled to compensation for idle hours. An employee is considered completely relieved of duties if told "in advance that he may leave the job and that he will not have to commence work until a definitely specified hour has arrived."[35] In short, employees are off duty if they can spend the idle time pursuing their own interests.

Employees who must remain "on call" or who must remain so close to the employer's business that they cannot use their off-duty time to their own benefit are often considered to be "working while on-call."[36] Employees who are working while on-call are entitled to compensation under the FLSA. Simply because an employee is on-call, however, does not necessarily entitle that person to compensation. Determining

[34] *Dole v. Enduro Plumbing, Inc.*, 1990 WL 252270 (C.D. Cal 1990); *Donovan v. 75 Truck Stop, Inc.*, 1981 WL 2333 (M.D. Fla. 1981).
[35] 29 C.F.R. § 785.16(a).
[36] *Id.*

whether payment is owed depends on a fact-specific examination of the control that the employer exerts on the employee during the period in question.

On-call employees who are permitted to use the bulk of their on-call time to perform activities for their own benefit are generally not entitled to compensation. The greater the ability of the employee to use on-call time to pursue personal interests and to dictate his or her own schedule, the less likely it is that this on-call time will be considered compensable work time. Courts generally consider the frequency of callbacks and the effect of the callback on the employee time. For example, where callbacks occur throughout the on-call period, making it nearly impossible for the employee to pursue personal interests, courts generally determine that the time is compensable.[37]

Compensation for on-call time must be included in the regular rate of pay calculation aspart of the hours worked.[38] On-call compensation must equal or exceed minimum wage. If the employee's duties while on-call are significantly different from the employee's duties during regular working hours, then compensation may be paid at a different rate, as long as statutory minimums are met.[39]

C. Independent Contractor Status Considerations

One common dispute in wage and hour law arises from situations in which the status of an employee is disputed. When determining whether an employee is a true "employee" or an independent contractor, courts examine several different factors. These factors include: (1) the degree of control exercised by the alleged employer; (2) the extent of the relative investments in equipment and material; (3) the worker's opportunity for profit and loss through managerial skill; (4) the skill and initiative required by the work; (5) the permanence of the relationship; and (6) the extent to which the service rendered is an integral part of the alleged employer's business.[40] Courts examine the entirety of the circumstances when examining these factors. When the status of an employee is in doubt, employers should always err on the side of caution and make certain that they are in compliance with state and federal wage laws.

D. Government Contracts

In addition to the FLSA, other state and federal laws affect the administration of wages in the construction industry. State laws vary, and each company should be aware of the relevant law in each jurisdiction where it employes workers. On the

[37]*See Pabst v. Okla. Gas & Elec. Co.*, 228 F.3d 1128 (10th Cir. 2000); *Renfro v. City of Emporia*, 948 F.2d 1529 (10th Cir. 1991).
[38]29 C.F.R. § 778.223.
[39]*See Townsend v. Mercy Hosp. of Pittsburgh*, 689 F. Supp. 503 (W.D. Pa. 1988).
[40]*See, e.g., Tritchler v. County of Lake*, 232 F.3d 897 (9th Cir. 2000).

federal level, several labor law[41] statutes can affect public construction work and those performing such work. These include:

(1) *Contract Work Hours and Safety Standards Act (CWHSSA).*[42] The CWHSSA covers laborers and mechanics on contracts exceeding $100,000 for public works of the United States or the District of Columbia. This law requires overtime wages beyond a 40-hour week and specifies health and safety requirements.[43]
(2) *Copeland (Anti-Kickback) Act.*[44] The Anti-Kickback Act is intended to protect the wages of any person engaged in the construction or repair of a public building or public work (including projects financed at least in part by federal grants or loans). This act prohibits employers from exacting kickbacks from employees as a condition of employment and requires contractors and subcontractors to submit weekly payroll reports and statements of compliance.
(3) *Service Contract Act of 1965 (SCA).*[45] Contractors performing any "service contract"[46] shall pay their employees not less than the FLSA minimum wage. Contracts in excess of $2,500 are subject to wage and fringe-benefit determinations. These wage determinations are set by the U.S. Department of Labor (DOL).[47]
(4) *Davis-Bacon Act (Davis-Bacon).*[48] This act requires contractors to pay mechanics and laborers a "prevailing wage"[49] on federal construction projects

[41]Although not normally considered to be a labor law, the Miller Act, 40 U.S.C. §§ 3131–34, provides for a statutory payment bond that provides a payment guarantee to certain individuals or parties furnishing labor for the construction, alteration, or repair of public buildings or work of the United States. See **Chapter 14** for a more detailed discussion of the application of the Miller Act to those furnishing labor.
[42]40 U.S.C. §§ 3701–8.
[43]It is the policy of the United States government that overtime not be utilized in the performance of a government contract, whenever practicable. *See* FAR § 22.103–2.
[44]18 U.S.C. § 874; 40 U.S.C. § 3145. Both statutes invoke potential criminal sanctions. Violation of the anti-kickback statute (18 U.S.C. § 874) carries a potential fine of up to $5,000 or five years imprisonment, or both. 40 U.S.C. § 3145 contains an express cross-reference to 18 U.S.C. § 1001 of Title 18, the False Statements Act.
[45]41 U.S.C. §§ 351–58. Violations of the SCA provide for the debarment of the contractor absent unusual circumstances (41 U.S.C. § 354(a)) and contract cancellation (41 U.S.C. § 352 (c)). There has been no authorization to proceed with a private right of action to enforce the SCA or DBA. *See Univs. Research Ass'n, Inc. v. Coutu*, 450 U.S. 754 (1981).
[46]The definition of a "service contract" does not encompass construction, alteration, or repair of public works of the United States, including painting or decorating. *See* FAR § 22.1003–3. The SCA does, however, cover support services such as grounds maintenance and landscaping, as well as the operation, maintenance, or logistical support of a federal facility. *See* FAR § 22.1003–5 and 29 C.F.R. § 4.130. (These types of contracts may include activities normally considered to be "construction.")
[47]This is termed as the "prevailing wage" determination by DOL. *See* FAR § 22.1002–2. This wage determination typically includes multiple classifications of workers and varying rates. A major area of risk for the contractor involves the classification of certain activities and wage rates.
[48]40 U.S.C. §§ 3141–48; FAR Subpart 22.4; 29 C.F.R. Part 5.
[49]40 U.S.C. § 3142(b).

performed in the United States that exceed $2,000. Violation of Davis-Bacon may result in a debarment of the contractor if the Comptroller General of the United States finds that the contractor "disregarded their obligations to employees and subcontractors."[50]

Of these four federal statutes, the Davis-Bacon Act traditionally has been the most frequent basis or source of DOL or agency actions involving contractors or their subcontractors on covered construction projects. Davis-Bacon applies to construction activity performed on "the site of the work." Generally, construction activity does not encompass manufacturing, supplying materials, or performing service/maintenance work.[51] The "site of the work" usually is limited to the geographical confines of the construction job site.[52] Transportation of materials to and from the project site is not considered to be construction for the purposes of this act.[53] Davis-Bacon also may apply to construction work performed under a nonconstruction contract—for example, an installation support contract. If the contract requires a substantial and segregable amount of construction, Davis-Bacon applies.[54]

The prevailing wage is the key to Davis-Bacon labor standards. "Wages" include both basic hourly rates for various classifications of labor needed for the project plus fringe benefits. DOL wage determinations are not subject to review by the Government Accountability Office, the agency boards of contract appeals, or the United States Court of Federal Claims.[55] Laborers and mechanics employed by a contractor or subcontractor at any tier are covered. Working foremen who devote more than 20 percent of their time during a workweek to performing duties as a laborer or mechanic are also covered.[56]

Many Davis-Bacon disputes involve issues regarding the proper classification of work to a particular craft (wage rate) and accurate record keeping. Employees who "work with the tools" part of the time and also perform work as laborers can lead to alleged violations and enforcement questions.

Enforcement of the Davis-Bacon Act may begin with either the DOL or the contracting agency. The contracting officer must withhold contract payments if the contracting officer believes that a Davis-Bacon violation exists or if requested to do so by the DOL. If an alleged Davis-Bacon violation is not resolved at the local level, the DOL resolves the dispute.

[50]40 U.S.C. § 3144(b)(1).

[51]FAR § 22.402.

[52]*L.P. Cavett Co. v. U.S. Dep't of Labor*, 101 F.3d 1111 (6th Cir. 1996); *Ball, Ball and Brosamer, Inc. v. Reich*, 24 F.3d 1447 (D.C. Cir. 1994).

[53]*See United States v. Kasner*, 2007 WL 765245 (W.D. Wis. 2007); *Cf.* 29 C.F.R. § 3.2(b); 29 C.F.R. § 5.2(j).

[54]DFARS § 222.402–70. These DoD regulations contain specific tests to assist in the determination of whether the DBA (repair) or SCA (maintenance) applies.

[55]*Abhe & Svoboda, Inc. v. Chao*, 508 F.3d 1052 (D.C. Cir. 2007); *Am. Fed'n of Labor—Congress of Indus. Orgs., Bldg, and Constr. Trades Dep't*, Comp. Gen. B-211189, 83–1 CPD ¶ 386 (1983).

[56]FAR § 22.401; 29 C.F.R. § 5.2(m).

IV. EMPLOYEE BENEFITS: ERISA

Although all employers generally must deal with providing health and pension benefits to employees, this area of the law is particularly relevant to many small to medium-size companies in the construction industry, many of which lack the in-house expertise or resources of large employers. State laws dealing with employee benefits vary and generally deal with the fringes of employee benefit issues given the strong federal regulation in this field. Further, the federal laws and regulations themselves represent some of the most complex and specialized areas of legal practice today. As such, this section focuses primarily on the relevant federal laws regarding employee benefits.

The Employee Retirement Income Security Act of 1974 (ERISA),[57] as amended, is an independent legal specialization on its own. ERISA, which preempts inconsistent state laws and regulations, however, is also key to various other areas of the law affecting construction industry employers, such as tax law, corporate law, bankruptcy law, and labor and employment law. ERISA covers every "employee benefit plan"[58] unless there is a specific exemption in the statute.[59]

Under ERISA, an "employee benefit plan" means any "employee pension benefit plan" or an "employee welfare benefit plan, or a plan which is both an employee welfare benefit plan and an employee pension benefit plan."[60] An employee "pension benefit plan" means: any "plan, fund, or program" established by an employer or union, or both, to the extent that, by its express terms or as a result of the surrounding circumstances, it "provides for retirement income" or "results in deferral of income" to termination of employment or beyond.[61] An "employee welfare benefit plan" is "any plan, fund or program" established by an employer or union, or both, to the extent that it is established for the purpose of providing for its participants or their beneficiaries (through insurance or otherwise) for medical, sickness, accident, disability, death, or unemployment or vacation benefits, or apprenticeship or training programs, or day care centers, scholarship funds, or prepaid legal services, or any benefit allowed by Taft-Hartley Act § 302(c) (other than pensions).[62]

Generally speaking, under ERISA's framework, if an employer inititates practically any benefit program for the employee other than those dealing with wages and vacation pay, it is typically presumed that such benefit program is an "ERISA plan," unless the plan clearly falls into one of the exemptions to coverage under the statute. If the goal of the plan is to defer or allocate income for retirement, the plan is generally an ERISA pension plan. Any other type of benefit is generally considered to constitute an ERISA welfare plan.

Although the generalized statements just listed sum up the ERISA coverage of many employee benefit plans, whether a specific plan is an ERISA plan or not is

[57] 29 U.S.C. § 1001 *et seq.*
[58] 29 U.S.C. § 1003(a).
[59] 29 U.S.C. §§ 1003(b), 1051, 1081(a), 1101(a).
[60] 29 U.S.C. § 1002(3).
[61] 29 U.S.C. § 1002(2)(A).
[62] 29 U.S.C. § 1002(1).

sometimes difficult to determine. One point to remember is that the employer's intent as to whether it believed its plan would be governed by ERISA is generally irrelevant. For instance, an employer's oral promise to provide retired employees with grocery vouchers was considered to fall under ERISA.[63] An employer's purchase of life insurance policies for two employees was considered to be an ERISA "plan."[64] Further, "severance pay" plans that pay after retirement may constitute ERISA pension plans.[65] Such informal plans can be problematic for employers—as if they are considered to be ERISA plans, the employer has already violated the law by failing to maintain the plan in writing and to perform the other requisite provision of notices and record keeping required by ERISA.

In sum, even with the best knowledge and intentions, it is often difficult to comply with the myriad requirements imposed on employee benefit plans by ERISA. Certainly such compliance cannot come by accident. Although this chapter (or even this entire book) is not long enough to detail all of the legal nuances that are intertwined with such plans, construction industry employers should be aware that the potential penalties for noncompliance with ERISA are great and that informal gifts of retirement benefits and the like may expose them to liability under ERISA—despite employers' good intentions. If any such plan is implemented, ensuring compliance with ERISA must be the first priority.

V. EMPLOYMENT DISCRIMINATION

Numerous laws are applicable to construction industry employers that prohibit discrimination against certain protected classifications. The key federal antidiscrimination statutes pertaining to the operations of construction employers are described next. Employers must be familiar with these provisions in order to comply with the law and avoid potential liability. As mentioned previously, many state and local governments have enacted laws that are similar in scope to the federal statutes. In fact, some of the state and local legislation may protect additional classifications that are not protected on the federal level (e.g., sexual orientation and marital status).[66] Accordingly, construction industry employers should survey the employment-related laws in each location where they have employees or anticipate competing for work.

A. Equal Employment Opportunity Commission

The U.S. Equal Employment Opportunity Commission (EEOC) is the agency responsible for enforcement of various federal laws that prohibit employment discrimination,

[63] Musmeci v. Schwegmann Giant Super Markets,332 F.3d 339 (5th Cir. 2003).
[64] *See Heidelberg v. Nat'l Found. Life Ins. Co.*, 2000 WL 1693635 (E.D. La. 2000).
[65] *See* 29 C.F.R. § 2510.3–2(b)(1)(i).
[66] *See, e.g.*, City of Atlanta Ordinance Sec. 94–112.

including Title VII of the Civil Rights Act of 1964,[67] the Age Discrimination in Employment Act, the Americans with Disabilities Act, and the Equal Pay Act. The EEOC's responsibilities entail acceptance and investigation of complaints, known as charges of discrimination, filed by employees and other affected persons. The EEOC also has the authority to file lawsuits where it determines that there is cause to believe discrimination has occurred. Congress also has provided for private rights of action for the individuals who claim to be aggrieved by discriminatory conduct. Many state and federal civil rights laws, however, require individuals to file a charge of discrimination with the EEOC or state agency and to observe other technical requirements before initiating such litigation.

B. Title VII of the Civil Rights Act of 1964

Title VII of the Civil Rights Act of 1964,[68] as amended by the Civil Rights Act of 1991[69] (Title VII), prohibits employers from discriminating against employees or applicants on the basis of race, color, religion, sex, or national origin, unless the employer can establish that discrimination based on one of the foregoing factors is permitted as a bona fide occupational qualification. Title VII also prohibits employers from retaliating against an employee based on activity that is protected under this statute, such as making complaints regarding discrimination or participating in an investigation involving allegations of discrimination. Employees who claim that they have been intentionally discriminated against in violation of Title VII are entitled to a jury trial and may be awarded relief, including compensatory and punitive damages.

One of the most widely utilized areas of Title VII jurisprudence involves claims of sexual harassment in the workplace. The term "sexual harassment" contemplates conduct on the part of a coworker, supervisor, or patron that is based on sex and is sufficiently severe or pervasive to alter the terms, conditions, or privileges of the victim's work environment. Sexual harassment continues to be an ever-developing area of law and can expose employers to significant liability. Moreover, in light of the historical male dominance of positions in the construction industry, prevention of sexual harassment should be a top priority for construction employers. To avoid complaints of sexual harassment, employers are encouraged to implement, disseminate, and adhere to workplace rules and policies prohibiting harassment and to act promptly to investigate and correct any harassing conduct.

C. Americans with Disabilities Act

Congress enacted the Americans with Disabilities Act[70] (ADA) to ensure that qualified individuals with disabilities receive the same employment opportunities as those

[67]42 U.S.C. § 2000e *et seq.*

[68]*Id.*

[69]42 U.S.C. § 1981a.

[70]The ADA has three titles that prohibit varying forms of discrimination by both public and private entities. Title I prohibits discrimination by private-sector employers. *See* 42 U.S.C. § 12111 *et seq.*

provided for individuals without disabilities. Accordingly, the ADA prohibits employers from discriminating against a qualified employee or applicant based on a disability. Nonetheless, employers are not required to employ or retain disabled individuals if they are incapable of performing the essential functions of their jobs. In such circumstances, the individual is not considered to be qualified for the position in question. Under the ADA, the term "disability" has three alternate meanings: (1) a physical or mental impairment that substantially limits one or more of the major life activities of an individual; (2) a record of such an impairment; or (3) being regarded as having such an impairment.[71] Homosexuality, bisexuality, temporary disabilities, and current use of illegal drugs are examples of conditions that are excluded from the coverage of the ADA. Where mitigating measures (such as medicine, eyeglasses, or prosthetic devices) would remove or reduce the impact of an individual's impairment, the ADA may not recognize the impairment as a disability within the meaning of the act.[72]

In order to be protected under the provisions of the ADA, an employee must be an individual with a disability, and qualified to perform the responsibilities of the job he or she holds or desires, with or without reasonable accommodation.[73] Covered employers must provide reasonable accommodations to qualified employees (including applicants) with disabilities absent undue hardship to the employer. In determining whether an employer has an undue hardship, the ADA allows consideration of the nature and cost of the accommodation in relation to the size, financial resources, nature, and structure of the employer's operation as well as the impact of the accommodation on the specific facility providing it.

In addition to the defense of undue hardship, employers can utilize qualification standards and tests to screen out individuals who are incapable of performing the requirements of a particular position, if such qualification standards are job-related or reveal conditions that would pose a direct threat to the health and safety of the employee in question or his or her coworkers. The ADA provides:

> It may be a defense to a charge of discrimination . . . that an alleged application of qualification standards, tests, or selection criteria that screen out or tend to screen out or otherwise deny a job or benefit to an individual with a disability has

[71]Under the three-part definition of 42 U.S.C. § 12111 *et seq.*, an employer can be held liable for a violation of the ADA if it acts in a manner that indicates that it regards an employee as having a disability, even if the employee, in fact, does not suffer from a disability within the meaning of the act (i.e., an impairment that substantially limits a major life activity). *See* 42 U.S.C. § 12102(2).

[72]*See Sutton v. United Air Lines, Inc.*, 527 U.S. 471 (1999).

[73]The ADA does not comprehensively define the term "reasonable accommodation" under varying circumstances; however, the ADA provides that "reasonable accommodations" may include any of these: (1) making existing facilities used by employees readily accessible to and usable by an individual with a disability; (2) job restructuring; (3) modifying work schedules; (4) reassigning employees to other vacant positions; (5) acquiring or modifying equipment or devices; (6) adjusting or modifying examinations, training, materials, or policies; and (7) providing qualified readers or interpreters to disabled employees. *See* 42 U.S.C. § 12111(9).

been shown to be job-related and consistent with the business necessity, and such performance cannot be accomplished by reasonable accommodation . . . .[74]

Moreover, to rely on the direct threat defense, employers must demonstrate the applicability of these factors: (1) a significant risk of substantial harm; (2) a specific risk that is identifiable; (3) a current risk rather than a speculative or remote risk; (4) an objective medical assessment of the risk; and (5) whether the risk could reasonably result in an undue hardship.[75]

The ADA covers employers, including state and local governments, with 15 or more employees who are engaged in an industry affecting commerce. The definition of an employer also includes individuals who are agents of the employer, such as managers, supervisors, foremen, or others who act on behalf of the employer, including agencies used to conduct background checks on prospective employees. Employers can be held responsible for their agents' violations of the ADA. Under the ADA, employees are entitled to the same remedies afforded under Title VII.

D. Age Discrimination in Employment Act

The Age Discrimination in Employment Act[76] (ADEA) protects employees and applicants who are over age 40 from discrimination on the basis of age. The ADEA also provides guidelines for settlement or waiver of ADEA claims. These guidelines discuss steps that employers must take to implement a layoff as a result of a reduction in force and require employers to justify any decrease in benefits offered to employees within the protected age classification in comparison with employees who are not protected by the ADEA. Employers can defend allegations of age discrimination if the discrimination is the result of a bona fide occupational qualification. In such circumstances, the employer must demonstrate that the individual's age is simply incidental to other factors that are reasonably necessary to the normal operation of the particular business.[77] The ADEA provides for back pay, liquidated or double damages, and other relief, including trial by jury.

E. Discrimination and Public Contracting

In addition to the above-mentioned laws that prohibit certain covered employers from discriminating against employees based on protected classifications, federal statutes and executive orders further regulate the workplace of employers who perform services under contracts with the federal government. In particular, Executive Order 11246, the Rehabilitation Act of 1973, and the Vietnam Era Veterans Readjustment Assistance Act of 1974 (VEVRAA) all have detailed record-keeping and affirmative action requirements that federal contractors must satisfy. Penalties for failure to

[74] 42 U.S.C. § 12113(a).
[75] *See* 42 U.S.C. § 12113(b); *Sch. Bd. of Nassau County, Fla. v. Arline*, 480 U.S. 273 (1987).
[76] 29 U.S.C. § 621 *et seq.*
[77] 29 U.S.C. § 623(f)(1).

follow these laws and regulations could lead to revocation of federal contracts and debarment from participation in future federal contracts.

F. Office of Federal Contract Compliance Programs

The Office of Federal Contract Compliance Programs (OFCCP) has been delegated the responsibility of ensuring that federal contractors do not discriminate against individuals based on certain classifications and that federal contractors abide by affirmative action requirements of creating, maintaining, and implementing affirmative action plans. The OFCCP also has authority to initiate compliance evaluations to ensure that covered entities are in compliance with their nondiscrimination and affirmative action obligations. The OFCCP can conduct compliance evaluations by any one or combination of these methods: (1) desk audit of the contractor's written affirmative action plan; (2) on-site review to investigate unresolved problem areas identified in the contractor's written affirmative action plan (including examination of personnel and employment policies); (3) off-site review of records; (4) compliance check (visit to the facility to determine whether information submitted is complete and accurate); and (5) on-site "focused review" (restricted to analysis of one or more components of the contractor's organization or employment practices).[78]

The OFCCP has the authority to implement enforcement proceedings and regulations for Executive Order 11246 and the VEVRAA.[79] Moreover, although the DOL has enforcement authority with regard to the Rehabilitation Act of 1973, it has adopted regulations that parallel those implemented under Executive Order 11246 and the VEVRAA.

G. Executive Order 11246

Executive Order 11246 prohibits employment discrimination based on race, color, religion, sex, or national origin by contractors and subcontractors operating under federal service, supply, use, and construction contracts and by contractors and subcontractors performing under federally assisted construction contracts. Additionally, all contracts and subcontracts covered under Executive Order 11246 must include a clause pledging not to discriminate because of race, color, religion, sex, or national origin and to take affirmative action to ensure that applicants are employed and that employees are treated during employment without regard to those protected classifications. These dual obligations are contained within an Equal Employment Opportunity clause that all contracting federal agencies are required to include with their contracts with private employers. The Executive Order applies to companies doing business with the federal government under contracts or subcontracts that exceed $10,000. Moreover, employers with 50 or more employees, and that have federal contracts worth at least $50,000, are required to prepare and maintain written affirmative action plans.

[78]41 C.F.R. § 60–1.20.
[79]38 U.S.C. § 2012.

H. Uniform Service Employment and Re-Employment Rights Act

Given the various conflicts around the world and the increasing number of soldiers and sailors who have been called away to significantly long-term military operations overseas, it is likely that increased scrutiny will be applied to employers of these military servicemen and women when these employees reenter the U.S. workforce. The Uniform Services Employment and Re-Employment Rights Act (USERRA)[80] clarifies and strengthens the earlier Veterans' Re-employment Rights (VRR) statute in order to assist the reentry of military personnel into the workforce after duty. In essence, USERRA protects the civilian job rights and benefits for veterans and members of reserve components. USERRA also protects service member rights by providing enhanced enforcement mechanisms.

USERRA establishes five years as the total time that an individual service member may be absent from work for military duty and retain the right to reemployment. There are important exceptions to the five-year limit, however, including initial enlistments lasting more than five years, periodic National Guard and Reserve training duty, and involuntary active duty extensions and recalls, especially during a national emergency. USERRA clearly establishes that reemployment protection does not depend on the timing, frequency, duration, or nature of an individual's service as long as the basic eligibility criteria are met.

USERRA protects disabled veterans, requiring employers to make reasonable efforts to accommodate the disability, similar to what is required under the ADA. Service members convalescing from injuries received during service or training may have up to two years from the date of completion of service to return to their jobs or apply for reemployment.

USERRA requires that returning service members be reemployed in the job that the employee would have achieved had the employee not been absent during military service, with the same seniority, status, and pay, as well as other rights and benefits determined by seniority. This aspect of the statute is often referred to as the "escalator" principle.[81] USERRA further requires employers to undertake reasonable efforts, including retraining, to enable returning service members to update or upgrade their skills as they pursue reemployment. USERRA further provides for alternative reemployment positions if the service member cannot qualify for the "escalator" position. USERRA also provides that while an individual is performing military service, he or she is deemed to be on a furlough or leave of absence and is entitled to other nonseniority rights accorded other employees on nonmilitary leaves of absence.

USERRA requires that service members provide their employers with advance written or verbal notice for all military duty unless giving notice is impossible, unreasonable, or precluded by military necessity. An employee should provide his or her employer notice as far in advance as is reasonable under the circumstances. Additionally, service members are able (but are not required) to use accrued vacation or annual leave while performing military duty.[82]

[80]38 U.S.C. § 4301 *et seq.*

[81]Nancy Bloodgood & Brian L. Quisenberry, *From War to Work*, 53 Fed. Law 38, 40 (2006).

[82]*See* 38 U.S.C. § 4301 *et seq.*

If an employee complains that his or her USERRA rights have been violated, the service member may have his or her claim referred to the Department of Justice for consideration of representation in the appropriate federal district court, at no cost to the claimant. Federal and Postal Service employees may have their claims referred to the Office of Special Counsel for consideration of representation before the Merit Systems Protection Board (MSPB). If violations under USERRA are shown to be willful, the court may award liquidated damages against the employer. Individuals who pursue their own claims in court or before the MSPB may be awarded reasonable attorney and expert witness fees if they prevail.[83]

VI. FAMILY AND MEDICAL LEAVE ACT

The Family and Medical Leave Act[84] (FMLA) requires employers to provide unpaid leave (up to 12 weeks per year) for an eligible employee with a serious health condition or for the birth or adoption of a child, and for the care of a child, spouse, or parent who has a serious health condition. The FMLA prohibits employers from discriminating against an employee as a result of a request for leave and requires reinstatement to the employee's original position upon expiration of leave. Employers must also provide access to continued health insurance, at the employee's expense, during the term of the employee's leave period. The DOL has enforcement and regulatory authority of the FMLA.

The FMLA allows eligible employees of a covered employer to take job protected, unpaid leave, or to substitute appropriate paid leave if the employee has earned or accrued it, for up to a total of 12 workweeks in any 12-month period. The FMLA also provides for employees in certain circumstances to work a part-time schedule or take leave on an intermittent basis rather than all at one time, if such a schedule is medically necessary.

The FMLA defines an "eligible employee" as an employee of a covered employer who: (1) has been employed by the employer for at least 12 months; (2) has been employed for at least 1,250 hours of service during the preceding 12-month period before commencement of the requested leave; and (3) is employed at a work site where 50 or more employees are employed by the employer within a 75-mile radius. The FMLA defines an employer as "any person engaged in commerce or in any industry or activity affecting commerce who employs fifty or more employees for each working day during each of twenty or more calendar workweeks in the current or preceding calendar year."[85] Employers covered by the FMLA also include any person acting, directly or indirectly, in the interest of a covered employer to any of the employees of the employer, any successor in interest of a covered employer, and any public agency.

[83] *Id.*
[84] 29 U.S.C. § 2601 *et seq.*
[85] 29 U.S.C. § 2611(4)(A).

An eligible employee can take FMLA leave only as a result of a birth or adoption, or an employee's serious health condition or a serious health condition affecting a close family member of the employee. The FMLA defines a serious health condition as an illness, injury, impairment, or physical or mental condition that involves either inpatient care or continuing treatment by a healthcare provider.[86] The FMLA defines "family member" as parent, spouse, or child.[87]

Because both the ADA and the FMLA regulate circumstances in which an employee's health condition impacts his or her work performance, there is a substantial amount of overlap and interplay between the two statutes. The FMLA's leave-of-absence provisions, however, are wholly distinct from an employer's responsibility to provide a reasonable accommodation pursuant to the ADA.[88] For example, under the ADA, if an employee is a qualified individual with a disability within the meaning of the ADA, the employer must make reasonable accommodations barring undue hardship to the employer. Yet a covered employer under the FMLA also must afford an eligible employee his or her FMLA rights for a serious health condition. Although the ADA's notion of a "disability" and the FMLA's notion of a "serious health condition" are distinct concepts, both could apply to protect an employee in the same circumstance.[89]

For example, a reasonable accommodation under the ADA may be accomplished by providing an employee with a reduced work schedule or an alternative part-time position with no health benefits for an undetermined amount of time. Under the FMLA, however, the employee would be required to work only on a reduced leave schedule (or in a part-time position) until the equivalent of 12 workweeks of leave was utilized, with group health benefits maintained during this period at cost to the employee. Under the ADA, if determining whether an employee's leave entitlement exceeding 12 workweeks would amount to an "undue hardship," the regulations permit an employer to consider the amount of FMLA leave already taken. Nonetheless, employers must apply the ADA's undue-hardship analysis to each individual case to determine whether leave in excess of 12 workweeks would pose an undue hardship.

VII. UNION ACTIVITY

The National Labor Relations Act (NLRA) governs protected concerted activity in the workplace of covered employers. The NLRA provides employees of covered employers with the right to discuss wages, hours, and work conditions, and to organize a union and collectively negotiate the terms and conditions of their employment, or to refrain from such conduct. The NLRA also prohibits certain conduct on the

[86]*See* 29 C.F.R. § 825.114.
[87]29 U.S.C. § 2612(a)(1)(C).
[88]29 C.F.R. § 825.702(a).
[89]29 C.F.R. § 825.702(b).

part of employers, including threats, interrogations, promises of benefits to avoid unionization, or spying that is intended to interfere with the employees' rights. The National Labor Relations Board (NLRB) has the authority to administer and enforce the NLRA.

After some decline, unions have again assumed a more pro-active stance in organizing employees and adding to their membership rolls. Unions have adopted aggressive strategies to "organize every working woman and man who needs a better deal and a new voice."[90] This enthusiasm seems to have rejuvenated union activists. Many unions have scaled back traditional organization drives and moved toward a new culture of innovation. This section provides insight into the innovative tactics adopted by union organizers and discusses strategies for employers to use in order to reduce the likelihood of a successful union campaign.

A. Union-Organizing Tactics

1. Salting

One familiar tactic that continues to be popular is "salting."[91] This is the practice of sending a union member or sympathizer to apply for a job at a nonunionized workplace. These individuals, who are either volunteers or paid by the union, are known as salts. The NLRA expressly protects union organizers from discharge based on union affiliation, and the United States Supreme Court has extended the same protection to job applicants.[92] In *Town & Country,* a union filed a complaint with the NLRB after the company refused to interview 10 of 11 job applicants when it learned that they were union members. Despite the fact that these 10 individuals were paid union organizers, the Court held that they fell within the NLRA's definition of "employee" and could not be discriminated against because of their union affiliation. As a result, salts often openly disclose their union membership or advocacy when applying for work, which has the effect of putting the employer on notice that the salts fall within the legal protection of the NLRA.

2. Intermittent/Partial Strikes

In an intermittent strike, workers halt production or work as part of a plan to force the company to accede to certain worker demands. They generally occur when a union salt has found his or her way into an operation. As a rule, however, intermittent strikes by union employees aimed at forcing an employer to bow to their contract requests are unlawful. This principle extends to employees' refusal to work mandatory overtime. The courts and the NLRB grant a small amount of leeway

[90]44 Daily Labor Report D26 (Mar. 6, 1996).

[91]*See Tualatin Elec., Inc. And Int'l Bhd. of Elec. Workers, Local No. 48*, 1993 WL 361183 (N.L.R.B. 1993) (explaining the origin of the term as derived from the use of the term "salt" as in "salting a mine," which is the artificial introduction of metal or ore into a mine by subterfuge to create the false impression that the material was naturally occurring).

[92]*Nat'l Labor Relations Bd. v. Town & Country Elec., Inc.*, 516 U.S. 85 (1995).

to intermittent strikes, though, and usually do not find unlawful behavior where the intermittent strike is a one-time incident or occurs several times for unrelated reasons. More latitude is granted to intermittent strikers by the courts when the strikers are nonunionized.

A related tactic is the partial strike, in which employees refuse to perform a certain aspect of their jobs. It too is illegal, but courts generally hold that a partial strike exists only when employees blatantly refuse to perform some essential duty while remaining on the job.

One final strike-type approach is dubbed work-to-rule, in which employees perform only those tasks absolutely required by their employment while refusing to do things they previously did out of goodwill or because the employer requested it. The NLRB has not ruled on the legality of this tactic, meaning it will recur until and unless it is found illegal in a future ruling.

3. *Nontraditional Picketing*

Picketing and passing out handbills at job sites have been favorite traditional union activities for decades. More recently, however, unions have attempted nontraditional picketing methods to facilitate union organization. Nontraditional picketing is nontraditional in location rather than in style. It moves the picketing from the work site to corporate headquarters or a corporate executive's home. Although nontraditional picketing lessens the union's chance for contact at the picket line with workers whom the union seeks to organize, the union may receive heightened media attention due to its chosen location. If the media attention is sympathetic in nature, it may aid the union in organizing workers.

Another nontraditional method of organizing is high tech in nature, via the Internet. Through the Internet, unions may be able to reach people in their homes with electronic handbills. Additionally, the Internet can be used to organize boycotts and protests.

B. Appropriate Employer Responses to Union Activism

Employers may legally ban nonemployee union organizers and sympathizers from company property, unless there is no other reasonable means for the union to reach employees. Typically, because there are other reasonable ways to get the union's message to employees, a nonsolicitation/distribution policy prohibiting access to employees (including work areas, cafeteria, and parking lots), if strictly enforced, will generally suffice to keep out nonemployee union advocates. To be effective, however, any nonsolicitation/distribution policy must be applied in a nondiscriminatory fashion—that is, it must be invoked to prohibit any acts that violate it, not just those that arouse management concern. For example, in *Lucile Salter Packard Hospital v. NLRB,*[93] the defendant hospital had a nonsolicitation policy that prohibited

[93]97 F.3d 583 (D.C. Cir. 1996).

nonemployees from distributing literature or soliciting on hospital property. Despite this rule, the hospital allowed several outside groups (including an insurance company and credit union) access to the hospital to solicit employees. When a union representative attempted to distribute union literature on hospital grounds, she was denied access pursuant to the hospital's nonsolicitation policy. Her union filed an unfair labor practice against the hospital, insisting that the policy was discriminatorily applied to keep out union activists. The union prevailed.[94]

As the foregoing case makes clear, a company that permits employee solicitations by certain organizations, including charities, on work time cannot legally hide behind its nonsolicitation policy to bar union solicitations during the same hours. As a result, a nonsolicitation policy is only as good as the supervisors who enforce it. Therefore, it is important that contractors ensure that their supervisors understand the importance of consistent application to all employment policies.

In addition to the policy against solicitation and distribution, uniform application of these policies also will assist employers in combating union organization: (1) no loitering—a policy that requires nonworking employees and nonemployees to leave the premises unless conducting business with the company; (2) restrictions on access to employee names and addresses—access to employee information should be restricted to upper-level management of the company, so as to avoid requests for employee information by the union before establishment of the bargaining unit; and (3) limited use of company bulletin boards—bulletin boards should be limited to company use only, so as to avoid sharing bulletin board space with union activists. Again, keep in mind that these policies must be enforced strictly and uniformly.

Additionally, one of the simplest tactics to avoid unionization, and salting in particular, is to explain in some detail during the interviewing process the terms and conditions of employment. Wages or fringe benefits that fall below what the union is accustomed to may discourage the casual or underfunded salt from completing the application process. Moreover, a thorough application procedure, requiring references and other information, can reveal salts who, despite the legal protections, attempt to evade being discovered during the hiring process. As with all employment policies, contractors need to ensure that all individuals involved in the hiring process, including nonmanagement personnel, strictly follow any and all prerequisites to the letter to avoid a union claim that nonunion applicants were not asked for references or otherwise faced less of a hurdle to employment. Even employees who are not members of management can significantly damage the company's chances of prevailing in a union campaign. For example, statements made by a receptionist regarding the company's position on hiring union members when taking applications from salts could result in an unfair labor practice charge regarding the company's hiring practices. Only people who fully understand company policies and the legal consequences should comment on them to prospective employees.

[94]*Id.* at 592.

VIII. UTILIZING EMPLOYEE BACKGROUND INVESTIGATIONS

An important tool used in making an informed hiring decision is the background investigation. Prospective employers generally rely on factors such as credit history, prior employment history, criminal background, and driving record in considering whether a particular applicant is qualified to fill a vacant position. To obtain pertinent background information, some employers traditionally have relied on consumer reporting agencies.

The Fair Consumer Credit Reporting Act (FCRA) (formerly known as the Consumer Credit Reporting Act),[95] places significant restrictions on an employer's use of certain personal information about prospective or current employees obtained through consumer reporting agencies. The requirements outlined in the FCRA specifically apply to circumstances in which a employer seeks to rely on such information to make employment decisions (e.g., hiring, promotion, reassignment, and retention).[96]

The act applies only if the employer relies on information obtained from a consumer reporting agency and does not apply to information employers obtain through their own background investigations.[97] The act, as currently amended, is designed to ensure that individuals: (1) are aware that consumer reports may be used for employment purposes and agree to such use and (2) are notified promptly if information in a consumer report may result in a negative employment decision.[98]

Employers face a variety of penalties if they violate the act, including criminal prosecution (if false pretenses are used to obtain a consumer report).[99] Generally, actions to enforce rights under the FCRA must commence within two years of the discovery of the alleged violation and no later than five years after the violation, or the aggrieved party is statutorily barred from pursuing a claim under the act.[100]

Under the FCRA, a "consumer reporting agency" is defined as:

> any person which, for monetary fees, dues, or on a cooperative nonprofit basis, regularly engages in whole or in part in the practice of assembling or evaluating consumer credit information or other information on consumers for the purpose of furnishing consumer reports to third parties, and which uses any means or facility of interstate commerce for the purpose of preparing or furnishing consumer reports.[101]

A "consumer report" is "any written, oral, or other communication of any information . . . bearing on a consumer's creditworthiness, credit standing, credit

[95]15 U.S.C. § 1681 (2003) *et seq.*

[96]*See* 15 U.S.C. § 1681b (a)(3)(B).

[97]Simply obtaining such firsthand information is not sufficient to bring an employer that is otherwise not engaged in the business of providing "consumer reports" within the definition of a "consumer reporting agency" under the act. *See* 15 U.S.C. § 1681a.

[98]15 U.S.C. § 1681b.

[99]15 U.S.C. §§ 1681o and 1681q.

[100]15 U.S.C. § 1681p.

[101]15 U.S.C. § 1681a(5).

capacity, character, general reputation, personal characteristics, or mode of living."[102]

If an employer seeks to rely on a consumer report obtained from a consumer reporting agency to assist it in making an employment decision, it must also:

(1) Certify to the consumer reporting agency that the employer has or will comply with the specific requirements of the FCRA, including the disclosure requirements, and that the employer will not use the report in violation of any applicable federal or state equal opportunity law.

(2) Inform the employee or applicant that the employer may request a consumer report in a clear and conspicuous manner before obtaining the report, and in a separate document (i.e., the disclosure cannot be a statement in a handbook or on the employment application). The employer also must obtain the employee's or applicant's written permission to obtain the report.

(3) If the employer plans to take an adverse employment action based, in whole or in part, on information contained in the consumer report, the employer must provide the employee or applicant with a copy of the report and a written statement of his or her rights before taking the adverse action. At least five days is an appropriate waiting time, but individual circumstances should be considered. The definition of an adverse employment action is broad, and can include hiring an applicant with a better credit history. If the consumer report has any bearing whatsoever on the employer's decision, disclosure is required under the FCRA.

(4) After the employer takes an adverse employment action based on the consumer report, it is required to provide the employee or applicant with notice of the adverse action, which must include information about the consumer reporting agency and the employee's or applicant's rights.[103]

The FCRA imposes additional responsibilities on employers who seek to rely on consumer reports obtained through personal interviews with neighbors, friends, or associates of the individual reported on or with others with whom the individual being reported is acquainted or who may have knowledge concerning any such items of information. Such detailed reports are described as "investigative consumer reports" under the FCRA.[104] In addition to the responsibilities required for employers who rely on consumer reports, when seeking to rely on investigative consumer reports, employers must:

(1) Inform the employee or applicant in writing, within three days of the request, if the employer actually orders such a report.

[102] 15 U.S.C. § 1681a(d)(1).
[103] 15 U.S.C. § 1681b.
[104] 15 U.S.C. § 1681a.

(2) Inform the employee or applicant that he or she can request a disclosure of the nature and scope of the investigation upon written request. This must be done before the employer actually gets the report. If the employee or applicant makes a written request regarding the investigative consumer report, the employer must promptly provide a complete disclosure of the nature and scope of the requested investigation.[105]

The principal step in avoiding liability under the FCRA is to determine whether the information being obtained reveals factors that are appropriate for consideration with regard to a current employee or applicant. For example, an employer may wish to use consumer reports only after determining whether an employee or applicant is otherwise qualified instead of using consumer reports as a preliminary screening device to narrow the field of candidates. An employer exposes itself to considerable liability under the FCRA however, if it does not maintain confidentiality over the information obtained. Furthermore, in order to obtain a consumer report, an employer may require an applicant to reveal otherwise unknown information pertaining to a protected classification, such as the applicant's date of birth. Accordingly, if the consumer report does not reveal information pertinent to the position sought, the employer could face liability based on the fact that the information obtained has the effect of excluding certain protected classifications from the position in question.

If, after considering the appropriateness of utilizing a consumer report, the employer determines that it is a useful tool to assist it in making difficult employment decisions, care must be given to ensure detailed compliance with the FCRA's requirements. Employers should also include a disclaimer when requesting protected information, as well as take other measures that are appropriate, to ensure that the information is not used for impermissible purposes.

➢ POINTS TO REMEMBER

- Always adhere to the Form I-9 requirements and implement policies to ensure that the record-keeping requirements as to the employee's immigration status are in line with state and federal laws.
- Never threaten an employee or discipline or discharge for union activity.
- Prepare and maintain a safety manual outlining employee responsibilities and workplace safety requirements under the Occupational Safety and Health Act.
- Never discharge or discipline an employee for raising legitimate safety concerns regarding the work environment.
- Ensure that employees are being compensated at a rate not less than minimum wage or rates established by your contract with a public agency.
- Ensure that employees are compensated at a rate not less than one and one-half times minimum wage for all hours worked in excess of 40 during the workweek.

[105] 15 U.S.C. § 1681b.

- Ensure that all employees are treated equally with regard to the terms, conditions, and privileges of employment.
- Never ask applicants questions related to previous workers' compensation injuries or medical history.
- Always ask applicants whether they can perform the essential functions of the position sought, with or without reasonable accommodation.
- Ensure that any employer obligations to provide an employee with a leave of absence satisfy the requirements of both the Americans with Disabilities Act and the Family and Medical Leave Act, to the extent that these statutes are applicable.
- Always ensure that all employment policies are strictly and uniformly applied.
- Never promise to reward those who oppose the union.
- Never seek or ask employees to seek information about union meetings or who is for or against the union.
- Even an informal or oral promise of benefits can expose an employer to liability under ERISA; always ensure that employee benefit plans are strictly drafted and maintained in accordance with the ERISA requirements.
- Always disclose to applicants and employees that background checks will be relied on and obtain permission before seeking background investigation materials from a credit reporting agency.

20

BANKRUPTCY IN THE CONSTRUCTION SETTING

I. INTRODUCTION

When bankruptcy strikes a construction project, all members of the project delivery system are affected. A party filing for bankruptcy in the construction setting can either liquidate its assets (Chapter 7 under the Bankruptcy Code)[1] or reorganize (Chapter 11 under the Bankruptcy Code)[2] its business and get a fresh start by eliminating certain debts and obligations. The bankruptcy process can be complicated and frequently has a negative impact on a construction project. A basic understanding of the law can benefit the construction professional when a project faces the bankruptcy of a key participant.

The Bankruptcy Code is a federal statute. Thus, many of the cases that interpret its provisions are decisions of the United States bankruptcy courts and federal courts of appeal. The characterizations of property interests are usually matters of law; however, for this reason, results can vary from jurisdiction to jurisdiction.

This chapter addresses the effects of bankruptcy on three vital areas of a construction project: (1) the status of contracts with the debtor, (2) the status of materials in the debtor's possession at the time of bankruptcy, and (3) the status of contract funds and alternate sources of funds.

II. THE PLAYERS

To understand how bankruptcy works in a construction setting, it is essential to be familiar with the key "players" in a typical liquidation or reorganization proceeding.

[1]11 U.S.C. § 701 *et seq.*
[2]11 U.S.C. § 1101 *et seq.*

Debtor. This is the party filing for bankruptcy and can include the owner, general contractor, subcontractors, or suppliers.[3]

Debtor in Possession. This is the party filing for bankruptcy (Chapter 11) that maintains control and continues operating its business through the reorganization process.[4]

Trustee. This is the "overseer" of a bankruptcy case who looks out for the creditors' interests and is either appointed by the court (Chapter 7) or may be a Debtor in Possession (Chapter 11).[5]

Bankruptcy Estate. This means the Debtor's assets that become part of "the estate" upon the filing of bankruptcy and can be used to pay off creditors.[6]

Secured Creditors. These are the entities whose claims or interests (i.e., a valid lien claim on property) are part of the Bankruptcy Estate.[7]

Unsecured Creditors. These are the entities whose claims or interests are not part of the Bankruptcy Estate.[8]

III. BANKRUPTCY CODE

Bankruptcy laws are contained in the United States Bankruptcy Code (the Code), which is codified in Title 11 of the United States Code. The Code provides for two types of business bankruptcies, Chapter 11 (Reorganization) and Chapter 7 (Liquidation), both of which may be encountered on a construction project.[9]

A. Chapter 11—Reorganization

Under Chapter 11, the Debtor remains in control of its business and property as a "Debtor in Possession" unless there is a good reason for a Trustee to be appointed by the court to control the Debtor's property. A Debtor in Possession has considerable

[3] 11 U.S.C. § 101(13).

[4] 11 U.S.C. § 1101(1); 11 U.S.C. § 1107; *In re Cybergenics Corp.*, 226 F.3d 237, 243 (3d Cir. 2000) (the debtor in possession acts and assumes the role of a trustee when one has not been appointed by the bankruptcy court).

[5] 11 U.S.C. § 322; 11 U.S.C. § 1101; *see In re Teligent, Inc.*, 307 B.R. 744, 747 (Bankr. S.D.N.Y. 2004) (the trustee takes the place of the debtor in bankruptcy including bringing legal action).

[6] 11 U.S.C. § 541.

[7] 11 U.S.C. § 506; *In re Taylor*, 289 B.R. 379, 383 (Bankr. N.D. Ind. 2003) (a creditor's collateral must be part of the bankruptcy estate to be a secured creditor); *see also* 11 U.S.C. § 101(10) for a general definition of "creditor" as such is defined in the Code.

[8] 11 U.S.C. § 506(a)(1) (unsecured claim mentioned); *see In re Kiwi Intern. Air Lines, Inc.*, 344 F.3d 311, 318 (3d Cir. 2003) (unsecured creditors must share according to their interests in any assets left over after secured creditors have received payment); *see also* 11 U.S.C. § 101(10) for a general definition of "creditor" as such is defined in the Code.

[9] 11 U.S.C. § 544; *see In re Holmes*, 298 B.R. 477, 484–85 (Bankr. M.D. Ga. 2003) (liquidation can also take place under a Chapter 11 reorganization plan and is not exclusive to Chapter 7).

discretion and authority to continue operating its business. The Debtor, however, must develop and obtain court approval of a plan of reorganization that includes payments to creditors. Once the plan is approved, the Debtor emerges from bankruptcy with a fresh start.

B. Chapter 7—Liquidation

The other form of bankruptcy often encountered on a construction project is a Chapter 7 liquidation. Under Chapter 7, the Debtor's assets are liquidated, creditors are paid a pro rata share of the proceeds of the liquidation, and the debtor corporation then ceases to operate. In a Chapter 7 liquidation, the bankruptcy court appoints a Trustee to handle the liquidation and winding up of the business.

Whether a Chapter 11 reorganization or a Chapter 7 liquidation, the basic goal of the bankruptcy process is to give the Debtor relief from debts that no longer can be paid in the ordinary course of business.[10] Under Chapter 7, a corporate Debtor is not discharged from its debts, but all assets obtained from the liquidation are distributed and the corporate Debtor ceases to exist.

From a creditor's standpoint, a basic premise of bankruptcy law is that creditors in the same class should be treated equally. As mentioned, in most bankruptcies, there are several different classes of creditors, including Secured Creditors and general Unsecured Creditors. Within each class, creditors should receive equal treatment so that no one creditor receives more than its fair share of money or other assets from the Debtor. Assets of the Debtor are protected in various ways to facilitate an orderly gathering and distribution of funds to all creditors.

C. Automatic Stay

A fundamental element of the bankruptcy system is the automatic stay, a rigidly enforced prohibition against taking any steps that are hostile to the Debtor or that affect the Debtor's property. Once a Debtor has filed for bankruptcy, all actions against the Debtor by its creditors are automatically brought to a standstill.[11] The automatic stay allows the Debtor in Possession or Trustee the necessary breathing space to determine what steps to take in order to reorganize or liquidate the company while holding off creditors.[12] The automatic stay also prevents any one creditor from obtaining more money or property than other creditors.[13]

[10] *Vill. of San Jose v. McWilliams*, 284 F.3d 785, 790 (7th Cir. 2002) (bankruptcy is to provide creditors with payment for what they are owed while giving debtors a chance to rebuild).
[11] *United States v. White*, 466, F.3d 1241, 1244 (11th Cir. 2006) (Chapter 7 debtors are provided with a break from the collection efforts of creditors through the "automatic stay"); *In re Jones*, 348 B.R. 715, 717 (Bankr. E.D. Va. 2006) (legal action against the debtor comes to a standstill in bankruptcy).
[12] *In re Jones*, 348 B.R. 715, 717–18 (Bankr. E.D. Va. 2006).
[13] *See generally* 11 U.S.C. § 362.

The automatic stay prohibits almost all creditor attempts to collect debts or property from the Debtor. The stay prohibits such actions as:

- Filing a lawsuit or demanding arbitration against the Debtor[14]
- Advancing a lawsuit or arbitration against the Debtor[15]
- Enforcement of a judgment against the Debtor[16]
- Seizing the Debtor's materials, tools, equipment, or supplies[17]
- Filing or foreclosing a lien against the property of the bankruptcy Estate[18]
- Filing or foreclosing a lien against the Debtor's property[19]
- Recovering the Debtor's property or claims[20]
- Reconciling amounts owed to the Debtor with amounts the Debtor owes creditors[21]
- Advancing a tax proceeding against the Debtor[22]

The stay is "automatic" because no court order is necessary to implement it. Instead, the stay is legally in effect from the moment the bankruptcy petition is filed.[23]

D. Sanctions for Violation of Automatic Stay

The automatic stay is enforced by the bankruptcy court, which may impose sanctions against a creditor for violating the automatic stay. Such sanctions may include imposition of administrative penalties and attorneys' fees. Any individual injured by reason of the violation may recover actual damages, costs, attorneys' fees, and, in some instances, punitive damages against the violator. Violation of the automatic stay in connection with a contract, such as terminating the contract for default, may be treated as a breach of contract by the party terminating the contract. The automatic stay does not, however, stop time or prevent a contract from expiring under its terms.[24] The automatic stay also does not preclude a defaulting party from having a contract canceled if proper notice is given.[25] Because the consequences of violating the automatic stay can be severe, before a creditor takes any action against a Debtor, the creditor must determine whether the planned action violates the stay.

[14]11 U.S.C. § 362(a)(1).
[15]*Id.*
[16]11 U.S.C. § 362(a)(2).
[17]*See generally* 11 U.S.C. § 362(a)(3).
[18]11 U.S.C. § 362(a)(4).
[19]11 U.S.C. § 362(a)(5).
[20]11 U.S.C. § 362(a)(6).
[21]11 U.S.C. § 362(a)(7).
[22]11 U.S.C. § 362(a)(8).
[23]*In re Best Payphones, Inc.*, 279 B.R. 92, 97 (Bankr. S.D.N.Y. 2002).
[24]*In re Margulis*, 323 B.R. 130, 133 (Bankr. S.D.N.Y. 2005).
[25]*In re Margulis*, 323 B.R. 130, 133 (Bankr. S.D.N.Y. 2005).

E. Relief from the Automatic Stay

A creditor is not, however, without an avenue for relief from the impact of the automatic stay, particularly with regard to efforts to obtain secured property held by the Debtor or to determine amounts owed by the Debtor in a litigation or arbitration forum. The United States Bankruptcy Code provides that a creditor may petition the bankruptcy court for relief from the automatic stay to allow the creditor to proceed against the Debtor or the Debtor's property.[26] A court may grant relief from the automatic stay when the property against which the creditor seeks to take action is of no value to the Bankruptcy Estate because the Debtor does not have any equity in the property. Creditors can request that the automatic stay be in some way altered.[27]

For example, a Debtor lacks equity in certain property when the value of the property is less than the security interest in the property. For example, a Debtor may have, prebankruptcy, pledged a piece of earthmoving equipment as security for a bank loan. If the amount of the bank's security interest exceeds the value of the equipment, the bank could move for relief from the automatic stay on the ground that the Debtor has no equity in the property. Because the Debtor has no equity, the property would have no value to the Bankruptcy Estate or to other creditors, and therefore the bank should be permitted to foreclose its interest.

Other circumstances may justify relief from the automatic stay "for cause."[28] For example, a common problem in a construction setting is the failure of the Debtor to maintain insurance on equipment. A creditor with a security interest in the equipment could move for relief from the automatic stay to allow foreclosure if the Debtor is not adequately protecting the security interest by maintaining insurance. Relief from the automatic stay may be justified if the Debtor fails to maintain or secure materials or equipment.

A creditor also may seek relief from the automatic stay to allow arbitration or litigation to proceed to determine the amount of a debt owed by the Debtor. In this instance, the creditor would not be allowed to proceed to judgment and collection; however, the alternate forum (which presumably would be more familiar with construction cases) would be permitted to determine the amount of the debt. The creditor's pro rata share of the proceeds of the bankruptcy liquidation or reorganization then would be calculated according to the amount of the debt established by the arbitration or litigation proceeding.

In order to obtain relief from the automatic stay, a creditor must file a motion with the bankruptcy court having jurisdiction over the Debtor's case. The court must take some action on the motion within 30 days after the motion is filed or the stay is automatically terminated as to the moving party.[29] Although the court may simply set the motion for a hearing within that 30-day time limit, the time limitation allows a creditor to shorten what otherwise could be a lengthy process.

[26]11 U.S.C. § 362(d).
[27]*In re Jones*, 348 B.R. 715, 718 (Bankr. E.D. Va. 2006).
[28]11 U.S.C. § 362(d)(1).
[29]11 U.S.C. § 362(e)(1).

F. Preferential Transfers

The Code allows a Trustee to void a prebankruptcy transfer of property or money by the Debtor to a creditor.[30] The purpose for this rule is to prevent Debtors from showing favoritism by paying one creditor at the expense of others.[31] Debtors that make payments 90 days before filing for bankruptcy run the risk that such will be viewed as suspect and rescinded by the Trustee.[32] Under the Code, Debtors are considered to be insolvent 90 days before filing; as such, the Trustee may question how creditors were paid during this prefiling period.[33]

If a transfer is voided, the creditor is required to return the money or property to the Bankruptcy Estate.[34] A transfer is voidable if it is a transfer of an interest in the Debtor's property:[35]

(1) Benefiting a creditor[36]
(2) For an antecedent debt of the Debtor[37]
(3) Made by an insolvent Debtor[38]
(4) Made within 90 days before the date of bankruptcy or one year before the date of bankruptcy if the creditor was an "insider"[39]
(5) The transfer allows a creditor to receive more than it would have received under a Chapter 7 liquidation[40]

A transfer that satisfies all of these five criteria is voidable even if the payment to the creditor was a lawfully made payment of a preexisting debt. Such transfers are termed "preferential" transfers, even though no intent to defraud other creditors is necessary. Simply stated, a preferential transfer is voidable to prevent favoritism among creditors even if the creditor's increased share was obtained before the Debtor filed for bankruptcy.

While the Code does not define "Debtor's Property," such includes all assets that could have been used by the Debtor to pay creditors, had the assets not been improperly transferred.[41] Assets not available to pay creditors usually are not considered part of the estate and are not subject to being reclaimed by the Trustee.[42]

[30] 11 U.S.C. § 547(b); *In re Paradise Valley Holdings, Inc.*, 347 B.R. 304, 308 (Bankr. E.D. Tenn. 2006) (transfers can be rescinded in Chapter 7 cases).
[31] *In re Lambert Oil Co., Inc.*, 347 B.R. 173, 177 (W.D. Va. 2006); *In re Pameco Corp.*, 356 B.R. 327, 337 (Bankr. S.D.N.Y. 2006); *In re Goss*, 378 B.R. 320, 326 (Bankr. E.D. Okla. 2007).
[32] 11 U.S.C. § 547(b)(4)(A); *In re GS Inc.*, 352 B.R. 858, 863 (Bankr. E.D. Ark. 2006).
[33] 11 U.S.C. § 547(f); *But see In re ML & Assocs., Inc.*, 301 B.R. 195, 200 (Bankr. N.D. Tex. 2003) (while debtors are assumed to be insolvent 90 days before filing for bankruptcy, such can be shown to not be the case).
[34] *In re Bridge Info. Sys., Inc.*, 460 F.3d 1041, 1044 (8th Cir. 2006).
[35] *See Wolff v. United States*, 372 B.R. 244, 251 (D. Md. 2007) (citing to 11 U.S.C. § 547(b)).
[36] 11 U.S.C. § 547(b)(1).
[37] 11 U.S.C. § 547(b)(2).
[38] 11 U.S.C. § 547(b)(3).
[39] 11 U.S.C. § 547(b)(4)(A) & (B); *see* 11 U.S.C. § 101(31) (defining "insider").
[40] 11 U.S.C. § 547(b)(5)(A).
[41] *In re Contractor Tech., Ltd.*, 343 B.R. 573, 580 (Bankr. S.D. Tex. 2006).
[42] *Id.* at 580.

Preferential transfers may include payments made by third parties to a Debtor's creditors where the Debtor controls the transaction,[43] payment to a supplier in exchange for dismissal of an involuntary bankruptcy petition,[44] exchanging an obligation for a preexisting duty,[45] and garnishment of funds owed to the Debtor.[46] A mechanic's lien that becomes effective before the Debtor's insolvency or bankruptcy filing is generally not considered a preferential transfer because the Code excludes statutory liens.[47]

G. Exceptions to the Preferential Transfer Rule

The Code provides that under certain circumstances, a Debtor can transfer assets without such being considered preferential.[48] The eight exceptions to the preferential transfer rule include:

(1) Transfers made for new value contemporaneously given to the Debtor[49]
(2) Payments made in the ordinary course of business or under ordinary business terms[50]
(3) Loans made to the Debtor for the purchase of property secured by new value with the property serving as collateral[51]
(4) Transfer benefiting the creditor whom gave new value to the Debtor and was proper[52]
(5) Secured interest in inventory or receivables[53]
(6) Valid statutory liens[54]
(7) Domestic support and consumer-type debt payments[55]
(8) Avoided transfers made to noninsiders to benefit insiders are considered avoided only as to the insiders[56]

[43]*In re Goss*,378 B.R. 320, 327 (Bankr. E.D. Okla. 2007); *see also Warso v. Preferred Technical Group*, 258 F.3d 557, 564 (7th Cir. 2001) (third-party transfers to pay debts may be avoided).
[44]*In re Ramba, Inc.*, 416 F.3d 394 (5th Cir. 2005) (supplier's dismissal of an involuntary bankruptcy petition in exchange for payment was not considered new value and was not allowed).
[45]*In re Payless Cashways, Inc.*, 306 B.R. 243, 253 (8th Cir. 2004).
[46]*In re Smith*, 333 B.R. 739 (Bankr. D. Md. 2005) (transfer of funds by writ of garnishment served on bank was found to be involuntary and voidable by the debtor); *In re Morehead*, 249 F.3d 445, 448 (6th Cir. 2001) (earned wages can be voided); *But see In re Cummings*, 266 B.R. 138, 146 (Bankr. N.D. Ohio 2001) (an enforceable lien prevented the avoidance of garnished proceeds).
[47]11 U.S.C. § 547(c)(6).
[48]11 U.S.C. § 547(c).
[49]11 U.S.C. § 547(c)(1)(A) & (B).
[50]11 U.S.C. § 547(c)(2)(A) & (B).
[51]11 U.S.C. § 547(c)(3)(A)(i)–(iv) & (B).
[52]11 U.S.C. § 547(c)(4)(A) & (B).
[53]11 U.S.C. § 547(c)(5).
[54]11 U.S.C. § 547(c)(6).
[55]11 U.S.C. § 547(c)(7) & (8).
[56]11 U.S.C. § 547(i).

1. "New Value" Exception

Perhaps the most important exception to the preferential transfer rule is the contemporaneous exchange of property for new value.[57] Under this exception, a transfer of new value going to the Debtor's estate will make up for any outgoing payments such that creditors will not be cheated out of available assets.[58] For example, a release of lien or bond rights in exchange for a payment of past-due amounts may be considered as an exchange for new value so that the payment is not a voidable preference.[59]

Although a waiver and release of lien or bond rights is not universally held to be an exchange for new value,[60] a contractor or supplier receiving a payment from an owner or another contractor in shaky financial condition can improve its chances of keeping the payment even if the payor files for bankruptcy within 90 days of the payment. The contractor or supplier receiving the payment should document that any lien or bond rights that are waived or released as a result of the payment are being given up expressly in exchange for the funds received. The waiver or release language also should be worded carefully to state that the waiver or release is contingent upon actual receipt of the funds.

In jurisdictions where a waiver and release are not considered new value, the contractor or supplier still may preserve lien or bond rights by entering into an agreement with the Debtor that the waiver and release are contingent upon the paying party not filing for bankruptcy within 90 days after the payment is made. Although the payment still could be treated as a voidable preferential transfer, the contractor or supplier would at least retain its lien or bond rights. Depending on the notice and filing requirements in the jurisdiction, however, it may be necessary to proceed with perfecting lien or bond rights during the 90-day preference period to avoid losing such rights for failure to comply with applicable time limits.

2. Ordinary Course of Business Defense

A second exception to the voidable preference rule relevant to the construction industry involves payments for work or supplies made in the "ordinary course of business" or under "ordinary business terms." The Trustee cannot rescind such payments.[61] A subcontractor receiving a payment shortly before a general contractor filed for bankruptcy must prove that such was made through normal business dealings.[62]

[57]11 U.S.C. § 547(c)(1)(A) & (B) & (c)(4); *see In re Hechinger Inv. Co. of Del., Inc.*, 489 F.3d 568, 574–75 (3d Cir. 2007) (the intent of the parties is a consideration in determining whether a transaction involves new value).

[58]*In re Inland Global Med. Group, Inc.*, 362 B.R. 459, 463 (Bankr. C.D. Cal. 2006).

[59]*In re Gem Constr. Corp. of Va.*, 262 B.R. 638 (Bankr. E.D. Va. 2000) (a subcontractor forgoing its rights against a surety that had posted a payment bond in exchange for payment by the general contractor was considered new value); *In re J.A. Jones, Inc.*, 361 B.R. 94 (Bankr. W.D. N.C. 2007) (release of lien rights could potentially constitute new value).

[60]*See In re Gem Constr. Corp. of Va.*, 262 B.R. 638, 649 & 653 (Bankr. E.D. Va. 2000).

[61]11 U.S.C. § 547(c)(2)(A) & (B); *In re ML & Assocs., Inc.*, 301 B.R. 195, 198 & 203 (Bankr. N.D. Tex. 2003) (check sent by general contractor to subcontractor less than 90 days before general contractor filed for bankruptcy was not a preferential transfer because the payment was made in the ordinary course of business and under ordinary business terms. *Id.* at 205–7).

[62]*See Id.*at 203 (citing 11 U.S.C. § 547(g)).

To invoke this exception, the creditor must establish that the "ordinary business terms" comported with industry standards.[63] In deciding if a payment was made in the ordinary course, courts will consider how long the transfer took, if the method or amount of payment deviated from prior transactions, whether any questionable payment or recovery practices were used, and if the creditor unjustly benefited from the Debtor's unfortunate circumstances.[64] In deciding if a payment was ordinary, the date the Debtor sent the check is more telling than when it was cashed or deposited.[65]

The rationale for this exception is that payments in the ordinary course of business neither drain funds from the Bankruptcy Estate nor treat other creditors unfairly because the Debtor has received something of value, that is, short-term credit. No hard and fast rule is applied when determining whether a transfer from one entity to another was made in the "ordinary course of business;" instead the courts examine the circumstances surrounding each individual case.[66]

Courts will look to the parties' past dealings to determine whether the payment at issue is consistent with those past dealings.[67] When there is no history between the parties to use as a baseline to determine what is normal, the courts will look for transactions that generally "raise a red flag."[68]

The ordinary course of business exception should apply to a progress payment made to a contractor or subcontractor in accordance with the contract payment terms or other terms established by the course of dealings between the parties. Payments made by a bankrupt subcontractor to a supplier that appear from the invoice to be late still may have been made in the ordinary course of business.[69] The closer that a payment is to the ordinary business dealings between the parties, the more likely it is that the court will treat it as an exception to the preferential transfer rule.

H. Discharge

At the conclusion of a bankruptcy case, the Debtor's plan is approved, the Debtor is discharged, and the automatic stay ceases.[70] The automatic stay is replaced

[63] *In re Matlock*, 361 B.R. 879, 884–85 (Bankr. W.D. Mo. 2007); *In re ML & Assocs., Inc.*, 301 B.R. 195, 206 (Bankr. N.D. Tex. 2003) (payments that are being analyzed to see if they were made following "ordinary business terms" are compared with industry standards and not the parties' normal practices).

[64] *In re GGSI Liquidation, Inc.*, 313 B.R. 770, 775 (Bankr. N.D. Ill. 2004).

[65] *In re ML & Assocs., Inc.*, 301 B.R. 195, 204–5 (Bankr. N.D. Tex. 2003).

[66] *In re GS Inc.*, 352 B.R. 858, 864 (Bankr. E.D. Ark. 2006).

[67] *Apex Auto Warehouse, L.P. v. Whitlock Corp.*, 245 B.R. 543, 549 (Bankr. N.D. Ill. 2000) (in determining if payments were made in the "ordinary course of business," courts will look at the parties' past dealings); *In re H.L. Hansen Lumber Co. of Galesburg, Inc.*, 270 B.R. 273, 279 (Bankr. C.D. Ill. 2001) (late payments were a deviation from the parties' past dealings and could not be saved by the "ordinary course of business defense").

[68] *In re GS Inc.*, 352 B.R. 858, 865 (Bankr. E.D. Ark. 2006).

[69] *In re GS Inc.*, 352 B.R. 858 (Bankr. E.D. Ark. 2006) (subcontractor that paid supplier for heating and cooling units approximately two months after delivery to a construction project was found to be in the ordinary course of business and not subject to be called back by the trustee despite the fact that the supplier's invoice stated payment was to be made within 30 days).

[70] *United States v. White*, 466 F.3d 1241, 1245 (11th Cir. 2006).

with a permanent injunction that prevents recovery of dischargeable debts but nondischargeable debts can be collected.[71] The discharge order does not specify which debts fit into each category.[72] Instead, such a determination usually is made in a different adjudication.[73]

The discharge is accomplished by a court order that bars the Debtor's liability on most claims.[74] Creditors can be penalized for purposely ignoring the injunction.[75] In a Chapter 7 liquidation, an individual is discharged from liability for prepetition debts.

Businesses cease to exist under a Chapter 7 liquidation rather than being discharged. Pursuant to Chapter 11, the order confirming the plan usually relieves the Debtor from obligations incurred before the confirmation.[76]

I. Nondischargeable Debts

Certain debts of an individual are not dischargeable.[77] Such nondischargeable debts include: debts for money, property, or services obtained through false statements or fraud; debts obtained through false financial statements upon which the creditor relied and which the Debtor provided with the intent to deceive; and debts arising from fraud while the Debtor was acting in a fiduciary capacity.[78]

In the construction setting, a construction trust fund statute usually treats all parties having possession of the funds as "trustees." Courts have reached conflicting results when deciding whether a construction trust fund statute creates a fiduciary duty on the "trustee" so that the breach of the fiduciary duty renders a debt nondischargeable.[79] Whether a Debtor's violation of a construction trust fund statute results in a nondischargeable debt depends on the state construction trust fund statute and the particular bankruptcy court's prior decisions interpreting the statute. Although federal laws controls the creation of a fiduciary duty, state laws are considered to determine whether such a relationship creates a nondischargeable debt.[80]

If the bankruptcy court determines that a debt is nondischargeable, the Debtor still is exposed to liability for suit and collection. Any debt owed by a Debtor on a

[71]*Id.* at 1246; *In re Diaz Rodriguez*, 357 B.R. 691, 697 (D. P.R. 2006) (the term "discharge injunction" is used, which has the same effect as a permanent injunction).
[72]*In re Jones*, 348 B.R. 715, 718 (Bankr. E.D. Va. 2006).
[73]*Id.*
[74]11 U.S.C. § 524(a).
[75]*In re Dynamic Tours & Transp., Inc.*, 359 B.R. 336, 342–43 (Bankr. M.D. Fla. 2006).
[76]11 U.S.C. § 1141(d)(1)(A).
[77]*In re Jones*, 348 B.R. 715, 718 (Bankr. E.D. Va. 2006).
[78]11 U.S.C. § 523; *see In re Webb*, 349 B.R. 711, 716 (Bankr. D. Or. 2006) ("false representation" stems from a party's statements while "false pretense" can arise from a party's actions).
[79]*In re Halversen*, 330 B.R. 291, 297–98 (Bankr. M.D. Fla. 2005) (Maryland Constructive Trust Statute does not create the type of trust whereby violation of the statute alone makes the debt nondischargeable); *In re Neal*, 324 B.R. 365 (Bankr. W.D. Okla. 2005) (under Oklahoma constructive trust fund statutes, a fiduciary duty exists only for lien claims that are nondischargeable).
[80]*Id.*at 370.

construction project should be investigated carefully to determine whether there is a basis for claiming fraud or a breach of a fiduciary duty.

IV. STATUS OF THE DEBTOR'S CONTRACTS

A. Executory Contracts

When a party files for bankruptcy, all "executory" contracts remain in full force and effect.[81] Under the Code, an executory contract is one in which there are material obligations remaining to be performed by both parties.[82] In determining whether a contract is executory, courts typically conclude that if the failure to complete performance by either party would constitute a material breach of contract, the contract is executory.[83] Generally, a construction contract is considered executory before substantial completion. Considerable punch list work remaining after substantial completion may, however, constitute a substantial remaining obligation on the contractor's part. Likewise, payment of retainage may be considered a substantial remaining obligation. Therefore, under certain circumstances, even if substantial completion has occurred on a construction project, the underlying contract still may be treated as an executory contract.

The Code offers many protections for the Debtor in Possession or Trustee to address the Debtor's executory contracts. For example, the Code prohibits termination of a contract solely because of the insolvency or bankruptcy of the Debtor. A contract clause that gives the right of termination for insolvency or bankruptcy is known as an ipso facto clause and is usually deemed invalid.[84] The Code does not recognize these clauses; instead it chooses to allow the Debtor to maintain profitable contracts as a resource for funding the Debtor's operations and obligations.[85]

B. Affirmance or Rejection

Under the Code, a Debtor in Possession or Trustee may affirm or reject an executory contract.[86] In a Chapter 7 bankruptcy, the Trustee has 60 days to affirm or reject

[81] *In re Aerobox Composite Structures, LLC*, 373 B.R. 135, 139 (Bankr. D. N.M. 2007); *In re Gencor Indus., Inc* ., 298 B.R. 902 (Bankr. M.D. Fla. 2003) (where it was determined that a "Mutual Release and Settlement Agreement" regarding licensing of a patent was not considered to be an executory contract because the parties did not have continuing obligations).
[82] 11 U.S.C. § 365; *In re Lewis*, 185 B.R. 66 (Bankr. N.D. Cal. 1995).
[83] *In re Lucre, Inc.*, 339 B.R. 648, 652 (Bankr. W.D. Mich. 2006); *see Kent's Run P'ship, Ltd. v. Glosser*, 323 B.R. 408 (W.D. Pa. 2005) (determining that a contract regarding an easement was not executory because the parties had no continuing duties whereby inaction would constitute a material breach).
[84] 11 U.S.C. § 365(e)(1)(A) – (C); *In re EBC I, Inc.*, 356 B.R. 631, 640 (Bankr. D. Del. 2006).
[85] *In re S. Pac. Funding Corp.*, 268 F.3d 712, 715–16 (9th Cir. 2001).
[86] 11 U.S.C. § 365(d)(2); *In re JZ, LLC*, 357 B.R. 816, 820 (Bankr. D. Idaho 2006) (the debtor's unexpired leases can also be accepted or declined).

an executory contract, or it is considered rejected.[87] Under Chapter 11, the Debtor in Possession can affirm or reject a contract at any time up until the court approves the Debtor's plan of reorganization. When a Chapter 11 Debtor neither declines nor accepts an executory contract, a "ride through" occurs and the agreement survives bankruptcy.[88]

A Debtor in Possession must use its "business judgment" when deciding which contracts to continue and which to abandon.[89] The ebtor should reject those contracts that are not profitable.[90] Those contracts that are rejected are considered to be breached, and the other contracting party then has an unsecured claim for damages.[91]

If the Debtor in Possession or Trustee decides to affirm a contract that is in default, any default first must be cured, and the other party must receive adequate compensation for damages incurred as a result of the default.[92] The Debtor also must give adequate assurance of future performance.[93] As a practical matter, curing existing defaults and giving adequate assurance of future performance can be insurmountable obstacles if the Bankruptcy Estate is in a condition sufficient to warrant filing for bankruptcy.

If an executory contract is affirmed, performance of contract obligations by both parties can continue in the normal course of business. An affirmed contract creates additional obligations on the part of the Debtor in Possession or Trustee and gives substantial rights to the other party to the contract.

All debts and expenses incurred by the Debtor's estate for an affirmed contract are considered as "administrative expenses" of the estate.[94] "Administrative expense" is a term of art applied to expenses and debts incurred by the Debtor during the administration of the Bankruptcy Estate.[95]

Typically, an expense or debt must be shown to benefit the Debtor's estate in order to be treated as an administrative expense.[96] The expenses a Debtor incurs in attempting to acquire assets, however, may not be administrative in nature.[97] Since administrative expenses are given a priority, a party to an affirmed contract may have some

[87]11 U.S.C. § 365(d)(1).

[88]*In re Hernandez*, 287 B.R. 795, 799 (Bankr. D. Ariz. 2002) (the "ride through" concept is not part of the Code and gets its origins from case law whereby contracts that the debtor or trustee does not act on survive as if bankruptcy had never been filed).

[89]*In re Helm*, 335 B.R. 528, 538 (Bankr. S.D.N.Y. 2006).

[90]*Id.*at 538; *see In re Globe Metallurgical, Inc.*, 312 B.R. 34, 38 (Bankr. S.D.N.Y. 2004) (debtors are allowed to utilize contracts that are money makers and shun those that are not).

[91]11 U.S.C. § 365(g)(breach); *In re Health Mgmt. Ltd. P'ship*, 332 B.R. 360, 363 (Bankr. C.D. Ill. 2005) (damages).

[92]11 U.S.C. § 365(b)(1)(A) – (C).

[93]*Id.*

[94]11 U.S.C. § 503.

[95]*In re HNRC Dissolution Co.*, 371 B.R. 210, 224 (E.D. Ky. 2007).

[96]*In re Vantage Invs., Inc.*, 328 B.R. 137, 142 (Bankr. W.D. Mo. 2005).

[97]*In re Lickman*, 273 B.R. 691 (Bankr. M.D. Fla. 2002) (debtor's attorneys' fees incurred in attempting to claim an inheritance in probate court were found to not benefit the bankruptcy estate and were not allowed as administrative expenses).

assurances that the debts and costs associated with the affirmed contract may be given preference in the ultimate distribution of the Debtor's assets if such benefited the Bankruptcy Estate.[98]

C. Assignment

Although the Debtor in Possession or Trustee may assign an executory contract,[99] adequate assurance of performance by the contract assignee must be given.[100] If applicable state law or nonbankruptcy federal law provides that the other party would have to consent to an assignment of the contract, an executory contract cannot be assigned without such consent.[101] Contracts that usually cannot be assigned (or assumed) frequently involve intellectual property rights.[102]

D. Minimizing the Impact on Executory Contracts

Because the Trustee (or Debtor in Possession) can affirm or reject a contract,[103] the other party to a contract with a Debtor is effectively denied the freedom to exercise its contract rights against the bankrupt Debtor. Most construction projects require quick decisions and aggressive actions to continue the work. If a party in bankruptcy continues to perform without difficulty, the bankruptcy may not have any impact. But if the Debtor in Possession or Trustee ceases performance and fails to promptly reject the contract, the impact on the progress of construction can be immediate and severe. Fortunately, the nonbankrupt party to a contract can take certain measures to minimize the impact of a Debtor's executory contract on a construction project.

1. Terminate Before Bankruptcy

In many cases, there are advance signs that a party on a construction project is about to file for bankruptcy. The problems inherent in having an executory contract with a Debtor can be avoided by terminating the contract before the bankruptcy filing. Such termination, however, must be based on a default for some reason other than insolvency or impending bankruptcy, or the contract must contain a termination for convenience clause.[104] For the termination to be effective, the terminating party

[98]*See generally*11 U.S.C. § 503; *In re Vantage Invs., Inc.*, 328 B.R. 137, 142 (Bankr. W.D. Mo: 2005); *In re Lickman*, 273 B.R. 691, 697 (Bankr. M.D. Fla. 2002).
[99]11 U.S.C. § 365(f)(1).
[100]11 U.S.C. § 365(f)(2)(A) – (B).
[101]11 U.S.C. § 365(c)(1)(A) – (B).
[102]*In re Wellington Vision, Inc.*, 364 B.R. 129 (S.D. Fla. 2007) (franchise agreement involving the use of certain trademarks could not be assigned or assumed without the owner's consent because of the Lanham Act). *But see In re Quantegy, Inc.*, 326 B.R. 467 (Bankr. M.D. Ala. 2005) (where license agreements involving trademark and manufacturing process patents could be assumed and assigned because language allowing such in limited circumstances appeared in the contract).
[103]11 U.S.C. § 365.
[104]*See In re J.E. Adams Indus., Ltd.*, 269 B.R. 808, 813 (N.D. Iowa 2001) (provisions that allow for termination because of default can be enforced even if bankruptcy has been filed).

must follow all applicable contract termination procedures. All required notice and cure periods must run in full before the bankruptcy filing. Otherwise, the termination cannot be made final because of the automatic stay.

The right to affirm or reject a contract does not apply if the contract has been terminated before the bankruptcy petition is filed or is in default.[105] After termination, there is no valid contract to be affirmed or rejected.

2. Exercise Contract Rights

A typical construction contract includes a clause allowing an owner or general contractor (or a subcontractor in the case of a sub-subcontractor's bankruptcy) to supply the necessary labor, materials, and equipment to complete the other party's work in the event of a failure of performance. If a contractor or subcontractor files for bankruptcy and ceases performance, the other party to the contract could use such a clause to supplement the Debtor's forces without violating the automatic stay and without violating the right of the Debtor in Possession or Trustee to affirm or reject the contract.[106] Typically, such a contractual clause would permit the costs incurred for supplementing the Debtor's forces to be backcharged to the Debtor in Possession or Trustee. Costs not recovered through backcharges become general, unsecured claims against the Bankruptcy Estate when a contract has been rejected, and administrative expenses of the Bankruptcy Estate when a contract has been affirmed.

3. Seek Relief from the Automatic Stay

If the Debtor is in default of its contract, the other party may petition for relief from the automatic stay to allow termination of the contract. Such a request limits the period of uncertainty as to whether the Debtor will affirm or reject the contract. The motion forces the Debtor in Possession or Trustee to take a position on the contract, and the other party to the contract is assured that the bankruptcy judge will act within a reasonable time on the motion. A party that can demonstrate that the contract is seriously in default and has few hopes for a cure would likely establish good cause for lifting the stay.[107] As a practical matter, if the Debtor's situation is hopeless, the Debtor in Possession or Trustee may not oppose lifting the stay to allow termination, thereby agreeing to the requested relief.

4. Seek a Time Limit on Affirming or Rejecting the Contract

If the Debtor continues performing the contract, and is not in default, there are no grounds for terminating the contract. As previously discussed, neither insolvency

[105]11 U.S.C. § 365(b)(1) (executory contracts that are in default must be cured or a promise given that they will be cured before they can be assumed); *In re New Breed Realty Enters., Inc.*, 278 B.R. 314 (Bankr. E.D.N.Y. 2002) (nonmonetary defaults that are incapable of being cured can prevent a contract from being assumed).

[106]*See generally* 11 U.S.C. § 362 (automatic stay); 11 U.S.C. § 365 (executory contracts).

[107]*See In re Cooper*, 296 B.R. 410, 412 (Bankr. E.D. Va. 2002) (injury to a secured creditor may be grounds for lifting the automatic stay).

nor bankruptcy standing alone is a ground for termination. The Chapter 11 Debtor that continues performance, therefore, has a prolonged period of time during which it may decide to affirm or reject the contract. As previously mentioned, an executory contract can be affirmed or rejected until the Debtor's reorganization plan is approved.[108] Even if a contract is not in default, however, the Code allows the nonbankrupt party to petition the bankruptcy court to set a "reasonable" time limit on the Debtor in Possession's or Trustee's choice to affirm or reject the contract even before the reorganization plan is confirmed.[109] Although the question of what is a "reasonable" time depends on the circumstances of each case,[110] where the contract is not in default, this request for a time limit to affirm or reject the contract is the only way to shorten the process.

V. STATUS OF MATERIALS AND EQUIPMENT

On a construction project, materials and equipment incorporated into the work can be a significant part of the project's value. During construction, materials and equipment needed for the project can be in the possession of suppliers, subcontractors, the contractor, or the owner. Materials and equipment may or may not be paid for, depending on whether the materials are installed, stored at site, stored off site, in transit, or still in the possession of a supplier. When any party in the chain of supply files for bankruptcy, the question of ownership of materials and equipment inevitably arises.

A. Property of the Debtor's Estate

Unless the Debtor in Possession or Trustee continues performance, construction materials and equipment intended for the project may be seized by the Debtor in Possession or Trustee and used for reorganization of the Bankruptcy Estate in the case of a Chapter 11 or in a liquidation of assets in a Chapter 7 bankruptcy. Problems with materials and equipment arise primarily when the Debtor in Possession or Trustee attempts to make materials and equipment part of the Bankruptcy Estate rather than allow materials and equipment to be incorporated into the project.

The Bankruptcy Estate consists of all of the Debtor's legal and equitable interests in property as of the date of filing the bankruptcy petition.[111] The extent and validity of the Debtor's interest in such property, however, is a question of state law rather than bankruptcy law.[112]

The ultimate use of materials and equipment intended for incorporation into the construction project depends on who has possession of materials, whether the

[108] 11 U.S.C. § 365(d)(2).
[109] *Id.*; *See In re Adelphia Commc'ns Corp.*, 291 B.R. 283, 292 (Bankr. S.D.N.Y. 2003).
[110] *See In re Teligent, Inc.*, 268 B.R. 723, 738 (Bankr. S.D.N.Y. 2001).
[111] 11 U.S.C. § 541(a)(1).
[112] *In re Jazzland, Inc.*, 322 B.R. 610 (Bankr. E.D. La. 2005).

materials are paid for, the terms and conditions of the contract, and the intentions of the Debtor in Possession or Trustee regarding affirming or rejecting the contract. If the Debtor has not yet been paid for the materials in a progress payment at the time of the bankruptcy filing, the materials will be considered property of the Bankruptcy Estate. If the Debtor rejects the construction contract, the materials may be used for a Chapter 11 reorganization or a Chapter 7 liquidation. If the Debtor in Possession or Trustee affirms the contract, the materials and equipment can be installed and paid for in a progress payment in the ordinary course of business.

B. Supplier's Right to Recover Goods

A supplier must exercise caution in recovering materials from a Debtor in Possession or Trustee. If the Debtor has not paid for the materials or equipment, however, the supplier of the materials or equipment can repossess them in certain circumstances. Under the Code, a supplier of goods can regain possession of materials delivered to a Debtor if the supplier demands their return within 45 days after the Debtor's receipt of the goods or not later than 20 days after the commencement of the bankruptcy case, if the 45-day period expires after the commencement of the case.[113] Although the automatic stay does not apply to the supplier's reclaiming of the goods in such a circumstance, suppliers must be extremely diligent in exercising the right to recover goods. The right to recover goods cannot be exercised if the Debtor was solvent at the time of delivery.

An owner, contractor, or subcontractor must be alert to the possibility that stored materials or equipment for which payment has been made may be seized by the supplier for nonpayment. The safe course is to make sure that the goods were delivered more than 45 days before making payment, although given the timing requirements in some contracts, this may not always be feasible. After the 45-day recovery period has run, with certain narrow exceptions, the automatic stay prohibits seizure. Even if the materials or equipment are covered by a security agreement giving the supplier a security interest in the goods, the supplier cannot repossess them without obtaining relief from the automatic stay from the bankruptcy court.[114]

The supplier may recover materials and equipment only where the goods are still in the Debtor's possession.[115] Materials and equipment that have been incorporated into the work are not subject to seizure. Stored materials and materials stored at the project site could be at risk, however, depending on the contract terms governing possession, risk of loss, and title to the goods.

C. Stored Materials

Special problems arise when materials have been paid for but have not been permanently incorporated into the work. The Debtor in Possession or Trustee could claim

[113] 11 U.S.C. § 546(c). The statute was amended in 2005 to lengthen the demand period. Previously, a supplier had only 10 days to reclaim goods under these circumstances.
[114] 11 U.S.C. § 362(d).
[115] *In re Nitram, Inc.*, 323 B.R. 792 (Bankr. M.D. Fla. 2005).

that stored materials are property of the Bankruptcy Estate because the Debtor has some legal or equitable interest, even though the materials have already been paid for in a progress payment.[116] The contrary argument is that the Bankruptcy Estate no longer has any legal or equitable interest in the materials or equipment because the Debtor received full value upon payment.

The owner, contractor, or subcontractor can obtain some protection against the claims of a Debtor in Possession or Trustee to stored materials by including appropriate terms in the contract. One alternative is to include a contract clause that treats stored materials as a bailment arrangement. As a "bailee," the Debtor does not have any property interest in the materials or equipment. Lacking a property interest, the Debtor in Possession or Trustee would have no right to use the property in a Chapter 11 reorganization or a Chapter 7 liquidation.[117] Another option, frequently seen in government contracts, is to include a clause that passes title to stored materials to the owner or general contractor upon payment. In light of such a clause, it would be unlikely that the Debtor in Possession or Trustee would prevail in claiming that the debtor's estate has a property interest in the goods already paid for by the non-bankrupt party to the contract.[118] Clauses affecting title to and possession of materials and equipment must be coordinated carefully with clauses on insurance and risk of loss because coverage may be affected when property ownership interests change.

D. Voiding Unperfected Security Interests

The Code gives a Debtor in Possession or Trustee the authority to void unperfected security interests in property.[119] A Debtor in Possession or Trustee obtains a secured interest in property in which the Debtor has any legal or equitable interest other than mere possession. If the Debtor in Possession or Trustee voids an unperfected security interest in property, the security interest given to the Debtor in Possession or Trustee takes priority and allows seizure of the goods. This provision of the Code is known as the strong-arm provision.

The strong-arm provision has been applied to construction materials to allow a Trustee to seize materials provided by a general contractor to a debtor-subcontractor for installation on the project. Even though the general contractor retained title to the materials, the general contractor's interest was treated as a mere security interest. The Trustee was permitted to void the general contractor's security interest because a Uniform Commercial Code (UCC) financing statement had not been recorded, a requirement for perfecting the general contractor's security interest. Thus, the Trustee was entitled to possession of the materials.[120]

To avoid these harsh results, if the jurisdiction where the project is located treats the owner's or contractor's interest in stored materials as a security interest, then

[116] *See In re A-1 Hydro Mechs. Corp.*, 92 B.R. 451, 457 (Bankr. D. Haw. 1988).

[117] *In re Warde Elec. Contracting, Inc.*, 308 B.R. 659, 664 (Bankr. S.D.N.Y. 2004).

[118] *In re Coated Sales, Inc.*, 112 B.R. 560 (Bankr. S.D.N.Y. 1988).

[119] 11 U.S.C. § 544.

[120] *In re A-1 Hydro Mechs. Corp.*, 92 B.R. 451 (Bankr. D. Haw. 1988).

the safest course of action is to file a UCC financing statement (commonly called a UCC-1) to protect the security interest from being voided under the Code's strong-arm provision.

VI. STATUS OF CONTRACT FUNDS

Once work begins, construction funds will be flowing through the system for payment to the contractor, subcontractors, laborers, and suppliers. If one of the parties in the payment process files for bankruptcy, any funds in the system could be claimed by the Debtor in Possession or Trustee as property of the Bankruptcy Estate. It is possible, although unlikely, that payment will be made downstream in the ordinary course of business. The Debtor in Possession or Trustee, however, may hold or seize funds intended to be paid to those downstream in the payment process. Unless those downstream entities gain control of the construction funds in the payment system, the funds may be absorbed in bankruptcy into the Bankruptcy Estate and used in a Chapter 11 reorganization or a Chapter 7 liquidation.

From the perspective of nonbankrupt participants in the construction project, the best result is for payments to be made downstream in the ordinary course of business. The filing of a bankruptcy petition, however, can set off an intense competition among the Debtor in Possession or Trustee, those downstream seeking payment for labor and materials already provided, those upstream asserting backcharges, and banks, guarantors, and sureties that have put money into the project to make up for the Debtor's financial shortcomings.

In an ideal world, all of the claimants are entitled to be made whole. The reality, however, is that only claimants with valid claims under the Code and applicable state law will be successful. Numerous Code provisions and state laws apply to the process of sorting out the parties' competing rights and interests in contract funds.

A. Unearned Contract Funds

Problems arise only for funds earned for work performed. The Debtor in Possession or Trustee cannot claim portions of the contract price for work that has not been performed any more than a nonbankrupt contractor, subcontractor, or supplier is entitled to payment for work not performed or for materials not supplied.

B. Earned but Unpaid Contract Funds

There are several legal theories to claim funds flowing through the payment pipeline when one of the parties on the project files a petition in bankruptcy. Most of these theories involve the principle that it is more equitable for the parties downstream that have spent time and money on the project to be paid the funds than for the Debtor to keep the funds for use in a reorganization or a liquidation.

1. Constructive Trust

One of the most important theories used in capturing contract funds is the constructive or "construction" trust doctrine. The constructive trust doctrine means that the court imposes a trust for the benefit of contractors, subcontractors, materialmen, and laborers. If contract funds are treated as being held in trust, a debtor does not have any property interest in the funds.[121] Whether a constructive trust exists depends on state law.

A constructive trust may be created by a builder's trust fund statute[122] or by terms of the owner/general contractor contract that create a retainage fund for the payment of subcontractors and suppliers.[123] At least one court has held that owner/general contractor contract terms obligating the general contractor to pay for labor and materials in performing the work create a constructive trust for payment of subcontractors and suppliers.[124]

2. Equitable Lien

Rather than treating the contract funds as being held in trust, certain courts have given claimants an equitable lien on the funds.[125] Third-party beneficiary entitlement to funds is another theory that may be applied. The Debtor's contract may contain terms that expressly make downstream claimants third-party beneficiaries of the contract's payment terms. In that event, the claimants could claim funds as beneficiaries of rights created under the contract.[126]

3. Joint Check Agreements

Some owners and general contractors routinely use joint check agreements. The wording of a joint check agreement usually will determine whether joint check funds can be seized by a Debtor that is a joint payee on the check. A joint check agreement that expressly states that joint check funds are the property of the ultimate recipient of the funds and not the property of the other joint payee (the Debtor) will make it very difficult for the Debtor in Possession or Trustee to claim that the funds are property of the Dankruptcy Estate.[127]

[121] *In re R.W. Leet Elec., Inc.*, 372 B.R. 846 (B.A.P. 6th Cir. 2007).
[122] *Id.*at 850.
[123] *In re La Follette Sheet Metal, Inc.*, 35 B.R. 634 (Bankr. E.D. Tenn. 1983).
[124] *Gold v. Alban Tractor Co.*, 202 B.R. 424 (Bankr. E.D. Mich. 1996).
[125] *Matter of Gebco Inv. Corp.*, 641 F.2d 143 (3d Cir. 1981).
[126] *Id.* at 147.
[127] *See Mid-Atl. Supply, Inc. of Va. v. Three Rivers Aluminum Co.*,790 F.2d 1121 (4th Cir. 1986); *T & B Scottdale Contractors, Inc. v. United States*, 866 F.2d 1372 (11th Cir. 1989) (joint check agreement specified that funds in construction account were to pay sub-subs and suppliers); *But see Ga. Pac. Corp. v. Sigma Serv. Corp.*, 712 F.2d 962 (5th Cir. 1983) (joint check funds held part of debtor's estate).

4. Setoff

Cases involving constructive trust or other equitable theories focus on the rights of claimants downstream in the payment process. Equally valid and often superior rights to funds exist for parties upstream of a Debtor. These rights include setoffs and backcharges for claims against the Debtor. Parties upstream also often have the right to insist that contract funds be paid to downstream claimants rather than be seized by a Debtor in midstream as property of the Debtor's estate.

The Code allows parties upstream of the Debtor a right to set off certain claims against the Debtor's right to payment of contract funds.[128] The right of setoff is available for mutual prepetition debts.[129] Setoff is available even though the mutual debts arise out of different transactions or contracts.[130] The Code recognizes the right of setoff, but it is available only if allowed under applicable state law.

Setoff is important for a party upstream of the Debtor when the upstream party wants to pay contract funds to claimants downstream of the Debtor and bypass the Debtor in Possession or the Trustee. A general contractor may pay off claimants with claims against a debtor-subcontractor and set them off against the Debtor's claim for contract payments arising out of the subcontract.[131] A project owner may set off amounts paid to a debtor/general contractor's subcontractors and suppliers against balances due to the debtor/general contractor.[132] Setoff is limited, however, to cases where the party paying off the claims has an independent obligation, such as under a payment bond or mechanic's lien statute, to make such payments. Setoff is allowed even though the claims of subcontractors and suppliers are not liquidated at the time of the petition of bankruptcy. It is sufficient that the claimants have claims that may be liquidated at some later point in settling up with the owner or general contractor.[133]

Although setoff is permitted in certain contexts, it is an action against the Debtor that is prohibited by the automatic stay.[134] Therefore, a party seeking to pay claims and set them off against amounts owed to the Debtor must obtain relief from the automatic stay before asserting a right of setoff.[135] Before making any payments to claimants, the safe course is first to develop a list of all outstanding claims and then to petition the court to allow the payment of those claims and to set off the amount of the payments against the Debtor's claims against contract funds.

[128] 11 U.S.C. § 553.

[129] *In re M&T Elec. Contractors, Inc.*, 267 B.R. 434 (Bankr. D. Col. 2001).

[130] *Westinghouse Credit Corp. v. D'Urso*, 278 F.3d 138 (2d Cir. 2002).

[131] *In re M&T Elec. Contractors, Inc.*, 267 B.R. 434 (Bankr. D. Col. 2001). In *M&T Elec. Contractors*, a first-tier subcontractor paid off claimants of a bankrupt second-tier electrical subcontractor. Because the first-tier subcontractor had provided a payment bond and was obligated to pay the debtor's claimants, the first-tier subcontractor was allowed to pay the debts and then set off the amounts of the payments against the debtor's claim for payments due under its purchase order with the bankrupt second-tier subcontractor.

[132] *In re Fulghum Constr. Corp.*, 23 B.R. 147, 152 (Bankr. M.D. Tenn. 1982). In *Fulghum Constr. Corp.*, the project owner had an independent obligation to pay claimants under the state mechanic's lien law.

[133] *Id.* at 151–52.

[134] 11 U.S.C. § 362.

[135] *In re Rand Energy Co.*, 256 B.R. 712 (Bankr. N.D. Tex. 2000).

5. *Recoupment*

Another right that can be asserted by parties upstream from the Debtor in the payment process is known as recoupment. Recoupment is available for claims arising out of the same transaction or contract, and is not subject to the Bankruptcy Code's setoff rules. Recoupment allows an owner, or a general contractor or subcontractor, to reduce the Debtor's claims to contract funds by deducting backcharges or other claims against the Debtor arising out of the same contract.[136] Recoupment is available for prepetition and postpetition obligations. An owner or general contractor can assert recoupment as a matter of right to reimburse itself for costs incurred in supplementing the Debtor's forces. The right of recoupment also applies to completion costs if the Debtor stops work and the owner or general contractor is forced to complete the work.[137]

6. *Surety Claims to Funds*

A bond surety often will be obligated to step in and pay off claimants on a payment bond provided by a contractor or subcontractor that petitions for bankruptcy. The surety then is held to have stepped into the shoes of those claimants. This entitles the surety to assert all of the constructive trust and equitable lien rights that the claimants would have had to recover contract funds.[138] Because the surety's rights "relate back" to the date of the surety bond, the surety's rights often are held to be superior to other claimants, including the Trustee in bankruptcy.[139]

VII. OTHER SOURCES OF FUNDS

Because of the difficulties and costs inherent in battling over construction project funds, parties on a construction project generally will seek to recover from any other available source of funds. An advantage to recovering from alternate sources of funds is that claims against parties other than the Debtor generally are not subject to the automatic stay. Alternate sources of funds and theories of recovery against those sources are reviewed in the following sections.

A. Performance and Payment Bond Claims

A claim may be asserted against a performance or payment bond surety even though the principal on the bond is in bankruptcy.[140] Because the automatic stay does not

[136]*In re Faust*, 353 B.R. 94 (Bankr. E.D. Pa. 2006).
[137]*In re Clowards, Inc.*, 42 B.R. 627 (Bankr. D. Idaho 1984).
[138]*In re Cone Constructors, Inc.*, 265 B.R. 302 (Bankr. M.D. Fla. 2001).
[139]*Id.*
[140]*In re Am-Haul Carting, Inc.*, 33 F. Supp. 2d 235 (S.D.N.Y. 1998) (automatic stay did not apply to debtor subcontractor's performance bond surety for claim by general contractor).

extend to actions against parties other than the Debtor, relief from the automatic stay is not needed to pursue a performance or payment bond claim.[141]

Sureties, however, sometimes seek an injunction against a bond claim under the court's equitable powers. A court may issue an injunction preventing a bond claim against a surety when the claims against third parties may frustrate the Debtor's reorganization efforts in a Chapter 11.[142] The injunction prevents such an adverse impact on the Debtor's estate. Once the plan of reorganization is approved or liquidation is completed, however, the injunction should be lifted to allow the suit to proceed against the surety. (See **Chapter 14 and Chapter15.**)

B. Mechanic's Liens

On private projects, contractors, subcontractors, laborers, and materialmen ordinarily have the right to file a mechanic's or materialman's lien under applicable state law to collect unpaid sums due on a construction project. The impact of bankruptcy on lien rights depends on the nature of the lien right that the claimant seeks to assert.

The automatic stay may prohibit steps to pursue mechanic's or materialman's liens against the owner's property, depending on the nature of the lien under the applicable state law. The Bankruptcy Code does not apply the automatic stay to any act to perfect a security interest. A filing of a claim of lien has been held as an act to "perfect" a lien.[143] Acts to create or enforce a lien, however, are subject to the automatic stay.[144]

If a mechanic's lien already was filed before the owner's bankruptcy, the automatic stay prohibits filing an action against the owner to foreclose the lien.[145] Although actually filing in state court to foreclose the lien is prohibited, the Code does allow the filing of a notice to temporarily satisfy a state law requirement for filing suit to perfect the lien.[146] The Code also tolls the running of the period for filing suit to perfect the lien until 30 days after relief from the stay is granted or the underlying bankruptcy case is discharged.[147] Some states allow perfection of a mechanic's lien by filing a proof of claim in the bankruptcy court. In these states, no other action is necessary to perfect the lien after the proof of claim is filed.

In the case of a general contractor bankruptcy, claimants downstream may file a lien against the owner's property without violating the automatic stay because the automatic stay extends only to the Debtor and the Debtor's property. Whether the claimants can file suit to foreclose on the lien depends on underlying state law. In states where the general contractor is treated as a necessary party to the foreclosure

[141] *Id.*at 242–43.
[142] *In re Calpine Corp.*, 354 B.R. 45 (Bankr. S.D.N.Y. 2006) (staying litigation by owner against payment and performance bond surety in another state's federal district court).
[143] 11 U.S.C. § 362(b)(3); *In re Premier Hotel Dev. Group*, 270 B.R. 234 (Bankr. E.D. Tenn. 2001).
[144] *In re Durango Ga. Paper Co.*, 297 B.R. 316 (Bankr. S.D. Ga. 2003) (notice of action requirement under Georgia lien law subject to stay).
[145] *Id.*at 321.
[146] 11 U.S.C. § 546(b).
[147] 11 U.S.C. § 108.

action, filing suit to foreclose a mechanic's lien violates the automatic stay, and the claimant must obtain relief from the automatic stay in order to foreclose on the lien.[148] In states treating the general contractor as a necessary party to a foreclosure action, the claimant should act immediately to obtain relief from the automatic stay to allow the foreclosure proceeding to go forward against the owner's property.

C. Guarantors

The parties on a construction project sometimes obtain guarantees of performance from the other parties with whom they contract. This may take the form of a personal guarantee by a corporate shareholder or a guarantee by a parent company of a subsidiary. A Debtor's guarantor is not treated as being the same entity as the Debtor and therefore is not protected by the automatic stay.[149] As with bond sureties, some courts will issue an injunction precluding an action against a guarantor where the action may frustrate the Debtor's reorganization.[150]

➢ POINTS TO REMEMBER

BANKRUPTCY CODE

- There are two types of bankruptcy in the business setting: Chapter 11 reorganization and Chapter 7 liquidation. A Chapter 11 Debtor is discharged from its debt and can continue to run its business with a court-approved plan of reorganization. A Chapter 7 Debtor has its business liquidated by a court-appointed Trustee.

AUTOMATIC STAY

- The automatic stay prevents any creditor from taking a step that is hostile toward the Debtor, including filing a lawsuit, terminating a contract, filing a lien, and so on.
- The automatic stay does not require a court order, and persons violating it are subject to sanctions.
- Relief from the automatic stay can be obtained "for cause," such as when the creditor has no equity in the property, fails to maintain insurance on the property, or fails to maintain or secure the property, or when proceedings are necessary to determine the debt owed by the Debtor.

[148] *Thompson v. Air Power, Inc.*, 448 S.E.2d 598 (Va. 1994) (if bankrupt general contractor is required to be joined, then lienor runs risk of violating automatic stay).
[149] *Reliant Energy Servs., Inc. v. Enron Can. Corp.*, 291 B.R. 687 (Bankr. S.D. Tex. 2030).
[150] *See In re Calpine Corp.*, 354 B.R. 45 (Bankr. S.D.N.Y. 2006).

PREFERENTIAL TRANSFERS

- Under certain conditions, a Debtor can have a "prebankruptcy" transfer of property to the creditor voided.
- A contemporaneous exchange of property for new value is not a voidable preference.
- When the Debtor receives valuable property in exchange for a transfer made in the ordinary course of business, the transfer cannot be voided.

DISCHARGE

- After bankruptcy, the Debtor is discharged of all its remaining debts, and a permanent injunction is placed against all creditors. Debts arising from fraud or false statement tend to be nondischargeable, leaving the debtor liable for them after bankruptcy.

STATUS OF THE DEBTOR'S CONTRACTS

- All construction contracts remain in effect after bankruptcy until substantial completion.
- The Debtor in Possession or the Trustee can affirm or reject any executory contract.
- To affirm a contract in default, the default must be cured and the Debtor must give adequate assurances of future performance.
- Debts and expenses incurred by the Debtor for an affirmed contract are treated as administrative expenses of the Bankruptcy Estate.
- Debtors can assign executory contracts if allowed under state law.

MINIMIZING THE IMPACT ON EXECUTORY CONTRACTS

These methods can minimize the impact of contracting with a Debtor:

- Terminate the contract before bankruptcy if nonfinancial reasons for default exist,
- Exercise protection rights under the contract,
- Seek relief from the automatic stay, and
- Seek a time limit on affirming or rejecting the executory contract.

STATUS OF MATERIALS AND EQUIPMENT

- Materials and equipment slated for incorporation into the project may be used by the Debtor in Possession or Trustee for the reorganization or liquidation of the Bankruptcy Estate.
- A party can recover goods in the Debtor's possession that have not been paid for and have not been incorporated into the project.
- Precautions must be taken to prevent a Debtor from claiming property in its possession that another party has already paid for.

VOIDING UNPERFECTED SECURITY INTERESTS

- This strong-arm provision of the Bankruptcy Code allows a Trustee to void an unperfected security or interest in construction materials.

STATUS OF CONTRACT FUNDS

- A party can use several methods to recover contract funds from a Debtor.
- The Debtor cannot seize contract funds that are held in constructive trust for the benefit of the nonbankrupt parties.
- Pursuant to specific contract provisions, parties may have an equitable lien on contract funds held by the Debtor.
- The use of joint check procedures can prevent the ultimate seizure of the contract funds by the Debtor in Possession or Trustee slated for a lower-tier subcontractor or supplier.
- Owners and general contractors can set off monies paid to lower-tier subcontractors and suppliers against contract funds owed to the debtor/subcontractor when the owner or general contractor has an independent obligation to pay the claims of lower-tier parties. The party can then recoup contract funds it owes to the Debtor in an amount representing backcharges or other claims it has against the Debtor.
- The surety of the Debtor that has made payments to the Debtor's claimants has the same rights against the Debtor as those bond claimants.

OTHER SOURCES OF FUNDS

- Parties can seek relief from a Debtor's surety unless the surety secures an injunction preventing the claim.
- Depending on state law, the filing of a mechanic's lien against a bankrupt owner may violate the automatic stay.
- The automatic stay prevents filing an action against a bankrupt owner to foreclose a valid lien.
- State law determines whether an action to foreclose a lien for monies owed by a bankrupt contractor violates the automatic stay.
- An action on the guarantee of an entity or person other than the Debtor does not violate the automatic stay.

21

RESOLUTION OF CONSTRUCTION DISPUTES

On a construction project of any complexity, disputes are often the rule, not the exception. Avoidance or quick resolution of disputes is often crucial to the project's economic success. In this regard, one point must be stressed at the outset: The key to quick resolution of disputes is the use of systems designed to collect, preserve, and organize information, including documents, throughout the project. Project documentation helps all parties avoid disputes and facilitates prompt resolution through negotiation. This chapter addresses presentation of claims and dispute resolution through alternative dispute resolution measures and litigation.[1]

It is always advisable to try to settle construction disputes before resorting to more formal dispute procedures. Anyone who has been involved in a typically costly and time-consuming legal battle will affirm the old adage: "A poor settlement beats a good lawsuit."

I. EARLY CLAIM RECOGNITION AND PREPARATION

Before there can be any preparation or prosecution of a claim, the claim must be recognized. Early recognition is required to ensure that notice requirements are met and that evidence needed to support the claim is preserved. Familiarity with the contract requirements is needed to recognize claims and to avoid unknowingly

[1]The basic discussion in this chapter on claim preparation and documentation also applies to claims on federal government construction projects. The specific procedures and statutes governing the submission and resolution of a claim on a federal government project are addressed in **Chapter 22** of this book. For a more in-depth discussion regarding federal government projects, *see* Thomas J. Kelleher, Jr., Thomas E. Abernathy IV and Hubert J. Bell, Jr., *Federal Government Construction Contracts—A Practical Guide for the Construction Professional* (John Wiley & Sons, Inc. 2008).

providing or accepting a nonconforming quantity, quality, or method of performance. Consequently, all job-site personnel should be familiar with the contract terms, including the plans and specifications, the general conditions, schedules, and special provisions, so they can evaluate the performance actually rendered or demanded, as compared to the performance specified in the contract. In-house educational programs to enhance this ability and better equip job personnel to identify and handle possible claim situations should also be considered.

Of course, asserting a claim for every minor incident or disagreement would be counterproductive. Filing claims with little merit or significance merely wastes resources, strains project relationships, and squanders credibility. Conversely, never filing a claim on the assumption that every dispute can be amicably resolved at the end of the project without ruffling feathers is naive. Part of an effective program for identifying claims requires targeting those incidents that are sufficiently meritorious and substantial to justify the cost of preparing and prosecuting a claim.

Once a determination is made that a claim merits prosecution, comprehensive preparation and organization should be promptly undertaken. The facts, evidence, and documents bearing on the claim should be assembled, organized, and reviewed when they are fresh and before they are lost or forgotten. This preparation should be undertaken with an eye toward resolving the claim in the formal setting of arbitration, litigation, or another dispute resolution process while still seeking early resolution through informal and less onerous means. If early resolution is not achieved, complete preparation at an early stage provides important insight for developing a claim strategy and a factual foundation that can be relied on, subject to revision, as prosecuting the claim continues.

The first step in claim preparation should be an exhaustive investigation of the claimant's own records and sources of information about the project and the claim. Project records generally are voluminous. Although the review of the records must be sufficient in scope to cover the documents relevant to the claim and anticipated defenses, it also must be sufficiently focused and specific enough to avoid inundating the claim preparation process with unnecessary and irrelevant documents. Certain categories of documents almost always merit some consideration, such as the contract documents, change orders, requests for information (RFIs), pay applications, daily logs or reports, bonds and insurance policies, and certain key correspondence. Some further organization of the documents may be required beyond that used during construction in order to make individual documents relating to the claim more readily accessible.

Although documentation is certainly critical in any construction claim, it is not everything. At arbitration or trial, the witnesses will bring life to the documentary evidence with their words, perceptions, and recollections. Those individual resources should not be overlooked in the claim preparation process, but they often are. The more remote that claim preparation is from the people actively involved in the field, the more likely there will be unpleasant surprises and inconsistencies as the claim is subjected to greater scrutiny in discovery or at trial. Consequently, the field personnel involved should be interviewed to confirm that management's secondhand understanding about the facts and circumstances of the claim is accurate and complete. To the extent possible, project personnel should be utilized to staff and assist in the claim preparation effort. At a minimum, field personnel should be given the opportunity to review the claim at various stages of preparation, and certainly before it is submitted, to confirm that they can vouch for its accuracy.

II. EARLY INVOLVEMENT OF EXPERTS AND ATTORNEYS

Construction claims often require the assistance of experts to help solve problems and to assemble and analyze the facts. Part of a program of prompt and cost-effective claim preparation requires considering the involvement of attorneys experienced with construction claims and other technical experts at an early stage. Of course, the use of outside assistance will depend on the size and complexity of the claim, but in most claims such early involvement will facilitate prompt resolution or better preparation for arbitration or trial and will be worth the investment. Scrimping on experienced and qualified legal and technical support for a claim can prove very costly in the long run.

The advice and guidance of an attorney experienced in construction claims and litigation often is desirable at the earliest stages of claim preparation. The construction attorney who ultimately would be charged with presenting the claim to judge, jury, or arbitrator should be consulted to ensure that the claim and supporting documentation and evidence are being assembled and preserved in a manner consistent with favorable resolution of the claim in a formal proceeding. The construction attorney need not take over the claim effort but should be consulted to ensure that the claim effort will not be wasted or undercut the claimant's position in any proceeding that may ensue. An experienced construction attorney also can suggest competent technical consultants in specialized areas, such as accounting and scheduling, and thereby help avoid the expense and frustration of relying on an individual who lacks the proper qualifications to testify in the case. Early involvement of an attorney does not presuppose a resort to litigation or arbitration. On the contrary, it should facilitate comprehensive claim preparation, which, it is hoped, contributes to the early resolution of disputes.

Early involvement of technical and accounting expertise likewise can enhance claim preparation efforts and, possibly the claimant's leverage in negotiation or persuasiveness at trial or arbitration. Construction is a complex process, involving a broad spectrum of scientific and technical disciplines. The immediate concern with technical qualifications must be balanced with the need to have a witness who is capable of persuasive testimony in the forum selected for resolution of claims. A technical expert respected in a particular field may be understood by a technically oriented arbitration panel but be incomprehensible to a lay jury.

Delay is a frequent subject of construction claims. Hence, scheduling analysis and scheduling experts often are involved in claims resolution. Beyond their scheduling expertise, scheduling consultants must have a detailed, working knowledge of the construction process so their analysis reflects the practical problems and difficulties experienced on the construction site, not merely a computer-generated abstraction. Likewise, it is essential that the scheduling expert involved in a claim be provided access to contemporaneous project documentation that accurately reflects the manner in which the work proceeded. Costly and complex "as-built" scheduling analyses presented in support of claims can be undermined severely if the dates and data used in the analysis conflict with those contained in project documentation, such as daily reports or monthly schedule updates.

Certified public accountants who are familiar with the construction industry and its financial and accounting practices also can contribute significantly to quantifying and proving the financial consequence of the technical problems that generated the claim.

If it is possible, and cost-effective, to involve an expert during the actual construction phase, when a claim is merely a probability, that option should be considered. At such an early stage, the expert may be able to suggest alternative construction methods or ways of reducing the impact of an injurious condition. The expert also can assist in identifying, compiling, and preserving evidence and creating demonstrative evidence for use during negotiation, mediation, arbitration, or trial. Further, testimony based on firsthand observation of the construction can be more credible and persuasive than testimony based solely on secondhand input. There are, however, some significant disadvantages to relying solely on the dual fact witness/expert witness to present expert opinion testimony at arbitration or trial. The expert's involvement in the project may rise to such a level that the expert develops a "personal stake" in the facts in dispute, compromising that person's credibility as the expert shifts from neutral observer to active and adversarial participant. The dual role of such an expert as both fact witness and expert witness also raises some difficulties for the attorney in discovery and at trial under the Federal Rules of Evidence. These concerns include the potential loss of the attorney-client privilege and work product protections over communications and information shared between the attorney and the expert. These risks can be weighed only on a case-by-case basis.

III. DEMONSTRATIVE EVIDENCE

Demonstrative evidence has the special advantage of presenting in pictorial form abstract, complicated, and extensive facts. It can clarify or explain oral testimony or documentary narrative in concrete terms. In addition, demonstrative evidence adds interest and avoids the tedium of a relentless one-dimensional recitation of facts. The simplicity and clarity demonstrative evidence can provide is particularly effective in large, highly technical, and complex construction claims. But the utility of demonstrative evidence should not be overlooked in smaller, more straightforward disputes.

Demonstrative evidence can range from photographs and videotapes to charts summarizing facts or making comparisons, such as a chart comparing as-planned manpower to as-built manpower in order to graphically depict an overrun. Charts and graphs often are used in connection with scheduling presentations, again usually comparing the as-planned schedule to the as-built schedule, with a focus on those problems that created delays. By displaying this information in an attractive visual way, in combination with other written and oral presentations, the claim can be advanced in a more compelling and persuasive manner. The goal of the demonstrative presentation is lost if it is not clear, understandable, and firmly supported by the facts.

Discussions of the importance and usefulness of demonstrative evidence as a means of persuasion generally are found in trial advocacy materials. But there is no need to hold such a powerful and effective tool in reserve until trial or arbitration. Demonstrative evidence should be developed and used to simplify the claim and persuade the other side as soon as possible.

IV. CONTEMPORANEOUS RECORDS

The importance of creating and maintaining contemporaneous project records cannot be overemphasized. The success of any claim presentation, whether formal or informal, largely depends on the quality of the documentary and visual evidence that can be relied on. Contemporaneous records engender integrity and trustworthiness essential to persuading a judge, jury, or arbitrators. Federal and state rules of evidence require contemporaneity in order to admit certain types of documents at trial. The "business records" exception to the hearsay rule is often the only avenue by which certain project records are admissible. Under the Federal Rules of Evidence, a project document qualifies as a "business record" when it: (1) contemporaneously records events or conditions; (2) is authored by a person with knowledge; and (3) is routinely kept in the regular course of the business.[2] Most states adhere to this evidentiary rule as well.

All projects need documentation systems in place at the start of the project that are maintained through completion. Documentation programs must be designed and implemented in a manner that accurately and comprehensively records job progress and problems in "real time." Documentation programs can take many forms and can be in paper or electronic format, or both. Common types of project records include RFIs, meeting minutes, daily reports, correspondence and e-mail, job-site diaries, progress schedules, progress and defective work photographs and video tapes, as-built drawings, and cost documentation. The documentation system should be designed to record project progress and events on a continual basis. Compliance with contractual notice provisions, the circumstances giving rise to requests for time extensions, extra work, changes, differing site conditions, and the grounds for or against termination should all be thoroughly documented when they occur. The properly implemented documentation program, faithfully followed, greatly enhances the likelihood of resolving disputes without arbitration or litigation and winning disputes that cannot be amicably resolved.

V. COMPONENTS OF A WELL-PREPARED CLAIM DOCUMENT

Simplicity to promote prompt understanding, while making the claim interesting and well supported, is the key to effective claim preparation and presentation. One way to synthesize the claim is to use a claim document—that is, a written synopsis of

[2] *See* Fed. R. Evid. 803(6).

the claim that can be presented to the opposition at the early stages of the dispute. As with essentially every other aspect of claim preparation, the claim document serves two alternate purposes. First, its immediate and primary goal is to bring about a prompt and satisfactory resolution of the claim. Failing that, the second purpose of the claim document is to provide a blueprint or plan for further claim prosecution.

The claim document provides an opportunity for the claimant to explain its grievance in a complete and comprehensive fashion. The process of preparing the claim document is an important step in developing a claim strategy, because it requires the claimant to refine and synthesize the claim from beginning to end. The claim document should be viewed as telling a story. It should have a clear and definite theme that can be communicated, understood, and remembered readily. The theme should be the strongest argument supporting the claimant's theory of recovery.

There will certainly be a considerable quantity of facts gathered in support of the claim. But trying to present and argue each and every one of these facts will simply overwhelm and confuse the reader, and the resulting claim document as a tool of persuasion will be a failure. When multiple and unrelated claims are presented in one document, the document must be structured to emphasize the strongest claim.

The primary communicative component of the claim document is the factual narrative. Although this narrative certainly will focus on the claimant's point of view, it should not be expressed in overly argumentative or combative terms. Instead, to the extent possible, the facts presented should be permitted to speak for themselves. The writing style should be clear and precise but should not read like technical specifications. It is, after all, a story, not simply a recital of a string of facts. The narrative should be comprehensive and logically organized, so it can be used as a resource throughout negotiations and further prosecution of the claim. If the complexity of the matter is such that the narrative is exceedingly long, an executive summary should be prepared.

The factual narrative often is followed by a written discussion of the applicable legal principles that support and illustrate the theories on which the claim is based. Assistance from an experienced attorney in construction claims generally is required to fashion the legal arguments and otherwise to ensure that the factual narrative is presented in a manner consistent with the applicable legal principles. The need for or extent of a legal discussion generally is geared to the expertise or experience of the ultimate decision maker for the opposition. For example, in federal construction contracts, certain theories of entitlement are so firmly established and recognized on all levels that no or only a limited legal discussion is required. In other situations, the legal discussion may be a crucial element in causing the other side to recognize its liability and exposure. A one-time owner may have no idea of what a differing site condition is or why the contractor should be paid for it. Claims against local governmental entities that contract regularly may be ruled on by elected officials who also require an education about construction law before they can be expected to recognize the need to settle a claim.

Pricing the claim and supporting such calculations is every bit as important as establishing liability for the claim. The claim document must recognize the importance of damages and include a specific dollar figure and fairly detailed cost analysis and breakdown. Supporting information and sources should be identified and appended if they are not too voluminous.

Finally, the claim document should be used to showcase and highlight the most persuasive documentary and demonstrative evidence. The most potent documents should be quoted or even reproduced in their entirety in the body of the narrative. Those documents that do not merit incorporation into the text, but which are referenced and support the claim, can be included in an indexed appendix that is cross-referenced with and organized like the factual narrative. In this manner, the narrative can be reviewed without having to sift through every bit of paper, but the reader will know that backup is readily available should further review be desired.

In addition to supporting documents, charts, graphs, drawings, and photographs, other demonstrative and visual evidence should be incorporated into the claim document to the extent practical. Similarly, consideration should be given to including relevant reports by experts as attachments to the claim document as exhibits, with appropriate references to and quotes from the reports in the narrative.

In certain situations, the nature of the claim or the character or capacity of the opposition may counsel against submitting an extensive claim document. The opposition may lack the financial resources or genuine interest in resolving the claim by negotiation, which is a primary goal of the claim document. Instead, the opposition may seek a one-way flow of information, it being willing to receive a detailed presentation of the claim but unwilling to explain or document any response, defense, or counterclaim until trial. The claimant must evaluate whether pursuing the race for disclosure by providing a claim document ultimately will eliminate the roadblocks to negotiation and settlement or simply better equip the opposition to defend the claim, without a commensurate benefit to the claimant. Generally, but not always, a sound, well-documented and prepared claim should be able to withstand and be improved through feedback from the opposition's scrutiny. Moreover, even if the decision is made that an extensive claim document should not be submitted, that conclusion should not stop the claimant from preparing a claim document for internal use to better synthesize the claim and prepare for whatever proceeding may follow. Of course, if formal claim submission is mandated by the contract, such a requirement should be followed, although the extent of the submission may vary.

VI. CALCULATING AND PROVING DAMAGES

The issue in construction disputes that generally receives the most attention and focus is liability. Does a differing site condition exist? Who caused the delay, and is it compensable? The issue of damages or costs flowing from the events that give rise to liability, however, is no less important. Too often, calculating and proving damages takes a backseat, with little precision or scrutiny applied until the eve of the trial. That approach can result in an entirely misguided claim effort, missed opportunities for settlement, and loss at trial or in other dispute forums. An early analysis of damages can help determine whether a claim really exists and the best means of preparing and positioning the claim for the affirmative recovery sought.

The problem of calculating and proving damages can be reduced substantially by initiating proper cost accounting at the project's inception. This may help provide an early indication of the unanticipated early work, delay, or disruption. Once a claim is identified, these cost accounting procedures can be used to capture the extra or unanticipated costs—for example, by cost-coding extra work. Accounting measures can be established to segregate and carefully maintain separate records. If such a procedure is followed, proof of damages can be reduced to little more than the presentation of evidence of separate accounts. Unfortunately, this ideal situation seldom exists; the problem is not recognized in time to set up separate accounting procedures, the maintenance of separate accounts is simply not possible because of an inability to isolate costs, or no attempt is made to establish the requisite procedures. These circumstances necessitate developing some sufficiently reliable formula that permits the court or arbitrators to allow its use as proof of damages. If settlement is being sought, the claimant likewise must convince the other side of the validity and reliability of its damage calculations.

The sections that follow offer specific approaches and alternatives for prosecuting claims of the types frequently encountered. Many of the principles and possible approaches are common to all construction claims.

VII. PURSUING NEGOTIATION AND SETTLEMENT

Although claims and disputes are a part of the construction process, they need not and should not dominate the process. When claims and disputes cannot be avoided, efforts should be redoubled to resolve them as quickly as possible. The complexity, time, and cost of arbitration, litigation, and other forms of alternative dispute resolution naturally cause the parties to favor negotiation and settlement. Formal dispute resolution such as litigation or arbitration is not a good business model because of the uncertainties inherent in the process. Therefore, an approach favoring prompt resolution should be part of a claims policy, and project personnel and management should be indoctrinated and trained along those lines. Although contract provisions regarding notice of claims and other technical requirements should be complied with, other lines of communication on the project should not be overlooked as a means of bringing a claim to quick settlement and avoiding the need to have the disputes process run its full course. It is far easier and less expensive to resolve problems in the field, where they arose, than in the courtroom. Even if early settlement is not achieved, negotiations force the claimant to seriously examine the merits of its claim and also reveal the strengths and weaknesses of the claim at an early stage.

Comprehensive and careful claim preparation greatly enhances the likelihood of early resolution and settlement. A party attempting to settle a claim should know its own case intimately and should anticipate as many of the opposing party's points as possible. Use of a well-prepared claim document, as previously discussed, is helpful both as a starting point and as a reference during settlement discussions. People with firsthand, detailed knowledge of the underlying facts and with the authority to negotiate are also an essential part of any negotiating effort. There is simply no substitute for involving the person who lived with the project's problems on a daily basis. Most important, successfully negotiating a claim to resolution often means compromise.

VIII. ALTERNATIVES TO LITIGATION

Alternative dispute resolution (ADR) is now well known in the construction industry as a means of resolving disputes when negotiations break down. ADR includes mediation, a combination of mediation and arbitration ("Med-Arb"), minitrials, dispute review boards, and several other methods of conflict resolution. In order to invoke one of these ADR processes on a private construction project, the contracting parties would need to include an agreed-on process in their contract. On public projects, such processes may be required by statute, regulation, ordinance, or terms of the contract.

A. Dispute Review Boards

In disputes involving state and local public projects, many state and local government authorities are requiring that disputes between the government authority and the contractor first be heard by a dispute review board (DRB) before litigation can commence. The goal is to create a mechanism whereby disputes that occur throughout the project are addressed without waiting until the end of the project and without interrupting the progress of work. The process is somewhat similar to presenting claims before the various boards of contract appeals on federal government projects, except that the parties participating in the DRB process have more control over the process and selection of the panel members who will decide the case.

Typically, the DRB is composed of three panel members, one selected by the government authority and one selected by the contractor; these panel members in turn select a third member. The parties present their case to the panel in a manner similar to the way in which they would present their case in arbitration, and the panel then renders a decision. The difference between the DRB and arbitration is that the DRB's decision is subject to a full appeal, unlike an arbitration decision, which may be overturned only under very limited circumstances.

B. Minitrials/Summary Jury Trials

The minitrial and summary jury trial are hybrids of the more formal arbitration and litigation procedures. In the minitrial, each of the parties presents its case to a mediator and to senior representatives from each party, who are generally unfamiliar with the dispute—usually corporate executives. After hearing the presentation, the corporate representatives attempt to negotiate the dispute with the assistance of the mediator. The benefit of this process is that representatives from the disputing parties themselves are serving as judge, jury, and negotiators. The goal is for the representatives to hear the evidence and, with the help of the mediator, reach a settlement based on the evidence presented, weighing the strengths and weaknesses of each side's case.

In a summary jury trial, the parties engage in an abbreviated trial before a judge and a mock jury. The entire trial generally lasts no more than a day. The jury then renders an advisory opinion, which is nonbinding on the parties. Again, the purpose of the procedure is to allow the parties to understand the strengths and weaknesses of their respective cases and reach a resolution without the protracted trial and trial preparation.

C. Mediation

Mediation is a popular method for attempting to resolve construction disputes. Although not a panacea for all project ills, mediation can provide a unique opportunity to resolve disputes without the costs of arbitration and litigation. Many standard-form contracts require mediation as part of the formal process for dispute resolution. For example, Article 16.01 of the Engineers Joint Contract Documents Committee (EJCDC) C-700 (2007 ed.) allows either the owner or the contractor to request mediation before arbitration or litigation. The ConsensusDOCS 200 (2007 ed.) provides for mediation before arbitration or litigation, but after direct discussions between the parties.[3] American Institute of Architects (AIA) Document A201 (2007 ed.) continues to require claims to go to mediation as a condition precedent to binding dispute resolution.[4] The language used in the AIA Document A201 (2007 ed.) provides:

> Claims, disputes, or other matters in controversy arising out of or related to the Contract, except those waived as provided for in Sections 9.10.4, 9.10.45 and 15.1.6 shall be subject to mediation as a condition precedent to binding dispute resolution.[5]

Mediation, unlike arbitration, is nonbinding. Rather, it relies on the parties' true desire to end the dispute and their willingness to compromise their respective positions to reach this goal. The mediator, therefore, does not "decide" who is right or wrong but facilitates the process of bringing about a mutually acceptable (or in many cases distasteful) solution to the problem. Also, mediation is much less costly than arbitration or litigation and is a relatively quick process.

Each mediation likely will be unique to the facts and circumstances of the underlying dispute; and an effective mediator will mold the process based on the parties' dynamics. There are, however, some basic components to a mediation.

1. *Preparing for the Mediation*

Frequently, once the mediator is selected, the parties will be required to sign a mediation statement (or agreement). This document is intended to memorialize the

[3]ConsensusDOCS 200, ¶ 12.4 (2007 ed.).
[4]AIA A201, § 15.3.1 (2007 ed.).
[5]*Id.*

parties' intent to preserve the confidentialty of the mediation process. Mediation usually is conducted in conjunction with a more formal dispute resolution process (arbitration or litigation). Further, an effective mediation can happen only when both sides are willing to compromise their positions. Such compromise may involve acknowledging responsibility (or no responsibility) for certain conduct. Taking a compromising position fosters settlement through mediation; however, it may not be conducive to the underlying dispute resolution process. The mediation agreement will allow the parties to compromise their positions without the concern that this may be used against them if the mediation is unsuccessful and litigation or arbitration follows.

Before the mediation begins, the mediator may ask the parties to submit written position statements. These statements are intended primarily to assist the mediator in understanding the issues and the parties' respective positions. A position statement is also a useful tool to help formulate a game plan for the mediation. Even if the mediator does not request a written position statement, taking some time to map out positions, frame the issues, and develop negotiating positions can only help to increase the likelihood of a successful mediation.

A successful mediation also will depend significantly on who attends the mediation for the parties. Although a qualified mediator will facilitate the settlement discussions, he or she is ultimately not the decision maker. Each party to a mediation needs to bring a person or persons within the organization with authority to settle the dispute. Without such a person, the mediation will likely fail. In addition to a decision maker, each party likely will want to have someone at the mediation with extension knowledge of the project, the disputed issues, and the claims being negotiated.

More and more, parties are now represented by counsel at mediations. This is most likely because the mediation process is intertwined with arbitration or litigation. Although not required in the mediation, a qualified construction attorney can assist a client in framing the issues for mediation, in articulating the claim(s) or defense against the claims, and in negotiating strategy.

In some mediations, particularly those involving multiple complex claims or construction issues, a party may consider bringing an opinion expert or a claims consultant. These persons should attend the mediation only if they can assist the mediator and the other parties in better understanding the topic on which they are speaking.

2. *Understanding the Mediation Process*

The actual mediation takes place in the form of a conference between the parties. This settlement conference usually is scheduled for a set time period (typically one or two days). Most often, the parties to the mediation will attend the mediation in person, although lesser involved parties may participate as needed by telephone or through video conferencing. The conference typically convenes with all parties in the same room particpating in a preliminary conference led by the mediator. After introductory comments and ground rules, the claimant will be asked to discuss its position, followed by the party (or parties) to whom the claim is being presented. After all sides have staked out their positions, the mediator may allow some further discussions or debate between the parties.

Once the preliminary session concludes, the parties typically will adjourn to their own rooms and the mediator will begin caucusing with individual parties. This is when the negotiations truly begin. Caucus sessions are fluid, and there are no set rules. Many times the mediator will attempt to bring party representatives, usually decision makers, back together for one-on-one negotiations. Whatever the process, the ultimate goal of an effective mediator is to facilitate discussions that will lead to a settlement and compromise of the claim.

3. Timing of Mediation

Because mediation is nonbinding, the timing of the mediation is a critical component in reaching a successful resolution of the claim. Some contracts require that mediation occur before, or as a condition precedent to, any formal dispute resolution process. Also, more and more courts are now either mandating or strongly encouraging these parties to engage in mediation before a dispute goes to trial.[6] Although such a provision can dictate the timing for mediation, in general, to have a successful mediation, the parties must agree that a compromise is in their respective best interests. This typically means that they must appreciate the strengths and weaknesses of their cases and those of their opponents, the legal costs of prosecuting or defending against the claim, and the costs in committing personnel to move forward with arbitration or litigation.

In some cases, the parties may understand these factors early in the dispute. In other cases, however, it may require one or both of the parties to proceed with a more formal dispute resolution procedure—arbitration or litigation—until they appreciate the factors more fully. Forcing mediation before the parties are ready may be a waste of time and money.

IX. ARBITRATION

Arbitration is a binding dispute resolution process that offers an alternative to the lawsuit. One of the contracting parties will commence (or demand) arbitration when the parties cannot resolve the claim between themselves amicably. The availability of arbitration, like any ADR procedure utilized, must be planned and provided for at the outset of the project, preferably in the contract itself. Although the parties are free to agree to arbitrate after a dispute has arisen, without mutual agreement, the parties must look to the courts. Generally, no particular form or special words are necessary to establish an agreement to arbitrate a dispute between the parties. There must, however, be a clear indication from the contract language that the parties intended the disputed issue to be subject to arbitration.

[6]See **Section X.F** of this chapter.

To avoid any dispute about the scope of the agreement to arbitrate, it is best to state expressly that all disputes arising from the contract will be arbitrated. The modified arbitration clause, contained in AIA Document A201 (2007 ed.), provides a simple and clear directive if the parties choose arbitration as the ultimate forum for resolving disputes, and while also referencing the required mediation:

> If the parties have selected arbitration as the method for binding resolution in the Agreement, any claim subject to, but not resolved by mediation, shall be subject to arbitration, which, unless the parties mutually agree otherwise, shall be administered by the American Arbitration Association in accordance with its Construction Industry Arbitration Rules in effect on the date of the Agreement.[7]

Under this contract, the parties can opt for arbitration by checking the box on the signature page. The default for the 2007 AIA documents is litigation as opposed to arbitration. The 2007 ConsensusDOCS likewise allow the contracting parties to decide between arbitration and litigation for resolution of contract disputes. Unlike the AIA documents, however, ConsensusDOCS do not specify that litigation is the "default" process if the parties do not check the box for arbitration. Ideally, the parties utilizing a contract document that allows opting for arbitration should state this affirmatively in the contract—they should check the box. If they do not, it could be argued that the parties' failure to check the box for arbitration would indicate that the parties did not elect arbitration as a means of dispute resolution.[8]

Arbitration is not new to the construction industry. Contract clauses providing for arbitration of disputes have been commonplace in the construction industry for many years. Although arbitration generally is perceived as a way to avoid the delays and problems associated with litigation, time has demonstrated the drawbacks as well as the advantages of relying on arbitration as a means of formally resolving construction claims. Many of the issues discussed below with respect to arbitration also may be equally applicable to other forms of ADR.

A. Time and Costs of Arbitration

On balance, and particularly for smaller claims, arbitration does provide a faster resolution. Given the right set of circumstances, however, arbitration can be as excruciatingly time consuming and expensive as litigation. A dispute over the existence, scope, or validity of an arbitration agreement itself can engender a protracted court proceeding and appeal before there is even a determination of whether and to what extent the parties should proceed with arbitration. In addition, arbitration of larger

[7]AIA A201, § 15.4.1 (2007 ed.).

[8]*See McAllister Bros., Inc. v. A & S Transp. Co.*, 621 F.2d 519, 522 (2d Cir. 1980) (*quoting United Steelworkers of Am. v. Warrior § Gulf Navigation Co.*, 363 U.S. 574, 582 (1960)) ("arbitration is a matter of contract and a party cannot be required to submit to arbitration any dispute which he has not agreed so to submit.").

claims involving multiple parties can match the delays and complexity of the most arcane court proceeding.

Although ultimately dependent on the parties involved and the scope and complexity of the issues, the cost of arbitration may be less than that of a comparable court proceeding. Generally, a shortened period for resolution will keep costs down, but there are certain costs that may be incurred in arbitration, just as in litigation. These costs include the examination and analysis of documents, legal research, the use of experts, the development of demonstrative evidence, and limited discovery.

Costs that can be avoided in arbitration usually involve certain trade-offs. For example, unless provided for by agreement or statute, discovery may be either unavailable or significantly limited.[9] The costs of discovery, which can be substantial, can thus be avoided, or at least diminished. The lack of discovery, however, means less preparation for and knowledge of the opposition's case. It also can possibly lead to a less focused hearing, which may then require more time and increased costs. In addition, certain costs are unique to arbitration, such as arbitrators' fees as well as administrative fees and the cost of meeting rooms, all of which can be substantial.

There are costs of arbitration that are not typically incurred (at least on the same order of magnitude) in litigation. First, if the arbitration is administered by an outside organization, as it frequently will be, that organization will charge a fee for such service. These filing or administrative fees typically are based on a percentage of the claim (and counterclaim).[10]

The cost for filing a lawsuit is usually much less than an arbitration filing fee. Second, the arbitrator (or arbitration panel) ultimately selected will most likely charge the parties a fee for such services. This is typically in the form of an hourly rate for preparation and study time, participating in hearings, and deliberations and award preparation. This fee can roughly triple if a three-member arbitration panel is required to resolve the dispute. Judges and juries, while compensated for their time, are not paid directly by the parties for their services. Although there may be a benefit to incurring these additional costs, such as a more expeditious resolution of the dispute, the prudent construction professional should understand and acknowledge the magnitude of these costs before agreeing to arbitration.

B. Selection of Arbitrators

The qualifications and fairness of each arbitrator are essential to the viability of arbitration as a means of resolving disputes. The selection of arbitrators, like the selection of jurors, can be a major factor in the success or failure of a claim. Some considerations must be addressed well in advance of any dispute at the contract drafting phase

[9]*See* W. Michael Tupman, *Discovery and Evidence in U.S. Arbitration: The Prevailing Views*, Arb. J. (Mar. 1980).

[10]The American Arbitration Association (AAA), for example, has a sliding scale for filing fees. *See, e.g.*, AAA Construction Industry Rules (Sept. 1, 2007).

of the project, when the terms of the arbitration clause are determined. Some basic but strategic considerations are the number of arbitrators and the manner in which they are selected.

The fact that arbitrators of construction disputes generally have more expertise in the construction industry than a judge or jury generally is cited as one of the major advantages of arbitration. Of course, as with a juror, an arbitrator's experience and background can predispose the arbitrator for or against a particular argument or party. For example, an owner in a dispute with an architect might fear choosing an architect as an arbitrator or placing an architect or engineer on the panel. An owner with a strong claim against an architect, however, actually might benefit from the heightened scrutiny another architect might apply to a fellow professional. Another scenario might bring an arbitrator to the panel who is considered by many, including him- or herself, to be a preeminent authority in a particular field applicable to the dispute. If the arbitrator rejects the testimony of one's expert, it is likely he or she will prevail in convincing fellow arbitrators to do the same. As with the selection of a jury, many factors go into the selection of an arbitration panel.

In terms of numbers, the basic choice is between a single arbitrator and a three-member panel. A single arbitrator costs less and probably will simplify scheduling hearings. A three-member panel probably would be more balanced. The three-member panel generally decides by majority vote—that is, two of the three panel members can render a decision over the objection or dissent of the third member.

Parties in an arbitration are free to agree on their own procedures for arbitrator selection. The procedure often is set forth in the contract's arbitration clause. Provided both sides agree, the procedures also can be established or changed when the claim is submitted to arbitration. The method for selecting arbitrators set forth in the American Arbitration Association (AAA) Construction Industry Arbitration Rules (Construction Rules) is most widely employed. The AAA's Construction Rules provide for selection of one or three neutral arbitrators, either by mutual agreement of the parties or by administrative appointment if no agreement can be reached. An alternative procedure that sometimes is used to select a three-member panel allows each party to appoint one arbitrator, with the two party-appointed arbitrators jointly selecting a third "neutral" arbitrator. As a practical matter, this process often makes the neutral arbitrator the swing vote that decides the arbitration. Many arbitrators decline the invitation to serve as neutrals on such panels. The prospect of having the respective positions of the parties reargued by one's fellow arbitrators during deliberations does not appeal to everyone, and may not best accomplish the goal of a truly neutral panel of arbitrators.

C. Informality and Limited Appeals in Arbitration

The emphasis on the technical expertise of arbitrators usually involves a substantial deemphasis of legal principles and evidentiary rules and procedures, including the right of appeal. This aspect of arbitration often is cited as a positive, but the contrary can be argued as well.

In arbitration proceedings, strict rules of evidence do not apply, and arbitrators are generally liberal in their acceptance of evidence. This permits an easier and faster presentation of records, correspondence, documents, photographs, and live testimony. In fact, the AAA Rules encourage arbitrators to accept any and all evidence that may shed light on the dispute. In the more relaxed environment of arbitration, substantive legal defenses such as statute of limitations, no-damage-for-delay clauses, and notice requirements may be given less weight than in a trial setting.

In addition to the arbitrators being granted considerable latitude in their conduct of hearings and rendering of awards,[11] the right of appeal by means of a challenge of an award is extremely limited in scope.[12] This limited scope of judicial review of arbitration awards is yet another trade-off. Although it certainly curtails a party's rights as compared to the scope of an appeal of a jury verdict, the limited scope tends to reduce the number and length of appeals from arbitration awards. This is, of course, in contrast to the court system with its lengthy and expensive appellate procedures.

D. Enforceability of Agreements to Arbitrate

In addition to the contract's arbitration clause, a party's right to arbitration may depend on its ability to go to court and enforce the agreement to arbitrate. At common law, the courts historically were jealous of their jurisdiction and very protective of a person's right of access to the courts. Today, however, the overburdened court system has made alternative dispute resolution a necessity, and most states have attempted to broaden the right to arbitration either by statute or by court decision.

In addition to state laws, construction arbitration agreements are often enforceable under the Federal Arbitration Act (FAA),[13] which is an expression of a strong federal policy favoring arbitration.[14] The FAA applies only if the arbitration clause

[11]*See* 9 U.S.C. §§ 1–16. The Federal Arbitration Act provides that a court may vacate an arbitration award only on these grounds: (1) where the award was procured by corruption, fraud, or undue means; (2) where there was evident partiality or corruption in the arbitrators, or either of them; (3) where the arbitrators were guilty of misconduct in refusing to postpone the hearing, upon sufficient cause shown, or in refusing to hear evidence pertinent and material to the controversy; or of any other misbehavior by which the rights of any party have been prejudiced; or (4) where the arbitrators exceeded their powers, or so imperfectly executed them that a mutual, final, and definite award upon the subject matter submitted was not made.

[12]*See Patten v. Signator Ins. Agency, Inc.*, 441 F.3d 230, 234 (4th Cir. 2006) ("The process and extent of federal judicial review of an arbitration award are substantially circumscribed. As a general proposition, a federal court may vacate an arbitration award only upon a showing of one of the grounds specified in the Federal Arbitration Act, . . . or upon a showing of certain limited common law grounds.")

[13]9 U.S.C. §§ 1–16 (1999).

[14]*Terrebonne v. K-Sea Transp. Corp.*, 477 F.3d 271 (5th Cir. 2007); *Davis v. O'Melveny & Myers*, 485 F.3d 1066 (9th Cir. 2007); *In re Elec. Mach. Enter., Inc.*, 479 F.3d 791 (11th Cir. 2007); *In re Mintze*, 434 F.3d 222 (3d Cir. 2006); *Albert M. Higley Co. v. N/S Corp.*, 445 F.3d 861 (6th Cir. 2006); *ING Fin. Partners v. Johansen*, 446 F.3d 777 (8th Cir. 2006); *Image Software, Inc. v. Reynolds and Reynolds Co.*, 459 F.3d 1044 (10th Cir. 2006); *Hill v. Peoplesoft USA, Inc.*, 412 F.3d 540 (4th Cir. 2005); *James v. McDonald's Corp.*, 417 F.3d 672 (7th Cir. 2005); *JLM Indus., Inc. v. Stolt-Nielsen SA*, 387 F.3d 163 (2d Cir. 2004); *Nat'l R.R. Passenger Corp. v. ExpressTrak, L.L.C.*, 330 F.3d 523 (D.C. Cir. 2003); *DiMercurio v. Sphere Drake Ins.*, PLC, 202 F.3d 71 (1st Cir. 2000).

in a contract "evidences" a transaction involving commerce, meaning interstate commerce. Transactions of the type generally involved in a large construction project often satisfy this interstate commerce requirement and come within the scope of the FAA.[15] If the interstate commerce requirement is met, the FAA must be enforced, even in state court, and it preempts and supersedes all contrary and inconsistent state law.[16] State law, however, incorporated into the contract by a choice of law provision, still may affect the manner in which the arbitration proceeds, even if the FAA were applicable.[17]

After the fact, many times parties to a contract or beneficiaries of a contract will attempt either to avoid or to enforce an arbitration clause contained in the contract. The United States Supreme Court has held that questions of arbitrability such as whether a party is bound by an arbitration clause and disagreements about the application of an arbitration clause are questions for the court to decide.[18] Procedural questions arising out of the dispute, however, are left for the arbitrator.[19]

Regarding arbitrability, a common question for the courts to decide is whether a nonsignatory to a contract who is seeking to enforce or avoid an arbitration provision is bound. The federal circuit courts generally have recognized five conditions under which a nonsignatory may be bound to an arbitration agreement: (1) incorporation by reference; (2) assumption; (3) agency; (4) veil-piercing/alter ego; and (5) estoppel.[20] Although each case depends on its respective facts, the majority rule regarding nonsignatories is that courts are willing to estop a signatory from avoiding an arbitration clause where the nonsignatory is seeking to enforce the contract clause but are reluctant to enforce an arbitration clause against a nonsignatory who is trying to

[15]*See, e.g., Garten v. Kurth*, 265 F.3d 136 (2d Cir. 2001) (contract involved interstate commerce where owner was in New York, contractor was located in Connecticut, architect did business in multiple states, and materials were transported from various states); *Del E. Webb Constr. v. Richardson Hosp. Auth.*, 823 F.2d 145 (5th Cir. 1987) (contract related to interstate commerce where contract involved persons from different states, employees of contractor traveled interstate, and materials used in construction were manufactured and moved in interstate commerce); *Am. Home Assurance Co. v. Vecco Concrete Constr. Co., Inc. of Va.*, 629 F.2d 961 (4th Cir. 1980) (contract was one involving interstate commerce where general contractor was a Delaware corporation with its principal place of business in Alabama, subcontractor was a Virginia corporation, and materials were shipped to Virginia from various states).

[16]*Perry v. Thomas*, 482 U.S. 483 (1987); *Southland Corp. v. Keating*, 465 U.S. 1 (1984); *Moses H. Cone Mem'l Hosp. v. Mercury Constr. Corp.*, 460 U.S. 1 (1983); *see also Doctor's Assocs. v. Casarotto*, 517 U.S. 681 (1996) (holding that a Montana law requiring that arbitration agreements be subject to specific first-page notice requirements that were not required of other contracts was invalid and preempted by the FAA).

[17]*Volt Info. Scis. Co. v. Bd. of Trs. of Leland Stanford Junior Univ.*, 489 U.S. 468 (1989); *see also Roadway Package Sys. v. Kayser*, 257 F.3d 287, 292 (3d Cir. 2001) (holding that courts must enforce the terms of parties' arbitration agreement and that "[w]hen a court enforces the terms of an arbitration agreement that incorporates state law rules, it does so not because the parties have chosen to be governed by state rather than federal law. Rather, it does so because *federal law* requires that the court enforce the terms of the agreement.").

[18]*Howsan v. Dean Witter Reynolds, Inc.*, 537 U.S. 79 (2002).

[19]*Id.*

[20]*Thomson-CSF, S.A. v. Am. Arbitration Ass'n*, 64 F.3d 773, 777 (2d Cir. 1995).

avoid it.[21] In most cases, where a nonsignatory brings an action, whether in contract or tort based on the contract or related to the transaction involving the contract, the courts have found that the nonsignatory is seeking a direct benefit and should be bound.

E. Special Problems Involving Multiple Parties to Arbitration

A recurring issue in the administration of construction arbitrations is the consolidation of a number of separate arbitrations and multiple parties on the same project into one proceeding. The potential for problems and the desirability of consolidation need to be considered when the arbitration clause is being drafted and the contract signed, long before any claim develops. Although the consolidation of court proceedings involving numerous parties is common, few construction contracts currently provide for such consolidated proceedings in arbitration. Even without contractual authorization, however, some courts have required consolidation as an expeditious means to resolve construction disputes.[22] The traditional and majority rule, however, appears to be that, without express contractual consent to multiparty arbitrations, courts will not require consolidation.[23] Notwithstanding the absence of specific contractual provisions authorizing consolidation, several state legislatures have enacted statutes to empower their respective state courts to address multiparty arbitration through either consolidation or joinder of parties.[24] The AAA's Construction Rules currently provide a mechanism for consolidation or joinder of related arbitrations if the parties' agreement or law permits consolidation or joinder.[25]

The arbitration clause contained in AIA Document A201 (2007 ed.), shows a change of position by this professional organization. Whereas the 1997 version

[21]*See Tracinda Corp. v. DaimlerChrysler AG*, 502 F.3d 212 (3d Cir. 2007); *Am. Bankers Ins. Group, Inc. v. Long*, 453 F.3d 623 (4th Cir. 2006); *R.J. Griffin & Co. v. Beach Club II Homeowners Ass'n*, 384 F.3d 157 (4th Cir. 2004); *MS Dealer Serv. Corp. v. Franklin*, 177 F.3d 942, 948 (11th Cir. 1999).

[22]See *Indep. Ass'n of Mailbox Ctr. Owners, Inc. v. Superior Court*, 34 Cal. Rptr. 3d 659 (Cal. Ct. App. 2005); *Ill. Farmers Ins. Co. v. Glass Serv. Co.*, 683 N.W.2d 792 (Minn. 2004); *Birmingham News Co. v. Horn*, 901 So. 2d 27 (Ala. 2004); *Travelers Cas. & Sur. Co. of Am. v. Long Bay Mgmt. Co.*, 792 N.E.2d 1013 (Mass. App. Ct. 2003); *New England Energy Inc. v. Keystone Shipping Co.*, 855 F.2d 1 (1st Cir. 1988); *Plaza Dev. Serv. v. Joe Harden Builder, Inc.*, 365 S.E.2d 231 (S.C. Ct. App. 1988); *Maxum Found., Inc. v. Salus Corp.*, 817 F.2d 1086 (4th Cir. 1987); *Kalman Floor Co. v. Jos. L. Muscarelle, Inc.*, 481 A.2d 553 (N.J. Super. Ct. App. 1984).

[23]*Lefkovitz v. Wagner*, 395 F.3d 773 (7th Cir. 2005); *Seretta Constr., Inc. v. Great Am. Ins. Co.*, 869 So. 2d 676 (Fla. Dist. Ct. App. 2004); *Phila. Reinsurance Corp. v. Employers Ins. of Wausau*, 61 Fed.Appx. 816 (3d Cir. 2003); *Matter of E. Coast Serv., Inc. v. Silverite Constr. Co.*, 623 N.Y.S.2d 1020 (N.Y. Sup. Ct. 1995); *Bateman Constr., Inc. v. Haitsuka Bros., Ltd.*, 889 P.2d 58 (Haw. 1995); *U.K. ex rel. U.K. Defense Procurement Office, Ministry of Def. v. Boeing Co.*, 998 F.2d 68 (2d Cir. 1993).

[24]*See* CAL. CIV. PROC. CODE § 1281.3 (West 1997) (consolidation); GA. CODE ANN. § 9–9–6 (1996) (consolidation); MASS. GEN. LAWS ANN. ch. 251, § 2A (West 1997) (consolidation); N.C. GEN. STAT. § 1–569.10 (2004) (consolidation); N.J. STAT. ANN. § 2A–23A–3 (West 1997) (consolidation); S.C. CODE ANN. § 15–48–60 (1996) (joinder); UTAH CODE ANN. § 78–31a-9 (1996) (joinder).

[25]AAA Construction Rule R-7 (Sept. 1, 2007).

imposed strict limits on consolidation and prohibited consolidation of the owner's claim against the architect in any arbitration between the owner and the contractor, the 2007 version of AIA A201 contains no such prohibition. Instead, AIA A201, Sections 15.4.3 and 15.4.4 (2007 ed.), provide:

> **§ 15.4.3** The foregoing agreement to arbitrate and other agreements to arbitrate with an additional person or entity duly consented to by parties to the Agreement shall be specifically enforceable under applicable law in any court having jurisdiction thereof.
>
> **§ 15.4.4** CONSOLIDATION OR JOINDER
>
> **§ 15.4.4.1** Either party, at its sole discretion, may consolidate an arbitration conducted under this Agreement with any other arbitration to which it is a party provided that (1) the arbitration agreement governing the other arbitration permits consolidation, (2) the arbitrations to be consolidated substantially involve common questions of law or fact, and (3) the arbitrations employ materially similar procedural rules and methods for selecting arbitrator(s).
>
> **§ 15.4.4.2** Either party, at its sole discretion, may include by joinder persons or entities substantially involved in a common question of law or fact whose presence is required if complete relief is to be accorded in arbitration, provided that the party sought to be joined consents in writing to such joinder. Consent to arbitration involving an additional person or entity shall not constitute consent to arbitration of any claim, dispute or other matter in question not described in the written consent.
>
> **§ 15.4.4.3** The Owner and Contractor grant to any person or entity made a party to an arbitration conducted under this Section 15.4, whether by joinder or consolidation, the same rights of joinder and consolidation as the Owner and Contractor under this Agreement.

Thus AIA A201 (2007 ed.) permits consolidating Owner claims against, and joining the Architect in, an arbitration between Owner and Contractor where: (1) the Owner/Architect agreement permits consolidation; (2) there are common questions of law or fact; and (3) similar procedural rules governing and processes. Yet, the Architect still controls its destiny where it has the right to consent or refrain from consenting, to such consolidation and joinder.

The ConsensusDOCS 200 takes a slightly different approach and requires that, for all necessary parties to the dispute resolution procedure, appropriate provisions "shall be included in all other contracts relating to the Work to provide for the joinder or consolidation of such dispute resolution procedures."[26] Article 16.02(C) of

[26]ConsensusDOCS 200, ¶ 12.6 (2007 ed.).

the EJCDC C-800 document (2007 ed.) prohibits joinder for nonparties to the contract, except where the inclusion is necessary to obtain complete relief, or a common question of law or fact exists.

F. Med-Arb

Med-Arb is a term of art meaning mediation followed by arbitration if the mediation is unsuccessful. In Med-Arb, the parties may elect to have the same person serve as both the mediator and arbitrator. Such an arrangement has its advantages in that the mediator becomes familiar with the parties, the project, and the dispute, thus ideally leading to a more efficient process and a more thorough understanding of the issues. The disadvantage in the mediator serving as the arbitrator is that any biases or preconceptions that the mediator acquires during mediation likely will carry over to the arbitration. Consequently, Med-Arb may be most useful in cases where the stakes are not that high and the parties want a quick resolution of the dispute.

G. Arbitration Agreements and Procedures

American courts recognize the rights of parties to contractually bind themselves to an arbitration process in order to resolve their disputes. The parties are free to select the method and manner of arbitration, and may agree to arbitrate before or after a dispute arises.

A number of organizations provide for the administration of formal arbitration proceedings that parties may choose to reference in their project contracts or in subsequent arbitration agreements. The suitability of one organization over another depends largely upon the nature of the parties, the underlying transaction, and personal preference.

H. American Arbitration Association Rules and Procedures

The American Arbitration Association (AAA) is probably the most widely used organization in the United States for administering the resolution of construction disputes through arbitration. The AAA is a public service, not-for-profit organization that offers a broad range of dispute resolution procedures, including arbitration. As previously mentioned, the AAA's Construction Rules provide a framework for the resolution of construction disputes of any size or complexity, whether national or international in scope. The applicable rules and procedures vary depending on the size and complexity of the dispute, with larger and more complex construction disputes being subject to a higher degree of subject-matter discovery and procedural management. The AAA's Construction Rules also include fast-track procedures for resolution of disputes less than a set amount. As mentioned in **Section IX.A of this chapter**, administrative fees are determined on a sliding scale in relation to the size of the claims in dispute. Arbitrators are selected from a large panel of industry professionals and construction lawyers, and are paid based on individual hourly or daily rates. All AAA rules and procedures can be found at www.adr.org.

Historically, the AIA, AGC, and EJCDC have favored arbitration administered by the AAA, but the recent trend, especially with the involvement of the state court systems to implement ADR programs and to certify mediators and arbitrators, is to let the contracting parties select their own means for arbitration including the entity, if any, administering these processes. AIA A 201 (1997 ed.) provides in pertinent part:

> The demand for arbitration shall be filed in writing with the other party to the Contract and with the American Arbitration Association, and a copy shall be filed with the Architect.[27]

Although a subtle change, the language in the 2007 edition does not limit arbitration to those filed with the AAA as the lone means for commencing an arbitration. Although the AAA is the default administrator unless otherwise agreed, the AIA has modified AIA Document A201 (2007 ed.) to provide in pertinent part:

> A demand for arbitration shall be made in writing, delivered to the other party to the Contract, and filed with the person or entity administering the arbitration.[28]

The ConsensusDOCS 200 Section 12.5 similarly sets the AAA as the default administrator but allows the contracting parties to mutually agree on other arbitration rules and administration. The EJCDC C-800 Guide to the Preparation of Supplementary Conditions (2007 ed.) has a fill-in-the-blank for the arbitration rules and arbitrator.

More and more, parties are relying on their own counsel to select an arbitrator or arbitrators and enter into an arbitration agreement pursuant to remedies and rules either provided for in their respective state's Uniform Arbitration Act or pursuant to terms and conditions that are acceptable to the parties. See **Section IX.K** of this chapter for more information concerning party-drafted agreements.

I. Center for Public Resources Rules and Procedures

The Center for Public Resources (CPR) serves as a vehicle and resource for self-administered arbitration of private and public business disputes in the United States and abroad. CPR rules and procedures emphasize flexibility in permitting sophisticated business entities to craft dispute resolution procedures appropriate for their particular dispute without the traditional assistance of outside organizations. Members of the CPR include multinational corporations, law firms, legal academics, and public institutions. As with the AAA, the CPR suggests contract language that parties may include in their arbitration clauses and provides a framework of rules and procedures the parties may employ in arbitrating a dispute. Additional information on this dispute resolution institute can be found at CPR's website at http://www.cpradr.org.

[27]AIA A201, § 4.6.2 (1997 ed.).
[28]AIA A201, § 15.4.1 (2007 ed.).

J. International Chamber of Commerce Rules and Procedures

Established in 1923, the International Court of Arbitration (the Court) organizes and supervises arbitrations held throughout the world under the Rules of Arbitration of the International Chamber of Commerce (ICC). The primary function of the Court is to provide for the resolution by arbitration of business disputes of an international character, although parties are free to invoke the ICC Rules in noninternational cases. The ICC Rules have a particularly international flavor and are designed to provide a mutually acceptable arbitration framework for adversaries that may not share common customs, culture, or language. The rules and procedures of the ICC vary considerably from those of the AAA and the CPR. One significant difference is that the arbitrator's fees under the ICC Rules are generally calculated on a sliding scale in relation to the amount in dispute. Before agreeing to an arbitration clause that binds one to an ICC arbitration, the importance of gaining an understanding of the rules and procedures that will apply cannot be overemphasized.

K. Party-Drafted Arbitration Agreements: Federal and State Law Considerations

Parties drafting arbitration clauses in private contracts must consider to what law will govern the enforcement and applicability of the agreement to arbitrate.[29] The Federal Arbitration Act was enacted in 1925, then reenacted and codified in 1947 as Title 9 of the United States Code. Its "purpose was to reverse the longstanding judicial hostility to arbitration agreements that had existed at English common law and had been adopted by American courts, and to place arbitration agreements upon the same footing as other contracts."[30] The FAA "was enacted to promote the enforcement of arbitration agreements and to make arbitration a more viable option to parties weary of the ever-increasing cost and delays of litigation."[31]

The United States Supreme Court has further ruled that the FAA was intended by Congress to create not merely federal *procedural* law but also a federal *substantive* law of arbitration.[32] Consequently, arbitration clauses subject to the FAA are to be liberally construed in favor of arbitration, notwithstanding any state substantive or procedural policies to the contrary.[33] This is true in both federal and state court proceedings.[34]

State arbitration statutes have been tailored closely to the Uniform Arbitration Act (UAA) as drafted by the National Conference of Commissioners on Uniform State

[29]See **Section IV.D** of this chapter discussing enforceability of arbitration agreements under the FAA.

[30]*Gilmer v. Interstate/Johnson Lane Corp.*, 500 U.S. 20, 24 (1991).

[31]*Dean Witter Reynolds, Inc. v. Byrd*, 470 U.S. 213 (1985).

[32]*Volt Info. Scis., Inc. v. Bd. of Trs. of Leland Stanford Junior Univ.*, 489 U.S. 468 (1989), *citing Moses H. Cone Mem'l Hosp. v. Mercury Constr. Corp.*, 460 U.S. 1 (1983).

[33]*Moses H. Cone Mem'l Hosp. v. Mercury Constr. Corp.*, 460 U.S. 1, 24–5 (1983).

[34]*See Buckeye Check Cashing, Inc. v. Cardegna*, 546 U.S. 440 (2006), *citing Southland Corp. v. Keating*, 465 U.S. 1 (1984).

Laws in 1955. The UAA was revised substantially in 2000 in the form of the Revised Uniform Arbitration Act.

X. LITIGATION

When there is no arbitration provision, the traditional avenue for obtaining relief is litigation. Rules governing court proceedings are quite complex and vary among federal, state, and local courts from jurisdiction to jurisdiction. For this reason, the discussion here presents only a thumbnail sketch of what is involved in going to court.

A. Profile of the Construction Trial

The trial of a construction contract dispute offers a unique challenge to both sides of the case. Complex antitrust cases are perhaps the closest parallel, since both involve numerous parties and virtual mountains of documentation. Certain problems, however, are unique to construction controversies.

In contrast to the typical civil dispute or crime where the wrong occurred at a single point in time, construction claims generally grow out of an accumulation of events developing over many months or years. To further complicate matters, a single activity constituting a breach of contract may have a ripple effect on the remainder of the project, expanding the effect of the breach and entitling the injured party to recover "impact" costs as an additional element of damages. The category of impact damages includes the cost of inefficiency and lost productivity resulting from out-of-sequence or accelerated work. Although at times difficult to prove, such damages often can be substantial. (See **Chapter 16**.)

An added difficulty in a construction case is the large cast of characters, which can include one or more prime contractors, subcontractors, suppliers, the owner, one or more sureties, design professionals, construction managers, lenders, and others. As soon as litigation becomes inevitable, the contractor and its counsel must carefully analyze the list of possible parties from the standpoint of liability, financial responsibility, jurisdiction, venue, access to facts, and other factual and legal viewpoints.

In a contract action, the most obvious parties are those that have direct contract relations, or "privity," with one another: owner and prime contractor, or prime contractor and subcontractor. Yet other parties may become involved. For example, assume that an owner-plaintiff sues a bonded prime contractor for defective work. The owner also would likely join the contractor's surety as a defendant. Assume that the prime contractor believes the defects are not its fault but the responsibility of one of its subcontractors. Since there was no "privity of contract" between the owner and the subcontractor, the owner could not sue the subcontractor directly. The prime contractor, by virtue of the subcontract, has a direct contractual relationship with the subcontractor and therefore may "join" the subcontractor as a "third-party defendant." This mechanism is commonly referred to as "third-party practice."

This scenario often also results in counterclaims by the prime contractor against the owner for delays, defective plans, or the like. Every party considering whether to file suit should take into account the defendant's likely counterclaim, which may make it impossible to withdraw voluntarily from litigation once suit has been filed.

The subcontractor may repeat the process. It may believe one of its subcontractors or suppliers is actually responsible for the defect and may, in turn, join that party as a "fourth-party defendant." What was once a simple, two-party lawsuit can mushroom into one involving numerous parties and their counsel, with the attendant increase in complexity, delay, and expense.

B. The Court System

A decision that must be made concurrent with the identification of parties to the lawsuit is the choice of the most desirable court or, in legal terminology, the forum. From the standpoint of strategy and convenience, it is generally a distinct advantage to be the plaintiff and to select the forum. The plaintiff tells its story first to the judge or jury. The plaintiff also has some latitude in determining where to initiate the lawsuit. Frequently, this allows a convenient or hometown location to be selected.

In order to evaluate the factors that bear on the selection of a particular court, it is helpful to have a basic understanding of the various court systems. Essentially, there are two court systems within the United States: one federal and a variety of state court systems. The federal judicial system is composed of the United States District Courts (approximately 90 in number), the United States Courts of Appeals (12 in number), and the United States Supreme Court.

The federal courts are courts of "limited jurisdiction." In contrast, state courts are courts of "general jurisdiction." The term "limited jurisdiction" means that some specific constitutional or federal statutory basis must exist before a party can have its case heard in federal court. In other words, jurisdiction is "limited" to those special instances prescribed by law. In contrast, a presumption exists that a state court has jurisdiction over a particular controversy, unless a showing is made to the contrary.

"Subject matter" jurisdiction refers to the power of a court to hear a particular type of case. The subject matter jurisdiction of the federal courts in construction cases typically has a "federal question" or "diversity of citizenship" basis. "Federal question" jurisdiction encompasses disputes arising "under the Constitution, laws or treaties of the United States."[35] The second basis for subject matter jurisdiction in the federal courts—and by far the more common in construction disputes—is the "diversity of citizenship" basis.[36] "Diversity" in this sense refers to controversies between citizens of different states or between citizens of a state and an alien (i.e., a resident of a foreign country). To satisfy the jurisdictional requirement, diversity of

[35]*See* 28 U.S.C. § 1331.
[36]*See* 28 U.S.C. § 1332(a).

citizenship must be complete—that is, all defendants must be from states different from all plaintiffs.[37]

"Diversity" cases are also subject to a $75,000 threshold limit on the amount in controversy, exclusive of interest and costs. This means that suits between residents of different states cannot be brought in federal court where the realistic recovery is less than $75,000.

The second facet of jurisdiction—"personal" jurisdiction—focuses on whether the defendant has sufficient contact with the forum state to give the court the right to resolve the dispute. The most common bases for personal jurisdiction include the defendant's physical presence within the jurisdiction or, in the case of a corporation, transacting business within the area of the court's control, such as performing work on a construction project in the forum state.

Venue also presents a preliminary problem in any lawsuit. The term "venue" refers to the proper geographic location for bringing suit, rather than which court system has appropriate jurisdiction to hear and decide the case.[38] Venue considerations apply in both federal and state courts, although they tend to be more complex under state law due to the characteristic focus of state statutes on the county of residence of the defendant.

C. Federal Rules of Civil Procedure

The Federal Rules of Civil Procedure, which govern all procedural aspects of federal trials from initial pleadings to judgment, are the product of years of study, analysis, and recommendations by attorneys, judges, and scholars. They represent an effort to provide a workable procedural system. Many states have adopted the Federal Rules in whole or in part for use in their own state court systems or have developed minor variations. Substantial variations in the rules of civil procedure do exist from state to state. This is in contrast to the standardization of the Federal Rules of Civil Procedure in all federal courts across the country. Although the Federal Rules of Civil Procedure apply in all U.S. District Courts, many of federal courts have their own "Local Rules" that expand on the federal rules. Although these Local Rules may vary in some particulars, the attorney familiar with the Federal Rules of Civil Procedure ordinarily will not feel handicapped in a district court outside of that attorney's usual geographic area of practice.

D. Discovery

Discovery under the Federal Rules of Civil Procedure and its counterparts at the state level is designed to eliminate courtroom surprise and foster settlement. Discovery is particularly important in construction litigation because most complicated cases are predicated on vast amounts of files, plans, specifications, engineering data, and other volumes of paper. Discovery is the statutory means by which a party is able to elicit

[37] *Exxon Mobil Corp. v. Allapattah Servs., Inc.*, 545 U.S. 546 (2005); *Owen Equip. & Erection Co. v. Kroger*, 437 U.S. 365 (1978).
[38] *See* 28 U.S.C. § 1391.

facts from the opposing party, and thereby narrow issues, pin down the opposition's contentions, and generally prepare for trial. The most important discovery devices for construction cases are described next.

1. Interrogatories

Interrogatories[39] are written questions prepared by one party to the lawsuit and directed to another party. Answers are generally prepared with the assistance of the party's counsel. This involvement of counsel in the preparation of written answers often restricts the information provided in response to the interrogatories. Nonetheless, interrogatories are a useful tool for initially obtaining information upon which subsequent discovery can be based. The party answering interrogatories typically is required to swear under oath (or verify) that the answers are true and correct.

2. Depositions

A deposition[40] is the oral examination of any person, whether or not a party to the action, whose knowledge and perspective are important to the case. The deposition allows the questioner a more effective method of obtaining information, since questions can be tailored to prior responses of the deponent, and the questioner has an opportunity to follow up spontaneously on new avenues of inquiry.

Depositions play a crucial part in the development of trial strategy because the testimony is taken under oath and recorded by a stenographer. Because there is an opportunity for cross-examination during the course of a deposition, a deposition transcript may be used at trial when a witness dies in the interim or is otherwise unavailable for trial. The deposition also may be used for the purpose of contradicting, or "impeaching," the courtroom testimony of the witness whose testimony contradicts prior deposition statements.

3. Requests for Admission

One party can require another party to the lawsuit to admit, in writing, the truth of certain facts, by means of a formal request for admission[41]. This saves both time and money, and allows the parties to direct their energies toward those issues or facts that are truly disputed.

4. Motion for Entry upon Land

This discovery rule provides a means for one party to request permission to enter onto another's property to inspect, survey, photograph, test, or sample the property

[39]Fed. R. Civ. P. 33(a).
[40]Fed. R. Civ. P. 30, 31 & 32.
[41]Fed. R. Civ. P. 36.

or any designated object or operation thereon.[42] The value of this procedure is readily apparent if the contractor has relinquished control over the site and subsequently needs to photograph site conditions or sample the soil in connection with a differing site condition claim.

5. *Production of Documents and Things*

This rule allows a party to compel another person, whether or not a party to the lawsuit, to furnish designated documents (including writings, drawings, graphs, charts, photographs, and other data compilations) and electronically stored records or tangible objects having a bearing on the case and that are in the possession of the other party.[43] A request for production to a party under this rule frequently accompanies a set of interrogatories, with the requested documents or objects to be produced or made available for inspection, copying, testing, or sampling at the same time that the answers are filed. Production of documents and objects at a deposition can also be required by means of a subpoena *duces tecum.*

E. Judge or Jury?

Another important decision associated with a trial is the choice between presenting the case to a judge or a jury. This choice lies with the parties.[44]

A number of factors enter into the decision. For example, a judge sitting alone may be more likely to admit disputed evidence, which normally would be excluded from consideration by a jury, on the rationale that it is better to let in doubtful evidence than to face reversal on appeal because relevant evidence was not admitted. In such a case, the judge is relying on experience and training to give the evidence only the weight it deserves. However, a judge is less likely to be swayed by sympathy factors and equities that frequently work in a party's favor before a jury.

The type of construction project also may have a bearing on this decision. A local jury may prove disastrous if the case involves construction of a public facility such as a school or hospital, particularly one funded by local property taxes or levies or one that is highly controversial. A similar result may occur if the party being sued is a local business concern—for example, a subcontractor or supplier—and the plaintiff is an outsider.

Ultimately, the determining factor may be that many construction cases are simply too complex for the average jury—although some trial attorneys will argue that most judges are likewise not familiar, through experience or background, with construction principles. In any event, the complexity of the matter and the relative advantage that may be gained by simplification (or by confusion) are factors to be weighed in deciding whether to proceed before a judge or jury.

[42]Fed. R. Civ. P. 34(a).
[43]*Id.*
[44]*See* Fed. R. Civ. P. 38.

F. Alternative Dispute Resolution in the Courts

Even if the parties have elected to settle their dispute through litigation, most U.S. District Courts "encourage" litigants to engage in a mediated settlement conference through their local federal court rules. The court will set forth the requirement for all parties to participate in the mediation in its initial Scheduling Order, which is issued after all known parties have appeared in the action. A party's failure to comply with the court's order can result in sanctions.[45] Although discouraged by the courts, the parties may opt out of mediation by making a motion to the court where they believe that mediation would be an act of futility.[46]

G. Trial

The final focal point of the litigation process is the presentation of evidence calculated to persuade the trier of fact of entitlement to relief. Evidence is presented through: (1) live testimony or by deposition if a witness is not available at the time of trial; (2) documentary evidence; and (3) demonstrative evidence. The trial attorney weaves the case by combining the facts with legal theories supporting the client's entitlement.

H. Live Testimony

Live testimony is, in most instances, the most persuasive type of proof. It is crucial, however, that witnesses be prepared well in advance of trial to ensure their knowledge of important facts as well as to bolster confidence. Live testimony is needed to explain and authenticate documentary and demonstrative evidence used in the trial. See **Sections III and IV** of this chapter.

The client must assist counsel in selecting the most effective witnesses. In most cases, effectiveness can be equated with personal knowledge of the facts. Often it is not the president of the company who has this firsthand knowledge; rather superintendents or inspectors, who are "in the trenches" on a daily basis, fighting the battle, ultimately wind up in the courtroom. Another important point to consider is that these individuals—who use the unique language of the construction industry in expressing themselves—frequently make excellent witnesses. They have the capacity to translate complicated construction problems into simple, everyday language that is frequently both colorful and persuasive.

Members of management usually are most effective in establishing broad overview points, such as assumptions about labor productivity that ultimately were translated into the bid and how those assumptions and expectations were frustrated. Management, including operation and finance personnel, also are frequently needed to prove damages claimed by a party.

[45]*See Lucas Auto. Eng'g, Inc. v. Bridgestone/Firestone, Inc.*, 275 F.3d 762 (9th Cir. 2001) (affirming district court's imposition of sanctions for defendant's failure to attend mediation due to an alleged "incapacitating headache," where defendant failed to notify parties in advance of his nonappearance).

[46]*Performance Chevrolet, Inc. v. Mkt. Scan Info. Sys., Inc.*, 2005 WL 1768650 (D. Idaho 2005); *Paragon Fin. Group, Inc. v. Bradley Factor, Inc.*, 2003 WL 23471548 (E.D. Tenn. 2003).

In addition to field and office personnel, "expert" witnesses often are used. The use of experts is commonplace in construction litigation, especially in the ever-increasing number of complex "delay damages-interference-impact" cases. It is common practice in such cases for an expert to be used to analyze the validity of the critical path method or similar schedule originally relied on by the contractor in submitting its bid, as well as in the preparation of an as-built schedule if needed to depict where and how the job went wrong. Testimony from such an expert is important to quantify the impact of breaches and delays and the assessment of damages.

One pitfall of using expert witnesses in construction cases is the failure to provide the expert with reliable source material, of the kind usually relied on, to use in forming opinions and conclusions. It is appropriate for the opposing party to inquire into the foundation, facts, and data on which an expert's opinion is based. If it is found that the facts and data are untrustworthy or not of the type an expert normally would rely on, then the testimony of the expert witness is subject to attack. As a part of its ruling on the admissibility of expert opinion testimony, the trial court will review the reasonableness of the witness's reliance on facts or data to determine if the expert has deviated from the recognized area of expertise by basing an opinion on untrustworthy material. It is then the responsibility of the court to rule whether the testimony can be received or should be excluded.[47]

Live witnesses, lay and expert alike, are subject to direct examination by counsel representing their side of the dispute and to cross-examination by opposing counsel. The primary function of direct examination is to place all pertinent facts into evidence for consideration by the trier of fact. To ensure that this is properly done, counsel generally discusses with the witness before trial the types of questions that will be asked, although the answers should be those of the witness.

On cross-examination, opposing counsel has a different goal and will take a different approach in an effort to discredit the witness or the testimony, by, for example, pointing out inconsistencies in the testimony or showing bias. A cardinal rule generally followed by attorneys during cross-examination is "Never ask a question you don't know the answer to." Given this, it is frequently possible for the construction attorney and the client to anticipate and prepare for most of the questions that will come up during cross-examination. Prior deposition testimony of witnesses who will testify at trial is a very useful tool in preparing for cross-examination.

The litigation process can be long, drawn out, and expensive. Consequently, it is no wonder that other forms of dispute resolution are finding their way to the forefront among parties attempting to resolve construction claims. This does not mean that litigation should never be used; it is often an effective means for addressing claims. This is not, however, typically the case, and other means of reaching a resolution of the claim always should be investigated and considered before turning to the courts.

[47]*See generally Kumho Tire Co., Ltd. v. Carmichael*, 526 U.S. 137 (1999).

POINTS TO REMEMBER

- Early claim recognition and preparation with the assistance of "experts" is important to ensure that the claim and supporting documentation and evidence are assembled and prepared in a manner likely to result in a favorable resolution.
- Preparation and use of demonstrative evidence can be effective tools in simplifying and advancing a claim in a persuasive manner.
- A well-drafted claim should be an exercise in storytelling that will educate, inform, and, it is hoped, persuade its reader.
- Early analysis of actual damages is essential in determining if a claim really exists and the best means for either advancing the claim or encouraging an early resolution and settlement.
- The advantages as well as the drawbacks of the various methods of dispute resolution should be considered at the time of contracting. If a particular form of dispute resolution is desired, it should be spelled out in the parties' contract.
- Procedures for selecting neutrals, as well as the consolidation of multiple parties, should be considered when drafting the construction contract's dispute-resolution clause.
- Generally, arbitrators are granted considerable latitude in conducting hearings and rendering awards, and the right of appeal is extremely limited.
- Enforceability of agreements to arbitrate may be governed by both state and federal laws.
- In the absence of an alternative dispute resolution provision, the traditional avenue for obtaining relief is litigation. Construction claims are inherently complex. They often involve multiple parties, and the governing laws vary widely among federal, state, and local jurisdictions.
- Selection of the court that will hear the claim depends on several factors, including the total dollar amount in controversy and the state citizenship of all the parties.
- The Federal Rules of Civil Procedure provide numerous important discovery devices designed to elicit facts and narrow the issues in preparation for trial.
- Parties to a complex construction claim must carefully consider whether to present evidence of the claim to a judge or to demand a jury trial.
- Whether witnesses were actual participants in the project giving rise to a claim or are hired "experts," it is crucial that all witnesses be prepared well in advance of a trial to ensure their effectiveness.

22

FEDERAL GOVERNMENT CONSTRUCTION CONTRACT DISPUTES

I. HISTORICAL OVERVIEW

Federal government construction contracts reflect policies contained in statutes and in the Federal Acquisition Regulation (FAR). In addition to containing numerous standard contract clauses, the FAR also contains extensive guidance to the federal agencies and their personnel regarding the award and administration of federal construction contracts.[1]

Disputes arising out of or related to the performance of a government construction contract are governed by the Contract Disputes Act (CDA).[2] The CDA and its implementing regulations set forth a comprehensive approach to the resolution of contract claims by contractors and the federal government.

The citations in this chapter are to the appropriate provisions of the CDA, other relevant statutes, and the applicable regulations, particularly the FAR, as well as to the cases. The CDA and the other cited statutes are found in West Publishing Company's United States Code Annotated. FAR citations including FAR Supplements issued by the various agencies are from Title 48 of the Code of Federal Regulations (C.F.R.).

Government contract case law is found in a variety of sources. Since 1921, selected bid protest decisions issued by the United States Government Accountability Office (GAO) have been published in the Decisions of the Comptroller General

[1]In addition to the FAR, many federal agencies have promulgated supplements to the FAR. For example, the Department of Defense publishes the Defense FAR Supplement (DFARS). Other regulations specific to a particular agency, such as the U.S. Army Corps of Engineers, were issued. These supplements and regulations can substantially alter a contractor's rights, obligations, and remedies on a government contract.

[2]41 U.S.C. §§ 601–13.

of the United States.[3] Beginning in 1974, Federal Publications, Inc., now part of the West Group, has published the Comptroller General's Procurement Decisions (CPD) service containing the full text of all of the GAO's bid protest decisions.[4] Court decisions regarding bid protests have been issued by the federal district courts,[5] the various federal circuit courts of appeals, the United States Claims Court (now the United States Court of Federal Claims),[6] and the United States Court of Appeals for the Federal Circuit. The case law involving claims and disputes arising out of or related to the performance of a contract basically consists of the decisions of the various boards of contract appeals (boards),[7] the former United States Court of Claims, the United States Claims Court, now the United States Court of Federal Claims (Court of Federal Claims), and the United States Court of Appeals for the Federal Circuit (Federal Circuit).[8]

The Court of Claims, which was abolished in 1982, had jurisdiction to entertain suits involving government contract claims, including claims under the CDA. When Congress abolished the Court of Claims, it created the Claims Court, now the Court of Federal Claims,[9] and granted to it all of the original (trial) jurisdiction of the Court of Claims.[10] At the same time, Congress also created a new United States Court of Appeals for the Federal Circuit.[11] The Federal Circuit reviews appeals from decisions of the boards and the Court of Federal Claims.[12] The Court of Federal Claims and the Federal Circuit view decisions of the old Court of Claims as binding precedent.[13]

[3]Formerly, the General Accounting Office. Each year the GAO selects decisions for publication that it believes would be of widespread interest. Typically, 10% of all of the GAO's decisions in a given year are included in this publication. *See also www.gao.gov.*

[4]Until 1974, the vast majority of the GAO's decisions on bid protests were not published and were not readily available.

[5]The U.S. Federal District Courts' jurisdiction over bid protests ended as of January 2001.

[6]28 U.S.C. § 1491 (1992).

[7]Previously, many agencies had agency-specific boards of contract appeals. These have been consolidated into two major boards: the Armed Services Board of Contract Appeals (ASBCA) and the Civilian Board of Contract Appeals (CBCA). The U.S. Postal Service and Tennessee Valley Authority (TVA) continue to maintain separate boards.

[8]These venues dispose of the vast majority of all contract-related issues. Congress, however, has also granted the Executive Branch extraordinary powers to be used in the course of procurements related to national defense. One of these laws, 50 U.S.C. §§ 1431–35, permits certain procuring activities to grant relief to contractors that may have no legal right to such relief—for example, an amendment without consideration. This avenue for relief is not a substitute for relief under the CDA and will be considered only after it is determined that the CDA does not provide an adequate remedy. The procedures related to extraordinary contractual actions are found in FAR part 50.

[9]28 U.S.C. § 171.

[10]28 U.S.C. § 1491. The United States Court of Federal Claims has the same basic jurisdiction, but broadened to include nonmonetary claims. See **Section** XIV of this chapter. *See also Alliant Techsystems, Inc. v. United States,* 178 F.3d 1260, 1268 (Fed. Cir. 1999); *Clearwater Constructors, Inc. v. United States,* 56 Fed. Cl. 303, 307–8 (2003).

[11]28 U.S.C. § 41.

[12]41 U.S.C. § 607(g)(1); 28 U.S.C. § 1295(a)(3), (a)(10).

[13]*S. Corp. v. United States,* 690 F.2d 1368, 1369 (Fed. Cir. 1982) (*en banc*); U.S. Court of Federal Claims General Order No. 33 (Dec. 4, 1992), 27 Fed. Cl. 25 (1992).

II. SCOPE OF THE CONTRACT DISPUTES ACT

Before enactment of the CDA, the government contract disputes process was a mixture of statutes, regulations, and interpretive case law. Reflecting a series of United States Supreme Court decisions,[14] the boards became the principal forum for the resolution of contractor claims, while the United States Court of Claims assumed the more limited role of an appellate court under the Wunderlich Act.[15] Except for the relatively unusual circumstance that could be characterized as a claim for breach of contract, nearly all claims arising under a contract had to be brought to the boards. However, the boards' jurisdiction was limited to "contract" claims, and any suit alleging breach of contract had to be filed in the United States Court of Claims. In addition, due to a decision of the United States Supreme Court in *S&E Contractors, Inc. v. United States,*[16] (*S&E*), the government had no right of appeal from an adverse decision by a board.

Attempting to improve the overall disputes process, the CDA created a comprehensive statutory basis for the disposition of contract disputes. The CDA applies to any express or implied contract that is entered into by an "executive agency" of the federal government for the "procurement of [the] construction, alteration, repair, or maintenance of real property."[17] The CDA also applies to "the executive agency contracts[18] for the procurement of property, other than real property, for the procurement of services and for the disposal of personal property, as well as for supplies."[19]

[14]*United States v. Wunderlich,* 342 U.S. 98 (1951); *United States v. Moorman,* 338 U.S. 457 (1950); *United States v. Holpuch,* 328 U.S. 234 (1946).

[15]Congressional reaction to the *Moorman* and *Wunderlich* decisions resulted in the passage of the Wunderlich Act, which limited the finality of board decisions. 41 U.S.C. §§ 321–22. The U.S. Supreme Court interpreted this statute in *United States v. Bianchi,* 373 U.S. 709 (1963) and *United States v. Grace & Sons, Inc.,* 384 U.S. 424 (1966). *Bianchi* and *Grace* establish that a court reviewing a board decision is confined to the record created during the board proceeding and cannot conduct an independent evidentiary hearing into issues not addressed by the board. Thus the boards are the primary fact-finding bodies, with significant emphasis placed on the development of a record that will support the board's findings with substantial evidence.

[16]406 U.S. 1 (1972).

[17]41 U.S.C. § 602(a).

[18]The term "executive agency" is defined in 41 U.S.C. § 601(2). It encompasses those entities that are commonly thought of as government agencies, such as the Department of Defense (DOD), the General Services Administration (GSA), the Department of Energy (DOE), the Department of Transportation (DOT), the Department of Veterans Affairs (VA), the Department of Homeland Security (DHS), and others. It also encompasses the U.S. Postal Service, the Postal Rate Commission, and various independent bodies and government corporations. The act also contains provisions covering the Tennessee Valley Authority. *See* 41 U.S.C. § 602(b).

[19]It is well established that the act applies to leases for real property. *See Forman v. United States,* 767 F.2d 875, 878–79 (Fed. Cir. 1985); *New London Dev. Corp.,* ASBCA No. 54535, 05–2 BCA ¶ 33,108; *Omni Dev. Corp.,* AGBCA No. 97–203–1, 05–2 BCA ¶ 32,982; *Henry H. Norman v. Gen. Servs. Admin.,* GSBCA Nos. 15070 *et al.,* 02–2 BCA ¶ 32,042. However, jurisdiction over a dispute outside of the terms of the lease, such as a decision to expand the area subject to the lease, has been rejected by a board. *See John Barrar & Marilyn Hunkler,* ENGBCA No. 5918, 92–3 BCA ¶ 25,074.

The comprehensive statutory basis in the CDA for the resolution of disputes made significant changes in the old process. It made the boards and their members more professional by requiring that all board members be full-time positions and have significant federal procurement experience. It also gives a contractor a choice of forums with respect to the appeal of a final decision by the contracting officer. A contractor may elect either to appeal to a board or to file a suit in the United States Court of Federal Claims. Furthermore, the CDA's provisions applied "notwithstanding any contract provision, regulation, or rules of law to the contrary."[20] Accordingly, it is not possible to agree by contract to limit the right of appeal to a particular forum.[21] The CDA effectively reversed the *S&E* decision by giving the government the right to appeal an adverse board decision to the Federal Circuit.

The CDA addresses the processing of claims from their initiation to final disposition and payment. It describes the manner in which claims are asserted and seeks to provide time frames for decisions on such claims by the contracting officer. The act also provides for a contractor's right to appeal the contracting officer's final decision or the lack of a final decision, as well as the time frames for appeals to each of the alternate forums (board or Court of Federal Claims) and any appellate review of those decisions. Since the CDA's primary focus is the disposition of a claim, this chapter begins with a discussion of an assertion of a claim and follows the processing of that claim through a decision and any appeal. It also reviews provisions of the CDA and related laws dealing with fraudulent or inflated claims,[22] small claims,[23] interest on amounts found due the contractor,[24] payment of claims,[25] and recovery of attorneys' fees by certain contractors pursuant to the Equal Access to Justice Act.[26]

Before the CDA, the typical disputes clause extended only to disputes "arising under" the contract. Typically, this meant that breach-of-contract claims were not subject to the disputes clause and were outside the jurisdiction of the boards.[27] The CDA extends to all disputes arising under or related to a contract. This broader formulation of the scope of the disputes process clearly includes claims for breach of contract. By its terms, however, and as its name implies, the CDA applies only to contract disputes. Tort claims or claims seeking specific performance by the government are not subject to the CDA.[28] Although the CDA governs any "express or implied contract," the Federal Circuit has held that the CDA did not give jurisdiction to the boards to hear claims based on an implied contract by the government to treat

[20]41 U.S.C. § 609(b).

[21]*OSHCO-PAE-SOMC v. United States,* 16 Cl. Ct. 614 (1989); *D&R Mach. Co.,* ASBCA No. 50730, 98–1 BCA ¶ 29,462; *Binladin Org.,* ENGBCA No. 5304, 89–3 BCA ¶ 22,188.

[22]41 U.S.C. § 604.

[23]41 U.S.C. § 608.

[24]41 U.S.C. § 611.

[25]41 U.S.C. § 612.

[26]5 U.S.C. § 504; 28 U.S.C. § 2412(d).

[27]*United States v. Utah Constr. & Mining Co.,* 384 U.S. 394 (1966).

[28]41 U.S.C. § 602(a); *Godley v. United States,* 26 Cl. Ct. 1075, 1080 n.3 (1992); *Malnak Assoc. v. United States,* 223 Ct. Cl. 783 (1980); *Pac. Legacy, Inc. v. Dep't of Agric.,* CBCA No. 641 (Dec. 3, 2007); *Innovative (PBX) Tel., Inc. v. Dep't of Veterans Affairs,* CBCA No. 12, 07–2 BCA ¶ 33,685; *Tyrone Shanks,* ASBCA No. 54538, 06–1 BCA ¶ 33,137.

bids fairly and honestly,[29] which is the basis for bid protest and bid preparation cost actions. The Federal Circuit held that the implied contract to treat bids fairly and honestly is not a contract for the procurement of "goods or services"; thus it did not fall within the definition contained in § 3(a) of the CDA.

Even though § 605(a) of the CDA provides that all contract-related claims shall be submitted to the contracting officer for a decision, the CDA also states that it does not extend to "a claim or dispute for penalties or forfeitures prescribed by statute or regulation which another federal agency is specifically authorized to administer, settle, or determine." Therefore, disputes arising under the Walsh-Healey Act,[30] the Davis-Bacon Act,[31] and the Service Contract Act of 1965[32] are not subject to the CDA. Those acts involve labor laws that are administered by the Secretary of Labor, and disputes related to the enforcement of such statutes generally are beyond the jurisdiction of the boards or Court of Federal Claims.[33]

III. THE FREEDOM OF INFORMATION ACT: A CLAIM PREPARATION TOOL

Often, agency files contain documents that may provide support for a contractor's request for a contract adjustment, for the contractor's understanding of its contract obligations, or the agency's responsibilities under the contract. Typically these materials are available during discovery that is allowed under a dispute resolution process. Document discovery in appeals or litigation occurs, however, ***after*** a claim has been submitted, a final decision has been issued, and an appeal or suit has been initiated. The Freedom of Information Act (FOIA)[34] may permit the contractor to obtain access to records maintained by a government agency before the submission of its claim. Contractors and their counsel should consider using FOIA requests to obtain additional information pertaining to the project. Agencies subject to the FOIA include: (1) any department or agency

[29]*Coastal Corp. v. United States,* 713 F.2d 728 (Fed. Cir. 1983). *But see Conoco Phillips v. United States,* 501 F.3d 1374, 1382 (Fed. Cir. 2007) (Court of Federal Claims had CDA jurisdiction over alleged government pre-award application of minority preference clause in a manner that affected the contract price.); *Weslayan Co. v. Harvey,* 454 F.3d 1375, 1378–79 (Fed. Cir. 2006) (ASBCA had CDA jurisdiction over dispute involving procurement of prototype for testing even though "final contract" had not been awarded); *LaBarge Prods., Inc. v. West,* 46 F.3d 1547 (Fed. Cir. 1995) (contractor entitled to assert a claim for reformation of contract based on government's allegedly improper preaward actions).

[30]41 U.S.C. § 35 *et seq.*

[31]40 U.S.C. § 276(a).

[32]41 U.S.C. § 351 *et seq.*

[33]*Herman B. Taylor Co. v. Barram,* 203 F.3d 808, 811 (Fed. Cir. 2000); *Emerald Maint., Inc. v. United States,* 925 F.2d 1425 (Fed. Cir. 1991); *Hunt Bldg. Co.,* ASBCA No. 55157, 06–1 BCA ¶ 33,213; *Schleicher Cmty. Corr. Ctr., Inc.,* DOTCAB No. 3067, 02–2 BCA ¶ 31,902. *But see Myers Investigative & Sec. Servs., Inc. v. EPA,* GSBCA No. 16587-EPA, 05–2 BCA ¶ 32,983 (GSBCA had jurisdiction over a dispute that centered on mutual contract rights even though matters reserved to the U.S. Department of Labor were part of the factual predicate).

[34]5 U.S.C. § 552.

of the Executive Branch; (2) government corporations; (3) government-controlled corporations; or (4) any independent regulatory branch.[35] The FOIA does not apply to the federal courts or to Congress.

Basically, information is made available to the public in three ways: (1) publication in the Federal Register;[36] (2) by sale to the public or availability for examination in public reading rooms;[37] or (3) upon request for documents that are reasonably described.[38] This third category of records is generally the best source of information pertaining to contract performance issues. To obtain these documents in the third category, it is necessary to submit a written FOIA request. Each agency's procedures governing the submission should be reviewed and followed carefully when making a FOIA request.[39] The regulations, found in the C.F.R., usually identify the FOIA official to whom a request should be sent and provide the appropriate address.[40] In addition, these procedures set forth time limits for agency responses and appeal procedures if the request is denied. Following these procedures usually will save time in processing the request. It should also avoid having a court decline jurisdiction over a suit to compel disclosure due to a failure to follow these rules.[41]

To ensure that a request for agency documents is processed under the FOIA, the written request should state that it is a "FOIA request" and acknowledge that the government may be entitled to be paid certain fees and costs for responding to the request.[42] In addition, it is necessary to provide a "reasonable description" of the desired records.[43] A "reasonable description" is one that enables a professional employee of the agency who is familiar with the subject area of the request to locate the record with a reasonable amount of effort.[44] Broad categorical requests that make it impossible for the agency to reasonably determine what is sought are not permissible.[45] If the agency denies the initial request, the person seeking disclosure may file an action in a United States district court to compel disclosure after exhausting the applicable administrative procedures (including any appeal process) set forth in the agency's FOIA regulations.[46]

Even after an appeal or suit is pending, contractors and their counsel should consider the potential for appropriate contemporaneous FOIA requests and formal

[35]5 U.S.C. § 552(f).

[36]5 U.S.C. § 552(a)(1); 1 C.F.R. Part 5.

[37]5 U.S.C. § 552(a)(2).

[38]5 U.S.C. § 552(a)(3)(A).

[39]5 U.S.C. § 552(a)(3)(B).

[40]*See www.gpoaccess.gov/cfr/* for a detailed listing of CFR titles.

[41]*Dettmann v. U.S. Dep't of Justice,* 802 F.2d 1472, 1476–77 (D.C. Cir. 1986); *Television Wis., Inc. v. Nat'l Labor Relations Bd.,* 410 F. Supp. 999 (W.D. Wis. 1976).

[42]If the requesting party is not sure of the cost (scope) associated with the FOIA request, it is possible to advise the agency to contact the party making the request before conducting a search that it is expected to cost more than a stated amount.

[43]H.R. Rep. No. 93–876, 93d Congress, 2d Sess. (1974).

[44]5 U.S.C. § 552(a)(3)(A). *See Jimenez v. Fed. Bureau of Investigations,* 910 F. Supp. 5 (D.D.C. 1996).

[45]S. Rep. No. 93–854 93 Congress 2d Sess. (1974). *See* 5 U.S.C. § 552(a)(3)(A)(i); *Krohn v. Dep't of Justice,* 628 F.2d 195, 198 (D.C. Cir. 1980); *Fonda v. Cent. Intelligence Agency,* 434 F. Supp. 498 (D.D.C. 1977).

[46]5 U.S.C. § 552(a)(6)(c); *Wilbur v. Cent. Intelligence Agency,* 355 F.3d 675, 677 (D.C. Cir. 2004).

discovery requests. For example, many projects involve federal agencies other than the contracting agency. An appropriate FOIA request to the noncontracting agency may provide quicker records access than the use of subpoenas on nonparty federal agencies.

IV. CONTRACTOR CLAIMS

The CDA states that "[a]ll claims by a contractor against the government relating to a contract shall be in writing and shall be submitted to the contracting officer for a decision."[47] Claims may be either monetary or nonmonetary. In that regard, the CDA further provides that monetary claims by a contractor in excess of $100,000 must be certified.[48] The existence of a claim in dispute and the proper submission of that claim to the contracting officer for a decision are prerequisites to the contractor's ability to invoke the CDA's disputes-resolution procedures. Until a claim is submitted, the contracting officer has no obligation to issue a final decision and the contractor has no right of access to either a board or the Court of Federal Claims. If a contractor initiates either a board or court proceeding before submitting a proper claim, the proceeding will either be dismissed or stayed, depending on the deficiency in the contractor's claim submission. Even though a dismissal would be without prejudice to the contractor's right to reinitiate the process,[49] the contractor must restart the process by submitting a proper claim to the contracting officer for a decision.[50] This costs time and money. Under the CDA, interest on amounts sought pursuant to a proposal or request for an equitable adjustment does not begin to accrue until the submission qualifies as a claim. Failure to understand and follow the CDA requirements can be very costly.

V. WHEN MUST A CLAIM BE SUBMITTED?

The CDA provides that "each claim" relating to a contract shall be submitted "within six years after the accrual of the claim."[51] The CDA does not define the term "accrual"; however, the implementing regulations have defined that term. For purposes of the CDA, "accrual of a claim" is defined as follows:

> [T]he date when all events, which fix the alleged liability of either the Government or the contractor and permit assertion of the claim, were known or should

[47] 41 U.S.C. § 605(a).
[48] 41 U.S.C. § 605(c).
[49] *Thoen v. United States,* 765 F.2d 1110, 1116 (Fed. Cir. 1985).
[50] *D.L. Braughler Co. v. West,* 127 F.3d 1476, 1480–81 (Fed. Cir. 1997); *Skelly & Loy, Inc. v. United States,* 685 F.2d 414, 419 (Ct. Cl. 1982).
[51] 41 U.S.C. § 605(a).

have been known. For liability to be fixed, some injury must have occurred. However, monetary damages need not have been incurred.[52]

The concept of claim "accrual" and a statute of limitations applies to claims by either the contractor or the government, except for government claims based on a contractor claim involving fraud. This definition still may stimulate questions of what constitutes an "injury" when no monetary damages have been incurred. Contractors should consider that the CDA distinquishes between a "claim" and a proposal, or a request for an equitable adjustment (REA). Submission of the latter may not satisfy the six-year submission requirement if it is later determined that the REA was not a "claim."[53]

Even if the six-year statute of limitations does not apply, a prolonged delay in the submission of a claim by a contractor may permit the government to assert an equitable defense of "laches." In such context, the claim may be denied if the prolonged passage of time substantially prejudiced the government's ability to defend against the claim.[54] Similarly, failure to submit a claim for relief due to a bid mistake until several years after award may be the basis for rejection of the claim on the grounds of waiver by the contractor, even though the government failed to demonstrate any prejudice and no statute of limitations had expired.[55]

VI. NOTICE REQUIREMENTS

Lack of timely notice of a potential claim can be a costly omission for contractors. Factual and objective notice is good business. It allows the agency the opportunity to address a problem before it becomes a costly dispute. To assist with the identification of notice requirements, the staff responsible for the daily project management should prepare a Notice Checklist for each project. That checklist should be based on a thorough review of the prime contract, subcontracts, and purchase orders. The checklist should identify the contract clause, the time requirements for notice, the subject matter of the notice, whether the notice must be in writing, and the stated consequences for failing to give timely notice.

Regardless of a contractor's familiarity with the standard clauses found in federal government construction contracts, each contract should be reviewed in detail, as contracts may contain special requirements regarding notification and documentation of delays, extra work, and so on. A Notice Checklist can take various formats. **Appendix 13.4A** in **Chapter 13** summarizes three of the standard government contract clauses with critical notice requirements and illustrates a format that can be adapted to any contract agreement.

[52]*Id.*
[53]*See Reflectone, Inc. v. Dalton,* 60 F.3d 1572 (Fed. Cir. 1995).
[54]*S.E.R. Jobs for Progress, Inc. v. United States,* 759 F.2d 1, 5 (Fed. Cir. 1985); *Sys Integrated,* ASBCA No. 54439, 07–1 BCA ¶ 33,575; *Houston Ship Channel Repair, Inc. v. U.S. Dep't of Transp., Mar. Admin.,* DOTCAB No. 4505, 06–2 BCA ¶ 33,381.
[55]*Turner-MAK (JV),* ASBCA No. 37711, 96–1 BCA ¶ 28,208.

VII. WHO MAY SUBMIT A CLAIM?

Generally, under the CDA, only a prime contractor may assert claims against the government. The CDA refers to "contractor claims"[56] and states that the term "contractor means a party to a government contract other than the government. . . ."[57] Similarly, FAR § 33.201 defines a "claim" as a demand or assertion "by one of the contracting parties." This requirement for "privity of contract" denotes a contractual relationship between the government and the prime contractor and is strictly enforced in government contracts.[58] Generally, subcontractors are not considered to be in privity of contract with the government. Accordingly, subcontractors may not assert claims directly against the government under the CDA.[59] Only in rare situations does privity of contract exist between a subcontractor and the government, such as when:

- The government utilizes an agent to enter into a contract "by and for" the government.[60]
- There has been an assignment of a subcontract to the government pursuant to a clause such as the Termination for Convenience clause.[61]
- The contracting officer consented to an assignment of a contract payment to a subcontractor.[62]
- The contract establishes a third-party beneficiary subcontractor status.[63]

Consistent with the strict requirement for privity of contract, a noncompleting performance bond surety is not a contractor under the definition set forth in the CDA.[64] A surety that expressly or implicitly assumes contract performance, however, is in privity with the government and may assert CDA claims.[65] Finally, unless the government is a party to the assignment transaction, the assignee of a contractor's rights under a government contract does not attain the status of a contractor and is not in the position to assert a claim under the CDA.[66]

Given the nature of most construction projects, subcontractor performance is often a key issue, and subcontractor claims are common. Although subcontractors

[56] 41 U.S.C. § 605(a).
[57] 41 U.S.C. § 601(4).
[58] *S. Cal. Fed. Sav. & Loan Ass'n v. United States,* 422 F.3d 1319, 1328 (Fed. Cir. 2005).
[59] *United States v. Johnson Controls, Inc.*, 713 F.2d 1541 (Fed. Cir. 1983); *Lockheed Martin Corp. v. United States,* 50 Fed. Cl. 550 (2001); *Doug Wiggs D/B/A/ Sloan Welding & Constr. Co. v. Envtl. Prot. Agency,* GSBCA No. 16817-EPA, 2006 WL 727772.
[60] *Kern-Limerick v. Scurlock,* 347 U.S. 110 (1954); *Deltec Corp. v. United States,* 326 F.2d 1004 (Ct. Cl. 1964); *Lockheed Martin Corp. v. United States,* 50 Fed. Cl. 550 (2001).
[61] FAR § 49.108–8.
[62] *D&H Distrib. Co. v. United States,* 102 F.3d 542, 546–47 (Fed. Cir. 1996).
[63] *State of Mont. v. United States,* 124 F.3d 1269, 1273 (Fed. Cir. 1997).
[64] *See Universal Sur. Co. v. United States,* 10 Cl. Ct. 794, 799–800 (1986).
[65] *United Pac. Ins. Co. v. Roche,* 380 F.3d 1352, 1355–56 (Fed. Cir. 2004).
[66] *United Pac. Ins. Co. v. Roche,* 380 F.3d 1352, 1357 (Fed. Cir. 2004); *HAM Invs., LLC,* ASBCA No. 55070, 06–2 BCA ¶ 33,406.

do not have the right to directly assert a claim against the government, subcontractor claims are routinely considered in the context of the disputes process. With the prime contractor's consent and cooperation, a subcontractor claim can be submitted to the contracting officer for decision and appealed through the disputes process. In that situation, the prime contractor acts as a "sponsor" for the subcontractor's claim.[67] But a prime contractor must recognize that only it can certify a subcontractor's claim when it is submitted under the CDA. See **Section X** of this chapter.

VIII. WHAT CONSTITUTES A CLAIM?

The CDA states that all claims "shall be in writing and shall be submitted to the contracting officer for a decision."[68] The CDA does not, however, define what constitutes a claim. Moreover, although it would seem that the definition of a claim would be uniform in federal government contracts, there is a marked lack of uniformity. There are differing definitions of a claim for the purposes of: (1) the Disputes clause,[69] (2) the disallowance of certain costs associated with the prosecution of a "claim" against the government,[70] and (3) contractor liability under the False Claims Act.[71]

FAR § 33.215 provides that, absent specific circumstances, each government contract must contain the Disputes clause set forth at FAR § 52.233–1. That clause defines a claim as follows:

> (c) Claim, as used in this clause, means a written demand or written assertion by one of the contracting parties seeking, as a matter of right, the payment of money in a sum certain, the adjustment or interpretation of contract terms, or other relief arising under or relating to this contract....[A] written demand or written assertion by the Contractor seeking the payment of money in excess of $100,000 is not a claim under the Act until certified....
>
> (d) (1) A claim by the Contractor shall be made in writing and, unless otherwise stated in this contract, submitted within 6 years after accrual of the claim to the Contracting Officer for a written decision....

In *Reflectone, Inc. v. United States,*[72] the Federal Circuit held that the FAR definition of a "CDA claim" distinguishes between "routine" requests for payment—that

[67] *Erickson Air Crane Co. of Wash., Inc. v. United States,* 731 F.2d 810, 813–14 (Fed. Cir. 1984). If a prime contractor refuses to authorize prosecution of a claim in its name, however, the claim will be dismissed. *Hamilton & Voeller, Inc.,* AGBCA No. 79–137, 79–2 BCA ¶ 13,992.
[68] 41 U.S.C. § 605(a).
[69] FAR § 52.233; *Reflectone, Inc. v. United States,* 60 F.3d 1572 (Fed. Cir. 1995).
[70] FAR § 31.205–33; *Bill Strong Enters., Inc., v. Shannon,* 49 F. 3d 1541, 1546–50 (Fed. Cir. 1995).
[71] 18 U.S.C. § 287; *United States v. Neifert-White Co.,* 390 U.S. 228, 233 (1968); *United States ex rel. Siewick v. Jamieson Eng'g, Inc.,* 214 F.3d 1372, 1375–76 (D.C. Cir. 2000).
[72] 60 F.3d 1572 (Fed. Cir. 1995).

is, a voucher or a progress payment invoice—and "nonroutine" claims for payment. Even though a routine request for payment is in writing and seeks payment of a sum certain, it is not a CDA claim until it is converted into a "claim." The CDA time limits for action by a contracting officer do not apply to and CDA interest does not accrue on a routine request for payment. A routine request for payment can be converted into a CDA claim by written notice to the contracting officer if it is disputed as to liability or dollar value or if it is not acted on in a reasonable time. In those circumstances, the conversion to a CDA claim occurs because the otherwise routine payment request is considered to be "in dispute."

A non-routine payment request does not have to be "in dispute" in order to be considered a CDA claim. The analysis, however, does not stop with the distinction between a "routine" and "nonroutine" payment request. For example, a termination for convenience settlement proposal is a basis for negotiation, not a claim. It can become a CDA claim after subsequent negotiations fail and the contractor impliedly requests a final decision.[73] Given this uncertainty surrounding nonroutine payment requests, contractors should not assume that a change order proposal or REA proposal will be treated as a CDA claim absent notice to the contracting officer and the submission of any necessary certification.

A. Written Submission to the Contracting Officer

As indicated in the Disputes clause quoted above, a CDA claim must be in writing and must be submitted to the contracting officer for a written decision. Accordingly, oral demands or assertions seeking compensation are not CDA claims. Similarly, a written demand or submission that is submitted to a person who is not the contracting officer is not a CDA claim. The CDA defines a contracting officer as "[a]ny person who, by appointment in accordance with applicable regulations, has the authority to enter into and administer contracts and make determinations and findings with respect thereto."[74] This definition also includes an authorized representative of the contracting officer, acting within the limits of that person's authority, usually published in an agency letter or memorandum. However, the fact that a government employee is designated as the "contracting officer's representative" does not necessarily mean that this person has the necessary authority to direct changes to the contract.[75] This lack of authority may have a direct adverse effect on the success of a contractor's claim.

Under some circumstances, the submission of a written claim to a subordinate of the contracting officer has been held to be ineffective for the purposes of the CDA.[76] A contractor should avoid the problem of misdirected claims by ascertaining

[73] *Rex Sys., Inc. v. Cohen,* 224 F.3d 1367, 1371–73 (Fed. Cir. 2000); *Ellett Constr. Co. v. United States,* 93 F.3d 1537 (Fed. Cir. 1996).

[74] 41 U.S.C. § 601(3).

[75] *Winter v. Cath-dr/Balti Joint Venture*, 497 F.3d 1339 (Fed. Cir. 2007).

[76] *Lakeview Constr. Co. v. United States,* 21 Cl. Ct. 269, 272–73 (1990). *But see Dawco Constr., Inc. v. United States,* 930 F.2d 872, 879–80 (Fed. Cir. 1991); *Blake Constr. Co. v. United States,* 25 Cl. Ct. 177, 181 (1992); *Gardner Zemke Co.,* ASBCA No. 51499, 98–2 BCA ¶ 29,997.

at the outset of performance the specific individuals, in addition to the contracting officer, who are authorized to receive claims. Moreover, if there is any doubt regarding whether the submission to the contracting officer's subordinate will be deemed appropriate under the CDA, the contracting officer should be copied on the submission. This should remove any doubt regarding when the contracting officer received the claim for the purposes of the CDA.

B. Elements of a Claim

In submitting a claim, the contractor need not use any particular wording or format.[77] The contractor must, however, give the contracting officer sufficient information such that the contracting officer has adequate notice of the basis of the claim and, if appropriate, the amount of the claim.[78] Therefore, there are basically two elements of any monetary claim: (1) a description of the facts and contractual basis for entitlement and (2) an explanation of the quantum requested.

The entitlement portion sets forth the factual and contractual basis supporting the contractor's right to the requested relief. For example, in a claim involving the government's alleged misinterpretation of a specification, the entitlement portion would describe how the government misinterpreted the specification as contrasted to the contractor's reasonable understanding of the contract's requirements. That section also would describe the effect on the contractor's performance resulting from the misinterpretation.

If the claim seeks monetary relief, the second part states quantum. At a minimum, the quantum portion of a monetary claim requires that the contractor state a "sum certain."[79] The contractor should describe the amount of money, time, or both to which it is entitled. The contractor also should attempt to relate cause (action or inaction by the government) and effect (expenditure of money, delay of progress, or both). The relation between cause and effect can be quite difficult, as it often encompasses issues of scheduling, cost accounting, and estimated cost or other impact. As with any contract (public or private), project documentation is often critical to establishing cause and effect. The evaluation of a contractor's claims for time or money can be significantly affected by the nature and quality of the contractor's contemporaneous records. For example, some government contracts expressly require that the contractor document on a daily basis the specific critical path activities affected by delays.[80] Even if there is no specific contract requirement for contemporaneous documentation of the extra time or expense incurred by the contractor, the absence of such documentation can create a substantial hurdle to recovery.[81]

[77]*Contract Cleaning Maint,, Inc. v. United States,* 811 F.2d 586, 592 (Fed. Cir. 1986).

[78]*H.L. Smith, Inc. v. Dalton,* 49 F.3d 1563, 1565 (Fed. Cir. 1995).

[79]FAR § 52.233–1(c); *Reflectone, Inc. v. Dalton,* 60 F.3d 1572, 1575 (Fed. Cir. 1995).

[80]*Commercial Contractors Equip., Inc.,* ASBCA No. 52930, 03–2 BCA ¶ 32,881.

[81]*Fru-Con Constr. Corp. v. United States,* 43 Fed. Cl. 306, 335–36 (1999); *Aleutian Constructors, J.V.,* ASBCA No. 49255, 01–1 BCA ¶ 31,392; *Centex Bateson,* VABCA Nos. 4613 *et al.*, 99–1 BCA ¶ 30,153 (number of RFIs and changes are not enough by themselves to demonstrate lost productivity).

There is no simple test to determine the degree of detail necessary to constitute a claim. In general, the boards and Court of Federal Claims have stated that the contractor must provide the contracting officer with adequate notice of the basis and the amount of the claim together with a request that the contracting officer render a final decision.[82] In short, the contractor should assert specific rights and request specific relief. Ultimately, whether a contractor's submission constitutes a claim depends on the totality of the circumstances and communications between the parties.[83] For example, the following submissions have been found to be claims:

(1) A letter in which a contractor specified various items that a government audit had disallowed but to which the contractor claimed entitlement. The letter was viewed together with a prior letter from the contractor giving a detailed breakdown of the additional amounts to which the contractor believed it was entitled and referring to the contractor's previous request for "funding of [a] back-wage demand."[84]

(2) A letter sent by a transportation services contractor at an Air Force base stating that the company viewed certain newly demanded bus service as beyond the contract's requirements and specifically seeking "compensation of $11,000.04 per year, to be billed at $916.67 per month."[85]

(3) A letter from the contractor's attorney to the contracting officer that "expressed interest" in a final decision with respect to the contractor's request for contract reformation and that stated the contractor was seeking a decision so that it could pursue its appeal routes under the CDA, if necessary.[86]

(4) Letters that, when taken together, showed the contractor protesting the payment of additional sums under a contract to purchase crude oil from the government and demanding that certain identified wire transfer payments comprising those sums be returned to the contractor.[87]

(5) A contractor's cover letter that attached a detailed argument by a subcontractor setting forth the basis of the subcontractor's disagreement with the government regarding the effect of a contract modification.[88]

Still, the claim must be made "by the contractor." Thus, a letter from the contractor's lawyer to the contracting officer has been held to be insufficient to establish

[82] *Transamerica Ins. Corp. v. United States,* 973 F.2d 1572, 1578–79 (Fed. Cir. 1992); *Isles Eng'g & Constr., Inc. v. United States,* 26 Cl. Ct. 240, 243 (1992); *D.N. Kelley & Son, Inc.,* DOTCAB No. 2444, 92–2 BCA ¶ 24,926; *Anchor Fabricators, Inc.,* ASBCA No. 40893, 91–3 BCA ¶ 24,231.

[83] *James M. Ellett Constr. Co., v. United States,* 93 F.3d 1537, 1546 (Fed. Cir. 1996); *Roxco, Ltd. v. United States,* 77 Fed. Cl. 138, 142–43 (2007); *Freeman Contracting, Inc. v. Dep't of Agric.,* CBCA No. 16, 07–2 BCA ¶ 33,593; *Ventas de Equipo, S.A.,* ENGBCA Nos. PCC-135 *et al.,* 00–1 BCA ¶ 30,913; *Bared & Co.,* ASBCA No. 47628, 95–2 BCA ¶ 27,710.

[84] *Contract Cleaning Maint., Inc. v. United States,* 811 F.2d 586, 592 (Fed. Cir. 1986).

[85] *Tecom, Inc. v. United States,* 732 F.2d 935, 937 (Fed. Cir. 1984).

[86] *Paragon Energy Corp. v. United States,* 645 F.2d 966, 976 (Ct. Cl. 1981).

[87] *Alliance Oil & Ref. Co. v. United States,* 13 Cl. Ct. 496, 499–500 (1987).

[88] *Clearwater Constructors, Inc. v. United States,* 56 Fed. Cl. 303 (2003).

a claim.[89] The contractor bears the burden to identify, specify, and perfect its claims. For example, in *Mingus Constructors, Inc. v. United States,*[90] the contractor had complained of malicious harassment by the government's project representative during the construction of a road and had stated an intention to claim the added costs resulting from the alleged harassment. These general complaints were set forth in various letters. At the conclusion of the project, the contractor executed a standard final release form and added a notation to the release that it did not apply to "claims" stated in the earlier correspondence. The Federal Circuit ruled that the contractor had never properly identified, specified, or perfected its claims. As no claims were ever properly submitted, the release, once executed, barred the contractor from making a valid claim.

Some decisions have emphasized the need for an explicit request for a final decision as part of the submission of a claim. In *Transamerica Insurance Corp. v. United States,*[91] however, the Federal Circuit expressly rejected a rule that a "claim" must include a specific request for a final decision. In particular, a formal demand for final decision has been held to be unnecessary if the claim clearly indicates that a decision was desired.[92]

Even though there is case law holding that there is no absolute requirement for a specific request for a final decision, there is a risk that the failure to make a specific request for a final decision will be viewed as evidence that there was no dispute between the parties sufficient to constitute a CDA claim. For this reason, correspondence from a contractor containing detailed cost information and, rather than demanding or requesting that the contracting officer issue a final decision, merely expressing a willingness to reach an agreement, may not constitute a claim.[93]

The passage of time may not convert a proposal into a CDA claim. In *Santa Fe Engineers, Inc. v. Garrett,*[94] the contractor's certified proposal had been pending for more than two years, during which time the parties met on several occasions to discuss the proposal and the contractor submitted additional information for the government's consideration. During that time, the contractor never requested a contracting officer's final decision. The Federal Circuit affirmed the board decision,[95] which had concluded that the failure to ask for a contracting officer's final decision was an indication

[89]*Constr. Equip. Lease v. United States,* 26 Cl. Ct. 341 (1992). In contrast, the bankruptcy trustee for a contractor in bankruptcy has been held to be the proper party to assert a pre-bankruptcy claim against the government. *In re Dodds,* ASBCA No. 51682, 02–1 BCA ¶ 31,844.

[90]812 F.2d 1387 (Fed. Cir. 1987).

[91]973 F.2d 1572 (Fed. Cir. 1992).

[92]*James M. Ellett Constr. Co. v. United States,* 93 F.3d 1537, 1546 (Fed. Cir. 1996).

[93]*Hoffman Constr. Co. v. United States,* 7 Cl. Ct. 518, 525 (1985). *See also Technassociates, Inc. v. United States,* 14 Cl. Ct. 200, 209–10 (1988) (a contractor's letters to the contracting officer seeking negotiations "on the future direction of the contract" did not constitute claims). In *D.H. Blattner & Sons, Inc.,* IBCA Nos. 2589 *et al.,* 89–3 BCA ¶ 22,230, however, the board held that a properly certified letter using the term "proposal" was not a claim. The Federal Circuit reversed this decision in an unpublished decision. *See Blattner & Sons, Inc. v. United States,* 909 F.2d 1495 (Fed. Cir. 1990).

[94]991 F.2d 1579, 1583–84 (Fed. Cir. 1993).

[95]*Santa Fe Eng'rs, Inc.,* ASBCA No. 36292, 92–2 BCA ¶ 24,795.

that the parties had not reached an impasse in their negotiations and that the matter was not sufficiently in dispute to constitute a CDA claim.

Whether a particular submission is a claim often depends on the totality of the circumstances. For example, in one case, a certified REA accompanied by a letter requesting a final decision constituted a valid claim even though the letter also suggested the possibility, or hope, of a negotiated resolution to the dispute.[96] In contrast, a letter sent to the contracting officer during negotiations on a proposal requesting that the matter be referred to an auditor did not qualify as a claim, because a contracting officer had not been asked to issue a decision.[97]

Finally, the contractor may not submit a money-based claim in an unspecified or open-ended amount.[98] A proper claim and request for a final decision exist only when the amount sought is set forth in a "sum-certain,"[99] or is determinable by a simple mathematical calculation or from the information provided by the contractor.[100]

C. Certification Requirement

The CDA currently requires that a claim in excess of $100,000 be certified.[101] The purpose of the certification requirement is to discourage the submission of unwarranted or inflated contractor claims, decrease litigation, and encourage settlements. Section 907(a) of the Federal Courts Administration Act has softened the rigid formalities of the CDA pertaining to claim certification while leaving intact the basic policy safeguards underlying the certification requirement.[102] This amendment to the

[96] *Isles Eng'g & Constr., Inc. v. United States,* 26 Cl. Ct. 240, 243 (1992).

[97] *G. S. & L. Mech. & Constr., Inc.,* DOTCAB ¶ No. 1856, 87–2 BCA ¶ 19,882; *see also Huntington Builders,* ASBCA No. 33945, 87–2 BCA ¶ 19,898, at 100,654–655 (letters to contracting officer that, when taken together, alleged defective specifications and requested a 30-day time extension to contract and release of monies withheld for liquidated damages did not constitute a claim because no specific monetary relief was requested for costs incurred as a result of contractor's having to comply with the allegedly defective specifications).

[98] *Metric Constr. Co. v. United States,* 14 Cl. Ct. 177, 179–80 (1988) (contractor's submissions made it clear that the contractor was seeking to recover extended home office overhead and third-party indemnification fees, but the submissions did not constitute a claim because the monetary amount was not specified); *Eaton Contract Servs., Inc.,* ASBCA Nos. 52888 *et al.,* 02–2 BCA ¶ 32,023 (no jurisdiction where contractor stated that claims "exceeded $2,000,000" but provided no total of all damages); *Fluor Daniel, Inc.,* EBCA No. C-9912312, 02–2 BCA ¶ 32,015 (no jurisdiction over "claim" for "in excess of $5 million").

[99] Disputes clause ¶ (c), FAR § 52.233–1; *CPS Mech. Contractors, Inc., v. United States,* 59 Fed. Cl. 761, 764–65 (2004). (A claim for a "sum certain" that reserved the right to include additional line items to "modify the presentation" was, in fact, deemed to be a predicate for negotiations rather than a CDA claim); *McElroy Mach. & Mfg. Co., Inc.,* ASBCA No. 39416, 92–3 BCA ¶ 25,107. (The sum certain requirement is determined when the claim is filed, not when the complaint is filed); *Morgan & Son Earthmoving, Inc.,* ASBCA No. 53524, 02–2 BCA ¶ 31,874 (same); *Sentra Health Sys.,* ASBCA No. 51540, 99–1 BCA ¶ 30,323. Jurisdiction over a sum certain Claim No. 1 not precluded by reservation of rights of future "Claim" No. 2.

[100] *Cubic Corp. v. United States,* 20 Cl. Ct. 610, 621 (1990); *Allstate Prods. Co.,* ASBCA No. 52014, 00–1 BCA ¶ 30,783 (claim for Prompt Payment Act interest amount "was easily calculable).

[101] 41 U.S.C. § 605(c)(1) and (c)(2).

[102] Pub. L. No. 102–572.

CDA changed the requirements for a proper CDA claim certification, and the consequences if the certification was somehow "defective" or improper, in five ways:

(1) Broadened the class of individuals who could properly certify a claim to include anyone authorized to bind the contractor with respect to the claim.
(2) Expressly stated that a "defective" certification would not deprive a board or court of jurisdiction over that claim.
(3) Required that a "defective" certification be cured before a board or court could render a decision on that claim.
(4) Excused a contracting officer from issuing a final decision on a defectively certified claim if the contracting officer advised the contractor in writing of the basis for the conclusion that the certification was inadequate within 60 days of the date of the contracting officer's receipt of the claim.
(5) Allowed CDA interest to accrue on a claim even though the certification was defective.

Still, a CDA claim over $100,000 must be certified. Consequently, the total absence of a required certification is not the equivalent of a "defective" certification.[103] If a claim is submitted without the required certification, and a final decision is issued, the claimant cannot cure such omission by making a retroactive submission. Instead, the claim and the certification must be submitted together.[104] The certification also must be separately signed. At least one board has ruled that the failure to sign the certification is "more akin" to a failure to certify.[105] Moreover, the failure to properly certify a claim most likely will delay its resolution and increase the cost of resolving the dispute. Therefore, it is still important to understand the monetary threshold for a claim certification, how it must be certified, who can certify it, and the relationship of that certification to other government contract certifications.

D. Monetary Threshold for Certification

The CDA requires a certification when the contractor asserts a claim exceeding $100,000. Therefore, it is not possible to bypass a certification requirement by breaking a claim into a series of separate claims each of which is $100,000 or less.[106] The test is whether there exists a single, unitary claim based on a common and/or related

[103]*Gulf Shores, LLC v. U.S. Dep't of Homeland Security,* DOTCAB No. 4530, 06–2 BCA ¶ 33,384; *Schnider's of OKC, Inc.,* ASBCA No. 53947, 03–1 BCA ¶ 32,160; *Westar Eng'g, Inc.,* ASBCA No. 52484, 02–1 BCA ¶ 31,759 (uncertified claim dismissed where board found that a request for contract interpretation was a veiled money-based claim in excess of $100,000).

[104]*Golub-Wegco Kan. City I, LLC v. Gen. Servs. Admin.,* GSBCA No. 15387, 01–2 BCA ¶ 31,553 (if a proposed claim and its certification are separately submitted, it is treated as a proper, new claim at the time of the certification); *J&J Maint., Inc.,* ASBCA No. 50984, 00–1 BCA ¶ 30,784 (certificate submitted with notice of appeal to board is ineffective as to previous defective claim but presents proper, new claim).

[105]*Haw. Cyberspace,* ASBCA No. 54065, 04–1 BCA ¶ 32,455.

[106]*Nussinow v. United States,* 23 Cl. Ct. 556, 560 (1991). *See also D&K Painting Co., Inc.,* DOTCAB No. 4014, 98–2 BCA ¶ 30,064; *Columbia Constr. Co., Inc.,* ASBCA No. 28536, 96–1 BCA ¶ 27,970.

set of operative facts, which the contractor, unintentionally or otherwise, divided into separate and distinct claims.[107] If the claims are distinct and independent, however, with one claim having no relationship to the operative facts of the other claim, each independent claim of $100,000 or less need not be certified.[108] Even if the contractor submits a single letter to the contracting officer that includes multiple claims totaling more than $100,000, certification is not required unless the claims arose from a common or related set of operative facts and are therefore truly a unitary claim.[109] In other words, do the contractor's claims arise from the same or different events or causes of action?[110]

The following examples illustrate application of this test. In one case, the contractor alleged that a differing site condition gave rise to three separate claims: one for additional paving costs; one for additional insurance, supervision, and maintenance costs; and one for loss of interest on funds spent to perform additional work. The court concluded that there was just one claim.[111] Similarly, when a contract was terminated for the convenience of the government, the court concluded that a contractor's demand for "pre-termination and post-termination items" constituted one claim because both items were directly related to the government's termination of the contract, and the resolution of both items depended on what, if any, liability the government incurred because of its action.[112] Similarly, in a case where the contract involved security guard services at five different locations in Boston, the court held that the contractor could not fragment its total dollar claim into separate claims based on each of the different locations. The rationale for this determination was that the amounts claimed from the various locations were based on the same operative facts (total hours of services performed for which a total number of dollars allegedly was due).[113] In another case, however, the Armed Services Board of Contract Appeal (ASBCA) determined that when 18 different claims arose from different causative events and were brought under different legal theories, such as differing site conditions and defective specifications, it was proper to separate the claims.[114]

E. Modification of Claim Amount

Sometimes a claim that initially does not exceed $100,000 (and therefore is not certified) increases in amount after the claim is submitted. In these circumstances, the

[107]*Nussinow v. United States,* 23 Cl. Ct. 556, 560 (1991); *Placeway Constr. Corp. v. United States,* 920 F.2d 903, 907 (Fed. Cir. 1990); *Lockhead Martin Aircraft Ctr.,* ASBCA No. 55164, 07–1 BCA ¶ 33,472; *Advanced Injection Molding, Inc .v. Gen. Servs. Admin.,* GSBCA No. 16504, 05–2 BCA ¶ 33,037.

[108]*Little River Lumber Co. v. United States,* 21 Cl. Ct. 527, 536 (1990); *Advanced Injection Molding v. Gen. Servs. Admin.,* GSBCA No. 16504, 05–2 BCA ¶ 33,037; *Gov't Bus. Servs. Group, LLC,* ASBCA No. 53920, 03–1 BCA ¶ 32,202.

[109]*Placeway Constr. Corp. v. United States,* 910 F.2d 835 (Fed. Cir. 1990); *Spirit Leveling Contractors v. United States,* 19 Cl. Ct. 84 (1989).

[110]*Gov't Bus. Servs. Group, LLC,* ASBCA No. 53920, 03–1 BCA ¶ 32,202.

[111]*Warchol Constr. Co. v. United States,* 2 Cl. Ct. 384, 391 (1983).

[112]*Palmer & Sicard, Inc. v. United States,* 4 Cl. Ct. 420, 422–23 (1984).

[113]*Black Star Sec., Inc. v. United States,* 5 Cl. Ct. 110, 116–18 (1984).

[114]*Zinger Constr. Co.,* ASBCA No. 28788, 86–2 BCA ¶ 18,920.

question arises whether the contractor still can proceed on the basis of the increased claim before the court or board, or whether it is necessary for the contractor to certify the claim in the increased amount and resubmit it to the contracting officer for a decision.

This question was addressed in *Tecom, Inc. v. United States.*[115] In *Tecom,* the contractor's claim was less than the certification threshold when it was submitted to the contracting officer. When the company filed its complaint before the ASBCA, however, the claim amount exceeded the monetary threshold for a certification. This increase was caused by two events that occurred after the contracting officer's decision: a reevaluation of the claim by the contractor and the government's exercise of an option to extend the contract for one year. Under these circumstances, the court held that it was not necessary for the contractor to certify and resubmit its claim.

Tecom stands for the proposition that a monetary claim properly considered by a contracting officer "need not be certified or re-certified if that very same claim (but in an increased amount reasonably based on further information) comes before a board or a court."[116] The Federal Circuit stated that it would be disruptive of normal litigation procedures "if any increase in the amount of a claim based on matters developed in litigation before the court [or board] had to be submitted to the contracting officer before the court [or board] could continue to final resolution on the claim."[117] In a footnote, however, the *Tecom* court noted that its decision should not be taken as an invitation to seek to evade the certification requirement.[118] Thus, a contractor that deliberately understates the amount of its original claim (with the intention of raising the amount on appeal on the basis of information that was readily available when the claim first was submitted) may well find its subsequent suit in the Court of Federal

[115]732 F.2d 935 (Fed. Cir. 1984).

[116]732 F.2d at 938.

[117]732 F.2d at 937–38 (*quoting J. F. Shea Co. v. United States,* 4 Cl. Ct. 46, 54 (parentheticals in *Tecom*). *See Modeer v. United States,* 68 Fed. Cl. 131, 137 (2005) (contractor claims for holdover rent, which would increase over time, were proper and need not be certified where "claim increases [were] based on new information available only after the claim was submitted to the contracting officer . . . as long as [the claim] arises from the same operative facts as the original claim and claims the same categories of relief"); *In re Todd Pac. Shipyards Corp.,* ASBCA No. 55126, 06–2 BCA ¶ 33,421 ("contractor can increase the amount sought in its proper CDA claim to the [contracting officer] when the increase is reasonably based upon further information developed in litigation before the Board"); *Whiting-Turner/A.L. Johnson Joint Venture v. Gen. Servs. Admin.,* GSBCA No. 15401, 02–1 BCA ¶ 31,708. *See also Glenn v. United States,* 858 F.2d 1577, 1580 (Fed. Cir. 1988). The contractor submitted a claim to the contracting officer in the amount of $31,500. Because the claim was less than $50,000 (then applicable monetary threshold for certification), the contractor did not certify it. The contracting officer issued a final decision denying the claim, which the contractor appealed to the ASBCA. Thereafter, the contracting officer issued a second final decision. In that decision, the contracting officer stated that he was withholding $66,570.32 from the contractor (consisting of the $31,500 that the contractor previously had sought to recover and an additional $35,070.32). Relying on its prior decision in *Tecom,* the Federal Circuit held that it was not necessary for the contractor to certify its $66,570.32 claim before bringing suit in the Claims Court. "Because Glenn was not required to certify his $31,500 claim before the CO,, he need not have certified the $66,570.32 resulting from the denial of his initial claim . . . and [the] additional setoffs."

[118]*Tecom, Inc. v. U.S.,* 732 F.2d 935, 938 n.2 (Fed. Cir. 1984); *Hydra Rig Cryogenics, Inc.,* ASBCA No. 36980, 90–1 BCA ¶ 22,532.

Claims or its board appeal dismissed for lack of any certification.[119] Dismissal occurs because the total absence of a certification is different from a defective (or inadequate) certification.

F. Certification Language

The CDA sets forth the language to be used in the certification. The contractor must certify

> "that the claim is made in good faith, that the supporting data are accurate and complete to the best of [the contractor's] knowledge and belief, and that the amount requested accurately reflects the contract adjustment for which the contractor believes the Government is liable." [120]

The CDA further requires that the person signing the certification state that "the certifier is duly authorized to certify the claim on behalf of the contractor."[121] The procurement regulations,[122] as well as the Disputes clause utilized in contracts covered by the CDA,[123] have the same elements for claim certification.

Under the pre-1992 statutory language, the boards, the United States Claims Court, and the Federal Circuit developed strict rules defining a defective or inadequate certification. These rules, in conjunction with the holding that the submission of a proper certification for any claim in excess of the monetary threshold was a jurisdictional requirement,[124] created extensive problems for claimants and their counsel. Although the 1992 Amendment eliminated the rule that the submission of a valid certification was a jurisdictional requirement that could not be waived, potential problems for a contractor regarding the form of the certification remain. To the extent that the cases interpreting the CDA as written before 1992 provide guidance regarding proper certification wording, it is possible that these decisions still will be relied on by the boards and the courts in determining whether a certification is "defective."

Before the 1992 Amendment to the CDA, there was a split in authority regarding the contractor's obligation to strictly track the statutory certification language in order to submit a valid certification. One line of cases took a very formalistic view and held that any deviation from the statutory language would be subject to strict scrutiny. In those cases, a contractor's attempt to deviate from the statutory

[119]Even the reduction of a claim below the applicable threshold will not eliminate the need for a certification if the claim, as submitted to the contracting officer, exceeded the certification threshold. *See Kaco Contracting Co.*, ASBCA No. 43066, 92–1 BCA ¶ 24,603; *Alaska Lumber & Pulp Co.*, AGBCA No. 82–107–1, 91–2 BCA ¶ 23,824.

[120]41 U.S.C. § 605(c)(1).

[121]41 U.S.C. § 605(c)(7).

[122]FAR § 33.207(a).

[123]FAR § 52.233–1.

[124]*See Envtl. Specialists, Inc. v. United States*, 23 Cl. Ct. 751 (1991); *Skelly & Loy v. United States*, 685 F.2d 414, 419 (Ct. Cl. 1982); *Paul E. Lehman v. United States*, 673 F.2d 352 (Ct. Cl. 1982); *Kaco Contracting Co.*, ASBCA No. 43066, 92–1 BCA ¶ 24,603; *Guilanani Contracting Co.*, ASBCA No. 41435, 91–2 BCA ¶ 23,774.

language by substituting alternate language usually was held to invalidate the certification.[125] A second line of cases held that substantial compliance was sufficient and the inadvertent omission of a few words in the certification and the omission of the claimed amount, which was stated elsewhere in the claim, were not fatal defects.[126] In general, any certification must simultaneously state all elements of the statutory requirements. An effort to satisfy the certification requirement by reference to multiple letters or by piecemeal submissions has not been deemed to be sufficient to satisfy the statutory requirement.[127]

In endorsing the substantial compliance approach, the General Services Administration Board of Contract Appeals (GSBCA) accepted a certification that omitted any reference to "knowledge or belief." The board held that this unqualified certification "more fully exposed [the contractor] to potential liability" for false statements than "if it had mimicked the words of the statute."[128] In contrast, the ASBCA has held that a failure to state that the contractor "believes" the government is liable invalidates the certification.[129] In addition, the following certifications have been held to be defective:

(1) A certification that varied from the language of the statute and the disputes clause by referring to "all data used" instead of the "supporting data" for the claim (thereby restricting the certification to "unidentified data [which the contractor] chose to use while the statute requires certification of all data that support the claim").[130]

(2) A certification that omitted the assertion that the supporting data was accurate and complete.[131]

(3) A certification in which the contractor stated that it would not assume any legal obligations that it would not have without the certification, that the data submitted was "as accurate and complete as practicable," and that the contractor was not demanding a "particular amount."[132]

[125]*Centex Constr. Co.*, ASBCA No. 35338, 89–1 BCA ¶ 21,259; *Liberty Envtl. Specialties, Inc.*, VABCA No. 2948, 89–3 BCA ¶ 21,982.

[126]*United States v. Gen. Elec. Corp.*, 727 F.2d 1567, 1569 (Fed. Cir. 1984); *Young Enters., Inc. v. United States*, 26 Cl. Ct. 858, 862 (1992); *P.J. Dick, Inc.*, GSBCA Nos. 11847 *et al.*, 93–1 BCA ¶ 25,263. In *Metric Constructors, Inc.*, ASBCA No. 50843, 98–2 BCA ¶ 30,088, the ASBCA held that a signed termination for convenience settlement proposal on Standard Form 1438 contained certification language sufficiently similar to the CDA to constitute a correctable certification. However, in *Keydata Sys., Inc. v. Dep't of Treasury*, GSBCA No. 14281-TD, 97–2 BCA ¶ 29,330, the GSBCA held that the 1992 Amendment did not authorize contractors to cure defective certifications resulting from fraud, bad faith, or "negligent disregard" of the certification requirements.

[127]*United States v. Turner Constr. Co.*, 827 F.2d 1554, 1560 (Fed. Cir. 1987); *Black Star Sec., Inc. v. United States*, 5 Cl. Ct. 110, 117 (1984); *TechDyn Sys. Corp.*, ASBCA No. 38727, 91–2 BCA ¶ 23,749; *Echo W. Constr.*, VABCA No. 3186, 90–3 BCA ¶ 23,106.

[128]*P.J. Dick, Inc.*, GSBCA Nos. 11847 *et al.*, 93–1 BCA ¶ 25,263.

[129]*C.F. Elecs.*, ASBCA No. 40777, 91–2 BCA ¶ 23,746.

[130]*Gauntt Constr. Co.*, ASBCA No. 33323, 87–3 BCA ¶ 20,221 at 102,412.

[131]*Raymond Kaiser Eng'rs, Inc./Kaiser Steel Corp., a Joint Venture*, ASBCA No. 34133, 87–3 BCA ¶ 20,140, at 101,940–41.

[132]*Cochran Constr. Co.*, ASBCA No. 34378, 87–3 BCA ¶ 19,993 at 101,280–81, *aff'd on recons.*, 87–3 BCA ¶ 20,114.

By contrast, in another case, a contractor whose certification did not contain the amount of the claim involved and did not have the words "the amount requested accurately reflects the contract adjustment for which the contractor believes the government is liable" still was found to be in substantial compliance with the certification requirement. The statement in which the certification was contained did have the remainder of the elements required by the CDA; and when the statement was read in its entirety and together with documents that accompanied it, all of the information and statements required by the statute were found to be present.[133] Notwithstanding the degree of flexibility allowed by the 1992 Amendment to the CDA and various decisions, the prudent course is to track the CDA language exactly when certifying a claim to avoid potential delays regarding an alleged defective certification.

G. Supporting Data

A contractor that certifies its claim by tracking the CDA language still may find itself confronted with an argument that it provided inadequate supporting data for its claim for purposes of the certification requirement. For the most part, though, neither the courts nor the boards have taken an overly stringent attitude with respect to the extent of the supporting data.

In *Metric Construction Co. v. United States,*[134] the government argued that the contractor's certification was defective because the contractor had failed to attach copies of the pertinent change order modifications to its claim. In rejecting the government's argument, the court observed that the certification requirement "was not intended, nor should it be so construed, to require a full evidentiary presentation before the contracting officer."[135] The court noted that the contracting officer had not denied the contractor's claim for lack of supporting data and that the data presented had assisted the contracting officer "in making a meaningful determination on the dispute before him.[136]

The Department of Energy Board of Contract Appeals took a similar position and cited *Metric* with approval in *Newhall Refining Co.*[137] In response to the government's argument that the contractors involved had not submitted accurate and complete supporting data when they certified their claims, the board noted that: on their face, the certifications met the CDA requirements; the claims were "articulated in a clear and concise fashion"; the contractors had notified the contracting officer of the basis for their claims before submitting them; and the contracting officer already had information relating to the claims. The board also noted that the claims before it involved a legal issue of contract interpretation, and it found "highly persuasive" the fact that the contracting officer had not requested additional information from the contractors. Under these circumstances, the board determined that the data submitted with the claims were "adequate."[138]

[133] *United States v. Gen. Elec. Corp.*, 727 F.2d 1567, 1596 (Fed. Cir. 1984).
[134] 1 Cl. Ct. 383 (1983).
[135] 1 Cl. Ct. 383, 391 (1983).
[136] *Id.*
[137] EBCA Nos. 363–7–86 *et al.*, 87–1 BCA ¶ 19,340.
[138] 87–1 BCA ¶ 19,340. *See also Santa Barbara Research Ctr.*, ASBCA No. 27831, 85–2 BCA ¶ 18,046.

These cases suggest that, when the contractor's certification language meets the CDA's requirements, and the contracting officer is provided with the needed information to render a final decision and has issued a final decision, the contractor probably does not have to worry about having its case derailed by a government assertion at the board or court that it failed to submit adequate supporting data. Yet supporting data that are adequate for the purposes of a CDA claim submission almost certainly will not be sufficient to prevail on the merits.

H. Who Can Certify the Claim?

Based on the CDA's current wording, questions of a person's authority to certify a claim should be minimal, so long as the certifier follows the CDA's language. This requires an express representation that the person signing the certification is authorized to do so.[139] In that context, the person certifying the claim need not have personal knowledge of the facts and data supporting the claim. Rather, the person certifying the claim can rely on data and facts developed by others within the contractor's organization.[140]

IX. OTHER CERTIFICATION REQUIREMENTS

The potential complexity of CDA certification is compounded by the fact that there are at least two other claim-related certifications that a contractor may be required to make under other statutes. Contracts with agencies of the Department of Defense (DOD) are subject to an additional statutory certification requirement.[141] This law requires that any REA to the contract terms or request for relief under Public Law 85–804 exceeding the "simplified acquisition threshold"[142] may not be paid unless it is certified by a person authorized to bind the contractor at the time of submission. This certification must state that the request is made in good faith and that the supporting data are accurate and complete to the best of that person's knowledge and belief. The Defense Federal Acquisition Regulation Supplement (DFARS) clause implementing this law potentially expands the scope of the certification.[143] The clause states that the REA include only the costs for performing the change, does not include any costs that have already been reimbursed or separately claimed, and that all claimed indirect costs are properly allocable to the change.

Another certification is required by the Truth in Negotiations Act.[144] There are significant differences between the Truth in Negotiations certificate and the two certificates previously discussed. Under the Truth in Negotiations Act, the certificate is

[139] 41 U.S.C. § 605(c)(7).

[140] *Fischbach & Moore Int'l Corp. v. Christopher,* 987 F.2d 759, 762 (Fed. Cir. 1993).

[141] 10 U.S.C. § 2410(a).

[142] Currently, $100,000. *See* FAR § 2.101.

[143] DFARS § 252.243–7002 (Mar. 1998).

[144] 10 U.S.C. § 2306(a). This statute applies only to the Department of Defense and the National Aeronautical and Space Administration (NASA). By regulation, the requirements of this statute have been extended to the civilian agencies. FAR §§ 15.804; 52.215–2; 52.215–22; 52.215–24; 52.215–25.

Table 22.1 Comparison of Certifications

	Contract Disputes Act	Department of Defense Contracts	Truth in Negotiations Act
Language:	Claim made in good faith	Claim made in good faith	Data in support of proposal are accurate, current, and complete
	Supporting data accurate and complete to best of contractor's knowledge and belief	Supporting data accurate and complete to the best of contractor's knowledge and belief	
	Amount requested accurately reflects adjustment for which contractor believes government liable	Amount requested accurately reflects adjustment for which contractor believes government liable	
	Certifier is duly authorized to certify the claim on behalf of the contractor		
$ Threshold:	Claim > $100,000	Claim or request for adjustment > $100,000	Price or adjustment in excess of $650,000 (subject to adjustment for inflation to nearest $50,000 every 5 years)
Certified by:	"Any person duly authorized to bind the contractor with respect to the claim"	"Any person duly authorized to bind the contractor with respect to the claim"	"The contractor" is anyone authorized to sign contractual documents
Date Required:	When submitted as a "claim" under CDA	Upon submission of a request for equitable adjustment or request for extraordinary relief	At time of agreement ("handshake") on price

not provided until the parties agree on a price—at the time of the handshake. The current threshold for a Truth in Negotiations certificate is $650,000 for contracts as well as subcontracts under such contracts and modifications to any contract.[145] The Truth in Negotiations Act provides a specific remedy when the data do not meet the requirements of accuracy, currency, and completeness. Under this act and the related regulations, the government is entitled to a price reduction if the data are found to be defective—that is, inaccurate, not current, or not complete.[146]

These certification requirements overlap to some extent; however, there are sufficient differences that confusion can develop. Table 22.1. compares the current requirements set forth in each of these statutes and their implementing regulations.

X. CERTIFICATION OF SUBCONTRACTOR CLAIMS

Submission, and CDA certification, of subcontractor claims can present problems for prime contractors. To varying degrees, the prime contractor may not fully agree with

[145] 10 U.S.C. § 2306(a); 41 U.S.C. § 2546; FAR § 15.403–4 (as of 2008).
[146] FAR § 15.407–1.

the positions asserted by the subcontractor. It is also likely that the prime contractor must, to some extent, rely on data and information developed outside of its organization. Notwithstanding these practical problems, both the boards and the courts have held that the certification of a subcontractor's claim must be signed by the prime contractor and contain all of the required elements.[147]

The prime contractor is not entitled to qualify its certificate by stating that it is "based on" a certificate provided by the subcontractor[148] or that it was "subject to review."[149] Absolute agreement with the subcontractor's claim is not essential, as the prime contractor may certify a claim with which it does not fully agree, if it concludes the subcontractor's claim is made in good faith and is not frivolous.[150] In *Arnold M. Diamond, Inc. v. Dalton,*[151] the prime contractor, which had previously rejected its subcontractor's claim, certified it on the order of a federal bankruptcy court. The Federal Circuit reversed an earlier board decision that refused to accept the contractor's certification as being in good faith. In the Federal Circuit's opinion, certification of a claim upon the direction of a federal bankruptcy judge satisfied the CDA's requirements. Even with these decisions, prime contractors should not consider certification of subcontractor claims as simply matters of form. In practice, contractors should review and certify subcontractor claims with the same diligence by which claims from the prime contractor's organization are reviewed and certified. Likewise, contractors should also consider obtaining indemnification from the subcontractor whose claims are being presented to mitigate the part of the risk of exposure to an assertion by the government that the subcontractor's claim was inflated or fraudulent.

XI. GOVERNMENT CLAIMS

The CDA also covers government claims. In this regard, the CDA provides that "[a]ll claims by the Government against a contractor relating to a contract shall be the subject of a decision by the contracting officer."[152] Although most government claims

[147]*United States v. Johnson Controls, Inc.,* 713 F.2d 1541, 1551 & 1557 (Fed. Cir. 1983); *Century Constr. Co. v. United States,* 22 Cl. Ct. 63, 65 (1990) (prime may not substitute subcontractor's name for itself); *Int'l Tech. Corp.,* ASBCA No. 54136, 04–1 BCA ¶ 32,607 n.1; *Lockheed Martin Tactical Def. Sys. v. Dep't of Commerce,* GSBCA No. 14450-COM, 98–1 BCA ¶ 29,717.
[148]*Cox Constr. Co.,* ASBCA No. 31072, 85–3 BCA ¶ 18,507.
[149]*Alvarado Constr., Inc., v. United States,* 32 Fed. Cl. 184 (1994).
[150]*United States v. Turner Constr. Co.,* 827 F.2d 1554 (Fed. Cir. 1987).
[151]25 F.3d 1006 (Fed. Cir. 1994).
[152]41 U.S.C. § 605(a). Some cases indicate that the boards may decline jurisdiction over government counterclaims where the counterclaims were never presented to the contractor nor given an opportunity to comment on it. *See Osborn Eng'g Co.,* DOTCAB No. 2165, 90–2 BCA ¶ 22,749; *Instruments & Controls Serv. Co.,* ASBCA No. 38332, 89–3 BCA ¶ 22,237. *But see Sec. Servs., Inc.,* GSBCA No. 11052, 92–1 BCA ¶ 24,704. In *Sec. Servs.,* the GSBCA concluded that the contracting officer had the discretion to either first negotiate the claim or issue a final decision. Similarly, a contracting officer's refusal to negotiate a government claim before issuing a final decision did not negate the finality of that decision. *See also Siebe N., Inc. v. Norton Co.,* ASBCA No. 34366, 89–1 BCA ¶ 21,487. The contractor cannot control the timing of the assertion of a government claim.

are the subject of a final decision, a government withholding of a "sum certain" due a contractor,[153] an assessment of liquidated damages,[154] or a default termination action[155] may constitute an appealable final decision even though no formal final decision is issued. The one clear exception to the requirement for a final decision arises where the government asserts a fraud claim against a contractor. Such a claim need not be the subject of a contracting officer's final decision.[156] Moreover, such claims are beyond the jurisdiction of the boards.[157]

All government demands are not government claims. For example, the ASBCA has refused to consider an appeal of a government demand that a contractor repair defective work. Although the government directed the contractor to perform the work, and the contractor disputed that it was required to do so without additional compensation, the board held that there was no government claim that could be appealed until the government either defaulted the contractor or the contractor did the work and submitted a claim.[158]

The question of when a claim is a government claim or a contractor's claim requiring certification to obtain a final decision has not been answered consistently by the boards and the courts. Several boards have held that the government's withholding of payment due a contractor is a government claim that does not require contractor certification.[159] Similarly, a demand for repayment of money allegedly paid to the contractor by mistake is a government claim, and no contractor certification is required.[160]

Although the assessment of liquidated damages has been characterized as a government claim,[161] the decisions addressing this question are not consistent. Some decisions have held that where a contracting officer assesses liquidated damages in a final decision, a government claim exists and no contractor certification is required.[162] One board has indicated that the burden of proof determines whose claim it is.[163]

In the context of an appeal of a termination for default, the ASBCA has ruled that it did not have jurisdiction over the contractor's claim for a partial remission of liquidated damages because the contractor had never submitted a claim seeking remission

[153]*Sprint Commc'ns Co. v. Gen. Servs. Admin.*, GSBCA No. 14263, 97–2 BCA ¶ 29,249. *But see McDonnell Douglas Corp.*, ASBCA No. 50592, 97–2 BCA ¶ 29,199.

[154]*Midwest Props., LLC v. Gen. Servs. Admin.*, GSBCA Nos. 15822 *et al.*, 03–2 BCA ¶ 32,344.

[155]*Armour of Am. v. United States*, 69 Fed. Cl. 587 (2006) (holding that Court of Federal Claims had jurisdiction over contractor's request that termination for default be converted to termination for convenience but did not have jurisdiction over contractor's claim for termination for convenience damages); *K&S Constr. v. United States*, 35 Fed. Cl. 270 (1996).

[156]*Martin J. Simko Constr., Inc. v. United States*, 852 F.2d 540 (Fed. Cir. 1988).

[157]*Turner Constr. Co.*, GSBCA No. 16840, 06–2 BCA ¶ 33,391; *Envtl. Sys., Inc.*, ASBCA No. 53283, 03–1 BCA ¶ 32,167.

[158]*H. B. Zachry Co.*, ASBCA No. 39209, 90–1 BCA ¶ 22,342.

[159]*Cal. Bean Growers Ass'n*, AGBCA No. 92–116–1, 92–1 BCA ¶ 24,628; *Triasco Corp.*, ASBCA No. 42465, 91–2 BCA ¶ 23,969; *Alaska Lumber & Pulp Co., Inc.*, AGBCA Nos. 83–301–1 *et al.*, 91–2 BCA ¶ 23,890; *TEM Assocs., Inc.*, NASA BCA No. 33–0990, 91–2 BCA ¶ 23,730.

[160]*PX Eng'g Co., Inc.*, ASBCA No. 40714, 90–3 BCA ¶ 23,253.

[161]*Whitesell-Green, Inc.*, ASBCA Nos. 54135 *et al.*, 06–2 BCA ¶ 33,323.

[162]*G. Bliudzius Contractors, Inc.*, ASBCA Nos. 42366 *et al.*, 93–3 BCA ¶ 26,074.

[163]*Equitable Life Ins. Soc'y of the U.S.*, GSBCA No. GS-7699R, 87–2 BCA ¶ 19,733.

of the liquidated damages.[164] The Claims Court required a contractor to certify a claim for the return of liquidated damages that the government withheld from payments due a contractor, apparently because acknowledged delays occurred that the contractor claimed were caused by the government."[165] Therefore, although the government generally withholds liquidated damages, a contractor's claim to recover the money withheld as liquidated damages must be certified.[166]

Even if the contractor is not required to submit or to certify a claim because it is considered to be the government's claim, the contractor should recognize that it likely needs to submit a claim in order to create the basis to recover CDA interest on the funds held by the government. Many government claims—such as the assessment of liquidated damages or deductive changes may result in the government withholding funds that are otherwise due under the contract. Even if the government's position eventually is determined to have no merit, the contractor is not entitled to receive CDA interest on those funds unless it submits a CDA claim.[167] When responding to a government claim and deduction of monies for liquidated damages or a deductive change order, the contractor's more prudent course of action is to submit a CDA claim and to certify it if it exceeds $100,000.

XII. CONTRACTING OFFICER'S DECISION

Once the contractor has submitted a claim meeting the CDA's requirements, the next step in the dispute resolution process is the issuance of a contracting officer's decision. The issuance of a valid contracting officer's decision, or the failure to issue such a decision within the time allowed by the CDA, is a prerequisite to bringing suit on the claim in the Court of Federal Claims or filing an appeal with a board of contract appeals.[168]

A. Time Allowed for Issuing the Decision

The CDA provides that, for claims of $100,000 or less, the contracting officer will issue a decision within 60 days of receipt of a written request from the contractor that a decision be issued within that period.[169] For claims over $100,000, the CDA provides that, within 60 days of receipt of a certified claim, the contracting officer will issue a decision or notify the contractor of the time within which a decision will be issued.[170] If the claim's monetary value requires a contractor certification, the

[164]*AEC Corp., Inc.,* ASBCA No. 42920, 03–1 BCA ¶ 32,071.
[165]*Warchol Constr. Co. v. United States,* 2 Ct. Cl. 384, 392–93 (1983).
[166]*Technocratica,* ASBCA Nos. 48060 et al., 06–2 BCA ¶ 33,316.
[167]*Id.*
[168]*England v. Sherman R. Smoot Corp.,* 388 F.3d 844, 852 (Fed. Cir. 2004). As noted in **Section XI** of this chapter, an assessment of monies by the government may constitute an appealable final decision even though the assessment was not labeled as a "final decision."
[169]41 U.S.C. § 605(c)(1).
[170]41 U.S.C. § 605(c)(2).

contracting officer has no obligation to render a decision on a claim accompanied by a defective certification so long as the contracting officer notifies the contractor in writing of the basis for the conclusion that the certification is defective within 60 days of the date of receipt of the claim.[171] The CDA states that contracting officer's decisions on claims in excess of $100,000 are to be issued "within a reasonable time" in accordance with agency regulations, considering such factors as the size and complexity of the claim and the adequacy of the information in support of the claim.[172]

The CDA also provides that, in the event of undue delay by the contracting officer in issuing a decision, a contractor may request the appropriate board or the Court of Federal Claims to direct that a final decision be issued in a specified period of time.[173] In making such a request, the contractor should be sure that it has provided the contracting officer with all information reasonably necessary for a proper review of the claim and the issuance of a decision.

Any failure by a contracting officer to issue a decision on a claim within the period required by the CDA or as directed by a board or the Court of Federal Claims may be deemed by the contractor to be a decision by the contracting officer denying the claim (a "deemed denied" decision), and such failure authorizes the commencement of suit in the Court of Federal Claims or an appeal to the appropriate board.[174] The fact that the contracting officer fails to issue a decision, however, does not mean that the government is barred from contesting the claim in subsequent proceedings. Failure to issue a decision is deemed a denial and not a default that precludes the government from contesting the merits of the claim at the board or court.[175]

A contractor should be aware, however, that even when a claim is properly submitted and the contracting officer fails to issue a decision, the Court of Federal Claims or a board still can stay the proceedings for the purpose of obtaining a decision on the claim.[176] It is reasonable to expect, however, that the Court of Federal Claims or a board will not exercise this option in the situation where the contracting officer involved has been directed to issue a decision but has failed to do so, or in a situation where the contracting officer gave no reason for the failure to issue a decision.

B. Contents of the Final Decision

The CDA requires that each contracting officer's decision "state the reasons for the decision reached and . . . inform the contractor of his rights as provided in [the CDA]. Specific findings of fact are not required, but, if made, shall not be binding in any subsequent proceeding."[177]

[171] 41 U.S.C. § 605(c)(6).
[172] 41 U.S.C. § 605(c)(3); *Eaton Contract Servs., Inc.,* ASBCA No. 54054, 03–2 BCA ¶ 32,273; *Fru-Con Constr. Corp.,* ASBCA No. 53544, 02–1 BCA ¶ 31,729. *But see Def. Sys. Co.,* ASBCA No. 50534, 97–2 BCA ¶ 29,981 (nine months to review $71 million claim not unreasonable).
[173] 41 U.S.C. § 605(c)(4).
[174] 41 U.S.C. § 605(c)(5).
[175] *Maki v. United States,* 13 Cl. Ct. 779, 782 (1987).
[176] 41 U.S.C. § 605(c)(5).
[177] 41 U.S.C. § 605(a).

FAR § 33.211 also sets forth the procedure that the contracting officer must follow if a claim by or against a contractor cannot be settled by mutual agreement. In preparing the final decision, the contracting officer is directed to include the following information:

(1) Description of the claim or dispute
(2) Reference to pertinent contract terms
(3) Statement of the factual areas of agreement and disagreement
(4) Statement of the final decision, with supporting rationale
(5) In the case of a final decision that the contractor is indebted to the government, a demand for payment in accordance with FAR § 32.610(b)

As a practical matter, the extent of the findings of fact and the rationale provided in the final decision can vary greatly, depending on the nature of the dispute, agency practices, and the specific contracting officer. Although the degree of detail and explanation may vary, the FAR requires that every final decision contain a paragraph that reads substantially as follows:

> This is the final decision of the Contracting Officer. You may appeal this decision to the agency board of contract appeals. If you decide to appeal, you must, within 90 days from the date you receive this decision, mail or otherwise furnish written notice to the agency board of contract appeals and provide a copy to the Contracting Officer from whose decision this appeal is taken. The notice shall indicate that an appeal is intended, reference this decision, and identify the contract by number. With regard to appeals to the agency board of contract appeals, you may, solely at your election, proceed under the board's small claims procedure for claims of $50,000 or less or its accelerated procedure for claims of $100,000 or less. Instead of appealing to the agency board of contract appeals, you may bring an action directly in the United States Court of Federal Claims (except as provided in the Contract Disputes Act of 1978, 41 U.S.C. 603, regarding Maritime Contracts) within 12 months of the date you receive this decision.

Although a final decision that fails to contain this paragraph is considered defective,[178] a defective statement of the contractor's appeal rights does not automatically toll (stop) the period for filing an appeal or suit. To excuse a late appeal, the contractor must demonstrate detrimental reliance on the defective statement of its appeal rights.[179] Since receipt of the final decision triggers the time periods for an

[178]*Fahey v. United States,* 71 Fed. Cl. 522, 528 (2006) (citing *Pathman Constr. Co. v. United States,* 817 F.2d 1573, 1578 (Fed Cir. 1987))
[179]*Fla. Dep't of Ins. v. United States,* 81 F.3d 1093 (Fed. Cir. 1996); *Decker & Co. v. West,* 76 F.3d 1573 (Fed. Cir. 1996); *Swanson Group, Inc.,* ASBCA No. 54863, 05–2 BCA ¶ 33,108 (reliance shown). *But see Am. Renovation & Constr. Co.,* ASBCA No. 54039, 03–2 BCA ¶ 32,296.

appeal or filing of a suit, the FAR directs the contracting officer to furnish the contractor a copy of the decision by certified mail, return receipt requested, or any other method that provides evidence of receipt.[180]

Notwithstanding the statutory requirement that the contracting officer act on claims within specific time frames, a contractor and its counsel must consider whether to petition the board to set a deadline for a final decision or to commence formal proceedings by filing an appeal or a suit on a deemed denied basis if the contracting officer fails to act on the claim.[181] One practical consideration is that a board or the court may stay the proceedings to await the issuance of a final decision,[182] particularly if there is an indication that the agency is attempting to comply with the CDA's requirements.

In order to provide the necessary foundation to appeal or to file an action from the lack of a final decision or petition a board or the court to set a deadline for the issuance of a final decision, certain basic documentation should be available to clearly establish the key events and their dates. This would include a letter notifying the agency that the matter is in dispute and that a final decision is requested. If the claim exceeds $100,000, the request for a final decision must be certified. If the proposal in excess of $100,000 has been certified previously and nothing has occurred that would require a new certification,[183] a basic request for a final decision is sufficient.

Once it appears that no final decision will be received, the contractor must decide whether to petition the appropriate board or the Court of Federal Claims to set a date by which the contracting officer is required to issue a final decision,[184] or just file an appeal or commence a suit. Regardless of whether a petition, appeal, or suit is filed, it is important to set forth the history of the efforts to obtain a final decision. Although this approach is more detailed than notice pleadings, a detailed event-by-event statement of the facts with supporting documents enables the board or court to quickly evaluate the reasons why an action was instituted before receipt of a final decision.

XIII. APPEAL DEADLINES

The CDA provides that a contracting officer's final decision on a claim (whether a contractor or a government claim) is "final and conclusive and not subject to review by any forum, tribunal, or Governmental agency unless an appeal or suit is timely

[180]FAR § 33.211(b); *Riley & Ephriam Constr. Co. v. United States,* 408 F.3d 1369, 1372 (Fed. Cir. 2005).
[181]*Boeing Co. v. United States,* 26 Cl. Ct. 257 (1992); *Fru-Con Constr. Corp.,* ASBCA No. 53544, 02–1 BCA ¶ 31,729.
[182]*Sarang Corp. v. United States,* 76 Fed. Cl. 560, 570 (2007); *Cont'l Mar.,* ASBCA No. 37820, 89–2 BCA ¶ 21,694.
[183]*Santa Fe Eng'rs, Inc. v. Garrett,* 991 F.2d 1579 (Fed. Cir. 1993).
[184]41 U.S.C. § 605(c)(4).

commenced as authorized by this [act]."[185] Once a contractor receives a contracting officer's final decision, it has two possible alternatives. The contractor can take no action, in which event the final decision becomes binding on both parties; or it can challenge the decision by taking an appeal before a board or filing an action in the appropriate court.

Under the CDA, within 90 days of the date of receipt of the contracting officer's final decision, the contractor may appeal the decision to the appropriate board.[186] Alternatively, within 12 months of the date of receipt of the decision, the contractor may initiate an action in the Court of Federal Claims.[187] With respect to either an appeal to a board of contract appeals or an action in the Court of Federal Claims, it is important to consider three basic points.

(1) There can be no appeal or suit unless there has been a valid contracting officer's final decision or the failure to issue such a decision within the period required under the CDA.
(2) Once a valid final decision has been issued, it is essential that a board appeal, or Court of Federal Claims suit, whichever the contractor wishes to pursue, be filed within the required time frame.
(3) The contractor should realize that, once it has elected either the appropriate board or the Court of Federal Claims as the forum in which to challenge the contracting officer's decision, it may not switch to the other forum.[188]

As noted, the contracting officer is directed to obtain evidence of the date on which the contractor received a final decision.[189] This regulatory directive reflects the requirement for strict compliance with the time limits set forth in the CDA for appealing to boards or for filing suit in the Court of Federal Claims. Neither a board nor the court can consider an appeal that is not timely presented to it,[190] as the periods for challenging a contracting officer's decision set forth in the CDA are jurisdictional and cannot be waived.[191]

[185]41 U.S.C. § 605(b). The government may not appeal a final decision of its own contracting officer. *See Zhengxing v. United States,* 71 Fed. Cl. 732, 737 n.17 (2006); *Vertol Sys. Co.,* ASBCA No. 52064, 00–2 BCA ¶ 31,081. A final decision favoring a contractor, however, can be rescinded and a new final decision denying the claim may be issued so long as it is done within the one-year CDA appeal period. *Daniels & Shanklin Constr. Co.,* ASBCA No. 37102, 89–3 BCA ¶ 22,060.

[186]41 U.S.C. § 606.

[187]41 U.S.C. § 609(a). *See also Roxco, Ltd. v. United States,* 77 Fed. Cl. 138, 139–41 (2007).

[188]*But see* 41 U.S.C. 609(a) and **Section XV** of this chapter *infra.*

[189]FAR § 33.211(b). When the government alleges that an appeal is untimely, it bears the burden of proving the date of the contractor's receipt of the final decision. *Brickwood Contractors, Inc. v. United States,* 77 Fed. Cl. 624, 630 (2007); *Alco Mach. Co.,* ASBCA No. 38183, 89–3 BCA ¶ 21,955; *Atl. Petroleum Corp.,* ASBCA No. 36207, 89–1 BCA ¶ 21,199.

[190]*Cosmic Constr. Co. v. United States,* 697 F.2d 1389 (Fed. Cir. 1982); *Renda Marine, Inc. v. United States,* 71 Fed. Cl. 782, 789 (2006); *L. C. Craft,* ASBCA No. 47351, 94–2 BCA ¶ 26, 929.

[191]*Cosmic Constr. Co. v. United States,* 697 F.2d 1389 (Fed. Cir. 1982); *States Roofing Corp. v. United States,* 70 Fed. Cl. 299, 300 (2006) .

Board rules typically provide that filing of the appeal occurs when it is mailed or otherwise furnished to the board.[192] "Mailing" has been interpreted as meaning deposit into the United States Postal Service system. Thus, an appeal submitted to a commercial carrier before the expiration of the 90-day appeal period but received after that period expired was untimely.[193] Sometimes contracting officers send out a copy of the final decision by facsimile, followed by a copy sent via certified mail. Unless the facsimile copy clearly indicates that it is an "advance" copy, the 90-day period has been calculated from the date of receipt of the facsimile.[194]

Often counsel participate directly in the transmission of a claim and may correspond with the agency regarding the claim. If the contracting officer sends the final decision to the contractor's attorney, that attorney may be treated by the boards or the court as the contractor's representative for the purpose of receiving the final decision. Accordingly, the time period for filing an appeal or suit would begin to run upon the attorney's receipt of the final decision.[195] Even if the real party in interest is a subcontractor, and the prime contractor is only sponsoring the subcontractor's claim, the period for an appeal or suit begins to run when the prime contractor receives the final decision.[196]

A contracting officer cannot waive the filing deadlines.[197] Reconsideration of final decision by a contracting officer, however, can have the effect of starting a new appeal period, which would allow the board to assume jurisdiction over a timely appeal of the second final decision[198] or, under certain circumstances, even the lack of a second final decision.[199] Relying upon post–final decision communications to extend the appeal period is very risky. Such communications do not revive appeal rights unless they clearly constitute a reconsideration of the final decision.[200] If both parties are interested in further negotiations after a final decision is issued, a safer course is to file an appeal or suit and then mutually seek a brief stay to explore a negotiated resolution

[192] *See* ASBCA Rule 1(a). The Civilian Board of Contract Appeals will accept a facsimile notice of appeal. CBCA Rule 1(b)(5)(ii).

[193] *Tiger Natural Gas, Inc. v. Gen. Servs. Admin.*, GSBCA No. 16039, 03–2 BCA ¶ 32,321.

[194] *See Brickwood Contractors, Inc. v. United States,* 77 Fed. Cl. 624, 630–33 (2007) (collecting cases addressing receipt of contracting officer's final decision by facsimile); *Leixab, S.A.*, ASBCA No. 51581, 98–2 BCA ¶ 29,962; *Mid-E. Indus., Inc.*, ASBCA No. 51287, 98–2 BCA ¶ 29,907. *But see AST Anlagen-und Sanierungstechnik GmbH,* ASBCA No. 51854, 04–2 BCA ¶ 32712 (where multiple copies of final decision were forwarded to contractor, and contracting officer did not inform the contractor which copy of the decision was legally effective, latter date of receipt is controlling for computation of the statutory period).

[195] *Riley & Ephriam Constr. Co. v. United States,* 408 F.3d 1369, 1374 (Fed. Cir. 2005); *AST Anlagen-und Sanierungstechnik GmbH,* ASBCA No. 51854, 04–2 BCA ¶ 32712.

[196] *Carothers Constr. Inc.*, ASBCA Nos. 44891 *et al.*, 93–3 BCA ¶ 26,069.

[197] *J. Leonard Spodek Nationwide Postal Mgmt,* PSBCA No. 5285, 05–2 BCA ¶ 33,086, 06–1 BCA ¶ 33,175; *Watson, Rice & Co.,* AGBCA No. 82–126–3, 82–2 BCA ¶ 16,009 at 79,359.

[198] *Arono, Inc. v. United States,* 49 Fed. Cl. 544 (2001); *Dawson Builders, Inc.*, ASBCA No. 53172, 01–2 BCA ¶ 31,618.

[199] *Westland Builders,* VABCA No. 1664, 83–1 BCA ¶ 16,235.

[200] Ongoing negotiations without clear evidence of an agreement to reconsider the decision will not prevent the appeal period from running. *Compare TLT Constr. Corp.,* ASBCA No. 52532, 00–1 BCA ¶ 30,805, *with Royal Int'l Builders Co.,* ASBCA No. 42637, 92–1 BCA ¶ 24,684.

unless there is a clear written record that the decision is being "reconsidered" by the contracting officer.

Although strict compliance is required with the appeal limitation periods set forth in the CDA, the limitation periods are not triggered when the contractor's right to proceed to either a board or the Court of Federal Claims arises because the contracting officer has failed to issue a decision on a proper claim within the time period required by the CDA and the claim therefore is deemed denied.[201]

Finally, in the case of a termination for default, the circumstances may be such that the time the contractor has to challenge the termination does not begin to run when the contracting officer issues the final decision terminating the contract but at a later date. This exception to the general basic requirement for strict compliance with the appeal deadlines reflects the continuing application of the doctrine set forth in *Fulford Manufacturing Co.*[202] Under the *Fulford* doctrine, when a contractor makes a timely appeal to an assessment of excess reprocurement costs, the propriety of the default termination can be challenged even though the default termination was not appealed.[203] The CDA has not altered the *Fulford* doctrine.[204] Thus, in most cases, the limitation periods set forth in the CDA do not "bar a contractor from contesting the propriety of a default termination in an action appealing a contracting officer's decision assessing excess reprocurement costs" if such an action is filed within 90 days (a board appeal) or 12 months (Court of Federal Claims) of the decision assessing excess costs.[205] Failure to seek review of a default termination final decision within the 90-day or 12-month period, however, bars a contractor from challenging the default termination if excess costs are not assessed.[206] The prudent course is to appeal the default termination timely and to preserve any claim that may be related.

XIV. CHOOSING A FORUM: BOARD OR COURT OF FEDERAL CLAIMS

When it appears that the agency will or is likely to issue an adverse final decision, the contractor and its counsel must carefully consider which forum (board or Court of Federal Claims) to select. The CDA gives the contractor the right to seek a *de novo,*

[201] *United Partition Sys., Inc. v. United States,* 59 Fed. Cl. 627 (*citing Pathman Constr. Co. v. United States,* 817 F.2d 1573 (Fed. Cir. 1987)).

[202] ASBCA Nos. 2143, 2144 (May 20, 1955), Cont. Cas. Fed. (CCH) ¶ 61,815 (May 20, 1955) (digest only) (timely appeal of the default action will also preserve right to contest excess cost assessment even though second final decision is not appealed in timely manner); *see also Deep Joint Venture v. Gen. Servs. Admin.,* GSBCA No. 14511, 02–2 BCA ¶ 31,914.

[203] *D. Moody & Co. v. United States,* 5 Cl. Ct. 70, 72 (1984); *see also Roxco, Ltd. v. United States,* 60 Fed. Cl. 39 (2004).

[204] *D. Moody & Co. v. United States,* 5 Cl. Ct. 70, 72 (1984); *Sw. Marine, Inc.,* DOTCAB No. 1891, 96–1 BCA ¶ 27,895; *Tom Warr,* IBCA No. 2360, 88–1 BCA ¶ 20,231.

[205] *D. Moody & Co. v. United States,* 5 Cl. Ct. 70, 72 (1984); *see also Am. Telecom Corp. v. United States,* 59 Fed. Cl. 467 (2004) (holding that the court lacked jurisdiction to entertain a contractor's claim seeking the conversion of a termination for default to a termination for convenience, but possessed jurisdiction to hear the contractor's challenge against the assessment of reprocurement costs).

[206] *D. Moody & Co. v. United States,* 5 Ct. Cl. 70 (1984).

or complete, review of a final decision in either forum.[207] When considering whether to elect to go to a board or to the court, there are six factors to consider:

(1) *Time and money.* Ordinarily, board proceedings are less time consuming and less costly than court proceedings. Often this perception reflects the fact that the formal rules of procedure of boards are not as extensive as the rules in the Court of Federal Claims. But some board judges issue extensive prehearing orders that mirror, to a great degree, orders issued by a federal district court or the Court of Federal Claims.

(2) *Judicial background and experience.* In accordance with the CDA, board judges must have at least five years experience in government contract law.[208] Typically, they have much more than that minimum level of experience in government contracting. There is no parallel specialized experience requirement for Court of Federal Claims judges. Board judges hear only government contract cases. Judges on the Court of Federal Claims hear a wide range of matters besides contract cases.

(3) *Case issues.* If the case involves a particular issue or contract provision, it is important to learn how that board or the court views that issue. For example, a board or the court may have recently issued a decision reflecting its views on the proof of delay and the use of a CPM to demonstrate delay. If the case warrants the investment, this type of research should be conducted as part of the forum selection process.[209]

(4) *Agency involvement.* If a case is appealed to a board, the agency will provide the government trial counsel. Accordingly, counsel representing the government may be the same person who advised the contracting officer when the claim was being denied. When a case is filed in the Court of Federal Claims, the Civil Division of the Department of Justice (DoJ) represents the government. Under certain circumstances, the DoJ can, in theory, settle a case over the contracting agency's objection. The DoJ trial attorney, however, typically will consult the agency as the court case progresses.

(5) *Hearing/trial location.* The boards and the Court of Federal Claims are located in the Washington, DC, area. In practice, the boards and the court can and often do hold hearings outside of Washington. Location usually depends on the convenience of and agreement by all hearing participants.

(6) *Representation by legal counsel.* Board practice permits an officer of the corporation to represent it. At the Court of Federal Claims, a corporation must be represented by an attorney admitted to practice before that court.[210]

[207]41 U.S.C. §§ 606, 609(a)(1) – (2). Once appealed, the final decision is accorded no finality. *See England v. Sherman R. Smoot,* 388 F.3d 844 (Fed. Cir. 2004).
[208]41 U.S.C. § 607(b)(1).
[209]For example, a comparison of the decisions of the ASBCA and the Claims Court (now the Court of Federal Claims) concerning the interpretation of essentially the same specification illustrates the value of this type of research. *Cf. W. States Constr. Co. v. United States,* 26 Cl. Ct. 818 (1992), *with Tomahawk Constr. Co.,* ASBCA No. 41717, 93–3 BCA ¶ 26,219.
[210]*Alchemy, Inc. v. United States,* 3 Cl. Ct. 727 (1983).

Once a contractor has elected either a board or the Court of Federal Claims as the forum in which to challenge a contracting officer's final decision, it may not switch to the other forum. In this regard, a contractor that is poised to proceed to either a board or the court should be aware of the Election doctrine. The term "Election doctrine" refers to the body of law related to the contractor's right to initially select the forum in which it will challenge a contracting officer's decision. The CDA does not allow the contractor to pursue its claim in both forums.[211] Thus, once a contractor makes a binding election to appeal a contracting officer's decision to the appropriate board of contract appeals, the contractor cannot change course and pursue its claim in the court.[212] The converse is also true.

A binding election takes place when a contractor files an appeal or initiates a suit in a "forum with jurisdiction over the proceeding."[213] This means that when a contractor timely initiates proceedings on its claim before a board, it has made a binding election to proceed before the board and it is barred from initiating suit in the court. Any suit it later files in the court on the same claim will be dismissed.[214] The filing of an appeal with the appropriate board, however, is not a binding election if it is determined by the board that the contractor's appeal was untimely and that the board was without jurisdiction; hence the subsequent filing of a claim in the Court of Federal Claims would not be barred.[215] The rationale is that a contractor's choice of forums in which to contest the contracting officer's decision is a binding election only if that choice is truly available, which it is not if resort to a board is untimely. In those circumstances, the untimely appeal to the board is an absolute nullity and the Election doctrine is not applicable.

XV. TRANSFER AND CONSOLIDATION OF CASES

The CDA provides that if two or more suits arising from one contract are filed in the Court of Federal Claims and with one or more boards, the court (but not a board) is authorized to order consolidation of the suits before it, or to transfer suits to or among the boards involved "for the convenience of parties or witnesses or in the interest of justice."[216] In deciding whether a case should be consolidated or transferred, the court will consider a number of factors, including: (1) whether the disputes in the different forums arise out of the same contract; (2) whether the cases present overlapping

[211]*Phillips/May Corp. v. United States,* 76 Fed. Cl. 671 (2007).

[212]*Zoeller v. United States,* 65 Fed. Cl. 449, 456 (2005).

[213]*Am. Telecom Corp. v. United States,* 59 Fed. Cl. 467, 471 (2004) (*citing Nat'l Neighbors, Inc. v. United States,* 839 F.2d 1539, 1542 (Fed. Cir. 1988); *Santa Fe Eng'rs, Inc. v. United States,* 677 F.2d 876 (Ct. Cl. 1982); *Tuttle/White Constructors, Inc. v. United States,* 656 F.2d 644 (Ct. Cl. 1981)).

[214]*States Roofing Corp. v. United States,* 70 Fed. Cl. 299, 301–02 (2006) (*citing Nat'l Neighbors, Inc. v. United States,* 839 F.2d 1539 (Fed. Cir. 1988)).

[215]*Nat'l Neighbors, Inc. v. United States,* 839 F.2d 1539, 1541–42 (Fed. Cir. 1988).

[216]41 U.S.C. § 609(d). *See Am. Renovation & Constr. Co. v. United States,* 77 Fed. Cl. 97, 102 (2007) (*citing Joseph Morton Co. v. United States,* 757 F.2d 1273, 1280 (Fed. Cir. 1985)).

or the same issues; (3) whether the plaintiff initially elected to initiate proceedings at the board; (4) whether substantial effort in the case already has been expended in one forum but not the other; (5) which proceeding involves the most money; and (6) which proceeding presents the more difficult and complex claims.[217]

XVI. ADR AND GOVERNMENT CONTRACT DISPUTES

There has long been a need for alternatives to the traditional manner in which government contract disputes are resolved. The Administrative Conference of the United States (the Conference), whose purpose is to promote efficiency and fairness in federal agency procedures, has been a major proponent of alternative dispute respolution (ADR) in government contracts. The Conference has strongly supported ADR and has several publications that discuss the government contract disputes dilemma and various ADR efforts.[218]

FAR § 33.214 states that the objective of using ADR procedures is to increase the opportunity for relatively inexpensive and expeditious resolution of issues in controversy. Essential elements of ADR include: (1) existence of an issue in controversy, (2) a voluntary election by both parties to participate in the ADR process, (3) an agreement on alternative procedures and terms to be used in lieu of formal litigation, (4) participation in the process by officials of both parties who have the authority to resolve the issue in controversy, and (5) contractor certification of claims in excess of $100,000.

ADR procedures may be used any time that the contracting officer has authority to settle the issue in controversy and may be applied to a portion of a claim. When ADR procedures are used after the issuance of a contracting officer's final decision, their use does not alter any of the time limitations or procedural requirements for filing an appeal of the contracting officer's final decision and does not constitute a reconsideration of the final decision. If the contracting officer rejects a request by a small business to use ADR, the contracting officer must, by regulation, set forth a written explanation for that decision and provide it to the contractor.[219]

The CDA states that the boards shall provide, to the fullest extent practicable, informal, expeditious, and inexpensive resolution of disputes. This is the authority for their use of ADR.[220] The boards have implemented ADR procedures and issued a Notice Regarding Alternative Methods of Dispute Resolution that strongly endorses the use of ADR and suggests several techniques.[221] Many of the procedures outlined

[217] *Am. Renovation & Constr. Co. v. United States,* 77 Fed. Cl. 97 (2007); *Glendale Joint Venture v. United States,* 13 Cl. Ct. 325 (1987).

[218] *See* 1 C.F.R. § 305.87–11, Alternatives for Resolving Government Contract Disputes, *available at www.law.fsu.edu/library/admin/acus/3058711.html; see also* DOD Directive 5145.5, *available at www.dtic.mil/whs/directives/corres/pdf/514505p.pdf.*

[219] FAR § 33.214.

[220] 41 U.S.C. § 605(d).

[221] *See* ASBCA Notice Regarding Alternative Methods of Dispute Resolution, *available at http://docs.law.gwu.edu/asbca/adr.htm*; CBCA Rule 54, *available at www.cbca.gsa.gov/CBCA%20Rules/Contract%20Appeals%20Cases.htm.*

by that notice come from the Conference's recommendations. The boards routinely provide a notice of the availability of ADR procedures when the docketing notice is sent to the parties. In addition, most boards will seek to assist in the resolution of claims before issuance of a final decision by making a judge available to mediate a claim or matter in controversy. Similar to the ADR procedure after an appeal is filed, this voluntary process requires a joint request by the contractor and the government.

The parties must jointly decide to use ADR, a board will not accept a unilateral request. The board may, however, take the initiative to suggest ADR as an option in dispute resolution. There are a number of ADR methods, and both the parties and the board may agree to the use of any of them, including a settlement judge, a minitrial, a summary trial with binding decision, or other agreed methods.

The Court of Federal Claims has formally approved the voluntary use of ADR. The court's rules mandate consideration of ADR at the required Early Meeting of Counsel[222] and in the Joint Preliminary Status Report,[223] and allow consideration at any pretrial Conference.[224] The adoption of a procedure for Alternative Dispute Resolution as an appendix to the court's rules "reflects the court's recognition of the increasing usefulness of ADR procedures in the resolution of claims."[225] The procedure recognizes ADR techniques that include, but are not limited to, mediation, minitrials, early neutral evaluation, and nonbinding arbitration. These may be conducted either by a settlement judge or by a third-party neutral. These techniques are voluntary, and both parties must agree to their use.

When both parties agree to utilize one of these ADR methods, they notify the presiding judge as early as possible in the proceedings or concurrently with the submission of a joint preliminary status report. If the presiding judge agrees, the case will be referred to the clerk, who will assign the case to another court judge, who will preside over the ADR method and who will exercise final authority, within the general guidelines adopted by the court, to determine the form and function of each method. If the ADR method utilized by the parties fails to produce a satisfactory settlement, the case will be returned to the presiding judge's docket. All representations made in the course of utilizing a method of ADR are confidential, and, except as permitted by Federal Rule of Evidence 408, may not be utilized for any reason in subsequent litigation.

General Order No. 44 established the ADR Automatic Referral program.[226] In this program, all Court of Federal Claims cases (except for bid protests) assigned to certain judges are automatically and simultaneously referred to certain other judges of the court for ADR. The program's stated goal is to reach a better understanding of the parties' differences and the prospect for settlement. The program allows the parties to

[222]U.S. Ct. Fed. Cl. R. App. A. ¶ 3(f), *available at www.uscfc.uscourts.gov/Rules/Rules%20-%20November% 2015,%202007.pdf.*
[223]U.S. Ct. Fed. Cl. R. App. A. ¶ 4(i).
[224]U.S. Ct. Fed. Cl. R. 16(c)(9).
[225]U.S. Ct. Fed. Cl. R. App. H (formerly General Order No. 13).
[226]U.S. Court of Federal Claims General Order No. 44 (June 21, 2007), Notice of ADR Automatic Referral Program and ADR Automatic Referral Procedures, *available at www.uscfc.uscourts.gov/adr/ADR_ Procedures.pdf.*

meet with the settlement judge to assist in settling the case or in narrowing the issues for resolution at trial.

XVII. RECOVERY OF ATTORNEYS' FEES IN GOVERNMENT CONTRACT CLAIMS

Generally, the FAR disallows the recovery of attorneys' fees and expenses, as well as claim consultants' fees and expenses associated with the preparation and prosecution of government contract claims.[227] With the passage of the Equal Access to Justice Act (EAJA),[228] however, Congress provided a statutory basis for certain eligible contractors to recover some or all of their legal costs and expenses of litigation with the government.

An EAJA application for recovery of legal fees and expenses must be filed within 30 days after the conclusion of the primary appeal or suit.[229] To recover its fees and expenses, the claimant must meet these criteria:

(1) Have a net worth of not more than $7 million. and
(2) Have no more than 500 employees;[230]
(3) Be the prevailing party in the litigation with the government.[231]

The government will not be held liable for the claimant's legal fees and expenses if it can demonstrate that its position in the litigation was substantially justified.[232] Even though the contractor recovers less than it claimed or prevailed on fewer than all of the issues, the claimant still may be deemed to be the prevailing party.[233]

The size and net worth criteria must be satisfied by the prime contractor. Even though the real party in interest is a subcontractor, it is not in privity of contract with the government and is not eligible to recover EAJA legal fees and expenses.[234]

[227] FAR § 31.205–33; *Plano Builders Corp. v. United States,* 40 Fed. Cl. 635 (1998). If the contractor can convince the board or court that these costs were incurred in aid of contract administration, rather than claim preparation or prosecution, these costs can be recovered to the extent they are reasonable and allocable. *See Bill Strong Enters., Inc. v. United States,* 49 F.3d 1541 (Fed. Cir. 1995); *Betancourt & Gonzalez, S.E.,* DOTCAB Nos. 2785 *et al.*, 96–1 BCA ¶ 28,033.

[228] 5 U.S.C. § 504; 28 U.S.C. § 2412.

[229] 5 U.S.C. § 504(a)(2); 28 U.S.C. § 2412(d)(1)(B); *SAI Indus. Corp. v. United States,* 421 F.3d 1344 (Fed. Cir. 2005) (holding that the first day of the 30-day filing period for EAJA application began the day after the expiration of the appeals period); *The Sweetwater v. United States,* 75 Fed. Cl. 214 (2007); *CEMS, Inc. v. United States,* 65 Fed. Cl. 473 (2005).

[230] 5 U.S.C. § 504(b)(1)(B); 28 U.S.C. § 2412(d)(2)(B). These requirements apply to corporations, partnerships, or unincorporated businesses.

[231] 5 U.S.C. § 504(a)(1); 28 U.S.C. § 2412(a)(1).

[232] *CEMS, Inc. v. United States,* 65 Fed. Cl. 473 (2005); *Omni Dev. Corp. v. Dep't of Agric.,* CBCA 609-C (2007).

[233] *C.H. Hyperbarics, Inc.,* ASBCA Nos. 53077 *et al.,* 05–2 BCA ¶ 33,111; *Herman B. Taylor Constr. Co. v. Gen. Servs. Admin.,* GSBCA No. 15361-C (12961), 01–2 BCA ¶ 31,491.

[234] *SCL Materials & Equip. Co.,* IBCA No. 3866–97F, 98–2 BCA ¶ 30,000.

The EAJA limits the amounts that can be recouped for legal fees to a maximum hourly rate[235] plus out-of-pocket expenses. Expert witness rates can be no higher than those paid by the government to its expert witness.[236] Reimbursement for paralegal services is based upon the market rates for those services rather than the actual cost incurred by the law firm.[237]

XVIII. FALSE OR INFLATED CONTRACT CLAIMS

Contractors competing for an award or performing a government contract need to recognize that the government agencies and the Congress have been and remain deeply concerned about fraudulent, inflated, or false claims, Government contractors are expected to "conduct themselves with the highest degree of integrity and honesty."[238] To combat and deter wrongdoing the federal government mandates that most of its contractors adopt and provide to each of its employees a written code of business ethics and conduct.[239] Unless a contractor is a "small business,"[240] contractors must also develop programs to promote awareness of the applicable ethical standards, and control systems such as "Hotlines" to detect and deter wrongdoing.[241] In addition to regulations requiring its contractors to develop and maintain codes of business ethics and conduct and control systems to detect and deter wrongdoing, the federal government has the ability to invoke several statutory remedies if the contractor submits a false or inflated claim including the civil False Claims Act[242] (FCA) and the CDA.[243]

Under the FCA, a party submitting a false claim to the government is liable for a civil penalty of between $5,000 and $10,000 and three times the damages that the government sustains[244]. The FCA is one of government's more effective laws to combat fraud and false claims. In the 21st Century, the federal government has averaged over $1 billion per year in recovery under the FCA. potential liability to the government is not limited to prime contractors. In *Allison Engine Co. v. United States ex rel. Sanders*,[245] the Supreme Court held that § 3729(a) (2) of the FCA does not require that the false record or statement be presented to the government. Rather, it is

[235] 5 U.S.C. § 504(b)(1)(A); 28 U.S.C.§ 2412(d)(2)(A). For actions or appeals awarded on or after March 29, 1996, the maximum rate for legal fees is $125 per hour.

[236] 5 U.S.C. § 504(b)(1)(A); 28 U.S.C. § 2412(d)(2)(A).

[237] *Richlin Sec. Servs. Co. v. Chertoff*, ___U.S. ___, 128 S. Ct. 2007 (2008).

[238] FAR § 3.1002(a).

[239] FAR §§ 3.1004(a), 52.203-13. Applies to contract awards made after 12/24/2007 exceeding $5 million and 120 days duration. Contracts for "commercial items" as defined in FAR Part 12 and those contracts performed "entirely" outside of the United States are exempt.

[240] *See* FAR Part 19 *and* 13 C.F.R. Part 121.

[241] If a contractor is exempt because it qualifies as a "small business," that firm must display agency Hotline posters pursuant to FAR § 52.203-14.

[242] 31 U.S.C. §§ 3729-30. There also is a companion criminal False Claims Act. *See* 18 U.S.C. § 287.

[243] 41 U.S.C. § 604.

[244] *See Morse Diesel v. United States*, 79 Fed. Cl. 116 (2007).

[245] __U. S. __, 128 S. Ct. 2123 (2008).

only necessary to establish that the defendant caused a false record or statement to be submitted for the purpose of getting a false or fraudulent claim paid or approved by the federal government. In that context, the Court noted that a subcontractor or other person could violate that section of the FCA if it submits a false statement to a prime contractor intending for the prime contractor to use that statement to obtain payment from the government.

While the CDA does not extend the jurisdiction of the boards to government claims against contractors for fraud,[246] the CDA does contain a potential sanction if the contractor's claim is determined to have been based on a misrepresentation of fact or fraud. The act provides:

> If a contractor is unable to support any part of his claim and it is determined that such inability is attributable to misrepresentation of fact or fraud on the part of the contractor, he shall be liable to the Government for an amount equal to such unsupported part of the claim in addition to all costs to the Government attributable to the cost of reviewing said part of his claim. Liability under this subsection shall be determined within six years of the commission of such misrepresentation of fact or fraud.[247]

This section of the act provides the government with an effective option to address false or inflated claims. For example in *Daewoo Engineering and Construction Co., Ltd. v. United States*,[248] the Court of Federal Claims concluded that the contractor had misrepresented the facts and supporting cost data related to a certified claim. As a result, the court entered judgment for the United States for the overstated value in the claim – $50,629,855.88. Contractors need to appreciate that the courts are clearly inclined to strictly interpret and enforce the various anti-fraud statues.[249] Even if a contractor obtains a general liability policy addressing errors and omissions in performance, it is unlikely that the coverage of that policy extends to the defense of a government suit based upon an alleged false claim.[250]

[246]*See Martin J. Simko Constr., Inc. v. United States*, 852 F.2d 540, 545 (Fed. Cir. 1988):*Warren Beaves, d/b/a Commercial Marine Servs.,* DOT BCA 1324, 83-1 BCA ¶ 16,232, at 80,648; *see also P.H. Mech. Corp. v. Gen. Servs. Admin.,* GSBCA 10567, 94-2 BCA ¶ 26,785.
[247]41 U.S.C. § 604.
[248]73 Fed. Cl. 547 (2007).
[249]*See Long Island Sav. Bank v. United States*, 503 F.3d 1234 (Fed. Cir. 2007). The Federal Circuit reversed a $435 million judgment in favor of claimant because of a false certification.
[250]*Zurich Am. Ins. Co. v. O'Hara Reg'l Cent. for Rehab.*, 529 F.3d 916 (10th Cir. 2008).

➢ POINTS TO REMEMBER

- Although the Contract Disputes Act (CDA) addresses the procedures for processing claims on a government contract, it is important to understand the obligations and rights under the standard clauses, such as the Changes, Differing Site Conditions, and Suspension of Work clauses.
- Compliance with the contract's notice provisions, as well as consideration of the six-year statute of limitations, is essential to preserving the right to recovery on a claim.
- Every request for an equitable adjustment (REA) is not necessarily a claim. A "claim" is a nonroutine written submission or demand that seeks, as a matter of right, the payment of money in a certain.
- Every claim exceeding $100,000 must be certified by an authorized representative of the prime contractor in order to be considered a "claim" and entitle the claimant recover CDA interest.
- Prime contractors must provide unqualified certifications of their subcontractors' claims. In that context, consider obtaining an appropriate indemnity agreement between contractor and subcontractor.
- A defective claim certification may delay action by the contracting officer, a board of contract appeals, or the Court of Federal Claims.
- The CDA specifies time frames for action by the contracting officer on all claims and provides a means for a contractor to compel consideration of the claim if the contracting officer is unreasonably slow in acting.
- Once a contracting officer's final decision is received, the contractor has 90 days to file an appeal at the board of contract appeals or one year to file a suit in the Court of Federal Claims. If these periods are allowed to pass, the final decision becomes, in almost all cases, final and binding.
- Carefully consider whether the appeal should be heard at a board of contract appeals or in the Court of Federal Claims. Once an election is made, it is, in almost all cases, binding on the parties.
- Whether the final decision is appealed to a board of contract appeals or suit is filed in the Court of Federal Claims, the claim receives *de novo* consideration. That means that no presumption of correctness is attached to the decision being appealed—even those portions favorable to the contractor.
- The government expects its contractors to conduct themselves with the highest degree of integrity and honesty. In addition to potential liability under the False Claims Act, contractors need to be aware of the government's potential remedy under the CDA for claims that found to unsupported due to fraud or a misrepresentation of fact.

INDEX

Abandonment, cost and damage proof, contractors, 472–473
Acceleration:
constructive, 239–242, 292–293
elements of, 240, 293
delay causes, compensable delays, 283–288
delay causes, excusable delays, 281–283
delay causes, nonexcusable delays, 282–283
directed, 292
owner caused, cost and damage proof, contractors, 473
Acceptance:
authority as element of constructive acceptance, 319–320
constructive, 318–320
contract provisions, 318–319, 321
AIA A201, ConsensusDOCS 200, EJCDC C-700, 321
failure to make, delay causes, compensable delays, 288
final completion, 323
formal, 318–319
generally, 317
goods and materials, Uniform Commercial Code (UCC), 122–123
limitations on finality of, 320–321
nonconforming work, design professional (architect/engineer) acceptance, 150–151
partial occupancy, 318–319
AIA A201, 318
ConsensusDOCS 200, 318
EJCDC C-700, 318
federal government contracts, 319
points to remember, acceptance of work, 333–334
post-acceptance facility operations, 329–333
project commissioning, 329–333
bonding considerations, 333
commissioning programs, 330–333
revocation of, 323–324
substantial completion, 322
types of, 318–319
Acts of God, delay causes, excusable delay, 282, 291
Actual authority, design professional (architect/engineer):
AIA A201 (2007 ed.), 138
AIA B101 (2007 ed.), 138
ConsensusDOCS 200 (2007 ed.), 138
ConsensusDOCS 240 (2007 ed.), 138
ConsensusDOCS 245 (2007 ed.), 138
generally, 137–138, 228–229
Actual authority, inspectors, 312
Adequate assurance of performance, Uniform Commercial Code (UCC), 128–129
Affirmative action, federal contracts, 566–567
Age Discrimination in Employment Act (ADEA), 566
Agency construction management. See Alternative contracting methods
Air quality, environmental regulation, 500–503
Alternative contracting methods, 10–35. See also Building Information Modeling (BIM)
construction management, 14–17
agency CM, 15–16
AIA documents, 15–16
CM/GC, 16–17
ConsensusDOCS, 15

Alternative contracting methods, (*Continued*)
program management, 16
safety liabilities and, 520–521
design-build contracts, 17–29
checklist for owners, 22–24
code compliance, 29
ConsensusDOCS 300 (2007 ed.), 18, 27
contractor liability issues, 24–25
design-builder's perspective, 17–19
interpretations clause, 28–29
owner's viewpoint, 19–21
patent defects, 29
performance specifications, 25–27
secondary design review, 28
shop drawings, 27–28
engineer-procure-construct (EPC), 31–33
limitations of liability, 32–33
liquidated damages, 33
liability issues for design professionals, 30–31
multiprime contracts and fast-track projects, 12–14
points to remember, alternative contracting methods, 35
private public partnership (PPP), 34, 41
traditional delivery systems compared, 11–12
Alternative dispute resolution (ADR):
federal contract disputes, 667–669
generally, 611–623
Ambiguities, contract interpretation:
constructive changes, 226, 237
American Arbitration Association (AAA), 617–623
American Bar Association Model Procurement Code. See Model Procurement Code (ABA)
American Institute of Architects (AIA): Design Professional (Architect/ Engineer); (See chapter 13 for a detailed list of the AIA construction documents released in 2007)
actual authority of architect, 137–138
agency construction management, 15–16
construction manager/general contractor (CM/GC), 16
design professional's obligations to owner, 22
LEED certification, 516
payment bonds, claimant defined, 416–417
payment obligations, 189–190
privity, contractual, 135
remedies to owner, 454
subcontract agreement, preparation, 186–187
subcontract termination, 195
surety's obligations, notice of claim on payment bond, 424
AIA A201 General Conditions:
acceptance, 317–318, 321–323
actual authority, of architect, 138
additional insured, CGL 527
apparent authority, of architect, 140
arbitration, 615, 620–621, 623
architect withholding approval on certificate for payment, 151
builder's risk, policy period, 540
changes clause, 224
compliance with permits, codes, and regulations, 29
contract changes and extras, 470
convenience termination costs, 459
copyright for design documents, 176
damages, consequential, 462, 487–488
default clause, 448–449
design-build contracts, 27–29
differing site conditions clause, 253–254
dispute resolution, owner and contractor, 155–156
electronic communications, 360–362
environmental concerns, stop work in affected area, 511
environmental risks, hazardous materials, 506
evidence of financial capability, 341
extra work, cost and damage proof, contractors, 470
final completion, 323
grounds for default, 448–449
inspection clause, 305, 307
inspection, prompt, 314
inspections and testing, 148, 308–309
insurance, standard contract clause, 530–532, 540

interpretations clause, "catchall clause", 28
mediation, 612
notice checklist, 390–392
notice of claims, 234–235
notice requirements, 258
patent defects, errors or omissions, 29
payment bonds, 406
qualifying the project participants, 341
recoupment costs, remedies to owner, 455–456
safety, 518, 520
shop drawing clause, 27
shop drawings and submittals, review and approval, 145
standard contract forms, 345, 349
substantial completion, 322
third-party beneficiary claims, avoided, 165–166
time and scheduling clause, delays and extensions, 278–279
Type I and Type II changed conditions, 257
unanticipated hazardous materials and substances, 506, 511
warranties, 304, 326–327
written directive, change order, claim for additional work, 231–232
Americans with Disabilities Act (ADA):
conditions not covered, 565
disability defined, 565
employers covered by, 565–566
generally, 564–566
reasonable accommodation, 565
undue hardship, 565
Anticipatory repudiation/adequate assurance of performance, Uniform Commercial Code (UCC), 128–129
Anti-Kickback Act, 560
Apparent authority concept:
contract changes, 230–231
design professional (architect/engineer), 139–141
inspectors, 312
Appeals:
arbitration, 617–618
federal contract disputes, 661–664
Arbitration: See also Dispute Resolution; Litigation
agreements and procedures, 622–624
AIA A201 (2007 ed.), 615, 620–621, 623
American Arbitration Association (AAA) rules, 622–623
arbitrator selection, 616–617
architect/engineer decision and, 155–158
Center for Public Resources (CPR) rules, 623
ConsensusDOCS (2007 ed.), 615, 621, 623
effect on sureties, 444–446
EJCDC C-800 (2007 ed.), 621–623
enforceability of agreements, 618–620
Federal Arbitration Act (FAA), 624
generally, 614–615
informality and limited appeals, 617–618
International Chamber of Commerce (ICC) Rules and procedures, 624
multiple parties, complications of, 620–622
rules of evidence and, 618
time and cost of, 615–616
Uniform Arbitration Act (UAA), 624–625
Architect/engineer. See design professionals, 134–179
Asbestos, environmental regulation, 501–503
inadvertent asbestos abatement, 502–503
Associated General Contractors of American (AGC):
Docubuilder® software, 336, 346
State Law Matrix, 53
White Paper on Reverse Auctions for Procurement of Construction, 68
Attorneys, dispute resolution, 605–606
Automatic stay:
bankruptcy, 580–581
relief from, 582, 591
sanctions for violation of, 581

Bankruptcy, 578–602
automatic stay, 580–581
automatic stay relief, 582, 591
automatic stay violation, sanctions for, 581
bankruptcy code, 579–580
debtor's contracts, status of, 588–592
affirmation or rejection, 588–590
assignment, 590

Bankruptcy, (*Continued*)
executory contracts, 588
exercise contract rights, 591
minimizing impact on executory contracts, 590–592
termination of contract before bankruptcy, 590–591
time limit, affirming or rejecting contract, 591–592
discharge, 586–587
funds, 595–600
constructive trust, 596
contract funds (earned and unearned), 595
equitable lien, 596
guarantors, 600
joint check arrangements, 596
mechanic's lien, 599–600
payment bond claims, 598–599
performance bond claims, 598–599
recoupment, 598
setoff, 597
surety's claims to, 598
generally, 578
material and equipment, status of, 592–595
property of debtor's estate, 592–593
stored materials, 593–594
supplier's right to recover goods, 593
voiding unperfecting security interests, 594–595
nondischargeable debts, 587–588
players, 578–579
bankruptcy estate, 579
debtor, 579
debtor in possession, 579
secured creditors, 579
trustee, 579
unsecured creditors, 579
points to remember, bankruptcy, 600–602
preferential transfers, 583–584
preferential transfers, exceptions to rule, 584–586
Uniform Commercial Code (UCC), 129–130
voidable preference rule, 583–586
voiding unperfected security interests, 594–595
Bid shopping, 88–90
Bidding, 55–67. See also Contracts
bid bonds, 86–88
extent of liability, 87
points to remember, bidding principles, 95–97
surety refusal to provide performance bond and, 87–88
bid enforcement, subcontractor, 184–185
bid mistakes, 73–78
contractor, 73
private contracts, 76
relief elements, 74–76
state statutes, 76
timely notice of, 74–75
withdrawal versus reformation, 77–78
bid shopping, 88–90
bid depository, 89–90
subcontractor listing, 89–90
doctrine of promissory estoppel, 91–94
damages related to, 94–95
elements of, 92–94
electronic bids, 66–67
lowest and best bid, 61–63
negotiated "best value," selection process, 63–65
past performance evaluations, 69–73
challenges to, 72–73
procedures, 70–72
private contracts, generally, 68–69
protests, 78–86
American Bar Association Model Procurement Code and, 79
Contract Disputes Act (CDA) and bid protests, 637
contracting agency, 80–82
courts, 84–85
federal projects, 79
Government Accountability Office (GAO), 79–84
state and local governments projects, 85–86
timing of, 79–80
responsible bidder, 56–57
challenges to determination, 57
qualifications to be, 56
responsive bidder, 57–61
competitive advantage, 60
defects in bid bond, 59
failure to acknowledge addenda, 57

minor informalities and, 61
reservations or conditions in bid, 58
unbalanced bid, 59
reverse bid auctions, 67–68
site investigations, differing site conditions, 266–268
Statute of Frauds, 94
subcontractor bids, and dispute avoidance, 180–183
subcontractor capability and low price, 180–184
BIM. See Building Information Modeling.
Bonds: See also Payment bonds; Performance bonds
bankruptcy, 598–599
bid, 59–61
evidence of financial capability, 341
payment bond claims, 406–428
performance bonds, 429–447
subcontracts, considerations, 207–208
Building Information Modeling (BIM), 363–366
collaborative uses of BIM, 364–365
legal implications, 365–366
Breach of implied warranty, differing site conditions, 272–273
Builder's risk insurance, 528: See also Insurance; Commercial General Liability (CGL) insurance
all risk vs. specified risk, 528
coverage trigger, 528
exclusions, 541
fortuitious loss, 541
policy period, 540
Building, green, 513–516
Building Information Modeling; See management and documentation
Buyer's remedies, goods and materials, Uniform Commercial Code (UCC), 131

Capital requirements, evidence of financial ability:
contractors, 181–182, 339, 341
generally, 339–342
subcontractors, 182, 341
Cardinal Change, described, 242–244
Center for Public Resources (CPR), 623
CERCLA. See Comprehensive Environmental Response, Compensation and Liability Act of 1980 (CERCLA)
Certificates of progress or completion, design professional, 151–155
Change orders:
Construction Change Directive, AIA A201, 232
constructive, 236–242
contractor, in subcontract, 200
excessive, delay causes, compensable delays, 244–245, 288
Federal Acquisition Regulation (FAR) clause, 220, 222–223
notification letter sample, 395–396
pricing changed work, 470–472
AIA A201 (2007 ed.), 470
ConsensusDOCS 200 (2007 ed.), 471
FAR, 471–472
written, contract changes, 231–236
Changed conditions. See Differing site conditions
Changes:
Cardinal Change, 242–244
changes clause, defined, 219
common law, departure from, 219–220
purpose, 219
claims under payment bonds, 419–422
constructive changes:
acceleration, 239–242, 293
defective plans and specifications, 237–238
generally, 236–237
misinterpretation of plans and specifications by architect/engineer, 142–144
misinterpretation of plans and specifications by owner, 238–239
cost and damage proof, contractors, 470–472
impact of multiple changes, 244–245
impossibility/impracticability, 245–247
points to remember, 247–248
recovery, 225–236
authority to order, 228–231
change to work required, 226–228
generally, 225–226
notice requirement, 234–236

Changes: (*Continued*)
written change orders, 231–234
written notice of claim required, 234–236
subcontractors and, 199–201
Changes clause, subcontract agreement, 199–201
Charge of discrimination, EEOC, 563–564
Checklists:
contractor response to termination notice, 453–454
contracts in foreign states, 383–385
delay claim documentation, 298–300
design-build, owner's, 22–24
environmental, materials encountered, 504–505
federal government project checklist, 206–207
labor affiliation, 203–204
notice, 386–392, 640
pre-bid/pre-proposal environmental considerations, 381–382
qualifying project participants, 336–342
qualifying the project site, 342–344
risk assessment, 50–53
subcontract agreement drafting:
changes clause, 199–201
default clause, 196–198
disputes clause, 204–205
indemnity clause, 201–203
payment clause, 192–193
scope of work, 189
Civil Rights Act, 564
Claim document: See also Dispute resolution; Federal contract disputes
components of, dispute resolution, 607–609
federal contract disputes, 643–647
Clean Air Act, 500
Clean Water Act, 499
Code compliance:
design-build contracts, 29
violation, contractor recognition, 148
Commercial General Liability (CGL) insurance: See also Insurance; Builder's Risk Insurance
addition insured status, 527
cost of defense, 526
coverage issues:
care, custody, and control, 538
completed operations, 537–538
continuing damages, 536–537
contractual liability exclusion, 538
diminution in value, 537
incorrect performance of work exclusion, 538–539
real property exclusion, 538
"your product" exclusion, 539
"your work" exclusion, 539–540
deductibles, 526
environmental liability, 527–528
excess/umbrella, 526
layers of, 526–527
occurrence vs. claims-made coverage, 526
subrogation, 527
"third-party" insurance, 525
Common law bond, performance bonds:
generally and as compared with "statutory" bonds, 434
subcontracts, 207–208
Communication, management and documentation, 352–353
Compensable delays: See also Delays
inspection related, 314
Compensable work time, 558–559
Competing for the contract. See Bidding
Comprehensive Environmental Response, Compensation and Liability Act of 1980 (CERCLA):
contribution, 498–499
generally, 495
joint and several liability, 498
potentially responsible party (PRP), 495–498
categories of, 496
described, 495
operator, 497
strict liability, 495–496
transporter, 497–498
Concurrent delay: See also Delays
apportionment of damages for, 289–290
described, 288–289
liquidated damages and, 488–490
net recovery for, 289–290
ConsensusDOCS: (See chapter 13 for a detailed list of the ConsensusDOCS construction documents released in 2007.)

acceptance, 317–318, 321–323
arbitration, 615, 621, 623
authority, 135, 138, 140
changes clause, 220–224
copyright for design documents, 175–176
damages, 462, 471, 486, 488
delays, excusable noncompensable (weather), 290
differing site conditions clause, 251–253
dispute resolution, owner and contractor,156
DocuBuilder® software, 338, 346
electronic communications, 360, 362
environmental risk, minimizing, 507–508, 511
generally, 346–349
inspection clauses, 305–306
inspection, costs, 308–309
inspection, prompt, 314
inspection, safety-related obligations, 307
inspection and testing, 148
insurance, contract requirements, 531–532
interpretations clause, "catchall clause", 28
management, construction, 15–16
mediation, 612
notice requirements, 258
patent defects, errors or omissions, 29
pay requests, adjustment, 152
payment bond claimant defined, 417
payment bonds, 406
payment obligations, 190
performance bonds, 443–444
privity, contractual, 135
qualifying the project participants:
contractor's statement of qualifications for a specific project, 338, 371–380
owner and contractor, 341
safety, 518–519
shared responsibility and risk, 173–175
shop drawings, 27, 145–146
standard contract forms, 346–349
termination, 194, 449, 457
third-party beneficiary claims, avoided, 166
time and scheduling clauses, delays and extensions, 278–279
tri-party agreement, 18, 29
Type I and Type II changed conditions, 257
warranties, express contractual, 325
weather, 290
written directives, change orders, claims for additional work, 231–232, 234
Consequential damages:
cost and damage proof, owner, 487–488
damage principles, 463–464
Construction insurance. See Insurance; Builder's Risk Insurance; Commercial General Liability (CGL) Insurance
Construction management. See Alternative contracting methods; See also Management and documentation
Constructive acceptance:
authority and, 150–151, 319–321
formal acceptance versus, 318–319
goods and materials, Uniform Commercial Code (UCC), 122–123
standard contract forms and, 318–319, 321–322
Constructive changes: See also Change orders; Changes
acceleration, 239–242, 293
defective plans and specifications, 237–238
generally, 236–237
informal directives for extra work, 237
inspection related, 310–313
misinterpretation of plans and specifications by design professional, 142–144
misinterpretation of plans and specifications by owner, 238–239, 313–314
working under protest, subcontractors, 200
Constructive trust, bankruptcy, 596
Consumer reporting agency, 574–575
Contract Disputes Act of 1978. See Federal contract disputes
Contract funds, bankruptcy funds, 595–600
Contract law, generally, 1–2
contract breach, 2
contract obligations implied, 2–4
Contract Work Hours and Safety Standards Act (CWHSSA):
federal government contracts, 559–560
requirements, 560

Contracting agency, bid protests, 80–82
Contractor: See also Prime contractor; Subcontractor
  authority to make changes, 200
  basic responsibility for differing site conditions, 249–250
  cost and damage proof, 470–485
    contract changes and extras, 470–472
    defective drawings or specifications, 480
    inefficiency claims, 480–485
    owner-caused acceleration, 479
    owner-caused delay and disruption, 473–479
    wrongful termination or abandonment, 472–473
  defenses of, terminations for default, 451–453
  design-build projects, 17–19, 25–29
  environmental concerns, potentially responsible party, 495–498
  implied obligations, 108–112
  implied warranties, 327–329
  inspection by, 316
  liability for design, 30–31, 170–173
  liability to, project design professional, 30–31, 159–167
  qualifications for project, 336–344, 371–380
  sponsorship of subcontractor claims, 214–215
  use of subcontractor's bid, 88–94
    enforcement of bid, 91–94
    letter of intent, 184
    obligations to award, 88–90, 185
Contracts: See also Bidding; Bonds; Payment bond; Performance bonds; Subcontracts; Uniform Commercial Code (UCC)
  acceptance of work, 150, 317–321
  alternative. See Alternative contracting methods
  ambiguity resolution, 106–108
  bankruptcy effect on, 588–592
  changes clause, 219–225
    AIA A201 (2007 ed.), 220, 222, 224
    ConsensusDOCS 200 (2007 ed.), 220–224
    EJCDC C-700 (2007 ed.), 220, 223–224
    FAR, 220, 222–223
  conditions on enforcement, 98–99
  construed against the drafter, 107
  construed as a whole, 101
  date of commencement, 277–278
  date of completion, 277–278
  defined, 1–2, 98–99
  delays, 109–110. See also Delays
  differing site conditions. See also Differing site conditions
  duty to request clarification, 107–108
  environmental concerns, management practices, 510–513
  environmental concerns, risk minimization, 503–510
  excusable delay clauses, 282–283
  federal contract disputes. See Federal contract disputes
  formation context, 103–106
  framework of, management and documentation, 345–350
  goods and materials, Uniform Commercial Code (UCC), 114–133
  implied contractual obligations, 108–112
    duty to cooperate, 109–110
    duty of good faith and fair dealing, 108–109
    warranty of plans and specifications, 111–112
  industry custom and usage, 105–106
  inspection clauses, 305–307
  insurance, 530–532. See also Insurance
  interpretation goal, 99
  language interpretation, 100–103
  merger clauses, 104
  obligations arising by operation of law, 112–113
  parol evidence, 103–104
  points to remember, interpretation of, 113
  pricing of extra work:
    AIA A201, 470
    ConsensusDOCS 200 (2007 ed.), 471
    FAR, 471–472
  prior dealings, 104–105
  standard forms:
    AIA, 349
    ConsensusDOCS, 346–349
    EJCDC, 345, 349–350
  subcontract administration, 180–218

termination for convenience, 456–458
termination for default, 447–448
terms defined in, 100–101
time, 276–277
Copeland (Anti-Kickback) Act:
federal government contracts, 560
federally financed projects, 560
requirements, 560
Correspondence:
e-mail, 356
letter log samples (incoming and outgoing), 404–405
management and documentation, 355–360
Cost accounting records:
cost and damage proof, 465
management and documentation, 366–367
Cost and damage proof, 461–493. See Liquidated damages
contractor's damages, 470–485
contract changes and extras, 470–472
defective drawings or specifications, 480
differing site conditions, 267–268
home office Eichleay overhead, 475
inefficiency claims, 480–481
owner-caused acceleration, 479
owner-caused delay and disruption, 473–479
subcontractor/supplier refusal to honor bid, 91–95
wrongful termination or abandonment, 472–473
damage principles, 461–472
betterment, 466
causation, 465
compensatory nature of, 461–462
consequential damages, 463–464
contract provisions, 470–472
cost accounting records, 366–367, 465
cost reasonableness, 471–472
direct damages, 462–463
dispute resolution, calculating and proving damages, 609–610
mitigation of damages, 465–466
punitive damages, 464–465
generally, 461
owner's damages, 485–492
consequential damages, 487–488
cost to complete, 486–487
delayed completion, 487
direct damages, 486–487
liquidated damages, 488–492
reduction in value, 487
termination for default, 456
points to remember, 492–493
pricing claims, methods for, 466–469
home office (Eichleay) overhead, 475–479
measured mile (differential studies), 483–484
modified total cost method, 468–469
quantum meruit method, 469
segregated cost method, 468
total cost method, 467–468
termination for convenience, contractor recovery, 459–460
Courts:
bid protests and, 84–85
Economic Loss Rule, 159–163
federal contract disputes, 664–666
litigation, 626–627
Critical path method of scheduling (CPM). See also schedules, project
concurrent delay, 288–290
demonstrate delay, use of, 280–281, 296–300
generally, 276
management and documentation, 298–300, 367–368
Customs and usages, contract interpretation, 105–106

Daily reports:
claim preparation, 607
management and documentation, 357–358
sample documents, 401
Damages, bidding: See also Cost and damage proof
bid bonds, 86–88
subcontractor refusal to honor bid, 91–95
Davis-Bacon Act:
application to construction projects, 560–561
disputes, related to, 637
enforcements of, 561
generally, 560–561
prevailing wages under, 560–561
site of work, 561
wage classifications, 561

Debts, nondischargeable, bankruptcy, 587–588
Default clause, subcontract agreement, 196–198
Default, terminations for. See Terminations
Defective performance remedies, subcontract administration, 212–213
Defective plans and specifications. See Plans and specifications
Delay claims analysis, claims, process, 296–300. See also Critical Path Method of Scheduling; Delays; Liquidated damages, scheduling, project.
Delayed performance remedies, subcontract administration, 213–214
Delays, 276–302, 473–479, 488–492. See also Acceleration; Critical path method of scheduling; Damages; Liquidated damages; Management and documentation; Schedules
  causes of, 283–288
    access problems, 284–285
    acts of God, 291
    compensable delays excusable, examples of, 283–288
    concurrent delays, 288–290
    defective drawings or specifications, 284
    excessive change orders, 288
    excusable delays, examples of, 282–283, 290–291
    failure to accept completed work, 288
    failure to coordinate prime contractors, 286
    failure to give timely orders for work, 286–287
    failure to inspect, 287
    failure to make timely payments to contractors, 287
    failure to provide plans, approve shop drawings, 285–286
    failure to supply materials or labor, 285
    improper site preparation, 284–285
    labor problems, 291
    suspensions of work, 287–288
    weather, 290–291
  checklist for delay claim documentation, 298–300
  concurrent delay, treatment of, 288–290, 489–490
    modern trend, 289–290
    traditional view, 289
  contract time, 277
  CPM schedules and claims for, 296–300
  delay damages, 473–479, 487–492
    generally, 473
    liquidated, 488–492
    overhead, 473–479
  early finish, right to, 281
  Eichleay (home office) overhead, 475–479
    original decision, 475–476
    subsequent limitations, 476–479
  excusable delays, generally, 281–283
  excusable versus non-excusable, 281–282
  final completion date, 280–281
  final payment waiving delay liability, 491
  float in schedule, use of, 280–281
  implied duties, no damage for delay clauses, 112–113
  implied duty not to hinder or delay, 276
  milestone completion dates, 277
  no damage for delay clauses and, 294–295
  no damage for delay clauses, exceptions to, 294–295
  no damage for delay clauses, statutory prohibitions, 295
  non-excusable delays, 281–283
  notice requirements, 293–294
  notification letter sample, 393
  owner's inspection, 287, 314
  owner-caused, cost and damage proof, contractors, 473–479
  payment bond claims and, 419–422
  points to remember, 300–302
  responsibility for, apportionment of, 289–290
  schedules, use in project management, 280–281
  standard contract forms, 345–350
  submittal review by design professional, 146
  substantial completion date and, 278, 490–491
  time of essence clause and, 277
  trade-to-trade clauses, 295–296
  typical contract time and scheduling clauses, 278–280

Demonstrative evidence, dispute resolution, 606–607
Design-build contracts. See Alternative contracting methods
Design professional, 134–179. See also American Institute of Architects
  authority of, 137–141
    actual authority concept, 137–138, 228–229
    apparent authority concept, 139–141, 230–231
    construction phase of project, 137
    design phase of project, 137
    implied authority concept, 138–139, 229–230
    interpret contract documents, 142–144
    notification of, by owner, 147
    ratification of design professional's authority, 141
  copyright on design documents, 175–178
  design-build projects and, 17
  generally, 134–135
  inspection and testing, 147–151, 315–316
    failure to inspect, 148
    job site accidents, 148–149
    means and methods, 149
    observation of defective work, 149–151
  liability of contractor for design, 30–31, 170–173
  liability to contractor, 159–167
    generally, 159
    intentional torts, 163
    negligence and "Economic Loss Rule," 159–163
    professional liability coverage, 166–167
    third-party beneficiary, 163–166
  liability to owner, 135–136
  limitations on design professional liability, effects, 168–170
  points to remember, 178–179
  qualifications for project, 340
  shared responsibility and risk (standard form contract efforts), 173–175
  shop drawings and submittals review and approval, 144–147
    notice of deviations to, 148
    review period, 147–148
  standard of care, 135–136
  statutes of repose, 167–168
  supervisory and administrative function of the design professional, 141–158
    inspection and testing, 147–151
    interpretation of the plans and specifications, 142–144
    issuance of certificates of progress/completion/payment, 151–155
    resolution of disputes between owner and contractor, 155–158
DFARS Regulation, claim certification requirement, 654
Differing site conditions:
  basic responsibility for, 249–250
  contract indications of conditions, express, 259–261
  contract indications of conditions, implied, 262–263
  defined, 249
  operation of clause, 258–266
    exculpatory clauses, 268–269
    generally, 258–259
    notice requirements, 258, 269–270
    site investigations, failure to conduct, 266–268
    Type I condition, elements of, 259
    Type I condition, examples, 260–263
    Type II condition, elements of, 263
    Type II condition, examples, 264–265
    Type II condition, performance of expected material, 265–266
  points to remember, 274–275
  recovery obstacles, 266–270
  recovery without clause, 271–273
    breach of implied warranty, plans and specifications, 272–273
    duty to disclose, 272
    misrepresentation, 271–272
    mutual mistake, 273
  site investigation:
    failure to conduct, 267–268
    generally, 266–268
  standard clauses, 250–257
    AIA A201 (2007 ed.), 253–254
    ConsensusDOCS (2007 ed.), 251–253
    EJCDC, 254–257
    Federal Acquisition Regulation (FAR), 251
    generally, 250–251

Differing site conditions: (*Continued*)
notice requirements, 258
site investigation, typical clause, 266
Type I and Type II conditions, 257
Direct damages:
claim resolution, 610–611
cost and damage proof, contractors, 468–472
cost and damage proof, owners, 486–487
damage principles, 462–463
Disability (Americans with Disabilities Act):
conditions not covered, 565
definition of, 565
generally, 564–566
reasonable accommodation, 565
undue hardship to employees, 565
Disability leave, 569–570
Discharge, bankruptcy, 586–587
Discovery, litigation, 627–629
Dispute arbitration. See Arbitration; See also Dispute resolution
Dispute avoidance (subcontract administration), 80–85, 208–216
pricing, 180–181
subcontractor characteristics, 181–184
Dispute clause, subcontract agreement, 204–206, 214–216
Dispute resolution, 603–632; See also Arbitration; Federal contract disputes; Litigation
arbitration:
agreements and procedures, 622
American Arbitration Association (AAA) rules and procedures, 622–623
arbitrator selection, 616–617
Center for Public Resources (CPR) rules and procedures, 623
effect on surety, 444–446
enforceability, 618–620
Federal Arbitration Act, 624
generally, 614–615
informality, 617–618
International Chamber of Commerce (ICC) rules and procedure, 624
limited appeals, 617–618
multiple parties, complications of, 620–622
party drafted procedures, federal and state law considerations, 624–625
time and cost of, 615–616
Uniform Arbitration Act (UAA), 624–625
alternatives to litigation (ADR):
arbitration, 614–615
dispute review boards, 611
generally, 611
med-arb (mediation/arbitration), 622
mediation, 612
minitrials/summary jury trials, 611–612
attorneys and experts, 605–606
claim document components, 607–609
contemporaneous records, 607
damages, calculation and proving, 609–610
design professional role, 155–158
demonstrative evidence, 606–607
early recognition and preparation of claim, 603–605
generally, 603
litigation, 625–631
alternative dispute resolution in the courts, 630
court system, 626–627
discovery, 627–629
generally, 625
judge or jury selection, 629
trial, 625–626, 630
negotiation and settlement, 610–611
points to remember, 632
Severin doctrine and subcontractor claims, 215–216
subcontractor's claims, assertion by contractor, 214–215
Doctrine of promissory estoppel, bidding, 91–95
Documentation. See Management and documentation
Drawings and specifications. See Plans and specifications
Duty to cooperate:
contract obligations, 109–110, 209
inspections, 314
Duty to coordinate:
multi-prime contractors, 286
subcontract administration, 109–110, 209–210

Duty to disclose, differing site conditions, recovery without clause, 272

Economic Loss Rule, design professional liability, 159–163
Eichleay formula (home office overhead) recovery, 475–479
  original decision, 475–476
  subsequent limitations, 476–479
Electronic communications on construction projects, 360–369
Employee background investigations, 574–576
Employee Retirement Income Security Act (ERISA), 562–563
Employer's "general duty" (OSHA), 554
Employment Discrimination, 563–569
  Age Discrimination in Employment Act (ADEA), 566
  Americans with Disabilities Act (ADA), 564–566
  discrimination and public contracting, 566–567
  Equal Employment Opportunity Commission (EEOC), 563–564
  Executive Order 11246, 567
  Office of Federal Contract Compliance Programs (OFCCP), 567
  Title VII of the Civil Rights Act of 1964, 564
  Uniform Service Employment and Re-Employment Rights Act (USERRA), 568–569
Employment issues, points to remember, 576–577
Enforceability, arbitration agreements, 618–620
Engineers Joint Contract Documents Committee (EJCDC). (See chapter 13 for a detailed list of the EJCDC construction documents released in 2007)
  acceptance, 317–318, 321–323
  arbitration, 621–623
  change order, 231–232
  changes clause, 220, 223–224
  compliance with permits, codes, and regulations, 29
  copyright for design documents, 176
  damages, consequential, 462
  defined terms, 100
  differing site conditions clause, 254–257
  environmental concerns, risk minimization, 506–507
  generally, 349–350
  inspection clauses, 305, 307
  inspection, defective work, 309
  inspection, prompt, 314
  inspection and testing, 147
  interpretation clause, "catchall clause", 28
  mediation, 612
  notice requirements, 258
  patent defects, errors or omissions, 29
  payment bond claimant defined, 417
  performance bonds, 443–444
  privity, contractual, 135
  shop drawing clauses, 27
  standard contract forms, 345, 349–350
    agreement between owner and contractor, cost-plus (C-525), 349
    agreement between owner and contractor, stipulated price (C-520), 349
    contractor's payment application (C-620), 349
    standard general conditions (C-700), 349
  subcontract agreement, preparation, 186
  termination for convenience clauses and costs, 457, 459–460
  third-party beneficiary, 164–166
  Type I and Type II changed conditions, 257
  warranties, express contractual, 325
  written directive, change order, 231–232
Environmental concerns: See also Comprehensive Environmental Response, Compensation and Liability Act of 1980 (CERCLA).
  air quality, 500–503
  asbestos and lead, 501–503
  greenhouse gas emissions, 500
  commercial general liability insurance and, 527–528
  contribution, 498–499
  differing site condition, 265
  environmental risks checklists, 381–382
  generally, 494
  green buildings, 513–516

Environmental concerns: (*Continued*)
joint and several liability, 498
management techniques, 342–344, 381–382, 510–513
mold, 510–513, 542–545
points to remember, 523
potentially responsible party ("PRP"), contractor as, 495–498
regulation and liability sources, 495–503
air quality, asbestos, and lead, 500–503
CERCLA, 495–499
storm water runoff, 499
risk minimization, 503–510
contract provisions and idemnification, 505–509
generally, 503–510
insurance, 509–510
pre-bid/proposal inquiry, 503–505
scope of work, 504
water quality, 499
Equal Employment Opportunity Commission (EEOC), 563–564
Equal Pay Act, 564
Equipment:
bankruptcy, 592–595
failure to provide, delay causes, compensable delays, 285
payment bond claims (repair and rental), 418–419
Equitable lien, bankruptcy, 596. See also Liens
Estimating, management and documentation, 350–352
Evidence, demonstrative, 606–607
Exculpatory clauses, differing site conditions, 268–269
Excusable delays, 281–288, 290–291. See also Delays
Excuse of performance by failure of presupposed conditions, Uniform Commercial Code (UCC):
allocation of deliveries, 130
commercially impracticable, 130
force majeure clause, 130
increased costs, 130–131
Executive Order 11246:
federal government projects, 567
federally financed projects, 567
Exempt employees (wage and hour laws), 555
Experts, dispute resolution, 605–606
Express warranties, described, 124, 325–327
Extra cost:
cost and damage proof, contractors, 470–472
impact of multiple changes, 244–245
notification letter sample, 393–394
Extra work:
cost and damage proof, contractors, 470–472
inspection related, 312–313
notification letter sample, 397
payment bond claims, 419–420
subcontractor, 199–201

Fair Consumer Credit Reporting Act (FCRA):
background checks, 574–576
generally, 574–575
violations of, 574
Fair Credit Reporting Act, 574–575
Fair Labor Standards Act (FLSA), 555–561
False Claims Act, definition of claim, 642
False or inflated contract claims, 670–671
Family and Medical Leave Act (FMLA), 569–570
Fast track projects, 12–14. See also Alternative contracting methods
Federal Acquisition Regulation (FAR):
best value tradeoff, 56–57
bid mistakes, 77–78
bid protests, 78–81
changes clause, 220, 222–223
cost and damage proof, contractors, 471–472
differing site conditions clause, 251
electronic bids, 66–67
federal contract disputes, 642–643
inspection of construction clause, 305–306
materials and workmanship clause, 306
notice checklist, 389–390
permits and responsibilities clause, 306
reasonable cost, demonstrating, 471–472
responsible bidder, 56–57
responsive bid, 57

termination for convenience clause, 458
termination for default clause, 448–449
use and possession prior to completion clause, 306
Federal Arbitration Act (FAA), 618–619, 624. See also Dispute resolution
Federal contract disputes, 633–672
alternative dispute resolution (ADR), 667–669
appeal final decision, filing of:
deadlines, 661–664
default termination issues (Fulford doctrine), 664
process to file, 663
subcontract claims, 663
waiver of late filing not allowed, 663
attorneys' fees recovery, 669–670
certification requirements:
CDA certification requirements, 654–656
comparision of certifications, 655
Department of Defense Appropriations Authorization Act of 1979 (DoD), 654–655
DFARS claim certification, 654
generally, 647–648
modification of claim amount, 649–651
monetary threshold for, 648–649
subcontractor claims, 655–656
supporting data, 653–654
Truth in Negotiations Act certification, 654–655
who may certify the claim, 654
claim(s):
contractor, 639
deadlines for submission, 639–640
differing definitions of, 642
elements of, 644–647
generally, 642–643
government claims, 656–658
notice requirements, 640
what constitutes, 642–643
who may submit, 641–642
written submission to the contracting officer, 643–644
Contract Disputes Act (CDA), 633–637
bid protests and, 637
breach of contract claims, 636
contracting officer's final decision, 658–662
contents of, 659–661
deadlines for, 658–659
contractor claims, 639
deadlines for claim submission, 639–640
disputes not covered by, 636–637
express or implied contracts, 636
false or inflated contract claims, 670–671
forum selection, 661, 664–666
Freedom of Information Act (FOIA), 637–639
generally, 635–637
government claims, 656–658
historical overview, 633–634
labor law disputes and, 637
laches doctrine defense, 640
notice requirements, 389–392, 640
points to remember, 672
requirement for written submission, 643–644
Severin doctrine, 215
sources of law relating to, 633–634
statute of limitations, 639–640
subcontractor claim certification, 655–656
subcontractor claims and, 204–206, 214–215
transfer and consolidation of cases, 666–667
untimely appeal, effect of, 661–664
who may claim, 641–642
Wunderlich Act, 635
Federal government: See also Federal contract disputes.
Bankruptcy Code, 579–580
bid protests, 78–86
contract disputes, See Federal contract disputes
convenience terminations, 458
courts and litigation, 626–627
default clause, terminations, 448–449
differing site conditions, 251
environmental regulation by, 495–498
inspection, generally, 304–317
low bid award, 61–63
notice requirements, 389–392
past performance evaluations, 69–73
payment bond claims (Miller Act), 407–412

Federal government: (*Continued*)
responsible bidder, 56–57
reverse bid auctions, 67–68
subcontract agreement, 206–207
suspension of work clause, delay causes, 287–288
Field order status chart, sample documents, 403
Finance, management and documentation, 341–342
Financial resources, of subcontractor, subcontract administration, 181–182
"flow-down" obligations, subcontract document, 204, 207
Formal acceptance, constructive acceptance versus:
construction projects, 318–319
goods and materials, Uniform Commercial Code (UCC), 122–123
Freedom of Information Act (FOIA), federal contract disputes, 637–639
Fulford doctrine, 664. See also Terminations

Green buildings, 513–516
design and construction issues affected by green building construction, 515–516
AIA B214 (2007 ed.), 516
generally, 513
Leadership in Energy and Environmental Design (LEED):
Accredited Professional (AP), 515
certification points, 515
certification and system, 513–514
classifications, 514
Green Building Rating System, 513
risk and liability, 516
United States Green Building Council (USGBC), 513
Greenhouse gas emissions, 500
Government Accountability Office (formerly General Accounting Office) (GAO):
bid protests, 82–84
challenges to responsibility determinations, 57
federal contract disputes, 636
generally, 79
relation to agency protests, 81
responsive bidder, 57–61

Hazardous material, and substances, regulation and liability sources, CERCLA, 495–496
Hybrid contracts, Uniform Commercial Code (UCC), 115–116

Immigrants (illegal/undocumented):
"acceptable documents", 551
Department of Homeland Security (DHS), 550, 552
employer "reasonable steps" for "safe harbor" protection, 552–553
Employment Eligibility Verification Form I-9, revised, 550–551
E-Verify system, 551–552
generally, 549–550
Immigration and Naturalization Service (INS), 550
Immigration Reform and Control Act (IRCA), 550
"no-match letter", 552
"notice of suspect document", 552
points to remember, 576–577
public contracting, related issues, 553–554
U.S. Citizenship and Immigration Service (USCIS), 550–551
U.S. Immigrations and Customs Enforcement (ICE or USCIS), 550, 552–553
Implied authority concept:
contract changes, 229–230
design professional (architect/engineer), 138–139
Implied obligations, contracts, 108–112, 209–210
Implied warranties, described, 111–112, 124–125, 327–329
Impossibility/impracticability, contract changes, 245–247
Indemnification:
environmental liabilities, 505–509
performance bonds, 432–434
safety, 521
Indemnity clause, subcontract agreement, 201–203
Industry custom and usage, 105
Inefficiency claims, cost and damage proof, contractors, 480–485

Insolvency, contracts for goods and materials, Uniform Commercial Code (UCC), 129–130
Inspection:
acceptable work, rejection of, 313
authority to change requirements during, 312
contractor, 305, 316–317
cost of, 308–309
design professional's role in, 147–151, 315–316
disruption caused by, 314
failure to conduct, delay causes, 287, 314
generally, 303–305
of goods, Uniform Commercial Code (UCC), 121
inspector's role in, 315–317
limitations on owner's, 311–314
obligation to ensure compliance, 316–317
owner's right, not duty, 309–310
points to remember, 333–334
rejection and correction, 311
scope of, 310–311
standard clauses, 305–307
Insurance, 524–548. See also Builder's Risk Insurance; Commercial General Liability (CGL) insurance
additional insured, CGL, 527
alternatives to traditional, 529–530
captive insurer provided, 530
company response to claim, 534–535
concurrent causes of loss, 541–542
contract requirements, 530–532
environmental concerns, risk minimization, 527–528
importance of, 524–525
mold claims and, 542–545
notice, immediate, 533–534
Owner Controlled Insurance Program (OCIP), 529
points to remember, 545–548
prompt action to protect coverage, 533–534
proof of, 532
reservation of rights, 534–535
routine coverage issues, 535–536
Builder's Risk insurance, 540–541
Commercial General Liability insurance, 525–526, 536–537
standard contract clauses, 530–531
AIA A201 (2007 ed.), 530–532
ConsensusDOCS 200 (2007 ed.), 531–532
subrogation by insurer, 527
types of, 525
Builder's Risk, 528
Commercial General liability, 525–528
Errors and Omissions, 528–529
waiver of subrogration, 531–532
Intentional torts, design professional liability, 163
Intermittent partial strike, 571–572
International Chamber of Commerce (ICC), 624
Interpretation, contracts: See also Contracts
ambiguities, resolution of, 106–108
construed as a whole, 101
contract duty to seek clarification, 107–108
custom and usage as aid to, 105–106
design professional (architect/engineer) role in, 142–144
gaps in contract, Uniform Commercial Code, 119–120
implied obligations and, 108–112, 209–210, 276, 283–287, 314
implied warranties and, 111–112, 124–125, 327–329
order of precedence and, 102–103
parties' conduct and, 103–104
parties' prior dealings and, 104–105
points to remember and, 113
terms defined and, 100–103
general terms, 101–102
party defined terms, 100
technical terms, 100
Investigative consumer reports:
disclosure regarding, 576
generally, 575–576

Job site logs, management and documentation, 357–359
Joint check agreements, bankruptcy, 596

Labor affiliation. See Union status
Labor, failure to provide, delay causes, 285
Labor problems, delay causes, 291

Laches doctrine defense, federal contract disputes, 640
Lead, environmental regulation, 501, 503
  inadvertent lead-based paint abatement, 503
LEED certification, 513–515; see also green buildings
Liability:
  Commercial General Liability insurance, 509, 525–528
  of contractor, as designer, 30–31, 170–173
  to contractor, design professional, 159–167
  environmental, 495–498. See also Environmental concerns
  limitation of, subcontract administration, 210–211
  payment bonds:
    extent of, 417–422
    waiver of, 427–428
  performance bonds, 430–442
    arbitration and surety's liability, 444–446
    extension of surety's liability to third parties, 435–436
    increase in surety's liability, 436–438
    surety's defenses to, payment bond claims, 425–428
    surety's defenses to, performance bond claims, 438–442
Licenses:
  design build contracts, 19, 25
  subcontractors, evidence of qualification, 182
Liens:
  equitable, bankruptcy, 596
  mechanic's liens and bankruptcy funds, 599–600
  partial lien waivers, 211
  payment bond claims and, 425, 427–428
Lines of communication, management and documentation, 352–353
Liquidated damages:
  benefit to contractors, 491–492
  cost and damage proof, owners, 488–491
  defenses to, 490–491
    concurrent delays, 489–490
    final payment, 491
    owner caused delays, 490–491
    waiver, 491
  EPC contract, 33
  generally, 488–489
  provision as penalty, 489–491
  provision, validity of, 489
  subcontracts and liquidated damage provision, 491–492
Litigation, 625–631: See also Arbitration; Dispute resolution; Federal contract disputes
  court system, 626–627
  discovery, 627–629
  generally, 625
  insurance company, litigation with, 535
  judge or jury selection, 629
  live testimony, 630–631
  profile of trial, 625–626
  trial, 630–631
Loss, risk of, contracts for goods and materials, Uniform Commercial Code (UCC), 120
Low bid award, described, 61–63

Management and documentation, 335–405
  Building Information Modeling (BIM), 363–366
    collaborative uses of BIM, 364–365
    legal implications, 365–366
  change orders, written directives required, 231–234
  checklists:
    contracts in foreign states, 383–385
    notice, 387–392
    prebid/preproposal environmental considerations, 381–382
    qualifying project participants, 337–342
    qualifying the project site, 342–344
    subcontract agreement drafting, 189, 192–193, 196–207
  contract framework, 345–350
  correspondence, 355–356
  cost accounting records, 366–367
  daily reports/jobsite logs, 357–358, 605, 607
  delay claims, 298–300
  differing site conditions, 269–270
  dispute resolution, contemporaneous records, 607

documentation and notice requirements, 353–360
electronic communications on construction projects, 360–363
electronic communications, industry forms, 360–369
AIA A201 (2007 ed.), 360–362
ConsensusDOCS, 360, 362
e-mail, 356
environmental concerns, 381–382, 510–513
estimating, 351–352
foreign state project concerns, 343–344, 383–385
lines of communication, 352–353
meeting minutes, 356–357
photographs/videotapes, 359–360
points to remember, 369–370
preserving electronically stored information, 368–369
project documentation system, qualifying, 353–355
project participants, qualifying, 337–342
contractor/subcontractor, 339–340
design professional (architect engineer), 340
owner, 339
project site, qualifying, 342–344, 381
project type qualifying, 336–337, 343–344
response to termination notice, 453–454
rights, responsibilities, and risk definition, 344–345
risk avoidance, 335–351
sample documents, 371–405
correspondence logs (incoming/outgoing), 404–405
daily report, 401
field order status chart, 403
notice checklists, 387–392
notice letters, 393–398
notice of backcharge work, 402
request for information, 399
telephone conversation memorandum, 400
schedules, 298–300, 367–368
standard contract forms, 345–350; See also American Institute of Architects (AIA); ConsensusDOCS; Engineers Joint Contract Documents Committee (EJCDC)
standard operating procedures, 352
status logs, 358–359
web-based project management systems, online collaboration and project management (OCPM), 363
Autodesk, 363
Bentley, 363
Meridian, 363
Primavera Contract Manager, 362–363
Sage Timberline, 352, 363
Materials: See also Uniform Commercial Code (UCC)
bankruptcy, 592–595
failure to provide, delay causes, compensable delays, 285
Mechanic's liens, bankruptcy funds, 599–600
Med-Arb, 622
Mediation, 612–614
Meeting notes, management and documentation, 356–357, 607
Merger clauses, 104
Miller Act: See Payment bonds; Performance bonds; See also Bonds
employee payment guarantee, 559–560
generally, 406–412
rights, waiver of, performance bonds, 435–436
Misrepresentation:
differing site conditions, recovery without clause, 271–272
surety's defense to liability, payment bond claims, 426–427
Model Procurement Code (ABA), bidding procedures, 79
Modified total cost method, pricing claims methods, 468–469
Mold:
insurance and, 542–545
management plan to limit liability, 512–513
Multiprime contracting, alternative contracting methods, 12–14

National Labor Relations Act (NLRA), 570–571

National Labor Relations Board (NLRB), 571
Negligence: See also Tort law; Insurance
  bid mistakes, relief elements, 74
  design professional liability, 159–163
Negotiation, dispute resolution, 610–611
No-damages-for-delay clause:
  generally, 294
  limits on enforcement, 294–295
  relationship to implied duties, 112–113
Nondischargeable debts, bankruptcy, 587–588
Non-exempt employees, 555–556
Nonresponsive bid, described, 57–61
Nonsolicitation/distribution policy, 572–573
Nontraditional picketing, 572
Notice of backcharge work, sample documents, 402
Notice requirements: See also Timely notice
  bid mistakes, 74–75
  changes, 234–236
  checklists, 354, 386–392
  delays, claims process, 293–294
  differing site conditions, 258, 269–270
  dispute resolution, 607
  federal contract disputes, 389–390, 640
  management and documentation, 354
  of breach, contracts for goods and materials, Uniform Commercial Code (UCC), 122–123
  payment bond claims, 422–423
  subcontract agreement, 199

Occupational Safety and Health Act (OSHA):
  compliance checks, 554–555
  general duty, 554
  generally, 554–555
  inspections, 554
  warrants, 555
Occupational Safety and Health Review Commission, 555
Office of Federal Contract Compliance Programs (OFCCP), 567
On-call time, 558–559
Oral agreements, subcontracts, dispute avoidance, 184
Order of precedence, contract interpretation, 102–103
Overtime wage, 558–559
Owners:
  acceleration caused by, cost and damage proof, contractors, 479
  bid mistakes, obligation if mistake suspected, 77
  cost and damage proof:
    actual damages, 486–487
    consequential damages, 487–488
    direct damages, 486–487
    generally, 485–486
    liquidated damages, 488–492
  damages for, termination for default, 456
  defective plans and specifications, constructive change, 237–238
  delay causes, 285–286
  design-build projects, checklist for, 22–24
  directed work, constructive change, 237
  environmental liability of, 496
  failure to coordinate multiple prime contractors, 286
  inspection of work by, 309–312
  misinterpretation of plans and specifications by, constructive changes, 238–239
  owner damages, 485–492
  qualifying the owner, 339
  ratification by, contract changes, 231
  remedies of, terminations for default, 454–456
  safety liability of, 521
  subcontractor's claims and, 214–215

Parallel prime contracting, alternative contracting methods, 10–12
Partial lien waivers, pay applications and, subcontract administration, 211
Partial strike (intermittent strike), 571–572
Parties' conduct, contract interpretation, 103–105
Pay applications, partial lien waivers and, subcontract administration, 211
Payment bonds, 406–428: See also Miller Act
  American Institute of Architects, AIA A201 and AIA A312, 406, 416–417, 419
  attorneys' fees and other costs, 422
  bankruptcy funds, 598–599

defined, 406
delay cost, 419, 421–422
design professional (architect/engineer) representation regarding, 139
distinguished from performance bond claims, 422
extra work cost, 419–420
generally, 406–407
lien rights, 425
Little Miller Act, 411–412
Miller Act, 406–412
contract for construction, alteration, or repair of "public buildings" and "public works", 409–411
"dummy" subcontractors, alter egos, and joint ventures, 416
factors determing if applicable, 408–409
qualifying for coverage, 413–417
rights, waiver of, performance bonds, 435–436
"subcontractors" and "suppliers", 413–417
"substantiality and importance" of relationship with prime contractor, 415–416
suppliers of customized materials, 414–415
pay if paid clauses and, 426
points to remember, 428
private projects, 412
required by statute, 406–412
statute of limitations on claims, 423–424
subcontract agreement, 207–208
surety's defenses to liability, 425–428
surety's response to notice of claim, 424–425
timely notice of claims against, 422–423
waiver of rights, 427–428
who may claim under, 413–417
work qualified for coverage, 417–419
Payment clauses, 190–193
pay-if-paid, 192
payment clause checklist, 192–193
pay-when-paid, 190–192
Payment, timely, failure to provide, delay causes, compensable delays, 287
Payment obligations, subcontract agreement, 189–194
Performance bonds, 429–447
alternatives (Subguard), 446–447
arbitration, effect on surety, 444–446
common law v. statutory, 434
distinguished from insurance, 431–432
distinguished from payment bonds, 422
extension to third parties, 435–436
fundamentals of suretyship, 430–435
generally, 429
indemnification, 432–434
liability:
defenses to (surety's liability), 438–442
increase in (surety's liability), 436–438
Miller Act, 435
points to remember, 460
subcontract agreement, 207–208
surety's entitlements to funds, 442–443
surety's obligations, 435
Permits, design-build contracts, 29
Photographs, management and documentation, 359–360
Plans and specifications: See also Spearin doctrine
adequacy of, 19–20, 111–112
defective:
constructive changes, 237–238
cost and damage proof, contractors, 480
delay causes, excusable compensable delays, 283–288
design-build contracts, 17–21
failure to provide, delay causes, 285–286
interpretation of, design professional (architect/engineer), 142–144
misinterpretation of, by owner, constructive changes, 238–239
Portal-to-Portal Act, 557–558
Postliminary time, 557–558
Preferential transfers, bankruptcy, 583–586
Preliminary time, 557–558
Prevailing wage (Davis Bacon Act), 560–561
Pricing claims methods, 466–469
modified total cost method, 468–469
quantum meruit method, 469
segregated cost method, 468
total cost method, 467–468
Pricing, subcontract agreement, dispute avoidance, 180–183

Prime contractor: See also Contractor; Subcontractor
bid mistakes, 73–76
bid shopping, 88–90
duty to cooperate, 109–110
duty to coordinate, 109–110, 209–210
enforcement of subcontractor bid, 91–94, 184
failure to coordinate or cooperate, delay causes, compensable delays, 94–95
federal contract disputes, See Federal contract disputes
subcontractor financing, subcontract administration, 211–212
Prior dealings, contract interpretation, 104–105
Private contracts, bidding, 68–69
Production of documents and things, 629
Project commissioning, post-acceptance facility operations, 329–333
commissioning programs, 330–333
Executive Order (EO) 13423, 331
Green Guide, 331–332
GSA Guide, 332
preventive maintenance, inspection, and service program (PMIS), 330–331
Project designer. See Architect/Engineer and design professional.
Proof of insurance, contracts, 532
Protests, bids. See Bidding; Government Accountability Office (GAO)
Public contracts. See Contracts
Punitive damages, damage principles, 464–465
Purchase orders, goods and materials. See Uniform Commercial Code (UCC)

Qualifications of subcontractor, subcontract administration, 182–183, 339–340
Quantum meruit, pricing claims methods, 469

Recordkeeping. See Management and documentation
Recoupment, bankruptcy, 598
Reformation of bid, bid mistakes, 77–78
Rehabilitation Act, 567
Rejection of goods and materials, Uniform Commercial Code (UCC), 122–123
Reputation, of subcontractor, subcontract administration, 181
Request for information (RFI):
dispute resolution, 607
environmental assessments, 503–505
sample document, 399
Resource Conservation and Recovery Act of 1976, 495
Responsible bidder, described, 56–57
Responsive bid, described, 57
Risk of loss, contracts for goods and materials, Uniform Commercial Code (UCC), 120

Safety: See also Occupational Safety and Health Act; Occupational Safety and Health Review Commission
construction managers, 520–521
contract provisions, 518–519
contractors, 518–519
design professionals, 519–520
indemnification, 521
OSHA, 517–518
owners, 521
points to remember, 523
responsible parties, 518–521
worker's compensation and, 521–522
Salting, 571
Salts (Union):
definition of, 571
employer response to union activism, 572–573
Sarbanes-Oxley Act, 8
Schedules, management and documentation, 280–281, 296–300, 367–368. See also Acceleration; Delays
Schedules, project:
CPM personal computer-based scheduling software, 280–281, 296–298
Scope of work, subcontract agreement, 188–189
Secondary design review, design-build contracts, 28
Segregated cost method, pricing claims methods, 468
Seller's remedies, contracts for goods and materials, Uniform Commercial Code (UCC), 131–132
Sentencing Guidelines, U.S., 8

Service Contract Act of 1965 (SCA):
  application, 560
  disputes related to, 637
Setoff, bankruptcy, 597
Settlement, dispute resolution, 610–611
Severin doctrine, subcontractor claims and, 215–216
Sexual harassment, 564
Shop drawings and submittals:
  design-build contracts, 27–28
  failure to review/approve, delay causes, 285–286
  logs, 358–359
  review and approval of, architect/engineer, 144–147
Site conditions. See Differing site conditions; See also Site investigations
Site investigations, differing site conditions:
  extent, 266–268
  failure to conduct, 267
  hidden conditions and, 266–268
  qualifying the site, 342–344
  typical clauses, 266
Site of the work, Davis Bacon Act, 561
Site preparation, defective, delay causes, 285–286
Spearin doctrine:
  constructive changes, defective plans and specifications, 237–238
  contract obligations, 3, 19–20
  defective plans and specifications, delay causes, 284
  design-build contracts, 19–20
  differing site conditions and, 272–273
  warranty of plans and specifications, 111–112
Specifications. See Plans and specifications; See also Spearin doctrine
Standard forms, management and documentation, 358–359
  ConsensusDOCS
    collaborative (300 Series), 347
    construction management (500 Series), 347
    design-build (400 Series), 347–348
    electronic communications, 362
    general contracting (200 Series), 346–347
    program management (800 Series), 348–349
    subcontracting (700 Series), 186, 348
Standard of care, design professional (architect/engineer), 135–136
Standard operating procedures, management and documentation, 352
State and local governments:
  authority to issue changes, 228–229
  bid protests, 85–86
  low bidder award, 61–63
  mistake in bids, 74–76
  payment bond claims, 406–412
  subcontractor claims and Severin doctrine, 215–216
  subcontractor qualifications, 182–183
State, preparing to work in a new state, 36–54
  AGC State Law Matrix, 53
  bonds to secure payment of taxes, 38
  information sources, 53–54
  lien/bond rights, preservation, 48–50
  points to remember, 54
  public construction awards, 40–41
  qualifying to do business, 36–37
  risk assessment checklists, 50–53
  state licensing and qualifications, 38–39
  state statutes and policies affecting contract terms and conditions, 42–48
Status logs, management and documentation, 358–359
Statute of Frauds, bid enforcement, 94
Statutes of limitation:
  contracts for goods and materials, Uniform Commercial Code (UCC),
    exceptions, 126–128
    generally, 126–127
    modifications of, 126–127
  federal contract disputes, appeal deadlines, 661–664
  federal contract disputes, general limitation, 639–640
Statutes of repose, design professional (architect/engineer) liability, 167–168
Statutory bond, performance bonds, 434
Statutory and regulatory law affecting construction, 8–9

Storm water runoff, environmental regulation, 499
Straight time pay, 556
Subcontract documents, 186
  AIA A401 (2007 ed.), 186
  ConsensusDOCS 700 Series (2007 ed.), 186
  EJCDC, 186
Subcontractor: See also Contractor; Prime contractor; Subcontracts
  bid shopping, 88–90, 184–185
  contractor sponsorship of subcontractor claims, 214–215
  default clause, express termination rights, 194–195
    AIA A401(2007 ed.), 195
    ConsensusDOCS 750 (2007 ed.), 194
  doctrine of promissory estoppel, 91–94
  federal contract disputes, claim certification, 655–656
  federal contract disputes claim submission, 642
  payment bond claims, 193–194
  proof of insurance, 532
  qualifying, 339–340, 371–380
  reliability, 181–182
  rights against the prime contractor, 185
  Severin doctrine and subcontractor claims, 215–216
Subcontracts:
  agreement preparation, 185–207
    changes clause, 199–201
    default clause, 194–198
    delay claims, 198–199
    disputes clause, 204–205
    federal government projects, 206–207, 640, 655–656
    "flow-down" obligations, 188
    generally, 185–187
    indemnity clause, 201–203
    liquidated damages provision, 492
    payment bond surety and, 193–194
    payment obligations, 189–193
    scope of work, 188–189
    standard forms, use of, 185–187
    union status, subcontractor, 183–184, 203
  bonding considerations, 207–208
  checklists for subcontract agreements:
    changes clause, 199–201
    default clause, 196–198
    disputes clause, 204–205
    indemnity clause, 201–203
    labor affiliation, 203–204
    payment clause, 192–193
    scope of work, 189
  dispute avoidance, 208–216
    defective performance remedies, 212–213
    delayed performance remedies, 198–199, 213–214
    duty to cooperate, 109–110, 209
    duty to coordinate, 109–110, 209–210
    liability limitation, 210–211
    pay applications and partial lien waivers, 211
    pricing, 180–181
    Severin doctrine, 215–216
    subcontract bidding, 91–94, 184–185
    subcontractor characteristics, 181–184
    subcontractor financing and, 211–212
    subcontractor's claims and owners, 214–216
  points to remember, 217–218
Substantial completion:
  acceptance of work, 322
  design professional (architect/engineer) liability for, 153
  effect on delay liability, 278
  final completion compared to, 323
  liquidated damages and, 490–491
Superfund. See Comprehensive Environmental Response, Compensation and Liability Act of 1980 (CERCLA); See also Environmental concerns
Suppliers, payment bond claims, 413–415
Surety: See also Payment bonds; Performance bonds
  bankruptcy and, 598–599
  payment obligations, subcontract document, 424–428
Suspensions of work, 287–288. See also Delays

Telephone conversation, memorandum, sample documents, 400

Termination rights, convenience, subcontract agreement, 198
Termination rights, default clause, subcontract agreement, 194–198
  alternatives to termination, 195
  consequences of improper termination, 195
  express termination rights, subcontractor default clause, 194–195
    AIA A401 (2007 ed.), 195
    ConsensusDOCS 750 (2007 ed.), 194
  implied termination rights, 195
Terminations: See also Fulford doctrine
  certification of cause, design professional (architect/engineer), 449
  checklist, contractor response to termination notice, 453–454
  ConsensusDOCS, 449, 457
  cost and damage proof, contractors, 472–473
  EJCDC, 457, 459–460
  for convenience, 456–458
  for default:
    contractor defenses, 451–453
    default clauses, 448–449
    grounds for, 448–451
    owner remedies, 454–456
  generally, 194–198
  points to remember, 460
  subcontractors, 194–198, 448–449
  wrongful, damages for, 472–473
Testimony, trials, litigation, 630–631
Testing, design professional (architect/engineer), 147–151. See also Inspection
Third-party beneficiary, project design professional (architect/engineer) liability, 163–166
Time is of the essence, delays, 277. See also Delays
Time of lawsuit, procedural requirements, payment bond claims, 423–424
Time of notice, procedural requirements, payment bond claims, 422–423
Timely notice: See also Notice requirements
  bid mistakes, relief elements, 74
  bid protests, 80, 82–83
  changes, 234–236
  environmental concerns, 511–512
  insurance, 533–534
  subcontract agreements, flow down, 188
Timely payments, failure to provide, delay causes, 287
Timely work orders, failure to provide, delay causes, 286–287
Tort damage, damage principles, 462
Tort law, 5–8
Total cost method:
  modified, pricing claims methods, 468–469
  pricing claims methods, 467–468
Trials, litigation, 625–626, 630–631
Truth in Negotiations Act certification, 654–655

Uniform Arbitration Act (UAA), 624–625
Uniform Commercial Code (UCC), contracts for goods and materials, 114–133
  acceptance and notification of breach, 122–123
  anticipatory repudiation/adequate assurance of performance, 128–129
  applicability to construction, 115–116
  Article 2, determining when it applies, 115–116
  buyer's remedies, 131
  contract formation under, 117–120
    gaps in contract, 119–120
    methods of acceptance, 117–118
    total agreement not required, 117
    withdrawal of offer, 119
    written contract requirement, 118–119
  excuse of performance by failure of presupposed conditions, 130–131
  generally, 114
  hybrid contracts defined, 115–116
  insolvency, 129–130
  inspection of goods, 121
  modification of obligations, 116–117
  oral agreements, 118
  points to remember, 132–133
  rejection of goods, 122
  revocation of acceptance in whole or in part, 123–124
  risk of loss, 120
  seller's remedies, 131–132
  seller's right to cure, 123
  subcontract document preparation, 184

Uniform Commercial Code (UCC), (*Continued*)
warranties:
disclaimers, 125
express, 124
generally, 124–125
implied, 124–125
indemnification, relation to, 128
limitations, 125–126
statute of limitations and commencement of the warranty period, 126–128
Uniform Service Employment and Re-Employment Rights Act (USERRA), 568–569
Veterans' Re-Employment Rights (VRR) Statute, 568
Union activity, 570–573
Union status: See also Salting
labor problems, delay causes, 291
salts, 571, 573
subcontract agreement, 203–204
of subcontractor, subcontract administration, 183–184, 203–204
U.S. Bankruptcy Code, 579–580
U.S. Court of Federal Claims:
ADR, 667–669
bid protests, 84–85
federal contract disputes, 664–666
predecessor courts, 633–634
U.S. Department of Defense Appropriations Authorization Act of 1979 (DoD), 654–655
U.S. Environmental Protection Agency (EPA), storm water runoff, 499

Video tapes, management and documentation, 359–360
Vietnam Era Veterans Readjustment Assistance Act of 1974 (VEVRAA), 566–567
Violation of bankruptcy automatic stay, sanctions, 581
Voidable preference rule, exceptions to, 584–586

Wage and hour laws:
basic calculation, 556–557
generally, 555–561
independent contractor and, 559
overtime calculation, 557–559
points to remember, 576–577
Waiting and on-call time, 558–559
Walsh-Healey Act, disputes related to, 637
Warranty: See also Uniform Commercial Code; Spearin doctrine
breach of implied, differing site conditions, recovery without clause, 272–273
contract obligations, plans and specifications, 111–112
express warranties, completed work, 325–327
implied warranties, completed work, 327–329
points to remember, 333–334
statutory warranties, 329
Uniform Commercial Code (UCC), 124–126
Weather:
delay causes, excusable delays, 290–291
weather clause, ConsensusDOCS 200 (2007 ed.), 290
inefficiency claims, cost and damage proof, contractors, 480
Wet or muddy conditions, as differing site condition, 260–261
Withdrawal of bid, bid mistakes, 77–78
Work orders, timely, failure to provide, delay causes, 286–287
Work time, 556–559
Worker's compensation statutes and safety:
generally, 521–522
liability without fault, 522
willful/intentional conduct, 522
Wunderlich Act and federal contract disputes, 635